BEEF CATTLE

A. L. NEUMANN
UNIVERSITY OF ILLINOIS

BEEF CATTLE

SEVENTH EDITION

John Wiley & Sons

NEW YORK · SANTA BARBARA · LONDON · SYDNEY · TORONTO

Library of Congress Cataloging in Publication Data:

Neumann, Alvin Ludwig, 1915–
    Beef Cattle.

    First-4th ed. by R. R. Snapp; 5th ed. by R. R. Snapp and A. L. Neumann.
    Includes index.
    1. Beef cattle.   I. Snapp, Roscoe Raymond,
1889–1953.   Beef cattle.   II. Title.

SF207.S5 1977   636.2'1'3   76–46616
ISBN 0-471-63236-8

Printed in the United States of America

10 9 8 7 6 5 4 3

This edition is dedicated to my wife, Lorena P. Neumann, who, I am thankful, is also an experienced scientific editor. Without her help, time from my full-time position as researcher and teacher would have been inadequate to accomplish this and the previous two revisions of Beef Cattle.

PREFACE

Efficient conversion of the feed resources used in beef cattle production continues to receive much attention from progressive and innovative cattle producers and researchers alike. Some concentrate their efforts on the pasture and range forages used mainly by beef cows and their replacements. Others attempt to reduce the need for increasingly costly feed grains and protein concentrates used both by the farmer-feeder who grows his own grains and by the commercial feedlot operator who buys his concentrates.

Both the fixed and the variable nonfeed costs of raising and finishing cattle for market continue to escalate. Increasing the size of the operation, with the hope that resulting economies of scale will offset the cost increases, no longer is a viable alternative in all situations. Some of the largest feedlots and ranches are, in fact, unable to compete effectively with smaller and medium-sized operations that make maximum use of family labor and homegrown feedstuffs and crop residues.

Total feed grain use for feeding all species of farm livestock as well as poultry is on the increase. Feed grains also have become a significant international trade item and are now one of the important generators of income to be used as payment for increases in foreign imports. In spite of the release of farm acreage formerly diverted in government crop control programs in effect over the past several decades, and in spite of extremely favorable crop yields in recent years, feed grain supplies are not far in excess of current needs. Thus it is unlikely that concentrates will ever again be as low-priced as in the recent past.

For all these reasons, and especially because of the urgent need in today's overpopulated world to promote the fullest use of every food resource, this seventh edition of *Beef Cattle* focuses on the factors that influence the costs of producing beef.

The chapter on principles of beef cattle breeding added in the preceding edition proved so popular that another chapter has been added to deal with the related subject of crossbreeding. The decade of the 1970s may well become known as the period when beef cow producers fully accepted the practice of crossbreeding which, while slow in coming to the beef industry, promises to yield widespread benefits in the form of improved reproductive and feedlot performance and higher-yielding carcasses. Dr. Keith E. Gregory and Dr. Larry V. Cundiff have made a valuable contribution in these two chapters, which are based on their experience and investigations at the U.S. Meat Animal Research Center in Clay Center, Nebraska.

Nutritionists have been particularly active in the areas of grain processing and preservation, seeking to reduce storage losses and costs of the increasingly popular high-moisture grains. The agronomic advantages of harvesting grain in this form are such that this fairly new technique will continue to grow. Acid treatments to lower the cost of storage of high-moisture grain in oxygen-limiting facilities and the reconstitution of dry grain which allows refilling such storage structures several times yearly are also promising developments. The basic research dealing with "protecting" high-value energy and protein sources in rations against rumenal degradation, thereby promoting more efficient utilization by the animal, is exciting, and feed-processing methods to enhance this protection are being actively researched. It is timely to at least introduce the possibilities offered by this new concept.

Several important subjects in the broad area of beef production are, unfortunately, in various stages of change or revision at this very time. These include the recent changes in carcass grading standards, the impending removal of several hormone-like compounds and other substances from the approved list of feed additives, and the uncertainty concerning the environmental pollution control regulations that are to be approved by the various states. I believe, however, that it is unfeasible to delay publication any longer while awaiting the ultimate resolution of these questions. Instructors and producers who are accustomed to relying on their own and nearby state experiment stations and extension specialists for research results and recommendations should continue to consult these sources for developments in these important areas.

A few instructors have indicated an interest in the addition of suggested laboratory exercises and reference lists for reading assignments at the end of each chapter, but these have not been included because an even greater number who use the book as a text say they prefer to prepare their own selected reference lists and laboratory exercises. In this way the assigned readings can more nearly meet local needs with the most current publications and the laboratory materials may be suited to the available animals and equipment and the instructor's skills. Many institutions using the book do not include a laboratory with this course but instead offer an all-species skills course to meet the broader needs of students studying the production aspects of all species of farm animals.

As in previous revisions, I have incorporated several ideas for strengthening certain subject matter areas at the suggestion of instructors who use this book as a text in their courses; I am grateful for their constructive suggestions and will continue to welcome them in the future. I specifically thank Dr. Leroy Biehl of the University of Illinois College of Veterinary

Medicine for reviewing the two final chapters, dealing with diseases and parasites of beef cattle. His years of experience in the field as a practicing veterinarian before joining the university staff lend his suggestions much substance and practicality.

I also acknowledge with thanks Dr. T. R. Greathouse of Colorado State University and Professors Frank C. Litterst, Jr., and J. K. Riggs of Texas A. & M. University for their constructive reviews of the final draft.

Finally, I acknowledge the generosity of the many farm journal editors, government agency personnel, farm machinery manufacturers, and breed association representatives who have so kindly granted permission to use their photographs and other material in this new edition. I have attempted to cite the source and to give proper credit for all such materials and assistance; any omissions are entirely unintentional and are deeply regretted.

A. L. NEUMANN

Urbana, Illinois

CONTENTS

VI. SPECIAL PROBLEMS IN BEEF PRODUCTION

PART I
ORGANIZATION OF THE BEEF CATTLE ENTERPRISE

CHAPTER 1
PROGRAMS AND AREAS
OF BEEF PRODUCTION

Beef cattle production differs from most other kinds of livestock production in that the operation is frequently divided into several distinct steps or phases. It is possible to carry on all of these phases on a single farm or ranch as successive steps of a continuous process. More often, however, one or two phases are carried on to the exclusion of the others, not only on individual farms, but also in agricultural regions. In addition there are several highly specialized forms of beef production that differ so much in management methods from those commonly followed that they deserve special mention.

BEEF PRODUCTION PROGRAMS

COMMERCIAL COW-CALF PROGRAM

The first step in the beef enterprise is the production of a calf and raising it to weaning age, for the calf, of course, is the raw material from which the finished beast is eventually made. The breeding herds in which calves are produced need little grain or other finishing feeds. Consequently, the raising of beef calves is generally, though not always, confined to those sections of the country where there is an abundance of land that is incapable of producing cash crops profitably. Most of the important breeding centers are located either in regions that receive insufficient rainfall to produce cultivated crops or in hilly areas where the land is too rolling to be farmed to advantage. More and more, however, cows are being found in combination with grain farming, and this fact accounts for the large increases in beef cow numbers in the Corn Belt, as discussed elsewhere.

Climate also plays an important part in determining the location of the cow-calf enterprise. In this respect, the southern and southwestern states have a decided advantage over those farther north. Because of the shorter and milder winters in these regions, the calves are ordinarily born 4 to 6 weeks earlier than in the North, and they may even be born in the fall, making them larger and heavier when they are marketed the following fall.

3

The raising of beef calves is more economically feasible on comparatively large farms or ranches because of economies of scale. It requires little more labor and investment in equipment to care for a herd of 75 to 100 cows than for a herd of 10 to 20. So long as the United States has extensive areas of government-controlled land or of privately owned grazing lands where large herds are cared for at the rate of only one man for about 250 cows, it will be difficult for the farmer who operates a half-section of high-priced tillable farm land to compete in the breeding of commercial beef calves for the open market as his principal enterprise. This does not mean that a breeding herd may have no place on a great many farms. It does mean that the breeding of commercial cattle must often be secondary in importance to some other enterprise that more efficiently utilizes the available land and labor. On such farms the breeding herd is a sort of by-product plant that uses the otherwise unmarketable products of the farm. The crop residues of corn and small-grain production are the best examples.

Commercial—that is, not purebred—cow herds can be grouped into four broad categories, depending on system of land management, available feeds and pastures, and the method of marketing the calf crop.

1. There are the large spectacular herds that sometimes contain a thousand cows or more, operated on ranches located principally in the Rocky Mountain area. These ranches usually include some deeded land situated along rivers or streams where the winter feed supply of hay or silage, if needed, is produced on irrigated meadows and cropland. The remainder of the ranch usually consists of extensive acreages of low-carrying-capacity, government-controlled land, such as Bureau of Land Management land, national parks, or national forests, that may be near or adjacent to the deeded land. The rancher has grazing privileges or permits for a given number of cows for the summer grazing season. The calf crop is sold either as calves at weaning time in the fall or as yearlings the following fall after having spent another grazing season on the range. Little if any finishing is done on these ranches because grain is not grown to any extent.

2. Then there are herds varying in size from 30 to 50 cows to very large herds operated on ranches usually owned by the rancher or leased from private owners. Most of these herds are found in the Plains region and the Pacific Coast area. The operation of these ranches varies considerably, depending on whether they are located in the northern or southern portion of these regions and on the feed-producing capabilities of the soil. Although most of these ranches sell either calves or yearlings to other ranchers or to feeders, some may feed out their own production. This practice is followed by more and more of the ranchers in the Southwest,

who are growing dryland grain sorghums or who have access to irrigation water for the production of feed grains. Still other ranchers vertically integrate their operations by contract-feeding their calves and yearlings in feedlots in the vicinity, thus maintaining ownership of their cattle from start to finish.

3. Farm-sized herds of 20 to 100 cows are typically operated on farms in the Corn Belt and adjacent sections. On these farms commercial cows may or may not be a secondary enterprise. The beef cow herd often has replaced a dairy herd, and sometimes tillable land incapable of maintaining high yields of cash crops has been seeded down to improved pasture which is utilized by a cow herd. In other instances, cows may utilize only the permanent pasture and aftermath and other cash-crop residues such as straw, corn stalk fields, and corn cobs. The calves from these herds are usually either sold in the fall or spring or are fed out on the farms where they are produced. The large increases in beef cow numbers in the Lake states and Appalachian area consist of herds that are in this category.

4. A fourth kind of operation consists of herds varying in size from a few to several hundred cows, typically found in the former Cotton Belt and Southeast as a whole. These areas are favorable to the cow-calf program because long-season grazing is possible on pastures now occupying acres once depleted by erosion and continuous production of cotton, peanuts, sweet potatoes, or corn. Winter oats, Bermuda grass, and long-season grasses such as tall fescue make it possible to reduce costly winter

Fig. 1. A top-quality commercial cow herd with calves, on typical mountain foothill range in the Western Range area. (The Record Stockman.)

feeding. The climate is such that winter or very early spring calving is common, so that calves are weaned and sold earlier than are calves from the other areas. Herds that do not market their production as fat slaughter calves usually sell them as stocker or feeder calves rather than as yearlings. Much of the increase in finishing cattle numbers that has occurred in southern Arizona-California and the Texas High Plains is based on the so-called Okie calves and light yearlings produced in these herds. This term formerly was derogatory in reference to the nondescript breeding of such cattle, but it now refers to cattle that are currently in strong demand because of their potential for compensatory growth and the tendency to improve in grade upon fattening.

Successful commercial cow-calf operations usually have several things in common, regardless of category:

1. Relatively low investment in land required per cow.
2. Low investment in buildings and other improvements.
3. Maximum utilization of pasture and low-sale-value roughages.
4. Minimum outlay for supplemental feed.
5. Low labor costs—often only family labor being used.
6. High percentage calf crops of high-quality, heavy-weight calves.
7. Minimal losses from diseases and parasites and comparatively low costs for veterinary services.

Beef cow numbers have been increasing rather steadily during the last decade except for occasional short periods in the declining phases of the cattle cycle, which is described elsewhere. Some but not nearly all of this increase has occurred in the Lake and Corn Belt states, where there are great quantities of corn and soybean crop residues that even yet are not fully utilized. Thus there is great potential for further increases in herd size and for establishing literally thousands of new herds in these states. There is a strong chance that cow numbers may be reduced in the Western Range as a result of efforts to change land use policies on federal and state lands. Such changes could entirely eliminate grazing on some lands, and almost certainly will result in fewer grazing permits for much of the nondeeded land.

THE STOCKER PROGRAM

A stocker is a young animal that is being fed and cared for in a way that promotes growth rather than improved condition. Stockers or stock cattle are of two kinds: heifers that are intended for replacements in the breeding herd, and steers and heifers that are intended for the market as feeders or for finishing by the present owner. With both kinds of stockers, the owner's main objective is to produce as much gain as is economical, consistent with normal growth and development. Necessarily, then, stockers are handled mainly by farmers or ranchers who have quantities of home-raised nonfinishing feed in the form of range or pasture, or harvested roughage such as hay, silage, or straw.

Because heifers for breeding purposes are in demand principally in the breeding centers where they were produced, few animals of this class are to be found outside such areas. In general, they are managed in much the same way as the breeding herd in which they were produced. Many cow men do not select their replacements until their heifers have been wintered or are about a year old, and so the surplus heifer calves are in fact unavailable for feeder purposes at weaning time.

Fig. 2. Stocker steer calves being wintered on legume-grass silage in Indiana. These calves were born and raised to weaning age on a western ranch but will spend a year on this Corn Belt farm before being marketed as finished steers. (Corn Belt Dailies.)

Stocker steers often are grown out in the region where they were bred and raised by allowing them to graze grassland like that used by their mothers. Or they may be shipped soon after being weaned, either to grazing areas that are not fully stocked with cows and young calves or to winter-wheat-growing sections where they ultimately will be grazed through the winter. Many Corn Belt feeders make a practice of buying their cattle as calves in the fall and carrying them on homegrown roughages through the winter, before raising them to a full feed on grain the next spring. In this way the by-products of grain farming are utilized. Any undesirable or unthrifty cattle are weeded out before the use of expensive feeds is begun and, probably most important, the cattle are bought when market conditions are especially favorable to the buyer.

Stocker cattle may be laid in at almost any time of year on a well-diversified farm or by a commercial feedlot operator. Hence an order may be placed in the hands of a commission firm or dealer with instructions to buy when the next big "bargain day" occurs. On the other hand, cattle that are to go directly into the feedlot must be bought within a rather short period if the plans for feeding and marketing are not to be disarranged. Feeder cattle often sell unusually high at the time the feeder wishes to start his feeding operations. Had he bought his animals 3 to 6 months earlier when the market was depressed and carried them along on roughage rations until he really needed them, he might have saved $1 to $5 per hundredweight in the cost price.

Commercial feedlot operators may feed a stocker or "warm-up" ration for only 50 to 75 days before shifting to higher-energy finishing rations. In fact, many commercial lots have two distinct operations going on simultaneously. Some of their purchased lightweight calves are first warmed up as stockers before being placed on finishing rations or before being sent back to contracted range—either native range or winter small-grain pastures. The stockers are then removed from the growing lot or from pasture as needed to replace finished cattle. During periods of high grain prices the commercial lots prefer to start their feeders on grain at heavier weights, and it may be more economical for them to grow the calves on stocker rations than to buy the heavier feeders. Such specialized stocker programs often are referred to as "backgrounding" programs.

THE FINISHING PROGRAM

The finishing or fattening program consists of feeding thrifty calves and well-grown-out older cattle, mainly yearlings, a ration that is moderately high to very high in energy, until they are sufficiently conditioned or finished to yield a carcass with enough intramuscular fat to be palatable and acceptable to the consumer. Such rations must contain considerable low-fiber, high-energy feeds such as farm grains, molasses, protein concentrates, or waste animal fats. Roughages are less important in this program but nevertheless must be available economically, whether they are homegrown or purchased.

Corn (*Zea mays*) and the sorghums make up the major portion of the high-energy feeds. Naturally, then, the Corn Belt and the sorghum-growing areas of western Kansas, the Texas Panhandle, and parts of western Oklahoma and eastern New Mexico are important cattle-finishing areas. Other important finishing sections are found in the irrigated valleys of southern California and Arizona, east central Colorado and the Platte Valley in Nebraska, where the cattle are fed large quantities of corn silage, alfalfa, and sugarbeet by-products, plus shipped-in grains. The Northwest—especially Washington, but including Montana and Idaho—where quantities of barley and wheat are being grown and fed to finishing cattle, is developing rapidly in this industry.

Other important cattle-finishing centers are in the productive Mississippi Delta area in the South, in parts of Florida where large quantities of citrus by-products are available, and in Louisiana where blackstrap molasses, a by-product of the sugar industry, is abundant.

For the two decades following World War II, the fastest-growing area of all was in California and Arizona where the large-scale feedlot with completely mechanized feed-handling facilities made for efficient operation. These large lots are near the fastest-growing population centers of

Fig. 3. Choice yearling steers on a finishing ration in northwestern Iowa. Shelter belt and board fence in background are sufficient protection from prevailing winds. (American Hereford Association.)

the United States and find a ready market for their product. They have a favorable climate for cattle finishing and are fairly close to sources of feeder cattle and grain. (Some inconsistencies in the axiom that fat cattle and grain farms go together will be discussed later.) During the mid-1960s the sorghum belt in the Texas Panhandle experienced a similar growth period, and only some uncertainty about the supply of irrigation water for sorghum production dims its prospects for future expansion.

In some sections of the Corn Belt, where the land is somewhat rolling and where there is considerable improved pasture, a common method of finishing cattle is to self-feed corn on grass during the summer and fall months. Such sections often produce a small percentage of their supply of feeders, the balance being purchased from the western or southern states. In the Corn Belt proper, however, where most of the land is tillable, cattle are finished almost entirely on harvested feeds and the feeding is done in drylot or confinement and usually during the winter months. A common practice is to buy, in the fall, yearling steers or heifers or heifer calves that are ready to go directly into the feedlot. They are accustomed to a full feed of grain as rapidly as possible and by April or May are usually carrying sufficient finish to grade high enough (low choice) to suit the majority of consumers. Cattle handled in this way interfere very little with the growing of crops. They arrive at the farm in the fall at about the time the corn is harvested and leave in the spring before the busy season begins. On the other hand, many of the steer calves finished in the Corn Belt are carried well into the summer and fall, either in drylot or on pasture. Haylage, an ensiled product made from either straight alfalfa or a combination legume-grass seeding harvested at a moisture level between that of hay and silage, is also finding favor.

The finishing of cattle for market has several important advantages in addition to the direct financial profit. One benefit is the fertility that remains on the farm in the manure. Farms where cattle have been fed for a number of years are much more productive, as a rule, than adjoining farms where grain and hay have been sold. Cattle feeding makes use of damaged or inferior grain and hay that would contribute little to the farm income if sold for cash. Farmers located at a distance from shipping points can greatly reduce their marketing expenses by marketing their farm products through cattle rather than in the form of hay and grain. Another potential source of income from feeding cattle is the gain made by hogs that are kept in the feedlot to utilize the partially digested grain found in the cattle manure. This practice is of course not possible in slotted-floor confinement systems of feeding. As mentioned earlier, cattle-finishing programs also provide a good market for a lengthy list of by-product feeds.

Cattle feeding, like other industries that derive their profits from buying or producing raw materials and selling them later as a finished product, involves some speculative risk. This risk is necessarily high when both cattle and feed are purchased, and there is always the temptation to expand or contract operations according to the prospect of favorable or unfavorable prices for finished cattle in the future. However, in a feeding program that is adapted to marketing the feeds grown on a given farm, the risk is probably no greater than that connected with a number of other major farm enterprises. Futures trading has been developed for feeder and finished cattle and even for grains and supplements, to provide protection against risk. This device is discussed more fully in Chapter 26.

THE BABY-BEEF PROGRAM

Strictly choice or higher grade, fat, young cattle, varying in age from 8 to 15 months and weighing 650 to 950 pounds, are called baby beeves. The

Fig. 4. Home-bred baby beeves, 16 months of age and almost ready for market as 1,000-pound slaughter cattle. These steers spent their entire lives on this Corn Belt farm. Note effective homemade feeding equipment. (Corn Belt Farm Dailies.)

finishing of baby beeves is a highly specialized form of beef production; to finish animals at so early an age there must be maximum use of palatable and highly nutritious feedstuffs. Baby-beef production is carried on throughout the country wherever grain is grown, but the program is naturally concentrated in areas such as the fringes of the Corn Belt where grain is grown in large quantities and permanent pastures also play an important role.

The production of baby beef implies the breeding, rearing, and finishing of the calves on the same farm. Since only calves of strictly beef type show a pronounced tendency to finish at so early an age, the breeding herd itself must demonstrate the characteristics that signify early maturity. Dairy crosses usually are unsatisfactory because such cattle tend to grow rather than fatten at the desired weights for baby beef.

Calves are provided with creep feed as soon as they are old enough to eat, which is usually at 4 to 6 weeks of age. Roughage has a minor place in the finishing ration because the calves have little capacity for quantities of roughage after consuming the desired amount of grain. However, the cow herd will utilize the available roughage materials.

Western steer calves carrying considerable flesh and showing excellent breeding can be finished into acceptable baby beeves if carefully managed and fed for 6 to 7 months, especially if they have previously been accustomed to grain from a creep feeder. Commercial feedlots that feed predominately heifer calves may be said to be in the baby-beef program because their heifers are sold after 90 to 120 days on feed and at about 800 pounds slaughter weight. This beef is often labeled "chain store beef" but, if the grade is at least choice, the carcasses are quite similar to the more typical homegrown baby-beef carcasses.

THE FAT-CALF PROGRAM

Much of the increase in beef cow numbers during and since World War II in the southeastern states can be attributed to the financial success achieved by those farmers who adapted the cow-calf program to a set of circumstances peculiar to these areas. In the gradual shift from dairying, large numbers of dairy cows of all breeds were bred to beef bulls. The resulting calves made very rapid gains because of the extra milk supplied by their dams and because of the longer lactation period. Calves weighing upwards of 600 pounds in slaughter condition were not uncommon at 8 to 9 months of age, and this increase often was made without grain feeding. Traditionally, consumers in the South and Southeast preferred beef with less finish than was desired by the majority of consumers elsewhere. Therefore consumer acceptance of this young milk-fed beef in

Fig. 5. Native grade cows with their 600-pound milk-fat calves ready for slaughter. These calves, sired by a registered Hereford bull, were dropped in January and February and sold for slaughter at weaning time in October. (University of Kentucky.)

the regions where it was produced was sufficient to ensure acceptable selling prices. Additional weight and condition were being added by creep-feeding and by earlier calving, with many fall calves being dropped by cows on winter rye and oat pastures.

The future of this program is now at the crossroads. As income levels have risen in the belt extending from the southeastern coastal areas to Texas, and as more people from other sections of the country, with their traditional food habits, have moved into this area, the demand for beef has shifted from calf or young, grass-fat beef to more mature, fed beef. This has caused lower prices for fat calves. Furthermore, with successive use of beef bulls on the dairy-bred cows and their female descendants found in the area, the milk production levels have dropped. As a result, the condition or finish of the weaner calf has been greatly reduced, generally to below desired slaughter finish. The improved conformation caused by the increase in beef breeding with each generation has placed these calves into a competitive situation with western calves for feeding in the Corn Belt and High Plains feedlots. Therefore they are no longer available for slaughter as fat calves, and the fat-calf program may be on the way out, although there is likely to be some demand for fat-calf beef for some time to come.

Calves in this program should by all means be marketed as slaughter calves, right off the cows or "off the teat," even if creep-feeding is required to ensure adequate finish. Once these calves lose their milk bloom, they sell at a distinct disadvantage as stockers or feeders. With age, their defects in beef conformation become more pronounced, and the final selling price after feeding in the manner usually practiced on straight beef-bred feeders is considerably lower. Brahman bulls, which are commonly used in the Gulf Coast area, sire calves that fit into the fat-calf program very well, although western feedlots are quite competitive for these calves as feeders when the fat-calf market drops.

Fewer and fewer cows of the dual-purpose breeds such as Milking Shorthorn and Red Poll are milked and handled as dairy cattle. The more common practice is for most, if not all, of the cows to nurse their own calves until they are 8 to 10 months old. Mature cows of good milking ability can easily suckle two calves, and extra calves are often purchased or calves from cows that are being milked are transferred to foster mothers. If labor is available, calves often are housed in drylot, apart from the cows, and turned in to nurse twice daily. In this method of managing dual-purpose cows the calves readily learn to eat grain, and the result is an ideal slaughter calf at weaning time, with extraordinary weights combined with enough beef type to satisfy the most discriminating buyer. When research now under way with the new exotic or continental European breeds such as the Charolais, Simmental, and Limousin is more conclusive, this program may well be the one where they will fit best.

When calves are in excess supply, as happened during the mid-1970s, packers buy calves, especially heifers, in lieu of slaughter cows for the purpose of producing boning and grass-fed beef. It probably is never profitable to produce calves specifically for this market. The price received for the slaughter calf is only a little more than that for cows, which of course is low in comparison with stocker calf prices. The fat-calf program should not be confused with this practice.

THE PUREBRED PROGRAM

The breeding of purebred or registered cattle is a highly specialized form of beef production, requiring a relatively large amount of capital for foundation cattle and equipment and a great deal of husbandry skill and sound judgment on the part of the manager before success is possible.

This phase of cattle breeding is better suited to persons of considerable experience than to beginners. This does not rule out the 4-H Club or

Fig. 6. A purebred Angus herd on a productive Iowa farm. Purebreds, if high enough in quality, may be used to increase volume of business without increasing numbers. (American Angus Association.)

Future Farmers of America youngster who starts a heifer project and conducts it under the supervision of an experienced parent or other adult leader. In this instance immediate financial gain is not so much the goal; rather, the training involved in the conduct of the project is uppermost. Indeed, a large number of successful herds were started from just such a project.

The opportunities offered the breeder of purebred cattle are almost unlimited. Recognition, fame, and financial reward are all within possibility of realization. In the United States the purebred cattle business, large though it is, still represents only a small part of the entire industry, and there is great progress yet to be made in expanding it. It is estimated that only about 3 or 4 percent of all beef females in the country are registered.

In the western states, where cattle are the principal source of income, ranchers have long recognized the value of purebred stock and have been willing to pay $500 to $1,500 per head for young purebred bulls in order to produce grade calves and yearlings for market. Farmers in the Corn Belt and elsewhere, who maintain small breeding herds as a minor enterprise, have possibly been slower to recognize the importance of superior breeding. They have now learned, however—mainly from their swine-breeding friends—that superior performance-tested bulls are a good investment, and in some respects these farmers are today leading many of the western breeders in the acceptance of performance-testing as a tool for herd improvement.

Despite the need for large numbers of purebred bulls to serve as sires in both purebred and commercial herds, most purebred herds at present are not sufficiently high in quality to warrant saving more than one-half to two-thirds of the bull calves dropped. Castration of mediocre calves gives the purebred breeder a source of feeder steers that can contribute to his income as well as raise the quality of commercial feeder cattle. Purebred bulls and surplus breeding females must sell for substantially more than market prices if the purebred program is to be profitable.

Although purebred beef cattle have increased noticeably in total numbers in recent years, they have not kept pace with the increase in commercial beef cows. Considering that probably not more than half of the grade beef cows are now bred to purebred bulls and that probably a similar proportion of the purebred beef bulls now used in grade herds are of mediocre quality, the future demand for young purebred bulls with production records or with progeny-tested sires appears unusually bright. Oddly, breeders who choose to start crossbreeding programs all too often settle for sires that are mediocre or that certainly have not been performance-tested.

No single part of the country any longer holds a monopoly on good beef bulls. Excellent purebred herds of all the major breeds are found throughout the country, and the day is past when commercial cow producers have to drive hundreds of miles to find good bulls. There is considerable evidence that bulls produced in the immediate vicinity of a farm or ranch, from cows and bulls adapted to the prevailing environmental conditions, will work out more satisfactorily for the buyer than will bulls shipped in from a distance. Data from ongoing genetic-environmental interaction studies show also that certain breeds probably should not be advocated indiscriminately, without due regard for the environment in which they must perform.

REGIONS OF BEEF PRODUCTION

There are five rather well-defined regions of beef production in the United States. Each differs from the others in the relative importance of the various programs that have been discussed above. These five regions may be subdivided into type-of-farming areas, each noted for a particular phase of cattle raising or for a distinctive method of managing cattle. The smaller subdivisions will not be discussed in detail here because they are mainly of local interest.

THE PLAINS AND WESTERN RANGE REGIONS

The Plains and Western Range regions are contained roughly within the area lying west of the 100th meridian. This section may be subdivided into the Northern and Southern Plains, Rocky Mountain, and Pacific Coast areas and certain smaller sections of more or less local importance.

In these regions the breeding of calves and the growing out of young cattle on grass are the dominant phases of beef production. Because of the vast acreage of grazing land, much of which is still in public domain, these regions are particularly well suited for these extensive, rather than intensive, forms of cattle raising. Most of the cattle produced in these regions are sold as stockers and feeders for further development or finishing.

Because beef production is a major enterprise with a majority of the ranchers in these areas, the cattle are well bred and are handled according to modern methods of ranch management. Formerly most of the pure-bred bulls used in improving the type and quality of the range cattle were purchased from Corn Belt breeders, but at present a large percentage of the bulls used in the Western Range are bred there. In fact, breeding good bulls for sale to commercial ranchers, as well as breeding superior young bulls and heifers to go into other purebred herds, is now an important phase of beef production in the Range and Plains states.

Until the years following World War II, the ranchers in these states finished few cattle, preferring to ship most of their calves and yearlings to the Corn Belt. This practice has changed appreciably because of four principal factors:

1. Higher freight rates on shipment of live cattle to Corn Belt markets and of dressed beef back to consuming centers in the West.
2. Substantial disproportionate increases in consumer population in the western states.
3. Simultaneous expansion of irrigation facilities and development of high-yielding grain sorghum varieties that may be combine-harvested.
4. Development of large-scale feedlot operations.

California, Arizona, western Texas, eastern Colorado and New Mexico, the Oklahoma Panhandle, western Nebraska and Kansas, and Washington show especially large increases in numbers of beef cattle finished, and this may be said about other more or less isolated spots throughout the area. By the mid-1960s the western states were feeding about ten times more cattle than just before World War II. At times up to one-half of all cattle on feed today may be found in the areas named above, demonstrating the

Table 1

Membership and Registration Data for Selected Beef Breed Associations, 1974[a]

Breed	Brahman	Angus	Polled Hereford	Hereford	Brangus	Shorthorn
Membership	156	41,094	38,936	24,829	1,180	3,500
Registrations	25,295	350,558	207,882	272,416	12,686	24,204
Transfers	21,955	231,028	113,718	111,211	9,957	16,219
Registrations, rank by states[b]						
1	Texas	Missouri	Missouri	Texas	Texas	Iowa
2	Florida	Iowa	Texas	Nebraska	Oklahoma	Missouri
3	Louisiana	Texas	Tennessee	Montana	Arkansas	Illinois
4	Arkansas	Nebraska	Oklahoma	S. Dakota	Southeast	Nebraska
5	Mississippi	Montana	Kansas	Oklahoma	Heart of America	California
6	Alabama	Tennessee	Illinois	Kansas	Corn Belt	Kansas
7	Oklahoma	Illinois	Kentucky	Colorado	Southwest	N. Dakota

8	Georgia	Oklahoma	Arkansas	N. Dakota	West Coast	S. Dakota
9	N. Mexico	Kansas	N. Dakota	Wyoming	Rocky Mountains	Texas
10	Missouri	S. Dakota	Georgia	Idaho	Hawaii & Alaska	Indiana

Breeds whose associations were unable to supply registration data by states:

	Active Breeders	Registrations	Transfers
Charolais	22,125	95,525	75,668
Santa Gertrudis	3,000	28,060[c]	13,751
Red Angus	976	8,272	5,932
Scotch Highland	140	564	300
Devon	151	1,303	856
Milking Shorthorn	981	4,676	2,586
Red Poll	366	2,269	1,753

[a] Adapted from *Annual Report of the National Society of Livestock Record Associations*, 1974–1975.
[b] Rank by region instead of state for Brangus breed.
[c] Classification of animals over 12 months of age.

Fig. 7. A commercial cow herd is the major enterprise on this Oregon ranch on the western slope of the Rocky Mountains. Yearling replacement heifers, being bred to calve as 2-year-olds, may be seen in the herd. (The Record Stockman.)

real competition that traditional Corn Belt feeders are receiving from the feeders in the western states.

Although these sections have seen spectacular increases in cattle feeding, it is highly possible that further increases will be smaller, for a number of reasons. Lowering irrigation water tables, rising energy costs for pumping water, continued rise in freight rates for both grain and cattle, shortages of suitable roughages, competition for space and land from other industries, stricter pollution control laws, and labor shortages may be important deterrents to further expansion. If the present high level of beef consumption per capita is to be maintained, a large portion of the increased numbers of fed cattle may well come from the Corn Belt, the nonirrigated sections where grain sorghum and winter wheat are grown in western Kansas and western Oklahoma, and the barley- and wheat-growing areas in the Northwest.

THE CORN BELT

The Corn Belt is noted for its broad stretches of prairie land and its exceedingly fertile soil, making it a region devoted largely to the growing of crops. Land is high-priced and farms are comparatively small to

Table 2

Relative Growth in Beef Cow and Total Cattle Numbers in Different Regions and Type-of-Farming Areas[a]

Region and Type-of-Farming Area	1967 (1,000 head)	1975 (1,000 head)	1975 as Percentage of 1967	Rank in Increase Area	Rank in Increase Region
Beef Cows					
North Atlantic	**337**	**427**	**127**		**4**
Lake and Corn Belt	**5,308**	**8,357**	**157**		**1**
Lake	775	1,439	186	1	
Corn Belt	4,533	6,918	153	4	
Southeastern	**8,315**	**12,218**	**147**		**2**
Appalachian	2,711	4,222	156	2	
Southeastern	2,721	4,225	155	3	
Delta	2,883	3,771	131	6	
Plains	**13,302**	**17,348**	**130**		**3**
Northern Plains	6,109	7,671	126	7	
Southern Plains	7,193	9,677	135	5	
Western Range	**7,651**	**8,359**	**109**		**5**
Rocky Mountain	5,583	6,148	110	9	
Pacific Coast	2,068	2,211	107	8	
United States Total[b]	**34,592**	**46,900**	**136**		
All Cattle and Calves					
North Atlantic	**5,081**	**4,503**	**88**		**5**
Lake and Corn Belt	**29,560**	**31,722**	**107**		**4**
Lake	9,741	10,232	105	8	
Corn Belt	19,819	21,490	108	7	
Southeastern	**19,612**	**23,090**	**118**		**3**
Appalachian	7,589	9,092	120	4	
Southeastern	6,156	7,503	122	3	
Delta	5,867	6,495	111	6	
Plains	**33,618**	**44,305**	**132**		**1**
Northern Plains	18,553	22,035	119	5	
Southern Plains	15,065	22,270	148	1	
Western Range	**19,376**	**23,351**	**121**		**2**
Rocky Mountain	11,585	15,251	132	2	
Pacific Coast	7,791	8,100	104	9	
United States Total[c]	**108,491**	**127,670**	**118**		

[a] Compiled from USDA, *Agricultural Statistics*, 1967 and 1975.

[b] Includes 2 and 4 thousand and 87 and 107 thousand for Alaska and Hawaii in the two years, respectively.

[c] Includes 8 and 8.9 thousand and 236 and 240 thousand for Alaska and Hawaii in the two years, respectively.

medium-sized. They are largely used for general farming, with the growing of grain and soybeans as the major enterprises. Intensive rather than extensive methods of feed or crop production prevail.

Beef production under such conditions necessarily assumes a quite different role from that in the Range region where grass is the main crop to be sold. Beef production in the Corn Belt is conducted according to intensive management methods except for most cow-calf operations. The finishing of western-grown and, in increasing numbers, southern-grown cattle, either in drylot or on a limited amount of pasture, represents the typical method of handling cattle in the Corn Belt and adjoining areas. As a rule the cattle are bought in the fall at public stockyards, feeder cattle auctions, or direct from the range. Yearling steers are commonly allowed

Fig. 8. Open-fronted, cold confinement slotted-floor cattle-feeding sheds are becoming quite common in the traditional Corn Belt cattle finishing region. Farmer-feeders situated near urban centers are apt to make the substantial investment required for this kind of facility. (Prairie Farmer.)

to run on pastures and corn stalk fields for 3 to 4 weeks, after which they are confined in drylots and placed on finishing rations. Usually 5 to 6 months are required to make satisfactory market animals out of these yearlings, by which time they will at least grade choice as carcasses on the rail. As discussed in greater detail elsewhere, yearling heifers are fed for shorter periods than are steers and are hurried along faster on more concentrated rations.

Summer feeding on grass of overwintered calves is commonly practiced in sections that are somewhat rolling though tillable and therefore have large areas in both tame and permanent pasture. Because thin cattle are scarce in the spring, steers that are to be fed on pasture are often bought in the fall and carried through the winter as stockers on feeds such as hay, silage, and a low level of concentrates, which tend to produce growth rather than fat. Thus two phases of beef production are represented in such a method of management. In the level areas where pasture crops are less important, calves may be finished out in drylot during the summer after being wintered on stocker rations, or they may go directly onto grain in early winter.

Cow-calf operations in the Corn Belt, although found on fewer farms than at the beginning of the twentieth century, are more important than is commonly supposed. Most of the native calves are raised and fed in small groups of 10 to 100 head, making the breeding industry less spectacular than feedlot operations involving larger droves of shipped-in feeders. Nevertheless, fully one-half of the cattle and calves slaughtered within the Corn Belt are native cattle that were bred in the region. Many of these above-mentioned slaughter cattle are discarded dairy cows and veal calves, but a large number are young cattle of beef breeding that are marketed as baby beeves or as finished yearlings.

Another aspect of breeding operations in the Corn Belt is the raising of purebred cattle, for which this region enjoys particularly favorable conditions. The abundant feed supply promotes relatively lower production costs and the proximity of the Plains and Southeast regions ensures a broad market for all surplus breeding stock.

THE APPALACHIAN AND GREAT LAKES AREAS

The Appalachian and Great Lakes areas are noted for their rolling topography, which in places has a mountainous character. Permanent pastures and woodlands occupy a considerable portion of the total area and, as always in broken terrain, there are wide variations in the character and fertility of the soil. In certain sections there are extensive outcrop-

Fig. 9. One of the many new cow herds in Virginia, being used to utilize the improved pastures that have displaced cash crops from these rolling hills. (American Angus Association.)

pings of limestone, as in western Virginia, where the best bluegrass of the country is found. Grass grown on such a soil is exceedingly nutritious in spring and early summer and is prized by cattle grazers. These areas are usually stocked with older steers, which finish satisfactorily during the summer on this grass alone. In other parts of these regions the soil is quite thin. These sections are used mainly by breeding herds or by young stocker cattle that have been purchased in the area. At the end of the grazing season these stocker cattle are either finished on feeds grown on the bottomland or are wintered on hay or silage and held over for another summer.

Despite the predominance of dairy cattle in the Appalachian and Great Lakes areas, beef cattle numbers have increased noticeably there since the mid-1930s. Especially is this true in Kentucky, Tennessee, Virginia, West Virginia, Minnesota, Wisconsin, and Michigan. Many plantations and farms in these states, which formerly produced principally dairy products, truck crops, cash grain crops, or tobacco, are now largely in grass that is being utilized by beef cattle. Many excellent purebred beef herds have been established and are a source of improved breeding stock for the rapidly expanding beef industry of these areas.

The Appalachian and Great Lakes states are well suited for the production of grass because of the rolling topography, making them natural beef cattle areas. Further increases in beef cattle population may be expected there as existing pastures are improved and as more land is reclaimed from scrubby timber or is taken out of cultivation and used for pasture and hay.

THE SOUTHEASTERN AND DELTA AREAS

Before the mid-1930s the Southeast and Delta areas were not important beef-producing areas. As a rule, few cattle were found on each farm and often these were noticeably lacking in type and quality. This is no longer true, although there are several factors that handicap beef production in these areas.

Rainfall is highly variable in this section and short droughts are rather common. The result is a rather unreliable pasture situation. Winter feed supplies, unless provided in the form of winter pastures, are sometimes rather expensive and poor in quality as a consequence of the uncertain

Fig. 10. Many cotton plantations and farms in the Gulf Coast region have been sodded with Coastal Bermuda and other adapted perennial grasses. Brahman cattle or Brahman crosses are popular in this region because of their adaptation to the environment. (National Cottonseed Products Association, Inc.)

moisture supply. Once-productive, tillable soils used for growing cotton and other cash crops are sometimes badly eroded and depleted of soil nutrients. Consequently, either forage production is low or investment in fertilizers is high. Wet winter weather in parts of the area makes grazing of winter oats and tame pastures difficult during some years and adds to the uncertainty of the winter feed supply.

Offsetting these disadvantages are the relatively lower investment in land and buildings, the short, mild winters and long growing seasons, and the multitude of forage and pasture crops that can be grown successfully. The increased production of improved Bermuda grass strains, ryegrass, winter oats, rye, corn, the newer hybrid Sudans, and hybrid grain sorghums of both the forage and grain types, is making these regions competitive in the production of cattle. Proof of the superior adaptation of the Brahman breed and its crosses in the hotter, more humid portions of these areas is responsible for much of the increase in cattle numbers. Application of soil conservation and soil-building practices that were an integral part of government crop control programs, the improvement of pastures, and the production of better-adapted forage crops come in for their share of credit.

THE BEEF CATTLE CYCLE

A study of the fluctuations in cattle population in the United States over a series of years shows that these fluctuations have not been haphazard but have followed a rather regular order. Periods of large cattle numbers have occurred about every 10 to 12 years. Also, periods of unusually low numbers of cattle have been separated by about the same interval (see Fig. 11). Because price is largely a function of supply, it follows that approximately the same amount of time—that is, about 10 to 12 years—intervenes between periods of very high prices or between periods of very low prices. This time interval, called the *cattle cycle*, is much used by agricultural statisticians and market forecasters in predicting the supply and price of cattle in the immediate future.

To paraphrase a livestock marketing specialist's description of the evolution of a typical cattle cycle, it begins with an increased demand for breeding stock to expand herds. Prices of breeding stock soar, making the producing or cow-calf enterprise especially profitable. Cows, yearling heifers, and heifer calves are retained while larger numbers of steers make up total cattle slaughter. Later, when calves from enlarged breeding herds reach maturity, total slaughter increases and prices break, often severely. Declines are sharpest for breeding stock and least for high-grade fed cattle. Thus the producing enterprise becomes relatively unprofitable,

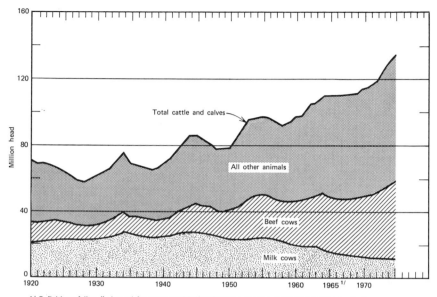

Million head

160

120

80

40

0

Total cattle and calves

All other animals

Beef cows

Milk cows

1920   1930   1940   1950   1960   1965 1/   1970

1/ Definition of "cow" changed from cows and heifers 2 years + to cows and heifers that have calved.

Fig. 11. Cycles in cattle numbers, 1920–1975. It is evident that "peaks" and "lows" in cattle numbers have occurred about every 10 to 12 years. (USDA.)

more cows are slaughtered, and there is a rush to expand the feeding business, using the now cheaper calves and yearlings. As the slaughter of cows and calves increases, cow herds are reduced and the calf crop becomes smaller. Eventually total slaughter decreases and prices turn upward, initiating a new cycle.

Having located our place on the cycle with reference to the last peak or low spot, we need only construct that part of the curve that represents the next 5 or 6 years to see what long-term supply and price changes are likely to occur during that period. However, it should be noted that cattle prices are not entirely determined by the cattle population, but also by the number of cattle marketed and by the consumer demand for beef. Consequently, any event that changes either of these factors has far more effect on cattle prices than does the total number of cattle in the country. For example, the severe droughts experienced in the Western Range and Corn Belt states during 1934 and 1936 resulted in a sharp reduction in cattle numbers after only 5 or 6 years of expansion instead of the usual 7 or 8. The decline was abnormally severe but was of short duration because of the onset of World War II.

Thus the cattle cycle may be greatly disturbed by unforeseen abnormal conditions. It should not be followed rigidly in planning a cattle enterprise on an individual farm but should be regarded as a rough guide. There are indications that the number of years required for the buildup to the peak of the cattle cycle has become less with each cycle—that is, the buildup appears to be occurring at a faster rate than formerly, as may be seen in Fig. 11. This no doubt is due to a number of reasons. Nowadays, heifers are nearly all bred to calve first as 2-year-olds rather than as 3-year-olds as in former years. Cows are being sold at younger ages as a result of more severe culling because of pregnancy-testing and perform-ance-testing. Both these practices would increase the rate at which numbers build up.

Anyone planning to add a beef cow enterprise to his other operations would possibly be better off to delay until the descending or liquidating phase of the cycle is in progress, rather than do the opposite. Cattle feeders could take advantage of the fact that feeder heifers would be relatively cheaper than steers in the liquidating phase and steers would be relatively cheaper in the buildup or ascending phase of the cycle.

These are only a few examples of how an understanding of cattle cycles in general, and of whichever phase of the cycle may exist at a given moment, can be used to advantage. The numbers cycle can be traced back to Civil War times when cattle numbers were first being recorded. There are few serious deviations from the cycle except for the already mentioned tendency for cycles to become shorter with time, and for wars and prolonged droughts to interrupt them.

CHAPTER 2
THE BEEF CATTLE
ENTERPRISE

The rancher usually does not have to be concerned with the question of whether he should be running cattle on his spread. He knows that range livestock, either a breeding herd or young stock or a combination of the two, is his program, and his only major decision, in certain parts of the country, is whether a combination of cattle and sheep, or possibly sheep alone, might be more profitable. The farmer, on the other hand, may first have to wrestle with the problem of whether he should engage in any livestock enterprise at all or whether he should give all his attention to commercial crop production. If he decides to divide his time between grain and animal production, the question of whether to cast his lot with cattle, swine, or sheep becomes a real one.

Beef cattle have some advantages over other kinds of livestock, and some of these advantages are of sufficient importance to warrant analysis and discussion. It should be noted that some of the data to be cited in support of certain statements are based on all cattle, both beef and dairy. Unfortunately, statistics are not always available for an accurate separation of these two types, but certain generalizations can be made with respect to the advantages of beef cattle, as briefly discussed below.

CATTLE AN IMPORTANT SOURCE OF CASH INCOME

Cattlemen operate farms and ranches to generate income. Philosophically it can be said that cattlemen use livestock to convert range forages, pastures, and crops into meat—a wholesome, high-quality, nutritious food for human consumption. But the cattleman's role in the economic picture is likewise not to be overlooked. In the last analysis there must be a profit factor; unless income above costs of operation is generated, the cattleman would have to find gainful employment elsewhere. Data such as those in Table 3 forcefully illustrate the impact that the sale of cattle and calves has on the economy of the United States. Such sales income accounts for 30 percent of all cash farm receipts. In almost half of the states the cattle industry ranks first, and in 46 states it ranks in the upper five, among all farm commodities in income generation. The cash receipts from the sale of cattle and calves exceed receipts from all other livestock enterprises combined.

29

Fig. 12. Purebred Polled Hereford heifers making use of Nebraska Sandhills range. The owner of this grass could have chosen from among several beef cattle programs, but crop farming is somewhat hazardous in this area because of limited rainfall and the type of soil. (Western Livestock Journal.)

CATTLE UTILIZE QUANTITIES OF CROP RESIDUE

Any system of general farming produces a large quantity of coarse, low-grade roughages that have low or no market value. In a typical Corn Belt cash-cropping system using corn, soybeans, and small grains, about 2 tons of roughage are produced for every ton of grain and soybeans harvested.

Table 3

Impact of Livestock Production on the Agricultural Economy of the United States in 1972[a]

	Annual Cash Receipts		Number of States in Which the Indicated Commodity Ranked		
	Millions of Dollars	Percent of Total	First	Second	Among the Top Five
Cattle and calves	18,247	30	23	9	46
Dairy products	7,156	12	11	11	41
Hogs	5,405	9	0	3	22
Eggs	1,818	3	0	2	15
Broilers	1,618	3	2	5	8
Turkeys	537	1	0	0	0
Sheep and lambs	352	0.5	0	0	4
All farm livestock products	35,596	58.5			
Total, all farm commodities	60,671	100.0			

[a] Wedin et al., *Journal of Animal Science* 41:671 (1975).

Fig. 13. Yearling steers on alfalfa-clover-timothy rotation pasture. Beef cattle provide the farmer with a strong incentive for staying with a good rotation cropping system, because cattle often make it possible for pasture crops to produce returns equal to those from cash crops. (University of Illinois.)

The roughage resulting from the corn and soybean crops is unsalable, and the small-grain straw is often hard to dispose of at a worthwhile price, especially if it is damaged by rain or the presence of weeds. All these roughages, properly fed, are good feed for beef cattle and may contribute returns to the farmer when fed or grazed instead of being left to weather and leach away, or even to be burned, before they are plowed under. Fantastically large quantities of corn and sorghum stover are being wasted which, if properly stored and fed or even grazed, would support millions of beef cows and stocker cattle during the winter months. The annual production of straws and stover in the United States is at least 250 million tons. When this amount is compared with the total production of wild and tame hay, namely 100 million tons, the importance of the roughages resulting from grain production is obvious. In the corn and sorghum production areas, where a system of general farming prevails, stover and straw constitute an even greater percentage of the total roughage supply than for the country as a whole, but comparatively little of this roughage is used efficiently in these areas.

A more efficient utilization of all farm products is one of the important problems of the general farmer. With respect to the economic use of these

Table 4

Item	Cattle on Feed (%)	Other Beef Cattle (%)	Dairy Cattle (%)	Swine (%)	Sheep and Goats (%)	Poultry (%)	Other Live-stock (%)
Source of All Feed Utilized by Different Classes of Livestock, 1971[a]							
Concentrates							
Corn	33.6	3.8	18.1	62.3	3.0	45.0	33.1
Sorghum	17.7	0.7	1.2	1.9	0.7	6.9	2.6
Other grains	13.7	1.1	6.4	6.1	1.0	9.6	9.8
High-protein feed	5.6	1.7	5.7	12.9	6.1	30.6	14.6
Other by-product feed	2.4	0.7	4.7	2.0	0.0	7.4	6.3
Total	73.0	8.0	36.1	85.2	10.8	99.5	72.7
Roughages							
Hay	14.2	14.6	22.1	0.0	4.8	0.0	11.7
Silage and stover	8.1	4.6	18.5	0.0	4.2	0.0	0.0
Pasture	4.7	72.8	23.3	14.8	80.0	0.0	15.6
Total	27.0	92.0	63.9	14.8	89.2	0.5	27.3
Combined total	100.0	100.0	100.0	100.0	100.0	100.0	100.0

[a] Compiled from USDA, *Agricultural Statistics,* 1971.

coarse roughages, beef cattle offer a usually satisfactory solution. Mature beef cows can be maintained effectively on rations composed of roughage alone, whereas steers being finished for the market may consume from 10 to 300 percent as much roughage as grain, depending on the level of grain being fed. Illinois studies summarizing the cost account data obtained from a typical cattle feeder in western Illinois for a 7-year period show that the ratio of roughage to concentrates fed was 141 to 100 (including straw used for bedding). In California and Arizona, roughages make up a much smaller portion of the rations fed feeder cattle, but this is usually because roughage is in short supply and high-priced. In other words, the situation is reversed from that in the more traditional feeding areas where roughage supplies are more economical and where, in fact, cattle are often being fed in order to provide a market for the vast quantities of roughage available. Only about 25 percent of the nutrients required in the life cycle of a typical 1,000- to 1,100-pound, grain-fed slaughter beast comes from grain and other concentrates. The remaining nutrients are derived from pasture or range and harvested roughages which have little or no cash value unless converted to beef or other ruminant animal gains.

CATTLE UTILIZE PASTURE CROPS

The importance of pastures and their utilization by beef cattle is taken for granted in the ranching sections of the country, but the importance of pasture crops in the nonranching areas is often overlooked. In the major general farming areas of the country—the Corn Belt and Lake states and portions of the Plains states—grain farming, pastures, and livestock production are highly developed. Beef cattle make it possible for the permanent and rotation pastures to contribute a fair share of income on the farms in this vast area. In other parts of the country real efforts are being made to reclaim and rebuild soils that no longer support cash crops. However, without beef cattle and other ruminants to convert the grass and roughage produced on these lands into income, such reclamation is economically unfeasible.

Pastures necessarily occupy a large fraction of the total farming area in this country. Permanent pastures have by no means disappeared from the Corn Belt and other farming areas, nor is it likely that they ever will, because even in the Corn Belt much of the land is better suited for pasturage than for anything else. Furthermore, farmers have found that a certain area of permanent grassland is almost indispensable to good animal management and over a series of years is likely to prove as profitable per acre as any other part of the farm. In addition to this permanent pasture there are acres of temporary pasture crops such as winter small grains, timothy, red and sweet clover, Sudan grass, and various mixtures of grasses and legumes, mostly grown as a part of the regular farm rotation. It is not surprising, then, that the total pasture area is considerable, even in the major farming states.

Just what percentage of this great pasture acreage is grazed by beef cattle can only be estimated. In some sections, such as central Missouri, southwestern Wisconsin, and southern Iowa, beef cattle are by far the most important kinds of livestock kept. Around Chicago, Omaha, Minneapolis-St. Paul, and other large cities dairy cattle are important. Making due allowance for the pasture used by horses, hogs, sheep, and dairy cattle, about 65 to 75 percent of the total pasture area of the north central states is utilized by beef cattle. This is equivalent to approximately 100 million acres, or 35 percent of the total area in farms.

CATTLE FURNISH A HOME MARKET FOR GRAIN AND HAY

Excepting wheat, comparatively little of the immense production of grain and hay is used other than as feed for farm animals. Most of these feeds

Table 5

Importance of Pastures on Farms and Ranches, by Region or Area, Percentage of Cropland Pastured, and Acreage of Various Types of Permanent Pasture per Farm[a]

Region or Area	Number of Farms (1,000)	Cropland per Farm (acres)	Cropland Used for Pasture (acres)	Cropland Used for Pasture (%)	Permanent Pasture per Farm (acres)	Woodland Pasture per Farm (acres)	Pasture in Federal and State Lands per Farm[b] (acres)	Total Pasture per Farm (acres)
Northeast	111	130	33	25	22	18	9	82
Corn Belt and Lake states	654	200	40	20	20	19	16	90
Northern Plains	214	470	53	11	335	6	11	405
Southeastern	396	150	62	41	55	49	71	237
Southwestern	184	282	95	34	971	97	352	1,515
Mountain states	142	260	36	14	874	44	991	1,930
Pacific states	93	251	40	16	401	58	419	919
United States (48 states)	1,794	232	49	21	250	35	160	494

[a] Compiled from USDA, *Agricultural Statistics*, 1969.
[b] Nondeeded federal and state lands.

Table 6

Percentage of Feeds Consumed by Beef Cattle and Other Classes of Livestock, 1971[a]

Item	Concentrates				Roughages			
	Corn[b]	Sorghum Grain	Other Grain	High-Protein Feed	Hay	Silage and Stover	Pasture	All Other
Cattle on feed	18.5	61.9	33.5	9.7	17.2	17.6	2.0	14.2
Other beef cattle	5.3	6.6	6.6	7.5	44.3	24.8	73.3	35.6
All beef cattle	**23.8**	**68.5**	**40.1**	**17.2**	**61.5**	**42.4**	**75.3**	**49.8**
Dairy cattle	12.2	5.0	18.7	12.5	32.7	53.4	10.5	17.3
Swine	34.6	6.6	15.2	22.7	0.0	0.0	0.6	14.4
Sheep and goats	0.3	0.4	0.4	1.7	1.0	1.4	5.0	2.2
Poultry and turkeys	17.6	17.1	16.7	38.0	0.0	0.0	0.1	10.1
Other livestock	11.5	2.4	8.9	7.9	4.8	2.8	8.5	6.2
Total	100.0	100.0	100.0	100.0	100.0	100.0	100.0	100.0

[a] Compiled from U.S. Department of Agriculture statistical data.
[b] Includes the corn in the silage fed.

are used on the farms where they are produced. Of the quantity sold, the larger portion is bought by feeders who do not raise enough for their own needs. Even in the heart of the Corn Belt more than 75 percent of the grain and hay is fed on the farms where it is grown.

Beef cattle undeniably constitute an important home market for these farm-grown feedstuffs. For the United States as a whole, beef cattle consume about 25 percent of the 5- to 6-billion-bushel annual corn crop. For the Corn Belt section this percentage is much higher, since corn is the principal feed used in the extensive finishing operations prevalent in this region. A still larger percentage, 68.5 in 1971, of the nation's sorghum crop is fed to cattle.

The demand for corn and sorghum for cattle feeding has an effect on the grain market that can hardly be overestimated. It is far more important than the total demand for commercial uses and export combined. Since the livestock industry, including the beef cattle enterprises, provides the principal outlet or market for feed grains, protein concentrates, and salable roughages, it is not surprising that the price of feed crops is closely related to the prices received by feeders and farmers for their livestock. This is conclusively shown in Fig. 14. Only in situations where heavy demand for export purposes may bolster prices above their value for feed do feed prices increase above what they are worth when converted to slaughter livestock. Another exception may be when government price support programs keep the normal supply and demand situation from functioning.

The commercial production of alfalfa in many of the irrigated valleys

Fig. 14. Relationship of prices received for all crops and for livestock and animal products. (USDA.)

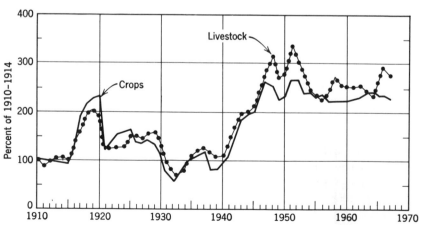

such as those of the Platte River in Nebraska, the Rio Grande in Colorado and New Mexico, or the Imperial Valley in California depends in large part on the cattle-finishing enterprise for its market. Dehydrated alfalfa has become such an important part of most commercially prepared fortified protein concentrates that it can almost be said that as the fat-cattle market goes, so goes the alfalfa market. The commercial production, under irrigation, of corn and forage sorghums, for sale as silage or as green chop to local cattle feeders, is becoming an important enterprise in such areas as Weld County, Colorado, and the Texas-Oklahoma Panhandle and southwestern Kansas, where cattle feeding is more highly concentrated than in any other part of the country. Naturally the price received depends on the profitability of the cattle-feeding enterprise at a given time.

BEEF CATTLE UTILIZE MILLING AND PROCESSING BY-PRODUCTS

The oilseed meals resulting from the processing of soybeans, cottonseed, and flax are the major components of the protein concentrates fed to beef cattle. Wheat bran, blackstrap molasses, beet pulp, citrus pulp, and cottonseed hulls are other important by-products of the milling and processing of farm crops that are used in beef cattle rations.

Farmers receive higher prices for crops from which these by-products are made because of the large quantity of such by-products being consumed by beef cattle and other farm animals. The total by-product production of at least 10 million acres is fed to beef cattle.

BEEF CATTLE REQUIRE A SMALL INVESTMENT IN BUILDINGS AND EQUIPMENT

The average investment in buildings and equipment per hundred dollars invested in livestock is generally lower for beef cattle than for other farm animals. The investment for buildings and other shelter for the cow-calf enterprise may range from no shelter at all to $15 to $50 per cow, depending on the section of the country. The increasing cost of labor has necessitated an increase in the investment in laborsaving equipment such as silo unloaders, auger bunks, self-feeders, and even complete confinement barns, but labor costs are lower for beef cattle than for other livestock enterprises, as will be discussed later.

Greater use of pole-type barns and open sheds instead of conventional barns, or the elimination of barns and sheds entirely, along with cheaper bunker-type or trench silos, is lowering the relative cost of buildings and equipment still further.

Although expensive barns, mechanical feed- and litter-handling equipment, extensive corrals, and the like are convenient and desirable from a laborsaving standpoint, they are not indispensable. Some of the best-known feeders of market-topping steers, and the owners of some of the better purebred beef cattle herds, use equipment that is no more expensive than that found on the average well-kept livestock farm.

BEEF CATTLE UTILIZE LABOR EFFICIENTLY

Beef cattle require little labor compared with dairy cattle and hogs or with cultivated crops occupying an equal area of land. Of the labor that is required, the larger part is needed during the winter and early spring when there is little demand for labor in the fields. Beef cattle thus tend to distribute the labor requirements of the farm throughout the year by (1) utilizing labor that otherwise would be unemployed during the winter and (2) lessening the summer demand for labor by using part of the tillable land for hay and pasture. A farm that would require the employment of an extra man during the spring and summer under a grain system is frequently operated by the owner alone under a livestock system in which beef cattle predominate. Moreover, the cattle furnish many hours of profitable employment for the owner and his tractor during the winter, which without cattle might not be profitably utilized.

Missouri workers found, in a study involving 132 cooperators, that only about 6 hours per cow per year of total labor was required in caring for a farm-sized beef herd. About two-thirds of this labor involved the feeding chores. The remainder was incidental labor such as checking cows, disease and parasite control, sorting, moving, and fence repair. Creep-feeding the calves added about 2 hours per cow. The variation in this study was great, ranging from a low of 3 hours to a high of 13 hours. Purebred operations require at least 50 percent more labor, especially if artificial insemination is used and if a show herd is fitted. These same workers reported that about 6 to 7 hours annually were required per head in feeder cattle programs, and Illinois workers reported that 2 hours or less are required when lots of more than 5,000 head are operated.

Table 7 shows the average monthly labor expenditure on a 295-acre Corn Belt farm where 40 to 79 cattle were fed annually over a 7-year period. Had cattle feeding not been practiced, it is likely that only a fraction of the 647 hours of labor expended during the winter would have gone into remunerative enterprises. The returns realized from this labor, therefore, may be counted as an almost clear gain to the farm income. In

Table 7

Monthly Expenditure of Labor in Feeding Beef Cattle[a]

	Labor Expended	
	For Cattle (hr)	For Total Farm (hr)
January	164	511
February	144	496
March	132	718
April	79	916
May	31	1,000
June	—	1,133
July	—	1,370
August	—	889
September	1	934
October	13	815
November	78	825
December	129	648
Total labor per year	771	10,255
Labor November 1 to April 1	647	3,198
Percent of labor coming in fall and winter	83.8	31.2

[a] Illinois Bulletin 261.

large commercial feedlots it is not uncommon for the labor requirement to be no more than one man per 2,000-head capacity.

Data such as those shown in Table 8 can be used to compare the various livestock enterprises on family-sized farms with respect to the magnitude of the various production cost items. Labor, building, and equipment costs for beef cattle tend to be relatively low, but interest and depreciation costs are high because of the greater investment in each animal unit.

DEATH LOSS IS LOW IN BEEF CATTLE

Beef cattle, unlike most other livestock, are subject to few ailments and diseases that are likely to result in death. Compared with sheep and hogs, beef cattle have an unusually low death rate. No beef cattle disease is at all comparable with hog cholera in mortality rate, nor are cattle an easy prey to internal parasites, as are sheep. Losses among calves and yearlings are, of course, higher than for mature cattle, but even here they are far below those for young pigs and lambs. Predator losses are certainly lower than they are for sheep.

Beef cattle are not, as a rule, seriously injured by improper methods of feeding. They do not go off feed easily, and usually show rapid recovery

Table 8

Percentage Distribution of Livestock Production Costs[a]

Livestock Enterprise	Feed	Labor	Buildings, Power, and Equipment	Livestock Expenses[b]	Taxes, Insurance, Interest	Miscellaneous
Dairy cattle (grade A)	45	30	9	5	9	2
Beef cow herds (calf sold)	41	23	6	10	18	2
Feeder cattle (steer calf)	61	15	6	6	10	2
Feeder lambs	56	14	6	10	12	2
Sheep flocks	48	26	7	8	9	2
Sow (2 litters raised)	68	11	7	7	5	2
Feeder pigs (bought)	72	9	5	8	4	2
Commercial laying flock (100)	58	28	7	1	5	1
Broiler hatching flock (100)	58	25	7	3	5	2
Broilers (1,000)	71	10	8	4	5	2
Turkeys (1,000)	73	10	5	2	8	2

[a] Adapted from USDA Technical Bulletin 1037.

[b] Includes veterinary and medicine costs, death losses, breeding fees, and marketing charges.

from the effects of improper rations or poor management methods. Two important exceptions are bloat and founder, discussed in other chapters. These points are of special importance to the ambitious young farmer whose experience with livestock is limited.

BEEF IS THE MOST POPULAR MEAT

A continuing increase in beef consumption is ample evidence of the overwhelming popularity of beef. Table 9, which shows consumption trends for meats and poultry, indicates that people are eating more total red meats than ever before and that the consumption of poultry has also increased. Even though dressed poultry usually sells for from one-third to one-half the price of beef per pound, it is not being used as a substitute for beef. The average consumer in California uses about 130 pounds of beef per year, suggesting that the ceiling on total beef consumption has not yet been reached in the United States. The average beef consumption in several other countries substantially exceeds that of the United States, as shown in Table 10, further illustrating the potential of the demand for beef in America.

VALUABLE AND ESSENTIAL BY-PRODUCTS FROM BEEF SLAUGHTER

The estimated value of hide and offal of all cattle slaughtered ranges from $2 to $4 per hundredweight of live cattle slaughtered. This amounts to

Table 9

Per Capita Meat Consumption in the United States[a]

	Pounds Consumed per Person					
Year	Beef	Veal	Lamb and Mutton	Pork	Total Red Meat	Poultry[b]
1910–1919 average	63.0	6.7	6.2	64.0	139.9	14.4
1950	63.4	8.0	4.0	69.2	144.6	24.7
1955	82.0	9.4	4.6	66.8	162.8	26.3
1960	85.2	6.2	4.8	65.2	161.4	34.3
1965	100.5	5.2	3.7	58.6	168.0	40.3
1970	113.7	2.9	3.3	66.4	186.3	49.7
1975[c]	119.5	4.0	2.0	55.0	180.5	48.5

[a] USDA-ERS.

[b] Not considered to be red meat, but shown for comparison purposes.

[c] Preliminary.

Table 10

	Worldwide Per Capita Meat Consumption (1964)[a]			
Country	Beef and Veal (lb)	Pork (lb)	Lamb and Mutton (lb)	Total (lb)
New Zealand	95	36	95	226
Uruguay	167	19	35	221
Australia	108	23	88	219
Argentina	149	17	11	177
United States	106	65	4	175
Canada	89	55	3	147
France	70	61	6	142
USSR	30	20	8	60
Mexico	24	12	3	39
Japan	5	7	2	15

[a] USDA, *Livestock and Meat Situation*, 1966.

about 8 to 10 percent of the cost of the cattle to the packer and, although costs are involved in the collection and processing of the by-products, much of this value is passed back to the cattle producer. The hide accounts for over half of this value, varying, of course, with the domestic and foreign demand for leather.

Medical scientists rely heavily on the by-products of beef slaughter. Many important pharmaceuticals have as yet not been synthesized commercially; thus natural sources must be relied upon. Insulin, required by an estimated 1 out of 20 Americans in the treatment of diabetes, is a prime example. Cortisone, prepared from bovine gall bladders, is used to treat arthritis. Still other pharmaceuticals derived from cattle slaughter include parathyroid extract, thyroxin, and adrenal cortex extract, all of which are in short supply. The list of useful by-products is literally endless. Even paunch contents and blood meal are being recycled as ration ingredients for cattle and other species of domestic livestock. Unfortunately, however, many by-products go unutilized in small local slaughter plants because of the relatively small quantities of product involved.

BEEF CATTLE MANURE IS A VALUABLE BY-PRODUCT

One of the important advantages enjoyed by the farmer who markets his crops through animals is the conservation of the fertility of his soil. Animals retain only a small part of the plant food elements contained in the feeds consumed, returning the greater part to the soil in the manure produced. (In beef cattle, for instance, the retention values for nitrogen, phosphorus, potassium, and organic matter are 25, 15, 10, and 70 percent respectively.) As a result the livestock farmer is able to maintain his land in a high state of fertility with the use of less—and ever higher-priced—commercial fertilizer than is needed by the farmer who sells his grain outright. Improved soil structure on cropped land in continuous corn or sorghum production is said by most soil scientists to be a very important advantage of selling grain production through livestock.

An estimated 1.5 billion tons of manure are produced annually by the livestock on the farms and ranches of the United States. The value of this manure in terms of the increased crop yield that would result if it were completely recovered and carefully used is enormous. The amount of manure produced by feeder cattle of various ages is shown in Table 11.

The nutrient composition and digestibility of the ration fed can cause the manure to vary, both as to quantity and fertilizer-element composition, but the average estimated values shown in Table 12 can be used to predict the value of the manure produced by a herd of cows or a drove of feeders. Obviously, for the manure to be of greatest value it must be properly collected and stored and must be spread and incorporated into the soil in a manner that prevents fertility loss. The distance manure must be hauled for spreading, the cost of truck or tractor fuel, and cost of

Table 11

Manure Obtained from Cattle Fed on Paved Floor in Open Shed and Adjoining Paved Lot[a]

			Average Manure per Head	
	Number of Lots Averaged	Average Days Fed	Total Period (tons)	Per Month (tons)
Calves				
Full-fed with silage	29	229	1.82	0.23
Full-fed with dry roughage	10	235	2.18	0.30
Full-fed with ear corn silage	8	231	2.16	0.28
Wintered without grain	5	132	1.59	0.36
Yearlings				
Fed over 140 days	4	150	2.24	0.45
Fed under 101 days	11	91	1.52	0.51
Dry beef cows	4	134	2.91	0.66

[a] Illinois Agricultural Experiment Station, mimeographed reports.

Table 12

Quantity and Quality Characteristics of Cattle Manure[a]

Source of Manure	Size of Animal (lb)	Manure Daily (lb)	Manure Daily (cu ft)[b]	Manure Daily (gal)[b]	Nitrogen (daily)	Nitrogen (yearly)	Phosphorus (daily)	Phosphorus (yearly)	Potassium (daily)	Potassium (yearly)
Feeder cattle	500	30	0.50	3.8	0.12	62	0.056	20	0.12	44
	750	45	0.75	5.6	0.26	93	0.084	30	0.19	66
	1,000	60	1.00	7.5	0.34	124	0.110	40	0.24	88
	1,250	75	1.20	9.4	0.43	155	0.140	50	0.31	110
Beef cow	1,000	63	1.05	7.9	0.36	131	0.120	44	0.26	95

[a] Derived from *Livestock Waste Facilities Handbook, 1975*, Midwest Plan Service, Iowa State University.

[b] Based on estimates of 88.4 percent water content and 60 pounds density per square foot.

commercial fertilizers that might be saved by substituting manure are items that determine the feasibility of using manure from commercial feedlots as fertilizer replacements. With commercial fertilizers costing $40 per acre for sorghum production on the irrigated High Plains of West Texas, it was estimated that cattle feedlot manure could profitably be hauled 40 miles if it could be obtained without cost at the feedlot, and 10 miles if it cost $2.25 per ton at the feedlot.

DISADVANTAGES OF THE BEEF CATTLE ENTERPRISE

The disadvantages of the beef enterprise should also be listed and discussed. Fortunately the list is short, but knowledge concerning these points is important in the choice of a farm or ranch enterprise.

A SPECULATIVE RISK IS INVOLVED IN STOCKER AND FEEDER PROGRAMS

In periods of rapidly declining prices there is a real possibility that purchased calves will sell 6 months or a year later, as feeder yearlings or as finished steers, for considerably less per hundredweight than the original cost. This of course applies only directly to the original weight bought, and this situation occurs only occasionally, but the possibility must be considered. In some of the finishing programs, especially those involving either heavy or low-grade feeders, the selling price may also be below the actual cost of the gains—that is, after conversion to cattle gains the feeds fed to such cattle may sell for less than their market value. It is even possible, as in 1973, that both these negative factors may occur at the same time, with disastrous results.

ANNUAL OUTLAY OF CAPITAL IS HIGH IN STOCKER AND FINISHING PROGRAMS

Most cattle feeders must buy their stockers or feeders and, since such cattle often remain on the farm or in the feedlot for 6 months to a year, it is necessary annually to invest as much as one-third to two-thirds of the final value of the finished cattle, aside from the investment in feed. Commercial feedlots may feed two or three droves per year; hence the interest charges on cattle on feed go on throughout the year. This requires a rather large supply of either available cash or credit. In any case, interest charges on this large investment must be reckoned with.

BEEF CATTLE PROGRAMS REQUIRE LABOR AND MANAGEMENT OF ABOVE-AVERAGE QUALITY

In order to reduce costs and losses from disease and death, and in order to utilize the latest research findings in the feeding and management of beef cattle, skilled labor and scientific knowledge are required. This is possibly even more true if the buying and selling of cattle are not delegated to a commission man or order buyer especially trained for the job. Knowing when fed cattle are ready to sell to best advantage will prevent the costly gains that result when cattle are held beyond the point where they are finished for their grade. Equally important is the ability to choose the proper feeding program to utilize the available feed supply, and to adapt the program to take advantage of seasonal trends in supply of feed and both feeders and fed cattle. The importance of such decisions means that exceptional managerial ability is essential to success. The alternative is to leave these managerial decisions to professional help, which is usually available on some sort of fee basis.

BEEF CATTLE PRODUCTION IS A YEAR-ROUND OPERATION

Except for a few short-term feeding programs, the production of beef cattle is a confining enterprise. For best results, daily inspection of cattle is desirable, and, of course, in practically all of the finishing programs daily chore work is necessary. True, the total hours required may be low, especially if laborsaving feeding systems are used, but still someone must be on hand. In ranching areas, where pastures are large and sometimes hard to reach, daily inspection is unfeasible and yet the cattle cannot be long neglected. Lengthy vacations and frequent weekend absences are not conducive to success with beef cattle.

EXTRA CAPITAL OUTLAY FOR POLLUTION CONTROL

It is interesting that, as cattle numbers on a farm or ranch are increased to take advantage of economies of scale and hence to increase efficiency through intensification, a problem has developed that must be considered a disadvantage for the beef enterprise. Accumulated cattle manure or waste imposes a potential environmental pollution hazard. Odors, flies, and hazardous runoff, and even the public nuisance aspects such as added traffic, noises, dust, and heavy bird and rodent populations, are all items that no doubt must be dealt with by cattlemen who concentrate cattle near population centers. Compared with some other agricultural uses of land— for example, production of cash grain and soybeans—the disadvantage of the added costs of environmental pollution control on intensive beef cattle farms or ranches is a formidable one indeed. Swine, poultry, and dairy farmers may be at an even greater disadvantage in this respect. The subject is dealt with further in a later chapter.

PART II
PRINCIPLES OF BREEDING, REPRODUCTION, AND FEEDING

CHAPTER 3
PRINCIPLES OF SELECTION FOR BEEF CATTLE IMPROVEMENT
LARRY V. CUNDIFF AND KEITH E. GREGORY[1]

The beef cattle industry in the United States is composed of several segments: (1) the purebred breeder or seedstock producer, (2) the commercial producer, (3) the feeder, (4) the packer, and (5) the retailer.

The purebred breeder of beef cattle maintains seedstock herds to provide bulls for the commercial producer. The commercial producer provides feeder stock to the feeder, who in turn provides the packer with finished beef cattle ready for slaughter. The packer slaughters the cattle and provides the retailer with either dressed carcasses or wholesale cuts from these carcasses. The retailer breaks down the dressed carcasses or wholesale cuts into retail cuts, trimmed and packaged suitably for his customers, the consumers.

There is an interdependence among these segments because each affects cost of production or desirability of product, or both. Both desirability and price of product are reflected in changes in consumption or use. Level of consumption is important to all segments. Consumption of beef depends primarily on how much it costs the consumers relative to other food items and on how well they like it. The profits to be realized by all segments of the beef cattle industry depend on continued improvement in both productive efficiency and carcass desirability.

Only traits that contribute to productive efficiency and carcass desirability are of economic importance to the beef cattle industry. These traits, frequently referred to as performance traits, are (1) reproductive performance or fertility, (2) mothering or nursing ability, (3) rate of gain, (4) economy of gain, (5) longevity, and (6) carcass merit.

THE PUREBRED BREEDER'S RESPONSIBILITY

The opportunity for genetically improving efficiency of production and desirability of product rests in the hands of purebred breeders and

[1] Larry V. Cundiff is Research Leader, Genetics and Breeding, and Keith E. Gregory is Director, U.S. Meat Animal Research Center, Agricultural Research Service, U.S. Department of Agriculture, Clay Center, Nebraska.

commercial producers who determine the matings made to produce beef and replenish breeding stock in our beef cattle populations. Improved breeding practices based on selection in purebred herds that provide seedstock to commercial producers who practice systematic crossbreeding will contribute substantially to increased productive efficiency and desirability of product.

Selection is the primary tool available to purebred breeders or seedstock producers for making genetic improvement. Most of the opportunity for selection in beef cattle is among bulls. In addition to benefits from systematic crossbreeding, level of performance in commercial beef cattle populations is determined primarily by the bulls available to commercial herds from the purebred segment of the industry. To fulfill his responsibility to the other segments of the beef cattle industry, the purebred breeder or seedstock producer should have a working knowledge of genetics or the science of heredity, along with an appreciation of all traits of economic importance to the industry. In addition, he should understand the procedures for measuring or evaluating differences in these traits and be able to develop effective breeding practices for making genetic improvement in them.

THE BASIS FOR GENETIC IMPROVEMENT

Differences among animals result from the hereditary or genetic differences transmitted by their parents and the environmental differences in which they are developed. With minor exceptions, each animal receives half its inheritance from its sire and half from its dam. The units of inheritance are known as *genes* and are carried on threadlike material, present in all cells of the body, called *chromosomes*. Cattle have 30 pairs of chromosomes. The chromosomes and genes are paired, each gene being at a particular place on a specific chromosome pair. There are thousands of pairs of genes in each animal, and one member of each pair comes from each parent. All cells in an animal's body have essentially the same makeup of chromosomes and genes.

The ovaries of females and the testicles of males produce the reproductive cells, which contain only one member of each chromosome pair, and it is purely a matter of chance which gene from each pair goes to each reproductive cell. In this halving process a sample half of each parent's inheritance goes to each reproductive cell, meaning that the genetic potentialities of an individual are determined at fertilization. The pairing of chromosomes restores the full complement when a reproductive cell from the male fertilizes a reproductive cell from the female, keeping the number of chromosomes constant over countless generations. Since the half of each of its parents' inheritance that each reproductive cell receives is strictly a matter of chance, some reproductive cells will contain more desirable genes for economically important traits than will others. This results in a superior individual and offers the opportunity for selection. The chance segregation in the production of reproductive cells and recombination upon fertilization is the cause of genetic differences among offspring of the same parents.

The genetic merit of a large number of offspring will average that of their parents. However, some individuals will be genetically superior to the average of their parents and an approximately equal number will be inferior. Those that are superior provide the opportunity for selection and genetic improvement. The basis for genetic improvement is differential reproduction, which is accomplished by permitting some animals to leave a greater number of offspring than others or some to leave offspring while others do not. This is what happens when selection is practiced.

Genes vary greatly in their effects. Some traits are controlled primarily by a single pair of genes, whereas other traits are affected by many genes. Examples of traits controlled primarily by a single pair of genes are dwarfism and color. Most of the economically important traits—carcass characteristics, growth rate, feed efficiency, and mothering ability—are affected by many genes. The thousands of genes present make countless combinations possible in any animal. Genes are too small to be individually identified, and their presence is evident only in outward effects such as differences in growth rate, feed efficiency, and conformation.

In traits controlled by a single pair of genes, one member of the pair must be dominant. The *dominant gene* has the capacity for covering or masking the effect of the other member of the pair, which is referred to as the *recessive gene*. For example, the gene for polled is dominant and masks the gene for horns when both are present. In another example, the gene for dwarfism is recessive to the gene for normal appearance. Thus, if N represents the gene for normal appearance and n represents the gene for dwarfism, individuals with the "genetic makeup" of NN and Nn are normal in appearance, but Nn individuals carry the gene for dwarfism and transmit this gene to approximately half their offspring. Dwarfs (nn individuals) can result from mating normal-appearing parents if each carries the gene for dwarfism ($Nn \times Nn$). Mating normal-appearing individuals that carry the dwarf gene ($Nn \times Nn$) results in noncarriers (NN), carriers (Nn), and dwarfs (nn) in a 1:2:1 ratio.

Among animals, all differences that are not genetic are classified as environmental. Even though every attempt may be made to provide a uniform environment, there are still random environmental differences among animals. For example, identical twins are exactly alike in their genetic makeup but differ in their performance because of random or chance environmental differences. All animals are not at exactly the same place at the same time, grazing the same area, and exposed to the same environmental elements. Some members of a group may contact infectious organisms while others do not. Another example might be injury to the udder of a cow, which would reduce her milk production and result in decreased weaning weight of her calf. There are many random environmental factors that may affect some members of a group and not others and thus affect the, expression of differences in economically important traits.

GENE FREQUENCY

The objective of selection for any performance trait is to increase in the population the number or frequency of desirable genes affecting that trait. This is accomplished by selecting animals that are above the herd average in genetic merit.

Differential reproduction is the basis for change in gene frequency and genetic improvement. Culling animals that are poor in economically important traits reduces the frequency of undesirable genes in a herd if the culled animals are replaced by animals that are superior in those traits and thus have a high percentage of desirable genes. Differential reproduction is the basis for continuous improvement in livestock, for the increase of desirable genes in one generation is added to those of the previous generation and the improvement tends to be permanent.

Gene frequency refers to the percentage of the available locations that a particular gene occupies in a herd or population. Since genes are paired in each animal, gene frequency includes both members of each pair and ranges from 0.00 to 1.00. For example, if a herd is free of dwarfism (*NN*), frequency of the dwarf gene in the herd is 0.00 and frequency of the gene for normal condition is 1.00—that is, it occupies every potential location. Conversely, in a herd of dwarfs (*nn*), frequency of the dwarf gene is 1.00 and frequency of the gene for normal condition is 0.00. In a herd where all animals carry the dwarf gene (*Nn*), the frequency is 0.5 for both genes; thus, combined frequencies of both members of a gene pair are 1.00.

KINDS OF GENETIC VARIATION

Genetic variation is caused by either additive or nonadditive effects of genes. When genes produce their effects in a manner comparable to adding block upon block, as in construction of a building, their effects are referred to as *additive gene effects*. It is the additive effects of genes affecting a trait that combine to determine an animal's *breeding value* for a specific trait. Parents transmit a sample of one-half of their genes to their offspring; therefore the average performance of an animal's progeny measures half of his breeding value relative to other progeny groups in a contemporary environment. The result of selection is to increase the frequency of desirable genes that produce additive effects. The proportion of total variation (genetic and environmental) due to additive gene effects is called heritability.

Nonadditive gene effects are caused by interaction of genes. These occur when specific pairs or combinations of genes produce favorable effects as a result of being present together in the individual. When specific pairs of genes produce a favorable effect, the result is referred to as *heterosis* or *"hybrid vigor."* Within a breed parents cannot consistently transmit these effects to their offspring because only half of their genes, one of each pair, is passed on to the next generation. However, systematic mating procedures involving different breeds can be used to restore favorable nonadditive genetic effects from one generation to another.

Traits vary in the degree to which they are controlled by these two kinds of genetic variation. For traits where most of the genetic variation is additive due to differences in breeding value and where environmental variation is relatively low, selection between and within breeds will be effective. For traits where most of the genetic variation is nonadditive, selection based on individual performance will be relatively ineffective. For the latter type of trait, the breeding program must be designed to make use of specific crosses that produce favorable gene combinations. This involves crossing lines or breeds to obtain favorable combinations of genes for the expression of these traits.

When both types of genetic variation are important, additive for some traits and nonadditive for others, then genetic improvement can be maximized by combining selection in purebred herds that provide seed-stock to commercial producers who practice systematic crossbreeding. A knowledge of the relative amounts of additive and nonadditive genetic variation that affect each economically important trait is fundamental to the development of an effective breeding program.

FACTORS AFFECTING RATE OF IMPROVEMENT FROM SELECTION

The factors affecting rate of improvement from selection are (1) heritability, (2) selection differential, (3) genetic association among the traits, and (4) generation interval.

HERITABILITY

Heritability is the proportion of the differences between animals, measured or observed, that are transmitted to the offspring. Thus it is the proportion of the total variation that is due to additive gene effects. The higher the heritability for any trait, the greater the rate of genetic improvement or the more effective selection will be for that trait. For traits of equal economic value, those with high heritability should receive more attention in selection than those with low heritability. Every attempt should be made to provide all animals from which selections are made with as nearly the same environment as possible, which will result in a larger proportion of the observed differences among individuals being genetic and will increase the effectiveness of selection. It is important to adjust for known environmental differences before making selections if the environmental factors can be evaluated. Adjustments can be made for differences in age, age of dam, season of calving, sex, and creep-feeding.

The average heritability estimates for some of the economically impor-

Table 13

Heritability Estimates of Some Economically Important Traits

Trait	Heritability (%)
Calving interval (fertility)	10
Birth weight	40
Weaning weight	30
Cow maternal ability	40
Feedlot gain	45
Pasture gain	30
Efficiency of gain	40
Final feedlot weight	60
Conformation score	
Weaning	25
Slaughter	40
Carcass traits	
Carcass grade	40
Rib-eye area	70
Tenderness	60
Fat thickness	45
Retail product (%)	30
Retail product (lb)	65
Cancer-eye susceptibility	30

tant traits of beef cattle are presented in Table 13. Of the total difference between the selected individuals and the average of the population from which they were selected, the percentage indicated in the table for each trait is actually transmitted to the offspring. For example, if the selected bulls and heifers were 30 pounds above herd average in weaning weight (selection differential), their progeny would be expected to average 9 pounds heavier than if no selection had been practiced for this trait (30% × 30 = 9).

These heritability estimates were obtained under carefully controlled environmental conditions from a large number of research herds, and adjustments were made for known major environmental sources of variation. The heritability of any trait can be expected to vary slightly in different herds, depending on the genetic variability present and the uniformity of environment. However, estimates from different research herds have been reasonably consistent. The heritability estimates in Table 13 probably represent average expectations for many herds, provided that the general environment is similar for all cattle within the herd. These estimates indicate that selection should be reasonably effective for most performance traits. But since these traits vary both in heritability and economic importance, the rate of improvement in them and the emphasis that they should receive will also vary considerably.

SELECTION DIFFERENTIAL

Selection differential is the difference between the selected individuals and the average of all animals from which they were selected. Selection differential is determined by the proportion of progeny needed for replacements, the number of traits considered in selection, and the differences that exist among the animals in a herd. If the average weaning weight of a herd is 450 pounds, and the individuals retained for breeding average 480 pounds, the selection differential is 30 pounds.

In beef cattle there are some rather severe limitations on selection differentials possible for the various traits. The relatively low reproductive rate of beef cattle usually necessitates keeping approximately 40 percent of the females for replacements to maintain the herd, and an even higher percentage to expand it. Most of the opportunity for selection is among the bulls because a smaller percentage of the bulls must be saved for replacement. Increasing the number of traits selected for, reduces the opportunity for selection for any one trait, making it important to select only for those traits of economic value that are heritable. Every effort should be made to obtain the maximum selection differentials possible for the traits of greatest economic importance and of highest heritability, ignoring traits that have little or no bearing on either efficiency of production or desirability of product.

GENETIC ASSOCIATION AMONG TRAITS

A genetic correlation among traits results when genes favorable for the expression of one trait tend to be either favorable or unfavorable for the expression of another trait. Genetic correlations may be either positive or negative. If the association is favorable, the rate of improvement in total merit is increased; conversely, if a genetic antagonism exists among traits, the rate of improvement from selection is reduced.

Available information indicates a favorable association between rate and efficiency of gain during postweaning growth periods from 7 to about 18 months of age. A major unfavorable genetic association that has been reported in beef cattle is a positive genetic correlation between outside fat thickness and marbling score. This means that when marbling, an important determinant of carcass grade, is selected for, excessive outside fat also may result. Estimates of genetic correlation between birth weight, preweaning gain, postweaning gain to 12 or 18 months of age, and eventual mature size have been high. Implications of these and other genetic associations will be discussed in a subsequent section on major performance traits of beef cattle.

GENERATION INTERVAL

The fourth major factor that influences rate of improvement from selection is the *generation interval*—that is, the average age of all parents when their progeny are born. Generation interval averages approximately 4.5 to 6 years in most beef cattle herds.

The progress made per generation in any trait is equal to the superiority of the selected individuals above the population average from which they came (selection differential), multiplied by the heritability of the trait. This can be put on a yearly basis by dividing by the average length of generation. For example:

$$\text{annual progress for a trait} = \frac{\text{heritability} \times \text{selection differential}}{\text{generation interval}}$$

If heritability of yearling weight is 50 percent, if the selected individuals (males and females) are 50 pounds heavier than the average of all animals, and if the generation interval is 5 years, then the rate of improvement per year in yearling weight would be (0.50 × 50)/5, or 5 pounds. It is evident that progress can be greater when the generation interval is shortened, which can be accomplished by vigorous culling of cows on the basis of production, but it also means that the herd would contain more young or nonproductive-aged cattle.

METHODS OF SELECTION

Selection may be based on (1) pedigree information, (2) individual performance information (mass selection), (3) family performance information, (4) progeny test, or (5) a combination of all four.

Pedigree information is most useful in selecting among young animals before their own performance or their progeny's performance is known. Pedigree information also may be used in selecting for characters that are measured late in life, such as longevity and resistance to cancer eye, or when selecting for traits expressed only in one sex, such as mothering or nursing ability (selecting bulls that are progeny of cows that have produced calves with a high average weaning weight). Pedigree information also is useful in selecting against defects such as dwarfism. When pedigree information is used, only the closest relatives should receive much consideration, since the most distant relatives can influence the heredity of the individual only through the close relatives (sire and dam). Pedigree information should be given less attention after information on an individual's own performance is available, especially in traits of higher heritability. When information on progeny is available, pedigree information is of even lesser value.

Selection on an individual's own performance (mass selection) will result in most rapid improvement when heritabilities are high. An example of such a trait is growth rate. The advantage of selecting on individual performance is that it permits a rapid turnover of generations, or shortens the generation interval.

Use of progeny test information results in the most accurate selection if the progeny test is adequate. Progeny tests are most needed in selecting for carcass traits (if good indicators are not available in the live animal), for sex-limited traits such as mothering ability (where individual performance information is not available on bulls), and for traits with low heritabilities. Progeny test information is more accurate than pedigree information and individual performance, provided the progeny test is extensive. Disadvantages are the less intense culling possible because of the small proportion of animals that can be adequately progeny-tested, the longer generation interval required to obtain progeny test information, and the decreased accuracy as compared to individual performance if not enough progeny are tested or if they are improperly evaluated.

If progeny test information is used, it is essential that the test be designed so that the results can be properly evaluated. The purpose of a progeny test is to obtain the best estimate of the relative genetic merit of the bulls being tested. Therefore it is necessary that cows be assigned to the bulls at random. Selection of cows for a particular bull tends to "stack the cards" either for or against him. It is recommended that cows be classified by their age and line of breeding before breeding assignments are made. Cows should then be assigned to breeding groups at random within line of breeding and age. The sires being progeny-tested can then be assigned at random to the different breeding groups. It is also important for the cows and progeny to be fed and managed uniformly until all the data are obtained. If these precautions are not taken, the progeny test results will be biased, and accuracy of selection can be seriously reduced.

When progeny test information is available, the same data can be used to select between individual progeny by different sires or between members of different half-sib families (half-brothers and -sisters). Effectiveness of selection based on half-sib family performance depends to a large extent on the intensity of selection possible, which is determined by the number of families that can be compared. Half-sib family selection is less accurate than selection based on progeny testing, but it has an advantage of permitting more rapid turnover of generations. This advantage is important for traits that cannot be measured on individuals before they begin reproduction. Thus half-sib family selection is especially useful for traits such as maternal performance and reproduction.

All four types of information should be used in selecting beef cattle. A good policy is to make initial selections on the basis of pedigree, individual performance, and half-sib family information and to determine the extent a bull or cow is used in a herd in later years on the basis of progeny test information. In traits with high heritability, individual performance should receive nearly all of the emphasis in selection.

ESTIMATED BREEDING VALUE

Breeding value is the value of an animal as a breeder determined by the average additive effect of all genes the animal possesses affecting a particular trait. It is most easily visualized as twice the transmitting ability of an individual—that is, as twice the difference between average performance of a very large number of progeny by a sire (or dam) and the population average when the sire (or dam) and other sires (or dams) are mated at random to cows (or bulls) in the population and all progeny are fed and managed alike. The difference is doubled because parents transmit only a sample one-half (one gene of each pair) of their genes to their offspring.

Breeding value also can be estimated from an individual's own performance relative to the average performance of contemporary animals in the population. The difference between an individual (I) and the average of contemporaries (C), multiplied by heritability (H), is the estimated breeding value (B); that is, $B = H(I - C)$. Similarly, breeding value of an individual can be estimated on the basis of information available on relatives such as the sire or dam, paternal or maternal half-sibs, or progeny by also taking into account their relationship to the individual and, in the case of half-sibs or progeny, the number of relatives.

Table 14 provides information on the accuracy of selection on the basis of each method of selection. Accuracy of selection depends on the correlation between estimated and actual breeding value. A correlation of 1.0 would reflect complete accuracy between estimated breeding value and actual breeding value. Accuracy of selection for each method is determined by the genetic relationship between the individual and the relative involved (for example, parent, half-sibs, or progeny), heritability of the trait and, in the case of half-sib and progeny information, the number of relatives with available unbiased records.

An individual's own performance, if available, is always the best single record that can be used to estimate the individual's breeding value because the genetic relationship is maximal (1.0). A record on the sire or dam is only half as valuable as an individual record because each parent transmits a sample one-half of its genes to the offspring. Therefore the genetic relationship between parent and offspring is 0.5. The genetic relationship between an individual (for example, sire) and one of his progeny is also 0.5. However, the individual transmits a different sample one-half of his genes to each of his offspring. Therefore, as the number of progeny increases, the accuracy of progeny-testing increases until, with a very large number of progeny, the correlation between actual breeding value of the sire and estimated breeding value based on average performance of progeny approaches 1.0. The same principle operates to increase the accuracy of half-sib family selection as family size increases. However, if the same number of records is available on each type of relative, then selection based on half-sibs is only half as accurate as selection based on progeny tests because the genetic relationship is just half as great (0.25 versus 0.50).

Comparative accuracies of each method of selection relative to individual selection (for example, comparative accuracy of progeny test = accuracy of progeny test/accuracy of individual performance) are also

Table 14

Accuracy and Comparative Accuracy (Relative to Individual Performance) of Selection Based on Pedigree Information, Half-Sibs and Progeny Tests

Type of Relative	Genetic Relationship	Number of Records	Accuracy[a] Heritability			Comparative Accuracy[b] Heritability		
			0.10	0.30	0.50	0.10	0.30	0.50
Individual	1.00	1	0.32	0.55	0.71	1.00	1.00	1.00
Sire (or dam)	0.50	1	0.16	0.27	0.35	0.50	0.50	0.50
Half-sibs	0.25	1	0.08	0.14	0.18	0.25	0.25	0.25
(paternal or maternal)		2	0.11	0.19	0.24	0.35	0.34	0.33
		4	0.15	0.25	0.30	0.48	0.45	0.43
		10	0.23	0.34	0.38	0.71	0.61	0.54
		20	0.29	0.39	0.43	0.92	0.72	0.61
		40	0.36	0.43	0.46	1.13	0.80	0.65
		100	0.42	0.47	0.48	1.34	0.86	0.69
Progeny	0.50	1	0.16	0.27	0.35	0.50	0.50	0.50
		2	0.22	0.37	0.47	0.70	0.68	0.67
		4	0.30	0.50	0.60	0.96	0.90	0.85
		6	0.36	0.57	0.68	1.15	1.04	0.96
		8	0.41	0.63	0.73	1.30	1.14	1.03
		10	0.45	0.67	0.77	1.43	1.22	1.08
		20	0.58	0.79	0.86	1.84	1.43	1.22
		40	0.71	0.87	0.92	2.25	1.60	1.30
		100	0.85	0.94	0.97	2.68	1.72	1.37

[a] Accuracy is the correlation between estimated and actual breeding value.

[b] Comparative accuracy is the accuracy of selection based on a particular type of relative expressed as a ratio to individual selection (for example, accuracy of progeny test/accuracy of individual performance).

shown in Table 14. About six progeny are required to make progeny test selection approximately as accurate as individual performance selection at the three levels of heritability. For example, when heritabilities are 0.10, 0.30, and 0.50, progeny test selection with six progeny per sire have comparative accuracies of 1.15, 1.04, and 0.96, respectively, relative to selection based on individual performance information. It is apparent that selections based on half-sib performance or on progeny tests have greater relative value for traits with low heritability.

Information from all sources including the individual, sire, dam, paternal half-sibs, maternal half-sibs, and progeny, if available, can be combined into a single estimate of breeding value for each animal in a herd. The information needed to do this is as follows:

1. An individual's own performance expressed as a deviation or ratio from the average of his contemporary average.
2. The same as (1) except for the individual's sire or dam.
3. The average performance of an individual's paternal half-sibs (that is, half-brothers and -sisters by the same sire) expressed as a difference or ratio deviation from the average performance of all individuals in the herd and the number of half-sibs. The individual's own record should be excluded from the average for the paternal half-sib group.
4. The same as (3) except for maternal half-sibs (that is, half-brothers and -sisters by the same dam).
5. The average performance of an individual's progeny expressed as a deviation or ratio from the average of all contemporaries and the number of progeny.
6. Heritability of the trait.
7. The genetic relationship between the individual and different relatives (that is, 0.5 for sire, dam, and progeny and 0.25 for paternal and maternal half-sibs).

This information can be combined into appropriate linear equations which when solved provide the best estimate possible of the individual's breeding value. The estimate is regressed back to average in inverse proportion to the amount of information available to maximize the accuracy of ranking every individual in the herd for breeding value in specific traits. The calculations are complex and difficult to make by hand. Thus computational details will not be shown (for computational details, see *Guidelines for Uniform Beef Improvement Programs*, Beef Improvement Federation, USDA, Extension Service, revised and in press, 1976), but they can be made relatively quickly by high-speed computers. A number of record-of-performance programs sponsored by breed associations or other beef cattle improvement associations provide estimates of breeding value based on individual, sib, and progeny information to their clientele for specific traits such as weaning weight or yearling weight.

Table 15 provides information on the accuracy of selection for estimated breeding value based on performance of the individual combined with different kinds and amounts of information that may be available on parents, half-sibs, and progeny. For traits that are highly heritable (for example, 0.50), information on the sire, dam, or maternal half-sibs adds very little to the accuracy of selection. Information on a large number of paternal half-sibs helps only a little, but information on progeny increases accuracy of selection substantially even for highly heritable traits.

Estimates of breeding value based on all sources of information are really most useful relative to individual selection for traits with a low heritability. When heritability is low (for example, 0.10), progeny information is again the most valuable source of information that can be added to that of the individual. However, if progeny information is not available, paternal and maternal half-sib data or even information on the sire and dam can effectively contribute to increased accuracy of selection for traits of low heritability. Use of estimated breeding value based on all available information should be encouraged as an appropriate selection criterion, especially for traits with a low heritability.

Reproduction traits have low heritability. Efforts have only recently been made to incorporate measures of reproduction into record-of-performance programs. As records on fertility traits become available, effectiveness of selection for fertility can be enhanced by use of estimated breeding value. Their use will be most effective in herds using a large number of sires to produce a large number of paternal half-sib families.

NATIONAL SIRE EVALUATION

In recent years, progeny testing has been extended to provide for evaluation of sires used in different herds. This was not possible in beef cattle for many years because artificial insemination was used to only a very limited extent. Progeny tests conducted within herds provided for good estimates of differences in transmitting ability between bulls used at the same time in the same herd, but it was not possible to compare between bulls used in different herds because the progeny were out of different sires and cows and raised in different environments common to each herd. To overcome this difficulty, a number of breed associations have developed national sire evaluation programs following guidelines of the Beef Improvement Federation based on widespread use in different herds of certain bulls designated as reference sires. When reference sires

Table 15

Accuracy and Comparative Accuracy (Relative to Individual Performance) of Selection Based on Estimated Breeding Value Combining Information That May Be Available on Parents, Half-Sibs, and Progeny

Type and Number of Relatives with Records						Accuracy[a] Heritability			Comparative Accuracy[b] Heritability		
Individual	Sire	Dam	Paternal Half-Sibs	Maternal Half-Sibs	Progeny	0.10	0.30	0.50	0.10	0.30	0.50
1	1	0	0	0	0	0.35	0.58	0.73	1.10	1.06	1.03
1	0	1	0	0	0	0.35	0.58	0.73	1.10	1.06	1.03
1	1	1	0	0	0	0.38	0.61	0.76	1.19	1.12	1.07
1	0	0	10	0	0	0.38	0.60	0.73	1.19	1.09	1.04
1	0	0	20	0	0	0.41	0.62	0.74	1.30	1.12	1.05
1	0	0	40	0	0	0.45	0.63	0.75	1.43	1.15	1.06
1	0	0	0	1	0	0.32	0.56	0.71	1.02	1.02	1.01
1	0	0	0	2	0	0.33	0.56	0.72	1.05	1.03	1.01
1	0	0	0	4	0	0.34	0.57	0.72	1.09	1.05	1.02
1	0	0	0	0	10	0.52	0.74	0.84	1.63	1.36	1.19
1	0	0	0	0	20	0.62	0.82	0.89	1.96	1.50	1.26
1	0	0	0	0	40	0.73	0.89	0.93	2.31	1.62	1.32
1	0	0	0	0	100	0.85	0.95	0.97	2.70	1.73	1.37
1	0	0	20	2	0	0.42	0.63	0.75	1.34	1.15	1.07
1	1	1	20	2	0	0.45	0.66	0.77	1.44	1.21	1.10
1	0	0	20	2	10	0.57	0.77	0.86	1.79	1.41	1.21
1	0	0	20	2	20	0.65	0.83	0.90	2.06	1.52	1.27
1	0	0	20	2	40	0.74	0.89	0.94	2.36	1.63	1.32
1	0	0	20	2	100	0.86	0.95	0.97	2.71	1.73	1.37
1	1	1	20	2	10	0.58	0.78	0.86	1.84	1.42	1.22
1	1	1	20	2	20	0.66	0.84	0.90	2.09	1.53	1.28
1	1	1	20	2	40	0.75	0.90	0.94	2.37	1.63	1.33
1	1	1	20	2	100	0.86	0.95	0.97	2.72	1.73	1.37

[a] Accuracy is the correlation between estimated and actual breeding value.

[b] Comparative accuracy is the accuracy of selection based on a particular type of relative expressed as a ratio to individual selection (for example, accuracy of progeny test/accuracy of individual performance).

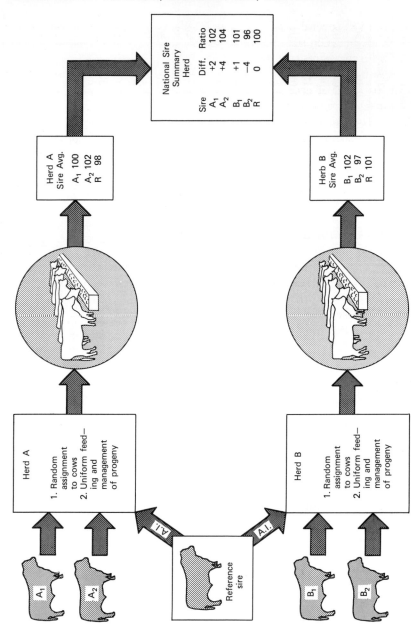

Fig. 15. National Sire Evaluation is possible by comparison with reference sires used through artificial insemination in different herds.

are used in several herds through artificial insemination, comparisons can be made between sires used in different herds through the tie provided by the reference sires common to each herd.

Results are reported in terms of "expected progeny difference" for various traits of economic importance. The expected progeny difference estimates the transmitting ability of sires evaluated in the program. Since each sire transmits one-half of his heredity to each offspring, the expected progeny difference estimates one-half of the sire's breeding value. The expected progeny difference is regressed back to the average in inverse proportion to the number of progeny produced by the sire, for maximum accuracy of evaluation. Thus the expected progeny difference provides the most accurate estimate available on how future progeny by various sires evaluated are expected to perform relative to the set of reference sires used in the breed.

Results of the sire evaluation program are generally published in a National Sire Summary. Effectiveness of selection and rate of genetic improvement in the breed can be enhanced if the outstanding sires are used extensively through artificial insemination or if the information is used as a paternal half-sib test to select outstanding sons of bulls that rank well in the National Sire Summary.

TYPES OF SELECTION

The three types of selection are (1) tandem selection, (2) selection based on independent culling levels, and (3) selection based on an index of net merit.

Tandem selection is selection for one trait at a time. When the desired level of performance is reached in a given trait, a second trait is given primary emphasis, and so on. This is the least effective of the three types and is not recommended. Its major disadvantage is that, by selecting for only one trait at a time, some animals that are at the same time poor in other traits will be retained.

Independent culling levels require that an animal reach specific levels of performance in each trait before it is kept for replacement. This is the second most effective type of selection, but it has one disadvantage: in requiring specific levels of performance in all traits, it does not allow for slightly substandard performance in one trait to be offset by superior performance in another.

Selection based on an index of net merit gives weight to the various traits in proportion to their relative economic importance and their heritability, and recognizes the genetic association, if any, among the various traits. The use of the index or some modification of it is the preferred type for most herds. It allows slightly substandard performance in one trait to be offset by outstanding performance in another. Also, by giving additional weight to traits of higher heritability or of greater economic importance, there can be greater improvement in net merit.

Differences in heritability of traits should be considered in selection because, obviously, if a trait has extremely low heritability, little genetic improvement in it can be expected, and emphasizing it will reduce the emphasis that can be put on traits with higher heritability that will give a greater response to selection.

Although increasing the number of traits reduces the selection differential for any one trait, it results in more rapid improvement in total genetic merit or net worth. Average reduction in progress in each trait as a result of considering several traits is approximately $1/\sqrt{n}$, where n is the number of traits selected for. For example, if four genetically independent traits are involved in selection, the selection differential for each will be approximately half what it would have been if only one trait were involved $(1/\sqrt{4}, = 1/2)$. This is based on the assumption that there are no genetic associations, favorable or unfavorable, among the four traits. It is obvious that considering all heritable, economically important traits simultaneously will result in faster improvement in genetic merit involving all traits.

Relative rates of improvement in some traits of economic value with different selection intensities or different percentages saved for breeding are considered in Table 16. These estimates are based on phenotypic evaluation (visibly obvious through performance) for the traits indicated, and assume that the percentage saved and used produce progeny that have an opportunity to be selected for the next generation—that is, the selected bulls from each generation are sired by bulls selected by the same criteria in the previous generation. Table 16 shows the advantages of saving bulls from among the top and selecting only for traits that have real economic values. Obviously, a closed-herd system must be used for the foregoing conditions to prevail.

MATING SYSTEMS

There are five fundamental types of mating systems: (1) random mating, (2) inbreeding, (3) outbreeding, (4) assortative mating, and (5) disassortative mating.

Random mating is the mating of individuals without regard to similarity of pedigree or similarity of performance.

Inbreeding is the mating of individuals that are more closely related than

Table 16

Estimates of Potential Progress in Ten Years when Different Intensities of Mass Selection Are Practiced for Specific Traits[a]

Item	Percentage of Bulls Saved					Assumptions
	1	10	20	50	70	
Weaning weight and no other traits (lb)	41.6	30.6	26.4	19.2	15.6	50% of heifers saved
Weaning weight and 1 other trait (lb)	29.4	21.6	18.7	13.6	11.0	h^2 = 0.3 in both sexes[b]
Weaning weight and 2 other traits (lb)	24.0	17.7	15.2	11.1	9.0	SD = 40 lb in both sexes[c]
Weaning weight and 3 other traits (lb)	20.8	15.3	13.2	9.6	7.8	
Postweaning daily gains and no other traits (lb)	0.44	0.30	0.25	0.17	0.12	50% of heifers saved
Postweaning daily gains and 1 other trait (lb)	0.31	0.21	0.18	0.12	0.08	h^2 = 0.5 in bulls
Postweaning daily gains and 2 other traits (lb)	0.25	0.17	0.14	0.10	0.07	h^2 = 0.3 in heifers
Postweaning daily gains and 3 other traits (lb)	0.22	0.15	0.12	0.08	0.06	SD = 0.29 lb in bulls / SD = 0.20 lb in heifers
Yearling weight and no other traits (lb)	147.4	103.2	86.4	57.6	43.2	50% of heifers saved
Yearling weight and 1 other trait (lb)	104.2	73.0	61.1	40.7	30.5	h^2 = 0.6 in bulls
Yearling weight and 2 other traits (lb)	85.0	59.5	49.8	33.2	24.9	h^2 = 0.4 in heifers
Yearling weight and 3 other traits (lb)	73.7	51.6	43.2	28.8	21.6	SD = 80 lb in bulls / SD = 60 lb in heifers
Yearling conformation score and no other traits (units)	1.39	1.02	0.88	0.64	0.52	50% of heifers saved
Yearling conformation score and 1 other trait (units)	0.98	0.72	0.62	0.45	0.37	h^2 = 0.4 in both sexes
Yearling conformation score and 2 other traits (units)	0.80	0.59	0.51	0.37	0.30	SD = 1 unit in both sexes
Yearling conformation score and 3 other traits (units)	0.70	0.51	0.44	0.32	0.26	

[a] Assumes that selection will be only for the criteria indicated, that when selection is for more than one trait, each trait is given equal emphasis, and that the traits are inherited independently. Generation interval is 5 years.

[b] h^2 = heritability.

[c] SD = standard deviation; it is an estimate of variation.

the average of the breed or population. Linebreeding is a form of inbreeding, and refers to the mating of individuals so that the relationship to a particular individual is either maintained or increased. This method automatically results in some inbreeding because related individuals must be mated to accomplish it.

Outbreeding is the mating of individuals that are less closely related than the average of the breed or population. The term "outcrossing" also is used to mean outbreeding when matings are made within a breed. Crossbreeding is a form of outbreeding.

Assortative mating is the mating of individuals that are more alike in performance traits than the average of the herd or group.

Disassortative mating is the mating of individuals that are less alike in performance traits than the average of the herd or group.

Inbreeding and outbreeding refer to similarity of pedigree of relationship, and assortative and disassortative mating refer to phenotypic resemblance.

Inbreeding adversely affects most performance traits or results in some reduction in general vigor. However, herds of reasonable size, where several sires are used, can be maintained closed to outside breeding for relatively long periods without any appreciable increase in inbreeding or decline in performance associated with inbreeding.

Within a closed herd where the mating is random as far as relationship is concerned, the rate of increase in inbreeding per generation is $1/8m + 1/8f$, where m is the total number of males used in each generation and f is the total number of females in the herd in each generation. Thus, in a 100-cow herd where 4 sires are used per generation with 100 cows in the herd per generation, the increase in inbreeding per generation is $1/8(4) + 1/8(100) = 1/32 + 1/800 = 0.031 + 0.0012 = 0.0322$, or 3.22 percent per generation. If generation interval is 5 years, 15 years on such a program would result in a herd with average inbreeding of 9.66 percent. This is not a rapid rate of inbreeding compared, for example, with the mating of half-brothers and -sisters which results in offspring that are 12.5 percent inbred. Offspring of sire-daughter, son-dam, and full brother and sister matings are 25 percent inbred.

Sire numbers per generation are of paramount importance in affecting rate of inbreeding. The rate of inbreeding can be reduced by deliberately avoiding close matings such as sire-daughter and half-brother and -sister. Whereas linebreeding will result in some loss of vigor, if the animal to which a herd is being linebred is truly outstanding the increase in performance as a result of intensifying the genes of an outstanding individual may more than offset any decline in performance due to inbreeding. Rigid selection accompanying linebreeding should also reduce some of the undesirable effects of inbreeding. When inbred or linebred herds are outcrossed, the loss of vigor that accompanies inbreeding is restored.

Linebreeding and inbreeding make the individuals in a herd more alike genetically and thus more uniform in their transmitting ability. A major advantage of linebreeding and inbreeding is that a breeder knows his own herd better than he knows someone else's and is likely to do a more effective job of selecting from within his herd. The effectiveness of linebreeding depends primarily on the genetic merit of the animal to which the linebreeding is directed.

Many breeders fear the consequences of inbreeding because it intensifies what is already present in the herd, including both bad and good traits. If an undesirable trait is present, inbreeding tends to bring it to light, but the genes responsible for the undesirable effect were already present. For example, if genes responsible for dwarfism are present in a population, inbreeding may increase the number of dwarf calves born, but it is not the cause of dwarfism. Inbreeding may be used to determine the presence of undesirable genes in a herd and, if accompanied by rigid selection, may effectively reduce their frequency.

The main disadvantage of linebreeding and inbreeding is that the foundation animals may not be truly superior. A genetic defect in the foundation animals can by chance rise to a high frequency and greatly interfere with the breeding program and materially reduce the value of the herd regardless of its genetic merit for major performance traits. Because it reduces genetic variation, inbreeding results in decreased heritabilities, and selection on individual performance is less effective. Since inbreeding makes individuals more alike in their genetic makeup, it increases the effectiveness of family selection.

Linebreeding and inbreeding should be practiced only in herds of outstanding genetic merit. The herds should be large enough so that the rate of inbreeding will be slow enough to provide opportunity for selection before genetic variation is reduced to the point where selection is not effective. All commercial producers and purebred breeders with small herds or herds of only average genetic merit should avoid linebreeding and inbreeding.

Outbreeding or outcrossing is recommended for all commercial producers and for secondary seedstock herds. Close matings should be avoided, but owners of secondary seedstock herds may profitably secure bulls from linebred herds. If sources of linebred bulls are changed periodically for use in secondary herds, the system is still outbreeding. If it becomes necessary to outcross linebred herds to correct a deficiency, breeders may find it advantageous to outcross to other linebred herds that are particularly outstanding in the trait that needs improvement. After such an outcross it may be desirable to resume a program of linebreeding.

Many breeders practice both assortative and disassortative mating. Assortative mating is practiced when superior cows are mated to superior bulls or when the poorer cows are mated to the unproved or less highly regarded sires. Disassortative mating is practiced when a breeder attempts to make "corrective matings"—that is, by mating cows that are mediocre or poor in one trait to bulls considered superior or outstanding in that trait. Assortative mating results in increased genetic variation in a herd, while dissassortative mating tends to reduce the genetic variation in a herd.

USE OF RECORDS

Record of performance is the systematic measurement of traits of economic value and the use of these records in selection, with the aim of finding the genetically superior individuals in all economically important traits so that they may be used for breeding. Records increase a breeder's knowledge of differences between animals and thus increase the accuracy of his selections.

The preferred measurements are those that give most accurately the breeding value or genetic merit of an animal relative to the others in a herd. Research on beef cattle breeding has demonstrated that appreciable genetic improvement can be made in most economically important traits by selection on the basis of differences in individual performance, as indicated by the estimates presented in Table 16. Such research has involved methods of measuring these traits and estimating their heritability, and developing selection procedures for traits that contribute to both productive efficiency and carcass merit. The systematic measurement of differences among animals in the economically valuable traits, the recording of these measurements, and using the records in selection will increase the rate of genetic improvement.

Performance records of animals should be adjusted to eliminate known environmental differences between animals, so that genetic differences will be a larger part of the total differences measured or observed. Adjustments should be made for differences in age, sex, age of dam, and any other "environmental" variable that can be measured or evaluated. Because any increase in environmental variation tends to obscure genetic differences and decrease the effectiveness of selection, every precaution should be taken to measure economically important traits as accurately as possible. For example, an effort should be made to equalize fill in animals before they are weighed, because errors in weighing decrease the accuracy of selection. Fill can be equalized somewhat by removing water and feed for 12 hours before weighing and by recording more than one weight. This applies to both initial and final weights.

Fig. 16. Postweaning performance traits being measured under uniform conditions in a within-herd, on-farm test, to provide information making it possible to estimate the breeding value of any selected bull in the drove. (USDA Meat Animal Research Center.)

Record of performance is useful primarily to provide a basis for comparing cattle handled alike within a herd and not for comparing differences between herds, because there are apt to be large environmental differences between herds due to location, management, and nutrition. Because it is difficult to adjust accurately for these differences, the evaluation of genetic differences is extremely difficult, even though genetic differences between herds do exist.

Average weaning weights of 500 pounds may be realistic in some environments and in some production programs, whereas 350-pound weaning weights may be reasonable under more adverse conditions. Yet beef cattle may provide the most desirable means of utilizing the land under both conditions. Furthermore, the genetic merit of a herd weaning 350-pound calves may be equal or even superior to that of a herd weaning 500-pound calves. Standards of performance expressed as deviations from individual herd or group averages are advisable for making comparisons within a herd, but comparisons between herds based on minimum standards of performance can be undesirable and misleading.

Minimum standards of performance for the various production and carcass traits have been considered in some record-of-performance programs. Because of the variation in environmental conditions and production programs, standards involving between-herd comparisons may tend to recognize herds carried under superior environmental conditions rather than those that are genetically superior.

Comparing animals within a herd that are subject to different environmental conditions, such as having part of the calves on nurse cows or other variations in feeding and management, is as objectionable as comparing the records of different herds. If variations in treatments exist, comparisons should be restricted to animals treated alike unless appropriate adjustments can be made for treatment effects.

All economically important traits that are heritable should be evaluated for all animals in a herd. An effective record-of-performance program should be compatible with practical management regimes. Cattle should be evaluated under the approximate environmental conditions in which their progeny are expected to perform.

From the standpoint of genetic improvement for the entire beef cattle industry, record of performance will have greatest impact in purebred or seedstock herds. Commercial producers can use records of performance to cull cows, to select replacement heifers, and to evaluate bulls on their progeny's performance where progeny groups are kept under comparable conditions. Since approximately 40 percent of all heifers must be saved for replacements just to maintain a herd, opportunity for selection among females is limited.

Commercial producers also can make effective use of performance records by selecting bulls on the basis of records from purebred or seedstock herds that are on a systematic record-of-performance program. In selecting herd bulls from their own herds as well as from other breeders' herds, purebred breeders should evaluate prospects on the basis of their records as compared with the herd average. Over a period of time the inherent productivity of any herd depends largely on the genetic merit of the bulls used.

The goals in record of performance are not greatly different from those that have always been sought by progressive breeders. The principal differences lie in a systematic record-keeping program and the use of these records in making selections. Record of performance up to slaughter requires no new or additional facilities except a scale and forms for keeping records.

The principal features of a good record-of-performance program are the following:

1. All animals are given equal opportunity.
2. Systematic, written records are kept of all economic traits on all animals.
3. Records are adjusted for known sources of variation such as age of dam, age of calf, and sex.
4. Records are used in selecting replacement stock and in culling poor producers.

5. Nutritional program and management practices are practical and compatible with those where progeny of the herd are expected to perform and are uniform for the entire herd.

Space does not permit sufficient detail to provide guidance for an individual record-of-performance program. Extension county agents can supply guides and record blanks prepared by their extension livestock specialists for this purpose. The various purebred associations also provide guidebooks and record forms designed to meet their special needs. Methods differ slightly in different areas, and breeders are advised to adopt those generally in use in their areas or for their breed.

Relative emphasis put on the different traits may vary in different herds, but the attention given each trait should be based primarily on its heritability and economic importance to the entire beef cattle industry. Keeping records does not change what an animal will transmit. Records must be used to locate and use the genetically superior individuals if genetic improvement is to be accomplished.

MAJOR PERFORMANCE TRAITS OF BEEF CATTLE

The major traits influencing production efficiency of highly desirable beef are (1) reproductive performance or fertility, (2) mothering or nursing ability, (3) rate and efficiency of gain, (4) longevity, and (5) carcass merit. Maximum production efficiency is not necessarily related to maximum performance levels in all of these traits. For example, maximum milk production and larger cow size associated with rapid growth rate are not desirable when feed supply for cows is limited, because reproduction suffers if additional nutrient requirements for lactation and maintenance are not met. Thus, with the possible exception of reproduction, it is not desirable to emphasize the same criteria of selection within all breeds for use in all production situations.

REPRODUCTIVE PERFORMANCE OR FERTILITY

A high level of reproductive performance, or fertility, is basic to an efficient beef industry because the percentage of the total beef cattle population composed of cows is high, requiring a major portion of the resources used in beef production. No single factor in commercial cow-calf operations has greater bearing on production costs than percentage calf crop. Also, a high level of reproduction is fundamental for making genetic improvements because increased calf crop decreases the percentage that must be saved for replacement and thus increases the selection differential possible for other traits. Both the male and female should be considered in selecting for reproduction, because reduced calf crops can result from reduced fertility of either.

Reproductive performance or fertility is a complex trait. A live calf at weaning is the product of a long sequence of events, each of which must succeed, from the time a cow is turned with a bull until her calf is weaned. The bull must have a high degree of libido and be physically capable of mating and producing sufficient viable sperm to maximize the probability of fertilization. The female must reach puberty as a heifer, or have a sufficiently early calf and short postpartum as a cow, to exhibit a fertile estrus during the breeding season. Ovulation, implantation, embryonic and fetal development, and parturition must occur without failure. The calf must consume colostrum and milk vital for early survival and ward off other hazards before weaning. With such a long chain of events involving interrelationships between the sire, dam, and offspring and their environment, the probability that any one event can succeed can be very high and still the probability that the product of all events will succeed (percentage calf crop weaned) can be somewhat low. A breakdown at any point in the sequence can be devastating.

Variation in reproduction is large. For example, percentage calf crop can easily range from 75 to 90 percent even among large herds. Results indicate that heritability of traits such as calf crop, pregnancy rate, and calving interval are low (10 percent), indicating that most of the variation is not caused by additive genetic differences between herds, but rather by differences in management, nutrition, herd health, and other environmental factors or by nonadditive genetic differences (for example, heterosis) and interactions between genotype and environment. So many random, or chance, environmental factors affect fertility from the time a cow is turned with a bull until her calf is weaned, that fertility in any given year reveals little of the real genetic differences among cows. Better measures of fertility are needed for cows and bulls. There are indications that certain components of fertility such as age at puberty and first service conception rate in heifers are more highly heritable than calf crop percentage.

The fact that heritability is low does not mean that detailed records should not be kept on reproductive traits. Records on reproduction are useful in identifying management problems that can be modified to improve reproductive performance in the herd. Even with low heritability, rigid culling to remove open cows or problem breeders can be useful in bringing about a more profitable reproductive pattern in a herd. There are reported instances where close culling for fertility has improved calf crop. Environmental effects influencing traits such as early calving date tend to have a permanent influence on reproduction in subsequent years. Most of the improvement is not additively genetic; that is, if the practice were discontinued, reproduction would decline in subsequent generations. However, selection pressure on reproduction is not wasted because reproduction is of overwhelming economic importance. Also, the return from culling open and other problem cows and keeping pregnant cows increases the number of calves weaned relative to cost of production and does provide greater opportunity to select for other economically important traits because of a larger total number of offspring available.

In herds where reduced calf crops are a problem, close attention to feeding, disease control, and management practices is definitely indicated. Reproductive diseases markedly influence fertility. Level of feeding, particularly level of energy, vitamin A, protein, and phosphorus is important. Many breeders and commercial producers can profitably give attention to these items in increasing calf crop.

CALVING DIFFICULTY AND BIRTH WEIGHT

A high degree of calving difficulty cannot be tolerated in commercial beef production. Not only is the expense of labor for assistance at calving a prohibitive factor, but calving difficulty can cut deeply into calf crop weaned as well, by reducing calf survival and postpartum conception in cows. For example, in one report, calf mortality was four times greater in calves experiencing difficult births (assistance given with a calf-puller or Caesarean birth) than in those not experiencing difficult births (20 percent versus 5 percent). Also, conception rate in the next breeding season was 16 percent lower in cows requiring assistance at calving than in those requiring no assistance. Cow age and calf birth weight are two of the most important factors influencing calving difficulty. One study showed that, on the average over all cow ages, calving difficulty increased 1 percent for each pound increase in calf birth weight. However, the association between calving difficulty and birth weight was strongest in 2-year-old first-calf heifers, reduced but still strong in 3-year-olds, and much lower in 4- and 5-year-old cows experiencing a low incidence of calving difficulty.

Pelvic size of cows and other physical measures of both cows and calves have been shown to be associated with calving difficulty, but these factors also have been associated with calf birth weight. Their association, independent of age and birth weight, has been too low to predict calving difficulty accurately or to serve as an appropriate selection criterion for calving ease. It appears that selection of heifers with larger pelvic area will result in females with larger pelvic area, but they will also be larger in size generally and will produce calves with proportionately heavier birth weights. Thus the net effect of selection for pelvic area on calving difficulty may be quite small.

Selection for smaller birth weight appears to be the most effective criterion for improving calving ease because it is the best single indicator of calving difficulty. Birth weight may even be a more sufficient selection criterion of calving ease than calving difficulty score because it can be measured more accurately and objectively on calves from cows of all ages in the herd. Calving difficulty is generally expressed to a high degree only in first- and second-calf females. Birth weight is affected by sex of calf and increases with advancing cow age. Thus it is necessary to adjust birth weight for sex of calf and age of dam. This can be done by expressing birth weight as a ratio to sex-age of dam-management group averages.

NURSING OR MOTHERING ABILITY

The ability of a cow to wean a healthy, vigorous calf is vital to efficient beef production. In the broadest sense, reproduction, calving ease, livability, maternal behavior, and milk production are all important components of mothering ability.

Increasing pounds of calf weaned per cow exposed to breeding can increase efficiency of production because certain fixed costs such as veterinary, labor, and bull service are on a per head basis. Feed costs per cow seem to be rather closely related to size of cow and level of milk production, but faster gains of calves decrease feed requirements of calves per unit of gain, and heavier weights at weaning reduce the amount of feed and time required for calves to reach a desired final slaughter weight—which takes on increasing importance if feed grains increase in price relative to other feed resources and forages used by cows. Thus increased weaning weight from increased milk production can be efficient, provided cows are converting economical low-quality feedstuffs into milk, a high-quality source of protein and energy, which can be used efficiently by the calf supplementary to that available from forages.

Milk yields, perhaps more than any other trait, must be synchronized or matched with feed resources available to maximize efficiency of production. Optimum milk yield does not mean maximum or minimum milk yield in most situations.

Increased milk yield increases weaning weight per calf, but increased weaning weight per calf from milk yield can be detrimental if weaning weight per cow is reduced as a result of impaired rebreeding performance. Rebreeding performance of cows can be reduced through longer postpartum intervals and reduced conception in cows producing high levels of milk if nutrient requirements to provide for increased milk production and maintenance of body weight are not met.

On the other hand, milk production can be too low. Milk production can be so low that survival and ability of calves to combat disease, parasites, and other environmental hazards can be reduced. When weaning weights of calves are low and calves are unthrifty, selection for improved nursing ability is indicated provided other health and disease considerations have been ruled out.

Weaning weight of a calf is used as a measure of nursing ability. The calf's own genetic impulse for growth is confounded with nursing ability by this procedure, but this is not a serious handicap since half of the growth impulse of the calf is transmitted by the dam.

Selection of bulls and replacement heifers that have heavy weaning weights relative to the herd average will lead to genetic improvement in nursing ability. In selecting for increased weaning weight, the breeder often selects not only for mothering ability but for the calf's own ability to grow. However, research information indicates that selection among cows for mothering ability should be reasonably effective. This can be accomplished by selecting cows on the basis of the weaning weights of their calves, since cows that wean calves heavier than the herd average in one year are more apt to produce calves heavier than average in succeeding years.

Differences in mothering ability can be evaluated about as accurately on the basis of 112-day calf weights as on the conventional weaning age of approximately 200 days. If calves are creep-fed, 112-day calf weights are perhaps preferable. Adjustment for differences in age of dam, sex of calf, and age of calf is necessary, since these factors influence weaning weight. In adjusting for differences in calf ages, it is recommended that average daily gain from birth to weaning be used for each calf (subtract actual birth weight, calculate average daily gain, and adjust to standard age for the group).

Mothering ability of cows may be compared within groups of the same sex of calf and of similar age of cows if numbers are large. This avoids an adjustment for differences in sex of calf and age of dam. The most accurate adjustment factors for sex of calf and age of dam are those developed in the herd in which they are used, provided the data are not biased by selection or management differences and the herd is large enough for reliable estimates to be made. Adjustment factors for smaller herds should be developed from herds with similar management regimes. Records are more accurate where the calving season is relatively restricted so that major differences in age and seasonal influences are avoided. Since weaning weight is used as a measure of mothering ability, it is important that all calves be treated the same (such as having all or none be creep-fed) so that the major variable is difference in nursing ability of the cows.

Indications are that selection for yearling weight may put as much pressure on nursing ability as selection for weaning weight. It has been observed repeatedly that the effect of age of dam on final weight is essentially the same as on weaning weight and that effects of age of dam on postweaning gain are negligible. This indicates that differences in preweaning gain associated with maternal environment are not compensated for in the subsequent postweaning period.

Selection of heifers on the basis of heavy weaning weight to increase nursing ability is less effective than selection of bulls with outstanding weaning weights. Results from several experiments have shown that heifers raised by cows providing a superior maternal environment may in turn provide a poorer maternal environment for their progeny than would be expected based on their own weight. The relationship for weaning weights has been greater between offspring and grand-dam than between offspring and dam even though their genetic relationship is only half as large (1/4 versus 1/2). Studies of effects of age of dam on weaning weight and on subsequent nursing ability of females have helped to clarify these relationships. Heifer calves raised by young cows (2- or 3-year-olds) or by old cows (10 years old or older) have lighter weaning weights than those raised by mature cows (5 to 9 years of age). However, heifers raised by young or very old cows have been superior in nursing ability to those raised by mature cows. Thus a superior maternal environment promoting a heavier weaning weight of a heifer calf can have a permanent negative effect on her subsequent maternal performance as a cow.

The negative influence that a superior maternal environment provided by the dam has on subsequent maternal ability of her female offspring may be caused by additional fat deposition in heifers raised by mature cows compared to those raised by young cows associated with greater milk production of mature cows. The possibility exists that fat deposition interacts with subsequent mammary and udder development in the growing female to reduce subsequent milk production; however, other mechanisms may also be involved. Whatever the underlying biological mechanism, results indicate that, if increased nursing ability is a selection objective, progress may be increased by selecting heifer replacements from the youngest heifers in the herd. Although selection intensity will be reduced when only heifers from the youngest cows are considered for prospective replacement, the generation interval will be shortened and progress will be made if intense sire selection for weaning weight is practiced and cows with the poorest maternal performance are culled.

GROWTH RATE

Growth rate is important because of its high association with economy of gain and its relation to fixed costs—such as veterinary, buildings, grazing fees, and labor—that tend to be on a per head or per unit of time basis. In most instances differences in growth rate have been measured in time-constant, postweaning feeding tests, and results indicate that differences in growth rate can be appraised rather accurately in this manner. A postweaning period of at least 140 days is required to measure differences in growth rate. This minimum length is based on rather uniform initial weights, condition, age, and previous treatments. Final weight at 12 to 18 months (standardized for age differences) is a more highly heritable measure of differences in growth rate than any individual component of final weight (that is, birth weight, preweaning gains, and postweaning gains).

Final weight at a standard age of 18 months fits the management programs of many purebred herds. Bulls can be carried on a relatively low level of concentrate feeding (4 to 5 pounds of concentrates plus full feed of roughage) their first winter and fed at a higher level of concentrate either on grass or in drylot during their yearling summer. By this procedure bulls are developed at a high enough level of feeding and over a long enough period for genetic differences in growth rate to be expressed, and a good appraisal of growth can be made. Bulls handled in this manner are in good sale condition at a desirable age and season. Postweaning gains are measured for approximately 350 days and gains made in this period can be added to 200-day weaning weight, appropriately adjusted for age of dam, to arrive at an adjusted 550-day weight.

Final weight and grade at somewhere near normal market age for a high percentage of slaughter cattle seems to be of most interest on an industry-wide basis. The use of postweaning gain alone as a measure of growth could foster poor milking ability because of compensatory gains, in that a poor feed supply in one period tends to be followed by a period of increased rate of gain.

An alternate program for measuring growth rate in bulls is to feed at a higher level and for a shorter period immediately after weaning. Bulls may be put on feed when they are weaned and full-fed for 5 to 6 months on a ration ranging from approximately equal parts of concentrates and roughage to two parts concentrates and one part roughage. In this program an adjusted final weight at 365 days can be used as a measure of differences in growth rate. For example, adjusted 365-day weight may be obtained by adding the gain made in a 165-day postweaning period to

200-day weaning weight, appropriately adjusted for age of dam. The postweaning feeding period may be intermediate to the two described above; for example, it may be 252 days with an adjusted final weight of 452 days computed and used as a basis for selection.

Research results indicate that a reasonably high level of feeding is desirable to appraise differences in growth rate most accurately. If a lower level of feeding is used, the period for measuring differences in growth rate should be longer. However, it is recommended that a relatively low level of feeding, promoting gains of 1.0 to 1.25 pounds per day, be used for heifers during their first winter. Gains of 1.0 to 1.25 pounds per day are adequate in most breeds to promote early sexual maturity so that the heifers can be bred at 14 to 15 months of age to calve as 2-year-olds. Research results indicate that full-feeding a high-concentrate ration during the first winter may interfere with reproductive performance and mothering ability. Because a high percentage of heifers must be kept for replacements, there is little opportunity to select among heifers for differences in growth rate. Hence, from this standpoint, little can be gained from the heavy feeding of heifers.

In selecting heifer replacements for differences in growth rate, it is suggested that long yearling age (approximately 18 months) be used, with adjustments in the same manner suggested for bulls (by adding the gain made after weaning to weaning weight, adjusted to a constant age, and appropriately adjusted for age of dam). This assumes that heifers are carried at a relatively low level of feeding during their first winter. If heifers are bred as yearlings, it may be desirable to make selections prior to 15 months of age. This can be done effectively with a 252-day postweaning period and adjusting final weights to 452 days.

Genetic correlations among measures of growth or size at different ages (for example, birth weight, weaning weight, 12- or 18-month weight, and mature weight) are high. Genetic correlations among these traits are higher than the phenotypic correlations. Thus 12- or 18-month weight is more highly heritable than any of its components (that is, birth weight, preweaning gains, and postweaning gains). However, other consequences are that selection of 12- or 18-month weight leads to significant increases in birth weight and mature size. Increases in birth weight contribute to increased calving difficulty associated with reduced survival of calves and reduced rebreeding performance of dams. Increases in mature weight of cows increase nutrient requirements for maintenance of the cow herd, which at least partially offsets the advantages of more rapid and efficient gains of the progeny slaughtered.

Hence recent studies have been conducted to evaluate genetic variation in shape of the growth curve, to assess the feasibility of increasing weights at market ages while minimizing changes in weight at birth and maturity.

Results from several studies have been encouraging. For example, degree of maturity (ratio of immature weight to mature weight) at weaning or yearling ages is moderately to highly heritable (40 to 50 percent). However, degree of maturity is not an effective selection criterion because it requires measurement of weight at maturity, which comes too late in life for effective selection to be practiced. Thus some research has been conducted to evaluate alternative selection criteria involving functions or indexes of immature weights that may be favorably related to degree of maturity for weight at market ages.

Encouraging results were reported in a study of selection criteria for efficient beef production with net effects of calf mortality, reproduction, and cow size included in the definition of economic efficiency. Results indicated that selection for heavier yearling weight (Y) but light birth weight (B) with an index $= Y - 3.2B$ would increase improvement in efficiency 6 to 7 percent more than selection for yearling weight alone. Adding this degree of selection against birth weight reduced expected increases by 55 percent in birth weight and by 25 percent in mature weight but only 10 percent in yearling weight.

Other results have indicated that selection for postnatal relative growth rate would have a similarly favorable effect on shape of growth curve reducing response in birth weight and mature weight relatively more than weight at market ages. Relative growth rate (RGR) was measured as

$$RGR = \frac{\ln W_{t2} - \ln W_{t1}}{t_2 - t_1}$$

which can be visualized more clearly as daily gain relative to average size over the time interval tested, or as

$$RGR = \frac{(W_{t2} - W_{t1})/(t_2 - t_1)}{(W_{t2} - W_{t1})/2}$$

where ln denotes the natural logarithm of weight (W) at time 1 (t_1, such as birth) or time 2 (t_2, such as weaning), respectively.

More research is needed to determine the most appropriate selection criteria for optimizing the shape of the growth curve. At least it appears that postnatal growth to weaning (for example, adjusted weaning weight–birth weight) or to yearling (for example, adjusted yearling weight–birth weight) ages should be emphasized rather than their respective final weights, to eliminate the effect of direct selection for heavier birth weight. Results have indicated that, although birth weight would still increase because of a positive genetic correlation with postnatal growth, the expected increase in birth weight could be reduced about 30 percent if all emphasis were directed to postnatal growth rate rather than weaning or yearling weight. The increase in mature weight should also be reduced because estimates of the genetic correlation between birth weight and mature weight have been higher than between birth weight and weaning or yearling weight.

EFFICIENCY OF GAIN

Efficiency of gain is one of the traits of greatest economic importance in beef cattle. Efficiency of gain is difficult to estimate because it requires individual feeding and adjustments for differences in weight, as increased weight is associated with higher feed requirements per unit of gain.

Present information indicates that genetic improvement can be made in efficiency of gain by selecting for it through rate of gain, because the fast gainers will also be efficient gainers. It is therefore recommended that breeders depend on differences in rate of gain as an indicator of efficiency of gain rather than incur the added expense of individual feeding. However, if a breeder desires to feed individually and adjust the records for differences in weight in order to measure differences in efficiency of gain, this is more accurate.

LONGEVITY

The longer animals remain productive in a herd, the fewer replacements will be needed, and thus the costs of growing out replacements to productive age will be reduced. However, the longer an animal remains in a herd, the longer will be the generation interval, which may reduce the rate of genetic improvement from selection. Breeders of purebred cattle or seedstock herds should be concerned with making genetic improvement in longevity, so that commercial beef cattle populations will be productive at older ages. Yet a fairly rapid turnover of generations in purebred herds is desirable for making a maximum rate of genetic improvement in other traits of economic value.

With the trend toward marketing cattle at younger ages and somewhat lighter weights, a higher percentage of the beef cattle population must be cows in order to produce the same amount of beef. This higher proportion of cows tends to make longevity of greater economic importance from an industry-wide standpoint. Longevity in bulls is important because it decreases the annual cost of bull service.

The major factors affecting longevity—or, more important, number of years spent in the breeding herd—are infertility, unsoundness of feet and legs, serious eye diseases such as cancer eye, udder troubles, and unsound mouth. Research shows that susceptibility to cancer eye is heritable, and selection against it should be reasonably effective; however, it is a trait that can be measured only late in life.

Selection for longevity must be confined primarily to indicators such as structural soundness and to pedigree information—that is, selection of close relatives of individuals that have had a long productive life. There is a certain amount of automatic selection for fertility and longevity, because animals that remain in a herd long enough to produce a large number of offspring tend to have a larger number saved for replacements.

CARCASS MERIT

Carcass merit is of fundamental importance to the beef cattle industry because desirability of product together with price is the major factor affecting consumption. In selecting for improved carcass merit, the factors that contribute to carcass desirability and their relative importance must be known. Research in many states indicates that the American public desires beef with a high percentage of lean as compared to fat and bone, and the lean must be tender, flavorful, and juicy.

Variation in composition of carcasses (relationship of lean to fat) is a major factor influencing differences in carcass value. The value of fat trim is negligible in today's market relative to that for retail product (closely trimmed, boneless steaks, roasts, and lean trim) from the carcass. It is not uncommon for carcasses of the same grade to range from 10 to 30 percent fat. However, differences in composition of this magnitude are due to environmental sources of variation such as age, length of time on feed, and energy content of the ration as well as to genetic sources of variation responsible for highly heritable differences in growth of lean, fat, and bone and differences in composition at a constant weight associated with degree of maturity.

Variation in retail product growth (retail product at a constant age) is much greater than variation in composition or proportion of retail product (retail product at a constant carcass weight). Heritability is also higher for retail product growth (60 percent) than for proportion of retail product (40 percent). One study indicated that genetic variation or opportunity to create genetic change was about eight times greater for retail product growth than for proportion of retail product.

Yearling weight (12- to 18-month final weight adjusted for age) is highly related to growth of retail product. The relationship to retail product growth is stronger if an index is used incorporating an accurate measure for less fat thickness along with yearling weight. Results indicate that an index for yearling weight and less fat thickness is also more highly correlated with postweaning feed efficiency during age or weight constant intervals than yearling weight alone. However, indications are that retail product growth or an index for yearling weight and less fat thickness may be negatively related to degree of maturity at market ages and, consequently, leads to an undesirable shape of growth curve. Thus the contribution to total production efficiency from increases in postweaning feed efficiency of growing animals, only part of which are marketed at slaughter, may be offset by subsequently greater maintenance requirements of cows with larger mature size. Therefore selection for retail product growth or an index for yearling weight and less fat thickness is clearly justified only in terminal sire strains or breeds—that is, in strains or breeds from which bulls are used in commercial production on crossbred cows of smaller size to produce progeny for slaughter.

It would be desirable if muscle development could be maximized in regions of the carcass yielding the more preferred and higher-priced cuts—the loin, rib, rump, and round as opposed to the chuck, brisket, plate, shank, and flank. Traditionally, visual appraisal of carcasses or live animals has been used to assess differences in conformation. Morphological differences in items such as thickness and bulge of the round are quite evident, both within and between breeds. It has been hypothesized that differences in conformation are associated with variation in distribution of muscle so that the amount of muscle could be increased in the round, loin, and rib relative to muscle in other lower-priced regions of the carcass. However, within- and between-breed studies have now indicated that it is not possible to alter the relative distribution of muscle in the carcass or to shift the proportion of muscle to regions of the high-priced cuts. Genetic correlations between proportion of retail product in one cut are very high with that in all other cuts, at a constant carcass weight. Even between breeds, differing widely in conformation (for example, Jersey crosses versus Charolais crosses), the percentage of retail product in different cuts relative to total retail product from the entire carcass does not differ significantly.

It is possible, however, to change the ratio of muscle to fat in the various cuts and the entire carcass. Genetic correlations between retail product and fat trim are strongly negative. Selection for retail product or against fatness can increase the percentage of carcass weight represented by retail product by increasing the muscle to fat ratio in various cuts and the entire carcass.

Workers at the Livestock Division, Agricultural Marketing Service, U.S. Department of Agriculture, have shown that the yield of closely trimmed, boneless retail yield from the primal cuts (round, loin, rib, and chuck) can be predicted from the fat thickness at the twelfth rib, rib-eye area at the twelfth rib, percentage of kidney and pelvic fat of the carcass, and carcass weight. Their prediction equation is as follows: Estimated percentage of boneless, trimmed retail cuts from round, loin, rib, and chuck (cutability) = 52.56 − 4.95 (single thickness of fat over rib eye, [twelfth rib], inches) − 1.06 (percentage of kidney fat) + 0.682 (area of rib eye [twelfth rib], square inches) − 0.008 (carcass weight, pounds). They have also developed a system that classifies cattle into five yield grades identifying differences in estimated yield of boneless, closely trimmed retail cuts from the primal cuts. These four cuts represent approximately 80 percent of the value of the carcass, and the relation between yield of boneless, trimmed retail cuts from the primal cuts and from the rest of the carcass is high. Cattle with a yield grade 1 are superior to those of yield grade 5 in terms of cutability.

In evaluating beef carcasses in progeny or sib tests, use of estimated cutability is recommended because it more accurately accounts for variation in retail product yield than yield grade. Recent results indicate that cutability estimated with the foregoing equation is a relatively poor predictor of differences between breeds that differ widely in size or carcass weight in relation to fat thickness. However, results indicated that the prediction equation ranked animals rather accurately within each of the different breeds, regardless of their average size. Also, within-breed estimates of the genetic correlation between actual retail product yield and estimated cutability have been high. Thus estimated cutability is a useful predictor of retail product yield for within-breed improvement.

Palatability of beef characterized by differences in tenderness, flavor, and juiciness and other qualitative factors associated with appearance of products over the counter (color, firmness, texture) can have an important influence on consumer acceptance and value of beef. Qualitative traits associated with appearance of products are influenced more by methods of handling during transportation, by slaughter methods, and by conditions during processing and distribution subsequent to slaughter than by genetic differences.

In cattle fed, managed, slaughtered, and processed uniformly, tenderness is probably the most important factor affecting palatability of beef. Results indicate that, under these uniform conditions, tenderness has a fairly high heritability. Therefore, if information can be obtained on tenderness, selection for this trait would be an effective way of ensuring that animals will produce tender lean. Tenderness is also related to youthfulness. Therefore selection of animals that will reach desirable market weights at young ages would be an indirect means to the same end.

In grading carcasses, meat quality is determined by marbling, texture, color, and firmness in relation to maturity. Maturity is evaluated from ossification changes that occur in the skeletal system with advancing age and from color and texture changes in the lean. Among cattle that have been uniformly fed and managed and that are of similar age, marbling is the major determinant of quality grade.

The amount of fat thickness on a carcass is not a factor in determining its quality grade. The observed quantity of outside fat on carcasses is not closely related to their marbling—that is, phenotypic correlations between fat thickness and marbling have been low among carcasses of cattle fed and managed in the same manner. However, on the average, estimates of the genetic correlation between marbling and retail product yield have been strongly negative, indicating that selection for one trait will reduce the other or that simultaneous selection for increased retail product yield and increased marbling would be ineffective.

The strong negative genetic correlation between retail product yield and marbling found within breeds is consistent with relationships that have been found between breeds representing diverse biological types differing widely in mature size, growth, and carcass composition at different times along their weight-age growth curves. Numerous experiments have shown that large, growthy breeds have less outside fat and lower degrees of marbling but greater retail product yield than smaller, early-maturing breeds when compared at the same age and especially when compared at the same carcass weight.

One experiment involving 1,123 steers by Hereford, Angus, Jersey, South Devon, Limousin, Charolais, and Simmental sires out of Hereford and Angus cows indicated further that, although marbling differed significantly between breeds, taste-panel evaluations of tenderness, flavor, and juiciness were well above minimum levels of acceptance for all breeds when the steers were fed a ration containing about 71 percent total digestible nutrients (TDN) (60 percent corn silage, 34 percent grain, and 6 percent protein supplement) for 7 to 9 months after weaning. Differences between breeds in tenderness, when compared at the same age, though significant, were small while differences in flavor and juiciness were not significant.

When comparing breed groups at equal ages, tenderness increased slightly as marbling increased. However, when comparisons were made at equal carcass weights, differences in tenderness were smaller and not significant even though differences in marbling were increased. Within breed groups, tenderness tended to decline as cattle were fed longer periods of time to increase marbling or fat in the rib eye. As time on feed was lengthened, it appeared that slight increases in tenderness associated with increased marbling were more than offset by decreased tenderness associated with older age. Thus the penalty of increased fat trim and cost of production associated with increased time on feed to meet present marbling requirements for the low choice grade are not justified in terms of improvement in eating quality, at least not in cattle fed moderate levels of energy (71 percent TDN) for 7 to 9 months prior to slaughter.

CONFORMATION AND ITS EVALUATION

Performance traits other than carcass merit and structural soundness should be measured directly or through the indicators that have been discussed rather than through conformation. Conformation is a performance trait to the extent that it contributes to carcass merit and longevity. Basically, the important conformation items are structural soundness, which may contribute to longevity, and carcass composition. Research indicates that differences in outside fat can be appraised, but with somewhat limited accuracy, by visual appraisal of the live animal.

However, if feasible, more objective measures of fatness are recommended for assessing differences in composition. Ultrasonic measurements of fat thickness or measurements taken with a thermistor probe or a small-gauge needle are reasonably accurate predictors of carcass fat content.

In bulls developed alike it seems reasonable to give independent scores for differences in (1) structural soundness, (2) thickness of natural fleshing or muscling, and (3) outside fat. A muscle score reflects differences in thickness of muscling in relation to length of long bones. Thus muscle score usually reflects weight in relation to height. With two animals of the same weight but differing in height, the animal with less height will ordinarily receive the higher score for muscling. Preliminary information indicates that mature size may be highly associated with long-bone length at yearling age. Thus, in two animals weighing the same as yearlings but differing in height, the one with the greater height may be expected to have a heavier mature weight and a lower muscling score at yearling age.

On the basis of these preliminary results, selection for heavy weights at yearling age, along with a high muscling score, should result in a growth curve with rapid early growth without excessive mature size. Thus the use of a muscling score at yearling age may be a factor in affecting mature size. Yearling weight or postnatal growth from birth to a year of age should be given major attention because of its great economic importance, and indications are that near-maximum yearling weight may be obtained along with a high muscling score at yearling age.

CENTRAL TESTING STATIONS

Central testing stations are locations where animals are assembled from many herds to evaluate differences in some performance traits under uniform conditions. Present and potential uses of central testing stations include (1) estimating genetic differences between herds or between sire progenies in gaining ability, grade, finishing ability, and carcass characteristics; (2) determining the gaining ability, grade, and finishing ability of potential sires as compared with similar animals from other herds; (3) determining gaining ability, grade, and finishing ability under comparable conditions of bulls being readied for sale to commercial producers; and (4) as an educational tool to acquaint breeders with performance testing.

In setting up a central testing station, its objectives should be clearly defined and procedures designed to accomplish the objectives. Because objectives and procedures vary with location, only general principles will be discussed here.

In beef cattle, nutritional level at one stage of life usually has carryover effects on performance at later stages. A poor feed supply in one period tends to be followed by a period of increased or compensatory gain when rations are improved. Conversely, a higher than normal plane of nutrition, such as that provided by creep-feeding, is likely to be followed by a period of subnormal gains on a normal feeding regime.

Because pretest levels of nutrition and management usually differ from farm to farm or ranch to ranch, performance at a central testing station is influenced by pretest environment. From one standpoint this is a serious disadvantage of central testing stations, because part of the observed differences at a station will be due to pretest conditions. It is almost always impossible to estimate the importance of these effects, but carryover herd environmental effects are less important than herd differences due to environment when all animals have been fed for a comparable period in the herds in which they were produced. If this is considered, central testing stations minimize herd environmental effects.

Bull buyers must decide from which herds to buy bulls and which bull or bulls to buy within a herd. If the bulls are raised and fed entirely on the farm or ranch where dropped, the buyer has the difficult task of deciding how much of the apparent superiority or inferiority of bulls in a specific herd is the result of feeding and herdsmanship rather than heredity. If the bulls have spent part of their lives under standard conditions that minimize these effects, the buyer's task is easier, whether he is buying commercial bulls or herd sires for a purebred herd.

Similarly, if progeny test groups of steers from different herds are being fed out to determine the transmitting ability of the sires for growth rate, feed efficiency, and carcass characteristics, sire comparisons are more accurate if all progeny are fed under standard conditions for the final feeding period.

Central tests have limited use for estimating genetic differences among herds. The larger the herd size, the greater the number needed to adequately sample the herd. The precision of the tests is greatly improved if five to eight progeny of each of two or more sires from each herd are tested each year. This permits assessment of within-herd differences to compare with between-herd differences. Furthermore, there should be an adequate sample of animals from each herd on test, or little real information on herd differences will be accumulated.

If central testing stations are used to estimate genetic differences between herds, it is recommended that samples of those completing the evaluation be used in topcross comparisons in commercial herds, so that additional traits can be measured and the precision can be increased. If the purpose is to evaluate individual potential sires, the number tested per herd or per sire is of no importance; but between-herd comparisons should be discouraged if numbers from each herd are small. Preferably, bulls should be entered in this type of test only if they meet rigid qualifications for preweaning rate of gain and conformation score.

If the purpose of the testing station is solely to develop bulls and make objective performance information available to prospective buyers—a service especially valuable to small breeders—then the number of bulls per herd or per sire is immaterial. To be most useful, however, large numbers should be fed at a single location, giving buyers an adequate number from which to choose. This is possible if commercial-type feedlots are used.

Influences of pretest environment on test performance can probably never be eliminated, but they can be minimized. Animals should be used whose pretest treatment was similar, and should be grouped within relatively narrow age ranges. Animals for a given test should be delivered to the station on a specific date and should undergo an adjustment period of 14 to 84 days on the test ration before beginning the official test. The test should run for an adequate length of time—140 to 182 days if a high-concentrate ration is used the entire time, and longer if the ration is high in roughage.

Influences of pretest environment can be minimized in appraisal of results if the final reports include both pretest and test gains. If test gain alone is used, cattle on a suboptimum pretest feeding level that did not permit full expression of their inherent ability to grow are likely to compensate with inflated test gains. Using both pretest and test gains avoids labeling an unduly high test gain as the animal's real gaining ability. This can be done either by averaging pretest and test gains or, if test starts immediately after weaning, by computing a final weight as a standard weaning weight (for example, 200 days) plus test gain. The animal's entire life must be accounted for; "loafing periods" of unequal length, which tend to influence subsequent gains, should not be omitted.

The problem of compensatory gain is not limited to central testing stations. Within a herd, the inherently fast-gaining calf whose mother was a poor milker is likely to have a low weaning weight with a correspondingly inflated postweaning gain. Comparisons, whether between herds at test stations or within a herd on an individual farm, should consider both preweaning and postweaning gains.

Central testing stations are most valuable if it is recognized that they can evaluate only a limited number of traits and that at best they are only one phase of a complete performance evaluation program. A primary measure of their effectiveness should be the impact they have on increased herd testing for all economically important traits. Central testing stations can cause difficulty in the maintenance of herd health, but proper precautions can minimize this problem.

HEREDITARY DEFECTS OF BEEF CATTLE

A large number of hereditary defects of possible economic importance have been reported in all breeds of beef cattle and also among the dairy breeds. Perhaps the hereditary defect most widely known is "snorter" dwarfism, which occurred at troublesome frequencies in some herds in the late 1940s and early 1950s. Discrimination against lines of breeding known to carry this defect has reduced its frequency. Snorter dwarfism, like most other hereditary defects, is inherited as a simple recessive—that is, it is caused by a single pair of genes that must be present together before the trait is expressed—and thus it results from the mating of parents that both carry the defective gene.

Other types of hereditary dwarfism are due to different genes. "Comprest" dwarfism seems to result from a gene with incomplete dominance, meaning that the carrier individuals are comprest and an extreme type of dwarfism segregates from comprest × comprest matings. The comprest condition in Herefords and the "compact" condition in Shorthorns are probably due to the same gene. Snorter dwarfism has been authentically reported in both the Angus and Hereford breeds, and "longheaded" dwarfism, which is also inherited as a simple recessive, has been reported in the Angus breed.

A practical means of testing a bull for a specific defective recessive is to breed him to females that are known carriers of the gene. Table 17 shows

Table 17

Testing Bulls for Hereditary Defects Inherited as Simple Recessives		
	Percentage of Carrier Bulls that Will Not Be Detected when Mated	
Number of Matings	To Known Carriers of a Specific Defect (%)	To Own Daughters or to Unselected Daughters of Known Carriers of a Specific Defect (%)
5	23.73	51.29
6	17.80	44.88
7	13.35	39.27
8	10.01	34.36
9	7.51	30.06
10	5.63	26.30
11	4.22	23.01
12	3.16	20.13
13	2.37	17.61
14	1.78	15.41
15	1.34	13.48
16	1.00	11.80
17		10.32
18		9.03
19		7.90
20		6.91
21		6.05
22		5.29
23		4.63
24		4.05
25		3.54
26		3.10
27		2.71
28		2.37
29		2.07
30		1.81
31		1.58
32		1.38
33		1.21
34		1.06
35		0.93

the percentage of bulls that are carriers of a hereditary defect that will not be detected with different numbers of test matings.

To determine whether a bull is a carrier of any genetic defect, bulls may be progeny-tested for undesirable recessive genes by either of two methods. Both methods test simultaneously for all recessives.

One method is to breed by artificial insemination to a large cross-section of the female population of the breed. The probability of detecting a carrier of an undesirable recessive is related to the frequency of the gene in the population. The probability of detecting a carrier equals $1 - (1 - \frac{1}{2} q)^n$ where q is the gene frequency in the female population and n is the number of progeny. This method provides for a relatively short generation interval and, if animals identified as carriers are culled, it will be effective in keeping undesirable genes at a low frequency.

The second method involves mating a sire to a group of his own daughters. On the average, it is a bit conservative to assume that half of the daughters of a bull with a defective recessive gene will be carriers of that gene (the formula given above applies with $q = 0.25$). The number of sire-daughter matings determines the precision of the test. The percentage of bulls that are carriers of a hereditary defect inherited as a simple recessive that will not be detected with different numbers of sire-daughter matings is also shown in Table 17.

Although many genetic defects are present in all breeds of beef and dairy cattle and in all classes of farm livestock and probably cannot be eliminated, it is possible to curb their effect. Increased frequency of genetic defects in a breed or population can be explained by a gene's producing an effect in carriers that causes them to be preferred to noncarriers; or, by chance, a defective gene may happen to be present in a line of breeding that is favored and used extensively by the industry. Seedstock producers are in a position to keep these defective genes from becoming a problem to commercial producers by closely observing operations and by realistically approaching a solution once a problem arises. This may require careful screening of herd bulls by progeny-testing and the prompt elimination of those proved to be carriers of a defective gene.

If an abnormal calf is born in a herd, the breeder should establish the most probable cause of the abnormality, which can be done only by complete records. A limited number of developmental abnormalities may occur that do not have a genetic basis. If a breeder decides that an abnormality has a hereditary basis, he should breed away from the source of the trouble, so that minimum damage will result to his herd and to others. This may be done by outcrossing to a linebred herd, after a careful study of the outcross so that the same or an equally undesirable defect will not be introduced. Another method involves the progeny-testing of bulls from his own herd, to ensure that future herd bulls are not carriers of the gene responsible for the defect. The latter procedure may be indicated if the genetic merit of the herd is particularly high. If only a small percentage of the animals in a herd are possible carriers of the genetic defect, the best course is to eliminate those animals from the herd, provided that their genetic merit is not superior to the remainder of the herd.

Although it is unwise to use sons of bulls or cows known to be carriers of defect-producing genes without first progeny-testing the sons, discrimination against lines· of breeding involving animals several generations removed from a known carrier is unjustified. Only one-half of the progeny of a carrier bull will be carriers when the bull is mated to cows that are noncarriers. Thus it seems more reasonable to handle such situations on an individual herd or bull basis than to discriminate against other herds descended from similar lines of breeding if they are not directly incriminated.

DEVELOPING BREEDING PLANS

Attaining the maximum rate of genetic improvement in all traits of economic value in beef cattle requires a clear perspective of objectives and a planned breeding program for accomplishing them. The different segments of the industry are interested in various specific traits, but the breeder must be aware of the demands of the entire industry. The commercial producer wants cows that have long productive lives and wean a high percentage calf crop of heavy, high-grading calves. The feeder demands rapid and efficient feedlot gains, and the packer and retailer are interested in cattle that will produce high-grading carcasses with a minimum of excess fat and the maximum yield of closely trimmed retail cuts from the wholesale cuts of greatest value.

The heritability, genetic association with other traits, and relative economic importance determine the attention each trait should receive in selection. Traits vary in heritability and economic value. The greater the number of traits selected for, the smaller the selection differential will be for any one trait. Traits of low heritability respond less to selection than do traits of high heritability. The opportunity for selection should be used for traits that will result in the maximum genetic progress for the traits of greatest economic value. Little can be gained and much can be lost by paying too much attention to traits of little economic value and traits of low heritability. While there are genetic differences between herds, evidence indicates that the large differences in feed resources and management programs between herds make it extremely difficult to compare the records of different herds. Thus comparisons should be among animals in the same herd.

Rate of improvement in most economically important traits of beef cattle is relatively slow, primarily because of the inherently low reproductive rate, the large number of traits of economic value, and the long generation interval. The low reproductive rate makes it necessary to keep a high percentage of the offspring, especially females, as replacements, and the large number of desired traits limits the selection that can be practiced for any one trait. However, most of the economically important traits have reasonably high heritability, fertility being the most notable exception. Though improvement is slow, it tends to be permanent, cumulative from year to year, and transmitted to future generations. Over a period of 15 to 20 years, production in a herd or breed that has been subjected to systematic selection for all economically important traits should be noticeably higher than in herds where no such effort was made.

A systematic record-of-performance program with selection based on differences in records is basic to any planned breeding program. The choice of breeding plans involves many considerations. In seedstock herds, if genetic merit is already high, if the herd is large, if it is not particularly deficient in some trait of major economic importance, and if it is relatively free of hereditary defects, a closed-herd program of linebreeding may be desirable. One advantage of a closed herd is that the breeder knows the differences in performance of his own cattle better than another breeder's and can better evaluate differences in their most probable genetic worth.

In large herds where a relatively large number of sires are used in each generation, a closed herd can be maintained for extremely long periods without any appreciable increase in inbreeding if an attempt is made to avoid the mating of close relatives such as sire-daughter, full brother-sister, and half brother-sister. In herds where as many as 8 to 10 sires are used per generation, the decrease in performance and the reduction in genetic variation as a result of inbreeding will hardly be noticeable.

If the herd is not large and only a small number of sires are used in each generation, the level of inbreeding will increase more rapidly, and performance and genetic variation in the herd will decrease. Decrease in genetic variation decreases the effectiveness of selection.

If the genetic merit of the herd is already high, it will be difficult to bring in an outcross that is genetically superior to some individuals in the herd. Also, in herds that are relatively free of genetic defects, the chance of increasing this problem with introductions of outside bulls is greater.

Whenever a genetic defect is troublesome in a herd, or when performance in some economically important trait is particularly low, perhaps an outcross is indicated. Although the outcross should be selected to correct the deficiency, the other traits of economic value should also receive major consideration. Minimum sacrifice in other traits is a primary objective when bringing in an outcross to correct some deficiency. Outcrossing for any reason in herds of superior genetic merit should be done only on a cautious and systematic basis, and only herds known to be outstanding in the trait of major interest and superior in all traits should be considered. Perhaps this may be another linebred herd. Certainly, records are as fundamental in making selections for outcrossing as they are in making selections from within the herd. After selecting the outcross, a comparison with sires in the herd in a properly conducted progeny test is desirable before extensive use is made of the sires brought into a herd for outcrossing.

In herds that are only average or below average in genetic merit, an outcrossing program may logically be the one of choice. However, since most of the opportunity for selection in beef cattle is in the bulls used, records in the herds where bulls are selected should be helpful in locating individuals that have superior genetic merit. Securing outcross sires from linebred herds is desirable.

Pedigree, individual performance, and progeny test information all have a place in a constructive breeding program. Young sires should be initially selected on the basis of pedigree and individual performance data. The extent to which they are used in a herd will depend on their rank with other sires based on progeny test information. After progeny test information is available, it should be used in making decisions among sires, and one must remember that an increase in generation interval is involved. However, in herds of superior genetic merit, where increased accuracy is of fundamental importance, there is justification for using progeny test information more extensively than in herds only average or below average in genetic merit.

Because generation interval affects rate of improvement from selection, it should be kept relatively short. If a bull is truly superior, he should sire sons that have genetic merit surpassing his own when he is bred to cows comparable in genetic merit to the population that produced him. The problem is to devise an evaluation program based on use of records that aid in locating such sons. Perhaps one handicap to continued improvement in some herds is the extensive use of an old sire without sufficient attention to locating sons to replace him. When the old bull passes out of the picture, the herd is left without sires that are superior to him. Continued improvement depends on use of herd bulls that are superior to the ones used in the previous generation.

Before a bull is used extensively in a herd, as would be the case with artificial insemination, it may be desirable to progeny-test him for genetic defects on 30 to 35 of his daughters. If a herd is following a linebreeding program, it should be determined that the bull to which the linebreeding is directed is not a carrier of genetic defects, and known carriers should be discarded.

The breeding of cattle of truly superior genetic merit is a great challenge. Many decisions must be made on breeding plans, and in selecting herd bulls and replacement females. One difficult decision seems to be "dropping" a bull that still has a good market for his progeny, even though the breeder may have determined that the bull is not contributing to the accomplishment of his goals and may be inferior to others in the herd.

The more a breeder knows about the animals in his herd and the more clearly he understands his objectives, the more often he should make correct decisions. Success in breeding superior beef cattle, like success in other ventures, depends primarily on the usefulness of the goals and the accuracy of decisions while working toward the goals. A complete record-of-performance program provides the basis for making correct decisions.

Goals can be attained only by those who have the objectivity to keep them in perspective and the dedication to remain steadfast in achieving them. They can be accomplished only by a planned breeding program based on the systematic use of records for selection on all traits of economic value.

The contributions to genetic improvement have been made and will continue to be made by those who have chosen goals based on utility. The successful breeders have not been faddists, but have exercised common sense and good judgment with the long-term outlook in mind.

CHAPTER 4
CROSSBREEDING FOR COMMERCIAL BEEF PRODUCTION
LARRY V. CUNDIFF AND KEITH E. GREGORY[1]

The commercial beef producer has the opportunity to improve efficiency of production and desirability of product through systematic crossbreeding. Crossbreeding provides for use of heterosis or hybrid vigor to increase reproductive performance, survival, mothering ability, growth rate, and longevity. Crossbreeding also can be used to combine and synchronize desired characteristics of breeds with market requirements and with feed and other resources available in different herds or production situations. These are benefits that can be derived from systematic crossbreeding in addition to those of selection discussed in the preceding chapter. A working knowledge of the genetic basis for and the benefits of heterosis, the differences between breeds, and the various systems of crossbreeding that can be employed, is needed to develop effective breeding programs for commercial beef production.

GENETIC BASIS FOR HETEROSIS

Nonadditive gene effects, which are caused by interaction of genes, occur when specific pairs or combinations of genes produce favorable effects as a result of being present together. Examples of nonadditive gene effects are portrayed in Fig. 17 for interactions between genes of a single pair or different degrees of dominance.

The first graph portrays a completely additive case of no dominance. The heterozygous genotype (Aa) is intermediate in phenotypic value to the homozygous genotype (AA or aa). Coat color in Shorthorns is an example of this form of inheritance. Red animals are homozygous (both genes of each gene pair are alike) for the red gene (RR), white animals are homozygous for the white gene (rr), and roan animals are heterozygous (Rr, genes of the pair are unlike).

---

[1] Larry V. Cundiff is Research Leader, Genetics and Breeding, and Keith E. Gregory is Director, U.S. Meat Animal Research Center, Agricultural Research Service, U.S. Department of Agriculture, Clay Center, Nebraska.

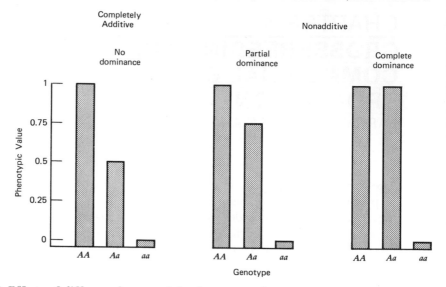

Fig. 17. Effects of different degrees of dominance on phenotypic value.

The second graph in Fig. 17 portrays a nonadditive case of partial dominance and the third graph portrays a nonadditive case of complete dominance. There are many examples of complete or nearly complete dominance in cattle where the heterozygote (Aa) has the same or nearly the same appearance as the homozygote for the dominant gene (AA), but is distinctly different from the homozygote for the recessive gene (aa). For example, polledness is dominant to horns, black coat color is dominant to red, and normal size is dominant to dwarfism. These are examples of qualitative traits whose inheritance is controlled by single gene pairs. Quantitative traits—such as growth rate, carcass merit, and so on—are controlled by many pairs of genes, each having a relatively small effect, part of which may be additive and part nonadditive.

The importance of nonadditive gene effects on quantitative traits has been studied in cattle through crossbreeding and inbreeding experiments. This is possible because pure breeds are more homozygous (AA or aa) than breed crosses, and inbred lines within pure breeds are more homozygous than ordinary purebreds. Figure 18 shows homozygosity and heterozygosity expected in pure breeds and inbred lines relative to crossbreds.

Inbreeding is the mating of animals more closely related to each other than the average relationship in the population. The primary effect of inbreeding is to make pairs of genes homozygous and to lower correspondingly the percentage of heterozygous gene pairs. This is why sire-daughter matings are often considered to test for deleterious recessives such as dwarfism. Inbreeding in offspring from sire-daughter matings is 25 percent relative to the pure breed involved. Thus the proportion of heterozygous "carriers" (Aa) resulting from sire-daughter matings is expected to be reduced 25 percent, and half of these are expected to emerge as homozygous recessives (aa), exposing any deleterious recessive traits for which the sire was a carrier.

One study showed that rather intensive inbreeding occurred during the formation of Shorthorns, and that inbreeding continued to accumulate slowly until, by 1920, the inbreeding coefficient for an average animal of the breed was 26 percent. This means that the average Shorthorn in 1920 had 26 percent fewer heterozygous gene pairs than the average of those animals used to form the breed. A study of pedigrees in the Hereford breed indicated that the average inbreeding coefficient increased by 8.1

Fig. 18. Effects of inbreeding on heterozygosity or homozygosity.

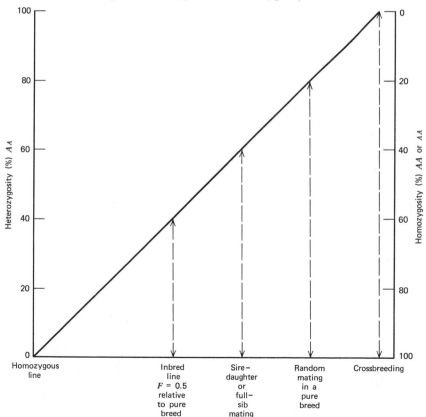

percent in 12.9 generations from 1860 to 1930. Thus, with inbreeding increasing 0.68 percent per generation from 1860 to 1930, inbreeding would be expected to be about 15 percent today relative to the Hereford breed in 1860. Similar rates of inbreeding have been found for other breeds.

Generally, it has not been possible to study the levels of inbreeding early in the formation of the breeds because records of early matings were too fragmentary. It is likely that considerable inbreeding occurred in cattle because of geographic barriers long before herd books and pedigree barriers were established. It is also likely that breeders used the Bakewell system of breeding, which encouraged mating of close relatives to increase prepotency, fix type, and form relatively uniform and distinct breeds of cattle. Even since breeds became established and reasonably large in numbers, inbreeding or homozygosity has slowly but inevitably increased with time. Although the exact extent to which breeds are inbred relative to their crosses is not known, it seems conservative to estimate that purebreds are on the average at least 20 percent more inbred or less heterozygous than their crosses. This level of inbreeding is assumed in Fig. 19 and Table 18 to demonstrate nonadditive effects of genes on performance of crossbreds, purebreds, and inbred lines.

Differences in inbreeding or proportion of heterozygosity would have no effect on average performance of crossbreds, purebreds, or inbreds if

Table 18

Average Expected Performance of Crossbreds, Purebreds, and Inbred Lines with Additive and Nonadditive Gene Effects[a]

Mating Type	Zygotic Frequency Relative to Crossbreds (%)[b]			Additive Gene Effect	Nonadditive Gene Effects	
	AA	Aa	aa		Partial Dominance	Complete Dominance
Crossbreds	0	100	0	0:5	0.75	1.0
Purebreds	10	80	10	0.5	0.70	0.9
Sire-daughter	20	60	20	0.5	0.65	0.8
Inbred line ($F = 0.5$)	30	40	30	.0.5	0.60	0.7
	40	20	40	.0.5	0.55	0.6
Homozygous line	50	0	50	.0.5	0.50	0.5

[a] Phenotypic values shown in Fig. 17 are assumed to compute the averages.
[b] Average expected frequency over a large number of gene pairs.

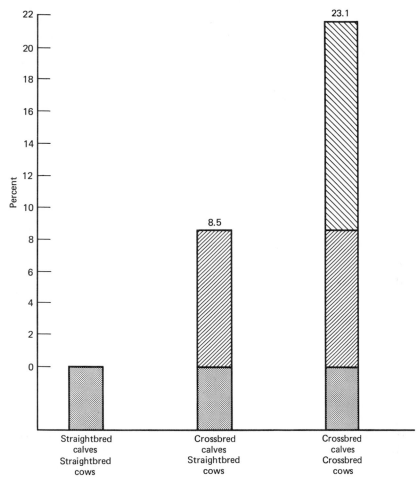

Fig. 19. Cumulative heterosis effects for pounds of calf weaned per cow exposed, Fort Robinson, Nebraska.

gene effects were completely additive (Table 18). However, with any degree of dominance, partial or complete, the mean performance of crossbreds is greater than that of purebreds. Similarly, performance of inbred lines would be expected to decline, on the average, relative to the pure breed or line crosses within a pure breed. This phenomenon is called *inbreeding depression*. It is due to nonadditive effects of genes and the reduction in heterozygosity in inbred lines that uncovers undesirable recessives that otherwise would be concealed or partially concealed by dominant genes. Its reverse, *heterosis*, is the difference in performance between crosses and the average of the parental breeds or lines used in the cross. Heterosis is caused by nonadditive effects of genes when heterozygosity is restored to the pure breed level by crossing inbred lines within a breed or when heterozygosity is restored to an even higher level by crossing breeds.

Explanation of the basis for heterosis could be expanded to include more complex nonadditive genetic models. However, this explanation provides sufficient understanding of the basis for heterosis to formulate the systems of crossbreeding to be discussed in subsequent sections. Fortunately, one need not depend entirely on theory, because a number of experiments have been conducted to obtain direct evidence concerning the importance of nonadditive gene effects on inbreeding depression and heterosis in cattle.

Considerable experimental evidence indicates that nonadditive genetic variation is important in beef cattle. A recent comprehensive study was made on 48 inbred lines from eight state agricultural experiment stations and one U.S. Department of Agriculture (USDA) station cooperating in the Western Region. Results indicated that, on the average, fertility (percentage of cows pregnant) declined 2 percent and 1.3 percent with each 10 percent increase in inbreeding of the dam and calf, respectively. Percent calf crop weaned declined 1.6 percent and 1.1 percent with 10 percent increments in inbreeding of the dam and calf, respectively. Results also have indicated that inbreeding has depressing effects on maternal ability and growth of calves. These results indicate that nonadditive gene effects are important and cause a decline in performance when homozygosity is increased at the expense of heterozygosity by inbreeding.

On the opposite side of the coin, crossbreeding experiments have indicated that heterosis is of significant economic importance in beef cattle. Although the results vary some for specific traits from different experiments and from crosses of different breeds, they generally show significant effects of heterosis on postnatal survival, preweaning and postweaning growth, age at puberty, and fertility and mothering ability of crossbred cows.

EFFECTS OF HETEROSIS

Results on effects of heterosis in beef cattle have been consistent with results in swine, sheep, and poultry, indicating that the level of heterosis (nonadditive genetic variation) is inversely proportional to heritability (additive genetic variation). Thus, in traits of highest heritability—such as postweaning growth rate, feed efficiency, and carcass composition—the level of heterosis is relatively low, whereas in traits with low heritability—such as livability and fertility—effects of heterosis are relatively large.

The following is a summary of results from a crossbreeding experiment conducted by USDA and Nebraska workers involving Herefords, Angus, and Shorthorns. The experiment has been conducted in three phases: Phase I dealt with individual heterosis expressed by the crossbred calf, phase II with maternal heterosis expressed by the crossbred cow, and phase III (in progress) is evaluating the level of heterosis restored from one generation to the next by alternative systems of crossbreeding.

In phase I, the three straightbreds and all reciprocal crosses among Herefords, Angus, and Shorthorns were produced during four calf crops. Heterosis or hybrid vigor was evaluated by comparing the crossbreds with the average of the straightbreds. Crossbred and straightbred calves were sired by the same bulls and were out of comparable cows. The study included 393 crossbred and 358 straightbred calves sired by 16 Hereford, 17 Angus, and 16 Shorthorn bulls. A series of economically important traits were studied, including those presented in Table 19. The effects of heterosis were significant for most of the traits evaluated.

There was no difference between crossbred and straightbred calves for percent calf crop born, but early postnatal survival was significantly greater in crossbreds. Crossbred calves were 4.6 percent heavier at weaning (7 months of age) than straightbreds. The combined advantages in survival and growth rate accounted for an 8.5 percent advantage in weight of calf weaned per cow exposed in favor of crossbred calves over straightbred calves.

There was essentially no difference between crossbred and straightbred steers in feed efficiency. The crossbred steers produced slightly fatter carcasses that graded higher when slaughtered at the same age. However, when adjusted for differences in carcass weight, there were no differences in carcass composition. Thus, if they had been slaughtered at the same weight, the carcasses would have been the same in composition.

In net merit (value of the boneless, closely trimmed retail meat, adjusted for quality grade, minus feed costs from weaning to slaughter), the advantage of the crossbred steers over the straightbred steers was $8.81 (1965) per carcass. This net merit difference is among the steers that lived to slaughter. The 3 percent advantage for the crossbreds in calf crop weaned was not involved in computing this difference.

The effects of heterosis on postweaning growth rate of heifers on lower levels of feeding were greater than in steers on a growing-fattening ration. There was a tendency for effects of heterosis to decrease with age after approximately one year of age. Age at puberty (first heat) was 35 and 40 days younger for crossbred than straightbred heifers on the moderate level of feeding associated with 2-year-old first calving and the low level of feeding associated with 3-year-old first calving, respectively.

The straightbred and reciprocal cross females produced at Fort Robin-

Table 19

Effects of Individual Heterosis in Herefords, Angus, and Shorthorns from Phase I of Fort Robinson Experiment[a]

Item	Crossbred Calves	Straightbred Calves	Heterosis Difference	Percentage
Number of matings	470	447		
Calves born (%)	89	89	0	—
Calves alive at 2 weeks (%)	86	82	4	—
Calves weaned (%)	84	81	+3	—
Birth weight (lb)	74.2	71.5	2.7	3.8
Weaning weight, 200 days (lb)	437.4	418.0	19.4	4.6
200-day weight/cow exposed (lb)	367.4	338.6	28.8	8.5
Steers				
Postweaning daily gain (lb)	1.845	1.794	0.052	3.0
452-day weight (lb)	912	883	29	3.3
TDN/unit gain	5.76	5.77	−0.01	−0.1
Retail product (lb)[b]	331	320	11	3.4
Retail product (%)	63.4	63.9	−0.5	−0.9
Carcass grade[c]	10.2	9.9	0.3	—
Retail product/unit TDN	0.1345	0.1338	+0.0007	0.5
Net merit ($)[d]	220.33	211.52	8.81	4.2

son in phase I of the experiment involving Angus, Herefords, and Shorthorns were retained to evaluate maternal heterosis for reproduction and maternal traits in phase II. Straightbred and reciprocal cross females of each pair of breeds were compared when they were mated to the same sires of a third breed. For example, to evaluate maternal heterosis in Angus-Shorthorn reciprocal crosses, the performance of Angus-Shorthorn and Shorthorn-Angus cows was compared with that of Angus and Shorthorn straightbred cows when the cows in all four groups were mated to the same Hereford bulls.

There were 570 matings of straightbred cows and 687 matings of crossbred cows accumulated over six breeding seasons to produce spring calf crops from 1963 through 1968. Approximately half of the cows were developed and managed to calve first as 2-year-olds and half as 3-year-olds, to evaluate the effects of maternal heterosis expressed from 2

Table 19 *(Continued)*

Effects of Individual Heterosis in Herefords, Angus, and Shorthorns from Phase I of Fort Robinson Experiment[a]

Item	Crossbred Calves	Straightbred Calves	Heterosis Difference	Heterosis Percentage
Heifers				
2-year-old management[e]				
Postweaning daily gain (lb)	1.173	1.100	0.073	6.6
550-day weight (lb)	853	805	48	6.0
Age at puberty (days)	321	356	−35	9.8
Weight at puberty (lb)	580	587	−7	1.2
3-year-old management[f]				
Postweaning daily gain (lb)	0.985	0.910	0.075	8.2
550-day weight (lb)	764	712	52	7.3
Age at puberty (days)	382	422	−40	9.5
Weight at puberty (lb)	528	534	−6	1.1

[a] From Gregory et al., *Journal of Animal Science* 24:21, 25:290, 25:299, 25:311; and Wiltbank et al., *Journal of Animal Science* 25:744, 26:1005.

[b] Pounds closely trimmed, boneless cuts from the carcass.

[c] Grade of 9 = high good, 10 = low choice (USDA grades).

[d] Net merit is value of retail product (dollars) minus feed costs from weaning to slaughter.

[e] Heifers managed and developed to calve as 2-year-olds were born in 1962 and 1963 and fed about 4 pounds of concentrate feed per head per day during their first winter.

[f] Heifers managed and developed to calve first as 3-year-olds were born in 1960 and 1961 and fed 1 pound of 40% protein supplement per head per day during their first winter.

through 6 years of age and 3 through 8 years of age in each management regime, respectively.

Calf crop weaned was 6.5 percent greater for crossbred than for straightbred cows (Table 20). This difference was due to higher pregnancy rates and first-service conception rate in the crossbreds. Differences in postnatal survival of calves were small and not significant in phase II when crossbred and straightbred cows were both raising crossbred calves.

Effects of maternal heterosis were 1.7 percent for birth weight, 3.6 percent for weight at 135 days, and 4.7 percent for weight at 200 days (weaning). These effects of maternal heterosis did reflect greater and especially more persistent milk production favoring crossbred cows over straightbred cows by 0.9 percent at 2 weeks postpartum, 7.5 percent at 6 weeks postpartum, 6.1 percent at about 14 weeks postpartum, and 38 percent at weaning at about 29 weeks postpartum. Actual weaning weight

Table 20

Effects of Maternal Heterosis in Herefords, Angus, and Shorthorns from Phase II of Fort Robinson Experiment[a]

Item	Crossbred Cows	Straightbred Cows	Heterosis Difference	Heterosis Percentage
Number of matings	687	570		
Conceived first service (%)[b]	63.2	56.6	6.6	—
Pregnant, end of breeding season (%)[b]	91.5	85.9	5.6	—
Pregnant in fall (%)[b]	89.7	84.5	5.2	—
Full term calf (%)[b]	87.2	81.1	6.1	—
Live calf born (%)[b]	86.2	80.4	5.8	—
Live calf 2 weeks (%)[b]	84.4	77.8	6.6	—
Live calf weaned (%)[b]	81.6	75.2	6.4	—
Calving to first estrus (days)	53.6	56.3	−2.7	—
Conception date, Julian date	156.3	159.1	−2.8	—
Gestation length (days)	284.7	283.5	1.2	—
Calving date, Julian date	76.0	77.6	−1.6	—
Calf weight				
Number of calves	555	420		
Birth weight (lb)	76.4	75.2	1.2	1.6
135 days (lb)	338.0	326.3	11.7	3.6
Weaning, 200 days (lb)	453.1	434.6	18.5	4.3
12-hour milk production (lb)				
2 weeks	6.79	6.73	0.06	0.09
6 weeks	7.55	7.02	0.53	7.5
June (approx. 14 weeks)	7.91	7.45	0.46	6.2
Weaning (approx. 29 weeks)	3.31	2.40	0.91	37.9
200-day weaning weight/ cow (lb)[b]	379.3	333.0	46.3	13.9
Actual weaning weight/ cow (lb)[b]	392.5	341.8	50.8	14.8

[a] From Cundiff et al., *Journal of Animal Science* 38:711, 38:728, 1974a,b.

[b] Based on all cows exposed to breeding: 687 cow-year-matings for crossbred females and 570 cow-year-matings for straightbred females.

was 14.8 percent greater per cow exposed to breeding for crossbred cows than for straightbred cows, on the average, over both management regimes due to combined effects of maternal heterosis on reproduction and maternal ability.

Preliminary results indicate that further advantages will be accrued through greater longevity and lifetime production of crossbred cows relative to straightbreds. The cows involved in phase II currently range from 11 to 14 years of age. Thirty-one percent of the original crossbred heifers assigned to breeding pastures to initiate phase II remained in the breeding herd during the 1974 breeding season compared to 16 percent of the straightbreds.

For growth, feed efficiency, and carcass traits, the heterosis effect was greater in the Hereford-Angus and Hereford-Shorthorn combinations than for the Angus-Shorthorn combination. For age and weight at puberty and for subsequent maternal performance, the effect of maternal heterosis was greatest for Hereford-Shorthorn reciprocal crosses.

When the advantages of individual heterosis on survival and growth of F_1 calves (phase I) and the advantage of maternal heterosis on reproduction and maternal ability of crossbred cows (phase II) are all combined, results from this experiment indicate that weight of calf weaned per cow exposed to breeding would be increased 23 percent (Fig. 19). More than half of the increased performance from heterosis was attributable to crossbred cows.

These results are generally typical of what has been obtained in other experiments. Results from an experiment in Virginia involving the same British breeds have varied some for specific traits, but the total effect of heterosis on pounds of calf weaned per cow exposed has been very similar. Results from USDA and Montana, Missouri, and Ohio experiments involving Charolais, Angus, and Herefords and from Iowa involving Herefords, Angus, Brown Swiss, and Holsteins have shown effects of heterosis similar to those summarized in Tables 19 and 20 for most traits. Differences in effects of heterosis between specific crosses have generally not been significant in these experiments. Results from experiments in USDA Southern Region involving crosses of Brahmans with Herefords and Angus have been similar in pattern to those involving British breeds, being inversely related to heritability, but effects of heterosis have generally been even higher in Brahman-British breed crosses than in British breed crosses.

COMBINING CHARACTERISTICS OF BREEDS

Additive genetic differences between breeds are responsible for differences observed between breeds in performance traits when they are managed in the same environment. Experiments have shown that there are substantial differences between breeds for traits such as milk production, growth, mature size, and carcass composition at various points along their growth curves. These differences are caused by differences between breeds in the frequency of genes affecting each trait that have emerged as a result of diverse selection goals (for example, beef versus dairy), natural selection for adaptation in different environments, and genetic drift (chance).

A very important advantage of crossbreeding is that it can be used to more nearly optimize gene frequency in a crossbred population and to combine desired characteristics in the crossbreds that would not be possible in any parent breed alone. This can be done very effectively because heritability of differences between breeds tends to be very high. *Heritability* is that portion of the differences observed between animals that is transmitted to their progeny. To the extent that breeds differ in average performance for a given trait, differences between animals of different breeds tend to be more highly heritable than differences of the same magnitude within a breed because progeny are regressing back to different breed means. This is partly a consequence of the inbreeding that has occurred in the evolution of breeds to increase the genetic likeness or uniformity of animals within breeds and the distinctness of differences between breeds.

Experiments involving crosses between Brahmans and Angus, Herefords, or Shorthorns in the Southern Region (Louisiana, Florida, Texas, Georgia, and South Carolina) were the first to focus attention on the opportunity to utilize additive as well as nonadditive genetic effects to combine desired characteristics of different breeds. Brahman-British breed crosses combine the heat and insect tolerance of the Brahmans, which evolved in the tropics, with the desired carcass characteristics, including carcass grade and tenderness, of the British breeds. This, combined with the substantial effects of heterosis on survival and growth of calves and fertility and maternal ability of cows, has resulted in more productive crossbreds that are better adapted for beef production in the Southern Region than either of the parent breeds.

Experiments involving Charolais top crosses at Louisiana, and heterosis experiments involving Charolais, Angus, or Herefords conducted by the USDA in Montana and by agricultural experiment stations at Ohio and Missouri, focused further attention on the opportunity that crossbreeding provides to combine characteristics of breeds. Effects of heterosis in these experiments have been similar to those in British breed crosses for growth and carcass traits. However, results have demonstrated that Charolais cattle are superior to Herefords and Angus in preweaning and postweaning growth rate and they produce carcasses containing a higher percentage of lean and less fat at comparable ages. While straightbred Charolais have graded one-third to two-thirds of a grade lower than Herefords or Angus, respectively, the grades of Charolais-Hereford and Charolais-Angus crosses have been only slightly lower than those of Hereford-Angus crosses. These experiments indicate that Charolais-British breed crosses do combine characteristics of both breeds in a desired direction; more specifically, to increase growth and percentage of retail product and reduce fat trim of the British breeds and to improve carcass grade of the Charolais. Results such as these stimulated importation of a number of new continental European breeds into North America via quarantine facilities in Canada, beginning in the late 1960s.

An extensive Germ Plasm Evaluation Program was initiated at the U.S. Meat Animal Research Center in 1969 which includes a number of these breeds and other domestic breeds formerly considered only for dual-purpose or dairy production. This program extends the evaluation of crossbreeding to a broad range of biological types represented by breeds varying widely in characteristics such as milk production, growth, mature size, and carcass composition at various points along their growth curves. The major objective is to characterize breeds representing different biological types in different feed environments and production situations for the full spectrum of biological traits relating to economic beef

Fig. 20. Straightbred Hereford and Angus cows with offspring sired by bulls of various breeds managed under uniform conditions to compare preweaning performance. (USDA Meat Animal Research Center.)

production. The program is being conducted in a series of cycles involving crossbreds by different sire breeds out of Hereford and Angus cows compared to control populations of Hereford-Angus reciprocal crosses.

The first cycle (cycle 1, phase I) involved breeding Hereford, Angus, Jersey, South Devon, Limousin, Simmental, and Charolais bulls (20 to 35 sires per breed) by artificial insemination (AI) to Hereford and Angus cows (ranging from 2 to 7 years old at calving) to produce three calf crops in March, April, and early May of 1970, 1971, and 1972.

Results on factors influencing calving difficulty and effects of calving difficulty on subsequent reproduction and prenatal and early postnatal mortality have been reported. A high degree of calving difficulty cannot

Fig. 21. Calving difficulty in Hereford and Angus cows according to sire breed of calf and age of cow. (U.S. Meat Animal Research Center, USDA, ARS; Laster et al., *Journal of Animal Science* 36:695, 1973.)

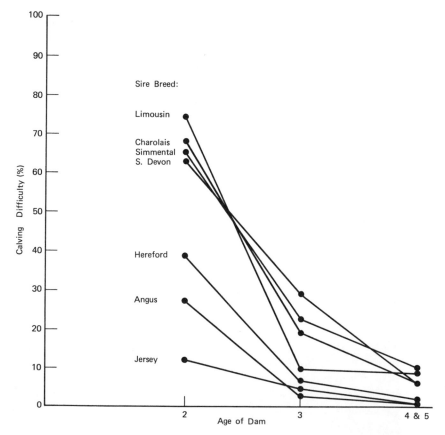

Fig. 22. Postweaning feedlot performance of F_1 steer offspring being measured under uniform conditions to study combining ability of various sire breeds with Hereford and Angus cows. (USDA Meat Animal Research Center.)

be tolerated in commercial beef production. Not only is the expense of labor for assistance at calving a prohibitive factor, but results from this experiment indicated that conception rate in the subsequent breeding season was reduced 16 percent by calving difficulty (85.3 percent conception rate for cows not experiencing difficulty versus 69.4 percent for cows experiencing difficulty). Calf mortality at or near the time of birth was four times greater in calves experiencing difficult parturitions than in those not experiencing difficulty (20.4 percent versus 5.0 percent).

Breeds siring the heaviest calves (Charolais, Simmental, Limousin, and South Devon) experienced significantly more calving difficulty than those sired by Hereford, Angus, and Jersey sires. The incidence of calving difficulty had a strong association with birth weight. On the average, calving difficulty increased 1 percent for each pound increase in birth weight. The association between calving difficulty and birth weight relative to sire breed was very strong in 2-year-old first-calf Angus and Hereford heifers and was also present to a lesser degree in 3-year-old cows (Fig. 21). These effects were substantial and must be taken into consideration when developing a crossbreeding program. The incidence of calving difficulty

Table 21

United States Meat Animal Research Center Germ Plasm Evaluation Program Breed Group Means for Calves Produced in Cycle 1, Phase I (1970–71–72 Calf Crops)[a]

Item	Breed Group[b]						
	HH + AA	AH + HA	JH + JA	SDH + SDA	LH + LA	SH + SA	CH + CA
Steers and heifers, preweaning							
Number	289	382	283	215	353	384	358
Birth weight (lb)	79.8	81.6	73.0	85.7	86.5	91.1	91.9
Adjusted 200-day:							
weight (lb)	459	475	452	478	485	501	505
ratio[c]	96.6	100.0	95.2	100.6	102.1	105.5	106.3
Steers, postweaning[d]							
Number	154	211	134	94	175	177	178
Daily gain (lb)	2.29	2.34	2.18	2.46	2.32	2.55	2.59
TDN/unit gain	6.63	6.75	7.07	6.73	6.56	6.61	6.62
Adjusted final							
weight (lb)	1,015	1,042	986	1,068	1,058	1,122	1,129
ratio (%)[c]	97.4	100.0	94.6	102.5	101.5	107.7	108.3
Dressing (%)	61.0	61.2	60.0	61.6	61.7	60.4	61.2
Retail product (%)[e]	66.1	65.1	65.1	66.2	69.8	68.4	69.3
Fat trim (%)	21.9	23.0	22.6	21.5	17.5	18.2	17.8

Bone (%)	12.1	11.9	12.3	12.3	12.6	13.3	12.9
USDA quality grade[f]	11.8	11.9	11.7	11.8	10.7	11.1	11.4
W-B shear (lb)[g]	7.0	7.1	6.6	6.6	7.5	7.6	7.0
Acceptance[h]	7.3	7.3	7.5	7.4	7.2	7.1	7.3
Heifers, postweaning							
Number	129	133	117	120	161	157	133
200-day average daily gain (lb)	1.04	1.16	0.99	1.21	1.12	1.20	1.18
400-day weight (lb)	610	640	581	661	641	688	685
550-day weight (lb)	686	721	660	740	720	776	779
Reached puberty (%)[i]	88.8	96.0	99.7	94.7	86.0	94.9	94.1
Age at puberty (days)	394	377	328	365	399	369	399
Weight at puberty (lb)	593	588	487	602	644	637	673
Pregnant (%)	82.6	87.4	88.2	83.9	81.5	84.5	80.6

[a] From Progress Report No. 2, U.S. Meat Animal Research Center, ARS, USDA, ARS-NC-13, March 1974.

[b] First letter denotes sire breed and second letter denotes dam breed where H = Hereford, A = Angus, J = Jersey, SD = South Devon, L = Limousin, S = Simmental, and C = Charolais.

[c] Ratio computed relative to Angus-Hereford and Hereford-Angus controls.

[d] Means are averaged by least squares over three slaughter dates after an average of 212, 247, or 279 days on feed postweaning. The finishing ration contained 60% corn silage (71.2% TDN, 12% crude protein).

[e] Retail product, % = actual yield of boneless, closely trimmed beef from carcass.

[f] USDA quality grade: 10 = average good, 11 = high good, 12 = low choice, and so on.

[g] Warner-Bratzler shear measures pounds of force required to shear one-half-inch cores of rib steaks cooked at 176.7° C to 65.6° C internal temperature (1,123 steers).

[h] Taste panel scores based on a 9-point hedonic scale, with higher scores indicating greater acceptability (72 per sire breed).

[i] First estrus was determined from weaning to an average age of about 15 months (end of AI) for the 1970 calf crop and to an average of about 16 months (end of AI plus clean-up) for the 1971-72 calf crops.

was much lower in 4- and 5-year-old Angus and Hereford cows, ranging from only 0.2 percent for Jersey-sired calves to 10.2 percent for Charolais-sired calves.

All male calves were castrated shortly after birth, weaned in October or November at approximately 200 days of age, fed out on a ration containing about 71 percent TDN, and slaughtered after an average of either 212, 247, or 279 days on feed. Detailed carcass data have been obtained in cooperation with Kansas State University. Results from the most recent report on the three calf crops produced in the first cycle of this program are summarized in Table 21 according to sire breed. All of the calves by Jersey (J), South Devon (SD), Limousin (L), Simmental (S), and Charolais (C) sires were F_1 calves out of Hereford (H) and Angus (A) dams; thus comparisons to Herefords and Angus should be made relative to the Hereford-Angus reciprocal crosses to account for the average effect of heterosis.

Substantial differences are indicated among sire breeds for preweaning and postweaning growth rate. Weaning weight ratios expressed relative to Hereford-Angus reciprocal crosses range from 95 percent for Jersey to 105.5 percent and 106.3 percent for Simmental and Charolais crosses, respectively. Similar results were found in average final weight. Limousin, Simmental, and Charolais crosses were relatively efficient in feed conversion in spite of heavier weights throughout the feeding period and produced carcasses with a higher percentage of retail product and less fat trim than Hereford-Angus reciprocal crosses. Although there were differences between the breed groups in marbling and USDA carcass quality grade that are important in terms of current market values, small differences were found in palatability characteristics evaluated by taste panel tests and by Warner-Bratzler shear determinations of tenderness.

Differences among breed groups in postweaning growth of heifers up to 18 months of age have been similar to those found in steers. As in the Nebraska heterosis experiment with Herefords, Angus, and Shorthorns, a sizable effect of heterosis for age at puberty has been observed in Hereford-Angus reciprocal crosses. Average age at puberty ranged from 328 days for Jersey crosses to 399 days for Limousin crosses.

It is too early to assess the relative merit of various biological types for the full spectrum of economic traits including fertility, maternal ability, and productive efficiency. Evaluations of reproduction and maternal performance are in progress. Intensive experiments are being initiated to evaluate nutrient requirements to support growth, maintenance, lactation, and reproduction among different biological types varying in mature size and milk production. A major objective is to develop an understanding relating to optimizing biological factors such as cow size and milk level, and to maximize reproduction and productive efficiency in different feed environments and production situations.

SYNCHRONIZING GENETIC RESOURCES WITH FEED AND OTHER PRODUCTION RESOURCES

Significant differences in characteristics such as growth, carcass composition, mature size, and milk production exist between breeds. At the same time, vast differences are noted between quantity and quality of feed supply from one region of the country to another. Stocking rates range from one cow per acre to one cow per section in the United States. Even within a region, large differences are noted between quantity and quality of feed supply from one operation to another. These differences often are determined by resources such as water, fertility, and variety of grasses, forages, crops, or crop residues available or supplied to provide the nutrient base for the beef operation.

Just as engines that vary in horsepower and performance require different quantities and quality of fuel for maximum efficiency, so do cattle that vary in performance capability require different quantities and quality of feed to provide their requirements for growth, maintenance, lactation, and reproduction, for maximum conversion of feed and other resources to beef. This has been demonstrated in recent experiments involving breeds varying widely in size and lactation potential.

In an experiment conducted under range conditions in Oklahoma (Table 22), Angus-Holstein heifers were compared to Angus heifers when

Table 22

Angus-Holstein Crossbred Heifers Compared to Angus Heifers under Range Conditions in Oklahoma[a]

Item	Angus	Angus-Holstein
Number of cows	27	23
200-day total milk production (lb)	1,753	2,505
Weight before calving (lb)	760	870
Weight after weaning (lb)	753	814
Weight change (lb)	−7	−56
Percentage rebreeding, during 90-day breeding season	63	13

[a] From Deutscher and Whiteman, *Journal of Animal Science* 33:337, 1971.

both groups were wintered on a relatively low level of supplement. Winter supplemental feed consisted of 2 pounds per head per day of cottonseed meal from November 15 to April 15 and 5 pounds per head per day of prairie hay from January 1 to April 15. Angus-Holstein heifers produced significantly more milk but lost significantly more weight in the subsequent breeding season than Angus. Of 23 Angus-Holstein F_1 heifers and 27 Angus heifers that calved in the spring as 2-year-olds, only 13 percent of the Angus-Holstein heifers rebred while 63 percent of the Angus rebred during a 90-day natural service breeding season.

A more recent experiment at Oklahoma involving Herefords, Hereford-Holstein crosses, and straightbred Holsteins fed different levels of supplement primarily during lactation has provided further evidence that reproduction is reduced if additional nutrients to support greater requirements for growth, maintenance, and lactation of larger cows with higher lactation potential are not met (Table 23). Two levels of supplementation were fed to each breed group through the winter beginning in the fall at calving. An additional group of straighbred Holsteins was fed a very high level of supplement. Average size, and particularly milk production, differed markedly among these breed groups. As level of supplementation increased, weight loss during lactation decreased and cows tended to exhibit estrus and conceive earlier. This trend has been more pronounced in the breed groups of larger size and higher lactation potential.

Holsteins represent an extreme biological type characterized by high milk production potential and relatively large size. Milk production has increased markedly from selection practiced for a long period of time and especially during the last 20 years when selection has been intensified by extensive progeny-testing and use of artificial insemination. This has been accompanied by changes in dairy herd management that include high levels of supplemental grain fed in relation to level of milk production. Similar relationships among nutrient intake, reproduction, size, and lactation exist, presumably to a somewhat lower degree among breeds that differ by a narrower range in milk production or size, or both. Experiments are in progress involving different biological types, to separate differential requirements for growth and maintenance from those for lactation while providing for maximum reproduction.

Although this subject is not completely understood, current results point to the need to synchronize genetic resources with feed and other production resources. Similar considerations are important in the growing-finishing segment of production in terms of optimal efficiency of growth and carcass composition to meet consumer and market requirements. When a crossbreeding program is being planned, it is important to choose breeds that are adapted to the feed and other resources available to promote maximum economy of production in the herd or specific operation.

IMPORTANT CONSIDERATIONS IN PLANNING CROSSBREEDING PROGRAMS

Substantial effects of heterosis can be utilized through crossbreeding as follows.

1. Results with Hereford, Angus, and Shorthorns indicate that heterosis can increase pounds of calf weaned per cow in the breeding herd by 23 percent.
2. It is important for crossbreeding programs to involve crossbred cows because more than half of this advantage is dependent on the use of crossbred cows.

Crossbreeding also can be utilized to combine desired characteristics of breeds and synchronize genetic resources with feed and other resources.

1. In certain situations, breeds of large size with lean carcasses can be mated to cows of small to medium size to increase efficiency of production—that is, to increase amount and value of retail product produced relative to feed costs of the calves and of the cows.
2. Advantages of mating sires of large size to cows of medium size are tempered by a high degree of calving difficulty when the cows are calving at 2 and 3 years of age.
3. To avoid calving difficulty and associated calf crop losses from calf mortality and reduced rebreeding, cows should be mated to bulls of other breeds that are similar in size until they are 4 years old or older. In cows calving when 4 years old or older, calving difficulty has not been a serious problem.
4. Maternal breeds of which the cow herd is made up should be well adapted to the climatic-feed environment and production situation.
5. Nutrient demands to support growth, maintenance, reproduction, and lactation of the cow herd need to be stabilized from one year to the next in most instances.

SYSTEMS OF CROSSBREEDING

It is possible for poultry breeders and plant breeders of many species—corn, for instance—to use static systems of mating that produce sufficient

Table 23

Performance of Hereford, Hereford X Holstein, and Holstein Fall-Calving Females as Influenced by Level of Winter Supplementation under Range Conditions[a]

Item	Hereford		Hereford X Holstein		Holstein		
	Moderate	High	Moderate	High	Moderate	High	Very High
2-year-olds							
Total winter supplement (lb)[b]	434	731	515	882	595	945	1,203
Precalving weight (lb)	885	904	988	995	1,151	1,090	1,116
Mid-lactation weight (lb)	753	788	813	882	946	917	954
Postlactation weight (lb)	985	983	993	1,055	1,152	1,156	1,200
Daily milk yield (lb)	12.0	12.9	17.3	19.3	23.5	24.5	28.4
Postpartum interval (days)[c]	71	62	82	68	83	71	65
Fraction conceiving[d]	12/12	12/12	13/13	13/13	8/11	9/11	11/11
Calf weaning weight (lb)[e]	507	500	550	563	604	621	634
3-year-olds							
Total winter supplement (lb)	353	691	343	763	390	780	1,189
Precalving weight (lb)	1,012	1,022	995	1,073	1,187	1,172	1,210

Mid-lactation weight (lb)	850	832	812	897	912	955	1,011
Postlactation weight (lb)	1,035	1,046	1,011	1,092	1,168	1,172	1,240
Daily milk yield (lb)	13.4	13.2	17.7	22.4	31.2	27.8	31.0
Postpartum interval (days)	72.7	68.3	79.3	66.2	126.7	78.5	57.7
Fraction conceiving	12/12	9/10	8/11	12/13	4/4	8/9	8/8
Calf weaning weight (lb)	601	592	645	641	723	736	730
4-year-olds							
Total winter supplement (lb)	221	565	263	564	313	626	891
Precalving weight (lb)	990	1,030	1,096	1,051	1,272	1,183	1,212
Mid-lactation weight (lb)	796	891	857	865	968	968	1,056
Postlactation weight (lb)	1,011	1,066	1,034	1,055	1,218	1,146	1,244
Daily milk yield (lb)	13.4	13.5	20.4	20.2	25.5	29.8	26.2
Postpartum interval (days)	58	51	69	58	83	91	63
Fraction conceiving	16/19	17/18	15/17	17/19	6/15	9/16	15/16
Calf weaning weight (lb)	574	576	625	659	732	699	692

[a] From Animal Science and Industry Research Reports 1972, MP-72; 1973, MP-90; and 1974, MP-92; Agricultural Experiment Station, Oklahoma State University and USDA, ARS.

[b] Soybean meal (44% C.P.), 60.1%; milo, ground, 30.3%; dehydrated alfalfa meal, 5.0%; dicalcium phosphate, 2.9%; Masonex, 1.3%; salt, 0.5%; plus vitamin A added at 10,000 IU/lb of supplement.

[c] Number of days from calving to first observed estrus.

[d] Fraction conceiving is number of cows conceiving per number of cows raising a calf during the breeding season.

[e] Adjusted for sex of calf and to an average weaning age of 240 days.

hybrids for complete use of heterosis in commercial production. Maximum use of hybrid vigor can be made in these species because only a small proportion of the total population is involved in seedstock production. Complete utilization of heterosis is more difficult in cattle because of their relatively low reproductive rate and long generation interval, which overlaps from one year to the next. However, this does not preclude utilization of a high level of heterosis in commercial beef production. Systems of crossbreeding can be used that restore significant levels of heterosis from one generation to the next, and provide for utilization of additive genetic variation between breeds to synchronize genetic resources with feed and other production resources.

There is no single, universal crossbreeding system of choice. The choice depends on resources specific to each operation, such as number of cows in the herd, number of breeding pastures, facilities, labor availability, and amount and quality of feed supply. Practically any system of crossbreeding is feasible in conjunction with artificial insemination, provided adequate facilities, labor, and technical expertise are available to accommodate a successful artificial insemination program. Since artificial insemination is not always feasible, discussion of crossbreeding systems will include breeding systems involving use of bulls in natural service. The following discussion is intended to be reasonably complete with regard to the basic systems of crossbreeding; however, it should be recognized that modifications often are necessary for optimum use of heterosis in any specific operation.

ROTATIONAL SYSTEMS

Rotational systems of crossbreeding have been used in commercial swine production for a number of years. The systems most commonly being implemented in commercial beef production are diagramed in Figs. 23 and 24.

In the two-breed rotation (Fig. 23), the program is initiated by mating cows of breed A to bulls of breed B. Heifers resulting from these matings are in turn mated to bulls of breed A. In the next generation heifers by breed A are mated to bulls of breed B, and so on, generation after generation. Thus a minimum of two breeding pastures are required for this system and it is necessary to identify heifers only by breed of their sire.

In the three-breed rotation the same pattern is involved except that a third breed is included in the rotation. Heifers sired by breed A are mated to bulls of breed B. Replacement heifers from these matings are bred to bulls of breed C, and the resulting heifer progeny kept for replacement

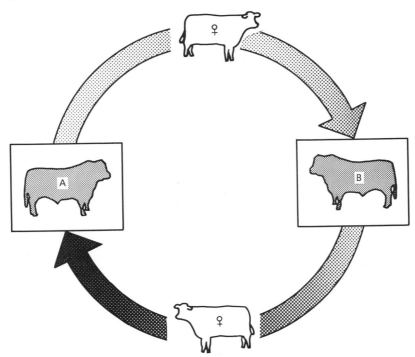

Fig. 23. A two-breed crossbreeding rotation system. Heifers sired by breed A are mated to bulls of breed B and heifers sired by breed B are mated to bulls of breed A. Pounds of calf per cow increased about 15 percent.

are in turn mated to bulls of breed A, and so on. In a three-breed rotation at least three breeding pastures are required, and again it is necessary to identify heifers according to the breed of their sire.

Rotational systems restore a substantial level of heterosis from one generation to the next. Tables 24 and 25 show the genetic composition and level of heterosis expected in each generation as two- and three-breed rotations continue over time. The level of heterozygosity in calves and cows fluctuates in rotational systems during the initial generations. However, once crossbred cows have entered the system this is hardly noticeable in terms of performance level, because heterozygosity of the calf and heterozygosity of the dam are both important, and low levels in one are offset by high levels in the other. Thus, on the average, the level of heterosis expected from a two-breed rotation is 67 percent of the maximum expected when an F_1 cow is mated to sires of a third breed, and levels of performance due to heterosis do not fluctuate greatly once crossbred cows have entered the system.

The level of heterosis sustained by a three-breed rotation is higher than

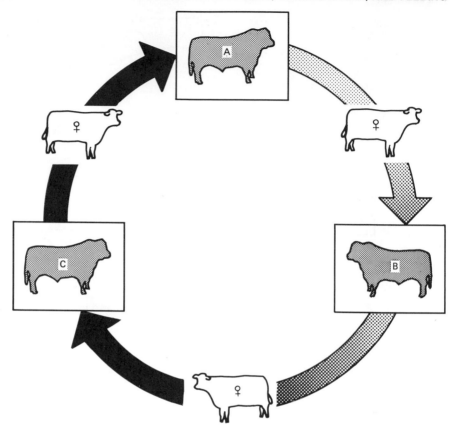

Fig. 24. A three-breed crossbreeding system. Pounds of calf per cow increased about 20 percent.

that in a two-breed rotation because the relationship between bulls and cows being mated is more remote. The three-breed rotation sustains an average level of 87 percent of the maximum heterozygosity realized in the second generation when the F_1 cow is mated to sires of a different breed. Once crossbred cows enter the system, performance due to heterosis is increased substantially and, in subsequent generations, performance level due to heterosis fluctuates very little.

Additive genetic composition, on the other hand, fluctuates greatly in two- or three-breed rotations, being highest for the most recent sire breed and lowest for the most remote sire breed. For example, in the eighth generation of a three-breed rotation and on the average over all generations, 57 percent of the genes of the cows are of the breed of their sire, 29 percent are of the breed of their grandsire, and 14 percent are of the breed of their great-grandsire, the latter being the same as the breed to which they are to be mated. Consequently, it is important for rotational systems to involve breeds that are reasonably comparable in characteristics such as birth weight, size, and lactation potential and well adapted to the feed and other production resources in the operation. This is necessary to avoid calving difficulty and associated calf crop losses, which can be a serious problem in females calving at 2 and 3 years of age if they have been mated to bulls of a breed that is substantially larger in size. Also, it is important to use breeds that are comparable in size and milk production to stabilize nutrition and management requirements in the cow herd. This is especially important if it is not feasible to feed and manage each group separately according to their size and lactation potential as well as age, during periods when supplemental feed is required.

Rotational systems have an advantage over some other systems in that all replacement heifers are provided from within the system. Opportunity to select replacement heifers intensely can be slightly greater in rotational systems than in straightbreeding systems because of the higher percentage calf crop weaned and the greater longevity of crossbred cows. However, genetic improvement within the herd is still largely determined by the sires selected within each breed. Bulls for use in rotational crossbreeding

Table 24

Genetic Composition and Level of Heterosis Expected in a Two-Breed Rotation

Gener-ation	Sire	Additive Genetic Composition (%)				Heterozygosity (%)		Calf Weight Weaned per Cow Exposed[a]
		Dam		Calf				
		A	B	A	B	Dam	Calf	
1	A	—	100	50	50	0	100	8.5
2	B	50	50	25	75	100	50	19.0
3	A	25	75	63	37	50	75	13.8
4	B	63	37	69	31	75	62	16.4
5	A	69	31	66	34	62	69	15.1
6	B	66	34	67	33	69	66	15.8
7	A	67	33	67	33	66	67	15.9
8	B	67	33	67	33	67	67	15.5
Average		50	50	50	50	67	67	15.5

[a] Based on heterosis effects of 8.5% in a crossbred calf and 14.8% in a crossed cow observed in phase I (Table 18) and phase II (Table 19) of the Fort Robinson Heterosis Experiment assuming that heterosis is directly proportional to heterozygosity. Experiments are in progress to determine the veracity of this assumption.

Table 25

Genetic Composition and Level of Heterosis Expected in a Three-Breed Rotation

Generation	Sire	Additive Genetic Composition						Heterozygosity (%)		Calf Weaning Weight per Cow Exposed[a]
		Dam			Calf					
		A	B	C	A	B	C	Dam	Calf	
1	A			100	50	0	50	0	100	8.5
2	B	50	0	50	25	50	25	100	100	23.3
3	C	25	50	25	12	25	62	100	75	21.2
4	A	12	25	62	56	12	31	75	88	18.5
5	B	56	12	31	28	56	16	88	88	20.4
6	C	28	56	16	14	28	58	88	84	20.1
7	A	14	28	58	57	14	29	84	86	19.8
8	B	57	14	29	29	57	14	86	86	20.0
		—	—	—	—	—	—	—	—	—
Average		33	33	33	33	33	33	86	86	20.0

[a] Based on heterosis effects shown in Tables 19 and 20.

should be selected just as judiciously as those selected for use in straight-breeding or any other system of crossbreeding.

STATIC TERMINAL SIRE SYSTEM

In a static terminal sire system (Fig. 25), straightbred cows of one breed (A) are mated to a second sire breed (B), to produce F_1 crossbred females (BA). The F_1 crossbred females are mated to a third breed (C) and the resulting progeny, both male and female, are marketed. The object of this system is to achieve maximum efficiency of production by selecting breeds used in the sequence of matings that complement each other to the highest degree possible.

The first two breeds (A and B) used in the sequence should be selected

Fig. 25. A static three-breed crossbreeding system. Male progeny of the first two matings and all progeny of the third mating are marketed. Pounds of calf per cow increased about 19 percent.

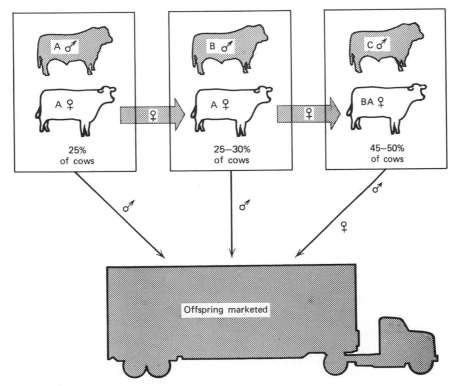

to synchronize cow size and maternal performance with feed resources available. They need not be comparable in both respects as long as the sequence is such that calving difficulty is not a problem. For example, one breed can be of larger size with lower milk production potential and the other of smaller size with higher milk production potential provided that the feed requirements of the resulting F_1 females are about the same as those of breed A or can otherwise be provided for economically in the operation.

Rate and efficiency of gain and carcass composition should be emphasized in sire breed C, the terminal cross, to maximize these characteristics in as many progeny marketed as possible. The primary advantage of this system is that pounds of beef marketed can be increased per unit of feed consumed by both the cows and the calves.

However, these advantages may be offset by considerable sacrifice of heterosis in this system if all replacements are produced rather than purchased. It is necessary to mate a high percentage of the cows (approximately 25 percent) as straightbreds to meet replacement requirements for straightbreds in the system. In addition, at least 25 percent of the cows in the herd must be straightbreds (A) mated to produce F_1 heifers to provide replacements for the herd mated to the terminal sire breed. Thus the advantages of greater fertility and increased maternal performance of crossbred cows are sacrificed in at least 50 percent of the herd, and the advantages of individual heterosis on survival and growth of F_1 calves are sacrificed in the straightbreds produced that make up 25 percent of the herd. Thus maximum heterosis is only realized in the terminal cross which comprises about 50 percent of the cows in the system, unless replacements are purchased.

Another disadvantage of the static terminal sire system is that very little selection pressure can be applied among female replacements entering the cow herd. Nearly all females (about 90 percent) produced by the straightbred cows are required as replacements if the number of cows in the herd is to be maintained. This can be altered by increasing the proportion of straightbred cows to crossbred cows, but this deducts from the benefits of heterosis and complementarity. The cost of sacrificing selection pressure in females is often overemphasized, however, since the impact of selection on the herd is determined primarily by the bulls selected to produce the heifers. Research has shown that 80 to 90 percent of the genetic improvement in a herd is attributable to sire selection. Selection pressure among sires can and should be just as great in a static terminal sire crossbreeding system as in any other system of mating. The characteristics receiving primary emphasis in selection should differ from one breed to the other and correspond to the characteristics that each breed contributes to the system.

A static terminal sire system does require at least three breeding pastures, as indicated in Fig. 25. However, to avoid calving difficulty in 2- and 3-year-old F_1 crossbred females, it may be necessary to use a smaller sire breed, requiring a fourth breeding pasture, or to make backcross matings to one of the maternal sire breeds (A or B) for the first and second calvings. The use of the fourth sire breed permits maximum heterosis, but the backcross matings would only sacrifice some of the individual heterosis in the calves produced and is probably easier to manage in most situations.

Assuming levels of heterosis shown in Tables 19 and 20 and assuming that the terminal sire breed increases weaning weight of calves by 5 percent, pounds of calf marketed per cow in the breeding herd should be increased about 18 to 19 percent over straightbreeding by this system. This is similar to the level of production shown for a three-breed rotation (Table 25). The sacrifice in heterosis is partly offset by the additional growth expected from use of a terminal sire, and the fact that a high percentage (42 percent) of the calves marketed are by the terminal sire. If technology were available to control sex of offspring, this system would have much more to offer; however, with the technology presently available, the static terminal sire system appears to have no more to offer than a three-breed rotation. Thus the system of choice will depend on personal preference, marketing considerations, and ease of management in specific operations.

ROTATIONAL-TERMINAL SIRE SYSTEMS

It is possible in some operations to combine most of the advantages of rotational systems with those of a terminal sire system. Rotational-terminal sire systems involve rotational mating of maternal breeds in a portion of the herd to provide female replacements for the entire herd, a portion of which are mated to a terminal sire breed as outlined in Fig. 26. In most herds, at least 45 percent of the cows will have to be mated in the rotational portion of the program to meet requirements for heifer replacements. The other cows in the herd (50 to 55 percent) would be mated to a terminal sire breed selected to excel in rate and efficiency of growth and carcass composition. The object of the system is to promote maximum efficiency of production in terms of pounds of calf or beef produced per unit of feed consumed by calves and cows in the herd.

To achieve maximum benefit from the system, the breeds used in the rotational portion of the program should be small to medium in size and selected to synchronize maternal performance with feed and other production resources available in the operation. Since a rotational crossing

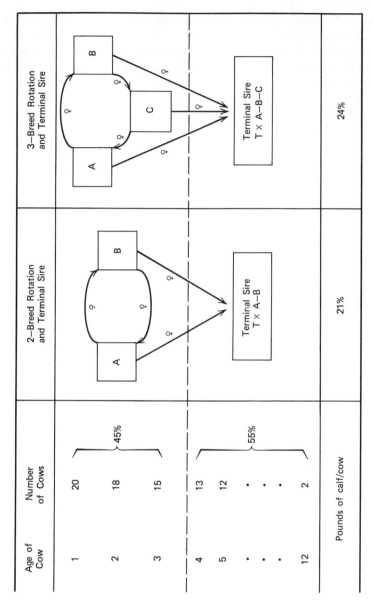

Fig. 26. A rotational and terminal sire crossbreeding system (100 cows calving and 20 yearling heifers are assumed).

system is employed, it is important for the maternal breeds to be reasonably comparable in terms of size and milk production to minimize calving difficulty and stabilize nutrition and management requirements in the cow herd.

It has generally been recommended that the rotational crossing portion of the program should involve the youngest cows in the herd. In many instances 1-, 2-, and 3-year-old females make up 45 percent of the breeding females in a herd. However, 4-year-old cows can be involved in the rotational crossing if they are needed to maintain or increase herd size. In this way cows are not mated to a terminal sire breed of large size until they are 4 years old or older, when the incidence of calving difficulty is low and not a serious problem.

Again, as in the static terminal sire program, nearly all of the heifers produced in the rotational portion of the program are required as replacements. This may be offset, however, by the fact that the heifers are all out of relatively young cows. Research at several experiment stations has shown that maternal performance of females raised by young cows is significantly greater than that of females raised by mature cows. This may be due to a tendency for heifer calves to fatten as a result of the favorable maternal environment associated with greater milk production of mature cows compared with younger cows. Fat deposition may in turn interact with mammary and udder development in the growing female to reduce subsequent milk production of females raised by mature cows relative to those raised by young cows. This interrelationship can be used to advantage if the youngest cows in the herd are employed in the rotational system to provide female replacements for the herd.

The terminal sire breed would be selected to excel in rate and efficiency of growth and to transmit superior carcass composition to his offspring. Female offspring from the terminal sire breed would not be kept for replacement, to avoid increased feed requirements for maintenance that would result if cows of large size were kept in the herd. Both the male and the female offspring from the terminal sire would be marketed; hence the name "terminal" sire breed.

Either the two-breed or the three-breed rotational-terminal sire system requires a high level of management. Cows must be identified by breed of sire and birth year. The two-breed rotational-sire system requires three breeding pastures. The three-breed rotational-terminal sire system can be run with four breeding pastures or, if bulls of the maternal sire breeds (A, B, and C) are replaced by bulls of the next breed in the sequence after only two years of use, then this system can be run with only two breeding pastures. This is because all females will be 4, 5, or 6 years old before bulls of the same breed as their sire are scheduled for use in the rotation. At these ages, the cows are scheduled to be mated to the terminal sire breed.

This is not possible in the two-breed rotational-terminal sire system because the females are still young (2, 3, or 4 years old) and still in the rotational portion of the herd when bulls of the same breed as their sire are scheduled for use in the rotation. Thus in herds with only 50 or 60 cows the three-breed rotational-terminal sire system is feasible provided two breeding pastures are available. In this system about 70 percent of the calves marketed are males and females from the terminal sire matings, and about 30 percent are steers from the rotational matings.

Assuming that the terminal sire breed increases growth rate by 5 percent and that the levels of heterosis in two- and three-breed rotations are as outlined in Tables 24 and 25, pounds of beef produced per cow in the breeding herd should be increased to levels of 21 and 24 percent over straightbreeding in the two-breed and three-breed rotational-terminal systems, respectively. This is greater than the increase produced by the static terminal sire system or the rotational systems described earlier. The three-breed rotational-terminal system provides for greater production than the two-breed rotational system because greater heterosis is sustained by the system.

SMALL HERD SYSTEMS

In herds of small size with only one breeding pasture, none of the systems described above is feasible. However, it is possible to utilize the principles of these systems to maintain a substantial level of heterosis. For example, a high level of heterosis can be restored from one generation to the next by using three sire breeds in rotation and replacing sire breeds every 3 years. Heterosis will approach the maximum level if replacements are kept from the last two of each three calf crops produced by each sire breed used in the rotation. A still higher level of heterosis will be realized if a fourth breed is included in the rotation. The feasibility of this is usually determined by the availability of desirable herd bulls from four breeds that are comparable in terms of size and lactation potential and compatible with feed and other production resources available.

Systems that are most commonly recommended for use in small single-sire herds involve purchase of crossbred females to provide for replacement requirements. This is quite often the simplest and easiest system to manage if a good source of crossbred females can be found. Breeders employing good management and sire selection for use in two- or three-breed rotations of breeds that are adapted to the feed resource base employed in the small herd are an excellent source for replacement females for small single-sire operations. Production level from heterosis will be the same as outlined earlier for two- and three-breed rotations, and

Fig. 27. Feedlot cattle with maximum heterosis result from mating crossbred females with a sire of a third breed possessing excellent growth rate and desirable carcass traits. (USDA Meat Animal Research Center.)

growth of the calves will be primarily determined by the performance level of the sire used in the small herd. As the cows become mature, a sire breed capable of transmitting rapid and efficient growth can be employed to increase production to the level characterized by the terminal sire systems discussed above.

CHAPTER 5
REPRODUCTION
AND MATING

The percent calf crop weaned is one of the most important single factors in determining profit or loss in the cow-calf program. The term *percent calf crop weaned*, as used in this book, refers to the percentage of all the cows and heifers of breeding age in the herd at breeding time that weaned calves. Figures obtained by calculating percent calf crop weaned by only those cows that actually calved may look and sound better, but they do not present the true picture of the reproductive performance of a herd, and they can be very misleading.

It is reliably reported that the percent calf crop weaned ranges from as low as 60 percent in most of the herds in certain areas having hazardous environmental extremes for cow-calf production, to not more than 90 percent in the best of herds throughout the country over a period of years. The national average is estimated at 80 percent. The influence of percent calf crop, weaning weight, annual cow cost, and selling price upon cost per hundredweight of calf weaned, and upon the value of the calf production per cow exposed, is shown in Table 26. Obviously all four of these variables are important in contributing to differences in calf costs and returns per cow, but the percent calf crop weaned is the single item that responds most to good management and thus offers the greatest potential for making the cow-calf program more profitable.

FACTORS THAT AFFECT REPRODUCTIVE PERFORMANCE

If only four out of five beef cows in the United States are weaning calves in any given year, obviously there are some difficulties and problems. An understanding of these problems should help to avert poor reproductive performance in a beef cow herd. Some of the causes of less than the ultimate or 100 percent weaned calf crop in a cow herd being managed by natural service techniques are the following.

1. Delayed puberty or sexual maturity in yearling first-calf heifers.
2. Failure of cows to reestablish early postcalving estrus cycles.
3. Too short a breeding season and poor choice of breed and calving season.

Table 26

Influence of Percent Calf Crop Weaned, Weaning Weight, Annual Cow Cost, and Calf Grade on Calf Costs and Gross Return per Cow

Percent Calf Crop	Weaning Weight (lb)	Pounds Calf Weaned per Cow	Cost/cwt Calf Weaned ($)		Value of Calf Production per Cow ($)		
			Cow Cost		Fancy-Choice[a]	Good	Standard
			$125	$175			
90	600	540	23.15	32.40	189.00	178.20	167.40
90	500	450	27.75	38.90	157.50	148.50	139.50
90	400	360	34.70	48.60	126.00	118.50	111.60
80	600	480	26.05	36.45	168.00	158.40	148.80
80	500	400	31.25	43.75	140.00	132.00	124.00
80	400	320	39.05	54.70	112.00	105.60	99.20
70	600	420	29.75	41.65	147.00	138.60	130.20
70	500	350	35.70	50.00	122.50	115.50	108.50
70	400	280	44.65	62.50	98.00	92.40	86.80

[a] Assumes selling prices or values of $35, $33, and $31 for fancy-choice, good, and standard grades, respectively.

4. Low conception rates from first or second services.
5. High percentage of embryonic mortality and abortions.
6. Calves lost at birth from dystocia or delayed and difficult calving.
7. Baby calf losses from trampling, drowning, freezing, starvation, and predators.
8. Calfhood diseases, particularly scours.
9. Bulls that are too immature for the number of cows to be served.
10. Too few bulls for the size of cow herd, and breeding pastures that are too large.
11. Infertile or unsound bulls.
12. Low adaptability of breeds or breed crosses to the particular environment.
13. Too high a percentage of inbreeding in cows and bulls.
14. Inadequate plane of nutrition of cows, calves, and bulls.
15. Incidence of specific reproductive diseases.
16. Choice of unsuitable mating system.
17. Lack of managerial skill in breeding and calving out cows.

Some special problems arise when artificial insemination is used rather than natural service. These problems are discussed later, in the section on artificial insemination.

Current data on average percent calf crop weaned in farm or ranch herds are difficult to obtain. Table 27, reporting results that are mainly from agricultural experiment station herds, shows the great variation from state to state and region to region. Wide differences between pregnancy status (usually determined by rectal palpation) and percentage of cows actually calving can be seen in all the data reported; these differences strongly suggest that embryonic mortality and abortions contribute to the fact that 2 to 4 percent of the cows fail to produce live calves. Differences between the percentage of cows calving and percentage of cows giving birth to live calves average about 5 percent and give some idea of the importance of dystocia during calving. Data reported in Chapter 4 on effect of breed of sire on calving losses illustrate the relative importance of this factor also. Apparently if calves are alive 2 weeks following birth, they stand an excellent chance of surviving to weaning age.

The great variation seen in number of cows weaning live calves, by regions, reflects largely environmental differences, including plane of nutrition. The extremely low figure reported for the South undoubtedly reflects the effect of extremes in temperature, rainfall, humidity, pasture condition, parasite and disease level, inappropriate choice of breeds or crosses, and possibly the level of skill of the operator. This region today

Fig. 28. The first goal of every cattle breeder should be the production of a healthy, vigorous calf, normally born and well started by a mother cow of sufficiently good inherent milk production to raise a calf for every cow in the herd. (American Hereford Association.)

has some highly efficient herds and regional differences have surely become smaller, but the data do indicate strongly that the factors that contribute to differences in percent calf crop are not the same throughout the country and the world.

THE REPRODUCTIVE ORGANS OF THE COW

An understanding of the physiology of reproduction, especially in the female, is extremely important because it clarifies the need for concentration of effort, skill, and resources during the breeding and calving seasons.

The reproductive system of the cow consists of the genital tract and certain related organs. The genital tract may be described as a tube

Table 27

Reproductive Losses in Beef Cattle and Stage at Which They Occur[a]

State or Area and Length of Study (yrs)	Cow Years[b]	Cycled during Breeding Season (%)	Pregnant (%)	Calving (%)	Cows Giving Birth to Live Calves (%)	Cows with Live Calves at 2 wks of Age (%)	Cows Weaning Live Calves (%)
Experiment Station Herds							
Montana (17)	4,753			85.6			81.0
Montana (19)	7,619			82.6			
South (3)	19,388			77.3			69.1
Virginia	612	98.0	88.0	84.0	76.0	73.0	72.0
Louisiana	121	98.0	77.0	74.0	71.0	63.0	63.0
Nebraska	300	98.0	92.0	90.0	84.0	78.0	77.0
New Mexico (31)	999		94.5	92.5	86.8		82.4
South Carolina	844		88.0		76.0		74.0
California (12) (supplemented)[c]	448		89.5	87.9	84.8		81.9
California (12) (unsupplemented)	456		75.0	73.7	71.2		66.2
Commercial Herds							
Plains (1)	38,300				90.5		79.9
West (1)	94,600				76.6		67.2
Northwest (1)	17,900				87.8		74.0
South (1)	65,500				77.2		45.7
Wyoming (11)	4,470				86.0		66.7

[a] Adapted from *Better Beef*, 1974.

[b] Average number of cows exposed annually, multiplied by years of study.

[c] Supplementation with energy and protein, depending on pasture quality, occurred during early lactation and rebreeding.

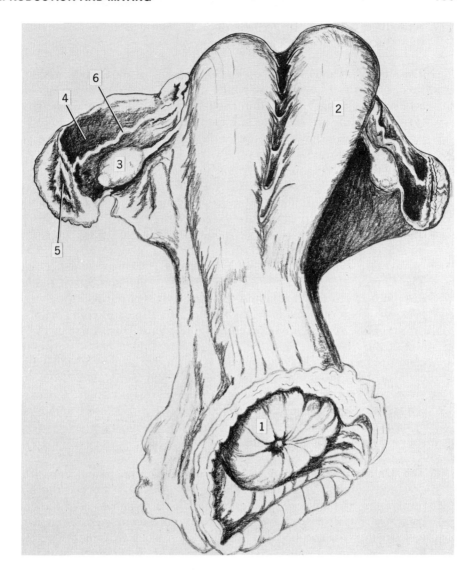

Fig. 29. The reproductive system of the cow. 1, cervix or os uteri; 2, right horn of the uterus; 3, ovary; 4, broad ligament; 5, the infundibulum of the oviduct; 6, the oviduct. (After Williams in *Diseases of the Genital Organs of Animals.*)

extending from the posterior end of the body, forward into the body cavity. This tube varies considerably in size and shape throughout its length and, inasmuch as each part performs a special function in the process of reproduction, each part is spoken of as an *organ* of the reproductive system. In accordance with such a definition the following reproductive organs exist in the cow:

The Vulva.   The vulva is the exterior opening of the female genital tract. It consists of two *labia*, or lips, which close the opening of the tract, and of an internal chamber just within the labia called the *vulvar cavity*. Into this cavity opens the *urethra,* the duct from the bladder. The vulva of course provides the external outlet for urine deposited into the vulvar cavity by way of the urethra.

The Vagina.   The portion of the tract just forward of the vulvar cavity is called the vagina. It is about 10 inches long and lies immediately below the colon. Its principal functions are to receive the penis or inseminating tube during service and to afford a passage for the calf from the uterus to the outside of the cow's body at birth.

The Cervix.   This is a constriction in the genital canal that marks the division between the vagina and the uterus. It is also called the *os uteri*, or neck of the uterus, and is really a cone-shaped part of that organ which projects back into the forward end of the vagina. During estrus or "heat" and at the time of calving the cervix is much dilated, but normally it is contracted so as to close the uterus. If, at time of service, whether natural or by artificial insemination, the cervix is closed so tightly as to make the passage of sperm cells into the uterus unlikely, conception of course will not occur. During pregnancy the cervix is tightly closed and is sealed against the entrance of bacteria from the vagina by a plug of mucus secreted by the mucous membrane of this region.

The Uterus.   The uterus is the portion of the genital tract that is designed to retain and nourish the embryo or fetus between the time of fertilization and calving. The uterus consists of a main portion or *body*, lying just ahead of the cervix, and two branches or *horns* at its forward end. In the cow the body of the uterus is relatively small, whereas the horns are long and large. For a short way they extend forward nearly parallel to each other, then spiral outward. The horns of the uterus are held in place by a tough, elastic membrane called the *broad ligament*, which forms a connection between the uterus and the abdominal walls.

The Oviducts.   The end of each horn of the uterus narrows down to a threadlike tubule, the oviduct or *Fallopian tube*, leading toward the ovary or female gonad. There are two oviducts, just as there are two ovaries and two horns. Although the ovaries are located quite near the end of the uterus, the oviducts are so tortuous that the total length of each is some 5 or 6 inches. At its outer extremity the oviduct broadens to form a funnel-shaped opening called the *infundibulum*, into which the ripened ovum or egg migrates when liberated from the ovary. The walls of the oviduct are covered with cilia or threadlike projections, which facilitate the passage of the egg into the uterine horn.

The Ovaries.   The ovaries are groups of specialized cells, actually outside the genital tract, that produce, at fairly regular periods, the *ova* or eggs, as the female sex cells are called. In the cow the ovaries lie loosely in the body cavity alongside the body of the uterus. They are oval in shape, about 1 inch in diameter around the thickest part. They may be felt with comparative ease by inserting the hand into the colon for about 15 inches by way of the rectum. Each ovary consists of a cluster of small egg sacs, probably several thousand in number, and each sac is called a *follicle*. Every female is born with a large number of follicles and still others develop throughout her lifetime. Each follicle contains an egg that eventually is theoretically capable of being fertilized and growing into a calf. The follicles remain in an undeveloped stage until puberty when, one at a time and at 3-week intervals, they begin to enlarge through an increase in the amount of follicular liquid within, until eventually the follicle wall is ruptured and the ovum is liberated. Probably fewer than a hundred ova are liberated through ovulation during the life of a cow. Stimulation of the follicles by injecting the cow with certain hormones may result in super or multiple ovulation, as will be discussed later.

THE ROLE OF REPRODUCTIVE HORMONES IN THE COW

From physiological studies it is known that reproduction is a complex series of processes that are carefully synchronized by substances called *hormones*. Hormones may be defined as chemical substances that are formed by endocrine (ductless) glands in one part of the body and carried by the blood or lymph to another part of the body or to an organ, where they modify the activity of that organ.

Briefly, the role of hormones in reproduction as generally understood today is as follows. The process is begun by the tiny *pituitary* gland, situated at the base of the brain, which secretes two *gonadotropic hormones*. Both are secreted into the bloodstream during the estrus cycle. The *follicle-stimulating hormone* (FSH) predominates early in the cycle, to bring about the growth and maturation of an ovarian follicle. Near the end of the cycle the *luteinizing hormone* (LH) is released at a concentration sufficient to cause rupture of the now ripened follicle and release of the ovum, this process being called *ovulation*.

In the late stages of the maturation process described above, the enlarging follicle itself begins to produce a hormone called *estrogen*, which, upon entering the bloodstream, stimulates the secretion of mucus in the vagina and acts upon the central nervous system to cause the cow to show signs of heat. Upon the rupture of the follicle and release of the egg, the production of estrogen ceases, and a substance known as *corpus luteum* is formed in the ruptured follicle from which the egg has just been released. The corpus luteum then begins to secrete a third hormone called *progesterone*, which (1) inhibits the production of further gonadotropin by the pituitary, thereby preventing the development of another follicle, and (2) prepares the mucous lining of the uterus to receive and nourish the fertilized egg. However, if the cow was not served or if conception did not occur, the corpus luteum begins to degenerate in 16 to 18 days and the secretion of progesterone ceases, whereupon the pituitary gland again begins the production of gonadotropin to start the process all over again.

Should the cow be served during estrus and the ovum be fertilized, the corpus luteum remains in the ovary throughout pregnancy and continues to produce progesterone, which prevents the development of more

Table 28

Time Interval Between Estrus and Ovulation in Beef Cows[a]

End of Estrus		Time of Ovulation		Interval Between End of Estrus and Ovulation	
Hour of Day	Number of Cows	Hour of Day	Number of Cows	Interval in Hours	Number of Cows
4– 5 P.M.	2	10–11 P.M.	1	1– 2	0
6– 7 P.M.	7	12– 1 A.M.	0	3– 4	0
8– 9 P.M.	7	2– 3 A.M.	0	5– 6	2
10–11 P.M.	7	4– 5 A.M.	0	7– 8	1
12– 1 A.M.	10	6– 7 A.M.	1	9–10	0
2– 3 A.M.	4	8– 9 A.M.	7	11–12	8
4– 5 A.M.	2	10–11 A.M.	5	13–14	8
6– 7 A.M.	0	12– 1 P.M.	5	15–16	11
8– 9 A.M.	0	2– 3 P.M.	11	17–18	6
10–11 A.M.	1	4– 5 P.M.	5	19–20	1
		6– 7 P.M.	1	21–22	2
		8– 9 P.M.	2	26	1
Total observations	40		38		40
Average time estrus ended, 10:30 P.M.		Average ovulation time, 1 P.M.		Average interval, 14.6 hours	

[a] Adapted from *Journal of Animal Science* 1:192.

ovarian follicles and thereby prevents the occurrence of estrus. Normally the corpus luteum atrophies within about 4 weeks following calving, and preparation for the resumption of the estrus cycle takes place. Should the plane of nutrition be too low at this time, resumption of the cycles mentioned is delayed, but in this case the delay is thought to be caused by failure of the triggering mechanism that initiates the pituitary secretions, and not by a retained corpus luteum.

Occasionally the corpus luteum does fail to atrophy at the normal time after calving, inducing temporary sterility. Such a "retained" corpus luteum is commonly termed an *ovarian cyst*. If the cyst is forcibly expelled from the ovary, which can be done by a veterinarian or an experienced herdsman, working through the rectum, the cow usually begins estrus within a few days and can then be served. The forced removal of an ovarian cyst may be a poor practice, however, because cows that require this treatment are apt to be chronic "problem breeders." Culling them will help ensure a larger calf crop and a shorter calving season. Nymphomania, an aberration in the estrus cycle expressed as almost constant heat, is usually though not always caused by a retained corpus luteum. Cows with this tendency should be culled because the trait appears to be heritable. Such "bullers" are sometimes used in the collection of semen with the artificial vagina or to assist in heat detection in cows to be artificially inseminated.

ESTRUS AND CONCEPTION

In practical cattle production the reproduction process begins with the first heat period of the young heifer. Ordinarily this condition, known as *puberty*, is first observed in heifers soon after they are 1 year old, although some heifers reach puberty while still nursing and, if a bull is running with the herd, are in danger of being bred when they are only 6 to 8 months old. If bulls are left with the cows year-round, some suckling heifers will almost certainly be bred prematurely unless heifer calves are weaned systematically when they reach 7 months or so in age. Creep feeding appears to hasten the advent of puberty and scanty feeding tends to retard it. Apparently Brahman females or Brahman crossbreds and some of the newer breeds containing some Brahman blood reach sexual maturity at a considerably later age than heifers of the British breeds. As a rule, the larger the breed average for mature size, the later the onset of puberty. This delayed puberty can be a problem in settling yearling heifers in certain breeds or breed crosses.

A few hours after the end of estrus, an egg is liberated by the rupture of a follicle, thereby presenting the conditions for conception and pregnancy. The interval between the end of estrus and ovulation varies from 6 to 26 hours (Table 28), whereas the time required for the sperm to travel from the vagina to the oviducts, where fertilization normally occurs, is only a few minutes. Consequently it is advisable to delay breeding until toward the close of the heat period, or even a few hours afterward if hand breeding or artificial insemination is practiced, in order to favor the presence of strong, vigorous sperm in the oviduct at the time the egg is liberated.

Because estrus is of short duration in cattle, seldom exceeding 18 hours, if natural mating is attempted, breeding should not be delayed because the cow may soon be out of standing heat. In the studies done at the Michigan station with beef cows, 4 P.M. was the earliest and 10:30 P.M. the average time of day at which estrus ended, indicating that cows that are to receive only a single service should be bred in mid- or late afternoon. If two services are possible, one should be made soon after the cow is observed to be in heat, usually in the morning, and another about 6 hours later or in the evening. Artificial breeding is most likely to be successful if insemination is done on the morning of the day after estrus if only one service is made. If hand mating in a breeding chute is practiced, bulls can be trained to serve cows the morning following estrus. Young bulls used in pasture mating may often breed cows only once, during the early part of estrus, and their semen may not remain viable long enough to ensure fertilization.

New research data suggest that much embryonic mortality or early abortion may be caused by the death of an embryo that resulted from fertilization by old or senile sperm cells. This could be the result of breeding too early in the heat period or, in artificial insemination, of using old (unfrozen) semen. As shown in Table 28, ovulation occurs, on the average, 14.6 hours after the end of estrus, although it may occur still later. Delaying artificial breeding until late in the day after estrus may very well result in conception, but can also result in embryonic mortality, in this case caused by senility of the ovum rather than the semen. There is evidence, mainly in other species, that sperm cells must be exposed for several hours to uterine or tubal secretions before they are capable of fertilizing eggs. This phenomenon, called *capacitation* or aging, could explain poor conception rate in either early or late services.

SIGNS OF ESTRUS

Cows vary greatly in their behavior during estrus. The condition can usually be detected by extreme nervousness in the animal and by her attempts to mount other members of the herd, which in turn often mount her. Examination usually discloses a noticeable swelling and relaxation of the vulva, which often appears slightly inflamed in light-skinned animals.

As a rule, estrus is accompanied by a slight mucous discharge. Rarely is there any loss of blood until a day or two after estrus, when a slight bloody discharge or "menstruation" sometimes occurs.

RECURRENCE OF ESTRUS

Unless fertilization takes place, heat periods normally recur at intervals of approximately 3 weeks. There is some variation in length of the estrus cycle even in the same individual. Usually, however, it is seldom shorter than 18 or longer than 21 days. Cows that are to be bred should be closely observed at least twice a day during the third week following their last heat period.

Estrus, of course, does not normally appear during pregnancy. Its occurrence at this time is often followed by abortion, although there are numerous instances on record where pregnant cows have been naturally served in apparently normal heat periods with no ill effect whatever on the fetus. Artificially breeding a pregnant cow that is in abnormal estrus nearly always results in an abortion or a mummified fetus because of injury to the cervical region or breaking of the cervical seal or plug. Good breeding records would prevent most mistakes of this nature.

The period between calving and recurrence of first estrus after calving is normally 6 to 10 weeks but may vary from as short as 4 weeks to an absence of cycling altogether during lactation. The length of this interval, which is extremely important, is influenced by numerous factors, most of which delay first postcalving estrus rather than hasten it. The gestation period of the cow ranges from 280 to 290 days, leaving only 85 to 75 days from time of calving until the cow must be pregnant again if the desired 365-day calving interval is achieved. Delayed first postcalving estrus almost certainly will reduce both the size of calf crop and weaning weights. In herds where bulls are left out for more than 3 months or even year-long, fair percent calf crops may be weaned but the calving season will be prolonged, making efficient management of the herd almost an impossibility.

Two-year-old cows, nursing their first calves, are most apt to show delayed postcalving estrus; for this reason, yearling heifers should be bred to calve about 3 weeks earlier than the main cow herd in order for subsequent calvings to occur early or at least no later than in the older cows. It is extremely difficult to move the calving date of beef cows forward in the year. Thus, if cows are delayed in calving, for whatever reason, they are apt to be late calvers for life. This is the basis for the practice, followed by some good stockmen, of automatically culling the open cows and even the late calvers. Use of short breeding seasons of 60 days or so guarantees identification of the cows that are late in cycling.

The role of adequate vitamin A, phosphorus, energy, and protein in initiating estrus and improving conception rate is discussed in Chapter 10.

AGE AT WHICH TO BREED HEIFERS

Because the process of reproduction, and especially lactation, imposes a heavy tax upon the mother, heifers should not be bred until they are reasonably mature. The growth of the fetus and care of the young appears to take precedence over everything else, even over the requirements of the mother's body for maintenance and growth. Whether delayed growth of the young cow, brought on by too early calving, is later resumed depends on the level of feed given the heifer after calving and during lactation and subsequent dry periods.

There is much evidence that gestation is less apt to stunt immature heifers than is lactation. This seems reasonable in view of the fact that the newborn calf contains only about 15 pounds of protein and 3 pounds of fat, whereas about 65 pounds of protein, 70 pounds of fat, and 90 pounds of carbohydrates are contained in the total milk production of the young mother during the first 4 months of lactation. The milk produced also contains many times more calcium, phosphorus, and other minerals than are present in the newborn calf. Thus it is obvious that the average daily demands on the mother during lactation are several times greater than those during gestation. Ranchers experiencing adverse range conditions often early-wean the calves from 2-year-old heifers when the calves are 4 to 5 months old, thereby reducing the deleterious effects of lactation. Actually, there is some evidence that deliberate stunting of young cows may reduce mature size and thus lower the maintenance requirements of the cow, without interfering with her productive efficiency as a mother cow.

Since a majority of farmers and ranchers want their calves to be born during not more than a 2- to 3-month period in the spring, replacement heifers must be bred to drop their first calves very close to either their second or their third birthdays. There is much difference of opinion as to which is the better age for first calving. Several experiments and surveys have been made, particularly in the range states, to study this question. Although the results do not agree in all respects, experiments such as the one summarized in Tables 29 and 30 (a study of both first-calving age and plane of winter nutrition) tend to justify the following statements regarding heifers bred as yearlings to calve as 2-year-olds.

1. Size at first breeding is more important than age, with a minimum weight of 600 pounds in average flesh being desirable in British breeds and 100 to 150 pounds additional for the larger breeds.
2. Total number of calves weaned and total weaned calf weight during a cow's lifetime favor breeding as yearlings if the heifer receives assistance, when needed, during calving.
3. When heifers are bred as yearlings, maturity is delayed 3 or 4 years, and they may never reach full mature size, especially if supplemental winter feed is inadequate.
4. Average weaning weights of the first two or three calves produced by heifers bred as yearlings will be slightly lower than weights of calves produced by heifers bred first as 2-year-olds.
5. Cow cost per 100 pounds of weaned calf favors breeding heifers to calve first as 2-year-olds.
6. More heifers will need assistance at first calving if bred to calve as 2-year-olds rather than as 3-year-olds.
7. Using small, refined bulls on yearling heifers is worthwhile if pasture

Table 29

Production Records of Cows, Bred First to Calve at Two and Three Years of Age, Through Fourteen Years of Age[a]		
Age at First Calving	2-Year-Old	3-Year-Old
Number of females started on test, fall 1948	60	60
Number remaining, March 1962	23	22
Reasons for removal from test		
Open or failing to calve in two successive years	16	16
Cancer eye	9	6
Spoiled udder	4	5
Crippled	2	2
Disease	1	2
Hardware disease	2	1
Accidental	0	2
Unknown	2	3
Heifers assisted at first calving	28	1
Average mature body weight, fall 1956 (lb)	1,148	1,178
Total number of calves weaned	533	482
Number of calves weaned per cow year	0.80	0.71
Total percent calf crop weaned	86.7	85.2
Average weaning weight, all calves (lb)	476	485
Average weaning weight, minus 2-year-old calf (lb)	482	485
Extra pounds of calf weaned	330	—

[a] Oklahoma MP-67:69.

Table 30

Summary of 8.5 Years' Results in Study of Beef Cows Wintered at Different Levels (1948–1956)[a]

Age at First Calving	2-Year-Olds			3-Year-Olds		
Lot Number	1	3	5	2	4	6
Level of Winter Supplement Fed[b]	Low	Medium	High	Low	Medium	High
Number of cows at start of experiment	15	15	15	15	15	15
Number remaining on test November 1956	14	14	10	14	12	13
Average weight changes of cows on test (lb)						
Initial weight 10/29/48	473	471	476	476	461	470
Average winter weight loss	−108	−98	−63	−112	−97	−67
Average summer gain	188	185	147	198	179	160
Final weight 10/30/56	1,103	1,165	1,165	1,182	1,128	1,223
Calf production records at 8.5 years of age						
Heifers assisted at first calving	6	8	4	—	—	1

calving is practiced and if experienced assistance is unavailable during the calving season, because the resulting calves tend to be somewhat smaller at birth. Using Angus bulls on first-calf Hereford heifers may be helpful, but does not necessarily insure against calving difficulty.

8. Level of supplemental winter feed has more effect on the weight of the young cow than on the weight of the weaned calf or percent calf crop weaned, provided that minerals, carotene or vitamin A and protein are all adequate.

9. The feed or pasture available during the nursing period affects weaning weight of the calf more than does the age at which its mother is first bred.

10. The number of open cows among second-calf heifers that calved first

Table 30 *(Continued)*

Summary of 8.5 Years' Results in Study of Beef Cows Wintered at Different Levels (1948–1956)[a]

Calves lost at first calving	1	1	2	—	—	2
Total number of calves weaned	91	93	75	82	71	73
Percent calf crop weaned[c]	93	95	89	97	89	90
Total number of calves weaned per cow	6.44	6.58	6.03	5.79	5.13	5.15
Average calving date	3/14	3/9	3/8	3/15	3/4	3/5
Average calf weights (lb)						
At birth (sex-corrected)	76	76	77	76	76	78
At weaning (age- and sex-corrected)	480	472	471	495	474	492

[a] Oklahoma MP, 48:46.

[b] Supplements fed: low level, 1 pound cottonseed meal pellets; medium level, 2.5 pounds cottonseed meal pellets; high level, 2.5 pounds cottonseed meal pellets and 3 pounds oats.

[c] Based on number of cows bred to calve each year. Calf losses not due to experimental treatment were not charged against the lot.

as 2-year-olds can be reduced by supplemental feeding during the breeding season when they are nursing their first calves.

The choice of age at which first to breed heifers thus depends on the ration available during the first winter as heifer calves, the level of management available during the calving season, and the quality and quantity of summer pasture available during the lactation period. Small ranch herds and farm herds tend to practice breeding as yearlings, whereas large spreads are more apt to breed heifers first as 2-year-olds. More information concerning the role of nutrition as it affects first-calf heifers is found in Chapter 10. In some commercial herds and many purebred herds, where both spring and fall calving are practiced, a compromise is often made by breeding at about 21 months and calving at about 30 months.

THE BREEDING SEASON

The time of the breeding season, of course, depends on when the farmer or rancher would most like the calves to be born. Because the average gestation period is about 283 days, mating should begin approximately 9 months and 10 days before the earliest date on which the calves are wanted. Although it is highly desirable to have all calves born early in the calving season and as close together as possible, it will be found that, because of delay of the estrus periods of some cows following calving and failure of others to conceive from the first service, the period of calving, even in the better-managed herds, usually extends over 2 or 3 months. Any greater irregularity in the span of calving time is usually regarded as unsatisfactory under ordinary circumstances, especially in a commercial herd. There are instances where low rainfall causes poor range conditions; hence the nutritional status of lactating cows is so poor that longer, or even year-round, breeding seasons are justifiable to insure any calves at all.

Usually bulls should be removed from the pastures within 4 months or less after the beginning of the breeding season. Leaving the bulls with the cows for 65 days will give almost every cow in the herd three opportunities to conceive, which should suffice if the bulls are fertile and active throughout the breeding season. A few cows will not begin cycling soon enough to have several opportunities to be served. Except in unusual circumstances, such as drought or injury, these late-cycling cows should be culled. Pregnancy tests in the fall will reveal open or nonpregnant cows, which should be sold for slaughter at weaning time or later if some cheap gains can be put onto the dry cow or if prices are almost certain to strengthen substantially during the holding period.

SPRING AND FALL CALVES

When possible, the commercial cattleman tries to have his calves born at a time when range and weather conditions are most favorable. This usually means that they should be born either in the spring, after the cold weather of winter but before the heat and flies of summer, or in the fall before cold weather sets in. The exact calendar dates depend somewhat on the

Table 31

Effect of Month of Birth on Survival, Growth, and Weaning Weight of Beef Calves[a]

Month	Number of Cows	Number of Calves	Percentage Calf Crop	Average Age Weaned (days)	Average Weaning Weight (lb)	Average Daily Gain (lb)	Percentage of Totals
January	130	128	98.5	275	558	1.72	1.24
February	714	676	94.7	250	503	1.69	6.52
March	3,625	3,474	95.8	221	466	1.75	33.53
April	4,366	4,183	95.8	198	433	1.79	40.37
May	1,548	1,447	93.5	180	402	1.79	13.96
June	335	326	97.3	163	374	1.81	3.15
July	20	20	100.0	130	355	2.10	0.20
September	11	10	91.0	336	596	1.53	0.10
October	16	15	93.7	284	517	1.61	0.13
December	90	83	92.2	304	557	1.57	0.80

[a] Charles R. Kyd, Missouri Cooperative Extension Service. Information to the author.

latitude and, in the western states, upon the altitude as well. Table 31 shows the results of one study of the effect of birth month on survival, growth, and weaning weight of beef calves. The light weights of summer calves are no doubt a reflection of both an early weaning age and poor environmental conditions for the young calf in summer. The fall calves were heavy when weaned but were much older. By far the greater number of the calves of the country are born in the spring. However, some farmers, especially in the central and southern states, find it more advantageous to have the calves dropped in the fall or even during winter. Below are the principal advantages claimed for spring- and fall-born calves, respectively.

Advantages of Spring Calves

1. Dry cows can be wintered more cheaply.
2. Calves are of good age by wintertime and can better withstand cold weather, if carried over as stockers.
3. Cows milk better while on grass than they do on dry winter feed.
4. Labor is saved when cows and calves run together on pasture.
5. Calves may be sold either at weaning time with no wintering or as yearlings with only one wintering.

6. Cows are bred while on pasture, when they are most likely to conceive.

7. A smaller investment in shelter and equipment is required.

8. Condition of cows at calving time is easier to control, because some winter feeding is usually practiced and this can be regulated to prevent unnecessary weight gain. Fall-calving cows are sometimes quite fat and will have more calving difficulties.

Advantages of Fall Calves

1. Cows usually are in better physical condition, nutritionally speaking, for calving in the fall than in the spring because of good summer grazing conditions; hence the calves are likely to be stronger.

2. Young calves escape the severe heat, pink eye, and flies of midsummer.

3. At weaning in the springtime the calves may be turned on grass to be sold in the fall as short yearlings with a sale weight at least 200 pounds heavier than in a spring-dropped calf.

4. Cows that freshen in the fall or early winter lactate longer than those that freshen in the spring. Spring grass stimulates milk flow in cows that calved the preceding fall, whereas changing to dry feed in the fall tends to diminish milk flow in cows that calved the preceding spring.

5. Cattle numbers are at their peak during the winter season when labor for their care is more apt to be available.

Fig. 30. Spring calves born on clean pasture are rarely affected by diseases or parasites. (Denver and Rio Grande Western Railroad.)

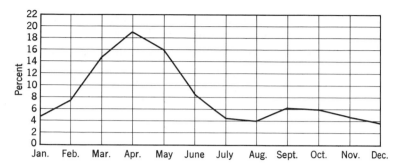

Fig. 31. Births of calves by months. Nearly 50 percent of all beef calves are born during March, April, and May. (USDA.)

6. Cows are bred in the winter when hand mating or artificial insemination can conveniently be used.

7. Creep-feeding, when practiced, is more conveniently managed.

8. Calves are weaned at a more favorable marketing time, whether sold as stockers or as fat slaughter calves.

The advantages of spring calves have the most weight under extensive rather than intensive methods of cattle production. Fall-born calves, on the other hand, are particularly well suited to farms where beef cattle are only one of several enterprises that contribute to the income from the farm. Farms once equipped for dairying may well utilize their equipment and buildings with a fall-calving program. In regions where winter small grain pastures are important, a fall or early winter calving system fits extremely well because it enables the cow herd to utilize such pastures to best advantage.

Calving in very early spring, such as in January or February, while seeming to be a compromise that would have the advantage of both seasons of calving, actually has most of the disadvantages instead. Feed costs are high, calf death losses are apt to be high, labor requirements are increased, and if calves are sold off the cows in the early fall, they are often just heavy enough to be discriminated against as to price. The result may well be less net income rather than the hoped-for increase.

In some of the Range region, notably the Sandhills section of Nebraska, the practice of splitting the calving season into a spring and a fall season is increasing. Apparently this is being done because the ranchers believe they can sell more pounds of calf per cow with this combination. The fall-dropped calves are usually sold the following fall, weighing 600 to 700 pounds. In marketing circles these calves are called "calf yearlings." Another reason some give for this practice is that the ranchers select from the fall-dropped calves their heifer replacements for the main herd, which calves in the spring. This procedure enables them to compromise on the breeding date for the first-calf heifers, since they will be calving at 2.5 years of age. The replacements for the fall-calving herd come from the spring-dropped calves for the same reason. Naturally this system is adaptable only to herds that are large enough to make dividing the herd practical. Purebred breeders use this system, along with longer calving seasons, to provide a wide variety of ages of cattle for show purposes and sale.

Researchers and extension service personnel in most states have collected data similar to those shown in Fig. 32. The data represent almost 25,000 weaner calves weighed in Webb County in South Texas near the Mexican border. Undoubtedly the extreme heat of the summer in this region accounts for the light weaning weights of the calves dropped between May and September. Corresponding data from other areas will probably differ in some respects, so that breeders in other sections should use data collected for their own regions, when possible, in choosing the ideal calving season. Differences in weight are not confounded by age differences because the weights shown are adjusted to a 205-day basis.

Fig. 32. Average 205-day weights of calves born in different months in South Texas. (Texas A. and M. University, Animal Science Department.)

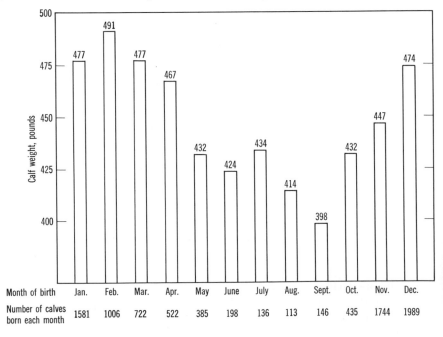

METHODS OF MATING

Three methods of mating are followed: hand mating, pasture mating, and artificial insemination. In hand mating the bull is kept separate from the cow herd. Whenever a cow is observed to be in heat during the breeding season, she is turned in with the bull, where she remains until she is served; or, as is often done in purebred herds, she is led into a level lot or into the breeding chute and held or tied while she is being served. The bull may or may not be managed by a herdsman. As a rule, only a single service is permitted in hand mating, and the cow is removed immediately after service. Two matings, one in the afternoon or evening and another the following morning, will almost certainly ensure that egg and sperm are both in best condition for fertilization. If two matings are attempted, a breeding chute is required, at least for the morning service, for the cow will have passed out of standing heat and will not readily submit to the second service.

In pasture mating the bull is allowed to run with the breeding herd throughout the breeding season. Data shown in Table 32 demonstrate that most cows that are easily settled will settle either from the first or second service. The table also shows that even after six estrus periods 27 percent of the cows were still not settled. These cows doubtless would be difficult to settle under any system of mating and should be culled, and

Table 32

Calculated Estrus Period at Which Cows Conceived During a Breeding Season with Pasture Matings[a]

| Estrus Period (20-Day Basis) | Number of Cows Conceiving | Percentage of Herd | | | |
| | | 100% Fertility | | 73% Fertility | |
		For Period	Cumulative	For Period	Cumulative
First	295	52	52	38	38
Second	155	28	80	20	58
Third	61	11	91	8	66
Fourth	30	5	96	4	70
Fifth	19	3	99	2	72
Sixth	3	1	100	1	73
Total: 120 days (4 months)	563	100		73	

[a] *Journal of Animal Science* 3:156. (Data from USDA Experiment Station, Jeanerette, Louisiana.)

such data often are the basis for recommending that bulls be left out no more than 90 or even 75 days.

Pasture mating saves the labor of inspecting the herd each day for cows that are in heat and driving them to the breeding pen for service. Moreover, it precludes the possibility of a cow's "going by" unbred because the herdsman failed to detect her estrus. The objections to pasture mating are the following.

1. All cows that run in the same pasture are bred by the same bull, or by any one of several bulls if more than one bull is present. This makes it impossible to be certain of the sire of each calf. This information is essential in herds on performance test or in purebred herds.

2. Bulls wear themselves out by repeatedly serving a cow while she remains in heat. Sometimes as many as six or more services are performed by several bulls, and should another cow be in heat the next, or even the same, day she might not be served.

3. If two or more cows are in heat at the same time in a single-sire herd, the bull may give all his attention to one of them, allowing the others to go unbred.

Fig. 33. Pasture mating of 2-year-old heifers. Labor requirements are low and conception rates are high in healthy cattle bred on lush spring pasture. (American Hereford Association.)

There is little doubt that in a herd of commercial cattle the disadvantages of pasture mating are more than balanced by the saving of labor and the greater certainty of getting all cows in calf. With purebred cattle, however, pasture mating is less widely used. Here, certain knowledge as to whether a cow has been served, the particular bull performing the service, and the date on which it occurred is so important, especially for sale cattle, that many purebred breeders resort to pasture breeding only to pick up any cows that were missed during hand breeding. Pasture mating is also used to some extent in large purebred establishments where bulls and pastures are numerous enough to permit dividing the herd into lots of 15 to 20 cows, each with a bull at the head.

ARTIFICIAL INSEMINATION

Artificial breeding has not been so widely practiced by beef cattle breeders as it has been by dairymen. This is changing rapidly, and only the matter of a generally reduced conception rate and the additional work involved are keeping the practice from becoming more widely accepted by beef cattlemen. Performance-tested beef bulls are now found in all important AI (artificial insemination) bull studs, and owners of small commercial herds should by all means avail themselves of this service if practicable. Of the 7.7 million cows bred by artificial insemination in 1965 in the United States, 0.5 million were beef cows. This figure represented about 2 percent of the total beef cow population. The use of recently introduced exotic breeds and larger dairy breeds as sire breeds, mainly in crossbreeding programs, has been responsible for increasing the percentage of artificially bred beef cows to about 5 percent in 1975.

Perhaps the most important reason for the limited use of artificial insemination by the practical cattleman is the difficulty of determining when beef cows are in heat, and the problems associated with sorting them out and restraining them for breeding. Beef cows that run together day after day are much less inclined to ride one another than are dairy cows that are penned at night and turned together after being milked in the morning. Because the breeding season for beef cows usually comes at the time of year when they are on pasture, often on a remote part of the farm or ranch, it is difficult not only to determine each day which cows should be bred but to separate these cows from the herd and drive them to the barn or corral to be inseminated. The daily sorting out of cows suspected of being in heat, driving them to the corral, confining them in the squeeze gate, and returning them to pasture cause a great deal of undesirable disturbance to both cows and young calves. Confinement or lot feeding of

---

open cows is feasible in many instances and eliminates some of the objections mentioned.

Use of vasectomized or surgically altered bulls in pastures or in drylot helps to locate cows that are in heat. A vasectomized bull is incapable of inseminating a cow because the vas deferens has been surgically cut, prohibiting the passage of semen from testicle to penis. Various other surgical procedures are applied to bulls that prevent penetration of the penis into the vulva. These "gomer" bulls are preferred as they are less apt to spread disease and vaginal infection from one cow to another. Chin ball markers and other devices often are worn by such bulls to help identify the cows in heat.

Added to the disadvantages mentioned is the fact that sometimes a relatively low percentage of cows have conceived after two and even three inseminations. Usually after 2 or 3 months of artificial breeding a cleanup bull is turned with the herd to settle the cows that are still open. As a result the next year's calves show much variation in age and weight, thereby complicating their feeding and management.

Artificial insemination is of great value in large purebred herds because it permits the mating of an outstanding sire to many more cows than he could handle by either hand or pasture breeding. Bulls owned in partnership can be used in several herds, often located hundreds of miles apart, by resorting to artificial insemination. However, purebred breeders should know the rules of their respective breed associations before using this method of breeding. Most beef breed associations permit a maximum of only four owners of a given bull, and only a few associations place no restrictions on numbers of breeders using one sire.

Artificial insemination can be employed to advantage in prolonging the usefulness of valuable sires which, because of accidents or advanced age, can no longer perform natural service. The development of a technique for freezing semen so that it can be stored almost indefinitely in a "semen bank" is one of the great advances in commercial animal husbandry. Semen from outstanding sires can thus be used to inseminate outstanding cows long after the sire's death, and enough semen can be stored to produce literally thousands of offspring. Here again, breed association rules should be known. Most purebred associations prohibit the registry of calves produced by stored semen from a bull no longer living.

There is evidence that blended semen—that is, semen from several sires mixed together—results in higher conception rates. Obviously it could be used only where sire identification is not important.

SYNCHRONIZATION OF ESTRUS

If all cows to be bred in a herd could be bred artificially in a 1- or 2-day period, and if a high conception rate could be assured without undue cost, artificial insemination would be readily accepted by the majority of cattlemen. Some promising research is under way on a workable method for achieving this goal.

Synthetic progesterone-like compounds, which have properties similar to those of the hormone produced by the corpus luteum of a pregnant cow, can be fed at low levels, causing cessation of heat in cows and heifers just as occurs during pregnancy. This makes it possible to halt the estrus cycles of all cows in a herd regardless of when their cycles have been occurring, and then cause the cycles to begin again simultaneously in all of the cows. All cows and heifers that are to be bred are fed the material in a small amount of supplement for 18 days; then the feeding of the compound is discontinued. Within 24 to 48 hours, 90 percent of the cows will be in heat. The best conception rate is being obtained by breeding all cows on both the second and third days following withdrawal of the hormone. Conception rates as high as 85 percent have been reported, although rates of 50 to 60 percent are more common.

A promising method of administering the synchronizing compounds is that of inserting a pessary, containing the absorbable compound, into the vagina. At the desired time the pessary is removed, by means of a protruding string. Ear implants which can be removed when desired also are under study.

Once synchronized, the cows will all continue to come in heat at about the same time. Synchronization will never have a place where natural breeding is practiced because too much bull-power would be required at one time. Even when artificial insemination is used, it can be a problem to provide enough cleanup bulls for natural breeding of the still-open cows.

Cows with calves less than 2 months old should not be synchronized until they are recycling; otherwise the chances of settling them will be lessened because normal ovulation seldom follows the artificially induced estrus. Getting a uniform consumption or absorption of hormone is difficult but absolutely essential. Research on use of injections of still other hormones, administered near the end of the quiet period to induce normal ovulation upon heat, holds promise of increasing the conception rate. Melengesterate or MGA, used to hold feedlot heifers out of heat, cannot be used satisfactorily to synchronize older breeding cattle.

If all of the problems associated with this practice can be solved, the number of bulls required in commercial beef herds could be reduced to the extent that purebred breeders who rely on sales to such breeders would be in serious financial difficulty.

MULTIPLE BIRTHS AND MULTIPLE OVULATION

The number of offspring from a given number of cows bred can be increased either by increasing the frequency of natural twinning or by

Fig. 34. Fifteen live calves resulted from the superovulation and artificial service of the purebred Simmental donor cow shown, although 15 synchronized recipient cows actually produced the calves after receiving the fertilized ova via embryo transplants. (Simmental Shield.)

causing more than the usual single ovum to be ovulated. If the resulting additional calves survived and were weaned at near-normal weights, and if the cow functioned normally for subsequent calvings, beef cow efficiency could be increased considerably.

However, twins occur infrequently in beef cattle—about 0.4 percent of all calvings—and the heritability of the tendency to twin is relatively low. Consequently it is doubtful whether selection for twinning results in its increase. Ranchers, in fact, generally prefer to have only single calves because, under extensive management, twins are seldom both weaned.

Today, highly valuable cows are being treated with combinations of FSH and LH hormones to induce *multiple ovulation*, with as many as 20 or more viable ova being produced at one time. Obviously the cow that produces these extra ova cannot produce this many calves, although up to sextuplets have been experimentally produced. A promising development is that of fertilizing the ova in the original or donor cow and then transplanting them to properly synchronized recipient cows, where they are implanted in the uterus and carried to term.

Thus the valuable cow serves the function of producing many ova, and consequently many offspring, per ovulation. Such multiple ovulation may be induced several times per year. When cows naturally produce twins, and especially when they produce hormonally induced multiple births, calves often are weak and light in weight and their survival rate is poor. Many such cows retain the placenta and rebreed slowly. Multiple ovulation would seem to have limited application except for embryo transplant purposes.

SEX CONTROL

From the beginning of animal husbandry history, cattlemen and researchers have been fascinated with the possibility of predicting or even of controlling the sex of calves. This would mean that certain sires and cows could be mated specifically to produce a herdsire or a number of replacement heifers at will. Or certain breed crosses could be made to produce heifers only, and thus hasten the process of establishing a herd. Terminal-cross sires could be used to produce males only, for feedlot purposes.

Applying certain chemical or physical processes to semen used in artificial insemination seems to hold the greatest promise for achieving these goals. For example, specialized centrifugation, to separate the semen into fractions that do or do not contain the sperm cells carrying the sex-determining Y chromosome, is being actively investigated. At this time there is no practical method of using this technique, but the possibility should not be completely ruled out.

Usually about equal numbers of both sexes of calves are dropped in any given herd in any one season, although males are slightly in the majority. For some unknown reason a greater number of male fetuses are aborted, but even so, the two sexes occur in almost equal numbers among all calves born.

NUMBER OF COWS PER BULL

The number of cows that can be successfully bred by a single bull during a short, intensive breeding season depends on, first, the age, health, and nutritional status of the bull and, second, the manner in which the cows and bull are handled. A yearling bull may be allowed an occasional service, but in no case should he be mated with more than 12 or 15 cows during a

Table 33

Effect of Bull-Cow Ratio on Length of Calving Period[a]

| | | | | | Length of Calving Period | | | |
| | | | | | Minimum Period | | Maximum Period | |
Cows per Bull	Number of Herds	Number of Cows	Average Calf Crop (%)	Average (days)	Number of Cows in Herd	Days	Number of Cows in Herd	Days
20 or less	5	756	95.4	77	51	53	455	99
21–30	36	3,682	94.5	102	101	41	110	219
31–40	12	1,223	93.1	118	34	46	80	212
Over 40	12	2,027	93.6	132	82	73	110	243

[a] Compiled from Mimeographed Reports of Kansas Beef Production Contest, 1946–1950 inclusive, Kansas Cooperative Extension Service.

breeding season of 2 or 3 months. Two-year-old bulls are capable of caring for 25 to 30 cows, and a bull 3 years old or over can handle 40 or 50 cows if hand mating is practiced. If pasture mating is followed, these figures should be reduced by about one-third.

As the size of the herd running in one pasture is increased, more bulls should be provided in relation to the number of cows, because there is always a tendency for a large herd to break up into small droves of 10 to 20 cows each. There should be enough bulls in the herd to make sure that there is little chance of any such drove of cows remaining long without a bull. Sometimes bulls running together in one pasture tend to bunch up, and it may be necessary to break up this tendency by riding the pasture daily for a while.

Table 33, compiled from published records of the Kansas Beef Production Contest, indicates that having more than 25 cows per bull is likely to result in a calving period that extends over 4 to 6 months or even longer, if the bulls are left with the herd until all the cows are settled. If they are removed earlier, some of the cows simply will not have been served and consequently will be found open at weaning time.

EVALUATION OF BULLS FOR BREEDING SOUNDNESS

A discussion of the physiology of the reproductive organs of the bull has been omitted, only because such information is perhaps less essential to the breeder or rancher than is similar information with respect to the cow. Most bulls are normal and fertile, and few anatomical abnormalities are seen. True, in small herds or single-sire herds an unsound bull may cause complete loss of a calf crop if commonsense observations are not made during the breeding season. A survey conducted in Texas found that approximately 16 percent of 1,369 bulls, representing all breeds and ages and all seasons of breeding, were placed in either the "questionable" or the "cull" category on the basis of semen evaluation—that is, examination for presence of a sufficient number of live, motile sperm cells. Such tests are indicative of the bull's soundness at the time of testing but do not necessarily reflect his ability to settle cows during a 60- to 90-day breeding season. Some bulls that produce a good semen sample may not have enough sex drive or libido to breed the cows, or they may be unable to follow cows or serve them because of leg or foot problems. On the other hand, the collection of a usable sample of semen is not always possible, and errors in testing can creep in by this means also.

In general, fertility testing of bulls should be left to a veterinarian or to specialists in physiology of reproduction who are technically trained in this procedure. Most small-scale breeders and even most ranchers cannot justify the cost of the specialized equipment needed to make the test themselves, nor will they be apt to examine enough bulls to become expert to the extent that the test will be reliable. Further information on the subject of evaluation of bulls for breeding soundness can be obtained from textbooks and bulletins devoted exclusively to the subject.

CHAPTER 6
PREGNANCY, PARTURITION, AND CARE OF THE CALF

Estrus and service of the cow, either naturally or by artificial means, followed shortly by ovulation, obviously must precede fertilization and pregnancy before a calf can be produced. Normally fertilization occurs in the upper end of the oviduct, just below the infundibulum, although there may be times when it does not occur until the egg reaches the uterus.

The length of time required for the fertilized ovum to move down the Fallopian tube to the horn of the uterus is not definitely known, but probably is about 10 days. During this time the fertilized ovum, now known as a zygote, is undergoing division or segmentation. While increasing little if any in size, the egg, by successive cleavage, divides first into 2, then 4, 8, 16, 32, and so forth, segments, finally reaching the blastocyst stage.

Soon after reaching the uterus the zygote becomes greatly enlarged by the absorption of fluids, and segmentation proceeds at a rapid rate. Also the cells begin to exhibit marked differences in size and shape, first assuming the appearance of well-defined layers; later, the differentiation of cells and tissues to form the different systems of organs takes place.

For as many as 30 days inside the uterus the segmented zygote lies free within the uterine cavity, but shortly thereafter it becomes implanted—that is, attached to the wall of the uterus. This implantation not only protects the embryo from sudden and violent displacement, but also affords a method for the transfer of nutritive material from the mother to the young, and the transfer of waste products from the embryo to the mother, thereby making possible growth and development. As soon as this exchange of materials begins to take place, the embryo is called a fetus.

THE FETAL MEMBRANES

The fetal membranes consist of three separate structures or parts: the *chorion,* the *amnion,* and the *allantois.* The chorion is the outer membrane and lies close to the mucous membrane of the uterus. The surface of the chorion is much greater than that of the impregnated horn. Consequently, it may extend into the nonpregnant horn, as well as into the body of the uterus. It has many blood vessels leading into the *placenta,* discussed below.

164

The amnion, the innermost of the three membranes, begins at the navel and surrounds the fetus like a sac, enclosing it entirely. It contains a liquid that protects the fetus from external injury. In the cow there are about 6 or 7 quarts of this liquid, called the *amniotic fluid*. During parturition the amniotic fluid serves to lubricate the vagina, thus aiding in the expulsion of the fetus. If the amnion does not break until the late stages of parturition and the birth canal does not benefit from this lubrication, birth may be slowed or made more difficult.

The allantois is a large membranous sac, between the chorion and the amnion, containing the fetal urine. The urine enters the allantois through a tube from the fetal bladder, called the *urachus*. All of these membranes, taken together, constitute the fetal membranes or "afterbirth."

THE PLACENTA

The placenta is the portion of the fetal membranes that unites the mother and the fetus. Although there is no direct vascular connection, the blood vessels of each lie quite close together, so that an interchange of materials can be made through their extremely thin but extensive walls. The capillaries of the fetal membranes, especially those of the chorion, become imbedded in the mucous walls of the uterus where they are in contact with the capillaries of the uterus.

This penetration of the capillaries of the fetal membranes into the walls of the uterus is by no means general over the entire surface of the impregnated horn. Rather, such contact is made only at specialized points on the uterine wall, which are known as cotyledons, illustrated in Fig. 35. These cotyledons are small prominences resembling scars or warts in the nonpregnant cow, somewhat oblong in shape, with their long axis at right angles to the long axis of the uterine horn. Because there are 40 to 60 cotyledons in each horn, the cow is said to have a multiple placenta.

During pregnancy these cotyledons greatly enlarge, and numerous follicles or depressions are developed on each one's surface. Into these follicles the villi of the chorion and other portions of the fetal membrane are inserted, making an extensive and extremely close attachment between the fetus and the mother. These groups of villi of the fetal membranes are called the *fetal cotyledons*. Since each fetal cotyledon surrounds and dovetails into a maternal cotyledon, it follows that the fetal and maternal cotyledons are present in equal numbers. Between the cotyledons the chorion is free of the walls of the uterus.

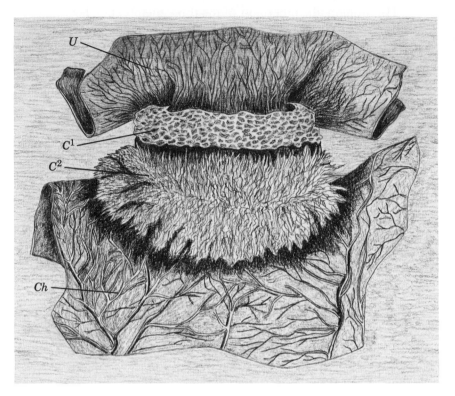

Fig. 35. Cotyledon of cow, showing relation of the maternal and fetal circulations: *u,* uterus; *Ch,* chorion; C^1, maternal, and C^2, fetal portion of cotyledon. (After drawing by Colin.)

THE UMBILICAL CORD

The membranous portion that unites the fetus and the placenta is the *umbilical cord*. Practical stockmen usually refer to it simply as the navel. Its sheath is composed of amniotic membrane within which are found two umbilical arteries, two umbilical veins, and the urachus, the tube leading from the urinary bladder into the allantois. Interspersed between these vessels is a gelatinous mass called the *Whartonian gelatin*. The umbilical arteries carry blood from the fetus to the capillaries of the fetal placenta, while the veins carry it back. These blood vessels have strong muscular walls that contract forcefully when ruptured at calving time, thereby preventing excessive bleeding from the calf's navel. The umbilical arteries are only loosely attached to the umbilical opening in the calf, and their ends retract within the calf's abdominal cavity upon being ruptured at birth. This retraction prevents the entrance of disease-producing bacteria through the severed vessels and also serves to check the flow of blood. The umbilical veins, however, are attached firmly to the umbilical opening or ring and do not retract when ruptured but remain open for a time and are occasionally the avenue for infectious bacteria such as those responsible for *navel ill*. Treatment of the navel with iodine at birth is recommended to prevent bacterial infection if calving takes place anywhere but on clean pasture.

POSITION OF THE FETUS IN THE UTERUS

In the early stages of development the embryo floats freely in the amniotic fluid, occupying no distinct position. With the growth of the fetus, however, it becomes fixed in position, usually with the anterior end, or head, toward the cervix. As the end of gestation approaches, the weight of the fetus causes it to rest upon its side on the floor of the cow's abdomen. As the cow's rumen or paunch occupies the entire left side, the fetus must arrange itself on the right. Advanced pregnancy can be diagnosed in a standing cow by "bumping" the cow with the heel of the hand just above the right flank. The fetus can be felt as it returns to its original position after being pushed inward.

MULTIPLE PREGNANCY

The cow usually is uniparous; more than one fetus is seldom formed in the uterus in a given pregnancy. However, twin calves occur occasionally, and triplets and quadruplets are not unknown, as discussed in the preceding chapter. When only one fetus is present, it usually occupies one horn of the uterus; with twins, each horn usually contains a fetus. As the fetal membranes of calves are large and extensive, filling, even in a single birth, nearly the entire uterus, those of twins are almost certain to be in close contact and to become fused. If the fetal membranes of twins of opposite sexes are fused and a more or less common circulatory system is established, development of the reproductive organs of the female fetus is arrested. Apparently the hormones of the male are dominant over those of the female; or, as seems more likely, the male sex cells are the first to appear in the development of the two fetuses. Heifers that are born twin to a male are called *freemartins* and fewer than 1 percent of such females are fertile. The condition is irreversible.

SIGNS OF PREGNANCY

The gestation period of the cow is approximately 283 days but varies with breed. Angus cows have shorter gestation periods by about one week, whereas Brahman cows have periods about one week longer. It is believed these differences in gestation length are largely responsible for the differences in calf birth weights seen among the various breeds. Breed of the calf's sire also can influence gestation length, for reasons that are not clearly understood, and breed differences are still under study.

Long before the end of the gestation period certain changes are observable in the pregnant female that indicate the existence of the developing fetus. Because a diagnosis of pregnancy is often of great importance, every cattleman would like to know how to determine, as early as possible, whether his cows are pregnant and *pregnancy testing* is becoming a management tool of many progressive cowmen.

Unfortunately none of the signs of pregnancy that can be observed by the layman is infallible during the first half of the gestation period. However, there are certain changes generally observed in pregnant cows that give good reason for suspecting pregnancy in any bred female that exhibits them. As gestation progresses, more obvious signs of pregnancy appear, although even these are sometimes misleading owing to the presence of certain diseases that produce changes similar to those caused by a developing fetus. Only direct examination made after the second month of gestation discloses beyond doubt the presence or absence of pregnancy. Among the many signs of pregnancy are the following important ones.

1. Cessation of estrus or heat. After seeing a bull serve a cow that has an identifying number, the cowman may easily make a record of it and determine when estrus should again occur if the cow fails to conceive. If pasture breeding is practiced, the experienced cowman checks the herd periodically during the breeding·season to see if the cows are "passing over." If identifiable cows do not return or show estrus, he can be assured that the cows are being settled. The artificial inseminator keeps records of estrus in individual cows to facilitate his work.

2. A noticeable enlargement of abdomen and udder. An enlargement of the abdomen is usually a good sign but not necessarily foolproof. As parturition approaches, the udder fills and teats firm up. First-calf heifers usually exhibit udder development sooner than mature cows do. Obviously these changes occur late in pregnancy.

3. "Pregnancy testing" by internal examination. Manual examination or palpation of the reproductive tract by way of the rectum and colon may

be made to verify pregnancy beyond a doubt. Experienced persons can detect a fetus as young as 2 months by this method, but examination after the third month is more reliable. The effectiveness of rectal palpation as a pregnancy test is possible because the uterus and ovaries lie just beneath the colon and can easily be felt through the wall of the large gut. In early pregnancy (2 to 6 months) the presence of the fetus can be felt beneath the floor of the colon. When gently pressed by the hand, the fetus slips away as though it were floating in a liquid, which of course it is, and it returns immediately to its original position when pressure is released. As the gestation period advances, the uterus is pulled down within the abdominal cavity owing to the weight of the fetus and accompanying fluids. In this stage it is usually impossible for the examiner to feel the fetus. However, careful exploration will disclose a large, thick, firmly stretched band—the posterior end of the uterus—which passes downward and forward into the abdominal cavity. The blood vessels supplying the fetus will be enlarged enough so that they can also be felt at this stage. The ovaries themselves are examined in early pregnancy by experienced persons, and presence of the corpus luteum of pregnancy is used to verify a tentative diagnosis of pregnancy.

IMPORTANCE OF PREGNANCY TESTING

Pregnancy testing or palpation as a routine practice at the end of the breeding season can be an important tool in the efficient operation of a beef cow herd, because carrying nonpregnant cows for a full year without any return is one of the largest drains on profits. Table 34 shows the breeding and calving record of a grade herd in Colorado over a 5-year period. All females, including breeding-age heifers, were examined for pregnancy in the fall before winter feeding started and, except for the last year, all open females were sold. Prompt disposal of the open females increased the annual net return of the remaining cows by approximately $8 per head by reducing the total winter feed bill. Higher cow costs would increase this figure to easily $20 to $25 in the mid-1970s. Experimental work has indicated, although not conclusively, that heritability plays a role in regularity of breeding, and vigorous culling of slow breeders or nonbreeders has been shown to increase the breeding efficiency of a herd materially. A cow-culling plan based on pregnancy status alone should be tempered with judgment. The high rates of open cows in the study just referred to were caused by poor nutrition brought on by severe droughts. Often the better-milking cows are the most apt to be open in such a case, and automatic culling might be unwarranted.

Table 34

Effect of Pregnancy Testing on Percent Calf Crop Dropped in a Colorado Herd[a]

Year	Cows Examined	Number Found Open	Percent Open	Percent Calf Crop Dropped	
------	---------------	-------------------	--------------	Without Pregnancy Testing[b]	With Pregnancy Testing[c]
1952	343	62	18.1	80.2	98.6
1953	352	24	6.7	92.4	99.1
1954	406	22	5.4	92.9	98.2
1955	469	28	6.0	92.1	97.9
1956	539	94[d]	17.4	82.0	99.3

[a] Personal communication to the author from Dr. Lloyd C. Faulkner, Colorado State University.

[b] Calculated on the basis of number of breeding-age females exposed to the bull.

[c] Based on the number of cows remaining in the herd after selling the cows declared open upon pregnancy testing.

[d] Not all the open females were sold in 1956 owing to the large number found to be open, but calving percentage is calculated as if they had been sold.

DETERMINATION OF CALVING AND BREEDING DATE

It is often desirable to know approximately the date on which a cow will calve in order to be prepared to give special attention should it be needed. If the date of service or the date the bull was turned in with a herd of cows is known, one need only count back 3 months and forward 10 days to determine the approximate calving date or the beginning of the calving season. For example, if a cow is served on July 21, counting back 3 months to April 21 and forward 10 days gives an approximate calving date of May 1.

It is often desirable to have calves dropped shortly after a certain date—for instance, if cattle are to be fitted for show, in which case full-aged animals may have a slight advantage over short-aged ones. The junior calf class, for example, usually consists of calves born after January 1 of the year they are shown, and it is obvious that a calf born on January 3 would be further along in development than one born on February 20. To determine when to start breeding so as to have the calves bunched after a certain base date, count forward 3 months and back 10 days from the base date. Because of the natural variation in length of gestation period, going back only 5 days instead of the usual 10 will give a safety factor. Thus, if it is desirable for calves to be born on or soon after January 1, determine the breeding date by counting forward 3 months to April 1, then back 10 or 5 days to March 20 or 25, depending on how safe one wishes to be. Gestation tables, found in many publications, are of course handy to use but may not always be available when needed; thus the preceding rules of thumb are a convenient substitute.

CARE OF PREGNANT COWS

If possible, cows in advanced pregnancy should be separated from the rest of the herd to avoid injury and possible abortion from riding, butting, and fighting. A good plan, where adequate lots and pastures are available, is to sort off and move all heifers and cows within 3 or 4 weeks of calving to separate calving pastures or corrals. This will save much time when cows are checked during calving season, because the other cattle require less attention. A good plan, where adequate lots and pastures are available, is to remove all bred cows from the herd at about the middle of the gestation period.

The pregnant cow needs exercise and, except during cold or stormy weather, there is no better place for her than out of doors where she can move about freely. A windbreak and a dry place on which to lie are ideal when available, but the majority of commercial beef cows do not enjoy such luxuries.

EMBRYONAL MORTALITY

It is well known that every pregnancy does not result in a live calf at parturition. Certain infectious reproductive diseases that are described in Chapter 27 account for a high percentage of these lost embryos or fetuses in infected herds, but the losses in "clean" herds are greater than generally supposed. British studies in 1973 reported that 20.3 percent of all pregnancies in a large sample of 280,215 cows failed to result in a live calf at birth. Since these cows all were bred by artificial insemination, it was possible to study the records and try to explain the losses. It was found that the lowest rate of embryonal mortality occurred among cows bred late in standing heat with fresh semen. Losses were greater if frozen semen was used, if the cow was bred after standing heat, if dairy-breed semen was used, or if aged (unfrozen) rather than fresh semen was used.

Data such as those in Table 35 show the extent of losses of potential

Table 35

Sources of Variation in Reproductive Performance in Experimental Beef Cow Herds at Four USDA Experiment Stations[a]

Item	Miles City, Montana[b]	Front Royal, Virginia[c]	Jeanerette, Louisiana[c]	Clay Center, Nebraska[d]
Number of cows exposed	5,658	882	462	2,563
Percent of cows failing to conceive	16.6	11.9	15.7	—
Percent of calves lost during gestation	2.4	2.7	10.1	—
Percent of calves lost at birth	6.4	8.5	3.7	5.5
Percent of calves lost, birth to weaning	4.1	3.6	8.0	—
Percent weaned calves (net calf crop)	70.5	73.2	62.8	—

[a] *The Profit Brand*, American Breeders' Service, 5(6), 1973.
[b] Bellows, National Association of Artificial Breeders' Conference on Beef Artificial Insemination, 1971.
[c] Wiltbank et al., *Journal of Animal Science* 20: 409.
[d] Preliminary report, Germ Plasm Experiment, U.S. Meat Animal Research Center, Clay Center, Nebraska, January 1973.

calves from all causes up through weaning in the United States. The data are a compilation of results from U.S. Department of Agriculture experimental herds which, it may be assumed, were free or nearly free of the more common reproductive diseases. Several environments are represented and beef breeds were used, with natural service being the main method of breeding. No explanation is at hand for the much higher embryonal loss at the Louisiana station, but it is recognized that higher environmental temperatures may be responsible to some extent. More Zebu breeding was used in this herd as well, but cattle with such breeding are known to be reproductively more efficient in warmer climates than British breeds. On the average, losses were about as great before calving as at calving, and from calving to weaning. Preliminary data with cattle and extensive data with other species indicate that most noninfectious embryonal mortality occurs during the first third of the pregnancy period, and that it is caused by an as yet unexplained but unsatisfactory uterine environment.

SIGNS OF PARTURITION

During the last few days of the gestation period certain changes in the pregnant animal will indicate to the experienced observer that parturition is not long away. The more important signs are the following.

1. A relaxation of the pelvic (sacrosciatic) ligaments that permits the muscles of the rump to drop inward, causing a noticeable falling away or sinking about the tailhead and pinbones, and a general softening or loosening of the flesh in this region.
2. An enlargement and thickening of the vulva, which appears swollen and somewhat inflamed.
3. A noticeable distension of the teats and an enlargement of the udder, as well as an abrupt change in its secretion, from a watery material to thick, milky colostrum.

These changes usually begin some 3 or 4 weeks before birth occurs, but become more and more pronounced as the time of parturition draws nearer. As a rule, they appear sooner in heifers than in older cows, and the latter sometimes calve with little or no "notice."

LABOR

The act of birth is accomplished through much exertion on the part of the mother. The first real sign of labor is a noticeable uneasiness on the part of the animal several hours before calving. The cow will, if possible, leave the herd and move to a more isolated part of the pasture. She turns her head, glancing nervously to the rear, and frequently lies down and gets up at short intervals.

The preliminary contractions of the uterus, which undoubtedly are responsible for the uneasiness described, are extremely important, because they move the fetus from a lateral, recumbent position on the floor of the abdomen to a longitudinal, upright attitude immediately in front of the pelvic girdle, from which position it can be easily expelled. The uterine contractions also bring about a dilation or enlargement of the cervix, through which the fetus must pass. In fact this restriction, the cervix, which normally separates the vagina and the uterus, is practically obliterated, and the two organs form one continuous passage to the exterior. Enlargement of the cervix is largely caused by the pressure exerted by the uterine walls upon the fetal fluids. These fluids within their elastic membranes are forced into the rear part of the uterus and, by transmitting the pressure from the contracting uterus equally in all directions, serve as an elastic dilator, first of the cervix and later of the parts beyond. Assisting the uterine muscles in this and the subsequent steps of parturition are the large muscles of the abdominal walls, as well as the diaphragm. These muscles, contracting in unison with those of the uterus at intervals of 1 to 2 minutes, exert a tremendous force upon the fetus. Normally they bring about its expulsion, unassisted, within 30 or 40 minutes after the contractions first begin.

The effect of this great and rhythmic pressure upon the fetus is to cause it and its enveloping membranes to be moved through the dilated cervix into the vagina. The chorion or outer membrane, being firmly attached to the uterine walls except near the fundus, does not withstand much stretching and is soon ruptured by the increasing pressure. This step in the birth process is highly essential to the well-being of the young, as the continuation of close contact between the chorion and uterus maintains the oxygen supply and nourishment of the fetus throughout labor. When the chorion ruptures, the allantois with its liquid contents—that is, the fetal urine—is forced into the vagina to appear at the vulva, where gradually it forms a semitransparent, balloonlike sac holding 1 to 2 pints of fluid. This sac is called the first water bag. Usually it increases in size until it ruptures from the pressure of the liquid within.

Following the escape of the allantoic fluid, the pressure exerted by the uterus and the abdominal muscles is applied directly to the fetus, now suspended only in the amniotic fluid. Should this pressure be applied before the fetus is in proper position or before the genital passages are sufficiently enlarged, birth will be more difficult. Therefore the longer the water bag remains intact, the better.

Within minutes following the rupture of the allantoic sac or first water bag, the amniotic membrane appears with the contained fetus. This membrane is glistening-white and contains a rather viscid, slimy, opalescent liquid. Within the membrane a portion of the fetus, usually the front feet, can plainly be seen. With each labor contraction more and more of the amniotic liquid is forced out to form a sac, the second water bag, which eventually bursts, thereby lubricating the genital passage.

The rupture of the amniotic membrane is followed by violent strainings on the part of the cow, which soon force the head and then the shoulders of the fetus through the pelvic canal. At this stage the cow often rises to her feet, and the calf completely emerges under the influence of gravity.

INDUCED PARTURITION

Several synthetic hormone-like compounds are now available that can be injected for the purpose of initiating parturition. This would appear to be a promising new management tool, for it would then be possible to plan the use of labor and equipment more efficiently. An even greater potential would be the possibility of shortening the gestation period to reduce the birth weights of calves sired by larger breedsires and thus reduce calving losses. Unfortunately, at present much research is still needed to reduce the incidence of undesirable side effects, principally retained placenta, lowered calf survival rate, and reduced milk production.

RENDERING ASSISTANCE

Assistance should not be given except when it is actually necessary. Some herdsmen and even veterinarians are prone to rush in and "take" the calf by force at the first appearance of the emerging fetus. This practice is not a good one, because there is likely to be injury to both cow and calf in the form of torn membranes and strained ligaments. If no progress has been made toward delivery within about 2 hours after the beginning of labor, a careful examination should be made, by a veterinarian if possible, to determine whether an abnormal presentation is causing the delay.

While unnecessary aid is to be discouraged, one should not make the opposite mistake of permitting the cow, and especially a first-calf heifer, to labor until she is completely exhausted before aid is furnished. Occasionally a calf has an abnormally large head or unusually heavy shoulders or hips that greatly delay or totally halt its passage through the pelvic cavity, even in a normal presentation. Anything other than a presentation where both front feet and the head appear almost simultaneously is considered abnormal. In a posterior presentation (hind feed first), the hips of the calf frequently cause trouble, mainly because the small size of the allantoic sac in the region of the hindquarters results in insufficient dilation of the cervix.

In either event assistance should be given by fastening small ropes or chains well above the calf's pasterns, to avoid injuring the soft hoofs, and pulling backward and downward each time the cow labors. No more force should be exerted than is necessary to overcome the obstruction. Under no condition should traction be applied except when the cow labors, unless she is exhausted to the point where she refuses to labor. In that event the calf must be removed entirely by traction. Use of mechanical calf-pullers by anyone but an experienced person usually proves fatal to the calf and results in permanent injury to the cow.

So long as there is no tension on the umbilical cord there is no cause for hastening the act of birth, at least in a normal presentation. However, when the calf's head has advanced as far as the eyes, the cord is pressed

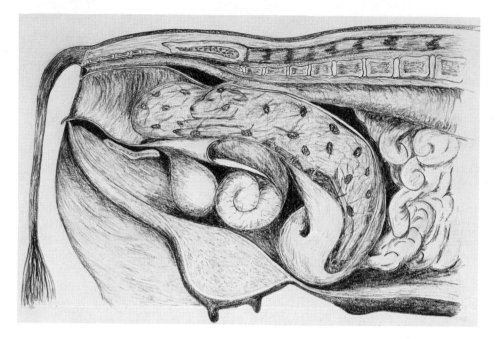

Fig. 36. The normal position of the fetus at the time of birth. (After Skelett.)

against the floor of the cow's pelvis in such a way that the placental circulation is jeopardized. At this point it is highly advisable to rupture the amnion, if it is still intact, to permit respiration and to prevent the calf from drowning in the amniotic fluid. Because respiration is unlikely to be effective so long as the chest is in the viselike grasp of the genital passage, birth should be hastened by traction in such situations. In a posterior presentation there is always considerable danger of the calf's suffocating through rupture or strangulation of the umbilical cord, and assistance in hastening parturition is more likely to be necessary in posterior than in anterior births. A posterior presentation can usually be anticipated if the feet first appear with the pads or soles of the hoofs topside.

Some other abnormal presentations occasionally encountered are a breech presentation (that is, a posterior presentation with the hind legs tucked under, not extended); one or both front feet turned back in an otherwise normal presentation; one with the front feet normally presented, but the head turned back. In all these situations the calf must be pushed back into the abdominal cavity and the limbs or head straightened. This is difficult to do because of the counteracting force of the labor contractions and because of the space limitations. Drugs or hormones are sometimes used to cause the cow to cease her straining and allow the corrections to be made. Generally only experienced herdsmen or veterinarians should attempt these adjustments, but when such help is unavailable an amateur will have to make the best of the situation.

BIRTH BY CAESAREAN SECTION

In some instances where it becomes evident that a valuable cow cannot deliver her calf, even with assistance, an operation known as a Caesarean or "C" section is performed. This surgery must be done by a veterinarian or it is almost certain to end in loss of both calf and cow. As already mentioned, excessive condition or fat can interfere with even a normal birth; if an abnormal presentation is found in such an overfinished animal, a Caesarean operation may be the only solution.

It consists of removal of the calf through a large incision in the right side just above the flank. Because the opening must be made through the skin, several membranes and muscle layers, and finally the wall of the uterus itself, it is obvious why only a skilled surgeon can do the work. The cow usually is immobilized for the operation by spinal anesthesia or a spinal block.

If the surgery is skillfully done and infection is held to a minimum, the cow should be left in good condition for future calvings. Usually the calf is saved if the operation takes place at the normal calving time or before prolonged labor has occurred. Ordinarily a Caesarean section is not attempted more than a few days in advance of what is believed will be the normal calving time.

Another surgical procedure being used on young cows, mostly by veterinarians, is that of "splitting the pelvis." The procedure is applied to a standing cow that has been immobilized by spinal block. Briefly, it consists of making a small opening through the skin just below the vulva and inserting a long, chisel-like instrument forward, below the vagina, until it is positioned posteriorially against the pelvis, which in a young cow is still cartilaginous. The chisel is gently driven forward, cutting apart the floor of the pelvis so that it spreads and allows passage of the calf. The cartilage heals back without permanent damage. Postoperative care should be taken to prevent infection and injury such as slipping on ice. Obviously only experienced persons should attempt this procedure but it has advantages over the "C" section.

CARE OF THE NEWBORN CALF

After the calf is born, all membranes should be cleared from its nostrils to facilitate breathing. Normally the mother cow takes care of this process. If parturition has taken a long time and it is suspected that the calf breathed while still inside the cow, all amniotic liquid should hastily be cleared from the nasal passages and throat to prevent drowning. This can usually be accomplished by holding the calf head-downward for an instant to permit the material to drain out of the lungs and air passages. Artificial respiration should then be given at once and continued until breathing is established. Many calves that appear to be dead can be revived by this method. Tickling the nostrils with a straw or stem often will stimulate the onset of breathing.

Because the umbilical cord is relatively short (12 to 15 inches), it always ruptures during the act of birth—usually about the time the forequarters pass through the vulva. If sanitary precautions have been taken in preparing clean quarters for the parturient cow, or if the cow has calved on clean pasture, there is little likelihood of any trouble from navel infection. However, some cattlemen make a practice of applying either tincture of iodine or formalin to the navel stump to destroy any pus-forming bacteria that may be present or the even more dangerous soilborne tetanus organisms found on some farms. Others dust the stump with antiseptic powder to hasten its drying and sloughing off.

As soon as possible after birth, the cow and calf should be left alone. The cow apparently derives satisfaction from drying her calf by licking it, and this may be nature's form of artificial respiration. Also it apparently stimulates the functioning of the calf's circulatory system and the cow's milk letdown.

Usually the calf stands and nurses of its own accord. The taking of nourishment a few minutes after birth is not absolutely necessary. If at the end of 5 or 6 hours the calf has not nursed, however, it should be given assistance in finding the udder. Seldom is more than one lesson necessary. Most authorities agree that the colostrum, or first milk, acts as a mild laxative, but perhaps its most important function is to provide protective antibodies that reduce the danger from various respiratory and gastrointestinal infections. A supply of frozen colostrum will save an occasional calf in cases where a cow does not claim her calf or cannot nurse it because of injury or some other reason. The colostrum should of course be warmed, but not boiled, before feeding.

It is important that young animals receive plenty of sunshine and exercise, which are necessary for health and rapid growth. Calves born on pasture naturally get sufficient exercise in following their mothers back and forth over the pasture. Fall- and winter-born calves in the nonrange area do not have this advantage and should be provided with a good-sized, well-drained lot in which to scamper and run on fine days when the ground is dry or frozen. In addition to this lot, a small paved corral or escape creep-pen on the sheltered side of the barn is ideal as a sunning place when the larger lot is wet and muddy.

EXPULSION OF THE PLACENTA

After the calf is born, the outer fetal membranes and the chorion are still attached to the walls of the uterus by means of the cotyledons. With parturition, however, the exchange of nutritive elements between the maternal and fetal placentae immediately stops, and there is a shrinkage of the villi of the cotyledons. The attachments between the chorion and the uterus are thus loosened, a process that is greatly hastened and facilitated by the contractions of the uterus, and through which the uterus returns to its normal size. One by one the fetal cotyledons separate from the cotyledons of the uterus, and the freed membranes are gradually forced out through the vulva.

RETENTION OF THE PLACENTA

Normally the fetal membranes are expelled 2 to 6 hours after parturition. If they remain longer than 24 hours, it is likely that an abnormal condition exists. Cattle are more susceptible to this condition than are other species of domestic animals, and it is not uncommon to encounter herds in which nearly 20 percent of the cows are troubled with this problem during some calving seasons.

There are various possible causes of retained placenta. In all probability many cases result from an infection that causes inflammation and enlargement of the maternal cotyledons. Such infection may have been present before parturition, in association with contagious abortion and other reproductive diseases (see Chapter 27), or it may have occurred while the cow was given assistance during labor. Also, retention is likely to accompany a failure of the uterine walls to contract promptly because of general weakness on the part of the animal or its exhaustion from an especially long labor. Cows that are thin and half-starved, as well as cows that are in high show condition, are likely to be troubled with retained placenta. A deficiency of carotene, or of vitamin A itself, in the ration of the pregnant cow has been strongly linked with retained placenta. Such a deficiency is most apt to occur in cows being wintered on cured grass, without supplementation, following one or more seasons of severe drought. Hormonal imbalance also has been implicated but treatments with specific hormone injections are hazardous.

A retained placenta is an ideal medium for the development of putrefying bacteria. An almost infallible signal of retained placenta is a stringlike portion of the unexpelled membranes hanging from the vulva, where it comes in contact with the tail and hindquarters of the cow. It of course becomes heavily laden with all sorts of bacteria, which quickly spread into the interior. Decomposition and putrefaction begin in a remarkably short time. Except in cold weather, an obnoxious odor, warning that attention is urgent, appears within 48 hours after parturition.

REMOVING RETAINED PLACENTA

There is considerable difference of opinion among veterinarians as to the proper time to remove a retained placenta. Some recommend removal as soon as 24 hours after calving. Others advocate waiting another day in order that the attachments may be partly loosened by the process of decomposition. Still other authorities advise that removal not be attempted at all, some preferring to treat the condition with sulfa drugs and/or antibiotics to help the cow combat the resultant infection.

DYSTOCIA

The term *dystocia* is applied to cases of prolonged and difficult birth. Death of the calf may or may not result and aftereffects often are noted in the cow. Some possible causes have been identified and studied by investigators at the Meat Animal Research Center in Clay Center, Nebraska. They conclude that the following items are most important and that their effects can be influenced by managerial decisions and husbandry.

1. *Size of calf* is by far the most important item that contributes to dystocia. Choice of sire breed, choice of sire within breed, and plane of nutrition in late stages of pregnancy are under the control of the cowman. Gestation length strongly influences calf birth weight but is not practically controllable at this time.
2. *Cow size*, independent of cow breed or age, is next in importance, and it should be noted that cow weight and cow size are not synonymous. Small cows, regardless of breed, and especially small 2-year-old cows, simply cannot bear calves as easily as larger cows, and the problem is complicated if sires that produce large calves are used.
3. *Cow age* is important because 2-year-old cows have reached only 75

percent of mature size but drop calves that are 90 percent as heavy as those of mature cows. Checking first- and second-calf heifers every 3 or 4 hours around the clock during the calving season is warranted.

4. *Cow breed* is important because breed is associated with calf weight, calf shape, and size and shape of the cow's pelvic opening. Also certain breeds are better adapted to harsh environments than other breeds and hence may be in stronger condition for calving under adverse circumstances.

The Clay Center studies have shown that shape of calf and pelvic area, as measured by a technique called *pelvimetry*, do not strongly influence dystocia, independent of breed, cow size, calf weight, and so on.

The aftereffects of dystocia should not be overlooked. The investigators just referred to found that cows that required assistance at calving had a 16 percent lower conception rate in subsequent breedings than cows requiring no assistance. They also found that each 21-day delay in rebreeding reduced weaning weight, at a constant date, by 40 pounds.

Abnormal presentations account for only about 5 percent of dystocia cases and are uncontrollable. Sex of calf is a factor, with bull calves causing difficulty twice as often as heifers, but, again, management cannot control the problem. Crossbreeding, per se, increases birth weight 2 or 3 pounds over the average of the parent breeds and so it can be a factor, whereas inbreeding, within a breed of course, reduces birth weight. The benefit of the reduced birth weight in the inbred calf is probably more than canceled by the resulting reduction in calf vigor.

The plane of nutrition of the dam during the last 4 months of gestation can influence birth weight of the calf and therefore the incidence of dystocia. This is discussed in Chapter 9, but suffice it to say here that adequate protein, phosphorus, and vitamin A, plus enough energy to just maintain flesh condition (about 1 pound gain per day and equal to weight loss at calving) is a safe course to follow.

With so many possible adverse factors, it is not surprising that a herd of 10 or more cows seldom escapes all of them. In other words, a perfect calf crop (100 percent) is seldom realized. Instead, the crop is likely to be somewhere between 70 and 95 percent.

In general, the size of the calf crop varies inversely with the size of the herd, because of the smaller amount of individual attention given to the animals of larger herds. However, studies by the U.S. Department of Agriculture, both in the Corn Belt and Range areas, indicate that good care and management on the part of the owner are much more important in determining the percentage of calves raised than is the size of the breeding herd or even the ratio of bulls to cows. In many instances great variation existed in the size of the calf crop on practically adjoining ranches "with no perceptible difference in range, feed, water facilities, quality of animals, or animal losses."

Undoubtedly such differences are largely due to careful culling of nonbreeding cows, the amount of attention given to the conditioning of bulls before the breeding season, the amount of time spent in systematic inspection of the herd for cows in heat, and the attention given the cows during both breeding and calving periods. Nevertheless, even with the same system of management, much variation in the percentage of calves raised was found on the same farm or ranch from one year to another.

Weather conditions, particularly as they affect the feed supply of the cow herd and the exposure to which the young calves are subjected during the 2 or 3 weeks following birth, are largely responsible for these yearly variations. These can be overcome to a great extent by providing sufficient emergency feed supplies and adequate shelter facilities to meet such emergencies if winter calving is practiced.

The influence of size of calf crop on the net cost of raising calves to weaning age is well illustrated in Table 36. It should be noted that 68 percent of these Texas ranches realized a calf crop under 80 percent. In sharp contrast to these figures are the records of more than 100 small breeders in the Corn Belt states, many of whom weaned calves from 90 percent or more of their cows (Table 37). In view of the gradual decrease in size of calf crop from east to west, it seems probable that feed

Table 36

Influence of Size of Calf Crop in Net Cost per Calf[a] (North Central Texas Ranches)

Percent Calf Crop (by Groups)	Number of Ranches in Each Group	Number of Calves	Net Cost per Calf (4-Year Average) ($)	Percent of Total Calves
30–40	1	590	43.92	1.4
40–50	2	585	49.21	1.3
50–60	14	11,880	36.15	27.4
60–70	18	8,020	38.88	18.5
70–80	22	10,848	28.03	25.6
80–90	22	9,749	25.46	22.5
90–100	5	1,695	22.15	3.9
Totals and averages	84	43,367	31.95	100.0

[a] *California Bulletin* 458.

Table 37

Percentage of Calves Raised in Corn Belt Herds[a]

State	Number of Farms	Average Number of Cows per Farm	Average Percent Calf Crop
Indiana	6	16.5	96.2
Illinois	13	21.0	89.3
Missouri	33	22.1	90.2
Iowa	76	31.1	86.5
Minnesota	12	21.7	85.0
South Dakota	6	30.2	86.4
Kansas	46	50.4	81.5
Nebraska	38	27.0	78.9
Total	230	31.5	84.9

[a] USDA Report 111.

conditions and available shelter were the chief causes for the variations noted. The data in both of these tables were obtained several years ago when many breeders, especially large ranchers, gave less attention to their herds than they do now.

CHAPTER 7
PRINCIPLES OF FEEDING BEEF CATTLE

Except for the increasingly large number of feeder cattle that are fed high-concentrate rations in drylot, beef cattle, being ruminant animals, are usually found on farms and ranches that produce large quantities of harvested roughage and/or pasture or range. The prevalence of roughage may result from the operator's choice, from necessity because perpetually low rainfall makes growing of other crops difficult, or from erosion or soil infertility.

Ruminant or cud-chewing animals have specially adapted digestive systems that enable them to utilize roughages or feeds that contain comparatively high levels of crude fiber or cellulose and related compounds. An understanding of these adaptations is valuable in determining feed or nutrient requirements of beef cattle, and knowledge of nutrient requirements is essential for proper ration formulation.[1]

SIGNIFICANT FEATURES OF RUMINANT NUTRITION

Monogastric animals—those having one simple stomach, such as the pig and man—have a relatively low-capacity alimentary tract consisting of stomach, small and large intestines, and accessory glands. In these animals digestion is largely enzymatic in nature and little provision is made for digesting roughages; therefore their diet must consist mainly of concentrates or feeds low in crude fiber. In contrast, ruminant animals have compound stomachs and a much more complex digestive system, and much remains to be learned about their anatomy and function.

Using tools such as the artificial rumen and various fistulae (for example, esophageal, rumen, and abomasal) in the live animal, researchers are shedding much light on the so-called darkest spot in animal nutrition, the rumen. By means of fistulae or semipermanent surgical openings, samples of feeds in various stages of digestion may be withdrawn at intervals to follow the progress of physical and chemical changes.

---

[1] The nutrient requirements of beef cattle, established through worldwide research, are regularly reviewed by the Sub-Committee on Beef Cattle Nutrition of the National Research Council. Recommendations published by the Sub-Committee in *Nutrient Requirements of Beef Cattle* (No. 4, 1975) have been used for reference in this chapter.

Also, specific chemical substances may be introduced through these openings to study their effects on the digestive process.

The most successful cattle feeders today are those who know and take advantage of the following unique characteristics of the ruminant animal.

The Four-Compartment Stomach.   The *rumen* or paunch, the first compartment of the compound stomach, may be likened to a huge fermentation vat where much of the carbohydrate fraction of the diet is converted to volatile fatty acids and absorbed directly into the bloodstream. The rumen constitutes about 80 percent of the total stomach capacity in adult cattle and may hold as much as 50 or 60 gallons. Connected with the paunch are the second and third compartments, the *reticulum* or honeycomb and the *omasum* or manyplies, which constitute 5 and 7 or 8 percent of the total stomach capacity, respectively, in mature animals. All three of these compartments have a common opening or passageway called the *esophageal groove*, through which materials may pass freely. The function of the reticulum is not well understood, but it is known that the omasum is the site where much water is absorbed from the paunch contents before it passes into the fourth compartment, the *abomasum* or true stomach. The abomasum holds about 7 to 8 percent of the total stomach contents and its function is similar to that of the stomach in monogastric animals.

Symbiotic Microorganisms of the Rumen.   The rumen ordinarily provides an ideal environment as to temperature, moisture, and nutrient supply for microbial life, and literally billions of bacteria—up to 100 billion per gram of dried rumen contents—and somewhat fewer protozoa live in the rumen, to the mutual benefit of both the microorganisms and the host animal, the ruminant. This mutual benefit or support is known as symbiosis. The breakdown of cellulose and related compounds by the enzymes produced by these microorganisms accounts for the higher feeding value of roughages when fed to ruminants.

Volatile fatty acids (VFA) are produced as a result of the mainly anaerobic microbial fermentation of carbohydrates in the rumen. The most important of these acids, in terms of amount produced, are acetic, propionic, and butyric, in that order. It is estimated that up to 80 percent of the digestible carbohydrate portion of beef cattle rations may be converted to these acids and absorbed directly from the rumen, to be metabolized and used as energy in meeting maintenance requirements or stored as fat. Thus fatty acids are a major source of energy in ruminant rations as contrasted with the situation in monogastric animals, where carbohydrates are largely absorbed as glucose after digestion in the stomach and small intestine.

Recent research in Great Britain and in the United States has shown that if the level of acetic acid resulting from the fermentation of carbohydrates in the rumen can be reduced while the level of propionic acid is increased, the energy of the ration will be more efficiently used. This shift of the acetic-propionic acid ratio in favor of propionic acid reduces the energy losses that occur in metabolism at the cellular level. Promising leads have been uncovered that may make it possible quantitatively to affect the VFA levels. A new feed additive, known commercially as Rumensin and approved by the Food and Drug Administration in 1976, has been shown to improve feed efficiency by at least 10 percent. It accomplishes this by favorably elevating the propionic acid level by as much as 45 percent, mainly at the expense of acetic acid.

Bacterial Synthesis of Protein in the Rumen.   As bacteria and the other microorganisms living in the rumen multiply, they build or synthesize the protein required for the next generation of organisms, using whatever source of nitrogen is available from the feeds consumed by the host. The bacterial protein thus synthesized in the rumen undergoes digestion in the abomasum or true stomach and intestine, and is absorbed by the host in the form of amino acids, regardless of the source or quality of the protein or nitrogen originally consumed in the ration.

Thus microorganisms play a significant role in protein as well as carbohydrate utilization in the ruminant. By contrast, the nonruminant animal has specific requirements for about 10 of the amino acids, the building blocks of protein. These specific amino acids are called the essential amino acids, and their balance—that is, presence in required proportions—in a feed protein determines the *quality* of the protein for the nonruminant.

When finishing rations containing mainly concentrates are fed, the passage rate through the first three compartments of the compound stomach, referred to as the rumino-reticular portion, may be so rapid that bacterial protein synthesis cannot occur at a rate sufficient to meet the host animal's dietary needs for certain of the essential amino acids. In experiments still under way, it is being shown that, under these rather artificial circumstances for the ruminant, there may indeed be a requirement for certain amino acids. Further research may well demonstrate that quality of protein is after all important in the feeding of ruminants.

The microflora of the rumen are not specific in their requirements as to source of nitrogen or protein; therefore low-quality proteins are well utilized by the ruminant. Furthermore, nonprotein nitrogenous (NPN) compounds, such as urea, ammonium salts, and amides, can make up a substantial portion of the total nitrogen or protein requirement of the ruminant. Such nonprotein nitrogenous compounds as urea are almost always a cheaper source of nitrogen, per pound of protein equivalent, than are the usual protein concentrates, and thus it is possible to formulate more economical protein supplements for cattle than for nonruminant animals. An adaptation period of 2 to 4 weeks is usually required before utilization of NPN is equal to that of natural protein sources because some shifts in the microfloral populations must take place and these require time. A majority of the commercially prepared protein concentrates designed for beef cattle now contain some urea, generally about 3 percent, or about 8 percent protein equivalent. More about urea and protein equivalent is found in Chapter 17.

Bacterial Synthesis of the B-Complex Vitamins in the Rumen. Just as the microflora of the rumen are able to synthesize protein, they also synthesize many of the vitamins required by the host. Once the rumen becomes functional and a bacterial population is established, B-complex vitamins such as riboflavin, niacin, pyridoxine, biotin, folic acid, and B_{12} are synthesized in the rumen at a rate sufficient to meet the needs of the host. The rate of synthesis of some of these vitamins can be altered by varying the level of certain nutrients in the ration. For example, if cobalt, one of the essential mineral elements, is deficient in the ration, vitamin B_{12} synthesis is too low for maximum performance on the part of the host. Vitamin K, one of the fat-soluble vitamins, is also synthesized in the rumen, but such important fat-soluble vitamins as A, D, and E must be supplied in the ration. Again, rapid passage of the diet through the tract may adversely affect the amount of B vitamins synthesized, as with amino acids; thus here is a case where a drastic change in feeding practice may be responsible for changes in earlier established nutrient requirements. This area is currently being actively researched in many laboratories.

Important Role of Saliva. The salivary glands of cattle, located beneath the tongue, produce 15 to 20 gallons of saliva per day. This saliva, a highly alkaline material, is mainly secreted during rumination or cud-chewing, but some appears during the act of eating. The saliva serves several functions but perhaps the most important is that of buffering or neutralizing the large amount of acid produced in the rumen by bacterial enzymatic fermentation.

When the total production of volatile fatty acids is extremely high, as often happens when high-concentrate rations are used in cattle-finishing programs, not enough saliva is secreted to neutralize the increased acid production. An abnormal condition, known as rumen parakeratosis, occurs. The problem is an inflammation and keratinization of the hairlike papillae of the rumen wall. Some researchers believe that lactic acid, an intermediate step in the formation of propionic acid, is the agent that causes the difficulty. Whatever the cause, it is believed that this condition interferes with absorption of nutrients through the rumen wall, and anything that would increase saliva flow, such as feeding even small amounts of coarse roughage (as little as 2 pounds of hay daily) to ensure rumination, would prevent the problem. The addition of buffering agents such as sodium bicarbonate has not been highly successful, but this also is a promising research area at present.

Saliva also contains urea, which is continuously recycled and thus provides the rumen microflora with another nitrogen source for the synthesis of protein. Because much of the salivary urea was originally absorbed as ammonia from the rumen, the recycling serves to increase the utilization of dietary nitrogen and is said to exert a sparing action on the protein or nitrogen in the diet.

Ruminant saliva contains minute quantities of certain enzymes, but they are considerably less important than are enzymes in nonruminant digestion.

FACTORS AFFECTING MICROBIAL ACTIVITY IN THE RUMEN

A more thorough understanding of the relationship between rumen microflora and the host animal has led to the concept that, in order to feed ruminants adequately or, specifically, to feed beef cattle, the nutrient requirements of the microflora should first be met. It is quite well established that the bacterial flora of the rumen consist of at least several dozen forms and that the relative distribution as well as total numbers present can be altered by changes in the ration. Certain forms are known to predominate if finishing rations that are high in readily available carbohydrates are being fed, whereas still other forms prevail when cattle are grazing lush pastures. Changing from one type of ration to another results in bacterial population shifts, but these shifts take place rather slowly. As a result, digestive disturbances may occur if the type of ration is changed too suddenly. Undoubtedly this fact partly explains the poor performance, often amounting to actual weight loss, that results when cattle are shifted from dry wintering rations to succulent spring pastures.

Protein or nitrogen level in the ration has a marked effect on the total digestibility of the dry matter of ruminant rations, and especially on the crude fiber. This level is extremely important, because differences in crude-fiber digestibility have an indirect influence on digestibility of remaining nutrients in the ration that may be encased within the cell walls of the fibrous portion of the plant.

Including a small amount of readily available carbohydrate such as ground corn or molasses in high-roughage rations has been reported to increase rumen microfloral activity and thereby increase the feeding value of high-fiber roughages. If large amounts of such easily digestible carbohydrates are fed, as in finishing rations, the reverse happens—that is, crude-fiber digestibility is reduced—but in this kind of ration the crude fiber content of the ration is relatively low.

Certain minerals, particularly phosphorus, sodium, potassium, sulfur, and cobalt, are essential for maximum microfloral activity. The relative availability or solubility of these minerals in the rumen seems to be involved, as well as the content of the ration, at least with respect to phosphorus.

Certain feeds, such as high-quality alfalfa and certain commercial fermentation by-products, reportedly contain as yet unidentified factors that stimulate the activity of the rumen microflora. Ashing the alfalfa and feeding the ash as a supplement to low-quality roughage has been reported to increase the concentration of rumen bacteria and improve crude fiber digestion. This suggests that the mineral matter of alfalfa may be responsible for its unique qualities.

Bulk and density of a ration undoubtedly play a part in microfloral activity and therefore in digestibility of crude fiber. Bulk provides for distention of the rumen and enhances normal rumination and physiological function of the rumen itself. Indirectly, bulk is involved in rate of passage of material from the rumen, and it is logical to assume that material passing too rapidly from the rumen (finely ground or chopped roughages, for instance) is not subjected to the normal cellulytic bacterial action. Hence the already mentioned synthesis of bacterial protein and certain vitamins may not occur at rates adequate to meet the dietary needs of the animal itself. Dense or high-concentrate rations that are low in fiber and therefore in bulk are not so dependent on rumen microflora as are the bulkier rations. The low-fiber rations are undoubtedly digested and utilized by cattle in a manner not much different from that of the monogastric animal. It is interesting that cattle on either an all-ground roughage ration or an all- or high-concentrate ration seldom ruminate or chew their cuds. Their saliva production is abnormally low as well.

WASTAGES IN THE RUMEN

The products of fermentation and microbial action in the rumen are unfortunately not all used to advantage by the host. Ingested or ration protein and nonprotein nitrogen compounds are partially converted to ammonia and volatile fatty acids in the rumen. Although much of the ammonia is utilized by microorganisms and converted to bacterial protein, variable amounts of ammonia are absorbed from the rumen into the bloodstream and either used in the synthesis of amino acids or wastefully excreted in the urine. Some of the absorbed ammonia is reexcreted into the rumen by way of the saliva, where it may be further utilized by the microflora.

The extent of loss of protein or nitrogen accruing from excessive ammonia production in the rumen is not known, but undoubtedly it is affected by the level of protein in the ration, solubility of protein or nitrogenous material in the ration, type of carbohydrate in the ration, and concentration and character of microflora in the rumen. A constant intake of protein, and especially of urea, throughout the day, such as happens in self-feeding, as contrasted to once or twice daily feeding, improves the utilization of ammonia by the rumen microorganisms. Less soluble forms of nonprotein nitrogen—biuret, for example—are being investigated because of their promise in reducing the wastage from excessive ammonia production.

The microbial fermentation in the rumen results not only in the production of volatile fatty acids from carbohydrate material used as energy, but also in the formation of the gases, methane and carbon dioxide. Both gases are normally eliminated by way of the esophagus and serve no useful purpose, but rather represent a loss in energy from the ration.

Rations that are high in soluble carbohydrates and proteins, such as finishing rations, are now being treated experimentally with certain compounds—formaldehyde, for example—so as to "protect" them from breakdown in the rumen. Being low in fiber, these diets pass rapidly from the rumen to be more efficiently digested and utilized in the abomasum and small intestine. This "rumen bypass" technique was first developed by using the experimental abomasal fistula, and commercial application of the findings made through its use probably will come in the area of ration additives that will, in effect, make nonruminants out of cattle.

Heat is also produced as a result of the fermentation in the rumen. Only in very cold weather, especially in the absence of shelter, do cattle derive any benefit from this heat production. On the contrary, elimination of this heat may impose a burden upon the animal and thus be responsible for wasting still more energy because of restlessness and rapid breathing. A more efficient body heat elimination system in Brahman cattle, due to a larger skin surface, is said to explain this breed's superior adaptation to hot, humid climates.

NUTRIENT REQUIREMENTS OF BEEF CATTLE

The nutrient requirements of beef cattle closely parallel those of the microflora found in the rumen, at least from a qualitative standpoint.

Because the requirements of both must be simultaneously supplied by the cattle ration, it is rather difficult to assess the requirements separately. For practical ration formulation it is unnecessary to do so. Quantitative nutrient requirements for rumen microorganisms have not been determined, and more information is needed before separate requirements can be established.

The nutrient requirements of beef cattle are discussed in the remainder of this chapter from a broad viewpoint under these headings: (1) feed capacity and bulk, (2) energy, (3) protein, (4) minerals, (5) vitamins, and (6) water. Specific requirements for different age groups and different feeding programs are discussed in the appropriate chapters.

FEED CAPACITY AND BULK

Cattle should be fed to capacity under all ordinary feeding conditions. Regardless of whether the aim is maximum daily gain in a steer finishing program or only maintenance of dry cows. performance is generally more favorable if the complete ration, or at least one item in the ration, is fed according to appetite. This procedure enables cattle to consume enough feed to satisfy their hunger and thus prevents the uneasiness and wasteful excess activity associated with a limited ration. Limited feeding of high-concentrate rations to growing cattle and dry cows is possible, but it requires some special techniques to overcome the so-called boss cow syndrome. Alternate-day feeding works well with cows but is not so satisfactory for calves.

Condition and age both affect feed capacity, and there is much variation between animals of the same condition or age. Cattle on finishing rations voluntarily consume daily an amount of feed equal to 2.25 to 3 percent of their live weight (air-dry basis), with the higher intake levels occurring in the early part of the finishing period. Older cattle such as cows in good condition, and fleshy individuals such as mature bulls and fitted show cattle, consume less, even as low as 1.5 percent of their live weight. Thin, growthy yearling or older steers may consume up to 4 percent of their weight daily for short periods of time. In general, the lower the fiber content or bulk of a feed, the lower the level of voluntary intake. Such feeds also will be higher in energy content in practical rations.

Certain additions can be made to the ration to cause increased consumption, especially if a major portion of the ration is unpalatable. An example is the addition of molasses to a wintering ration consisting largely of ground corn cobs or similar low-quality roughage. Such an addition to an already palatable ration increases intake only temporarily and serves later merely as a replacement for an equal weight of ration being fed prior to the addition.

Evidence has been presented to suggest that acceptance or palatability of a ration, or the level of voluntary consumption, is related to the digestible nutrient content of the ration—that is, the more digestible the ration, the greater the daily consumption, and vice versa. This line of reasoning would relate daily feed intake to the quality of the ration as well as to the size, age, or condition of cattle. There is a growing body of thought that all animals, when given full access to a palatable ration, will eat only enough feed to meet their energy needs; that differences in energy needs or intake, under hormonal control, are responsible for the genetic differences seen in rate of gain among animals.

Feeds high in water content, such as succulent spring pasture, winter small-grain pastures, and high-moisture silages, are apparently consumed at a lower level of dry matter or air-dry feed intake because of their high water content. The effect of moisture content would appear to be more physical than metabolic.

Environmental temperature and humidity apparently can affect voluntary feed intake, especially of finishing rations. Feed intake and daily rate of gain are somewhat higher in moderate winter temperatures than in summer, and cattle fed in the low-humidity environment of West Texas maintain feed intake well during the summer, despite high daytime temperatures. Differentials between day- and night-time summer temperatures also seem to favor the maintenance of high feed intake levels at the higher elevations.

A certain minimum amount of bulk or roughage is required to maintain feed intake at a constantly high level in the ruminant; otherwise bloat and other digestive disturbances will be frequent. The practical minimum roughage level in finishing rations in most parts of the country is from 0.3 to 0.5 pound per 100 pounds live weight daily, or 10 to 15 percent when expressed as percentage of the ration. In areas where roughage nutrients are not cheaper than concentrate nutrients, rations containing less than 10 percent or even no roughage are fed. This practice is discussed in greater detail in Chapter 15. Apparently roughages can provide the necessary bulk, even though finely chopped, if fed in pelleted form, as discussed in Chapter 24 dealing with preparation of feeds.

ENERGY

The energy requirements of beef cattle have been studied by various methods with the result that requirements are expressed in a number of ways. Specific requirements as used in this text are given in terms of *total digestible nutrients* (TDN) and in terms of metabolizable energy. The daily requirement for energy and other nutrients is given in most feed requirement tables, as is the required percentage nutrient composition of the total ration for all the various nutrients.

The total digestible nutrient content of a feed, as determined and expressed in older feed tables, has been defined as the sum of all the digestible organic nutrients—protein, fiber, nitrogen-free extract, and fat (the latter being multiplied by 2.25 because its energy value is approximately 2.25 times that of protein and carbohydrate). The TDN, as a measure of the energy value of feeds, is subject to some criticism, mainly because feeds are not digested in the test tube exactly as they are in the cow. There is also the criticism that the TDN method assumes an equal caloric or energy value for digestible protein, nitrogen-free extract, and crude fiber, and a constant energy value for all ether extract. It is now known that this assumption is invalid. Perhaps the most serious shortcoming is that it does not allow for correction of differences in net energy that occur when the ratios or proportions of roughage and concentrates vary. Roughages are higher in energy value when they make up most of the ration, as in maintenance rations, than when they are included in high-concentrate finishing rations at low levels. Within the normal ranges of roughage : concentrate ratios, this under- or overevaluation of roughage is probably not serious, practically speaking, but it can cause problems when a single value for a feed is used under all circumstances. As an example: if one were computing least-cost rations and gave too high a value to a roughage component of a finishing ration, the computer might select a combination of ingredients that would in fact not perform, in the feedlot, as expected because a roughage ingredient was overestimated as to its energy value.

In terms of precision, at the other end of the scale as an expression of the energy value of feeds and energy requirements for cattle is *net energy* (NE). It represents that fraction of the energy in a feedstuff that remains after accounting for or subtracting the losses of energy found in the feces, urine, and gases resulting from its digestion, and the metabolic losses resulting from the heat production at the cell level. Net energy is usually expressed in megacalories or Mcal (1 million calories) per kilogram or, roughly, per 2.2 pounds of feed. The 1975 revision of *Nutrient Requirements of Beef Cattle* prepared by the Subcommittee on Beef Cattle Nutrition of the National Research Council further divides the energy requirements into two fractions, one for maintenance and the other for gain or production. These two requirements or values for net energy are then added together in computing the daily requirements of cattle of various sizes or weights, being fed for a wide variety of rates of gain. The net energy values assigned to feeds are based on both actual feeding trials and derived values.

At present, more feeds need to be tested, singly and in combinations, and cattle of various initial weights need to be fed to achieve daily gains over a wider spectrum. Such feeding trials are currently under way in many places and data for use in the NE method are being accumulated for later revisions of requirement and feed tables. Space limitations unfortunately do not permit the inclusion here of the extra tables and detailed explanations necessary for using the NE system for balancing rations. Extension livestock specialists in many states have published bulletins and circulars that explain the system. The most up-to-date information is available in the National Research Council publication referred to above.

A third method of expressing energy requirements and energy values of feeds is in terms of *metabolizable energy* (ME). This method accounts for the fecal, urinary, and gaseous losses in digestion but does not account for heat losses incurred as a result of the metabolism of the digested energy. Metabolizable energy values are subject to somewhat the same limitations or errors as those of total digestible nutrients (TDN) when a combination of feeds that vary in fiber content are fed as a total mixed ration. Single values derived from feeding a feed alone are unlikely to be the same for that same feed when fed in combination.

The metabolizable energy requirements and the values in the feed tables in the Appendix and in appropriate chapters of this book are not actual ME values that have been derived from feeding trials. Rather, they are values transformed or back-computed from net energy requirements based on the efficiency of energy utilization for maintenance and production or weight gain. Only those roughage : concentrate ratios that are normally fed to cattle for the various functions were used in converting NE to ME values. Likewise the TDN requirements and feed values in the tables, except for the Morrison requirements, are transformed and, in this instance, each Kcal (1,000 calories) of ME is assumed to equal the energy in 1 gram of TDN. Under most practical circumstances, cattlemen have a certain size and weight of cattle to feed for a specific function or purpose and they have at hand all or most of the feeds, at least the roughages, that they will be wanting to feed. Certain supplemental concentrates, usually protein supplements and nearly always mineral supplements, will be purchased. By using the requirement tables and starting with expected daily feed consumption or intake, rations can be formulated, from the feeds at hand or purchased, by matching the requirements for energy, protein, minerals, and vitamins, and the values for the available feeds as found in the feed tables. Both arithmetical and algebraic methods may be used to combine these feeds appropriately or, in many instances, electronic computers can make the calculations much more expeditiously, especially when least-cost aspects are included in the computations.

A deficiency of energy, due to a simple lack of sufficient total feed, is undoubtedly the most common deficiency in beef cattle rations. The results of low energy intake are slow growth or even loss in weight, stunting, late recycling, failure to conceive, and increased disease and mortality. An energy deficiency is usually accompanied by deficiencies in all other nutrients but especially in protein. Overstocking of pastures and ranges, especially in periods of prolonged drought, is the principal cause of energy deficiency. Such overstocking often results in poor performance or even death, caused by consumption of toxic or poisonous plants and because of lowered resistance to disease and parasites. Cattle on finishing rations may receive rations adequate in total amount or weight to meet appetite needs but still not perform up to their capabilities in terms of gain, owing to a lack of sufficient energy in the ration.

PROTEIN

Protein requirements are usually expressed on the basis of percent of both the total and the digestible protein in the ration. In tables appearing in the appropriate chapters, requirements are also given in terms of daily requirements for total and digestible protein. The digestible protein values are approximately equal to 60 percent of the total protein for high roughage rations and to 75 percent for high-concentrate rations, owing to the differences in the digestibility of roughages and concentrates.

Ration nitrogen, from which bacterial protein can be synthesized in the rumen, can be supplied in nonprotein nitrogenous forms such as urea, as previously mentioned. Under ordinary circumstances up to one-third of the total protein requirement may safely be supplied in nonprotein nitrogenous forms in rations fed to stocker and feeder cattle and up to one-fourth in rations fed to pregnant and lactating cows. Some protein concentrates consisting almost entirely of nonprotein nitrogen are being successfully fed today, but in these situations the feeder must be alert to the danger of toxicity, as is further discussed in Chapter 17. Supplemental NPN, expressed as urea equivalent, should not exceed 1 percent of the total ration, and somewhat reduced nitrogen utilization may be expected when considerable nitrogen is supplied by NPN in high-roughage rations or when new cattle are being fed.

Symptoms of protein deficiency in beef cattle are poor growth, depressed appetite, reduced milk flow and light calf weaning weights, delayed onset of first estrus in heifers, delayed and irregular postpartum estrus in cows, and loss of weight in extreme cases.

MINERALS

The requirements for calcium and phosphorus are quite well established, and numerous determinations for the content of these minerals in feeds have been made. The requirements for these macrominerals and their sources are given in appropriate National Research Council (NRC) tables in other chapters. Requirements for other minerals, except for salt, are not so well established and thus are not included in the NRC tables. All of the established mineral and vitamin requirements of several categories of beef cattle are listed together in Table 38.

Phosphorus.   Phosphorus deficiency is the most common mineral deficiency in cattle and is most likely to occur in cattle being wintered on all-roughage rations or on mature, cured grass. Phosphorus deficiency is usually associated with use of feeds grown in phosphorus-deficient soils. Figure 37 shows that this mineral is deficient in the feeds raised in large areas of the country.

In the early stages of phosphorus deficiency or when a borderline

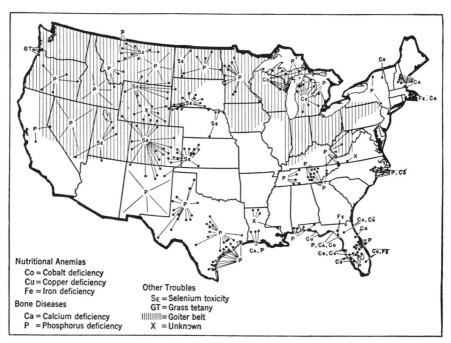

Fig. 37. Areas deficient in important minerals needed for the adequate nutrition of livestock. (USDA.)

Table 38

Mineral and Vitamin Requirements of Beef Cattle, Expressed as Percentage of Dietary Dry Matter or Amount per Kilogram or Pound[a]

Item	Growing and Finishing Steers and Heifers (kg)	(lb)	Dry, Pregnant Cows (kg)	(lb)	Breeding Bulls and Lactating Cows (kg)	(lb)	Possible Toxic Levels[b] (kg)	(lb)
Minerals								
Sodium, %	0.06		0.06		0.06			
Calcium,[c] %	0.18–1.04		0.18		0.18–0.44			
Phosphorus,[c] %	0.18–0.70		0.18		0.18–0.39			
Magnesium, %	0.04–0.10		—[d]		—[d]			
Potassium, %	0.60–0.80		—[d]		—[d]			
Sulfur, %	0.10		—[d]		—[d]			
Iodine, μg	—[e]	—[e]	50–100	23–45	50–100	23–45	100	45
Iron, mg	10.0	4.5	—[d]	—[d]	—[d]	—[d]	400	180
Copper, mg	4.0	1.8	—[d]	—[d]	—[d]	—[d]	115	50
Cobalt, mg	0.05–0.10	0.02–0.04	0.05–0.10	0.02–0.04	0.05–0.10	0.02–0.04	10–15	4.5–7.0
Manganese, mg	1.00–10.0	0.40–4.50	20.0	9.0	—[d]	—[d]	150	70
Zinc, mg	20–30	9–14	—[d]	—[d]	—[d]	—[d]	900	410
Selenium, mg	0.10	0.04	0.05–0.10	0.02–0.04	0.05–0.10	0.02–0.04	5.0	2.3
Vitamins								
Vitamin A,[c,f] IU	2200	1000	2800	1275	3900	1750		
Vitamin D, IU	275	125	275	125	275	125		
Vitamin E, IU	15–60	7–27	—[d]	—[d]	15–60	7–27		

[a] Adapted from *Nutrient Requirements of Beef Cattle*, Subcommittee on Beef Cattle Nutrition, National Research Council, 1975.

[b] The toxic levels indicated are at best estimates and depend upon length of period of intake, availability in the feedstuff, and possible interaction with other minerals consumed.

[c] See Tables 45, 46, 73, 74, 114, and 115 for more details.

[d] Unknown. It is suggested that the level for the growing and finishing animal be used.

[e] Very small but unknown.

[f] May be vitamin A or its precursor, carotene (400 IU/mg carotene).

deficiency exists, feed intake is decreased, gain is reduced, and milk production falls off, with a consequent reduction in suckling calf gains. Efficiency of feed utilization is reduced in the feedlot, and if the deficiency is prolonged, blood phosphorus levels fall. Pica, or depraved appetite, results from severe prolonged deficiency and is usually accompanied by bone alterations resulting in lameness and stiff joints. In breeding cattle, phosphorus deficiency results in reduced percent calf crop weaned because of a combination of problems, discussed in Chapter 10.

As mentioned earlier, phosphorus may be one of the most critical nutrients for normal bacterial action in the rumen. Therefore requirements of both the microorganisms and the host animal must be supplied in the ration. The relative availability or solubility of the various phosphorus supplements fed varies among the sources of phosphorus and even among batches of the same source. Dicalcium phosphate varies least in availability.

Calcium.   A deficiency of calcium is most apt to occur when high-concentrate rations, such as finishing rations, are fed. This is especially true if the limited roughage portion of the ration is nonleguminous. If cows, especially those nursing fall or winter calves, subsist principally on mature, weathered grasses or hay or cereal straw, they are almost certain to respond to calcium supplementation. Heavy corn silage rations are often borderline for calcium unless supplemented with legume hay. Calcium deficiency symptoms are not specific, except in extreme cases when fractures may occur because of depletion. Poor growth rate, inefficient feed conversion, and low ash content of the bones are usually the result of low calcium intake. Calcium deficiency is often associated with energy, protein, and phosphorus deficiencies.

It will be noted from the nutrient requirement tables that the ratio of calcium required to phosphorus required is generally about 1 : 1. Increasing the ration calcium content so that the Ca : P ratio is as high as 7 : 1 has been reported to prevent urinary calculi in feedlot cattle that experience this difficulty. Ordinarily, though, a Ca : P ratio of 2 : 1 would seem to be a practical upper limit for calcium.

Salt.   The salt or sodium requirement of full-fed beef cattle is met by including 0.5 percent salt in the total ration. Salt, of course, may be and often is satisfactorily fed free-choice rather than as part of a mixed ration.

Most stockmen believe that salt is essential to the growth and health of all kinds of livestock. Cattle obviously exhibit a strong liking for it and soon show signs of restlessness if salt is not provided, but recent studies suggest that the requirement for sodium, per se, is quite low. Feeding the usually recommended level of salt in cattle finishing rations also may result in manure that could in time cause a salinity problem in cropland soils if heavy annual manure applications are made.

The form in which the salt is fed has an effect on the amount consumed. Cattle that have free access to granulated, flake, or loose salt will consume approximately twice as much as they will when salt is furnished in the form of compressed blocks. Weathering loss is a factor to be considered. At the Kansas station the flake form of salt weathered 24 percent per month, whereas the blocks weathered only 11 percent. In the states with a heavier rainfall, greater losses may be expected. Cattle accounted for the disappearance of 604 pounds and weathering for 141 pounds of block salt exposed in pastures from April 30 until October 21, or 175 days, in a study at the Illinois station. The total disappearance of salt per 2-year-old steer was approximately 3 pounds per month. Little difference was observed at the Illinois station between the salt consumption of yearling cattle wintered on good roughages in the drylot and that of the same cattle on good pasture the following summer.

The use of salt to control free-choice protein supplement intake usually results in the consumption of considerably more salt than is required. Cattle may consume as much as 2 pounds of salt daily apparently without harm, provided they have free access to an abundant supply of drinking water to ensure elimination of the excess sodium and chlorine by way of the urine. Frozen waterers and stock tanks are to be avoided, because there is extreme danger of toxicity if water is unavailable for any length of time.

Magnesium.   The requirement for magnesium has been studied mainly as it relates to "grass tetany" or "grass staggers," a condition occurring mostly in lactating cows in the spring but sometimes in fall- or winter-calving cows on lush pasture regrowth or small grain pastures. The magnesium requirement of such cows has been determined to be at least 0.18 percent of ration dry matter. It often is necessary to provide supplemental magnesium in the form of magnesium oxide in mineral mixes. It also is available in some liquid pasture supplements. Grass tetany is further described in Chapter 11.

Potassium.   Requirements for potassium are reported to be 0.6 to 0.8 percent of ration dry matter. Breeding cattle and stockers on pasture are unlikely ever to be in a deficiency status. However, because feed grains often contain less than 0.5 percent of potassium, finishing cattle on high- or especially all-concentrate rations are apt to require supplementation with this mineral. Excessive potassium, on the other hand, has been found to interfere with magnesium absorption, resulting in increased incidence of phosphatic urinary calculi.

Sulfur.   When urea supplies a considerable portion of the nitrogen in the ration, sulfur, either in organic or inorganic form, has been shown to be beneficial as a mineral additive. Sulfur, or the sulfur-bearing amino acid methionine, is known to facilitate bacterial synthesis of protein. Practical rations containing considerable roughage have not always been improved by the addition of sulfur, indicating that the sulfur content of such rations is probably already adequate. A sulfur requirement of 0.1 percent of ration dry matter has been established and the requirement is sometimes expressed as a 15 : 1 nitrogen : sulfur ratio. When dietary urea is high, a level of 3 parts of inorganic sulfur to 100 parts of urea is indicated.

Trace Minerals.   Trace or microminerals are those found in only minute amounts in soils and plants. The content of such minerals in feedstuffs is closely associated with the amounts present in the soils in which the feedstuffs are grown. Figure 37 indicates areas where it has definitely been established that certain trace mineral deficiencies may occur. It will be seen that areas whose soils are most subject to leaching, either because of heavy rainfall or permeability, are apt to experience a greater incidence of trace mineral deficiencies in plants grown in such soils.

Iodine deficiency is usually evidenced by the production of goiterous calves that are either born dead or die soon after birth unless iodine is administered. The use of iodized salt containing 0.01 percent of stabilized potassium iodide in the pregnant cow's ration prevents deficiency symptoms. Iodized salt should be used as a general practice, if available, for the slight difference in cost over noniodized salt will usually be offset by the insurance provided.

Iron is undoubtedly required by beef cattle, but minimal quantitative requirements are not well established, nor have consistent responses been obtained from its use in practical rations. It may therefore be assumed that feedstuffs commonly consumed by beef cattle contain adequate iron. Calves on an exclusively milk diet become anemic because of an iron deficiency which persists until they begin consuming other feedstuffs. Milk replacers used as the sole diet for suckling calves should be fortified with iron.

Copper requirements for beef cattle are very low, but the need for copper may become critical if excessive molybdenum and sulfates occur in the ration. Copper deficiency can be prevented by adding 0.25 to 0.5 percent of copper sulfate to the salt, fed free-choice. This ensures a copper content of 4 to 8 parts per million in the total air-dry ration. Excess molybdenum apparently interferes with copper metabolism. Therefore, in areas where molybdenum toxicity is evident, additional copper should be fed. Although copper deficiency is almost exclusively a local problem, calves being fitted for show or sale by the use of nurse cows well beyond the normal time for weaning sometimes may develop copper deficiency. Generally poor performance along with intermittent to severe diarrhea and stunted growth are symptoms of copper deficiency. When copper deficiency is complicated by molybdenum toxicity, depigmentation of the haircoat is common. Apparently forages grown on muck soils such as those in parts of Florida are abnormally low in copper.

Cobalt is required by rumen microflora to ensure adequate vitamin B_{12} synthesis, as this element is an integral part of the B_{12} molecule. The requirement has been set at 0.07 to 0.10 milligram per 100 pounds body weight daily. It is generally supplied in the ration, when needed, in the form of cobalt sulfate added to the salt or mineral mixture at the rate of 1 ounce per 100 pounds of salt.

Hay and pasture forage rations containing 0.01 to 0.07 part per million of cobalt have resulted in cobalt deficiency, whereas if the cobalt content is in the neighborhood of 0.10 part per million, performance is normal. Thus the requirement is apparently no more than 0.10 part per million, expressed as content of the diet.

Symptoms of cobalt deficiency, like those of many other mineral deficiencies, are general rather than specific. Severe deficiency results in reduced feed intake, emaciation, weakness, and even death. Perhaps of more importance is the borderline deficiency, which may not be recognized and thus is not corrected. Such borderline deficiencies may result in generally reduced performance and inefficient feed conversion.

Manganese and *zinc* are undoubtedly required by beef cattle, but since forages contain 50 to 150 parts per million of manganese and 10 to 100 parts per million of zinc, the likelihood of a deficiency occurring under practical conditions is remote.

Although *selenium* is toxic at higher levels, there now appears to be a minimum dietary requirement for this mineral. The level required is related to the vitamin E content of the diet. Cows on diets containing at least 1 part per million of selenium produce normal calves, but cows on diets containing less than this amount produce calves that are afflicted with white muscle disease, a form of muscular dystrophy. Affected calves can be drenched or injected with selenium supplements. Selenium also can be added to cow supplements, but the level at which toxicity occurs is so low that at present this method is usually not recommended.

As indicated earlier, trace mineral deficiency is principally a geographical problem, and consultation with soils experts and nutritionists in specific areas is necessary to determine whether supplementation is needed. Trace-mineralized salt is available in most feed stores. The cost per unit of trace mineral may be quite high and its feeding may be unessential. This is especially likely if the rations being fed contain average to good quality roughages, grown in areas where the soils are adequate in trace mineral elements.

The elimination of most or all of the roughage in the newer high-concentrate rations has reawakened interest in the trace mineral question, and active research may yet show a greater need for trace minerals under such circumstances.

Toxic Minerals.   Some minerals, although they may actually be required in minute amounts, may produce harmful effects if ingested in excess of these requirements.

Fluorine, contained in either undefluorinated or raw rock phosphate and mine washings or deposited upon the forage grown in areas subjected to fluoride-containing smoke, is toxic to beef cattle if consumed in excessive amounts. The usual fluorine content of the mineral supplements fed to beef cattle is shown in the Appendix tables.

Symptoms of fluorine toxicity are mottling and erosion of the tooth enamel and a softening and thickening of the bones with a resulting decrease in breaking strength. The maximum level for fluorine in the ration has been recommended at not more than 65 parts per million for feeder cattle and not more than 30 parts per million for breeding cattle kept on such rations for extended periods of time.

Selenium toxicity is a local problem occurring primarily in North Dakota, South Dakota, and the Rocky Mountain states. Cattle consuming feeds containing 8.5 parts per million of selenium for a considerable length of time show symptoms of chronic toxicity. Death occurs if feeds consumed contain 500 to 1,000 parts per million of selenium. Characteristic symptoms of selenium toxicity are loss of appetite, sloughing of hoofs, loss of hair from the tail, and eventually death. No effective treatment has been determined other than removal of animals from the affected areas.

Molybdenum is apparently an essential mineral, but the requirement is low, since more than 10 to 20 parts per million in forages results in toxic symptoms. Excess molybdenum interferes with copper metabolism as previously mentioned. Until more information is available, molybdenum should not be added to beef cattle mineral supplements.

VITAMINS

Vitamin deficiencies in general are less apt to occur in beef cattle than are mineral deficiencies. This is because, first, as mentioned earlier, most of the vitamins required by cattle are synthesized by rumen microorganisms and, second, the feeds usually consumed by cattle are fair to excellent sources of those vitamins that must be supplied in the ration.

Vitamin A and Carotene.   Requirements for vitamin A are expressed as vitamin A in the requirement tables in the appropriate chapters whereas the feed composition tables in the Appendix show the requirements in terms of provitamin A or carotene content. Approximately 2.5 milligrams of carotene or 1,000 IU of vitamin A are required per pound of feed consumed for normal growth in young cattle, and for all finishing cattle. Dry pregnant cows require about 3.0 milligrams of carotene or 1,275 IU of vitamin A per pound of feed consumed, whereas nursing cows and bulls require 4.5 milligrams of carotene or 1,750 IU of vitamin A per pound of feed consumed.

Vitamin A, which is synthesized in the body from carotene obtained from the ration, may be stored in the body during periods of high carotene intake such as occur during the summer on lush pasture. Carotene reserves may be drawn upon during winter, and the amount stored of course affects the level of vitamin A or carotene necessary in the winter ration. Another factor affecting the overall practical problem of supplying vitamin A is the instability of carotene, evidenced by losses or destruction of the carotene in feeds through oxidation during storage. For example, hay may lose up to three-fourths of its carotene content in one winter storage period. Carotene and even added supplemental synthetic vitamin A itself may be destroyed in processing feeds with steam, high pressure, and so on, or when mixing them with certain minerals or organic acids.

Vitamin A deficiency in beef cattle is rather uncommon when good quality roughages are fed. However, when a combination of poor hay and concentrates low in carotene content, such as old corn, small grains, grain sorghums, or molasses, make up the ration, vitamin A deficiency may occur. A little-understood interrelationship between nitrate content of feeds and poor utilization of carotene apparently is causing considerable vitamin A deficiency today.

Common but not highly specific vitamin A deficiency symptoms in feedlot cattle are a marked reduction in feed intake, reduced gain, and poor feed conversion. These symptoms may become pronounced in cattle that are almost ready for market at the onset of hot weather.

Other vitamin A deficiency symptoms are night blindness (inability to adjust to sudden bright lights), incoordinated gait, and convulsive seizures in severe cases. Excessive lacrimation may also occur, but this symptom can easily be confused with watery eyes brought on by pink eye (see Chapter 27) or by irritation from cinders and dust during movement or shipment of the cattle. Diarrhea, from severe to intermittent, in both young and older cattle may result from vitamin A deficiency, but here, too, other causes are possible and may cause confusion. A characteristic deficiency symptom in feeder cattle is anasarca or generalized edema. The swellings are localized in the brisket area and in the knee and hock joints. Lameness usually accompanies the more severe cases.

Sexual activity declines in bulls suffering from vitamin A deficiency; spermatozoa decrease in numbers and motility, and increase in number of abnormal forms. In breeding cows, estrus may continue, but cows conceive less readily. Vitamin A deficiency in the pregnant cow may result in abortion if it is sufficiently severe. Calves may be born dead or weak, and retained afterbirths are common in deficient cows. A suspected vitamin A deficiency can be verified by analysis of liver tissue for vitamin A content. A value that is below 4 micrograms of vitamin A per gram of fresh liver is considered critical, and from 4 to 10 micrograms are considered suboptimal.

When supplemental vitamin A is required, it may be added to the ration as a dry, stabilized vitamin A premix or it may be injected intramuscularly or intraperitoneally in an aqueous solution. A single injection of 1 million IU will prevent deficiency symptoms for 2 to 4 months in all cattle.

Vitamin D.   A deficiency of this vitamin is extremely unlikely under practical conditions, because beef cattle receive sufficient vitamin D from exposure to direct sunlight or from the consumption of sun-cured hay. In areas where days are extremely short or where cloudiness persists during most of the year, and especially during haying season, it is conceivable that a deficiency, especially in borderline form, may exist, but clear-cut cases of vitamin D deficiency are seldom reported. A daily requirement of 125 IU has been established in controlled experiments, but needs are more than met by normal feeding and management practices.

The disease known as rickets, which results from poor calcification of bone, is the chief symptom of vitamin D deficiency produced under controlled experimental conditions. Posterior paralysis may result from fracture of vertebrae in severe cases. Poor performance may result from borderline deficiency. Adequate vitamin D is essential for efficient utilization of calcium and phosphorus and for bone formation.

Vitamin E.   Muscular dystrophy or white muscle disease in calves between the ages of 2 to 12 weeks is the chief symptom of a deficiency of vitamin E. Certain limited geographical areas have reported this condition but it is not widespread. As mentioned in connection with mineral requirements, selenium level in the ration is involved in this disease.

The quantitative requirement for vitamin E, expressed as tocopherol, is tentatively estimated to be less than 40 milligrams of tocopherol per 100 pounds of body weight daily. The oral requirements for vitamin E, expressed as international units, ranges from 7 to 27 units per pound of diet. In affected areas, losses due to muscular dystrophy in calves may be reduced by feeding 2 to 3 pounds of grain during the last 60 days of pregnancy or by oral administration of tocopherol to both cow and calf shortly after parturition. Vitamin E functions as an antioxidant and facilitates the absorption and storage of vitamin A.

Vitamin K. This vitamin is synthesized by the rumen microflora in adequate amounts under normal feeding conditions. Moldy clover hay and clover pasture sometimes contain dicoumarol, a compound that prevents normal blood clotting. Therapy with vitamin K is usually effective in combating this condition.

The B Vitamins. Although requirements for most of the B vitamins such as riboflavin, thiamin, and biotin have been demonstrated for the young calf before it has developed a functioning rumen, attempts to improve the rations of cattle more than 8 weeks of age with B-vitamin supplementation have not been generally successful. These vitamins, like vitamin K, are apparently synthesized by the microflora of the rumen at a

Table 39

| | | Daily Water Consumption | | Consumption per 100-lb Live Weight |
Water Consumed by Cattle Full-Fed in Drylot	Period	(lb)	(gal)	(lb)
Calves				
Iowa	Feb. 4–14	26	3.0	4.8
	Mar. 5–15	27	3.1	4.4
	May 4–14	51	5.9	6.7
	July 3–13	65	7.6	7.1
Ohio	Apr. 15–Aug. 18	57	6.6	6.5
Yearlings				
Iowa	Feb. 4–14	29	3.3	3.4
	Mar. 5–15	25	2.9	2.7
Illinois	Aug. 5–19	92	10.7	8.8
	Aug. 22–Sept. 6	86	10.0	7.7
2-year-olds				
Iowa	Feb. 4–14	36	4.1	3.4
	Mar. 5–15	33	3.8	3.0

Table 40

Total Daily Water Intake[a] as Affected by Temperature and Level of Feed Intake (Dry Matter Basis)[b]

Temperature (Fahrenheit)			40°	50°	60°	70°	80°	90°
Gallons of Water per Pound of Dry Matter Intake			0.37	0.40	0.46	0.54	0.62	0.88
Body Weight (lb)	Expected Daily Gain (lb)	Dry Matter Daily (lb)	(gal)	(gal)	(gal)	(gal)	(gal)	(gal)
Cattle on Maintenance Rations[c]								
400	0.0	5.4	2.0	2.2	2.5	2.9	3.3	4.8
800	0.0	8.8	3.3	3.5	4.0	4.8	5.5	7.7
1,200	0.0	11.8	4.4	4.7	5.4	6.4	7.3	10.4
Wintering Weanling Calves								
400	1.0	9.9	3.7	4.0	4.5	5.3		
500	1.0	11.7	4.3	4.7	5.4	6.3		
600	1.0	13.5	5.0	5.4	6.2	7.3		
Wintering Yearling Cattle								
600	1.0	14.4	5.3	5.8	6.6	7.8		
800	0.7	16.2	6.0	6.5	7.4	8.7		

[a] Total water intake includes both the water drunk and that contained in the feed.
[b] *Journal of Animal Science* 15:722–740.
[c] Animals that are neither gaining nor losing weight.

rate sufficient to meet the needs of the animal. An example of how dietary nutrients may affect bacterial synthesis of B vitamins has been mentioned, namely the relationship between cobalt intake and B_{12} synthesis. Other such relationships may be discovered in the research being conducted in this field at the present time. As already mentioned, extremely rapid passage rate of high-concentrate rations through the rumen may well result in insufficient synthesis of the essential B vitamin, niacin.

WATER

Water, because of its abundance and universal use, is seldom regarded as a feed, and yet it is one of the most essential nutrients for all animal life. The amount of water required by cattle, exclusive of the water contained in the ration, varies with the character of the feed, the amount of dry matter consumed, and the air temperature. Data on the water consumed on different rations and by cattle of different ages are extremely meager, but enough are available to permit a rough estimate of the daily water requirements of cattle full-fed in drylot. It can be seen from Table 39 that as the season advances from winter to summer, the amount of water consumed per 100 pounds live weight increases rather sharply. Similar data shown in Table 40 indicate that, in addition to temperature effect, increasing levels of feed intake result in the intake of more water.

Water should be kept slightly above freezing temperature during the winter by the use of insulated tanks and by tank heaters during extremely cold weather. Electric, oil, or gas heaters are all suitable, and the choice is usually based on cost of equipment, ease of operation, and freedom from mechanical difficulties, Heaters save labor in keeping the tanks ice-free, but cattle appear to thrive as well on ice-cold water as on water at moderate temperatures. When pond or stream waters are the only source of drinking water for cattle, ice must be broken at least every other day in extremely cold climates. If salt is used to regulate intake of supplement, ice that forms on water sources must be broken daily.

PART III
THE COMMERCIAL
COW-CALF PROGRAM

CHAPTER 8
THE COMMERCIAL
COW-CALF PROGRAM

The commercial cow-calf program, as the term is used in this discussion, refers to an operation consisting almost exclusively of grade or nonregistered mother cows and their suckling calves up to weaning time, the replacement heifers, and, of course, the bulls. Sometimes the steer calves and the surplus heifer calves are overwintered and summered before being sold as yearlings. In this case a stocker program has been added and the operation is no longer exclusively a cow-calf program. The weaner calves, sold at 6 to 8 months of age, are the principal production or source of income for the cow-calf man. Sale of cull cows and other surplus older cattle contribute income, of course, but this is incidental to the main objective—namely, the maximum return from the sale of heavy, high-quality calves. Ownership of calves all the way to finished weight is being tried by some ranchers, who have the calves custom-fed by nearby commercial feedlots, often with mixed success.

FACTORS TO CONSIDER IN CHOICE OF PROGRAM

Before starting a commercial beef cow herd—or any beef cattle program, for that matter—one should be certain that he has the necessary skills or can learn them, and that the program is the one that will best utilize the feed production capabilities of the farm or ranch. Among the important considerations involved in choosing a program are the following.

1. Kind and amount of pasture or range to be utilized.
2. Homegrown supply of grain and roughages, including cash crop residues.
3. Season during which labor is least needed for other work.
4. Local market demands for feeder or slaughter cattle, or both.
5. Proximity to market outlets and surplus feed supplies.
6. Climate, including temperature range, rain and snowfall, and relative humidity.
7. Available equipment, shelter, and stock water.
8. Extent and availability of financial resources.

9. Training, skill, and experience of the operator.
10. Personal likes and dislikes.

No single program is best suited to all conditions, and each has its advantages and disadvantages. The cow-calf program is growing in popularity, as shown by the increasing numbers of new herds being established, especially east of the Mississippi River. Some believe that this is largely due to the periodic high prices that must be paid for stockers and feeders obtainable from the range states, but it is doubtful if this is the main explanation.

ADVANTAGES OF THE COW-CALF PROGRAM

Among the many reasons for the growing popularity of the cow-calf program over other forms of beef cattle production are the following advantages.

1. Beef cows can produce more pounds of valuable product (calves) from poor to average pastures and low-grade harvested roughages.
2. This program is less speculative; that is, there is less risk of losing large amounts of money because of rapidly declining prices.

Fig. 38. The commercial cow and calf program maximizes returns from pastures or range such as the Sandhills area of northwestern Nebraska, as shown by research at agricultural experiment stations and nearly a century of practical experience on the part of ranchers. (The Record Stockman.)

3. A beef cow herd is a good stabilizer. The man with a successful cow herd is less likely to be an "inner and outer" trying to outguess the cattle market to his downfall. (There is probably reason to wonder whether stability, as used here, is cause or effect.)

4. The man with a combination of pasture, roughage, and grain to market through cattle can produce feeder cattle that best suit his needs by virtue of having under his control such management details as breeding and weaning dates, bred-in performance, and herd health.

5. The man with a cow herd can utilize labor and equipment that may already be on the farm and that would not otherwise be used—for example, a former dairy farm, or a feeder farm, well equipped with barns, silos, and fence and operated by family labor.

6. A cow herd can and usually does increase in value over the years by being graded up in quality through the use of better and better bulls, the keeping of only the very best heifers, and by growth in numbers.

Fig. 39. Income from a productive cow herd makes possible the clearing of land and application of seed and fertilizer, which are required to establish improved pastures such as this one in Georgia. (American Hereford Association.)

7. This program is perhaps the most satisfying because it covers the gamut of experiences with cattle. (No doubt this is the main reason why the beef-breeding project is among the most popular of all projects for 4-H and FFA Club members.)

8. Tax incentives are favorable to the cow-calf program. An investment credit of 10 percent for purchased breeding stock and depreciation schedules that reduce taxable income are encouragements to new breeders. Treatment of income from sale of home-raised brood cows on a capital-gains basis also is attractive. Income tax laws, of course, are constantly undergoing revision, and so this advantage may be only temporary.

DISADVANTAGES OF THE COW-CALF PROGRAM

The cow-calf program has some real disadvantages and these must also be considered in choosing a cattle program.

1. The cow-calf program is a longtime program. Returns come slowly, and the program is not well adapted to tenant farming. Concerning this point someone has said, "Many tenants cannot wait that long and most landlords and bankers won't."

2. The program is inflexible; that is, it cannot readily be changed in size or method of management to adapt to unforeseen difficulties or to take advantage of unexpected higher cattle prices. The cowman seldom is able to capitalize on drastic price rises because, even though his cows may increase in value, he cannot sell them and stay in business. Culling heavily at such times does give the cowman advantages over those engaged in other beef enterprises.

3. A better grade of labor and management is required than is needed for certain other programs.

4. Losses due to calving difficulties, sterility, and disease are high compared with other cattle programs.

5. Average to excellent quality pastures and harvested roughages, and concentrates can be converted to more pounds of gain with some of the other cattle programs. Dairying might be especially attractive in this case.

6. The volume of business or gross income is small for the large investment in cattle and land; thus the minimum-sized herd for profitability represents a comparatively large dollar investment.

When costs of land, fences, water developments, horses and vehicles, buildings, and feed storage and harvesting equipment, if required, and the investment in the cows and bulls are added together, an investment of $1,500 to $2,000 per cow unit in the herd may be expected. These figures have a way of being surprisingly similar throughout the country. Exceptions occur near urban and recreational areas, of course. The uniform values for cow land throughout the country provide evidence that there are no "bargains" in range or pasture land suitable for beef cows.

Investment costs and nonfeed costs associated with operating a cow herd—such as labor, taxes, interest on operating capital, veterinary costs, and depreciation or replacement costs—can be used in constructing budgets for estimating the economic feasibility of operating a cow herd. Such budgets are available from agricultural economists in most states, but rapidly changing prices, especially land prices, make it difficult for these budgets to remain current.

Unfortunately, few state budgets are identical with respect to how certain charges are made. Table 41 is an attempt to adapt three budgets, prepared in 1975 in three states of widely different type of country. The annual cow costs for western Colorado, Indiana, and North Carolina are remarkably similar, with pasture and feed costs being especially comparable.

Table 41

Estimated Annual Costs of Carrying a Beef Cow in Selected States, 1975[a]

Item	Western Colorado	Indiana	North Carolina
Feed Costs			
Pasture or land charge[b]	$ 92.20	$ 64.00	$ 94.61
Harvested roughage	40.80	30.00	22.17
Supplements	1.22	21.40	5.50
Total	134.20	115.40	122.28
Nonfeed Costs			
Interest[c]	60.29	69.60	47.92
Labor	24.67	22.50	14.55
Miscellaneous[d]	22.87	22.50	15.32
Total	107.83	114.60	77.79
Grand total	$242.03	$230.00	$209.07

[a] Compiled from adaptations of reports of agricultural extension economists.

[b] Consists of monthly or annual pasture rental or interest on land investment.

[c] Interest on capital requirements and depreciation of cattle, buildings, and equipment.

[d] Consists of taxes, insurance, veterinary services, marketing costs, utilities, and management charges.

It is generally conceded that, if a beef cow herd is the major source of income, from 200 to 250 cows is a practical minimum-sized herd. Obviously, if the enterprise is a supplementary one, no set number is required, but fewer than 25 cows as the final cow herd size is probably not economically feasible.

THE CHOICE OF THE FOUNDATION BREEDING STOCK

Choosing the animals that are to be the foundation stock for the breeding herd is a matter of prime importance. Often insufficient attention is given to their selection. The young breeder, especially, in his eagerness to start operations, is likely to take too little time to consider properly just which animals will best suit his needs. He grows impatient at what seems to be a loss of time, and to get started he purchases the animals that are immediately available even though he may know they fall short of the kind he really wants. Such a procedure is to be avoided, for usually a herd that is established hastily in this way is found to be so unsatisfactory that it is soon replaced either by cattle from another source or by some project entirely different from a beef-breeding herd.

Once a herd is established, it is usually best to raise one's own replacement heifers. This is because of the disease control problem present when outside animals are added, and because of the desire to make progress in improvement in performance and type or beef conformation. Decisions must be made almost every time a calf crop is weaned in order to choose the best heifers to keep, but since only a few cows are replaced each year, these decisions are not nearly so important as those involved in laying the foundation.

CHOOSING THE BREED

In some ways it is unfortunate that we have several breeds of beef cattle of the same general type, because many men waste considerable time and effort in attempting to decide which breed is best for their particular conditions.

The choice of breed, where straightbreeding is used, is, for practical purposes, unimportant to all but those cowmen operating in the Gulf Coast region. It is a well-established fact that the Brahman, the Santa Gertrudis, and the so-called exotic breeds that trace their foundation to an infusion of Brahman breeding, are better suited to this region, at least where performance is concerned. Otherwise all of the principal British and continental European breeds are well adapted to the climate and feed conditions of the remainder of the country. Some special problems that may arise from using breeds of larger mature size and heavier milk production in certain range and forage situations are discussed in Chapter 9, dealing with cow herd management. So much more variation exists between the individuals of any one breed than between the best, or even the average representatives, of any two of them that it is useless to argue over their respective merits. One thing is certain: if a man has a distinct preference or liking for a particular breed, that breed in all probability is the one for him to use because it is unlikely that he will ever be entirely satisfied with any other. In the absence of such a preference, however, he should probably choose the breed that is most popular in the community in which he lives. The very fact that it is prominent is a good indication that it is well suited to the prevailing environment.

Literally countless experiments have been conducted that were designed to compare British breeds of beef cattle. Because the environments in which the tests were conducted varied widely, and because the test animals in the herds studied were only a fractional sample of the breeds in question, it is important, in evaluating results from breed comparison studies, to consider all or most of these studies together, rather than only one or two examples.

Dr. D. D. Dearborn, while at the University of Nebraska, reviewed breed studies in the United States and other parts of the world and prepared a summary of his review, as shown in Table 42. Included are limited data from studies of the three most common dairy breeds. It is regrettable that only the Charolais is included to represent the continental European breeds. Understandably, no effort was made to weight economically the various performance traits, and no attempt was made to rank the breeds on the basis of overall merit. A numerical value of 100 was assigned to each of several important traits for the Hereford breed, with other breed values then shown as deviations from the Hereford value. As shown in Table 42, a carcass grade of C− was found for the Hereford breed and the deviations from it are expressed as portions of one-third of a grade. No details were learned concerning certain traits for the dairy breeds.

The following statements by U.S. Department of Agriculture specialists effectively summarize the various aspects of the question of choice of breed.

1. Differences in preweaning and postweaning gain are relatively small among the three British breeds—Angus, Hereford, and Shorthorn— and the polled types of the last two.
2. The Charolais and the new breeds based on Brahman-European

Table 42

Average Breed Performance Expressed as Percentage of Hereford Breed Average Performance, with Number of Studies Considered[a]

Breed	Trait						
	Male Fertility	Female Fertility	Calf Crop Weaned	Weaning Weight	Postweaning Gain	Carcass Cutability	Carcass Grade
Hereford	100 (4)[b]	100 (4)	100 (5)	100 (13)	100 (9)	100 (7)	12.0 (C−)[c] (9)
Angus	98.8 (2)	102.5 (2)	102.9 (4)	98.9 (2)	92.6 (4)	96.6 (3)	+0.44 (3)
Shorthorn	96.5 (3)	98.0 (3)	92.3 (3)	100.2 (5)	98.6 (4)	95.2 (3)	+0.30 (3)
Charolais	91.8 (1)	99.5 (1)	92.1 (1)	123.0 (3)	110.5 (1)	105.0 (1)	−0.45 (1)
Brown Swiss		82.6 (1)	96.9 (1)	122.3 (1)	103.8 (2)	100.7 (2)	−0.10 (2)
Friesian				144.9 (1)	117.5 (1)	115.9 (1)	−0.80 (1)
Jersey					84.8 (1)	118.8 (1)	−1.3 (1)

[a] Adapted from Proceedings, Symposium on Range Beef Cow Production, Chadron, Nebraska, 1969.
[b] Numbers in parentheses indicate number of studies evaluated.
[c] Numerical values of 12.0, 13.0, and 14.0 were assigned to carcasses grading low, average, and high choice, respectively, and values of 9.0, 10.0, and 11.0 were assigned to low, average, and high good carcasses, respectively.

crossbred foundations grow faster, both before and after weaning, than the British breeds. (The same is true of crosses of these breeds with British breeds.)

3. If breed differences exist in efficiency of growth (feed consumed per unit of gain), they have not been established. Similarly, breed differences in fertility and longevity have not been clearly defined in most cases.

4. Among the British breeds, differences in meat palatability and tenderness are small.

5. The Charolais and its crosses, the Brahman and its crosses, and new breeds based on Brahman-European crosses produced carcasses with less external fat and higher yields of trimmed preferred retail cuts than British breeds. As compared with British types slaughtered at the same weights, these crosses ordinarily have less marbling and do not grade as high by USDA quality grade standards.

6. Brahman cattle and breeds based on Brahman-European foundations have greater heat tolerance than European types and greater resistance to many insects and some diseases. As compared with British types, animals of these breeds are slower to reach sexual maturity, but brood cows are excellent mothers and have longer productive lives. The lean meat from the Brahman, and from breeds with part Brahman foundations, has been found in several experiments to be somewhat less tender than that of British breeds.

The more recent introductions of breeds from continental Europe and their place among the breeds already in use in the United States is discussed in Chapter 4. Although these breeds are larger and gain faster up to the same slaughter weights compared with common performance for the British breeds, they usually show less marbling and thus grade lower. If fed to heavier weights, or if slaughtered at comparable physiological ages, then they ordinarily grade as well as cattle of the British breeds.

Breeders of each breed, whether producing purebreds or commercial cattle, should strive to improve the breed of their choice in the characteristics in which weaknesses are most apparent. In each breed, breeding stock can be found that is acceptable or even very strong in those points in which the breed needs improvement.

CROSSBREEDING

The value of crossbreeding is discussed in some detail in Chapter 4. The Nebraska tests with three British breeds demonstrated the magnitude of

Fig. 40. F_1 cows resulting from the use of Brahman bulls on Hereford cows are proving to be excellent mother cows throughout the Gulf Coast region. (American Brahman Breeders Association.)

the heterotic effect from both the first cross and the backcross. Other tests are under way throughout the country with three- and four-breed crossing. In the South, when the cross involves the Brahman or some of the newer breeds with Brahman ancestry such as the Brangus, Beefmaster, or Charbray, even greater heterotic response occurs than when British breeds are crossed. The ideal combination of breeds to use in a crossbreeding program depends to a large extent on the environment in the locality where the herd will be run and on the market demands in the area. Guidance in this matter can be obtained from beef cattle specialists in the state or area in question.

CHOICE OF ANIMALS FOR THE COMMERCIAL HERD

The following items should be carefully considered in choosing the animals that are to form the foundation of a new herd or are to serve as additions or replacements for a herd already in existence: (1) freedom from disease, (2) individuality, (3) performance records if available, (4) age, and (5) cost.

Freedom from Disease.   The most important item in determining profits is percentage of calf crop, and the best insurance against poor calving percentage is a healthy herd. High selling prices due to extra quality and weight or low feed costs cannot offset low calving percentage and large death losses.

Important contagious reproductive diseases of breeding cattle that must and can be guarded against are brucellosis (contagious abortion), leptospirosis, and vibriosis in females and trichomoniasis in bulls. These diseases can be minimized by demanding, at the time of purchase, proof of negative results from tests conducted by qualified veterinarians on all breeding animals composing the foundation stock or additions to the herd. Replacements that are produced in one's own herd are likely to be free of these diseases if the herd in general is healthy; however, annual tests should be made on these additions as well. Compliance with federal and state regulations concerning these diseases is becoming mandatory in all parts of the country and, it is hoped, will continue until all herds are disease free.

Calfhood vaccination gives good protection against brucellosis, and approved vaccines are now in use for leptospirosis and vibriosis. In some states, vaccination for brucellosis is no longer encouraged because the incidence of the disease is now so low that the chances that cattle will contract it are believed to be almost nonexistent. The problem of vaccinated heifers' showing a positive reaction to a test is thus eliminated. Many cattlemen and even veterinarians are concerned about the long-range aspects of this relaxation of vaccination requirements, however, and they continue to vaccinate.

Breeding animals should be bought subject to negative tests for brucellosis, leptospirosis, and vibriosis in any case. Tests made more than 30 days before delivery of the animals to the buyer should not be accepted. All herd additions should be isolated from the rest of the herd for 60 to 90 days, and a retest should be made before they are released from quarantine.

Many other diseases affect breeding cattle—for example, tuberculosis, anthrax, and anaplasmosis—but these are comparatively uncommon. Nevertheless, the breeder should be aware of them and should consider the area of origin when buying foundation cattle. More details are available concerning these and other diseases in Chapter 28.

Individuality.   As used here, individuality includes all those characteristics in an animal that may be noted from "eyeball" or visual inspection. Size, body type, quality, bone and set of legs, breed and sex character, and temperament all come under this heading. The extent to which visual appraisal of individuality is used to determine acceptance or rejection of breeding animals varies from using such an appraisal as the sole means of selection to completely ignoring the appearance of the animals. Some of the characteristics of breeding cattle are due to a combination of inheritance and environment, as mentioned in Chapter 3. This fact complicates selection. Often the first calves produced by the foundation stock or herd replacements bear little resemblance to their sire or dam and do not measure up to expectations. This is usually because environmental factors, such as degree of finish and previous treatment of introduced cattle, led the prospective breeder astray. Some of the characteristics mentioned, however, are highly heritable and can thus be expected to be passed on to offspring with considerable regularity.

Characteristics such as breed character or breed type can contribute to the sales appeal of commercial cattle. These traits must be appraised visually, since they cannot be accurately measured with scales or calipers. Unfortunately, as experienced breeders know, these traits are no more reliably passed on to offspring than others previously mentioned—that is, these traits have low heritability. In addition, research indicates that these characteristics are not so closely associated with productivity as once believed. Observation of many near relatives is a good, though not perfect, guide when selecting young breeding cattle, insofar as the "hard to measure" characteristics are concerned.

Breeders and feeders alike often express a preference for a certain shade of red or a specific color pattern in their feeders or breeding cattle. Buyers have been known to pay as much as several dollars per hundredweight more for yellow-colored Hereford feeders, for instance, and shade of red may mean as much as several hundred dollars' difference in the selling price or cost of a bull. South Dakota workers have gathered data that indicate that shade of red in Herefords is not correlated with performance and that neither a premium nor a discount should be expected when selling or buying Herefords that are either extremely light or extremely dark colored.

Performance Records.   With respect to grade females that are to serve as a foundation for a commercial herd, records of their performance or productive ability are seldom available. This is especially true if these females are bought in the range areas where herds are usually large and individual records are not kept on each cow or calf in the herd. This is not to imply that western ranch-bred females are not desirable foundation material. To the contrary, more good surplus females are apt to be found there than elsewhere because of the general awareness of economically important characteristics over a period of many years. In addition, cow numbers are relatively stable in these regions and proportionately fewer heifer calves or yearlings are needed for herd replacements or buildup.

More and more ranchers and operators of farm herds are participating in performance-testing programs. Weaning weight and weaning conformation and body type scores are the three principal kinds of information recorded for heifer calves. Breeders usually practice performance testing mainly to aid in selecting their own herd replacements, but the surplus females with good weights and weaning scores should make suitable foundation material for many prospective purchasers. If yearling heifer weight and type score can be obtained in addition to the information obtained at weaning time, so much the better. However, this information is unlikely to be available for most grade yearling heifers excepting those sold in dispersal sales.

The breeding ability of the sire or dam of prospective foundation females is another good guide in making selections. Usually a sizable number of a sire's progeny can be seen on one ranch or farm, and one should inspect as many of them as possible in order to evaluate the sire. An unselected sample of progeny should be seen—those from poor cows as well as those from the better cows. Calves that are creep-fed or otherwise pampered are not of much help in evaluating a bull's breeding worth, although cattle with a tendency to mature or fatten too early can sometimes be sorted off this way.

The breeders of some localities have established reputations as good producers of foundation females. A new breeder would do well to ask for expert help, even when buying his females from such high-reputation areas, because not all herds in the area are outstanding, and prices are usually higher. Such an area is a good place to buy females, however, because more herds can be seen in less time and more good cattle will be offered for sale. The feedlot performance of large numbers of steers or heifers from a herd over a period of years is a good practical guide in seeking foundation females.

As for the bull or bulls to be used in a new commercial herd or as a replacement, nothing less than an excellent purebred bull on which performance data are available should be acceptable. Such data, to be complete, should include weaning weight, corrected for the age of the calf and for the age of the dam, weaning conformation and body type score, gain data for at least a 120-day postweaning period, and a 12- or 15-month weight and body type score.

Records alone, of course, mean nothing. Obviously bulls with poor records should not be offered for sale, but some breeders, in their eagerness to make use of a "performance-tested" bull, mistakenly think that a bull with records, even though only average, should be better than a bull with no records.

Age.   Age groups of females available for foundation material can be grouped according to numbers available, in descending order, as follows: (1) weanling heifer calves, (2) yearling heifers (usually bred), and (3) older cows with or without calves at side. From a quality and performance standpoint the older cows, having withstood annual cullings, may be best. However, they may be for sale because of advanced age and reasons related to age, such as bad disposition, eye problems, poor udder, and slow breeding tendency. Cows that are culled for such reasons are unlikely to be profitable additions to a herd. If mature cows are for sale owing to a forced reduction in cow numbers on a ranch or farm because of drought, settling of an estate, or the like, then such females are often very desirable foundation stock.

Western ranchers like to offer their surplus heifer calves and yearling heifers for sale in droves without undue culling or sorting. Therefore, if such females are bought to start a commercial herd, considerably more, even twice as many, should be bought than are needed to start the herd. The drove can be culled to size after the calves have been fed and handled in their new surroundings for 2 or 3 months or even until spring, and the rejected females can be fed out for slaughter or sold as feeders. Heifer calves offered for sale are usually of better quality than yearling heifers, especially if the above procedure is followed. Often the yearling heifer droves include late-dropped calves from the year before, or heifers that were intended as replacements in the original owner's herd but are now for sale because their subsequent development was disappointing. If yearlings are bought in the fall, some are apt to be pregnant and some not, depending on the management program followed by the seller. If the heifers are open, it is probably too late to breed them for calving during the next spring. If bred, they may be bred for late calves. Thus it is evident that buying yearlings can result in future problems and generally proves to be a disappointing way to start a herd. The effect of age on the future productive life of females is discussed in Chapter 5.

Age is also important in selecting bulls. Obviously a 4- or 5-year-old bull that has sired two or three crops of good-doing calves of the right type is a much safer bull to buy than a younger, unproven bull or a bull calf that has yet to demonstrate his own performance or his breeding ability. However, not many such bulls are offered for sale, and often they are in great demand if of excellent quality. If the herd is already in production, a good plan is to buy a yearling bull to use on yearling heifers or 2-year-old cows while the older bull is still in use. Thus he can be progeny-tested— that is, offspring sired by him can be compared with calves sired by the main herd bull. Young bulls should be used sparingly, as is discussed in Chapter 5. Great skill is required in selecting weanling bull calves as future herd sires, unless one has performance data on the calves themselves, as well as progeny data on the sire and dam of the calf, or unless one can see many close relatives in more mature stages of development.

Cost.   Naturally cost is an important consideration in selecting commercial breeding cattle, but it should not crowd out all others. Often price becomes the all-important factor in deciding which animals to purchase. As a matter of fact, price should be the last consideration within reasonable limits. If closely culled grade females are bought, a 25 to 50 percent premium over feeder cattle market price is not unreasonable. If young heifers are bought, the added weight at disposal time as old cows will often offset the premium paid at purchase time. If a drove of females is bought in which no culling was permitted by the seller, naturally one would reasonably expect to pay a somewhat lower premium amounting to perhaps only 10 percent above market price, or no premium at all. Locally grown cattle of equal quality have an advantage with respect to price because of lower transportation and incidental costs. There is also less likelihood of losses due to shipping fever and similar diseases.

Good bulls can usually be bought for the equivalent of 2 to 3 choice grade finished steers or 5 or 6 weaner calves. Producers of superior performance-tested bulls do and should expect somewhat higher prices because of the labor and expense involved in participating in such tests. Furthermore, not nearly all of the bulls tested will prove to be superior; hence the testing costs must be added to the bulls actually sold as superior or tested bulls.

SOURCES OF FOUNDATION FEMALES

Sources of good female breeding stock for the foundation of a commercial beef herd vary with the section of the country. Anyone wishing to start a herd in the range area has a relatively simple job with respect to this point because ranchers in the neighborhood are likely to be able to supply his wants. On the other hand, good females are harder to come by in areas where there are few herds or where most of the herds are small or are themselves expanding in numbers. Of course, even a buyer in these areas may secure surplus range area females through dealers or commission men or by direct purchase. Extra costs such as transportation and commission and the fact that little sorting can be done mean that this source may be more expensive. In the nonrange areas, then, more thought and time are required to obtain the right kind of females.

Some sources of females and methods of procurement are the following.

1. The occasional complete or intact herd available owing to circumstances beyond the control of the owner or operator.
2. The normally surplus heifer calves and yearling heifers in the range areas.
3. Cow and calf pairs in the range areas, especially in periods of extended drought when forced herd reductions are necessary.
4. Dispersal sales held to settle estates, divide partnerships, and so forth.
5. Regularly scheduled auction sales, especially those held in states that have closely supervised health requirements.
6. Private treaty purchase from neighbors.

SOURCES OF BREEDING BULLS

Purebred herd bulls for use in establishing a commercial herd may be obtained from several sources, and the source is not so important as the bull himself. Satisfactory sources of breeding bulls, in descending order of choice, are the following.

1. A proven but still sound bull, purchased from another breeder.
2. A young performance-tested bull, purchased privately from a breeder or in a sale of performance-tested bulls.
3. A young bull from a reputable purebred breeder, after inspecting the sire, dam, and other near relatives.
4. Purebred bull sales where a screening committee has rigidly culled the sale offering.

METHODS OF ESTABLISHING A COMMERCIAL HERD

Because the average farmer or rancher seldom establishes more than one cow herd in a lifetime, it behooves him to make his foundation choices carefully and wisely. Seeking advice from an experienced neighbor, a reputable, professionally trained dealer or commission man, or a county agent or vocational agriculture teacher is especially advisable if the buyer is inexperienced. Some time-tested methods of starting a herd are as follows.

1. Buy a complete herd with a good reputation, or the top half of such a herd, when a farm or ranch is changing hands. This method requires considerable capital, but returns come quickly, and ordinarily the quality is high unless some topping out has occurred. By all means the

lower end of such a herd should not be purchased alone even if it seems to be a bargain.

2. Buy twice as many heifer calves as will ultimately make up the cow herd from a herd with a good reputation. The better half of the drove may be selected for foundation females on the basis of performance as shown by weight (or size if scales are unavailable), beef type or conformation, disposition, and uniformity after being fed a good growing ration for 3 to 4 months or until early spring. The remaining half of the heifers may be disposed of through any of several heifer feeding plans discussed later. Returns come more slowly from such foundation females, but the initial investment is comparatively low and considerable culling may be practiced.

3. Buy yearling heifers in the fall, again buying extras to permit culling before making final choices, after the cattle have spent several months on the purchaser's farm. As yearling heifers from the range area will probably already have been exposed unless guaranteed open, a pregnancy test should be made to reveal those not bred. The open heifers, with the culls, can then be sold or fed out for the spring market. This method has the advantage over the purchase of heifer calves that the first calf crop may be obtained one year earlier. Initial costs per head are higher and there is less opportunity for culling, however.

4. Buy cow and calf pairs, preferably from areas where numbers are being reduced for reasons other than simple culling—for example, where extended drought is forcing a drastic reduction in cow numbers. Such cows will ordinarily be bred when offered for sale in late summer or early fall. Steer calves and undesirable heifer calves at side can be sold or handled in any one of the many programs discussed elsewhere. Naturally, such cows will not be the best cows on the farm or ranch from which they came, but the calves at side will tend toward the average of the herd of origin and will offer evidence of the quality of the herd. If such cattle are from a drought area, the buyer should take into consideration the lack of flesh or condition in the cows and the lack of weight and bloom in the calves resulting from the poor rations of their mothers. The cattle are probably better cattle than they appear to be to a beginner breeder.

5. Grading up the native beef females, or even dairy cows of the heavy breeds such as Holstein and Brown Swiss, available in the area or presently on the farm. This method takes several years, but results are often quite good. Using at least three successive beef bulls of excellent quality and bred-in performance will result in good beef-type cows, especially if a performance-testing program is begun from the start and only the best heifers are retained. Especially good foundation females

of this type are heifer calves produced in dairy herds where some of the cows have been artificially bred to beef bulls. This avoids the problem of too much milk being produced, as may be the case in the straightbred dairy females. The disease problem is greatly reduced by this method if the usual recommended procedures for disease control are followed. The initial outlay of capital is again comparatively low; consequently this method is well suited to young men just starting in the business. The selection of the herd bull is of the utmost importance in this method.

CHAPTER 9

BREEDING HERD
MANAGEMENT
IN WINTER

Herd management throughout the year should have as its aim the production of a 100 percent crop of uniformly high-quality, heavyweight calves. Items that contribute to the accomplishment of this aim are early sexual maturity of females, high conception rate, high calf livability, early rebreeding after calving, heavy and prolonged milk production, and longevity. Quality of management and plane of nutrition contribute to each of the items mentioned and are important determinants of profit or loss.

The cow-calf program is a low-gross-income enterprise at best; therefore cash expenditures must be kept at a minimum to ensure a net profit. As the largest item of cash expense is for winter feed, it is the item that requires the closest scrutiny. It has been correctly said that one cannot starve a profit out of cows, but it is equally true that supplying nutrients above the minimal requirement is uneconomical, especially when it applies to harvested or purchased feeds.

HERD GROUPINGS WITH SIMILAR NUTRIENT REQUIREMENTS

Sorting the herd into groups with similar feed requirements is the first step in planning an efficient feeding program for the winter. Usually the following groups of cattle are found in a typical well-managed commercial herd during the winter.

1. Dry cows.
2. Wet or nursing cows with late calves, unless only spring calving is practiced.
3. Weaner calves, usually heifers.
4. Bred or open yearling and 2-year-old heifers.
5. Bulls.

In actual practice these five groups do not always need to be fed separately. If the herd is not large, the weaner calves, the bred yearlings if 2-year-old first calving is practiced, and the first-calf and oldest dry cows

229

Fig. 41. Many ranchers wait until after the wintering period to decide which heifers to keep for replacements, in order to get more information concerning their growth rate and type. High condition in calves at weaning time, owing to heavy-milking mothers, can mislead a breeder in selecting replacements. (American Angus Association.)

can all be fed together. The ration should be such that the weaners and yearling-bred heifers will gain enough to ensure maximum growth and development and such that the suckled-down first-calf heifers and old cows will mend to some extent before their next calving and lactation periods. The other main grouping would then contain the medium-aged dry cows and the open yearlings and bred 2-year-olds if heifers are bred to calve first as 3-year-olds.

If the herd has either a fall calving, an early spring calving, or a year-round calving program, the cows that are nursing calves should be handled separately, especially if creep-feeding is practiced, or they may be grouped with the first large group mentioned above. The bulls should be fed separately if only a spring-calving season is practiced, but of course they will be running with the wet cows if fall calving is used.

FACTORS THAT DETERMINE WINTER FEEDING AND MANAGEMENT PLANS

The nutritional plane and feeding and management plans depend on a number of factors, as follows.

1. Condition of cows at the end of the fall grazing season.
2. Age of cows.
3. Calving season—early spring versus late spring, for example.
4. Probable mineral and vitamin stores at the end of the fall grazing season.
5. Condition, amount, and species of forage if cows are to be wintered on cured range feed.
6. Probable weather conditions.
7. Labor available for feeding.
8. Availability of supplemental feeds.
9. Breed of cattle.
10. Level of feed intake.

The four nutrients of greatest importance in wintering beef cattle are energy, protein, phosphorus, and vitamin A or its precursor, carotene. It will be shown how the ten factors above relate to these nutrients and how they affect the choice of a feeding program for the herd in wintertime.

CONDITION OF COWS

This factor is extremely important, because it tells much about the type of pasture or range the cows have just experienced during the summer and fall grazing seasons. It is also a good indicator of the kind and amount of feed that will probably be available for winter, especially if it is a range operation and cows will be wintered on "set-aside" or ungrazed range. Obviously if cows have "summered hard" on drought-stricken or over-grazed range or pasture, the feeds reserved for winter range also will be poor. Unless considerable culling or herd reduction has occurred to adjust cattle numbers to feed supply, the amount of winter feed will be short. On the other hand, if the cows are in fair or better flesh, they can safely lose as much as 200 pounds up through the calving season as a result of deliberate or forced restriction of energy during the winter. Such fleshy cows may be expected to have good liver vitamin A stores, but adequate protein and phosphorus must be supplied. Very thin cows require

Fig. 42. Cows being wintered in northern New Mexico with only the shelter provided by the small mountains and draws. Often a high-energy mineral-vitamin-fortified cubed supplement is fed in these areas as insurance against deficiencies. (National Cottonseed Products Association, Inc.)

Table 43

Effect of Winter Gains of Cows on Weights of Calves[a]

High, Medium, and Low Winter Gains

	Number of Cows	Average Winter Gain (lb)[b]	Calves at Weaning	
			Average Weight (lb)	Average Age (days)
8 high winter-gaining lots	62	110.4	384.4	161.0
8 medium winter-gaining lots	50	66.9	370.2	162.3
8 low winter-gaining lots	61	21.2	362.4	160.1

Winter Gains versus Winter Losses of Cows

	Cows			Calves		
	Average Winter Gain (lb)	Average Weight at End of Winter (lb)	Average Weight after Calving (lb)	Average Birth Weight (lb)	Average Weaning Weight (lb)	Average Age at Weaning (days)
11 highest winter-gaining cows	42.6	1,064.4	844.2	65.6	379.1	170.7
11 lowest winter-gaining cows	−62.7	981.3	781.3	64.6	326.9	177.1

[a] Montana Special Circular 7.
[b] Up to time of calving.

complete supplementation of their rations, including energy. A winter weight loss of about 100 pounds, up through calving, is the approximate weight change most preferred by ranchers for their mature cows.

AGE OF COWS

First-calf heifers and especially those that calved first as 2-year-olds are the major concern. Inasmuch as nutrient requirements for maintenance and lactation have priority over nutrients for growth and reproduction, this age group usually will be suckled thin. Most of the nonpregnant cows are found in this category. The winter dry period gives them an opportunity to make up for loss in growth; hence the ration should by all means contain sufficient energy, phosphorus, and protein. These first-calf heifers should not lose weight during winter, as is permissible with mature cows, but rather should gain up to 100 pounds before their next calving. Older cows, especially heavy milkers and cows with tooth problems, will also be suckled so thin that they will need to gain flesh before calving again in the spring.

CALVING SEASON

The period from 30 days before through 90 days after calving is the most critical period of the year for the beef cow from a nutritional standpoint. Most of the growth of the developing embryo occurs during the last 30 days of gestation. Adequate protein, vitamin A, phosphorus, and energy will ensure a strong calf with greatest chances of normal birth and survival. The early reestablishment of the estrus cycle and conception depend greatly on the plane of nutrition during this 4-month period, as discussed in Chapter 5. The relationship between the nutrient content of the harvested roughage and the cured forage available to the dry cows and the date when the calving season begins, in light of the above, is obvious. Thus the time to begin supplemental feeding, if required, is determined largely by the dates of the calving season. Fall-calving cows have nutrient requirements at least 50 percent higher than those of dry cows. Unfortunately, their voluntary intake is not 50 percent higher and they must be fed more highly concentrated rations to provide the additional nutrients.

VITAMIN AND MINERAL STORES

This point is discussed separately from the subject of condition because differences in condition may not necessarily be associated with differences in vitamin and mineral reserves. The quality and species of forage available in late fall and winter have more to do with these reserves than quantity of feed during the entire grazing season. Thin cows may have had access to late-fall weeds, small-grain pastures, or browse, which ensured good liver vitamin A reserves. Judgment and close observation of range conditions may tell one more than condition of cows with respect to this point.

CONDITION, AMOUNT, AND SPECIES OF WINTER FORAGE

If cured range is to be the chief source of winter feed, the species of forage plants that are most prevalent will determine the supplemental feed needs. In programs where hay or silage is to be fed to cows in winter, the question is simplified because such feeds are usually either legume or nonlegume and the matter of balancing rations containing these feeds is simple by comparison. Some range plant species retain their feed value longer during the dormant season—for instance, black grama grass retains some green growth throughout the winter as do the tall fescues and some of the browse plants—but others, notably the native tall grasses, may be devoid of carotene by February or March, as well as quite indigestible and thus low in all available nutrients (Table 44).

WEATHER CONDITIONS

Either snow cover or extremely wet, muddy conditions may necessitate a complete dependence on supplemental feeding of the entire ration in some localities. The length of time during which such conditions prevail will of course greatly influence winter feed needs and the best time for the calving season. Temperature can affect the nutrient requirement itself, with energy and protein requirements for maintenance being most

Table 44

						Percentage Composition of Dry Matter				
Analysis of Prairie Grass at Different Seasons[a] (Three-Year Average)										
	Percent Dry Matter	Ash	Protein	Fat	Fiber	Nitrogen-Free Extract	Ca	P	Caro-tene[b]
Grass[c]									
November	82.39	5.01	2.53	1.74	40.02	50.40	0.253	0.046	14
January	94.85	5.92	2.57	1.57	40.86	48.74	0.309	0.039	Trace
May	52.29	6.39	9.68	2.40	32.02	49.08	0.308	0.126	407
August	54.71	6.21	5.06	2.23	35.42	50.66	0.346	0.078	112
October	63.09	5.18	3.23	1.62	37.24	52.24	0.244	0.048	16

[a] Oklahoma Cattle Feeders' Day Report.

[b] Parts per million.

[c] Averages of the four predominant grasses: big bluestem, little bluestem, Indian, and switch.

involved. Cows are adaptable to wide temperature ranges, however, and supplemental feeding adjustments for this reason are not so critical as some believe. However, under extreme conditions, such as temperatures of −20° F or lower, energy requirements may increase by one-third. The absence of shelter, especially from high winds, in such instances can cause more serious problems.

LABOR AVAILABLE FOR FEEDING

This factor probably has more to do with how and how often cows are fed than with what they are fed. Cows can be fed as infrequently as once weekly with success. Except possibly for checking the water supply and fences, little labor is required in managing cows during the winter until calving begins.

AVAILABILITY OF SUPPLEMENTAL FEED

If distances between the source of supplemental feeds, when not home-grown, and location of the cow herd are great, concentrates are more apt to be used in supplementation programs than roughages, simply because of convenience and the comparative cost of transportation.

BREED OF CATTLE

The nutrient requirements, expressed on a percentage basis, of various breeds of mature beef cattle do not differ markedly during the dry period, but breed may influence the condition of cows in the fall and thus have a bearing on supplemental feed needs. One breed of cow may summer better under extremes of weather, forage, and terrain than others, or one may travel or forage over a wider area in winter. Some breeds, especially some of the dairy × beef crosses, suckle down to a greater degree and thus may require a higher-energy winter ration. The same problem may occur with some of the newer European exotic crossbred cows unless adjustments are made in stocking rates to provide the added forage for the larger mature size, and the sometimes stronger lactation tendencies, of such cows.

In summary, it is evident that numerous factors prevent making hard and fast recommendations concerning the feeding of the cow herd during winter for each and every farm or ranch condition. Knowledge of the nutritional requirements of cows and replacement cattle and the composition of available forages is essential for the development of sound supplemental feeding programs.

LEVEL OF FEED INTAKE

The amount of feed a cow will consume is of interest because, if supplemental feeding is required, the total amount of feed required for an individual or herd must be determined, whether homegrown or purchased feeds are fed. Cows of approximately the same skeletal size and body type will consume about the same amount of total air-dry feed regardless of condition, if fed free-choice or all they will clean up. However, because cows of the same skeletal size will weigh less or more depending on whether the cows are thin or fleshy, it is necessary to use different factors for determining feed capacity for cows of varying condition. Cows that are fleshy, average, or thin, with respect to condition, will consume daily (on an air-dry basis) an amount of feed that is equivalent to approximately 1.75, 2.00, and 2.25 percent of their live weight, respectively. Thus cows that vary in weight because of differences in condition and not body or skeletal size or type per se and, accordingly, weigh 1,100, 1,000, and 900 pounds, respectively, will each consume about 20 pounds of air-dry feed daily.

NUTRIENT REQUIREMENTS OF BREEDING BEEF CATTLE

The nutrient requirements for breeding beef cattle, as recommended by the National Research Council, expressed on the basis of daily nutrient needs and nutrient concentration in the ration, are shown in Tables 45 and 46. Note that the daily nutrient requirements, with energy expressed both as metabolizable energy and as total digestible nutrients, during the full 11- to 14-month cycle for the producing cow are divided into three separate but overlapping periods.

First is a 4- to 5-month postpartum period during which requirements are highest because of the added demands for both lactation and rebreeding. Then follows a 4- to 5-month period that simultaneously includes, at first, the 2 to 3 months of late lactation and midpregnancy, followed by 2 months or so during which the cow is dry and in advancing pregnancy. Finally there is the period corresponding to the last 3 to 4 months of pregnancy, during which the cow is dry and preparing for parturition. The range in length of the various periods makes it possible

Table 45

Nutrient Requirements for Beef Cattle Breeding Herd (Daily Nutrients per Animal)[a]

Weight[b] (kg) (lb)	Daily Gain (kg) (lb)	Minimum Dry Matter Consumption[c] (kg) (lb)	Total Protein (kg) (lb)	Digestible Protein (kg) (lb)	Metabolizable Energy[c,d] (Mcal)	Total Digestible Nutrients[c,d] (kg) (lb)	Calcium (gm)	Phosphorus (gm)	Vitamin A (1000 IU)
Cows Nursing Calves, Average Milking Ability,[e] First 4 to 5 Months Postpartum, Including Rebreeding									
350 772		8.2 18.1	0.75 1.50	0.44 0.97	15.9	4.4 9.7	24	24	19
400 882		8.8 19.4	0.81 1.78	0.48 1.06	17.0	4.7 10.4	25	25	21
450 992		9.3 20.5	0.86 1.89	0.50 1.10	18.1	5.0 11.0	26	26	23
500 1102		9.8 21.6	0.90 1.98	0.53 1.17	19.2	5.3 11.7	27	27	24
550 1213		10.5 23.1	0.97 2.13	0.57 1.25	20.3	5.6 12.3	28	28	26
600 1323		11.0 24.2	1.01 2.22	0.59 1.30	21.3	5.9 13.0	28	28	27
Cows Nursing Calves, Superior Milking Ability,[f] First 4 to 5 Months Postpartum, Including Rebreeding									
350 772		10.2 22.4	1.11 2.44	0.65 1.43	21.0	5.8 12.8	45	40	32
400 882		10.8 23.8	1.17 2.57	0.69 1.52	22.1	6.1 13.5	45	41	34
450 992		11.3 24.9	1.23 2.71	0.72 1.58	23.2	6.4 14.1	45	42	36
500 1102		11.8 26.0	1.29 2.84	0.76 1.67	24.3	6.7 14.8	46	43	38
550 1213		12.4 27.3	1.35 2.97	0.79 1.74	25.3	7.0 15.4	46	44	41
600 1323		12.9 28.4	1.41 3.10	0.83 1.83	26.4	7.3 16.1	46	44	43
All Pregnant Cows Nursing Calves in Late Lactation and Including Dry Midpregnancy Period, 4 to 5 Months									
350 772		5.5 12.2	0.32 0.70	0.15 0.33	10.8	3.0 6.6	10	10	15
400 882		6.1 13.4	0.36 0.79	0.17 0.37	11.9	3.3 7.3	11	11	17
450 992		6.7 14.8	0.39 0.86	0.19 0.38	13.0	3.6 7.9	12	12	19

500	1102	0.4[a]	0.9[a]	7.2	15.9	0.42	0.92	0.20	0.44	14.1	3.9	8.6	13	13	20
550	1213	0.4	0.9	7.7	17.0	0.45	0.99	0.22	0.48	15.1	4.2	9.2	14	14	22
600	1323	0.4	0.9	8.3	18.3	0.49	1.10	0.23	0.51	16.1	4.4	9.8	15	15	23

Dry, Pregnant, Mature Cows, Last 3 to 4 Months of Pregnancy

350	772	0.4[a]	0.9[a]	6.9	13.9	0.41	0.90	0.19	0.42	13.2	3.6	8.0	12	12	19
400	882	0.4	0.9	7.5	15.4	0.44	0.97	0.21	0.46	14.3	4.0	8.7	14	14	21
450	992	0.4	0.9	8.1	16.5	0.48	1.06	0.23	0.51	15.4	4.2	9.4	15	15	23
500	1102	0.4	0.9	8.6	17.9	0.51	1.12	0.24	0.53	16.4	4.5	10.0	15	15	24
550	1213	0.4	0.9	9.1	19.0	0.54	1.19	0.25	0.55	17.5	4.8	10.7	16	16	26
600	1323	0.4	0.9	9.7	20.3	0.57	1.25	0.27	0.59	18.5	5.1	11.2	17	17	27

Pregnant Yearling Heifers, Last 3 to 4 Months of Pregnancy

325	716	0.4[a]	0.9[a]	6.6	14.5	0.58	1.28	0.34	0.75	12.6	3.5	7.7	15	15	19
		0.6	1.3	8.5	18.7	0.75	1.65	0.42	0.92	16.2	4.5	9.9	18	18	23
		0.8	1.8	9.4	20.7	0.85	1.87	0.50	1.10	20.1	5.6	12.3	22	20	26
350	772	0.4	0.9	6.9	15.2	0.61	1.34	0.35	0.77	13.2	3.7	8.1	15	15	19
		0.6	1.3	8.9	19.6	0.78	1.71	0.45	0.99	16.9	4.7	10.3	19	19	25
		0.8	1.8	10.0	22.0	0.88	1.94	0.51	1.12	21.1	5.8	12.9	22	21	28
375	827	0.4	0.9	7.2	15.9	0.63	1.39	0.36	0.79	13.7	3.8	8.4	15	15	20
		0.6	1.3	9.3	20.5	0.81	1.78	0.46	1.01	17.7	4.9	10.8	19	19	26
		0.8	1.8	11.0	24.2	0.96	2.11	0.55	1.21	22.1	6.1	13.5	22	22	31
400	882	0.4	0.9	7.5	16.5	0.65	1.43	0.38	0.84	14.2	3.9	8.7	16	16	21
		0.6	1.3	9.7	21.4	0.84	1.85	0.48	1.06	18.5	5.1	11.3	19	19	27
		0.8	1.8	11.6	25.6	1.01	2.22	0.57	1.24	23.0	6.4	14.0	22	22	33

Table 45 (*Continued*)

Nutrient Requirements for Beef Cattle Breeding Herd (Daily Nutrients per Animal)[a]

Weight[b] (kg)	(lb)	Daily Gain (kg)	(lb)	Minimum Dry Matter Consumption[c] (kg)	(lb)	Total Protein (kg)	(lb)	Digestible Protein (kg)	(lb)	Metabolizable Energy[c,d] (Mcal)	Total Digestible Nutrients[c,d] (kg)	(lb)	Calcium (gm)	Phosphorus (gm)	Vitamin A (1000 IU)
\multicolumn{16}{Bulls, Growth and Maintenance (Moderate Activity)}															
300	661	1.00	2.2	8.8	19.4	0.90	1.98	0.55	1.21	20.4	5.6	12.3	27	23	34
400	882	0.90	2.0	11.0	24.2	1.03	2.27	0.62	1.36	25.2	7.0	15.4	23	23	43
500	1102	0.70	1.5	12.2	26.9	1.07	2.35	0.62	1.36	27.0	7.5	16.5	22	22	48
600	1323	0.50	1.1	12.0	26.4	1.02	2.24	0.60	1.32	26.4	7.3	16.1	22	22	48
700	1543	0.30	0.7	12.9	28.4	1.08	2.38	0.60	1.32	27.7	7.7	17.0	23	23	50
800	1764	0.00	0.0	10.5	23.1	0.89	1.96	0.50	1.10	21.0	5.8	12.8	19	19	41
900	1984	0.00	0.0	11.4	25.1	0.99	2.18	0.55	1.21	22.8	6.3	13.9	21	21	44

[a] Adapted from *Nutrient Requirements of Beef Cattle*, Subcommittee on Beef Cattle Nutrition, National Research Council, 1975.

[b] Average weight for a feeding period.

[c] Dry matter consumption, metabolizable energy (ME), and total digestible nutrients (TDN) requirements are based on all- or almost all-roughage rations.

[d] Approximately 0.4 ± 0.1 kg (0.88 ± 0.22 lb) weight gain/day over the last 3 to 4 months of pregnancy is accounted for by the products of conception. These nutrients and energy requirements include the quantities estimated as necessary for conceptus development.

[e] 5.0 ± 0.5 kg (11.0 ± 1.0 lb) milk/day. Nutrients and energy for maintenance of the cow and for milk production are included in these requirements.

[f] 10.0 ± 0.5 kg (22.0 ± 2.2 lb) milk/day. Nutrients and energy for maintenance of the cow and for milk production are included in these requirements.

Table 46

Nutrient Requirements for Beef Cattle Breeding Herd (Nutrient Concentration in Diet Dry Matter)[a]

Weight[b] (kg)	(lb)	Daily Gain (kg)	(lb)	Minimum Dry Matter Consumption[c] (kg)	(lb)	Total Protein (%)	Digestible Protein (%)	Metabolizable Energy[c,d] (Mcal) (kg)	(lb)	Total Digestible Nutrients[c,d] (%)	Calcium (%)	Phosphorus (%)	Vitamin A (1000 IU) (kg)	(lb)
Cows Nursing Calves, Average Milking Ability,[e] First 4 to 5 Months Postpartum, Including Rebreeding														
350	772			8.2	18.1	9.2	5.4	1.9	0.86	52	0.29	0.29	2.4	1.1
400	882			8.8	19.4	9.2	5.4	1.9	0.86	52	0.28	0.28	2.4	1.1
450	992			9.3	20.5	9.2	5.4	1.9	0.86	52	0.28	0.28	2.4	1.1
500	1102			9.8	21.6	9.2	5.4	1.9	0.86	52	0.28	0.28	2.5	1.1
550	1213			10.5	23.1	9.2	5.4	1.9	0.86	52	0.27	0.27	2.5	1.1
600	1323			11.0	24.2	9.2	5.4	1.9	0.86	52	0.25	0.25	2.5	1.1
Cows Nursing Calves, Superior Milking Ability,[f] First 4 to 5 Months Postpartum, Including Rebreeding														
350	772			10.2	22.4	10.9	6.4	2.0	0.91	55	0.44	0.39	3.1	1.4
400	882			10.8	23.8	10.9	6.4	2.0	0.91	55	0.42	0.38	3.1	1.4
450	992			11.3	24.9	10.9	6.4	2.0	0.91	55	0.40	0.37	3.1	1.4
500	1102			11.8	26.0	10.9	6.4	2.0	0.91	55	0.39	0.36	3.1	1.4
550	1213			12.4	27.3	10.9	6.4	2.0	0.91	55	0.37	0.35	3.1	1.4
600	1323			12.9	28.4	10.9	6.4	2.0	0.91	55	0.36	0.34	3.1	1.4
All Pregnant Cows Nursing Calves in Late Lactation and Including Dry Midpregnancy Period, 4 to 5 Months														
350	772			5.5	12.2	5.9	2.8	1.9	0.86	52	0.18	0.18	2.7	1.2
400	882			6.1	13.4	5.9	2.8	1.9	0.86	52	0.18	0.18	2.7	1.2
450	992			6.7	14.8	5.9	2.8	1.9	0.86	52	0.18	0.18	2.7	1.2

Table 46 (Continued)

Nutrient Requirements for Beef Cattle Breeding Herd (Nutrient Concentration in Diet Dry Matter)[a]

Weight[b] (kg)	(lb)	Daily Gain (kg)	(lb)	Minimum Dry Matter Consumption[c] (kg)	(lb)	Total Protein (%)	Digestible Protein (%)	Metabolizable Energy[c,d] (Mcal) (kg)	(lb)	Total Digestible Nutrients[c,d] (%)	Calcium (%)	Phosphorus (%)	Vitamin A (1000 IU) (kg)	(lb)
500	1102			7.2	15.9	5.9	2.8	1.9	0.86	52	0.18	0.18	2.7	1.2
550	1213			7.7	17.0	5.9	2.8	1.9	0.86	52	0.18	0.18	2.7	1.2
600	1323			8.3	18.3	5.9	2.8	1.9	0.86	52	0.18	0.18	2.7	1.2
Dry, Pregnant, Mature Cows, Last 3 to 4 Months of Pregnancy														
350	772	0.4[a]	0.9[a]	6.9	13.9	5.9	2.8	1.9	0.86	52	0.18	0.18	2.8	1.3
400	882	0.4	0.9	7.5	15.4	5.9	2.8	1.9	0.86	52	0.18	0.18	2.8	1.3
450	992	0.4	0.9	8.1	16.5	5.9	2.8	1.9	0.86	52	0.18	0.18	2.8	1.3
500	1102	0.4	0.9	8.6	17.9	5.9	2.8	1.9	0.86	52	0.18	0.18	2.8	1.3
550	1213	0.4	0.9	9.1	19.0	5.9	2.8	1.9	0.86	52	0.18	0.18	2.8	1.3
600	1323	0.4	0.9	9.7	20.3	5.9	2.8	1.9	0.86	52	0.18	0.18	2.8	1.3
Pregnant Yearling Heifers, Last 3 to 4 Months of Pregnancy														
325	716	0.4[a]	0.9[a]	6.6	14.5	8.8	5.1	1.9	0.86	52	0.23	0.23	2.8	1.3
		0.6	1.3	8.5	18.7	8.8	5.1	1.9	0.86	52	0.21	0.21	2.8	1.3
		0.8	1.8	9.4	20.7	9.0	5.3	2.1	0.95	58	0.23	0.21	2.8	1.3
350	772	0.4	0.9	6.9	15.2	8.8	5.1	1.9	0.86	52	0.22	0.22	2.8	1.3
		0.6	1.3	8.9	19.6	8.8	5.1	1.9	0.86	52	0.21	0.21	2.8	1.3
		0.8	1.8	10.0	22.0	8.8	5.1	2.1	0.95	58	0.22	0.21	2.8	1.3

375	827	0.4	0.9	7.2	15.9	8.7	5.0	1.9	0.86	52	0.21	0.21	2.8	1.3
		0.6	1.3	9.3	20.5	8.7	5.0	1.9	0.86	52	0.20	0.20	2.8	1.3
		0.8	1.8	11.0	24.2	8.7	5.0	2.0	0.91	55	0.20	0.20	2.8	1.3
400	882	0.4	0.9	7.5	16.5	8.7	5.0	1.9	0.86	52	0.21	0.21	2.8	1.3
		0.6	1.3	9.7	21.4	8.7	5.0	1.9	0.86	52	0.20	0.20	2.8	1.3
		0.8	1.8	11.6	25.6	8.7	5.0	2.0	0.91	55	0.19	0.19	2.8	1.3

Bulls, Growth and Maintenance (Moderate Activity)

300	661	1.00	2.2	8.8	19.4	10.2	6.3	2.3	1.04	64	0.31	0.26	3.9	1.8
400	882	0.90	2.0	11.0	24.2	9.4	5.6	2.3	1.04	64	0.21	0.21	3.9	1.8
500	1102	0.70	1.5	12.2	26.9	8.8	5.1	2.2	1.00	61	0.18	0.18	3.9	1.8
600	1323	0.50	1.1	12.0	26.4	8.6	5.0	2.2	1.00	61	0.18	0.18	3.9	1.8
700	1543	0.30	0.7	12.9	28.4	8.5	4.8	2.0	0.91	55	0.18	0.18	3.9	1.8
800	1764	0.00	0.0	10.5	23.1	8.5	4.8	2.0	0.91	55	0.18	0.18	3.9	1.8
900	1984	0.00	0.0	11.4	25.1	8.5	4.8	2.0	0.91	55	0.18	0.18	3.9	1.8

[a] Adapted from *Nutrient Requirements of Beef Cattle*, Subcommittee on Beef Cattle Nutrition, National Research Council, 1975.

[b] Average weight for a feeding period.

[c] Dry matter consumption, metabolizable energy (ME), and total digestible nutrients (TDN) requirements are based on all- or almost all-roughage rations.

[d] Approximately 0.4 ± 0.1 kg (0.88 ± 0.22 lb) weight gain/day over the last 3 to 4 months of pregnancy is accounted for by the products of conception. These nutrients and energy requirements include the quantities estimated as necessary for conceptus development.

[e] 5.0 ± 0.5 kg (11.0 ± 1.1 lb) milk/day. Nutrients and energy for maintenance of the cow and for milk production are included in these requirements.

[f] 10.0 ± 0.5 kg (22.0 ± 2.2 lb) milk/day. Nutrients and energy for maintenance of the cow and for milk production are included in these requirements.

to adjust the total requirement to the variation in number of months between calvings because, in real situations, every cow does not produce a calf every 12 months. Requirements during early lactation are shown separately for beef-type cows with average lactation levels and for heavier-milking cows such as dairy and dual-purpose × beef crossbred cows.

Obviously, daily nutrient requirements do not change as abruptly from one period to the next, as shown in the requirement tables, and ration changes, when made, should be made gradually. In practice, if cows calve in late winter or early spring, the change from early to late lactation will occur in about midsummer when pasture forages, especially cool-season grasses, tend to enter their summer dormancy. It is to be hoped that all cows have been rebred by at least the end of the first 3 months of the postpartum period of higher nutrient requirements, or otherwise the available pasture nutrients may be inadequate to meet all their needs.

The requirements for bred heifers are shown only for the last 3 to 4 months of pregnancy because this corresponds roughly to the winter period prior to first calving in the spring and is the period when supplemental feeding most likely will be practiced. Fall-calving heifers usually are on pasture during this stage of pregnancy, and the question of the adequacy of their nutrient intake from pasture is determined by the quality and quantity of the available forage. The nutrient requirements of yearling heifers are given in the tables in Chapter 11, which provide the necessary requirement data for the bred heifers up to the late stages of pregnancy covered by Tables 45 and 46.

The daily nutrient requirement values for bulls provide for enough nutrients for the lighter, and presumably younger, bulls to make adequate growth gains to reach mature size at normal ages. Older bulls that are light because of extreme weight loss on summer pasture would gain considerably on the amounts of nutrients shown, which is desirable in order to precondition them for the onset of the breeding season.

Morrison's feeding standards for breeding beef cattle are shown in Table 47. Rations that meet the needs of the various weight and age groups of breeding cattle, according to the requirements or standards indicated, can be computed exactly from information concerning the composition of feeds as found in the Appendix tables.

WINTER FEEDING IN FARM REGIONS

The winter ration of beef cows should consist largely of the common farm or ranch roughages. Indeed it usually is possible to maintain dry cows satisfactorily during the winter on roughages alone in the farm regions, but in range areas, cured range may need supplementation with concen-

Table 47

Morrison's Feeding Standards for Breeding Beef Cattle[a]

Class of Cattle and Weight (lb)	Dry Matter (lb)	Digestible Protein (lb)	Total Digestible Nutrients (lb)	Requirements per Head Daily				
				Calcium (gm)	Calcium (lb)	Phosphorus (gm)	Phosphorus (lb)	Carotene (mg)
Pregnant cows								
900	13.1–18.4	0.65–0.70	6.9– 9.7	20	0.044	17	0.037	55
1000	14.2–20.2	0.70–0.80	7.5–10.5	20	0.044	17	0.037	55
1100	15.2–21.5	0.75–0.85	8.0–11.3	20	0.044	17	0.037	55
1200	16.3–22.8	0.80–0.90	8.6–12.0	20	0.044	17	0.037	55
Cows nursing calves								
900–1100	22.0–27.0	1.20–1.40	12.0–15.0	30	0.066	24	0.053	90
Growing cattle[b]								
700	14.2–16.5	0.87–0.98	8.9–10.2	17	0.037	15	0.033	40
800	15.9–18.3	0.90–1.00	9.5–10.9	16	0.035	15	0.033	45
900	17.3–19.7	0.93–1.03	10.1–11.5	16	0.035	15	0.033	50
Mature bulls[c]								
1400	17.2–19.0	1.19–1.31	11.0–12.2	14	0.031	14	0.031	84
1600	18.6–20.6	1.28–1.42	12.3–13.5	16	0.035	16	0.035	96
1800	20.4–22.6	1.40–1.54	13.5–14.9	18	0.041	18	0.041	108

[a] Taken by permission of The Morrison Publishing Company, Ithaca, N.Y., from *Feeds and Feeding*, 22nd edition, by Frank B. Morrison.
[b] Upper limits of range should be adequate for pregnant yearling or 2-year-old heifers.
[c] Assumed to be the same as for dairy bulls as shown in Morrison's tables.

trates or with hay or silage if it can be purchased in the locality. The roughage sources vary so widely between the farm and range regions as to warrant discussing them separately.

In general, the harvested roughages found in the farming regions are higher in nutrient content, especially in energy or TDN, than the level required for feeding dry cows most economically, if full-fed. Thus either limited feeding of these roughages must be practiced or the ration must be a mixture of the better-quality roughage and a lower-grade roughage. For instance, a ration of legume hay can be balanced with access to a corn stalk field to achieve the most economical balance.

The roughages available in the farm regions can usually be grouped as follows.

Legume Hays.   No better feed than clover or alfalfa hay can be recommended for breeding cows, but an exclusive ration of legume hay for animals in average condition supplies more nutrients than are actually needed. If enough legume hay is fed to furnish from 20 to 30 percent of the total air-dry feed requirement or about 5 to 6 pounds, the remainder being supplied by nonlegume roughages, the ration will be sufficiently well balanced for dry cows in ordinary condition. Soybean straw (obtained

Fig. 43. Legume-grass hays, stockpiled as big bales at the edge of pastures or near windbreaks, make excellent feed for dry pregnant cows. Adjustable self-feeding racks reduce wastage and labor requirements. (Farmhand.)

from threshed or combined beans) is high in crude fiber because many of the leaves are lost during combining. However, it is excellent for breeding cattle when fed rather liberally. Catching the straw and seed pods behind the combine with a commercially available forage wagon such as a "Foster wagon," which dumps its 1.5-ton loads wherever desired, will increase the amount utilized, per acre, by at least 50 percent.

Although it is true that good-quality legume hays, when fed as the sole ration, may supply more nutrients than needed, especially protein and carotene, it still may be the most economical feed to use if homegrown and if other roughages are unavailable. Mixed hay, such as smooth brome-alfalfa, composed of about one-half grass and one-half legume, is excellent winter feed for cows, meeting their nutrient requirements almost perfectly when self-fed or when fed ad lib.

Grass Hays.   Grass hay composed principally of brome, orchard, or tall fescue and similar tall-growing bunch grasses is the principal grass hay fed to beef cows in the farm region. Improved wheatgrasses, found on the mountain meadows of the Northwest and sometimes on irrigated meadows, are especially nutritious if not overly mature when cut. Marsh hay, grown in the North Central states and elsewhere, may be of good quality but varies greatly with stage of cutting and state of the weather during curing. Coastal Bermuda, found in the southern region, has a composition

Fig. 44. Prairie or meadow hay plus a protein supplement is a common feeding program for wintering stocker steers in the Nebraska Sandhills. (The Record Stockman.)

similar to most grass hays. "Prairie hay," cut from hay meadows in localized regions such as the Flint Hills of eastern Kansas, consists mainly of big and little bluestem and represents a very common grass hay used in this section when snow covers the cured range or when a prairie fire destroys the reserve pasture set aside for cows. It also is preferred by stockyards and show cattle fitters. The principal deficiency of native hays is protein, but carotene and phosphorus may also be deficient if the hay is unduly mature when cut. Grass hays are often cut when overly mature in order to get more tonnage, but actual pounds of digestible protein and energy may in fact be reduced because of increased lignification and consequent lowered digestibility.

Straw. Next to corn stover, the straw of small grain crops is the most abundant low-grade roughage material or crop residue feed in the country. The value of straw for feeding purposes is determined largely by the stage at which the grain is cut and the amount of damage done by rains before baling or stacking. Good bright oat straw, cut with a windrower when a little short of being fully ripe and stored in well-constructed stacks or bales after combining, has considerable feeding value, ranking only a little below grass hays in the amount of total digestible nutrients. This method of harvesting is not widely practiced except in the North Central states. Oat straw obtained after a direct-cutting combine is much less valuable than early-cut straw because it is cut when much riper and correspondingly higher in fiber. Moreover, most of the leaves and chaff, which are the best parts of the straw, are lost during combining and subsequent raking. Barley straw is somewhat less valuable as a feed than oat straw, and wheat and rye straws are lowest in value.

The main problem with straw is that cows simply will not consume enough to meet their nutritional requirements for energy. Canadian workers at the University of Alberta, in attempting to increase intake of barley straw, found that cows ate 23 percent more straw in the form of pellets than in loose form, and their intake also increased by about 14 percent when the crude protein level of the total ration was raised from 5.0 to 6.5 percent by the addition of supplemental protein. Chemical treatment of straw with 4 percent sodium hydroxide shows promise as a means of delignifying straw and similar roughages to improve digestibility and hence feeding value for cows. This is further discussed in Chapter 18.

Corn, Sorghum, and Soybean Residue Feeds. By far the greatest reservoir of under- or un-utilized cow feed in the United States is the residue left in the fields after harvesting the more than 100 million acres of corn, sorghum, and soybeans grown annually. This feed resource consists mainly of the stalks of these crops, but not to be overlooked are the leaves of all three crops, the husks and cobs of the corn, and the pods of the soybeans. In all instances, because of inefficiencies in the threshing process, some grains or seeds are intermingled with the material leaving the combine harvester. Since these three crops are nearly all harvested with combines during the months of September to November, there is ample time before winter sets in to graze the fields containing this valuable residue feed—or, if fall plowing is desirable, the residue from some or all of the acreage may be harvested and stored as silage or in piles, big-package bales, or stacks.

From the standpoint of designing the most practical programs for utilizing these feeds, no doubt the most acceptable plan is a combination of grazing some of the acreage early and storing some of the residue feed in harvested form for use later in winter. After-harvest fall weather in most of the farm regions is extremely variable from year to year, being subject to frequent rains, much damp and foggy weather, and even early snowfalls. This causes great variation in the feeding value of the residue feeds and makes the chances of storing them, in bale or stack form at least, very uncertain in some years. At the other extreme, storage as silage becomes problematical if the fall is an unusually dry one, because the residue is then too dry for silage-making unless much water is added.

From a nutritional standpoint, these residue feeds should be thought of mainly as an energy source and, because they are fairly high in crude fiber, they contain only enough net energy to serve as maintenance feeds for dry, pregnant cows unless considerable supplementation is practiced. Their chemical composition is comparable to that of most grass hays and thus they generally must be supplemented with protein, phosphorus, and calcium, and possibly vitamin A and trace minerals, when used to winter dry, pregnant cows. Fall-calving cows or weaner calves are not as well suited for utilizing these feeds. The discussion that follows gives more detail concerning how best to utilize these crops.

Grazing Corn and Sorghum Stalk Fields.   It is estimated that approximately 90 percent of the corn and sorghum crops are harvested for grain and the stalks left standing in the fields, with the remaining 10 percent being harvested as entire-plant silage. As mentioned earlier, stalks can furnish a considerable amount of pasture for cattle during the late fall and early winter months, particularly for dry cows because they are capable of consuming and utilizing large quantities of coarse roughage and their nutrient requirements are low. Ordinarily stalk fields may be relied upon as the major feed source of the breeding herd during October, November, and December. The Iowa data summarized in Table 48 show that the full potential of corn stalk fields is not nearly being realized on most Corn Belt farms. This experiment was conducted during a single winter that was

Table 48

	Performance of Heifer Calves and Dry Beef Cows on Corn Stalk Fields[a]	
Age	Heifer Calves (140 days)	Dry Cows (112 days)
Acres of stalk fields	24	24
Number of females	12	12
Fall weight (lb)	439(11/11)	1,041(11/30)
Spring weight (lb)	588(3/31)	1,139(3/22)
Winter gain (lb)	149	98
Supplemental feed per head (lb)		
Protein supplement	138	—
Oats	264	—
Hay	140	—
Salt and mineral	17	Free choice
Feed cost per head	$15.12	$0.90

[a] *Results of Cattle Feeding Experiments,* Iowa State University AS-183, 1966.

exceptionally open or free of snow, and cow numbers were small, but the fact that the 12 cows in the Iowa study produced 12 calves with only a salt-mineral-vitamin supplement lends further support to the fact that the energy and protein requirement of dry bred cows is low as long as adequate trace-mineralized salt, phosphorus, and vitamin A are provided.

To use stalk fields for wintering cows, spring rather than fall plowing must be practiced, and some cash-grain farmers prefer not to postpone plowing, but this is a choice they will have to make. As a compromise, a stalk field may be reserved near the barns for cow feed, and the remainder of the corn ground may be plowed in the fall.

Cowmen with the most experience in utilizing this type of forage have found that each acre of corn or sorghum stalk fields provides about 60 cow days of grazing, and best utilization of a field is obtained if only a portion is grazed at a time. By using an electric fencer, it is a simple matter to allow access to perhaps only one-fourth of the field at a time. This prevents the cows from selectively grazing the more nutritious portions of the crop from the entire field at once.

Sorghum stalk fields are well utilized by dry cows in late fall and early winter. A sorghum stalk field adjacent to wheat pasture being grazed in winter makes an almost ideal combination because cattle on high-moisture wheat pasture crave a dry roughage such as the sorghum stalks left after combining the sorghums for grain. Such a combination of feeds will provide adequate nutrition even for lactating, fall-calving cows and stocker calves. If cows are to be maintained solely on stalk fields for longer than 6 weeks, a protein supplement or some legume hay should be provided.

Commercially prepared protein blocks, containing enough salt (about 20 percent) to limit protein intake to the desired level, are quite popular but expensive. Liquid supplements provided in lick tanks are even more popular because of the convenience and dealer service aspects of their use. Such supplements are discussed in greater detail in Chapter 17.

Sorghum stalk fields tend to stay green after combining and, in fact, often may even produce tillers or suckers and immature seed heads before frost, especially in the Southwest. Although this does add extra feed, as compared to corn stalk fields, it also may be the source of a problem in their utilization. The sorghum regrowth may contain dangerous levels of hydrocyanic or prussic acid, especially from regrowth after short droughts or light frosts. Before turning in the main herd, it would be wise to test the field first for a few days with less valuable cattle. Feeding some supplemental roughages to dilute or reduce the level of intake of the sorghum would reduce the danger of toxicity. Such sorghums are safe if harvested as hay or stalklage.

Stalk Field Grazing Combined with Stack Utilization.   Ordinarily, when stalk fields are grazed, only about one-half of the residue feed is actually utilized unless cows are left on the fields too long—that is, until they are finally forced to consume the very poorest of the stalks simply to survive. Unfortunately, weight losses of the cows are then so great that dystocia may occur and rebreeding problems almost certainly will follow. Another management problem arises when grazing alone is practiced if the fields are snow-covered during much of the winter or even for periods of 2 to 4 weeks, as is more likely to happen in some locations. Wet fields also can cause problems in southerly portions of the farm region. Harvesting part of the stalk fields and storing the feeds for supplementary or emergency use would seem to offer a solution. Equipment for harvesting crop residues is still being developed, but some reasonably efficient, though expensive, models are already available.

Iowa workers conducted an experiment in wintering dry cows to compare several combinations of grazing either corn stalk fields plus stacks or dumps, soybean residue plus dumps, and grain sorghum residue plus dumps. Protein supplementation of the various residue feed combinations was included in the study, which is summarized in Table 49.

The corn stalk stacks were made immediately following harvest in November by flail-chopping the center two rows of the six rows covered in one pass through the field with a six-row combine. This procedure thus resulted in two rows of chopped stalks plus the husks and cobs and a bit of grain from all six rows. Perhaps 75 percent of the material on the ground that was exposed to the 8-foot-wide combination chopper-stacker (Fig. 45) was actually picked up and deposited in stacks at the end of the field. The

Table 49

Performance of Dry, Pregnant Cows on Crop Residue Grazing with Supplementation (98 days, 12 December 1973–20 March 1974)[a]

Field Treatment	Number of Animals	Acres per Head	Supplement Fed	Additional Feed	Average Initial Weight (lb)	Weight Change (lb)	Post-partum Weight (lb)	Supplement Consumed Daily (lb)	Calf Birth Weight (lb)
Corn stalk grazing	18	1.12	Liquid 32[b]	6.5 lb hay daily	992	−16[c]	850	3.7	71
Corn stalk grazing + stalk stacks	16	1.04	None	None	960	−34	823	0	70
Corn stalk grazing + stalk stacks + supplement	18	1.12	Liquid 32	None	917	−50	792	2.5	80
Corn stalk grazing + stalk stacks + supplement	16	1.07	4 lb ear corn on alternate days	None	972	−39	840	1.9	74
Corn stalk grazing + husklage dumps + supplement	14	1.08	Liquid 32	Stacked corn residue	939	−34	801	2.9	69
Soybean residue grazing + dumps + supplement	9	2.41	Liquid 32	None	984	+11	882	2.0	66
Grain sorghum stubble grazing + dumps + supplement	13	0.98	Liquid 32	None	952	+27	897	2.1	78
Mean of all treatments	—	—	—	—	947	−27	836	—	73

[a] Iowa Beef Cow-Calf Research Report, 1974.

[b] A 32 percent crude protein (nonprotein nitrogen equivalent) liquid (molasses) supplement fortified with calcium, phosphorus, and vitamins A, D, and E.

Fig. 45. Corn stalk stacks make suitable feed for dry cows if properly supplemented and if moisture level is low enough to prevent molding of the stalks. It is important to locate stacks on well-drained sites, out of the fields. (Farmhand.)

remaining four rows of unchopped stalks provided the residue grazing. Stack-feeder panels were used to self-feed the stacks one at a time, while the remaining stacks were protected by an electric fence.

The piles or dumps used in the experiment were formed by attaching a Foster dump-wagon behind a combine, catching the material that emerged from the combine during the harvest, and dumping it at the ends of the rows. The corn residue, commonly called husklage, consisted mainly of cobs and husks. Soybean hulls and stalks plus a few beans made up the soybean residue dumps, whereas the sorghum field dumps consisted of tailings from combine-threshing the sorghum fields.

The acreage per cow fed, as shown in Table 49, is a combination of chopped and unchopped area. Snow covered the fields in December, temporarily ending field grazing and forcing the cows to rely on the stacks or dumps where available. Cows without a source of such emergency feed were given added feed. Cows that grazed sorghum and soybean fields, along with the dumps, experienced actual weight gains while all cows on corn stalk fields lost weight, though not at unduly high levels. Wastage of husklage dumps was severe, necessitating supplemental feeding of hay and stalk stacks to cows on this treatment. All calf birth weights and cow postpartum weights were normal. Mineral supplements were provided free-choice to all groups, and intake levels were within the normal range. There were no evident differences in performance between cows on ad lib consumption of a liquid protein supplement and cows on 2 pounds of ear corn daily, although the ear corn would be more economical.

There were many variables in this study, and weather conditions prevailing in central Iowa are not typical of the entire area where corn, sorghum, and soybeans are grown. The results demonstrate, however, that the acreage of any of these crops required to winter a dry cow is surprisingly low. When a combination of grazing and harvesting is used, each acre provides 75 to 100 cow days of feed, almost twice as much as might be expected from grazing alone. Supplementing the residue feeds presents no insurmountable problems. Thus if all the residue feeds of corn, sorghum, and soybeans were used most efficiently—which is of course an unreasonable expectation—adequate roughage would be available from these crops alone to winter the entire U.S. beef cow population. Changing from conventional hay or silage programs to total reliance on residue feeds should, however, be approached cautiously and gradually.

Management of fields and stacks is an extremely critical factor in cow wintering programs that involve heavy use of crop residues. Experiment station research and practical experience by producers over several years' use of the newer methods of harvesting and feeding residue feeds has pinpointed some problems that may be encountered.

1. Corn stalk stacks and big bales may be much lower in feed value than estimated, owing either to weathering before harvest, a moisture level above 25 percent with resultant molding, or low nutrient content caused by drought, early frost, or poor harvesting technique. Ideal harvesting weather is seldom encountered, and labor and machinery conflicts usually delay residue harvest.

2. Supplementation with available energy, protein, phosphorus, and vitamins A and E is often required to support normal digestion and metabolism of the stalk portion of the ration. Deciding when or whether to begin supplementation is important; obviously it should be begun before cows are in an advanced stage of malnutrition.

3. Insufficient intake may occur because of extreme coarseness of stalks, moldiness, and weather damage, or even because of bad teeth in older cows.

4. Poorly located stacks or bales and inattention to management of feeder panels and electrical fencing devices can result in poor cow performance.

5. Choosing a too-early calving season may impose lactation and rebreeding stress at a time when the poorest feed is available. A heavy parasite load and excessive mud and cold increase the energy needs still further.

6. Failure to sort cattle into groups with widely different nutrient requirements results in poor performance of some cattle. Mature, dry, pregnant cows utilize residue feeds best, whereas weaner calves, bred heifers, and first-calf heifers need better-quality feed.

Corn and Sorghum Residue Silage. Ensiling the stover left after combining corn or sorghum would be the most efficient way to utilize these residue feeds, provided harvesting and storage costs were not prohibitive and storage losses were not unduly high. Unfortunately, there is still no commercial harvester available that will, in one pass through the field, harvest the entire plant and separate the grain from the stalk portion. Satisfactory stover silage, often called stalklage, can be made, but inputs in machinery costs and labor are high.

Because most grain growers who practice combine field-shelling prefer that the grain moisture content at harvest time will have been reduced to

Fig. 46. Corn stalks can be harvested with equipment especially designed to harvest the stalks and chop them finely enough for storage, as stalklage, in a single efficient operation. It may be necessary to add water to the chopped stalks. (John Deere.)

20 to 24 percent or lower in the field, stalk dry matter at harvest time is usually too dry to ensile without adding moisture. If silo-filling is done immediately behind the combine, 15 to 20 gallons of water must be added per ton of material as it comes from the field, to ensure good packing and fermentation. The final product should contain at least 50 to 60 percent moisture if a conventional tower silo is used. For small bunkers or trenches, an even higher moisture level is preferred. With oxygen-limiting silos, moisture levels as low as 40 percent are satisfactory. If corn grain is harvested either as high-moisture shelled corn or as ear corn (26 to 32 percent grain moisture), then little or no moisture need be added to the chopped stalks to make satisfactory stalklage, especially if silo-filling occurs at grain harvest time or shortly afterward.

Mixing fresh manure, scraped from a paved cattle feedlot floor, with an equal weight of chopped cornstalks direct from the field at grain harvest time resulted in satisfactory stover silage, in preliminary Illinois studies. Cows consumed it well and maintained weight without any supplemental feeding except minerals.

Other methods of preserving stovers—such as treatment with propionic acid, formaldehyde, or anhydrous ammonia—are under investigation and appear promising. The anhydrous ammonia method supplies the needed supplementary nitrogen. Any new harvesting and storage methods for preserving the forage must be sufficiently economical to compete with simple grazing of stalk fields, because cows can harvest at least half the crop themselves with little more investment than that required to fence them in and provide a water source. The trend toward minimum, or even zero, tillage in grain production tends to favor cow wintering programs that rely on crop residues because the need for fall plowing obviously is lessened or eliminated.

CORN AND FORAGE SORGHUM SILAGES

In the Corn Belt and wherever corn and forage sorghums can be successfully grown, corn and sorghum silages made from the entire plant are excellent, though not always economical, winter feeds for the breeding herd. Because they are moist, succulent, and contain considerable grain, they tend to stimulate a good flow of milk in fall-calving cows and in wet cows that calve in late winter or early spring. Their high yield per acre enables the farmer with a limited acreage to winter a considerably larger number of cattle from a given size of farm. Their physical nature makes for ease and economy of storage, especially if stored in larger bunker or trench silos, and they lend themselves to mechanical feeding methods.

Corn and sorghum silages fed alone are not well-balanced rations for cows because they are low in protein in comparison with their energy content. This can be remedied by using a legume roughage or a protein supplement along with the silage. The oilseed meals, clover hay, and alfalfa are all higher in protein, and any of them can be used to supply the nitrogenous material or protein that the animals require. The addition of 10 pounds of urea per ton of silage at silo-filling time will provide the required supplemental nitrogen at a very economical figure. Ammoniated molasses with added minerals is available in commercial form (Pro-Sil) for adding to corn or sorghum silages at silo-filling time to make them complete feeds for beef cows as well as other cattle.

Corn silage and, to a lesser extent, sorghum silage, because of their grain content, are usually considered higher-cost feeds than straw, stover, and other residue feeds, which have little cash sale value. Consequently these silages usually are fed to commercial cows in limited amounts, along with some less valuable, dry roughage that is fed according to appetite. Sometimes no silage is fed until the cows start to calve, when feeding is begun at the rate of 20 to 30 pounds per head daily. Purebred cows, which are best kept in fairly good flesh to enhance the sale value of their offspring, will stay in about the desired condition on 5 to 8 pounds of legume hay and a full feed of corn or sorghum silage, which for an average-sized cow is about 40 pounds. When fed no hay or other dry roughage, cows will eat about 5 pounds of silage per 100 pounds body weight. Allowing only 25 to 30 pounds of these silages per day to dry cows, by feeding twice this amount every other day, has been a satisfactory method of controlling intake to only what the cow needs, as shown in Illinois studies. The "boss cow" becomes a real problem if limited silage is fed daily, but the foregoing method solved this problem.

From the standpoint of the acreage required to produce the winter feed for beef cows, corn or sorghum silage can be justified as a beef cow feed. For example, eight cows consuming 40 pounds of silage daily for 4 months can be wintered on the forage produced on one acre if the corn silage yields 20 tons per acre, which is an average yield. Few if any other crops can do as well.

HAY CROP SILAGE OR HAYLAGE

Silage made from mixed grasses and legumes is even better than corn silage, nutritionally speaking, as a winter feed for dry cows. Its high protein, carotene, and mineral content makes it an ideal source of nutrients for cows during both gestation and lactation. It is sufficiently low in total digestible nutrients or digestible energy to permit it to be full-fed. West Virginia station tests of corn silage and legume grass silage,

Table 50

Comparison of Corn Silage and Legume-Grass Silage for Wintering Beef Cows[a]		
	Corn Silage	Legume-Grass Silage
Average initial weight (lb)	807	789
Average final weight (lb)	1,164	1,143
Average gain per head (lb)	357	354
First winter (yearling heifers)		
Silage	20	20
Alfalfa hay	6	6
Cracked corn	1	3
Soybean meal	0.75	—
Average next 4 winters (2- to 5-year-old cows)		
Silage	31.4[b]	27.7[b]
Legume-grass hay	6.5	6.5
Average winter gain (cows plus newborn calves) (lb)	94.2	79.8
Average birth weight of calves (lb)	66.2	67.3
Average weaning weight of calves (corrected to 180 days of age) (lb)	368.5	377.8

[a] West Virginia Agricultural Experiment Station, Mimeographed Report.

[b] Silages were fed in amounts that provided equal amounts of dry matter to both groups of cows.

summarized in Table 50, disclosed no significant difference between these feeds for pregnant and lactating beef cows when they were fed in quantities that furnished the same amount of dry matter. Making silage from grasses and legumes not needed for pasture during the early summer is an excellent way to extend the pasture season, because good-quality hay-crop silage is similar in physical nature and digestible nutrients to fresh pasture forage. Because of its high protein content, no protein concentrate or legume hay is required; consequently, additional roughage if fed may well be straw or stover. It also makes an excellent supplement for a fescue or Bermuda pasture deferred for winter use.

MISCELLANEOUS HARVESTED ROUGHAGES

A variety of roughages that may be important in localized areas but are not of widespread interest are also being utilized by beef cows.

Oat Hay and Oat-Vetch Hay. Oats, either seeded alone or with vetch, makes excellent hay for cows, especially if cut in the soft dough stage.

Oat Silage.   If the oat crop is cut for silage at the same stage as for hay, the silage is equally good as feed for cows. In fact, because the hazard of unfavorable hay-curing weather is eliminated and less oat grain is shattered, an acre of oats cut for silage will winter more cows than if cut for hay. Neither of these feeds requires supplementation for dry cows other than a salt-mineral mix. Access to some straw or a cured grass pasture will keep the cows on the silage ration more contented.

Sugarbeet-Top Silage.   A full feed of this silage plus 2 to 4 pounds of legume hay makes a good cow wintering ration but this type of feed is naturally confined to areas that grow sugarbeets commercially. An all beet-top ration is apt to be too laxative, and limestone should be fed to counteract the diarrhea-producing agent, oxalic acid, which this silage contains.

Cannery Silage.   In the vicinity of commercial canneries, pea vines and other vegetable residues can often be bought for ensiling purposes, or the silage can be bought directly from the company stack. Feeding some hay or even a pound or two of grain may be necessary if thin cows are fed this type of silage, because the silage is sometimes too wet or unpalatable, and

Fig. 47. Haylage and hay crop silage are excellent beef cow wintering feeds but may need to be diluted with lower-quality residue feeds or used by limited-feeding. (Badger.)

intake may not be adequate to meet all needs. Corn cannery waste is usually fed to other types of cattle, which utilize it better than do cows.

Turnips, Mangels, and Rutabagas. These root crops are more popular in Great Britain and northern Europe than in the United States. They usually contain only about 10 percent dry matter, and extremely large quantities must be consumed by cows if their energy requirements are to be met. In fact, because of the high water content of such feeds it is recommended that some form of dry roughage be fed in addition to the roots. If the dry roughage is nonleguminous, a half pound of high-protein concentrate should be added per day.

Prickly Pear. This plant of the cactus family is sometimes used as an emergency feed in the Southwest. The spines must first be removed by singeing, and a completely fortified supplement is required to balance the ration. It should always be considered as an emergency feed only. Nevertheless, thousands of cows have literally been saved from starvation by the prickly pear during extreme droughts such as those of 1953, 1957, 1964, and 1971.

Corn Cobs. Corn cobs have been used successfully as a source of energy for cows during the winter season. Because of low palatability and almost complete absence of protein, minerals, and vitamins, they must be fortified with complete supplements that furnish these essential nutrients. The cobs must be finely ground to make them more palatable and to facilitate thorough mixing with other feed materials. One part of ground ear corn, two parts of ground corn cobs, and one part of ground alfalfa hay fed free-choice through a self-feeder should maintain beef cows in satisfactory winter condition. If they increase noticeably in weight on this mixture, the amount of ground cobs should be increased. If they eat less than 2 pounds of the mixture per 100 pounds live weight, about 1 pound of 50 percent molasses feed per head should be added to the mixture daily. Less and less ear corn is being harvested, and so supplies of cobs are diminishing. Seed corn companies often sell cobs at low cost.

Cottonseed Hulls. This product of the cottonseed processing industry is comparable to oat straw, corn stover, or prairie hay in energy content but is virtually devoid of digestible protein. Hulls are palatable and intake is not a problem; in fact, hulls sometimes are added to less palatable feeds to improve intake. Cattle-finishing feedlots generally pay more for hulls than they are worth as cow feed.

Gin Trash. Although admittedly a poor-quality roughage, this by-product of the cotton ginning process may be used to help carry cows through a winter, especially in emergencies. If supplemented with a high-protein, high-energy supplement that is also fortified with minerals and vitamin A, a ration consisting largely of gin trash will support dry cows. From 1 to 2 pounds of fortified supplement will be needed for cows in average flesh. If extremely thin, cows require an additional 2 pounds of an energy concentrate such as molasses, sorghum, or corn. Feeding cows alfalfa hay one day and gin trash on the next will result in a balanced ration.

WINTER PASTURE

Fall-sown small grains, annual ryegrass, crimson clover, and fescue are examples of winter pastures that are utilized for wintering beef cows as well as other livestock. There is a limit to how far north such a program can be reliably used, but evidently a line drawn through Virginia, northern Kentucky, southern Illinois, southern Missouri, Oklahoma, and the Texas Panhandle marks approximately the northern boundary where winter pastures can reasonably be depended upon. Fall and winter rainfall and some warm weather are essential for adequate growth in such pastures, but too much rain is a severe handicap because of the problem of grazing the wet fields. Tall fescue, a perennial grass, is grown in all but the western end of this winter pasture belt because it forms a matted turf that helps support the weight of the cattle on the unfrozen, wet fields used for winter grazing.

Western Winter Pasture Area. In the western portion of the region outlined above, winter wheat and oats are the main crops used. The crop may or may not be intended for grain production in addition to whatever pasture it provides. The following conclusions from the Oklahoma station point out the value of and the problems involved in winter pasture utilization.

1. The pasture value of winter small grains is so high that livestock farmers may profitably use them entirely for pasture, without taking a grain crop. This permits grazing as late as May or early June.
2. The protein content of small-grain and annual ryegrass forage, when young, green, and succulent, is high—about 30 percent or more (dry-matter basis) as compared with around 42 percent in the usual high-protein supplement.
3. The carotene (provitamin A) content is exceedingly abundant. This is an important point, for winter rations in the Southwest are often

seriously lacking in carotene. The forage is also high in minerals. Fiber is low—about the same as in alfalfa leaf meal.

4. Grain yield is not seriously affected by grazing until plants reach the jointing stage of growth.

5. The forage yield is about tripled if grains are completely pastured out instead of taking off the cattle when continued grazing begins to affect grain yield. (March 1 is about the end of the grazing period if a grain crop is desired.)

6. Forage production of the different varieties of the same crop tested differed enough to make it worthwhile to choose a variety specifically for pasture. (An extension forage crop specialist may be consulted for this purpose.)

7. A good mixture for both early fall and late spring pasture would include either barley or rye, winter oats, and annual ryegrass.

8. On a low-phosphate soil, cows show a definite preference for pasture grown on plots where phosphate fertilizer has been applied.

9. It is more important to have plenty of succulent, rapidly growing forage available to animals than it is to worry about possible differences in the palatability of pasture crops.

10. The protein, vitamin, and mineral content of small-grain and annual ryegrass forage cut at the stage of most rapid growth is so high that it apparently would be a valuable supplement for other feeds if dehydrated.

Eastern and Southern Winter Pasture Area.   In the eastern and southern portion of the winter pasture belt, three major kinds of winter pasture are available: (1) permanent winter forages such as fescue, (2) self-reseeding annual forage crops, usually grown in connection with permanent summer pastures such as crimson clover in Bermuda pasture, and (3) annuals such as small grains or ryegrass planted on a prepared seedbed as in the winter wheat belt. An experiment concerned with evaluating various winter pastures was conducted by the Georgia station and is summarized in Table 51. On the basis of this and other tests conducted by the Georgia workers, the following conclusions may be made.

1. A seeding mixture of 2 bushels wheat, 25 pounds annual ryegrass, and 15 pounds crimson clover is a satisfactory pasture mixture for winter-calving beef cows in the Southeast.

2. A complete fertilizer, applied at the rate of 500 to 600 pounds per acre in a split application, is generally required for maximum production. Additional nitrogen, to the extent of 100 pounds per acre, in three topdressings is often a good investment.

Table 51

Performance of Beef Cows on Winter Pasture[a]

Treatment	Drylot	Fescue Pasture	Crimson Clover Pasture	Winter Temporary Pasture
Number of cows	10	10	20	20
Total acres grazed	—	6	28	20
Date on test	1/18	1/18	2/9	1/18
Date off test	5/9	5/9	5/9	5/9
Days on test	112	112	84	112
Average cow weight on (lb)	958	955	920	957
Average cow weight off (lb)	896	916	1,015	1,099
Average daily gain, cows (lb)[b]	0.21	0.48	1.92	2.02
Number of calves born	6	6	15	20
Average birth weight, calves (lb)	65.8	68.2	68.0	68.1
Average calf weight off test (lb)	151.5	173.6	166.0	255.3
Average daily gain, calves (lb)	1.00	1.39	1.77	1.93
Daily milk production, 10th week of lactation (lb)	8.40	11.40	13.00	15.90
Average daily ration (lb)				
Cane silage	52.3	29.4	33.2	13.2
Oat straw	5.7	—	—	—
Cottonseed meal	1.5	—	—	—
Mineral	0.30	0.26	0.34	0.23
Gain per acre (lb) (cow and calf gains)	—	214	180	422
Cost				
Pasture per acre ($)	—	25.38	12.05	28.69
Pasture per head ($)	—	15.23	16.87	28.69
Supplemental feed per head ($)[c]	23.85	8.63	7.31	3.86
Total winter feed cost per head ($)	23.85	23.86	24.18	32.55
Cost per head daily (cents)	21.3	21.3	28.8	29.1

[a] Georgia Agricultural Experiment Station, Mimeographed Series N.S. 241, 1965.

[b] Corrected for an average loss of 120 pounds due to calving.

[c] Feed prices per ton: cane silage $5.24; oat straw, $12; cottonseed meal, $88.

3. Fescue is a highly dependable permanent winter pasture for beef cows, but it is improved by the addition of crimson clover.

Although Bermuda grass is dormant during wintertime, it is being used as a year-round pasture in the Southeast and thus needs to be considered among the feeds for wintering beef cows in the nonrange area. The results of an Oklahoma test conducted at the Fort Reno station with spring-calving mature Hereford cows are shown in Table 52. Of interest is the fact that the cows in all three lots on improved Bermuda grass (Midland) produced heavier calves than were produced by cows on native pasture—which in this instance was an excellent stand of bluestem, buffalo, and grama grasses, usually considered to be superior cured forage for beef cows in wintertime, as well as good summer range. All cows received supplemental cottonseed meal and an injection of 1 million IU of vitamin A in early January. Half of the cows in each Bermuda grass lot received 36 grams of monosodium phosphate daily in the cottonseed meal as a supplemental source of phosphorus. The lot on native pasture received a 2:1 salt:bonemeal mixture free-choice. During the summer the cows on Bermuda grass pasture were rotated among four pastures at weekly intervals, and it should be noted that 200 pounds of nitrogen were applied per acre in three equal applications during the grazing season. The supplemental phosphorus proved beneficial in the lots fed less than 3 pounds of supplemental cottonseed meal, itself a good phosphorus source.

WINTERING FALL-CALVING COWS

Feeding fall-calving cows calls for special application of the knowledge concerning the interrelationship of nutrition and reproduction. Cows that calve in September and October, for instance, need to be rebred mainly during December and January. To promote satisfactory rebreeding by having the cows in a rising plane of nutrition at this time, as well as to ensure optimum milk flow, some kind or combination of feeds must be supplied that is considerably higher in nutrient content than that usually provided by pastures and stalk fields at this season.

Note that, in the NRC recommendations included in Tables 45 and 46, the energy and protein requirements for lactating cows are almost double those for dry cows. Relying on the same feeds for the fall-calving cows that are used for wintering the dry, spring-calving cows will unquestionably result in lighter calves and a serious reduction in size of calf crop the following year. Either silages made from corn or sorghums—which of course contain considerable grain—should be fed or, if hay or stalk fields or stacks are used, 4 to 5 pounds of grain should be added. If the

Table 52

Supplemental Winter Feeding of Spring-Calving Cows on Bermuda Grass Pasture (December 3 to April 15)[a]

Pasture[b]	Bermuda		Bermuda		Bermuda		Native
Level of Cottonseed Meal (lb)	0		2		3		2
Phosphorus Supplement	−	+	−	+	−	+	+
Number of Cows	8	8	8	8	8	8	16
Cow Data							
Initial weight (lb)	1,102	1,073	1,084	1,096	1,093	1,117	1,066
Weight change (lb)							
To postcalving	−137	−123	−129	−107	−99	−101	−101
To weaning	−6	+43	+35	+25	+89	+86	+39
Average 24-hour milk (lb)	12.0	12.3	15.1	13.0	13.8	13.5	13.7
Plasma phosphorus (mg/100 ml), 4 cows per treatment							
12/3/64	5.9	5.5	4.7	5.6	5.3	5.6	—
4/15/65	4.1	6.5	5.3	5.9	4.0	4.5	—
Calf Data							
Average birth weight (lb)	75.6	73.0	80.1	89.1	83.1	75.4	73.0
Average weaning weight (lb)[c]	471	481	483	490	500	478	444

[a] Oklahoma M.P. 79, USDA-ARS cooperating.

[b] Each group of 16 cows grazed 140 acres year-round.

[c] Weaning weights adjusted to 205-day steer equivalents.

roughage found most abundantly on the farm is low in quality, necessitating considerable supplementation, it becomes debatable whether a fall-calving program should be used. The good small grain pastures found in the South and Southwest and the annual downy brome (cheat) pastures on the West Coast are exceptional situations. In these instances, the pasture forage, if sufficiently abundant, certainly is adequate and suitable for fall-calving cows.

WINTER FEEDING IN RANGE REGIONS

Balancing the winter rations for beef cows in the range regions is quite different from that in the farming regions. The nutrient requirements of the cows do not differ if the calving season is similar, except possibly for some cows in areas at very high altitude, such as the North Park section in Colorado where cows are wintered without shelter and as much as one-third more energy may be required for maintenance. The main differences lie in the roughage sources available and in the time of year at which new spring growth begins.

The extensive nature of the cow-calf program in the range regions and the large size of ranches makes it possible or, more accurately, necessary to set aside a portion of the range for winter use, making a year's growth of grass and browse, in cured form, available for wintering dry cows and replacements. The variety of grasses, annual browse plants, and perennial weeds or forbs found throughout the range area seems almost unlimited, and the nature of the available cured feed can vary tremendously. The types of range found in various parts of the range area may be classified and described as follows.

Tallgrass Range.   Located largely in the Plains states from Oklahoma northward; consists of big and little bluestem grasses, Indian grass, switch grasses, and others, but few browse plants. Intermediate rainfall and variable snow result in considerable wintertime deterioration of the forage. The average carrying capacity for year-round grazing is 10 to 15 acres per cow or animal unit.

Shortgrass Range.   Bounded on the east by the tallgrass range, on the west by the Rocky Mountains, and extending from Texas and New Mexico to Canada. The grama grasses, especially blue and side oats, dominate in the southern portion, with the wheatgrasses being introduced into the northern part. The prevailing low winter moisture conditions preserve the quality of the forage quite well. Only limited browse is available. From 20 to 30 acres are required per animal unit for year-round grazing.

Semidesert and Sagebrush Ranges.   Intermingling types of range that dominate the southwestern area of the country, extending from New Mexico westward to California and northward into Wyoming, Utah, and Nevada. Black grama grass, numerous weeds, and browse species such as chamiza, winter fat, and Apache plume make this a good cattle-wintering area. But because of scant rainfall during the growing season, stocking rate is quite low if year-round grazing is necessary. In general, from 75 to 150 acres are required per cow annually for year-round grazing. Increased carrying capacity is possible when open-forest range (see below) is available for summer use.

Pacific Bunchgrass Range.   Located from Washington to California, largely in the coastal area but extending eastward 200 to 300 miles in places. Once covered with permanent grasses and browse, this area is now characterized by annual grasses such as wild oat grass and downy brome, and by filaree, burclover, and other forbs. Good fall and winter rainfall ensures abundant feed for this season, but summer forage is dry and mature and very low in quality. Acreage required per animal unit during the fall and winter growing season is 4 to 6 acres. Very little of the area is used for year-round grazing because of the extremely low nutrient content of the mature grass. Obviously, a fall-calving season is indicated for this unusual reversal of seasonal feed supply.

Open-Forest Range.   Scattered throughout the West, principally at high altitudes and suited only for summer range. Most of the permanent grasses found in the West also occur in open-forest range, as do sedge and various annual grasses. The season is short and the available forage must be shared by domestic livestock and wild game, principally elk and deer. Because most of this range is federally owned and controlled, stocking rates and management practices are made a part of lease arrangements.

It is apparent that winter feed, in the form of cured range, varies greatly in the range regions, not only with respect to quantity but also nutritive quality and palatability. In sections that consist largely of grasses, especially the tallgrass and shortgrass types, supplemental protein and phosphorus are always required, as is evident from a comparison of Table 44 with the requirement tables, Tables 45 and 46.

SUPPLEMENTATION OF FORAGE FOR DRY, PREGNANT COWS

The question of whether supplemental energy is required has been studied extensively by the Oklahoma station workers. The results of one longtime study are presented in Chapter 5. Another study, summarized in Table 53, shows the results of restricting the supplemental energy level of spring-calving cows considerably below the lowest level used in the first-mentioned Oklahoma study. One lot received supplemental energy well beyond what commercial cowmen customarily feed but at a level often fed by purebred breeders. The winter feeding program began each year in early November and ended about mid-April, by which time green feed was available and calves averaged about 1 month in age. Cottonseed cake or meal and ground milo were fed as a supplement to cured tallgrass

Table 53

Longtime Performance of Cows Wintered at Different Levels (Seven Calf Crops)[a]

Wintering Level	Low	Moderate	High	Very High	Very High–Moderate[b]
Number of heifers started	30	30	30	15	15
Gain first winter, as calves (lb)	−9	97	140	280	268
Fall weight as 8-year-olds (lb)	1,146	1,182	1,262	1,372	1,172
Spring weight as 8½-year-olds (lb)[c]	926	1,011	1,139	1,248	971
Winter weight change (lb)	−220	−171	−123	−124	−201
Percent of cows still in herd	83.4	86.7	80.0	73.3	80.0
Average calving date	3/14	3/9	3/2	3/5	3/4
Average calf birth weight (lb)	75	77	79	75	77
Percent calf crop weaned	86.3	88.3	85.9	83.7	85.7
Average pounds calf weaned per cow[d]	442	465	481	452	451
Average daily milk production (lb)	11.2	12.7	12.1	10.6	10.2

[a] Oklahoma Agricultural Experiment Station MP-76, 1965.
[b] Fed at the "very high" level during the first three winters, then at moderate rate.
[c] Weights of cows at 8½ years of age, weights taken about April 15, or about 6 weeks after calving.
[d] Corrected for sex and age differences of the calves.

range at rates that would accomplish the following winter weight change patterns (from fall weight up through calving):

Lot 1 (low). No gain the first winter as calves, with a loss of approximately 20 percent of fall weight during subsequent winters as bred females.

Lot 2 (moderate). Gain of 0.5 pound per head daily the first winter as calves, with a loss of 10 percent of fall weight during subsequent winters as bred females.

Lot 3 (high). Gain of 1 pound per head daily during the first winter as calves, then less than a 10 percent weight loss from fall weight during subsequent winters as bred females.

Lot 4 (very high). Self-fed a 50 percent concentrate mixture during the first winter as calves and during subsequent winters as bred females.

Lot 5 (very high–moderate). Treated the same as Lot 4 for the first three winters, and as Lot 2 in the following years.

The Lot 1 cows had to be fed wheat straw in drylot for about the first 2 months to ensure the desired weight loss. The summary table shows the performance of all lots for the first six calf crops, and it would appear that the level of energy fed did not affect calf birth weight, calving percentage, or calf weaning weight; thus each increment of increase in energy level fed unnecessarily increased the cost of the ration, at least of the last two levels. Other items of interest are the fact that the lower energy levels resulted in a later calving date and a lighter mature cow weight, and the very high level resulted in the loss of a greater number of cows from the test. Individual yearling weights of cows (not shown) were lower for the low-level cows for the first 5 years, indicating that maturity was delayed by the stress of lactation. The Oklahoma workers conclude their discussion of results as follows:

The data presented for the individual yearly calf crops indicate that rather than select a level of wintering for the lifetime of the cow, consideration should be given to the life cycle feeding approach in which higher levels are used during growth and development of the female followed by lower levels after the cow has reached maturity since the major influence of the various levels on cow productivity occurs during the first three calf crops.

This conclusion fortifies the recommendation made earlier for sorting the cows in such a way that younger cows can be fed additional supplements if it is believed advisable after observing the condition of the cows.

It should be mentioned that drought is a common but unpredictable occurrence in the range areas. It is generally conceded by experienced cowmen and range management specialists that it is more economical to meet this situation by adjusting cattle numbers by culling cows heavily in the fall or by carrying over some steer calves from the previous fall to partially stock the range and sell as yearlings than it is to try to offset the feed shortage by heavier supplemental feeding. Drought-damaged range will be further damaged by overgrazing unless stocking rate is reduced.

ENERGY VERSUS PROTEIN SUPPLEMENTATION

A range must be carefully evaluated to determine what type of supplemental feeding program, if any, is called for. Energy intake may be limited because of a shortage of forage, unpalatability of species in the pasture, low digestibility, poor range pasture layout, snow cover, and combinations of these factors. If it is unfeasible to adjust cattle numbers to meet the shortage of energy, a choice between several types of feed must be made, and economics or cost per unit of energy generally will determine the choice. In the range areas, concentrates such as ground milo or corn or a 12 to 20 percent protein commercial range cube may be the preferred energy supplement, but if irrigated alfalfa meadows are accessible, this feed may be chosen because of its additional value as a protein, carotene, and phosphorus source.

An experiment of several years' duration conducted in the semidesert type of range in New Mexico is summarized in Table 54. This particular range provided considerable winter and early spring grazing during some years in the form of browse, forbs, and grasses. The supplements were fed at the rate of 1 pound daily before calving and 2 pounds daily from the

Table 54

Response of Range Cows to Supplemental Feed[a]		
Supplement Fed	Ground Milo	Cottonseed Meal Pellets
All years		
Weight loss of cow in calving season (lb)	46	24
Weaning weight of calves (lb)	387	389
Production per cow (lb)	321	311
Average-rainfall years		
Weaning weight of calves (lb)	424	430
Calves born following year (%)	91.7	90.6
Drought years		
Weaning weight of calves (lb)	366	366
Calves born following year (%)	87.4	84.1

[a] New Mexico Cattle Growers' Short Course, 1964.

beginning of the calving season until June or July when summer rains started the growth of the perennial grasses. In this experiment energy was apparently the limiting nutrient. It is interesting to note how limited energy, and probably multiple protein-mineral-vitamin deficiency, during a drought year can affect calf crop percentage the following year.

UREA IN WINTERING SUPPLEMENTS

Another chapter discusses how urea can be successfully substituted for a considerable portion of the supplemental plant proteins in cattle-finishing rations. While research on the subject of urea supplements for cattle being wintered on cured range forage is more limited, the results generally indicate poorer utilization of this economical nitrogen source. This is not surprising, since it is known that readily available sources of energy and minerals must be present in the rumen in order for the microflora to utilize urea nitrogen. Apparently the poor utilization of urea in this situation also affects the utilization of the forage itself. The resulting combined protein deficiency and poor forage utilization is responsible for such results as are shown in Table 55, which is compiled from an Oklahoma experiment on this subject.

The Oklahoma workers started with a set of weaner heifer calves and carried them through four successive winters and until they had weaned two calf crops. The similarity in performance between the lots on the 20 percent natural protein and the 40 percent protein-equivalent urea-containing supplement indicates little of the urea was utilized. The rations consumed by both these lots were too low in available protein, as evidenced by the fact that the heifers on the 2 pounds of 40 percent cottonseed meal outperformed the other two lots.

Most commercial range supplements today contain some urea in order to reduce the price per ton. When such supplements are adequately fortified with phosphorus, vitamin A, trace minerals, and a source of readily available energy such as molasses, they can be expected to perform satisfactorily if not more than 50 percent of the natural protein is replaced by nonprotein nitrogen. Such supplements may usually be obtained in block, cube, pellet, or liquid form, and some are prepared with sufficient salt or other intake depressants to permit self-feeding. Dehydrated alfalfa meal added to urea-containing range supplements at a rate that assures the daily consumption of at least 0.5 pound of the alfalfa meal, generally provides the necessary trace minerals, carotene (vitamin A precursor) and other as yet unidentified factors that apparently enhance urea utilization.

Table 55

Effect of Level of Protein and Urea on Performance of Breeding Females Wintered on Tallgrass Cured Range[a]

| | | Supplements[b] | | 40% Protein[c] with Urea |
| | | 20% Protein | 40% Protein | |
Ingredients		Cottonseed Meal, Corn	Cottonseed Meal	Cottonseed Meal, Corn, Urea
Daily supplement fed (lb)		2	2	2
Winter weight				
change (lb)	Days fed			
Calves	162	− 9	6	− 13
Yearlings	171	− 15	− 8	2
2-year-olds	196	−141	−214	−146
3-year-olds	212	−311	−255	−321
Calf weaning weight (lb)				
2-year-old heifers		415	436	426
3-year-old heifers		347	405	381

[a] Oklahoma Agricultural Experiment Station MP-51, 1958.

[b] All supplements were equalized with respect to calcium and phosphorus.

[c] One-half of the protein equivalent was supplied by urea.

LIQUID SUPPLEMENTS FOR WINTERING COWS

The use of liquid supplements has expanded several-fold during the 1970s, the main reason probably being convenience. Most suppliers either sell or provide a lick tank and keep it filled from bulk tank trucks that call on each farm or ranch at regular intervals. Cowmen with off-farm employment find the convenience factor to be highly important.

The ingredient composition of the supplements varies from one firm to another, but most of the formulas include molasses, some kind of nonprotein nitrogen source (usually urea), a phosphorus source (usually phosphoric acid), other minerals, vitamins A, D, and E, and an ingredient that makes the liquid feed sufficiently unpalatable so that it may be fed free-choice without the cows' overeating. Some liquid supplements contain as much as 50 percent natural proteins. Prices, after allowing credit for the laborsaving possibilities, are usually competitive, on a per unit of nitrogen basis, with dry mixed supplements.

Experimental results from many research stations around the country, combined with reports of farm and ranch experiences, would seem to justify the following statements regarding problem areas concerning use of liquid feeds.

1. Urea and other nonprotein nitrogen sources in liquid feeds are not quite as well utilized as are natural proteins in dry form when used to supplement the low-quality roughages that customarily provide the energy for wintering beef cows.
2. Toxicity may pose a problem, especially if tanks run dry for several days before being refilled, or if the roughage source is depleted.
3. Some ration components tend to settle out, or even break down, in the lick tanks or in the bulk storage tanks.
4. Intake per cow may vary considerably, but overeating is more common than underconsumption.
5. Unless lick tanks are moved about over the range or pasture, serious overgrazing may occur in spots near the tanks.
6. Failure to observe the cattle daily, in the absence of the need to feed supplement daily, encourages slip-ups in management. This point applies to self-feeding programs of both liquid and dry supplements.

Manufacturers of liquid supplements themselves promote active research programs, and some of the problems just discussed no doubt will be alleviated in time. The advantages of this system undoubtedly ensure its continued use by certain cowmen.

VITAMIN A IN WINTERING SUPPLEMENTS

The importance of vitamin A or its precursor, carotene, for growth and reproduction is discussed in Chapter 7. With reference to the winter feeding program for cows, the type and amount of available forages during the fall-winter period will have much to do with whether supplemental vitamin A is required. Fortunately cows and heifers can accumulate considerable vitamin A reserves in the liver, but if no green feed is available from frost in the fall until June or July or until after calving and the onset of the breeding season, the resulting vitamin A deficiency is likely to cause retained placentae, calf losses at or after calving, and poor conception rate for the succeeding calf crop. In certain range areas drought occurs frequently enough to warrant the use of vitamin A–fortified supplements. A daily intake of 50,000 IU of vitamin A should meet the needs of all cows. Field research, although leaving something to be desired in control, has shown that a single injection of an aqueous carrier containing 1 million units of vitamin A, either intramuscularly or intrarumenally, about 60 to 30 days before the calving season, is equally effective in preventing the symptoms of vitamin A deficiency.

Table 56 gives the results of an Oklahoma study that compares favorably with field trial results. Ranges that provide some forbs or annual grass forage, because of intermittent spring showers, will provide adequate carotene under most circumstances. Stabilized synthetic vitamin A is very inexpensive and should be included in virtually all protein supplements, which must be processed for the addition of other ingredients anyway. The question whether to feed or to inject vitamin A appears to be a matter of relative cost and availability of labor. Heifers that are to be bred after wintering should be treated the same as bred cows with respect to supplemental vitamin A. If it is already evident in the fall that the winter feed for heifers and cows will be low in carotene—in stalk field programs, for instance—vitamin A injections should be given at calf weaning time, since the cows no doubt will be in the chute anyway, for pregnancy-testing. Vitamins D and E usually are included in commercial vitamin A supplements. In isolated instances they may prove beneficial, especially the vitamin E, and the very small added cost is justifiable as insurance.

FREQUENCY OF FEEDING PROTEIN SUPPLEMENT TO RANGE COWS

The experience of some ranchers and results of recent research show that beef cattle being wintered on cured range, or in drylot, for that matter, do not require their supplement daily. Table 57 summarizes an experiment

Table 56

Effect of Vitamin A Injection of Cows before Calving[a]		
Treatment	No Vitamin A	Vitamin A[b]
Number of cows	77	78
Average weight upon injection (lb)	866	863
Average weight of cows, 5/7/63 (lb)	792	792
Weight change (lb)	−74	−71
Number of calves born	73	75
Number of calves at 112 days	66[c]	71
Weight of calves at 112 days (lb)	209	223
Date cows calved following year	3/10	3/11
Number of cows open following year	15	9

[a] Oklahoma Agricultural Experiment Station MP-76, 1965.

[b] One million IU vitamin A injected into rumen 1/31/63.

[c] Three calves were lost in this group to predators.

Table 57

Effect of Interval of Feeding Cottonseed Cake to Beef Cows[a]			
Frequency of Feeding	Daily	Three Times per Week	Two Times per Week
Winter gain (lb)			
1958–1959	91	88	38
1959–1960	−60	−62	−37
1960–1961	−80	−121	−98
1961–1962	−167	−176	−170
Percent calf crop weaned	81	89	86
Calf weight per cow (steer equivalent) (lb)	361	399	374

[a] Texas Agricultural Experiment Station Bulletin 1025, 1964.

with cows being wintered on shortgrass range in the Trans-Pecos area of West Texas. It made no difference whether the weekly allowance of supplement was fed at a given rate each day, three times weekly, or twice weekly. If anything, there may have been an advantage when some days were skipped. It has been observed that if cows do not expect to be fed daily, they will range out farther from the feedground or water source and thus will consume more and better-quality forage. Stocking rate in large pastures may be increased by this management practice because of the more uniform grazing pattern. Moving the feeding sites to various locations in large pastures can of course accomplish somewhat the same purpose if the terrain permits transporting the supplement throughout the pasture area.

A more recent study by Oklahoma workers with a 30 percent protein supplement in which half the protein equivalent was supplied by urea reported the same performance from feeding three times weekly as compared with self-feeding the same supplement. No toxicity problems were encountered. They observed that the self-fed cows spent 32 minutes each day at the feeder, whereas the cows fed three times weekly used only 62 minutes of the entire week to eat their supplements, an indication that the latter cows spent more time foraging for grass. Both groups of cows actually performed poorly, giving evidence of the poor utilization of the urea nitrogen.

When alfalfa hay is used as a winter protein supplement, the quantity fed must naturally be higher because of the lower protein content of the hay. Nebraska workers fed heifer calves an allowance of 4 pounds of hay per day or an equivalent amount at twice-weekly or weekly intervals, with satisfactory results, as shown in Table 58. The summer compensatory gain of the heifers fed daily is evident.

A disadvantage of not feeding daily is that the herd will not be observed

Table 58

Effect of Interval of Feeding Alfalfa Hay to Stocker Heifer Calves[a]			
Feeding Interval	Daily	Twice Weekly	Weekly
Winter gain (lb)	59	62	76
Summer gain (lb)	272	248	250
Total gain (lb)	331	310	326

[a] Nebraska Feeders' Day Progress Report, 1960.

daily. If fences are in good condition and the water source does not need frequent checking, failure to see the herd daily is perhaps not serious, at least until beginning of the calving season. On very large spreads, much less labor is required if only the herds in one or two pastures need to be fed and checked daily.

SELF-FEEDING DRY PROTEIN SUPPLEMENTS

Another recent innovation that makes possible a saving in labor is the addition of sufficient salt to dry protein supplements to permit self-feeding. From 25 to 35 percent salt content in supplements fed as meal, cubes, pellets, or blocks, will usually keep supplement consumption constant at about 2 pounds per head daily. Several factors affect the degree of success of this feeding method, including amount of forage available, location of feedground in relation to water supply, hardness of blocks if that is the form used, number of "boss" cattle in the herd, and added cost of preparation of the supplements. Salt-containing blocks seem to be gaining in favor in the area where large pastures are common. Regardless of the form of the high-salt supplements, plenty of unfrozen water must be available or salt toxicity will be a problem. Apparently the additional water is required for elimination of the excess sodium chloride by way of the urine and for prevention of dehydration. Ranchers in areas where urinary calculi are common have reported that the excess salt consumption and resultant flushing of the kidneys reduces the incidence of this aggravating problem.

CREEP-FEEDING FALL AND EARLY-WINTER CALVES

The question whether creep-feeding is a profitable practice has been studied extensively. Most studies have been with spring-calving herds, in which case the pasture being utilized by both cows and calves would normally be ample in supply and good to excellent in quality. However, as has been mentioned, the roughage supply for cows in winter, especially in the range areas, is likely to be of lower quality and possibly in short supply. If cows calve in fall or winter, the added nutritional requirements for lactation, especially for the younger cows, are not likely to be met, resulting in reduced milk production and poor calf performance. Creep-feeding the calves to make up for the reduced milk supply provided by the mother cow would seem a practical solution to the problem.

The results of a 2-year experiment conducted by Oklahoma workers are summarized in Table 59, which shows that creep-feeding the calves is a better solution to the problem than extra feeding of the cow herself. Calves were dropped in late fall, beginning in late September. The cows

Table 59

Value of Creep-Feeding Calves versus Supplemental Cow Feed for Young Fall-Calving Cows[a]

Lot Number	1	2	3
Cow Feeding Level	Low	Low	High
Creep-Feeding	Yes	No	No
Number of cows raising calves	27	25	29
Average weight per cow (lb)			
Initial (fall)	1,034	1,012	990
Winter change (203 days)	−283	−299	−194
Change to weaning	−42	−59	−6
Yearly change	24	32	60
Average weight per calf (lb)[b]			
Birth	73	72	73
Spring	222	160	207
Weaning	413	320	375
Average birth date of calves	10/21	10/26	11/3
Supplemental feed per head (lb)			
Cow			
Cottonseed meal	360	360	456
Milo	—	—	676
Calf (creep-fed)	944	—	—
Total feed cost per head ($)			
Cow	36.30	36.30	53.26
Calf	24.38	—	—
Total	60.68	36.30	53.26
Selling value per calf minus feed cost ($)	62.28	55.22	57.03

[a] Oklahoma Agricultural Experiment Station Bulletin, B-610, 1963.
[b] Corrected for sex and age.

were wintered on tallgrass range with one of two levels of supplementation. The creep-feeding, when used, extended from mid-January until weaning time in July. The creep ration consisted of 55 percent ground milo, 30 percent whole oats, 10 percent cottonseed meal, and 5 percent molasses. The results show that both creep-feeding the calves and feeding the cows additional feed were profitable practices, but feeding the calf itself was the more profitable of the two. It should be noted that the data are a summary of the first two calf crops of cows that calved first as 2-year-olds. If only the data for the first calf crop had been shown, creep-feeding would have appeared even more favorable. By the same token, benefits from creep-feeding the calves of mature cows undoubtedly would be less.

The Oklahoma workers concluded that creep-feeding the calves from first-calf heifers calving in the fall should be profitable if the selling price from the resulting calves is not reduced by their extra weight and condition. This means that, in practice at the farm or ranch level, first-calf heifers would best be pastured or fed separately from the rest of the herd to take advantage of the potential profits that may accrue from creep-feeding.

The separate handling of first-calf heifers makes possible two other recommended management practices. To ensure quick rebreeding and high conception rate of cows nursing their first calves, added supplemental feed or the best pasture on the farm or ranch is recommended. Then too, because it is often desirable to breed these young cows to younger unrelated sires, pasturing these cows separately means they are more unlikely to be exposed to their own sires. Open heifers being bred for the first time may be handled together with these young cows.

WINTER MANAGEMENT OF BULLS

Winter is the proper time to condition bulls for the spring and summer breeding season of the following year. Bulls that have been running out on pasture with the breeding herd during summertime are likely to be thin by fall, especially if they are young, and should have sufficient grain during the winter to put them in proper flesh before the arrival of spring. Young bulls should be fed more liberally than old mature ones because their growth requirements must be met before any improvement in condition can take place. A grain mixture composed largely of feeding stuffs that are growth-promoting rather than fattening is found best for bulls. Crushed oats make a good basal ingredient, to which corn, grain sorghum, or barley may be added only if a marked improvement in flesh is desired. Bulls that are in a run-down condition or that show signs of lack of thrift should have 2 to 4 pounds of protein concentrate per day, depending on age and weight. Mature bulls already carrying sufficient flesh may be wintered largely on choice roughages such as legume or mixed hay.

Bulls can be kept strong and vigorous during the winter only if they are given sufficient exercise as well as plenty of feed. To accomplish this end they should have the run of good-sized lots or, better yet, be turned with a few pregnant cows into an adjacent pasture or stalk field whenever the weather permits. When two or more bulls are kept, it is highly desirable to run them together because they will take more exercise when together than when they are alone. Although the turning together of old bulls that are strangers to each other is attended with some risk, a young bull may be put with an older one with comparative safety.

The importance of having bulls in strong condition (that does not mean fat, however) at the opening of the breeding season can scarcely be overemphasized. Valuable time is often lost by the use of a bull with impaired sex drive (libido) due to being either too fat or too thin. Examination of several semen samples for numbers and normalcy of sperm cells usually determines the fertility of a bull, and often a small or late calf crop can be prevented by such tests.

Western ranchers realize, perhaps better than other cattlemen, the importance of conditioning their bulls before the breeding season. They are well repaid for their pains, as shown more than 40 years ago by the data gathered on 15 ranches in northern Texas over a 3-year period.

On 10 of the 15 ranches, the bulls were taken from the cow herds in the fall for conditioning and were returned about June 1 of the following year. On the other 5 ranches the bulls were kept with the cows during the entire year. The ranches on which the bulls were removed from the cow herds for conditioning had a 77 percent average calf crop for the 3 years. On ranches where the bulls were not removed the average calf crop was 64 percent.

Whereas the better feed conditions prevalent in the central states make conditioning of bulls less urgent than on the range or in the coastal regions, more attention to this point on the part of all cattlemen will result in a material decrease in the number of cows that fail to raise calves through no fault of their own.

WINTERING BREEDING HEIFER CALVES

Rations for the heifer calves that are to be kept for herd replacements should be composed principally of the better-quality roughages available on the farm or ranch and details are discussed in Chapter 11. Actual nutrient requirements of heifer calves are listed in Tables 73, 74, and 75 in Chapter 11.

BUILDINGS AND EQUIPMENT

One of the advantages claimed for the beef cow by her ardent supporters is that she requires a small outlay for shelter and equipment. Expensive barns are by no means necessary. In fact, in some instances they may be a positive disadvantage. An investment of much more than $100 per cow in buildings and equipment is an unrealistic financial outlay for a herd of beef cows kept solely to raise calves for the open market. The situation is of course quite different with respect to high-priced registered cattle. Even here the elaborate barns and lots or corrals so often seen in large purebred establishments are not indispensable, and sometimes large barns that were built at great cost have been discarded in favor of inexpensive open sheds in which the cattle seem to do better.

Dry cows prefer to live entirely in the open if the ground is well drained, affords a reasonably dry area where they may lie down, and is protected from high winds by hills or trees. Such a place is provided by rolling pastures that have some sheltered valleys protected by surrounding hills and thick stands of timber, where the cows may spend the entire winter in apparent comfort, even in near-zero weather.

Fig. 48. Dry pregnant beef cows can derive much of their early winter feed requirement, especially energy, from corn stalk fields. More snow than the light covering seen here would necessitate supplemental feeds, protein in particular. (Animal Science Department, Iowa State University.)

A deep shed that opens to the south or southeast offers ideal shelter for dry cows when protection is needed in areas affected by wet, cold climates. Here the cows may run together in groups of 10 to 50 head or more and can be fed with a minimum of labor. Cows that are approaching parturition and those with very young calves should be handled apart from the rest of the herd if shed space is limited. However, after the calves are 10 days old, both cow and calf may be returned to the main herd. In a small herd the calves may be kept apart from the cows and allowed to nurse twice daily if a trap or pen is provided at one end of the shed where a deep, dry bed is available for the calves.

The lots occupied by cattle need not be large. If possible it is best to have two lots—a small paved lot adjoining the shed with about 50 or 60 square feet of space per cow and a much larger lot or trap into which the cattle may be turned for exercise when the ground is dry or frozen. If pasture land or stalk fields are available, the larger lot is unnecessary. The small paved lot, however, is almost indispensable in the nonrange areas during the spring when the ground is soft and muddy. Pavement is much more desirable in the lot just outside the shed than under the shed itself. This combination keeps the cattle more comfortable, reduces the labor of feeding, and repays its cost many times over in the extra amount of manure saved.

CHAPTER 10
BREEDING HERD
MANAGEMENT
IN SUMMER

In theory the feeding and management of the breeding herd during the summer is relatively simple, because the herd is then running on pasture or range and requires little attention. In practice, however, some of the cowman's most difficult problems may arise during this season of the year.

First in importance is the difficulty of estimating the forage supply, which is determined largely by weather conditions and can neither be definitely predicted nor controlled. Should the weather be unfavorable for the growth of grass, there may be a critical shortage of feed long before the usual end of the grazing season.

Second, the breeding season for most beef herds comes during the summer, making it necessary to inspect the herd regularly to determine whether the cows are being settled. Noting the dates on which a few familiar cows are observed to be in heat and then following up with an inspection 19 to 20 days later soon reveals whether the cows are being bred. Farmers and ranchers are invariably busy with fieldwork at this time of year, however, and are prone to limit inspections to once or twice weekly, especially if the herd is pastured some distance from the farmstead or ranch headquarters.

Third, early summer is the best time to castrate bull calves that were born on pasture or were too young to castrate while the herd was still in drylot. If this job is put off until the urgent fieldwork is finished, the calves are big and hard to handle when it is finally done. The same is true of such jobs as dehorning, vaccinating, and branding, which may not be done before cows go to pasture if calves were dropped late.

A fourth summertime problem is the need to attend to such parasites as hornflies, faceflies, and deerflies. In addition, in the range areas water shortage can be as serious a problem as grass shortage, and this also goes for farm areas that rely on farm ponds, springs, or creeks.

PASTURE

Pasture is relied upon almost entirely as feed for the breeding herd during the summer. On most farms and ranches no feed other than

Fig. 49. An almost ideal pasture for a cow herd with spring-dropped calves. Addition of legumes to the predominantly grass pasture would improve both quality and quantity of forage. (American Angus Association.)

pasture is provided. As discussed elsewhere, pasture is considered to be so essential for the breeding herd that beef cows are not usually kept by operators who lack sufficient grass or other forage to carry them through the summer. Although it seems desirable to keep the breeding herd on pasture during the summer for as long as forage is plentiful enough to provide milk and nutritious feed for the calf, some successful operations are conducted completely in drylot. The year-round drylot system offers an opportunity for maximum use of equipment and labor and for actually harvesting more forage, either as hay, green chop, or silage, from high-yielding cultivated pasture seedings. The subject of drylot- or confinement-rearing of calves is treated in greater detail later in this chapter.

KINDS OF PASTURE

Pasture areas usually are occupied by perennial plants—that is, plants that do not die after ripening their seed but are dormant through the winter while continuing to grow year after year during the summer growing season. Kentucky bluegrass, white clover, fescue, bromegrass, Bermuda grass, and nearly all of the native grasses of the Western Plains region such as the bluestems, the gramas, wheatgrasses, and buffalo, are examples of only a few of the perennials available, and they together occupy by far the greater percentage of the total pasture area of the country.

Annuals—that is, plants that produce seed and die in one year—provide much forage for beef cattle and, when used to supplement perennial forages, they can increase both the performance of beef cattle and the carrying capacity of a farm or ranch. Such seeded annuals as straight Sudan, forage sorghum–Sudan hybrids, ryegrass, and small grains, when planted on productive soils, can carry as many as 3 to 5 cows per acre for the growing season, especially when fertilized and irrigated. Other annuals that reseed themselves, such as wild oats, lespedeza, filaree, and cheat, also are valuable pasture forages, but these usually are found in permanent pastures containing perennial species as well. The fact that these annuals grow most actively during seasons when the perennial grasses may be dormant makes them especially valuable.

Pastures can further be divided into temporary and permanent pastures. Temporary pastures may consist of only cultivated perennials, only annuals, or combinations of both. On the other hand, permanent pastures are more likely to consist of native perennials, and those in the nonhumid areas of the country are more apt to be spoken of as range than pasture.

ROTATION AND TEMPORARY PASTURE

The temporary pastures that consist of cultivated perennials such as alfalfa, clovers, and bromegrass as opposed to native perennials, are also referred to as rotation pastures and are found mainly in the Corn Belt where they are grown in a planned sequence with the cash crops of corn, wheat, and soybeans. They are usually preceded in the rotation by oats or wheat, used as a nurse crop. The acreages of rotation pastures and oats are decreasing because it has become more economical to purchase fertilizers in the Midwest, especially nitrogen, than to depend on legumes, with their atmospheric nitrogen-fixing bacteria, to serve this function. Recent advances in the price of fossil fuels, required in the manufacture of nitrogen fertilizers, suggest that the role of legumes in cropping systems will again become important. Continuous cropping with corn is now an accepted practice in the more fertile areas of the country and no doubt will continue to be until fertilizer costs become prohibitive. The fact that the production from the rotation pastures and the oat crop returns considerably less cash income, if sold, than corn or soybeans, has also had much to do with the shifts in acreage mentioned. The claim that rotation pastures are required to maintain yields of cash crops over the long range is being disputed, and only time will tell whether the soil tilth, water-holding capacity, and micro- and macro-mineral content can be maintained without them.

Rotation pastures are usually allowed to stand for 1 to 3 years as part of the general crop rotation when used. These rotation seedings may be harvested for hay, haylage, or grass silage the first year and pastured for a year or two thereafter before they are plowed under for corn. Or they may be left standing for only one year, in which case the first cutting is sometimes cut for hay or grass silage and the crop is pastured for the rest of the season. Or they may be grazed by beef cattle or other stock from early spring until late fall. Obviously the method of utilizing a particular field is determined largely by the need for additional pasture at the time it is ready for use.

The plant species most widely used in seeding rotation pastures consist largely of legumes, because the main reason for including temporary pasture in the crop rotation is to improve soil fertility. Consequently, alfalfa and red, alsike, and sweet clovers, either seeded alone or mixed with several grasses, are commonly used. A mixture of red clover, sweet clover, and timothy is a satisfactory seeding for a pasture that is to be used for only one year, but alfalfa usually should be included in seedings that are to be allowed to stand 2 years or longer.

It is apparent that rotation forage crops have a real place in cattle raising. However, it would appear that their greatest use is to supplement rather than displace permanent pastures. The danger of not securing a stand, and thus being without any forage, is too great for the farmer to rely exclusively on rotation pastures for maintaining a breeding herd.

The best plan in all but the range areas is to depend on neither permanent nor rotation pasture exclusively for maintaining the breeding herd, but to provide enough of each to ensure a continuous supply of nutritious, palatable forage throughout the summer. Some permanent pasture is especially advisable for use in early spring and late fall when rotation pastures often cannot be used without damage to the stand because of wet weather. The permanent pastures such as bluegrass and bromegrass, which form a thick, dense sod, are damaged much less seriously both by extremely wet weather and occasional periods of severe drought, when all pastures become little more than exercise lots.

Permanent pastures outside the range areas are usually ready for grazing earlier in the spring than rotation pastures, not only because they have a denser sod but because they are lower in moisture and thus less laxative. They are most productive during cool, moist weather and may be grazed heavily from the last of April to the middle of June. They are relatively dormant during July and August, but make considerable growth in the fall with the return of cool weather. The rotation pastures are especially useful in filling the "summer slump" or dormant period of the cool-season grasses usually found in the permanent pastures of the farm regions.

Most rotation pastures that are 1 or more years old are at their best in June, July, and August. Often they are so productive during the early summer that only a portion of the area is needed for grazing, making it possible to harvest the remainder of the field for winter reserve feed. The second crop comes on quickly, and this fresh growth combines with the more mature forage of the grazed area to make an ideal pasture for beef cattle.

If rotation pastures are spring-seeded—as in the case of Sudan grass and the forage sorghum–Sudan hybrids, which are seeded alone, or sweet clover, Korean lespedeza, or alfalfa, which are seeded in small grain—they are usually ready to be grazed by late July or mid-August and furnish excellent grazing until the first killing frost.

Cattle should be removed from all legume pastures immediately after a killing frost, because frozen legume forage may cause serious bloat and scouring. However, such pastures may be safely grazed 2 or 3 weeks later after the stems and leaves are dead and brown. Legume pasture is much less palatable and nutritious after being frozen than while still green. Fortunately, at this season of the year permanent pastures usually are in excellent condition and are an ideal supplement to the cured-legume forage remaining in the fields.

Fig. 50. Beefmaster cows and heavy calves on high-carrying-capacity, irrigated Coastal Bermuda grass pasture. Rotation grazing of this kind of pasture decreases the amount of land required per cow by about one-third. (The Cattleman.)

The length of the grazing season and the carrying capacity of temporary or rotation pastures vary with the latitude and with the species of
plants making up the seeding. They also depend on weather conditions
prevailing during the growing season and on soil fertility. In general,
however, an average of 5, 7, and 9 months' grazing are obtained in the
northern, central, and southern latitudes of the United States respectively.
Pasture specialists and plant breeders constantly strive to lengthen the
grazing season by developing strains or finding new species of grasses and
legumes that begin their development earlier in the spring and are
productive longer in the fall. Consulting with the pasture specialists in a
given locality will ensure obtaining the latest recommendations concerning
suitable pasture crop species.

The application of nitrogen to grass pastures to stimulate early spring
growth is a rather common practice in some areas. Data similar to those in
Table 60, summarizing a cow-calf study with a nitrogen-fertilized Midland
Bermuda pasture, are numerous and can be obtained from almost any
state agricultural experiment station in areas where either adequate
rainfall or irrigation water is available to elicit a response to fertilization.
Economic analyses, using current fertilizer and calf prices, must be made
for each locality before one may conclude that nitrogen fertilization pays.
Beyond a doubt, more early-season forage is produced, especially by the
cool-season grasses, and often an additional response is obtained in the fall
if split nitrogen applications are used. There are some management
problems. Nitrogen fertilization almost certainly results in loss of much of
the legume fraction in mixed seedlings because of the vigorous response

Table 60

Effect of Nitrogen Fertilization of Bermuda Pasture on Cow-Calf Performance[a]

Item	Nitrogen per Acre (lb)[b]		
	60	180	300
Number of cows and calves	20	20	20
Average cow gains, April 21–October 9 (171 days) (lb)	168	157	159
Average calf gains, April 21–October 9 (171 days) (lb)	304	325	312
Calf average daily gain (lb)	1.74	1.78	1.74
205-day weaning weight (lb)[c]	421	429	421
Acres per cow-calf pair	2.51	1.91	1.33
Calf weight gain (lb/acre)	168	225	316

[a] Data from Oklahoma Research Report MP-92, 1974.

[b] Applied in equal split applications in May, July, and September.

[c] Adjusted for age and sex.

of the grasses. Tall fescue is especially notorious in this respect. Adjusting pasture stocking rates to utilize the great flush of growth in early summer presents problems, often resulting in excess cattle being on hand in the summer.

Nitrogen fertilization of pastures growing on average to poor soils usually improves forage quality as well as quantity, but it usually is still too low in energy content for any cattle program other than the cow-calf program and alternative options for recovery of costs of fertilizer are few. Harvesting excess growth as hay or haylage if sufficient cattle are unavailable for stocking during the peak pasture growing season (generally May and June) is resorted to by cattlemen who have sufficient acreages to justify the necessary investments in harvesting equipment. Custom harvesting usually adds $15 to $20 per ton to the cost of harvesting excess hay.

Irrigation is being used more extensively as a pasture extender in midsummer and late fall. Mowing, rotational grazing, strip grazing, and the feeding of harvested pasture in drylot as green chop or even as silage or hay are still other pasture extenders.

NATIVE AND PERMANENT PASTURES

By far the majority of beef cows graze native pastures or ranges in the summer. These pastures or ranges vary greatly in type and quality depending on geographic location.

Some of the very best native or permanent grass range, such as that of the Flint Hills region of Kansas where big and little bluestem are the predominant grasses, may require only 4 or 5 acres per cow for summer pasture. In contrast, in some parts of the West where annual rainfall is less than 8 inches, as many as 150 acres may be required per cow for the season. Unfortunately much of the native range area has such a low rainfall that improvement by introducing new species of grasses is unpromising because stands cannot be obtained with dependable regularity. In these situations, reducing the competition for moisture and fertility from brush and scrubby tree growth is the principal hope for improving carrying capacity. Figure 51 shows the extent of the acreage in the various states where brush removal can improve grazing and increase carrying capacity. Much research is being done in brush control and, although 2,4,5-T spray is producing good results, still more promising compounds are in the testing stage.

Overgrazing and burning, whether intentional or accidental, are per-

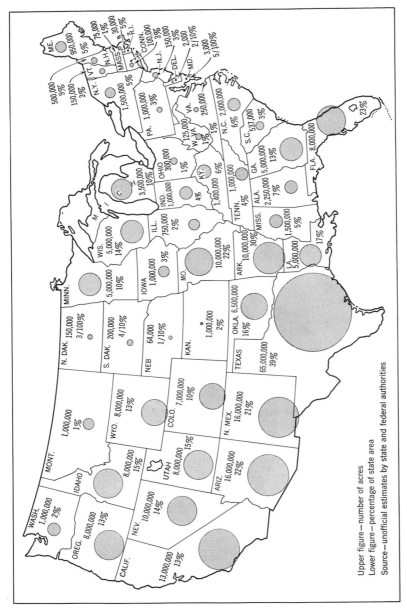

Upper figure—number of acres
Lower figure—percentage of state area
Source—unofficial estimates by state and federal authorities

Fig. 51. Estimated acreage where brush removal can improve beef gains in the United States. (Texas Livestock Journal.)

haps the two most destructive practices with respect to permanent range. In both instances stands of desirable grasses are injured, and erosion and water runoff further reduce forage production. Weeds and less nutritious annual grasses and brush compete for moisture and fertility. Burning reduces forage production in the long run, even though growth may appear to start a few weeks earlier as a result of the burning. Kansas agronomists checked the effects of burning bluestem range over a long period of years and found the following differences in yield of forage in pounds per acre annually:

Unburned plot	2,502
Burned April 30	2,161
Burned April 10	1,934
Burned December 1	1,926
Burned March 20	1,845

Conversely, controlled burning is a recommended range-improvement practice in some instances where certain poisonous and noxious weeds, such as snakeweed, and undesirable brush can be controlled by such burning. Consultation with local range-management specialists is suggested for determining local recommendations in this regard.

Such pasture management practices as contouring, pitting of the sod, building of stockwater ponds or tanks to help control grazing, use of drift fences, feeding supplements to prevent overgrazing in emergencies, rotation-deferred grazing to permit native grasses to set seed at least 1 year out of 3, rodent control, use of fireguards along highways, and other more localized improved management practices can enhance the productivity of permanent pastures. A discussion of all the peculiar problems of each range area is unfeasible in this text.

Permanent pastures for summer use in nonrange areas also vary widely in quality, carrying capacity, and length of grazing season, but at least in this situation lack of rainfall is ordinarily not the major concern. Soil infertility, poor choice of grass and legume species, overgrazing, erosion, and competition from weeds, underbrush, and scrub timber are the big problems here. Correcting soil mineral element deficiencies according to needs demonstrated by soil tests is the first step in pasture renovation. Seeding locally recommended mixtures of grasses and legumes according to proven procedures, followed by sound pasture management, makes possible carrying capacities as high as one cow and calf per acre for the summer pasture season in areas consisting largely of once-cultivated but now abandoned farmland.

GAINS MADE BY COWS ON PASTURE

The gains made by cows on pasture during the summer depend on a number of factors, including the abundance and nutritive value of the forage, the condition of the cows when turned onto pasture, the age of the suckling calves, and whether or not the calves are creep-fed. In most sections of the country the cows have calved before they are turned onto pasture and have lost considerable weight from parturition and lactation. In much of the range country, weight loss will have occurred throughout the winter. These losses from all causes often amount to 150 to 200 pounds per cow. Half or more of these losses usually is regained during the first 2 months on pasture while the grass is fresh and abundant, and the remainder is regained during the fall and early winter after the calves are weaned. Frequently the cows lose in weight during the middle of the summer when the pastures are scorched and dry and the demands of the nursing calves are greater. One advantage of creep-feeding calves on summer pasture is the reduction of stress on the cows. Early weaning of calves, especially from young and old cows, is another procedure for reducing stress on cows during summer drought.

CALVING ON PASTURE

There is no better place for a cow to deliver her calf than in a clean calving trap or pasture. Usually the danger from infection is much less than in the average cattle barn or drylot. However, if the cow requires assistance during parturition, it cannot be given so readily, especially if the pasture is a distance from the farmstead or headquarters. Consequently it is desirable to have a small calving pasture nearby in which cows well advanced in pregnancy may be kept. Under no circumstances should calving cows be allowed to remain in a pasture where there are swine.

SUPPLEMENTARY FEED FOR COWS ON SUMMER PASTURE

Ranges or pasture properly stocked will ensure the accumulation of enough reserve forage to prevent feed shortages during temporary droughts. However, in prolonged drought it is sometimes necessary to give the cows some supplemental feed if their milk flow and condition are to be maintained. Some cattlemen follow the practice of guarding their supply of grass against possible exhaustion by supplementing the pasture with other feeds from the very beginning, or soon after the beginning, of the grazing season. This is recommended on farms in the nonrange area where the pasture acreage is small and homegrown roughages are available, as it makes possible a greater carrying capacity. Table 61 shows how the pasture area was reduced 33 and 66 percent respectively at the Illinois station by using different amounts of supplementary feed.

One of the most satisfactory feeds for supplementing pastures is corn, sorghum, or grass silage. Silage is palatable, succulent, and usually economical. Seldom do cattle refuse to eat 15 to 20 pounds of silage per day, even when the supply of grass is abundant. On the other hand, dry hay or straw remains in the racks almost untouched as long as the supply of grass holds out.

Cowmen often make the mistake of delaying summer feeding too long. If it is to be a supplement to the grass and is to prolong the grazing period, feeding must be begun long before the grass is exhausted. The excuse so often heard—that the farmer hoped for a rain which would make feeding unnecessary—is invalid because a closely cropped pasture seldom recovers as quickly after a rain as one that has had better treatment. It is much easier to discontinue feeding when it proves unnecessary than it is to repair damage to grass and cattle resulting from delaying supplemental feeding.

If silage, hay, or green chop is unavailable, it may be necessary to feed some grain to the cows during the summer. Six to eight pounds of grain a day per cow is usually sufficient, or special emergency range pellets or blocks or liquid feeds may be fed as recommended by reliable manufacturers. If supplemental feeding on permanent pasture is required in most years, the choice of cattle enterprise should be examined. The farm or

Table 61

			Average Daily Supplement per Cow for Period Fed (lb)		Average Daily Supplement per Cow for Entire Period (lb)	
Pasture Acreage per Cow and Calf (Bluegrass)	Number of Days Pastured	Number of Days Supplement Was Fed	Cottonseed Meal	Corn Silage	Cottonseed Meal	Corn Silage
1.50	161	12	0.52	18.6	0.04	1.35
1.00	145	93	0.35	14.0	0.22	9.06
0.50	161	153	0.73	29.2	0.70	27.85

Effect of Supplementary Feed on the Pasture Area Required per Cow and Calf[a]

[a] Illinois Agricultural Experiment Station; average of three years' work.

ranch simply may be too small or otherwise not well suited to the cow-calf program.

SUPPLEMENTARY PASTURE CROPS

Since pastures consisting mainly of cool-season grasses frequently become dormant during the middle of the grazing season, the farsighted cowman takes steps, if possible, to provide forage crops with which to supplement his regular pastures, should a summer pasture shortage develop. Sudan grass or forage sorghum–Sudan hybrids are excellent pasture crops for summer and early fall. These crops should preferably be sown in ploughed and carefully prepared ground. They should be grazed with care because of the danger from prussic acid poisoning. This poisoning is most likely to occur when abundant new growth follows a period of drought or when second growth is grazed following earlier grazing or cutting for hay. Use of one or two "tester" animals is often resorted to, but the forage can be chemically analyzed to evaluate its safety, by means of readily available, low-cost testing kits.

SHADE AND WATER

Often too little attention is given to providing adequate shade and sufficient fresh water for cattle during hot weather. Many pastures are entirely without shade of any kind, and the cattle are exposed to the heat and glare of the summer sun. Tall-growing species of trees make the best shade for permanent pastures but in rotation or temporary pastures portable shades are preferable. The effect of shades, both natural and artificial, under humid conditions is illustrated by the Louisiana experiment summarized in Table 62. Roofs of the Louisiana shades were made of 6-inch-thick layers of hay, straw, or pasture clippings supported by wire netting. It should be noted that the breeds used in this study were Herefords and Angus, which are not so well adapted to the hot, humid weather of the coastal region as is the Brahman breed or its crosses. Thus the favorable response to shades reported here is unlikely to occur to the same extent if the latter breed or its crosses are used.

Water should be readily accessible to cows and calves in all pastures. Energy expended in walking great distances to water, or shade for that matter, is lost, from a productive standpoint, and thus should be kept to a minimum. Pasture areas that are more than 2 or 3 miles from water will not be well utilized and, still more serious, when water is not well

Table 62

| Effect of Type of Shade on Performance and Grazing Habits of Hereford and Angus Cows and Calves (Four-Year Summary)[a] | | | | | | | |
| Treatment Class | Abundant Natural Shade | | Scanty Natural Shade | | Artificial Shade | | Without Shade | |
	Cows	Calves	Cows	Calves	Cows	Calves	Cows	Calves
Number	255	255	110	110	40	40	112	112
Average weight (lb)	1,013	336	1,037	349	990	354	960	322
Average daily gain (lb)	1.29[c]	1.85[c]	1.00[b]	1.64[b]	0.84	1.78[b]	−0.05	1.18
Grazing habits (6 A.M.–7 P.M.)								
Grazing (hr and min)	4:16	3:05	4:13	2:45	3:58	3:06	3:47	2:42
Standing (hr and min)	4:16[c]	4:00[c]	4:30[c]	4:03[c]	4:55[b]	3:49[c]	6:04	5:06
Lying (hr and min)	4:28[c]	5:55[b]	4:17[c]	6:12[c]	4:07[b]	6:05[b]	3:09	5:12
Average maximum temperature (°F)	89.6	89.6	91.4	91.4	88.4	88.4	86.9	86.9
Average relative humidity (%)	63.1	63.1	62.4	62.4	62.8	62.8	64.0	64.0

[a] *Journal of Animal Science* 15:59.
[b,c] Values significantly different (5 and 1% level of probability, respectively) from those of the "without shade" treatment.

distributed, overgrazing and permanent damage to the range occurs near the watering sites.

Farm and ranch ponds should be fenced, and a tank with float control should be situated below the dam for best results. Water is cleaner and cattle do not damage the dam and pond itself by wading into the water. Some newly built ponds do not hold water at first, and purposely delaying fencing out the pond for a year or so will ensure trampling and sealing of the pond floor.

Metal or wooden tanks are most suitable for temporary pastures. Shades or covers over the tanks lower the temperatures of tank water with the result that somewhat less water is consumed during very warm weather.

In an Illinois experiment, conducted during June and July, water in tanks provided with shades 7 feet high averaged 11°F cooler during the day than water in unshaded tanks. Steers being full-fed corn on lush alfalfa pasture drank 9.4 percent more water in the lot in which the tank was unshaded. On such lush pastures, which are high in water content anyway, excess consumption of drinking water conceivably could reduce intake of dry matter. It has not been proved experimentally that water consumed by cattle must be absolutely clean for normal performance.

PROTECTION FROM FLIES

Flies annually cause millions of dollars of loss to cattlemen. The hornfly is the principal species that annoys grazing cattle, and the stable fly is the more troublesome to animals confined in or watering at barns and feedlots during the summer months. Both species are bloodsuckers, and the blood loss suffered by animals of all ages during a long period of heavy fly infestation constitutes a serious tax upon the herd. The annoyance, disturbance, and reduced amount of grazing done by the cattle in the daytime are often capable of retarding milk production, growth, and fattening. Fortunately, cattle may be protected from both hornflies and stable flies by dipping or spraying the animals with an effective insecticide such as a solution of chlordane or by using a systemic insecticide contained in a salt-mineral mix, as discussed in Chapter 28. Effective rubbers of various kinds are also available. Sheds, barns, and feeding equipment used by the cattle during the summer should also be sprayed with an effective residual spray to destroy house and stable flies, which breed in moist straw and manure around the farmstead or ranch headquarters.

SALT

Cattle on grass usually consume a somewhat larger quantity of salt than they do in the drylot. Offering either loose or block salt in a weather-protected self-feeder to the cattle at all times is strongly recommended. A mature cow requires about 0.1 pound of salt per day, or 3.0 pounds per month, during the summer grazing season. During winter months, salt consumption is about half as high.

If salt is provided as part of a salt-mineral mixture, the daily salt consumption level remains approximately the same as for plain salt. Hence the salt content of the formulas for the salt-mineral mixes should be adjusted according to season.

Location of the salt in a pasture can be effectively used to ensure more efficient range utilization. Successful ranchers often use salt to entice cattle into relatively inaccessible or isolated areas in mountain ranges that the cattle would otherwise not utilize.

With the exception of salt, phosphorus is the mineral most apt to be deficient in summer pasture or range. This deficiency is more likely to occur in areas where soils are deficient in phosphorus, as indicated in Fig. 37.

If cows were wintered on harvested forages of good quality, and especially if protein supplements such as soybean or cottonseed oil meals were fed, phosphorus-deficient or low phosphorus-content summer pasture or range is not so serious insofar as reproduction is concerned. On the other hand, if cows graze such mineral-deficient forage in summer and then are wintered on dry, cured forage of the same type, serious reductions in percent calf crop weaned and in weights of all ages of cattle are bound to follow.

New Mexico Experiment Station workers compared the production of two herds handled under similar conditions, except that one herd had year-round access to a mineral mixture composed of half salt and half steamed bonemeal (a common source of supplemental phosphorus in cattle rations). The pertinent data from their 7-year study are shown in Table 63. The annual consumption of mineral mix, including salt, was 51.7, 30.2, 34.5, and 73 pounds, respectively, for cows, calves over 7 months, yearlings, and bulls. Since half the mixture was salt, which was needed in any case, the consumption of 26 pounds steamed bonemeal resulted in a 9 percent larger calf crop weaned, and those calves weaned averaged 53 more pounds in weight. Improved performance for the other classes of stock was equally striking. Texas workers got similar responses from adding phosphorus to the drinking water in soluble form. Supple-

Table 63

Effect of Mineral Supplements on Weight and Production of Range Cattle (Results of 7 Years)[a]

Measure of Production	Cows Not Fed Minerals	Cows Fed Minerals	Gain from Mineral Feeding
Cows calving (%)	90.4	92.2	1.8
Calves died (%)	10.8	2.7	8.1
Cows weaning calves (%)	80.7	89.7	9.0
Average weight of calves (lb)[b]	408	442	34
Average production per cow (lb)	336	389	53
Gain of yearling steers (lb)	321	353	32
Gain of 2-year-old heifers (lb)	202	278	76

[a] New Mexico Agricultural Experiment Station Bulletin 359.

[b] Average weight of calves multiplied by percent of cows weaning calves.

mentation with calcium alone (the other important constituent of bone-meal) on summer pasture has not produced similar results.

Not all cattle respond similarly to phosphorus supplementation on pasture or range. Consultation with pasture specialists or animal nutritionists in the local area is recommended in all situations.

A special problem in mineral supplementation may present itself in early spring and again in late fall in areas where grass tetany occurs. This nutritional disease and its prevention and control are discussed in detail in Chapter 27. Let it suffice to say here that, in regions where spring calving is practiced and where pastures consist predominantly of cool-season grasses such as tall fescue, brome, or orchard grass or of winter small grains, one should be informed as to management programs that can be followed to prevent the likelihood of grass tetany.

BREEDING ON PASTURE

Cows that are to calve in the springtime must be bred during the early summer months. At this time they are, of course, on pasture, where their heat periods are likely to pass unobserved. It is advisable, under most

Fig. 52. Crossbred yearling heifers being bred on excellent legume-grass pasture. Combining the favorable effects of heterosis for reproduction and the flushing effect of the high-quality pasture should ensure a high conception rate and a high proportion of early calves. (Prairie Farmer.)

circumstances, to pasture-breed during the grazing season. It is, of course, understood that a bull used in pasture mating should be sound and mature. Also, the herd should be divided, where practical, so that there are not more than 25 or 30 cows with each bull. Rotation of bulls every 3 weeks is practiced by some breeders using single-sire herds. Purebred breeders do not use this method for obvious reasons.

Many breeders of purebred cattle seldom resort to pasture breeding. The importance of knowing the breeding dates and of being absolutely certain which bull has made the service makes hand mating or artificial insemination preferable. There is always risk that a bull running in a pasture may jump a fence and get with cows that are being kept for mating with another bull, or get with heifers that are too young to be bred. The man with valuable purebred cattle cannot afford to take such risks.

If artificial insemination is used on summer-pastured cows, the problems of heat detection and inconvenience are compounded. Use of small pastures through at least two heat cycles for each cow reduces the disadvantages of artificial insemination of spring-calving cows. A cleanup bull is recommended for breeding the unsettled cows.

CREEP-FEEDING CALVES ON PASTURE

In the commercial spring-calving cow-calf program, several factors must be considered before a suitable decision can be reached as to whether to creep feed. Among these factors are the following.

1. *Age, time, and method of marketing calves.* If lower-grade calves go direct to the killers, creep-feeding often pays. If higher-grading calves are to be kept over as stockers, it seldom pays. If calves are to be fed out at home, as baby beeves, starting them on creep eases the transition from suckling calf to full-fed weaner and eliminates stress.

2. *Quality of pasture and milking ability of cow.* If the mother cow has the potential to provide at least 2 to 3 gallons of milk per day, especially in late stages of lactation when the calf is large enough to consume this quantity, and if the pastures being used provide the necessary plane of nutrition for this level of milk production, creep-feeding seldom pays. Such a pasture, if available, would in itself provide excellent supplemental nutrients for the calf. However, most straight-British beef cows seldom are genetically capable of producing this volume of milk, nor will the pastures being used for beef cows in many instances support such a level of milk production because of low rainfall, soil infertility, lack of nutritious forage species, and other reasons.

Fig. 53. A well-located creep-feeder in use in a purebred, operation. (American Angus Association.)

3. *Age of cow.* Calves of first-calf heifers and older cows are most likely to need or to respond profitably to creep-feeding. Such cows also are apt to rebreed more easily and to go into the wintering dry period in better flesh if their calves are creep-fed.

4. *Cost of grain and selling price of calves.* Feed prices and cattle prices are usually closely associated, and so this item is highly important only when there is a sharp deviation in the above-mentioned association.

5. *Keeping heifers for replacements.* Creep-feeding of heifer calves destined to be kept as mother cows, whether in one's own herd or sold specifically for that purpose, is not a good practice. Recent experimental data show that weaning weights of calves that are suckling dams that were themselves creep-fed were reduced about 5 percent as compared with weaning weights of calves suckling comparable-aged dams that were not creep-fed themselves. Their dams gave less milk, a reduction generally believed to be caused by fatty infiltration of mammary tissue. Creep-feeding also obscures poor milk production in cows and thus interferes with selection for this trait. Weaning heifer calves at ages as

early as 4 to 5 months apparently would offset both of the disadvantages of creep-feeding just mentioned.

6. *Desirability of preconditioning calves.* It will become increasingly important to cow-calf producers to manage late-stage suckling calves so as to ensure a reduction in shrink at weaning weight, and in the incidence and severity of postweaning "shipping fever." Creep-feeding rations containing antibiotics, vitamins, and minerals—along with energy and protein, of course—for 4 to 8 weeks prior to weaning may well attract premium buyers to one's calves. The buyer will find that calves that have been taught to eat will make the transition to new feeds and surroundings with less stress, although this does not guarantee elimination of the problem. Details of preconditioning methods are presented elsewhere.

DRY COWS AND TWO-YEAR-OLD HEIFERS

Dry cows and heifers that will calve in the fall and yearling or bred heifers that will not calve until the next spring require comparatively little attention during the summer. Ordinarily the cows with calves are given the best pasture, whereas the dry she-stock is able to get along on somewhat less abundant and low-quality forage. It is also a common practice to take the dry cows to the pasture farthest removed from the farmstead or headquarters, as they do not require the daily attention needed by the unbred cows and those with calves. Only in very dry weather or pronounced overstocking is feed other than grass necessary.

YEAR-ROUND DRYLOT MANAGEMENT OF BEEF COWS

Pasture is not absolutely necessary for the well-being of cattle in summer, as was proved rather conclusively at the Illinois station 60 years ago. Ten heifers were kept in a small paved lot from the time they were weaned until they were $4\frac{1}{2}$ years old. This period covered four summers, during the last two of which the cows were suckling spring-born calves. During the entire 4 years the cows received nothing except corn silage supplemented with cottonseed meal at the rate of $2\frac{1}{2}$ pounds of cottonseed meal per 100 pounds of silage fed. The daily feed of silage for the mature cows was 40 pounds. These cows, kept constantly in the drylot, maintained their weight nearly as well as other cows that were on pasture each summer, and they produced calves that were in every way normal. The only trouble

Fig. 54. Crossbred cows in confinement during summer being fed on a fortified corn stalklage ration in Illinois. (Norris Farms—Prairie Farmer.)

experienced during the summer was a few cases of foot rot among the calves.

In an Iowa experiment, cows nursing calves consumed daily 114 pounds of an alfalfa-grass mixture, fed as green chop. These green-chop-fed cows gained an average of 7 pounds for the summer period as compared with a loss of 70 pounds for comparable wet cows on a permanent pasture. Weaned calf weights were 370 and 405 pounds respectively for the calves nursing the permanent-pasture and green-chop cow groups. However, because of the added charge made for the labor of harvesting and feeding the green chop, the heavier calves returned a smaller margin of profit per calf.

A more extensive test was conducted by the Texas Agricultural Experiment Station workers at the Rolling Plains Livestock Research Station at Spur, Texas. After six calf crops they concluded that the advantages and disadvantages of a continuous drylot system for cows are the following.

Advantages

1. Investment in land is lower. A small operator can enlarge his herd without buying more land.

2. Closer observation of cattle facilitates selection, physical attention, and record keeping.
3. Artificial insemination is easier to use and a more uniform calf crop is possible.
4. There is less hazard from drought through storage of reserve feed.
5. Drylot calves are easy to precondition, wean, and start on growing or finishing rations.

Disadvantages

1. More labor and equipment are required.
2. There is greater confinement of the operator; labor is required for feeding daily or every other day, or there must be a pasture where cattle can graze while the operator is away.
3. Disease problems may be more acute, especially for the calves.

The Texas experiment was started with two groups of 36 Hereford weaner heifer calves, with one group being continuously drylotted and the other run on pasture. Each group was further subdivided into three groups during the winter for comparisons of supplemental energy levels. Table 64 shows the supplemental levels of feed fed and the performance of the cows and calves through six calf crops.

Table 64

Effect of Method of Management and Energy Level on Performance of Beef Cows (1960–1965)[a]

	Level of Energy			Average, All Levels of Energy	
	Low	Medium	High		
Winter daily feed (lb)					
Cottonseed meal	1.25	0.75	—		
Sorghum grain	—	2.00	5.0		
Sorghum silage[b]	30.00	35.00	40.00		
Calf production	% lb	% lb	% lb	% lb	
Pasture	90 468	85 454	91 469	89 464	
Drylot year-round	85 483	85 488	92 466	88 478	
Average 345 calves	87 476	85 471	92 468	88 471	

[a] Texas Agricultural Experiment Station, Spur Technical Report Number 2.
[b] Silage was fed to the drylot cows only. Pasture cows grazed cured range plus the supplements indicated.

The drylot cows were fed 1 pound of cottonseed meal, 2 pounds of sorghum grain, and 50 to 55 pounds of sorghum silage or green chop after May 1, while pasture cows received no feed other than grass. Iodized salt was fed both groups year-round and a 50 : 50 bonemeal : salt mix was provided for 6 weeks in the spring. A creep-feed was provided for the drylot calves to compensate for the grass consumed by the calves on pasture. The drylot cows, in general, were heavier than the pasture cows, indicating that a still lower level of feed might be used. The different energy levels tested resulted in comparable results in both drylot and pasture cows. After six calf crops, the greatest difference evident was in the number of cows lost from the project. Only four cows were lost from the drylot whereas ten were lost from the pasture group. Percent calf crop weaned and weaning weights were so highly comparable that one must conclude that, with respect to performance, drylotting cows can be as successful as running cows on pasture. With respect to cost, the pasture cows, over the six years, had the advantage, as the summer feed for the drylot cows was more expensive. Using the low energy levels only, the pasture and drylot cows produced net returns per cow of $22.05 and $21.20 respectively for the last calf crop weaned. The Texas workers suggest that these data indicate that cows in drylot can produce calves as efficiently as cows on pasture. They add that if drylot cows could be grazed on more economical temporary pastures for 90 to 120 days during the summer, an additional $10 to $15 could be added to the net return per cow. In a later experiment still in progress, the Texas workers are finding that a combination drylot-pasture system of managing cows has both the advantages of lower total feed costs and more productive cows.

These experiments, together with the knowledge available concerning the wide use of soiling crops—that is, harvesting and feeding forage crops in a fresh state—in Europe, prove conclusively that pastures are not indispensable in raising beef cattle. However, the great amount of labor involved in caring for cattle under such conditions precludes any wide use of such methods in this country at the present time, unless automatic feeding systems, silo unloaders and auger bunks for feeding silage, or field choppers and self-feeding bunk wagons for soiling or feeding green chop, are installed and used year-round in relatively large operations. Drylot summertime rations for cows nursing calves should be compounded according to the requirements shown in Chapter 9 for nursing cows, although preliminary Illinois data suggest that the requirements for energy may be as much as 25 percent too high for drylotted cows.

Arizona workers have shown that, if calves are creep-fed in cow confinement programs, then cows with some dairy or dual-purpose breeding should be used and they should be bred to bulls of the larger exotic breeds in a terminal-cross program so as to produce calves that do not mature and fatten at too light a weight. Only such calves can respond maximally to the extra milk and creep-fed consumed.

WORKING THE SUCKLING CALVES

Because most calves are "worked"—that is, dehorned, castrated, marked, and vaccinated during summer and weaned in the fall—brief discussions of these jobs are appropriate in connection with summer management of the cow herd. This does not imply, however, that such chores can only be done then.

Dehorning.   Although a nicely shaped pair of horns undoubtedly adds to an animal's appearance in the show ring or purebred herd, the presence of horns on commercial cattle is considered objectionable by most cow-men. Horns are the cause of so much loss to the packer in damaged hides and bruised carcasses that polled or dehorned cattle normally sell from $1

Fig. 55. A purebred Polled Hereford bull being used to dehorn calves through heredity. If the bull is homozygous for the polled character, the horned cows will produce only polled calves. (American Hereford Association.)

to $2 per hundredweight higher than horned cattle of equal merit in other respects. This is particularly true when the animals are shipped some distance and are kept in crowded cars or trucks for several hours. Horns are also objectionable on the farm and in the feedlot. Cattle with horns require more shed room per animal as well as more space at the feed bunk or hay rack and in the truck or car when shipped. Among horned cattle there is always a tendency for some to be "bossy" and to keep the timid ones away from their share of shelter and feed. Dehorning tends to curb the aggressiveness of such animals, thus lessening feedlot disturbances. Even hornless cattle, however, establish a "peck order," with some being more aggressive or shy than others. Unfortunately, removal of the "boss cow" does not solve the problem, because another soon fills her position.

Dehorning by Breeding.   Using a polled bull results in a majority of hornless calves. If such a bull is "pure" polled, carrying no horn genes in his germ plasm, all his calves will be polled, even though all their dams have horns. If, however, the bull is heterozygous for the horn gene (the product, let us say, of a pure polled bull and a horned cow), only half his calves from horned cows will on the average be polled, and the rest will have horns. This explains why Angus bulls (which are almost invariably "pure" polled) seldom sire any but polled calves, whereas some polled bulls of the Hereford and Shorthorn breeds often sire a number of horned offspring. This problem is worsening because so many purebred breeders of horned cattle are in the transition stage from horned to polled cattle, and as long as they are still using their original horned cows and even their daughters, it will remain a problem.

Dehorning by Use of Chemicals.   A satisfactory way to prevent the growth of horns in newborn calves is to treat the horn button with a strong chemical. This is best done when the calf is 1 to 5 days old but can be delayed for as long as 2 weeks. Caustic sticks and pastes composed of such alkalies as potassium, sodium, and calcium hydroxides have been used for years but because they tend to burn the calves' heads, and their dams during the act of nursing, they have been largely replaced by improved preparations. One such product, marketed as "Pol," contains antimony trichloride, salicyclic acid, and flexible collodion. This material dries faster, adheres more firmly, and is water-resistant, making it much safer than the older product.

Because caustic, to be successful, should be used while the calf is very young, this dehorning method requires considerable labor when calves are born on pasture. Consequently it is used mainly on farms where the calves can easily be caught and treated. In a performance-tested herd, calves will usually be caught for tattooing or tagging anyway, and dehorning can be done at that time. Horn buttons can be detected at time of birth by an experienced observer.

The Bell Dehorner.   A heated dehorning iron is a popular tool for dehorning young calves on many ranches and farms. After the bell-shaped iron has been heated to the proper temperature, it is fitted over the horn button and held firmly against the head until the horn matrix has been destroyed. Care is necessary to make the burn deep enough to destroy the horn tissue but not so deep as to produce a bad sore. The operation is more painful and requires more skill than chemical dehorning but it may be used to dehorn calves up to 2 months of age. Fire irons and electrically heated irons are available from most commercial sources.

Dehorning Older Calves.   In large herds of commercial cattle, especially in the range area where cows and calves are ranging over a wide area, dehorning is usually postponed until the calving season is practically over, in order that all calves may be dehorned, castrated, vaccinated, and branded with only one working of the herd. At this time the horns of most calves are well developed but still soft and loosely attached to the skin and can easily be removed with a knifelike dehorning tool that separates the horn button from the adjoining skin with little blood loss. Such tools are called "gougers," "spoons," or "tubes," depending on their shape and the way in which they are used (see Fig. 56, *a* and *b*). Either a dehorning saw or some type of clippers is used in dehorning cattle older than 4 months (see Fig. 56, *c* and *d*).

The Barnes dehorner shown in Fig. 56 literally lifts the horn out by the

Fig. 56. Dehorning tools, listed in the order of age of calves for which they are best suited: (*a*) dehorning tube, 1 to 3 months; (*b*) dehorning spoon, 3 to 5 months; (*c*) Barnes dehorners (available in two sizes), 5 to 12 months; (*d*) Leavitt clippers, 12 to 24 months; (*e*) dehorning saw (not shown), over 24 months. (O. M. Franklin Serum Company, Denver, Colorado.)

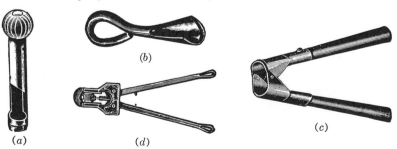

(*a*)   (*b*)   (*c*)

(*d*)

roots and crushes the blood vessels so that only a trickle of blood is produced. However, it is suitable only for dehorning calves and short yearlings.

If bleeding appears to be excessive, a commercial dust designed for the purpose should be applied to stop it. In extreme cases the cut end of the exposed artery can be picked up with tweezers and twisted until it is severed down inside the head, stopping the bleeding after a few minutes.

Equipment for Dehorning. Little equipment is necessary for dehorning calves. Often they are simply thrown to the ground and held firmly, or sometimes they are snubbed to a fencepost. However, if a number of calves are to be dehorned, it is advisable to construct or buy a regular dehorning chute. If herds are large and calves will be worked in readily accessible areas, special tilting tables are justified.

Castration. Bull calves are castrated for economic reasons. At the present time, steers sell higher per hundredweight than bulls whether sold as stockers, feeders, or fat cattle. Many believe that castration results in improved texture, tenderness, and flavor of the beef and produces a quieter disposition, which is an asset in the feedlot. It is noted elsewhere that if market discrimination against bulls were reduced, the feeding of bull calves for slaughter at 12 to 18 months of age might become a common practice because of their superior performance in the feedlot and higher yield grade or higher percentage of carcass lean and lower percentage of fat trim.

Age and Season for Castration. Castration ordinarily is best done when calves are 4 to 10 weeks old. Like dehorning, castration should be performed when weather conditions are favorable. The spring and fall months are considered the best, although winter castration is not objectionable if it is done on a mild, bright day and if adequate protection is furnished during the following night. As a rule, calves born during the winter and early spring are castrated just before being turned to pasture, whereas those born during summer and fall are allowed to go until they are brought in from the pasture in the fall. On the range, castration is performed at the summer roundup or "work" as it is referred to by ranchers.

Methods of Castration. Young calves are usually thrown to be castrated, but animals older than 4 months may be operated on better while standing. In throwing, the calf is placed on either side and the feet and legs are held or "hogtied." This position exposes the scrotum to the operator, who stands or kneels alongside the calf's rump. If not absolutely

Fig. 57. The entire family often helps at calf-working time on the range. Corrals with headgates, and sometimes tilting tables, are used on many ranches where less labor is available. (American Hereford Association.)

clean, the scrotum should first be cleansed with a mild antiseptic to remove any dirt or filth that might otherwise contaminate the wound. The operator's hands and the knife should always be clean.

Several satisfactory methods are used for surgical removal of the testicles, the more common methods being the following.

1. Removal of the lower third of the scrotum, exposing both testicles. In calves 3 months of age or less, the testicles can then easily be removed by working each one loose and simply pulling it from the opening in the scrotum. In older calves where bleeding may be more severe, the testicle should be worked loose, then the cord should be severed as high into the scrotum or as near the body as possible. This may be done either by scraping with a knife blade until the cord comes apart or by crushing the cord with an emasculatome or clamp.

2. A sharp-pointed knife may be inserted into the side of the scrotum all the way through to the opposite side. Then with one downward stroke the scrotum can be split to the bottom, exposing the testicles for removal as described above.

Larger calves and short yearlings can be castrated standing if well secured. Pulling the tail sharply upward and forward and holding it firmly against the backbone prevents the animal from kicking the operator. Either of the methods described above can then be used for removing the testicles. Both methods ensure good drainage, which is an important factor in trouble-free castration. Removing as much cord as possible with the testicle prevents stagginess.

Treatment after Surgical Castration. After castration, the scrotum should be examined to make sure that the incisions are sufficiently large and high for proper drainage. The calves should be put in a clean box stall or a small, grassy area with their mothers for a few hours until bleeding stops and all mothers have claimed their calves. The calves should have a clean, dry place to lie, shut away from muddy or manure-covered lots or sheds that have not recently been cleaned and rebedded. Daily observations should be made for a week or 10 days to make sure that any swelling due to faulty drainage is promptly relieved by reopening the incision and treating the wound with an antiseptic solution. If castration is done in fly season, repellents should by all means be used, but even then, both the castration and dehorning wounds should be observed closely.

Castration Pincers and Elastrators. A unique castration pincers called the Burdizzo or emasculator has been in use for some time. These pincers or "clamps" and others of similar design have blunt jaws that close with enormous force when sufficient pressure is exerted on the handles to lock them into the closed position. In using the clamps, the object is to crush or

Fig. 58. (*a*) Burdizzo castration pincers; (*b*) correct method of using to ensure crushing of spermatic cord. (O. M. Franklin Serum Company, Denver, Colorado.)

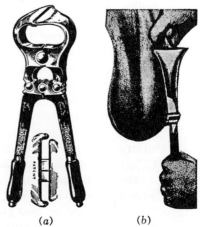

(*a*)   (*b*)

sever the spermatic cord and the blood vessels that supply the testicle so that the testicle degenerates for want of circulation. The advantage of the Burdizzo over the knife is that there is no blood loss and the skin is unbroken, eliminating the problem of blowflies.

A rather new castration instrument called the elastrator has been used with some success in castrating calves. It is a forceps-like instrument that slips a strong elastic band around the scrotum close to its attachment to the groin. The pressure exerted by the rubber band shuts off the blood supply to the scrotum and testicles, causing them to slough off. The virtual elimination of the screwworm fly has of course lessened the dangers from open-wound castration methods.

Short-Scrotum Castration.   The elastrator can be used to castrate calves physiologically without removal of the testicles. The testicles are forced up into the body cavity, or as close to it as possible, and the rubber band is placed on the scrotum below the testicles but as close to the body as possible. The body heat of the animal is such that the testicles produce no viable sperm cells, but the growth response from the testosterone and other male hormones produced by the testicles continues to be produced. Feeding trial and carcass study data evaluating the effect of short-scrotum castration are presented in another chapter. Such castration must be done before calves are 6 to 8 months old.

Marking.   It is highly desirable that all animals in the herd bear some mark or tag whereby each can be positively identified. In the range areas, marking or branding with a registered brand is required by law to establish ownership. In all herds, it is desirable to maintain individual animal identification, and it is a necessity if performance testing is practiced or if the herd is a purebred one.

The method used for marking depends on the object for which it is done. When the purpose is to establish ownership, as it is on the range and in poorly fenced pastures, permanency and ease of recognition are of paramount importance. Under such conditions branding with a hot iron is the best method. Although there have been objections to branding because of the pain inflicted and the damage to the hide, no perfect substitute for the iron has yet been devised. However, a brand should be no larger than necessary to permit easy identification at a distance of 30 or 40 feet, and no deeper than needed to destroy the hair follicles.

Other ownership marks sometimes used are ear notching or slitting and bobbing. Another mark of this sort is the slitting of the dewlap in a way that causes one or two "wattles" of skin to hang from the neck. The principal objection to such marks is that they detract from the animal's appearance and are easy to copy. They are often used to separate the sexes and in combination with brands.

Methods used to establish the identity of each animal in the herd include numbered neck straps and chains, ear tags, horn brands, numbered fire and freeze brands, and tattoos. The principal objection to neck chains is that they are frequently lost. Also, if they are adjusted to fit the cattle when turned onto pasture in the spring, the chains become either too tight or too loose later in the summer as the cattle gain or lose flesh. It is especially difficult to keep the neck chains of growing calves and yearlings properly adjusted. Cattle have been known to hang themselves on neck chains in brushy country or on fences and feedbunks.

Many kinds of ear tags are available at relatively low cost. They are usually easy to attach but are also easily lost and remain in place only a short time. Unnumbered plastic tags can be purchased that are attached by a wide variety of means. These are convenient in that any numbering or coding system can be applied with felt pens and renewed if necessary. One side may be used for individual identification, leaving the other for the number of dam or sire or both.

Horn branding is an excellent means of numbering mature horned cattle but is, of course, useless with calves and yearlings, and obviously the polled breeds cannot be horn branded. Horn brands must be reburned every few years. This is the common method of numbering cattle at sales of purebred Shorthorn, Hereford, and other horned breeds, where each animal's horn brand is made to correspond with the catalog number. Half-inch copper brands are the proper size for this purpose.

The best method yet devised for permanently identifying breeding cattle is tattooing the ear with indelible ink. The advantages of this method are that the mark is permanent and in no way disfigures the animal. Its only serious objection is that the animal must be caught and the ear examined at close range before the mark can be read. Precision and considerable skill are also required to produce readable tattoos. Tattooing outfits, including an initial letter and a set of numbers, may be purchased from any stockman's supply house. All purebred record associations now require that calves be properly tattooed before they are accepted for registration. Black ink is preferred for all but the black breeds, and in their case green ink seems more satisfactory.

A promising new method of marking cattle for individual identity is freeze-branding. The method is painless, eliminates danger of infection, does no damage to the hide, and results in an easy-to-read brand when well done. Essentially the method consists of destroying the hair-pigmentation cells in the skin by use of a copper or steel branding iron that has been held in an alcohol or gasoline bath chilled with dry ice to $-158°F$. The hair must be closely clipped and the chilled iron held in place for 40

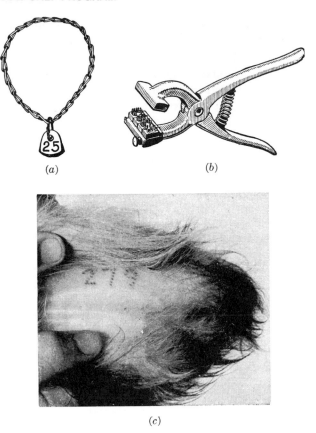

(a)　　　　　(b)

(c)

Fig. 59. Devices for marking cattle: (*a*) adjustable neck chain; (*b*) tattooing instrument; (*c*) a correctly tattooed ear. (O. M. Franklin Serum Company, Denver, Colorado.)

to 60 seconds. If the technique is successful, the hair and surface layer of skin will slough off in 4 to 6 weeks, to be replaced by permanently white hairs. At present the success rate is about 50 percent, but the method is new and improvements in the technique are certain to occur. This brand is not a substitute for fire brands in legally establishing ownership of cattle at present, but may become so in time if the method is perfected. Figure 60 illustrates the type of brand that may result from freeze-branding. Long hair may have to be clipped from the brand area once during the winter to facilitate easy reading.

Vaccination. Calves born on farms in areas that are infested with blackleg and malignant edema should be vaccinated against these diseases before or at weaning time (see Chapter 27). Vaccination at a much

Fig. 60. A 6-inch easily read, well-located freeze brand. This method of marking is painless and does not damage the hide. (North Dakota Agricultural Extension Service.)

younger age is advisable if an outbreak has occurred in the vicinity of the farm within the preceding 2 years. Vaccinating at the regular summer work is the practice on most ranches to reduce the number of times the cattle must be gathered.

Heifer calves that are to be kept or sold for breeding should be vaccinated for contagious abortion at 4 to 8 months of age unless the herd is "accredited" (see Chapter 27) as a result of annual clean tests. Although the value of calfhood vaccination is questioned by some breeders and veterinarians, cattle that are so vaccinated and later pass the blood test are regarded by most breeders as being much less likely to contract the disease than unvaccinated animals. While calfhood vaccination is more urgently needed in purebred than in grade herds because of the greater risk of exposure, the vaccination of grade heifers that are to be added to the breeding herd is generally recommended, since they too may be exposed to the disease through contacts with cattle on adjoining farms. Vaccination for abortion must be done by a veterinarian, who tattoos a "V" and the year of injection in the right ear. Other vaccinations such as those for leptospirosis and vibriosis should be done according to a veterinarian's instructions. State laws vary with respect to health certificates for vaccinated females that are to be shipped to another ranch or state and some states are reexamining their state health regulations concerning whether vaccinations are needed at all after the incidence reaches a certain acceptably low point in a given state or region.

Weaning. Beef calves ordinarily are weaned at 6 to 8 months of age. Usually the date is determined largely by necessary changes in the management of the herd. Spring-born calves are usually weaned about the first of October and fall calves are weaned in the spring when the cow herd is taken to an outlying pasture. Fall-born calves are often allowed to nurse until midsummer, since green pasture causes cows to increase their milk production if they are still being suckled when they are changed from drylot to pasture.

Calves that have been running with their dams should be removed from them once and for all. If they can be placed beyond earshot of one another, so much the better. The practice of turning them back on the second or third day to suck out the cows cannot be recommended, because it tends to prolong the period during which the calves pine for

Fig. 61. Pickups, jeeps, and even helicopters have replaced some of the cow horses and ranch hands as a laborsaving practice on many ranches. Much of the Western Range is so rough in terrain, however, that horses will never entirely disappear from use. (American Hereford Association.)

their mothers, and may give rise to digestive disorders. The question whether to move the cows or the calves from a pasture is a much argued one. Each farmer or rancher must, by experience, work out a system that seems best. Small weaning pastures with tight fences for either the cows or calves will save much labor in keeping the cows and calves apart and under control. Selling and delivering calves at weaning time of course eliminates many of these problems.

Preweaning treatments, usually initiated 3 to 4 weeks prior to weaning, are being tested experimentally to study their effectiveness in reducing the incidence of shipping fever. Such treatments include teaching calves to eat grain and drink from small waterers, and vaccinating to enable calves to build immunities to certain viruses and bacterial agents. Calf buyers are reluctant, as yet, to pay the extra costs of the practice. Seldom does the calf gain enough more as a result of the preweaning treatments to pay the cost incurred by the producer. Thus it appears that widespread use of the practice is at an impasse and the outcome awaits better documentation of the apparent advantages.

Early Weaning. Numerous studies during recent years have amply demonstrated that calves can be weaned earlier than the conventional 6 to 8 months of age. How early to wean depends upon why one is choosing to early wean in the first place. If the purpose is to reduce the stress upon 2-year-old first-calf heifers in order to enhance rebreeding performance, then weaning must be done at about 2 months. In a 42-day breeding season, researchers at the U.S. Meat Animal Research Center increased pregnancy rate by 26, 16, and 8 percent for 2-year-old, 3-year-old, and mature cows, respectively, by weaning their calves at 2 months instead of $6\frac{1}{2}$ months. The question then arises whether the resulting extra calves will have value enough to pay the higher cost of raising the younger weaned calf. After the cows are bred in such instances, the cows' ration can be reduced at least one-third; this saving could almost offset the early-weaned calf feed cost. Early weaning is more likely to occur during droughts or in confinement cow operations where high-priced harvested feeds are being fed. Early-weaned calves should not be fed rations that are overly high in energy content because very early maturity and excessive fattening may result.

PART IV
THE STOCKER AND
FINISHING PROGRAMS

CHAPTER 11
OPERATION OF THE
STOCKER PROGRAM

A discussion of the development of young cattle logically follows the chapters dealing with feeding and managing the cow herd. The question of finishing cattle for slaughter is treated in a separate section, and thus in the present chapter we discuss the subject of calves that are not to be finished immediately but are to be handled in a way that achieves maximum growth at the lowest possible feed cost.

This stocker or grower period, as it is called, consists of the period following weaning up to the time when the calves, if steers, are sold or put into the feedlot for finishing. The stocker program also applies to the replacement heifers in a cow-calf program that are handled so as to ensure normal growth and development. It is customary to refer to the animals on such a program as *stockers,* and they are considered as being on a growing, rather than a finishing, ration. *Feeders,* on the other hand, are older calves and yearlings carrying more finish and bloom, which therefore are placed on higher-energy finishing rations in order to take advantage of their extra condition.

Stocker cattle are found throughout the United States, but they tend to be concentrated mainly in the regions where cow herds are numerous and in the vicinity of cattle-finishing centers. Table 65 shows that two-thirds of all stockers are located in the Plains and Corn Belt and Lake regions. The fact that cowmen must keep almost half their heifers for replacements explains in part why stockers are concentrated in regions high in cow numbers.

The stocker program is sometimes used as the sole program on a particular farm or ranch, but as often as not it is either conducted on the same farm or ranch in connection with a cow-calf operation or it precedes the finishing program. In the latter circumstance, the program is often referred to as a "warm-up" or "backgrounding" program. When the stocker program is the only beef cattle program on a farm, it is usually managed in one of two ways. In *Plan One,* calves or light yearlings are bought in the fall to be wintered on roughage rations in drylot and sold in the spring to buyers who need "grass cattle" for the summer, or who need feeders to put on a summer finishing program. In *Plan Two* the operator may have summer pasture to utilize and lighter-weight calves are bought in the fall to be wintered as in the first plan, but usually at a slower rate of

Table 65

Regional Distribution of Stocker Cattle, 1 January 1974[a]		
Region	Number (1000)	Percentage of U.S. Total
North Atlantic	1,317	3.0
Corn Belt and Lake	11,963	26.6
Southeastern	7,254	16.2
Plains	17,871	39.8
Western Range	6,445	14.4
United States (48 states)	44,850	100.0

[a] Compiled from USDA *Agricultural Statistics*, 1974.

gain. Instead of being sold or placed on a finishing ration in the spring, they are grazed all summer, or until pastures mature and deteriorate in the fall, or until winter sets in.

Plan Three, limited to areas where winter small grains are grown, makes use of these winter pastures and, while still a stocker program in every sense of the term, has distinct features of its own and is separately discussed later in this chapter.

There is a tendency for more and more of the calves to be handled and fed according to the first plan, and even this program tends to be shortened in order to reduce the time from weaning to market by whatever segments of the industry are involved. This trend will undoubtedly accelerate because of rising nonfeed costs associated with the industry. Commercial feedlots in the West, Southwest, and western Corn Belt particularly are using an adaptation of this program on very lightweight calves, often starting with calves as light as 250 pounds. In effect, they use the program to lower their feeder costs and to have the condition of their feeders under their control. They usually use lower grades of calves from the Southeast, and heifers are seldom fed.

Plan Four, perhaps the most common type of stocker program, actually is a combination stocker-feeder program that is especially adapted to the Corn Belt and the irrigated sections of the Southwest where high-yielding corn and sorghum silage crops can be produced. The cattle feeder purchases steer calves or light yearlings of good to choice quality in the fall, or as early as August 1 if available. Sometimes the cattle are wanted early in order to utilize fall growth in small-grain stubble or stalk fields and similar aftermath. After most of these feeds are salvaged and gains have begun to slow down, these cattle should be brought to drylot. They are then fed a heavy feed of roughage, preferably corn or sorghum silage plus supplement or legume hay, until the cattle reach a stage of condition where gains are reduced below a level that permits economical conversion of roughages to gains. This point is usually reached after about 2 to 3 months with yearlings and 4 to 5 months with calves. Having reached feeder condition, the steers are finished out on a full feed of grain, either in drylot or on pasture, to be sold in late summer or fall. Some feeders add limited grain to the silage ration to improve rate of gain, but then the calves begin to fatten and they can no longer be said to be on a stocker program in the strictest sense.

If the farm or ranch also has a cow-calf program, the calf crop from the herd can be handled in the stocker program in the same manner as purchased calves. Ordinarily, however, a farm or ranch suited to the cow-calf program may as well be stocked to capacity with cows to increase the size of herd and thus reduce the fixed costs per calf raised. Keeping one's own calves for further grazing as yearlings means that, unless the operation is a large one, both the cow-calf and the stocker program will be reduced in size. If calves from one's own herd are fed out in drylot as baby beeves, that is another matter, as it would not be likely to reduce the maximum size of cow herd. Under certain semiarid range conditions, keeping one's steer calves over to run as yearlings provides a flexibility factor that allows for adjustment in cattle numbers to fit the almost certain fluctuation in grass or feed supply resulting from periodic droughts.

Lower grades of steers, heavy yearlings, and most heifers are less suited to the commercial stocker program, because such cattle usually should be placed on high-energy finishing rations soon after they are obtained. A warm-up on a rather high-energy stocker ration for about 50 to 60 days is an exception to this rule that is being practiced by southwestern and western feeders using "Okie" stockers. Another exception would be choice or prime grade heifer calves that have sufficient quality to justify carrying them to high choice or low prime condition.

ADVANTAGES AND DISADVANTAGES OF THE STOCKER PROGRAM

This program has the following advantages over other programs.

1. The program is adapted to an intensive type of farming–that is, a large volume of business can be done on either small or large farms that can produce large tonnages of roughages.
2. Returns come quickly, as early as 4 to 6 months, if not preceded or followed by other programs. In some instances this quick turnover permits feeding two to three droves of cattle per year.
3. If used in winter only, this program is completed by the time labor is needed for spring and summer farm work.

4. Feedlots that are poorly located with respect to a fat cattle market or a source of grain can grow cattle for feeders on a contract basis.

5. Stockers can utilize large quantities of harvested roughages and aftermath, thus cheapening the price of feeders if the same owner is also to finish them.

6. The stocker program is quite flexible because adjustments in numbers fed are easily made.

7. Death losses are lower than in the cow-calf program.

8. Little equipment is required for a farm-sized operation, but the use of mechanized equipment for handling silage or other roughage reduces labor requirements in the large commercial or custom feedlot.

9. Little capital is required if a grower contract can be arranged with a feedlot, thus not requiring the purchase and ownership of cattle.

Some inherent weaknesses of the stocker program are the following.

1. Much capital or available credit is required as compared with the cow-calf program, unless the program is a contract arrangement and someone else supplies at least the major share of the capital.

2. Buying and selling skills are extremely important in this program because shrink and other losses, on both ends, and mistakes in judging quality or in judging the health of the stockers can quickly offset the economical gains that may be made.

3. The stocker program may have conflicting labor loads; for example, roughage-harvesting time and the winter feeding period.

4. Commercial feedlots may not have available to them, for purchase, suitable roughages at prices that would result in low-cost gains. Storage of such feeds—silage, for example—also presents special problems in added investments.

5. The program carries above-average risk for the owner-feeder. Total gains made are not large in proportion to the weight purchased; hence profits must come both from a favorable price spread between buying and selling prices and from added weight made at feed and labor costs that are lower than the selling price. The latter is not difficult to achieve with stockers but, as mentioned, the total weight gain per head to which this factor may be applied is comparatively small. Thus the profits made because of economical gains can be lost when the cattle are sold if the selling price is much below the purchase price.

Large commercial feedlot operators who have a more or less continuous type of operation would do well to make most of their stocker purchases during the favorable fall season, carrying at least some of the cattle along on stocker rations until they are needed to replace finished cattle in the finishing pens. Sometimes early spring is a favorable time to buy stockers in the winter small-grain belt, but the cattle feeder should not depend on this situation without alternatives.

MARKET CLASSES AND GRADES OF STOCKERS AND FEEDERS

Throughout the year, many thousands of cattle are marketed daily in the large central markets and auction markets, and by direct sale or purchase on farms and ranches. These cattle are of every kind, displaying a wide range of combinations of the various characteristics such as sex, age, weight, size, conformation, breeding, and condition. There is fortunately a market for each kind of cattle and the variation in prices received reflects both the supply and the demand for each kind and the variation in suitability to the purpose for which each kind is purchased.

The need for standardized terms to describe the various kinds of cattle has long been realized. As early as 1902, H. W. Mumford of the Illinois station published the results of a lengthy study of the subject. In 1918 when the Bureau of Markets, now the Bureau of Agricultural Economics, of the U.S. Department of Agriculture (USDA) inaugurated its market-reporting service on livestock at Chicago, the market classes and grades of cattle suggested by Mumford were used to establish the terminology and classification of cattle for market-reporting purposes. It is a tribute to the pioneer workers in this field that the classes and grades of cattle have since been changed only slightly. The use of uniform descriptions of all classes and grades of cattle by producers, selling and buying agencies, and packers throughout the country has contributed much toward the orderly marketing of stocker, feeder, and slaughter cattle.

In recent years there has been some criticism of the grading system for stockers and feeders, because performance in the feedlot is not always related to grade—that is, the higher-grading stockers are not always the fastest and most efficient gainers. This justifiable criticism does not alter the positive features of the present system, which should continue to be useful until one that evaluates more than just the subjective features of feeder cattle can be developed. A number of experiment stations are pursuing this problem, but results to date are not consistent enough to warrant discarding the long-used system.

The official standards for live cattle as developed by the Department of Agriculture and as generally used today provide for segregation according to (1) use, such as stocker, feeder, slaughter, (2) class, determined by sex condition, and (3) grade, determined by the conformation and apparent desirability of the animal for its particular use. Table 66 shows the grouping of feeder cattle according to class or sex condition and grade.

Table 66

Market Classes and Grades of Stocker and Feeder Cattle[a]	
Class (Sex Condition)	Market Grade
Steer calves and yearlings	Prime, choice, good, standard, utility, inferior
Heifer calves and yearlings	Prime, choice, good, standard, utility, inferior
Cows	Choice, good, standard, utility, inferior
Bulls	Ungraded
Stags	Ungraded

[a] Based on USDA Consumer Marketing Service Regulatory Announcements 183, 1965.

The grade assigned to a stocker or feeder is based on a subjective or "eyeball" appraisal of the extent to which the animal meets certain established standards of conformation, quality, breeding, constitution and capacity, and condition or finish. Figure 62 shows photos of stocker and feeder calves that represent or correspond to the various grades. In practice, the upper three grades are further divided into three subgrades. For example, in the choice grade, market men speak of low, average, and high choice, although these subgrades are not a part of the standard government grading system.

It should be mentioned that there is presently a considerable body of opinion that believes there is a serious contradiction between the USDA grades assigned to stockers and feeders and the body-type scores that are being given to such cattle by the official graders of the Cooperative Extension Service who are working in the various performance testing programs. Body-type scores being used range from 1 for the lower-set, more compact, early-maturing cattle to a score of 5, representing the most extreme of the British breeds of cattle in scale and skeletal structure. The higher type scores are being assigned to the apparently more desirable kinds of stockers and feeders, but such calves would, by definition, receive a lower USDA grade unless some changes are made in the description of the various grades by the U.S. Department of Agriculture. It is felt that the sorting of cattle according to body-type score more accurately distinguishes them with respect to their feedlot performance potential and ultimate carcass value than does USDA grading. Several purebred breed associations subscribe to this view.

A brief description of the characteristics that typify each USDA grade of stocker and feeder follows.

Prime stocker and feeder cattle approach the ideal in beef type or conformation, possessing all the characteristics, except high degree of

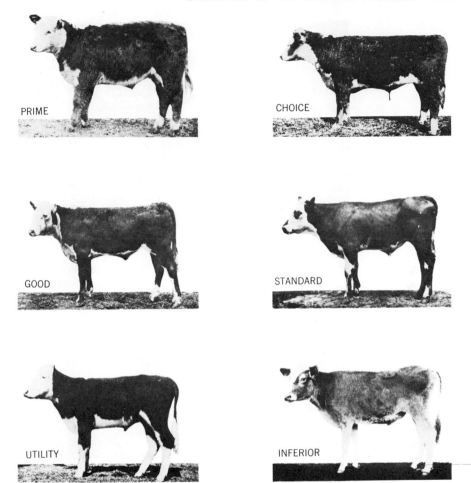

Fig. 62. U.S. grades of stocker and feeder steers. In planning a feeding program, the choice of grade depends on many factors, but mainly on the amount and quality of the feed supply. (USDA.)

finish, that will result in prime slaughter cattle if they are fed long enough. Thick natural fleshing, symmetry and balance, and quality are developed to the highest degree. A drove of such cattle is uniform in type and color and, because only a small proportion of the stocker and feeder cattle supply is desirable enough in conformation to make this grade, they command a premium price when either sold or purchased. In the trade such cattle are usually spoken of as "reputation brands," and they are ordinarily not found in central markets, but rather are either sold at feeder cattle auctions or are contracted for on the farm on ranch. Care

Fig. 63. A drove of high-choice and prime calves. Dehorning would have increased the selling price of these calves by 5 to 10 percent. Note uniformity, good heads, width of body, and straightness of lines. (American Hereford Association.)

must be exercised or too many of the early-maturing body-type 2 calves may be found in a drove of prime calves.

Prime calves usually carry considerable bloom and finish, often as a result of creep-feeding, and should be handled in a program that makes the fullest use of these characteristics. It seldom pays to place prime grade calves on a stocker program; they should rather be put on full feed as soon as practicable. Prime grade home-raised calves make ideal feeders for the baby-beef program, in which case they are probably finished as rapidly as possible. Prime grade feeders should be fed to high-choice or prime slaughter grade in order to command the selling price needed to offset the premium purchase price. It is worth repeating that the correlation between grade and performance in the feedlot is low, or even negative in some instances. Therefore, by all means a man who feeds prime grade cattle should be almost certain of a premium market for his cattle, as such cattle can be very risky because of high costs and the long feeding period required. Such cattle seldom are sold on a "yield grade" basis because they actually may be discounted on the market for carrying excess finish. Only a highly specialized trade demands such cattle; thus the market for them is rather unpredictable.

Prime grade heifers are seldom available for feeding purposes because heifers of such quality are usually retained as herd replacements or are sold at higher prices than they would bring as feeders, to add to or start other breeding herds.

Choice stocker and feeder cattle, especially those from the middle to lower end of the grade, are the most numerous kind of cattle in American feedlots. Choice grade cattle resemble those of prime grade but lack some of the eye appeal. Whereas they are moderately wide, muscular, and low

Fig. 64. Yearling feeder steers heading into some railroad shipping pens in Wyoming—destination: an Indiana farmer-feeder's farm where they will first utilize corn stalk fields and fall aftermath for about 50 days. Cow herds are being replaced by such steers in some of the better ranching areas of the West. (Union Pacific Railroad.)

set, they may have more scale and frame and some signs of coarseness and thus may lack the refinement and quality of prime grade feeders. Choice grade feeders usually show evidence of being "good-doing" cattle, with considerable thickness and depth, but they may be somewhat lacking in straightness of lines and in balance. They are more uneven as to type, weight, and breed characteristics. Cattle feeders who choose this grade are often able to top out part of a drove for further feeding to prime grade. Cattle of choice grade are usually fed for at least 5 months if put on feed as yearlings and at least 7 or 8 months if started on full feed as calves. Many body-type 3 and 4 calves will be found in this grade.

Good grade stockers and feeders have the appearance of thrifty cattle with moderate thickness, but they are usually quite rangy, upstanding, and lacking in muscling and balance. Good grade cattle usually show the color markings of the principal beef breeds. Many of the yearling heifers sold as feeders fall in this grade. Many cattle of this grade are thinner than those in the upper two grades. An alert feeder buyer often spots a potential low choice slaughter cattle end on a drove of good grade feeder cattle, which could well be fed longer to upgrade them into choice.

Many steers originating in the winter small-grain pasture and bluestem areas are cattle of good grade, although cattle of the upper two grades come from these areas as well. Number 1 "Okie" cattle, a term used to describe many of the stockers and feeders with British breed markings originating in the Southeast and Coastal regions, fit into the good grade. If they appear to have a cross or two of Brahman breeding they may be called "No. 2 Okies" and probably belong in the standard grade. Good grade stockers and feeders often may be upgraded by fattening them so that they will grade low choice on the rail.

Standard feeders are lacking in beef conformation. They are long, shallow, light muscled, and narrow and often show evidence of mixed dairy or Brahman breeding. Frequently they are the result of crossing British beef bulls on cows of either dairy or native breeding. Such cattle may not have made a profit for the breeder, but they may be quite profitable for the feeder. Because of their growthiness and size and because they are often older for their weight as a result of poor nutrition, they often gain quite well. Gains must be very economical and the purchase price must be low, because such cattle usually do not grade better than standard or good on the rail and thus will never sell very high. The feeding period must be short because these cattle soon reach the point where the gains cost more than the selling price warrants. Expert judgment is required in feeding standard grade cattle, because speculative risks are high. Heifers of this grade are rarely available in large numbers, because they will have been more profitably marketed as slaughter calves.

As slaughter cattle, standard grade feeders usually grade standard if

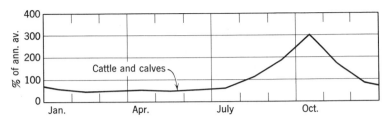

Fig. 65. Seasonally heavy shipments of stockers and feeders occur in the fall months of
September and October. (USDA.)

young, or commercial if mature. It is not uncommon, though, to find
these feeders being graded good on the rail after finishing. They produce
carcasses of high cutability but their dressing percent is low.

Utility and *inferior* feeders, at the bottom of the quality scale, can best be
described by saying that they lack beef characteristics. They are nondescript in breeding and color and are often unthrifty. Although some cattle
of this grade may make good gains for short periods, the low selling price
and the rather numerous stunted or "dogied" individuals that perform
poorly make this grade of feeder a very risky proposition. Carcasses from
cattle of this grade generally grade utility.

It is seldom advisable to feed anything but the good and choice grades
of cattle in strictly stocker feeding programs. Because the lower grades
seldom sell well enough to someone else as feeders for further feeding, a
grower program usually does not pay. The man who finishes the lower
grade of cattle usually does best not to delay unduly long in getting his
cattle fed out and marketed.

BUYING STOCKERS AND FEEDERS

Stocker programs should be planned to take advantage of the seasonal
fluctuations in supply, and therefore price, of such cattle. Figure 65 shows
that the fall months are the heaviest in movement of stocker and feeder
cattle, whereas the spring and early summer months are the lightest.
Figure 66 shows that prices react to the supply situation and that stockers
and feeders can be bought much more favorably in the fall months.

SOURCES OF STOCKER AND FEEDER CATTLE

Principal sources of stocker and feeder cattle and the relative importance
of each source vary specifically from area to area and even from

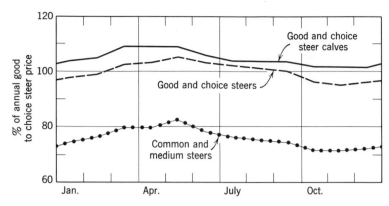

Fig. 66. Seasonality in prices of stocker and feeder cattle at Kansas City, one of the more important central markets for such cattle. (USDA.)

community to community. The principal sources in general are (1) public (terminal) markets, either from dealers or commission firms, (2) local dealers, (3) direct from ranchers or farmers, (4) auctions, and (5) contract arrangements.

A survey of 8 counties in Illinois, which included 123 cattle feeders and 270 different lots of cattle, supplied some interesting information concerning the source and area of origin of feeders. The counties surveyed are typical of much of the central Corn Belt area. The survey showed that the cattle feeders bought their stockers and feeders from the following sources: public stockyards, 46 percent; local dealer, 36 percent; direct from grower, 12 percent; auctions, 4 percent; other, 2 percent. Ranchers in the Northwest sell more of their feeders to local dealers who, in turn, sell them to cattle feeders. Southwestern ranchers and farmer-breeders with smaller herds in Missouri and Arkansas sell mainly through public stockyards and auctions. Even here, dealers play an important role, as they will have purchased the feeders from the rancher at the public stockyards or auctions and immediately offered them for sale to prospective feeder-buyers.

Outside the Corn Belt the public market has less relative importance as a feeder cattle source. Buying in auctions and direct from neighbors is more common where the number of feeders fed per farm is smaller and cattle feeding is relatively less important. Since completion of the study referred to above, the trend toward less selling through public stockyards has continued and, in fact, accelerated. There is greater use of auction sales and of direct dealing.

A recent regional publication from the Kansas station describes a study of buying methods and kinds of cattle purchased by operators of stocker programs in western Kansas and northeastern Nebraska. These two areas are representative of the growers who grow out stockers that ultimately are to be fed out in two distinctly different kinds of cattle finishing operations. Western Kansas has numerous large-scale commercial feedlots typical of the newer feedlot industry in the Southwest, whereas northeastern Nebraska is more representative of the farmer-feeder system of cattle finishing found throughout the Corn Belt.

Table 67

Percentage of Feeder Cattle Purchased for Use in Grower Programs, by Buying Method, Grade, Kind of Cattle, and Age and Sex[a]

	Northeastern Nebraska	Western Kansas
Buying Method		
Owner-buying, all methods	**46**	**22**
At auction	35	19
Direct from cowman or grower	10	3
On contract from other owner	1	0
Order-buyer buying, all methods	**54**	**78**
At auction	16	60
At terminal markets	1	3
Direct from cowman or grower	30	11
On contract from other owner	7	4
Grade Purchased[b]		
U.S. choice	75	53
U.S. good	45	70
Standard and commercial	4	13
Kind of Cattle[b]		
Straightbred British	78	53
Crossbred	81	73
Dairy	7	6
Okie	7	38
Age and Sex of Cattle[b]		
Calves	**81**	**75**
Steers	55	62
Heifers	45	38
Yearlings	**38**	**50**
Steers	69	74
Heifers	31	26

[a] Based on North Central Regional Publication No. 218, Kansas Agricultural Experiment Station, 1974.

[b] Percentages do not add to 100 percent because one operator may have used cattle of several kinds.

Selected data from this publication are shown in Table 67. Items of particular interest: auctions were important sources of stockers and especially when order buyers bought for the western Kansas operators; owners did more of their own buying in the Corn Belt area and these operators bought higher-grade cattle of mainly straight British breeding, whereas the western operators bought more Okie cattle, probably from the Southeast; the western Kansas feedlots bought and fed, for resale to neighboring finishing lots, relatively more yearling cattle than calves and more steers than heifers.

METHOD OF BUYING

There is an old saying, "Well bought is half sold." Certainly this adage applies to the buying of stocker and feeder cattle. An inexperienced feeder should not attempt to buy his own cattle, but should place an order with a dealer, commission firm, or order buyer who knows his business and, preferably, also knows the feed situation of the particular farm or feedlot, the capabilities of the feeder, and how much risk the feeder can afford to take. The commission charge is small compared with the savings usually made. This does not mean that ranchers or dealers who have stocker or feeder cattle for sale are dishonest, but naturally they place their own or the seller's interests uppermost, and the prospective buyer should also follow the good business practice of having a knowledgeable person represent him on the other side of the transaction.

The methods of buying stockers and feeders are as follows.

1. *Buying direct from rancher or neighbor* by oral or written contract. Cattle for fall delivery are usually contracted for during the summer. Price, pencil shrink, weighing conditions, sort, if any, date of delivery, method of transportation, and arrangements for paying for the cattle should all be agreed upon in advance and preferably verified with a written, signed contract. Beginners should by all means avoid driving out to the range areas, shopping for bargains from ranch to ranch. Much time and money can be wasted and cattle are not often bought satisfactorily in this manner. Many experienced feeders believe that the increasing use of this practice by cattle feeders has upset normal channels of buying and selling feeder cattle, sometimes to the detriment of both buyer and seller.

2. *Placing an order with a commission firm or dealer.* In this arrangement the buyer makes his wants known to a dealer, preferably 3 or 4 months in advance of desired delivery. The commission firm or dealer locates the cattle, negotiates for them, and makes all necessary arrangements for

shipping. The cattle are paid for upon delivery, with the commission firm or dealer acting as go-between. Ranch or delivery-point weight, less the customary 2 to 4 percent pencil shrink, is the weight actually paid for by the buyer. Freight, insurance, feed costs in transit, commission, and miscellaneous costs are added to the original negotiated price. Sometimes the commission firm or dealer pays for and assumes ownership of the cattle in order to compute more simply the incidental costs and values, in case shipments from a particular ranch are to be divided and sold to more than one buyer. Some commission firms also pay for cattle that were not specifically ordered and assume ownership themselves, hoping to sell the feeders again at a profit. Cattle bought through a commission firm are usually not seen by the buyer until delivery. The buyer thus depends on reliability of the firm that he deals with.

3. *Buying direct from a dealer.* Some feeders prefer buying direct from a dealer at his yards, because they can see what they are buying and the "asking price" usually covers all costs. Naturally the asking price is as high as the demand allows and it may seem unduly high compared with prices published in many livestock reports because freight and other incidental costs are included. Sometimes ranch weight and sometimes dealer's yard weight is used to determine price. This buying method is convenient because dealers are often located in the midst of the feeding territory, although every terminal market also has numerous dealers.

4. *Buying in person at feeder cattle auctions.* The number of feeder cattle auctions is increasing, especially in areas that are developing rapidly in the cow-calf business, such as the Southeast, the Corn Belt, and the South. A special once-a-year kind of auction is sponsored by cattlemen's associations, with large numbers of stocker and feeder cattle being brought together and sorted or graded into lots that are uniform in sex, grade, and weight. They are then auctioned in truckload or smaller lots to the highest bidder.

When a good-sized crowd is present, bidders may become overenthusiastic and actually pay more than the market on that day. At auction sales cattle may sell with more fill than cattle bought elsewhere, and cattle that appear uniform when purchased, as a result of the sorting, may soon grow uneven because they came from several different farms or ranches. On the other hand, because buying costs, freight costs, and profits to middlemen may be lower or absent, delivered or on-the-farm costs are also often lower than when cattle are bought by other methods.

Buying in "community sale" auctions is done by many smaller feeders. Buying at auctions requires great skill, especially if cattle must be bought

Fig. 67. Well-managed auction sales with convenient truck and rail loading facilities are gaining in prestige and popularity as a source of stockers and feeders. (Schnell Livestock Auction Market, Dickinson, North Dakota.)

"by the head" rather than by weight. It is difficult to put together a uniform lot of cattle and the chance of buying diseased, stale, or overexposed cattle is great. It should be said that most auctions of this kind operate under the highest standards, and in some localities they are the only source of stocker and feeder cattle. If the yards cover more than 100,000 square yards in area, they are "posted"—that is, they come under the supervision of the U.S. Department of Agriculture Packers and Stockyards Act, making them subject to enforcement of prescribed rules of sanitation, scales inspection, and management in general.

TRANSPORTATION OF STOCKERS AND FEEDERS

Railroads once hauled most of the stocker and feeder cattle from the range states to destinations east of the Mississippi River, because rail shipment was cheaper than other methods for such distances. The cattle were delivered to terminal markets, dealers' yards, and sometimes to holding pens on railroad sidings near the feeders' farms or feedlots.

Trucks are now used to move most of the cattle, regardless of the distance and the source of the shipment. Double-decked trucks, usually with a relief driver, have become highly competitive with railroads for long hauls, because they can carry heavier loads than formerly and they have appreciably reduced the time spent on the road.

Fig. 68. Gathering a drove of choice yearling steers off of Kansas Flinthills bluestem range. They will be trucked to a Corn Belt farmer-feeder's lot in Iowa. (Drovers Journal.)

Railroads are required by law to unload, feed, water, and rest cattle every 28 hours, except when the owner gives written permission for a "36-hour release," which is permissible if the haul takes less than 36 hours. This helps ensure delivery of the cattle in better condition with less shrink, because they will not be delayed for the additional minimum 5 hours otherwise required for a rest stop. Truckers are not required by federal law to unload and rest and feed the cattle. However, some buyers insist that their truckers stop and rest the cattle, to reduce stress when hauls are quite long, such as from Florida to Arizona. Studies have not conclusively shown that the loss from stress is greater in unrested cattle. They seldom eat or drink anyway during shipment.

SHRINK FROM SOURCE TO FEEDLOT

Shrink may be of two kinds in stocker and feeder cattle. *Excretory shrink* is the loss in weight from excretion of manure and urine, and this shrink can be regained in a short time. *Tissue shrink* is actual loss of flesh and body water, and this type of shrink is regained much more slowly. Length of the trip, previous feeding, and temperature affect both amount and type of shrink. Most of the shrink on a short trip on a hot day with a heavy fill at loading time is excretory shrink. "Sappy," milk-fat calves, shipped on a week-long trip, may incur heavy tissue shrink, especially if they do not eat or drink adequately at rest stops, as they seldom do.

It is difficult to generalize as to what is normal shrink for stocker and feeder cattle. Calves may shrink up to 10 percent or more, whereas yearling cattle may not shrink over 4 or 5 percent. The time required to return to pay weight is perhaps as important as pounds of shrink in transit, because all feed fed during the shrink recovery period is added cost. This period may vary from less than a week for older, heavy cattle to as much as a month for calves or stale cattle. An outbreak of shipping fever during this shrink-recovery period of course drastically increases the time required to return to pay weight.

HOMEGROWN STOCKERS AND FEEDERS

Even in the heavy cattle feeding areas in the Corn Belt there are large numbers of brood cows in comparatively small farm herds, and this is also true throughout the Southeast and in the Great Lakes region where cow numbers have increased greatly during the last 30 years. Calves or yearlings produced in these herds are often fed out on the farm where they are produced, but still others are offered for sale locally. There are both advantages and disadvantages to buying local or native feeder and stocker cattle.

Advantages of native or homegrown stockers and feeders:

1. Native stockers and feeders are more likely to escape shipping fever and similar diseases, especially if bought direct on the farm where produced.
2. Delivered price may be lower because of lower freight and buying costs or because of less demand for local calves.
3. Shrink, especially tissue shrink, may be lower. Of course, there is no shrink at all, except that resulting from weaning, in homegrown cattle that are fed on the farm where they are produced.
4. Native stockers and feeders do not have to adjust to sudden changes in weather, altitude, and the like.
5. Native cattle are somewhat accustomed to the feeds produced in the area.

Disadvantages of native or homegrown stockers and feeders:

1. They are likely to be less uniform in quality, condition, and age.
2. Native or homegrown stockers and feeders may be of lower quality

than those purchased in the traditional range areas, but this is not always true.

3. Such cattle are likely to be fleshier and thus do not make such large compensatory gains on roughages as shipped-in cattle are apt to do.

4. Several purchases must be combined to meet the needs of cattle feeders who want to feed more than a car or truckload of cattle.

5. Native stockers and feeders may not be available in large enough numbers when wanted.

PAY WEIGHT

Pay weight is the weight at the producer's farm or at the loading point, less whatever allowance for shrink, if any, the buyer can negotiate for. The method for determining pay weight should be agreed upon in advance in order to prevent misunderstanding. The most common practice in the range area is to allow the buyer a 3 percent shrink—that is, the calves or yearlings are weighed at the producer's ranch or at the loading point, and then 3 percent of the weight is subtracted. Such shrink is generally referred to as "pencil shrink." Final settlement is based on the calculated shrunk weight. If cattle must be driven some distance to weigh and load, no pencil shrink is expected by the buyer. If cattle are lotted overnight without feed or water, only 1 or 2 percent or possibly no shrink is deducted. The delivered weight is subtracted from pay weight to determine how much shrink occurred to the weight of feeder cattle actually paid for.

DISEASE PROBLEMS AND DEATH LOSS IN STOCKER AND FEEDER CATTLE

Losses due to disease and accident are highest in stocker and feeder calves and lowest in older, heavy cattle. A survey of reports on 113 droves of cattle fed by Farm Business Farm Management Service cooperators in Illinois showed that death losses of calves were 2.43 percent; yearlings, 0.85 percent; 2-year-olds and over, 0.43 percent. Recent studies of losses in the High Plains commercial feedlots show extreme variations, ranging from as low as 0.5 to as high as 25.0 percent loss. Items that seem to contribute most to the variation found were: source of cattle (that is, area and method of putting loads together), distance hauled, skill of feedlot pen riders and helpers, weather during the first 2 weeks after arrival, and original condition of the cattle. Even the length of the exhaust pipe on the hauler's truck had an influence; the lower decks of calves were adversely affected by the carbon monoxide emitted by short exhaust pipes or stacks. Most of these losses are caused by exposure to certain viruses as yet not well described, and to stress, which weakens the calves' resistance. Exposure occurs somewhere between the producer's farm or ranch and the feedlot. Shipping fever or "stress fever" is the principal disease that causes losses. This and other diseases are more fully discussed in Chapter 27.

Results of one of many similar experiments recently conducted on the effect of high-level feeding of antibiotics and sulfa drugs are shown in Table 68. Low levels of 75 milligrams daily of any of several broad-spectrum antibiotics have been fed to stocker calves, but usually this level does not prevent shipping or stress fever. It improves gains over a period of 120 to 150 days by about 5 to 10 percent. A higher level of 350 to 500 milligrams daily over the first 3- or 4-week period is apparently required to reduce the incidence of shipping fever and death loss. It should be noted that the gains reported in Table 68 are unusually high. This is

Table 68

Effect of High Levels of Antibiotic and Sulfa Drugs in Conditioner Supplements (28 days)[a]

	Control	350 mg Aureomycin	350 mg Sulfa-methazine	Aureomycin plus Sulfa-methazine
Number of animals	20	20	20	20
Average initial weight (lb)	441	416	427	421
Average final weight (lb)	500	493	489	494
Gain per animal (lb)	59	77	62	73
Average daily gain (lb)	2.11	2.75	2.21	2.61
Daily feed (lb)				
Corn	2.0	1.9	1.9	1.8
Soybean meal	2.0	1.9	1.9	1.8
Corn silage	17.8	18.9	17.7	18.3
Corn cobs[b]	0.9	0.6	0.8	0.6
Pounds of feed per pound of gain				
Corn	0.9	0.7	0.9	0.7
Soybean meal	0.9	0.7	0.9	0.7
Corn silage	8.4	6.9	8.0	7.0
Corn cobs	0.4	0.2	0.4	0.2

[a] Indiana Research Progress Report 250, 1966.

[b] Three pounds of ground corn cobs per head daily were fed only the first 11 days.

because "off-truck" empty weights were initially used in the study in order to reduce the variability in results that would be attributable to differences in initial fill.

Responses to sulfa drugs in the ration are usually negative, as in this trial, but some trials have been reported where responses were obtained from sulfa drugs dissolved in the drinking water. Many veterinarians prescribe the use of "electrolytes," sold under various trade names, in the drinking water to ensure rehydration to offset the loss of fluids during weaning, shipment, and adjustment to new feed and surroundings. It is not completely understood how these chemicals function, but evidently they are involved in the maintenance of the proper acid-base balance in the cellular fluids. Stocker calves that are extremely tired and stressed often drink but do not eat for some time after arrival, and perhaps medication by way of the drinking water is indicated in this situation.

Vaccines that are administered once in the summer as calves are last worked, and again when they are shipped at weaning time, appear helpful as a means of reducing the incidence of the shipping fever problem. Another promising method of pretreatment is early weaning of the calves and getting them accustomed to grain and hay. The injection of tranquilizers, vitamins, or antibiotics and a number of other practices, as preventives, are not especially promising in most situations. More research on the subject of pretreatment, conditioning, or "backgrounding," though it is expensive and difficult to conduct, is badly needed.

The most promising new technique in reducing the harmful effects of weaning and shipping stress is the practice of giving calves readily available sources of energy and other nutrients in their rations, in combination with oral or injected antibiotics and sulfa drugs. In the past, the observation that calves readily consumed grassy hay upon being unloaded led to the general use of this practice in feeding tired, newly received calves. It now appears that the calves consume such a low level of hay, under these conditions, that neither the energy nor the protein requirements are being met. Limiting the amount of roughage offered and supplementing a highly palatable concentrate mix containing grains, molasses, protein concentrates of natural origin, a phosphorus source, additives such as antibiotic-sulfa combinations, and vitamin A, is a program being followed by many progressive feeders and feedlot operators. If cases of shipping fever still are found, these are sorted off and handled with therapeutic treatments according to veterinarians' recommendations.

California workers, studying the effect of ration on newly received, stressed calves over a period of 28 days, used incremental increases in percentage concentrates in the total ration all the way up to 90 percent. They obtained increasingly favorable responses with each step-up in

Table 69

Effect of Various Supplements Fed to Stressed Calves on Incidence of Respiratory Disease and Performance (Two Tests, 18 and 24 Days)[a]

Treatment	Soybean Meal	Simple Grain Mix	Complex Grain Mix	No Anti-biotic	Anti-biotic + Sulfa
Number	60	64	60	94	90
Initial weight (lb)	464	472	466	466	469
Gain (lb)[b]	63	63	66	57	70
Average daily gain (lb)	3.00	3.00	3.10	2.70	3.37
Number treated for respiratory disease	17	21	8	34	12
Severe cases of respiratory disease	10	8	2	17	3

[a] Nebraska Beef Cattle Report, 1973.
[b] From empty weights off the truck.

energy, in terms of reduced incidence of sick calves and in costs of treatment, and in time and feed required to recover shrink.

Nebraska workers, using rations and systems more likely to be employed by farmer-feeders, obtained similar responses as shown in Table 69. They used a basal ration of either well-eared, green, chopped corn or oatlage, both of which contained about 40 percent grain on a dry basis. They added either soybean meal or a simple or complex grain mix, with and without an antibiotic-sulfa combination, as supplemental feed at the rate of 2 or 5 pounds per head daily. Addition of grain to the already grain-containing roughages failed to improve gain further, but the oral antibiotic-sulfa additive improved performance and reduced incidence of severe cases of respiratory disease. The Nebraska calves, having been shipped over a shorter haul, were not as highly stressed as the southeastern calves used in the California study.

"LAID-IN" COST OF STOCKERS AND FEEDERS

The buyer of stocker and feeder cattle is of course interested in buying his cattle as cheaply as possible without sacrificing quality and bred-in performance. The rancher or breeder is just as interested in selling as high as possible. However, the buyer always pays more than the grower receives. There are numerous items of expense from ranch to feedlot, and these are all added to the purchase price by the time the cattle are "laid in" or safely unloaded at their new home. Not all of the following items are included in every case, although most of them are. Charges naturally vary somewhat from area to area.

Items of expense in procuring stocker and feeder cattle:

1. *Freight.* Freight varies with distance and method of hauling, but it ranges from 15 cents per hundredweight for locally bought cattle to as much as $3 per hundred for cattle requiring a long haul.

2. *Commission and buying expense.* This charge applies only when a buyer has placed an order with a commission firm or order buyer. It may range from 10 to 20 cents per hundredweight depending on services rendered, but generally averages out at about $1 per head.

3. *Transit insurance.* Insurance is usually figured by the head, but it can be broken down to a hundredweight basis and varies from 3 to 6 cents per hundredweight depending on distance hauled.

4. *Feed and labor of feeding en route.* Feeding cost varies greatly depending on how many rest stops must be made, but it may amount to 10 or 15 cents per hundredweight. Trucked cattle are seldom fed on the road.

5. *Trucking from railroad or dealer's yard to feedlot or farm.* The size of this item depends on distance hauled, and it may vary from a low of 15 cents to $1 per hundredweight if the haul is as much as 200 miles.

6. *Personal expenses on the buying trip.* Some buyers who go to the West to buy cattle may mark the trip off as vacation expense, but usually one or more special buying trips are necessary, even if only to the local feeder yard. In any case, this expense must be borne by the newly acquired stocker or feeder cattle. A reasonable estimate of this item is from a few to 25 cents per hundredweight.

The items listed above usually add a total of $1 to $4 per hundredweight to the purchase price of stockers and feeders and must be added in calculating the laid-in cost. Locally purchased cattle, hauled in the buyer's truck, naturally have little of this type of cost added, but the feeding enterprise must carry its share of the depreciation and maintenance costs of the owner's truck.

Shrink and death loss occurring on the farm or in the feedlot have already been mentioned, but actually these items can well be added to the laid-in price of the feeders before final calculations of profit or loss are made. The same may be said for financing costs, or interest and carrying charges on borrowed money used to buy the feeder cattle. For example an 8 percent loan for 1 year on a long-fed calf costing $150 and weighing 400 pounds at time of purchase adds $3.00 per hundredweight to the cost of the calf.

FEEDING AND MANAGEMENT OF STOCKER CATTLE

In planning the management of stocker cattle, it should be remembered that such animals are kept for three reasons: (1) to ensure a supply of the right kind of cattle for the finishing lot at the proper time; (2) to utilize farm roughages, otherwise unmarketable; or (3) to "cheapen down" the feeder cattle. No matter which of these objects is uppermost in the owner's mind, economy in feeding and management is of utmost importance.

GAINS OF STOCKER CATTLE

Stocker cattle should be fed with the minimum outlay for feed consistent with normal growth and efficient feed conversion. Increase in condition much beyond that represented by normal growth may not be desirable. In fact, a noticeable increase in condition is usually undesirable if the stockers are to make the most efficient use of the finishing ration that subsequently will be fed.

The amount of gain necessary to account for normal growth depends, of course, on the age of the animal. Table 70 shows the results of an early experiment that demonstrates the levels of daily gain that constitute normal growth without fattening for annual periods up to 1, 2, and 3 years of age. Growth is most rapid in young animals and gradually decreases as the animals approach maturity. Thus calves will make greater stocker or growth gains than yearlings and yearlings greater gains than 2-year-olds when all are on the plane of nutrition that permits normal growth but little or no improvement in condition. In fact, even yearling

Table 70

Age of Cattle		Group 1 (Maximum Growth Without Fattening)	Group 2 (Growth Distinctly Retarded)
Yearly Gains Made by Cattle on Different Planes of Nutrition[a]			
		(lb)	(lb)
30 to 360 days	Weight gained	409.7	241.6
	Daily gain	1.24	0.73
360 to 720 days	Weight gained	320.8	235.6
	Daily gain	0.89	0.65
720 to 1,080 days	Weight gained	126.9	121.2
	Daily gain	0.35	0.34

[a] Missouri Research Bulletin 43.

steers grow so slowly that it is seldom advisable to carry them on nonfattening rations. Because they require a large amount of feed for maintenance, gains that represent only slow growth in older cattle are inefficient and costly.

As far as practicable, the stocker ration should prepare the cattle for making the maximum use of the finishing ration, or of grass if they are to be grazed without grain. The amount of gain made on grass or in the feedlot varies inversely with the amount of gain that is made during the stocker period. This inverse relationship—that is, the tendency for lightly wintered cattle to gain faster and for more heavily wintered cattle to gain slower in a subsequent grazing or finishing program—is the result of a physiological phenomenon called compensatory gain.

California workers have reported the results of a comprehensive study that explains, in part at least, the phenomenon of compensatory gain. In order to correct for the large differences in fill that occur when cattle are fed on rations of widely differing roughage content, they slaughtered some of their animals at the beginning of the test and others at the end of each of the three periods in a three-phase study. They then determined the empty-body composition of the slaughtered animals and calculated the energy gain made in each period from known amounts of feed. The report is too long and involved to review in detail in this text, but it is suggested for further study. In essence, the California group found that, even after correcting for differences in fill and body composition, compensatory gain did in fact occur. They concluded that this might be explained by an improvement in feed utilization and an increase in feed intake by those animals previously fed on the lower-energy rations.

The amount of gain desired for weaner calves during the winter, or on a grower program at any time for that matter, depends largely on the way the cattle are to be handled the following summer or subsequent feeding period. If they are to be fed concentrates on pasture, they should be wintered better than if they are to be only grazed. Also, if they are to be grazed only until midsummer and then full-fed for the late-fall market, they should be wintered at a higher level than would be advisable if they were to be grazed the entire summer and fall.

Table 71 shows the effect of level of winter gain on the gains made the following summer when cattle are handled under three different summer management programs. Total gains for winter and summer are also shown. In fertile areas where the yields of roughages such as corn and sorghum silage are high, wintering gains can often be made more economically than summer gains resulting from finishing rations fed on pasture or even in drylot. Therefore it seldom pays to skimp on wintering or grower rations for cattle that are to be finished immediately afterward.

Although it is true that the smaller the gain made by young cattle

Table 71

Effect of Level of Wintering Gain on Summer and Total Gain[a]

Daily Summer Gain and Total Winter plus Summer Gain (Summer Programs = 100 Days)

	Winter Gain (160 Days)		Pasture Only		Full-Fed on Pasture		Full-Fed in Drylot		All Summer Programs	
	Average (lb)	Total (lb)	Average Daily Summer Gain (lb)	Total Summer + Winter Gain (lb)	Average Daily Summer Gain (lb)	Total Summer + Winter Gain (lb)	Average Daily Summer Gain (lb)	Total Summer + Winter Gain (lb)	Average Daily Gain— 100 days (lb)	Total Gain— 260 days (lb)
Number of Steers	60		20		20		20			
Lot										
1	0.92	146	1.42	288	2.70	416	2.35	381	2.17	363
2	1.25	200	1.35	335	2.00	400	2.12	412	1.82	382
3	1.51	231	1.10	341	1.55	386	2.00	431	1.55	386
4	1.64	262	1.05	367	1.60	422	1.90	452	1.51	413
5	1.65	263	0.92	355	1.82	445	2.12	475	1.62	425

[a] Illinois Cattle Feeders' Day Report.

during the winter, the greater the gain on pasture or in drylot the following summer, the winter and summer gains are not exactly inversely proportional, as is often believed. For example, if one lot of steers gains twice as much during the winter as a second lot, its summer gains are not limited to half those of the second lot, but will probably be 70 to 90 percent as much. Consequently, it usually happens that the cattle that make the largest stocker gains also make the most economical and the largest total gain for the entire period. Although the effect of the stocker gain on the finishing gain varies from year to year and between different droves of cattle, a good rule for the practical cattleman is that, for every additional pound that stocker calves gain during the growing period, they will gain 0.5 pound less in drylot or on grass.

Recommendations for level of stocker gain for stockers that are to be handled subsequently according to the more common steer management plans are shown in Table 72. Recommendations for feeding replacement heifers are found in the two preceding chapters.

The level of stocker gain to aim for is complicated not only by the matter of its effect on compensatory gain in a subsequent finishing period but also by the relative costs of concentrates and roughages. This latter point explains in large part why commercial feedlot managers in the West and Southwest ordinarily feed for higher gains in their stocker or

Table 72

Recommended Gains for Stocker Cattle (120–150 Days)				
	Calves		Yearlings	
Method of Feeding and Management the Following Summer	Total Gain (lb)	Average Daily Gain (lb)	Total Gain (lb)	Average Daily Gain (lb)
1. Grazed entire summer; sold as feeders or started on feed in late fall (November)	115	0.75	100	0.66
2. Grazed until August 1; full-fed until late fall or early winter (November or December)	150	1.00	115	0.75
3. Pastured only until about June 1; then full-fed on pasture and marketed late fall (November)	185	1.25	150	1.00
4. Full-fed during summer in drylot or pasture; marketed in early fall (September)	275	2.00	200	1.50

backgrounded calves. Purchased roughage energy is relatively high in cost when compared to concentrate energy in these regions, whereas in the Corn Belt and wherever silage can be grown economically, homegrown silages supply roughage energy at a lower cost than grain or concentrate energy. In fact, the stalk portion of the grain crop may have no value at all unless it is fed to some kind of cattle, and stockers make efficient use of this valuable by-product of grain production. A third point that must be mentioned is that of length of subsequent finishing period. Stockers that have gained at a higher rate usually are fatter when started on finishing rations and thus require a shorter feeding period to reach satisfactory weight and grade. This in turn results in slightly more efficient conversion rates while on the finishing ration and reduces the nonfeed costs which are usually on a fixed cost-per-day basis.

NUTRIENT REQUIREMENTS FOR STOCKER CATTLE

National Research Council recommendations for the daily nutrient needs and for percentage nutrient concentration in the rations for different rates of growing gains by stocker steer calves and yearlings and replacement heifers are shown in Tables 73 and 74. Morrison's feeding standards for the same classes of cattle are shown in Table 75.

As already mentioned, the feeding of stocker cattle is not likely to prove profitable unless the ration consists largely of farm-grown roughages. The presence of much grain in such rations is unjustified. Calves usually require 3 to 5 pounds of grain, or about 1 percent of body weight, daily to ensure satisfactory gains unless they are fed almost a full feed of corn or sorghum silage. Heifer replacement calves or steer calves that will be grazed the next summer can get by on only a full-feed of excellent-quality legume or mixed hay or, at best, no more than 1 to 2 pounds of concentrates. In the absence of some kind of legume roughage, it is highly advisable to feed a high-level protein supplement to supply the protein requirements of the growing animals. Usually 1 pound of such feeds per head daily is sufficient. If 4 or more pounds of good legume hay or its equivalent in silage is fed, protein concentrates may be dispensed with entirely. Urea may be added at the rate of 10 to 20 pounds per ton at silo-filling time to corn and sorghum silage with reasonable success in meeting the protein requirements.

The importance of supplying adequate protein in the stocker ration is shown by results of the feeding experiments summarized in Table 76. The addition of 0.50 pound of cottonseed cake to a full feed of prairie hay increased the daily gain to 0.58 pound, whereas the addition of another 0.50 pound of cake further increased the gain by 0.37 pound. Little further increase in daily gain was obtained when the supplement feeding rate was increased to 1.5 pounds. This test shows that calves should have from 0.75 to 1 pound daily of high-level protein supplement when they are consuming nonlegume roughages if they are to make satisfactory gains. On the other hand, results secured at the Indiana station, reported in Table 77, in which 2 to 2.5 pounds of protein-mineral-vitamin concentrate were fed to calves wintered on oat straw, corn cobs, and soybean straw, indicate that relatively high levels of nitrogenous supplements, as well as minerals and vitamins, must be fed with such low-grade roughages if the feed nutrients in such roughages are to be efficiently utilized.

Complex protein-mineral-vitamin supplements are beneficial when added to low-quality roughages as just shown. However, such supplements are not required when such high-quality roughages as legume-grass or corn silages are fed, as shown by results of Illinois studies reported in Table 78. Had the complex supplements been fed at levels that would supply only the usually recommended level of protein concentrate, the results would not have been so overwhelmingly in favor of simple high-energy concentrates from a cost standpoint.

ROUGHAGES USED IN STOCKER PROGRAMS

The quality of the roughage to be utilized has much to do with the age of cattle chosen to be fed. Yearling steers can make good use of corn or sorghum stover, cobs, and straw. Calves, on the other hand, should be fed a limited amount of such materials. If possible, corn, sorghum, or grass silages and legume hay should furnish at least 50 percent of the dry matter of the roughage ration for calves. The remainder may well consist of straw or other low-quality roughage.

It is important that the cattle be given all they will eat of some component of the ration; otherwise their hunger is unsatisfied and they are restless and waste some of their energy in moving about. As a matter of economy, the better practice is to limit the quantities of legume hay and silage to the amounts actually required to produce the desired gains, and to keep a supply of cheaper feed such as grass hay, cottonseed hulls, or straw before the cattle at all times. Stocker cattle consume the equivalent of about 2.5 percent of their live weight daily, in total air-dry ration, if they are in average flesh. Thin stockers consume slightly more, up to 3 percent, and, conversely, fleshier stockers consume less.

Corn and Sorghum Silage.   Silage made from high-yielding corn and sorghum varieties is the most satisfactory feed for stocker cattle. The

Table 73

Nutrient Requirements of Stocker Cattle (Daily Nutrients per Animal)[a]

Body Weight (kg)	(lb)	Average Daily Gain (kg)	(lb)	Daily Dry Matter per Animal (kg)	(lb)	Total Protein (kg)	(lb)	Digestible Protein (kg)	(lb)	Energy ME[b] (Mcal)	TDN[c] (kg)	(lb)	Calcium (gm)	Phosphorus (gm)	Carotene (mg)	Vitamin A (1000 IU)
Growing Steers																
150	330	0.00	0.00	2.7	5.9	0.21	0.46	0.11	0.24	5.6	1.5	3.3	5	5	15.0	6.0
		0.25	0.55	3.1	6.8	0.34	0.75	0.22	0.48	7.1	2.0	4.4	8	7	17.0	6.8
		0.50	1.10	3.2	7.0	0.39	0.78	0.26	0.57	8.4	2.3	5.1	12	10	17.5	7.0
		0.75	1.65	3.2	7.0	0.43	0.95	0.29	0.64	9.0	2.5	5.5	17	13	17.5	7.0
200	440	0.00	0.00	3.3	7.3	0.26	0.57	0.14	0.31	6.8	1.9	4.2	6	6	18.5	7.4
		0.25	0.55	4.5	9.9	0.45	0.99	0.27	0.59	9.3	2.6	5.7	8	8	25.0	10.0
		0.50	1.10	4.9	10.8	0.54	1.19	0.35	0.77	11.2	3.1	6.8	13	10	27.0	10.8
		0.75	1.65	5.0	11.0	0.56	1.23	0.36	0.79	12.5	3.5	7.7	18	14	28.0	11.2
300	660	0.00	0.00	4.5	9.9	0.35	0.77	0.19	0.42	9.3	2.6	5.7	8	8	25.0	10.0
		0.25	0.55	6.1	13.4	0.54	1.19	0.32	0.70	12.6	3.5	7.7	11	11	34.0	13.6
		0.50	1.10	7.7	16.9	0.77	1.69	0.47	1.03	15.9	4.4	9.7	14	14	43.5	17.4
		0.75	1.65	8.0	17.6	0.89	1.96	0.57	1.25	18.2	5.0	11.0	17	15	44.5	17.8

Growing Heifers

150	330	0.00	0.00	2.7	5.9	0.21	0.46	0.11	0.24	5.6	1.5	3.3	5	5	15.0	6.0
		0.25	0.55	3.2	7.0	0.36	0.79	0.23	0.51	7.3	2.0	4.4	8	7	17.5	7.0
		0.50	1.10	3.2	7.0	0.39	0.86	0.26	0.57	8.4	2.3	5.1	12	10	18.0	7.2
		0.75	1.65	3.3	7.3	0.44	0.97	0.30	0.66	9.3	2.6	5.7	17	13	18.5	7.4
200	440	0.00	0.00	3.3	7.3	0.26	0.57	0.14	0.31	6.8	1.9	4.2	6	6	18.5	7.4
		0.25	0.55	4.6	10.1	0.46	1.01	0.28	0.62	9.5	2.6	5.7	8	8	25.5	10.2
		0.50	1.10	5.0	11.0	0.56	1.23	0.36	0.79	11.4	3.2	7.0	13	10	28.0	11.2
		0.75	1.65	5.4	11.9	0.60	1.32	0.38	0.84	13.5	3.7	8.1	18	14	30.0	12.0
300	660	0.00	0.00	4.5	9.9	0.35	0.77	0.19	0.42	9.3	2.6	5.7	8	8	25.0	10.0
		0.25	0.55	6.2	13.6	0.55	1.21	0.32	0.70	12.8	3.5	7.7	11	11	34.5	13.8
		0.50	1.10	8.2	18.0	0.82	1.80	0.50	1.10	16.9	4.7	10.3	15	15	45.5	18.2
		0.75	1.65	8.6	18.9	0.95	2.01	0.61	1.34	19.6	5.4	11.9	17	15	47.5	19.0

[a] Adapted from *Nutrient Requirements of Beef Cattle*, Subcommittee on Beef Cattle Nutrition, National Research Council, 1970.

[b] ME (metabolizable energy) requirements for growing cattle were calculated from net energy values for maintenance and gain, as published in appropriate tables, for the weights and rates of gain shown.

[c] TDN (total digestible nutrients) requirements were calculated from metabolizable energy by assuming 3.6155 Kcal of ME per gm TDN.

Table 74

Nutrient Requirements of Stocker Cattle (Nutrient Concentration in Ration Dry Matter)[a]

Body Weight (kg)	(lb)	Average Daily Gain (kg)	(lb)	Daily Dry Matter per Animal (kg)	(lb)	Total Protein (%)	Digestible Protein (%)	Energy ME[b] (Mcal/kg)	TDN[c] (%)	Calcium (%)	Phosphorus (%)	Carotene (mg/kg)	Vitamin A (1000 IU) (kg)	(lb)
Growing Steers														
		0.00	0.00	2.7	6.0	7.8	4.2	2.06	57	0.19	0.19	5.5	2.2	1.0
		0.25	0.55	3.1	6.8	11.1	7.1	2.28	63	0.26	0.23	5.5	2.2	1.0
150	330	0.50	1.10	3.2	7.1	12.2	8.1	2.61	72	0.38	0.31	5.5	2.2	1.0
		0.75	1.65	3.2	7.1	13.3	9.0	2.82	78	0.53	0.41	5.5	2.2	1.0
		0.00	0.00	3.3	7.3	7.8	4.2	2.06	57	0.18	0.18	5.5	2.2	1.0
		0.25	0.55	4.5	9.9	10.0	6.1	2.06	57	0.18	0.18	5.5	2.2	1.0
200	440	0.50	1.10	4.9	10.8	11.1	7.1	2.28	63	0.27	0.20	5.5	2.2	1.0
		0.75	1.65	5.0	11.0	11.1	7.1	2.50	69	0.36	0.28	5.5	2.2	1.0
		0.00	0.00	4.5	9.9	7.8	4.2	2.06	57	0.18	0.18	5.5	2.2	1.0
		0.25	0.55	6.1	13.4	8.9	5.2	2.06	57	0.18	0.18	5.5	2.2	1.0
300	660	0.50	1.10	7.7	17.0	10.0	6.1	2.06	57	0.18	0.18	5.5	2.2	1.0
		0.75	1.65	8.0	17.6	11.1	7.1	2.28	63	0.21	0.18	5.5	2.2	1.0

Growing Heifers

Weight (lb)	Weight (kg)	Gain	Gain					ME[b]	TDN[c]					
150	330	0.00	0.00	2.7	6.0	7.8	4.2	2.06	57	0.19	0.19	5.5	2.2	1.0
		0.25	0.55	3.2	7.1	11.1	7.1	2.28	63	0.25	0.22	5.5	2.2	1.0
		0.50	1.10	3.2	7.1	12.2	8.1	2.61	72	0.38	0.31	5.5	2.2	1.0
		0.75	1.65	3.3	7.3	13.3	9.0	2.82	78	0.52	0.39	5.5	2.2	1.0
200	440	0.00	0.00	3.3	7.3	7.8	4.2	2.06	57	0.18	0.18	5.5	2.2	1.0
		0.25	0.55	4.6	10.1	10.0	6.1	2.06	57	0.18	0.18	5.5	2.2	1.0
		0.50	1.10	5.0	11.0	11.1	7.1	2.28	63	0.26	0.20	5.5	2.2	1.0
		0.75	1.65	5.4	11.9	11.1	7.1	2.50	69	0.33	0.26	5.5	2.2	1.0
300	660	0.00	0.00	4.5	9.9	7.8	4.2	2.06	57	0.18	0.18	5.5	2.2	1.0
		0.25	0.55	6.2	13.7	8.9	5.2	2.06	57	0.18	0.18	5.5	2.2	1.0
		0.50	1.10	8.2	18.1	10.0	6.1	2.06	57	0.18	0.18	5.5	2.2	1.0
		0.75	1.65	8.6	19.0	11.1	7.1	2.28	63	0.20	0.18	5.5	2.2	1.0

[a] Adapted from *Nutrient Requirements of Beef Cattle*, Subcommittee on Beef Cattle Nutrition, National Research Council, 1970.

[b] ME (metabolizable energy) requirements for growing cattle were calculated from net energy values for maintenance and gain, as published in appropriate tables, for the weights and rates of gain shown.

[c] TDN (total digestible nutrients) requirements were calculated from metabolizable energy by assuming 3.6155 Kcal of ME per gm TDN.

Table 75

Morrison's Feeding Standards for Stocker Cattle[a]								
			Requirements per Head Daily					
Body Weight (lb)	Dry Matter (lb)	Digest-ible Protein (lb)	Total Digest-ible Nutrients (lb)	Calcium (gm)	(lb)	Phosphorus (gm)	(lb)	Caro-tene[b] (mg)
Growing Cattle, Fed for Rapid Growth								
300	7.2– 9.0	0.67–0.77	5.1– 6.2	18	0.040	13	0.029	20
400	9.1–11.4	0.76–0.87	6.2– 7.2	20	0.044	15	0.033	25
500	10.7–13.0	0.81–0.92	7.2– 8.4	19	0.042	15	0.033	30
600	12.4–14.7	0.84–0.95	8.1– 9.3	18	0.040	15	0.033	35
700	14.2–16.5	0.87–0.98	8.9–10.2	17	0.037	15	0.033	40
800	15.9–18.3	0.90–1.00	9.5–10.9	16	0.035	15	0.033	45
900	17.3–19.7	0.93–1.03	10.1–11.5	16	0.035	15	0.033	50
1,000	18.6–21.0	0.95–1.05	10.6–12.0	15	0.033	15	0.033	55
Wintering Calves, to Gain 0.75 to 1 lb per Head Daily								
300	7.0– 8.3	0.52–0.58	3.9–4.6	16	0.035	12	0.026	17
400	8.7–10.3	0.63–0.70	4.8–5.7	16	0.035	12	0.026	25
500	10.3–12.1	0.71–0.78	5.7–6.7	16	0.035	12	0.026	30
600	11.7–13.9	0.79–0.88	6.5–7.7	16	0.035	12	0.026	35
Wintering Yearling Cattle, to Gain 0.50 to 0.75 lb per Head Daily								
600	11.6–13.3	0.67–0.75	6.3–7.2	16	0.035	12	0.026	35
700	12.9–14.8	0.76–0.83	7.0–8.0	16	0.035	12	0.026	40
800	14.2–16.3	0.83–0.90	7.7–8.8	16	0.035	12	0.026	45

[a] Taken by permission of The Morrison Publishing Company, Ithaca, N.Y., from *Feeds and Feeding*, 22nd edition, by Frank B. Morrison.
[b] 1 mg carotene = 400 IU vitamin A.

palatability of such silage ensures a sufficient consumption of feed to produce rapid and efficient gains. Silage is usually economical enough to permit liberal feeding without unduly increasing feed costs. In fact, when yield per acre, ease of feeding, and the amount required to produce a given gain are considered, corn and sorghum silages are usually the cheapest feeds that can be fed to young stocker cattle. It is not uncommon to find reports of experiments where 1 ton of stocker gains are made per acre of corn or sorghums harvested and fed as silage. When grain prices are high, silages are even more competitive, even though each ton of silage contains 5 to 6 bushels of grain. Replacing some of the costly grain with the low-sale-value stalk portion of the silage accounts for a substantial reduction in feed cost of gain.

Weanling calves, after the stress of weaning, should be fed as much

Table 76

Importance of Adequate Protein in the Rations of Stocker Cattle[a]					
		0.5 lb Cottonseed	0.75 lb Cottonseed	1 lb Cottonseed	1.5 lb Cottonseed
Concentrate Fed	None	Cake	Cake	Cake	Cake
Number of trials averaged	5	2	3	3	2
Average initial weight (lb)	408	418	433	385	363
Average final weight (lb)	433	543	596	564	548
Average winter gain (lb)	25	125	163	179	185
Average daily gain (lb)	0.15	0.73	0.92	1.10	1.16
Average prairie hay eaten daily (lb)	10.6	12.7	13.7	13.2	11.6
Average summer gain (lb)	259	231	222	212	190
Average total gain (lb)	284	356	385	391	375

[a] Nebraska Bulletin 357.

silage as they will consume, which averages approximately 30 pounds or 4 to 5 percent of body weight daily if they are fed only a little hay or concentrates in addition. Corn and sorghum silages are low in both protein and minerals. Therefore approximately 1 pound of protein supplement and 0.10 pound of finely ground limestone or dicalcium phosphate should be fed daily per head unless 4 or more pounds of leafy legume hay are present in the ration. If limestone is used, a phosphorus source should also be included. The calcium needs can be met by adding 10 to 20 pounds of limestone per ton of silage at silo-filling time.

Legume-Grass Silage. Silage made from mixed grasses and legumes cut at the proper stage of maturity and carefully ensiled is an excellent feed for stocker cattle. Such silage has a high protein and mineral content and does not need to be supplemented with a protein concentrate. However, it is much lower in metabolizable or net energy than corn or sorghum silage, and smaller gains result from its use, unless grain is added as a preservative, at the rate of 150 to 200 pounds a ton, or unless it is supplemented at feeding time with some grain or other energy source. If corn or sorghum silage and legume-grass silage are both to be fed to a drove of stocker cattle, the legume-grass silage should be fed out first, since it usually is less palatable than corn or sorghum silage.

Table 77

Value of Low-Grade Roughages for Wintering Stocker Cattle When Fed with a Complete Supplement[a]

Roughage Fed	First Trial 112 Days			Second Trial 147 Days		
	Oat Straw	Corn Cobs	Corn Silage	Soybean Straw	Corn Cobs	Corn Silage
Average initial weight (lb)	485	479	478	480	478	481
Average final weight (lb)	589	671	721	595	698	806
Average total gain (lb)	104	192	243	115	220	325
Average daily gain (lb)	0.93	1.72	2.18	0.78	1.50	2.21
Average daily feed						
Roughage	12.4	12.8	31.0	13.3	13.4	37.0
Supplement A[b]	3.5	3.5	3.5	3.5	3.5	3.5
Minerals[c]	Free choice	Free choice	Free choice	0.06	0.05	0.04
Feed per hundredweight gain (lb)						
Roughage	1,337	742	1,423	1,707	890	1,671
Supplement A	377	202	160	449	233	158

[a] Indiana Mimeographs AH 47 and AH 59.
[b] Supplement A consisted of 2.25 lb soybean meal, 1 lb molasses feed, 0.18 lb bonemeal, 0.06 lb salt, and 0.01 lb vitamin A concentrate.
[c] Mineral mixture fed free-choice during both tests: 2 parts steamed bonemeal, 1 part iodized salt, plus 1 oz. cobalt sulfate per 100 lb of salt.

The addition of a low level of a dry roughage to a heavy feed of silage being fed to stockers is recommended. Although the reason for the beneficial responses obtained from such additions is not well established, the physiological well-being of both the rumen microorganisms and the host animal are apparently involved. Table 79 shows the results of one of several such trials. In this trial, dry matter intake was constant and the small amount of hay fed was comparable to the silage replaced with respect to species and stage of maturity when cut. Large commercial feedlots, which utilize fenceline bunks and automatic unloading wagons or trucks, find it economically unfeasible to feed a dry roughage in the form of long hay, but chopped hay or cottonseed hulls are usually substituted.

The beneficial effect of the hay is more pronounced in the first month on feed.

Legume-Grass Hay. Mixed hays containing up to 50 percent legumes such as alfalfa, red clover, or trefoil are excellent feed for stocker cattle, especially if the hays are cut before they are unduly mature and high in crude fiber, and if they are properly cured. In all but the western sections of the country, such hays are often lower in quality than the same crops would be if preserved as silage. This is because good hay-curing days are few in most parts of the country, especially for the first cutting. This first cutting also is usually much higher in grass hay than are later cuttings.

Energy supplementation of mixed-hay stocker rations is required, as it is with legume-grass silages, with 2 to 4 pounds of grain daily being the level usually fed. Protein supplementation may or may not be required, depending on the quality of the roughage. In general, if the hay consumed daily contains the equivalent of 4 pounds of good-quality legume hay, supplementation with a protein concentrate increases gains, but no more than if a similar amount of a high-energy concentrate such as corn or ground grain sorghum is added. Carotene or vitamin A, calcium,

Table 78

Comparison of Simple and Complex Supplements for Good-Quality Roughages Fed to Stocker Calves[a]

Type of Silage	Corn Silage		Legume-Grass Silage[b]	
Supplement Fed	Purdue Supplement A	Soybean Meal, Ground Shelled Corn	Purdue Supplement A	Ground Shelled Corn
Days fed	125	125	160	160
Average initial weight (lb)	333	320	400	400
Average daily gain (lb)	2.19	2.17	1.51	1.65
Average daily ration (lb)				
Silage	27.5	26.6	22.5	22.5
Protein supplement	3.5	1.25	3.5	—
Ground shelled corn	—	2.00	—	3.5
Minerals	0.08	0.11	0.06	0.07
Feed cost per cwt gain ($)	15.31	12.74	21.58	14.05
Winter feed bill per head ($)	42.09	34.61	49.85	36.95

[a] Illinois Cattle Feeders' Day Reports.

[b] Contained 150 lb of ground shelled corn preservative per ton of silage.

Table 79

	Steer Calves (133 Days)	
Roughages Fed	Legume-Grass Silage	Legume-Grass Silage + Hay
Number of calves	10	10
Initial weight (lb)	435	435
Average daily gain (lb)	1.68	1.80
Average daily ration (lb)		
Silage	28.7	22.8
Hay	—	2.2
Ground shelled corn	3.5	3.5
Mineral mixture	0.07	0.07
Feed cost per cwt gain ($)	13.26	12.45

Effect of a Small Hay Addition to a Silage Stocker Ration Fed to Calves[a]

[a] Illinois Cattle Feeders' Day Report.

and phosphorus supplements may be needed in very poor quality mixed-hay rations.

Haylage. Haylage is legume-grass or straight-legume forage that has been allowed to wilt in the windrow until it has been reduced to about 50 percent moisture. The forage is then usually chopped and stored in a conventional silo or, more commonly, in an oxygen-free silo such as the Harvestore. The forage undergoes normal silage fermentation and, except for the difference in moisture content, has a feeding value comparable to that of silage or hay that was properly made from the same crop. Supplementation of this feed thus presents no problems that were not discussed in connection with silage and hay.

The use of haylage is increasing despite the added cost in storage and harvesting equipment. This must be because of the convenience of harvesting and feeding, the reduction in field-harvesting losses, and the reduction in losses traceable to poor haymaking weather on the one hand or to high fermentation losses of unwilted silage on the other.

Grass Hay. Straight-grass hays, such as prairie hay, Coastal Bermuda, timothy, Sudan-sorghum or Johnson grass, may satisfactorily make up a major portion of the ration of stocker cattle, but fairly complete supplementation with protein, minerals, and vitamins is essential for normal growth. Added energy supplementation is needed if any improvement in condition of the stockers is desired.

Stalk Fields.   Stalk fields furnish much cheap feed for stocker cattle during the late fall and early winter. Many feeders buy their feeder cattle, especially if they handle yearlings, in October or November and run them on stubble and stalk fields until about the first of January, when they are put into the feedlot and started on finishing rations. Yearlings are better than calves for this method of handling, as they are better able to make use of the coarse roughage that stalk fields furnish. Stocker cattle, particularly calves, should never be maintained wholly on stalks because stalks are quite low in metabolizable energy and seriously deficient in protein. Unless there is access to a good bluegrass or legume pasture or winter small grains, cattle on stalk fields should be fed 4 to 6 pounds of legume hay or 1 to 1.5 pounds of protein concentrate per head daily. Protein blocks or liquid supplements that can be self-fed fit this situation nicely.

WINTER SMALL-GRAIN PASTURES

Winter wheat pastures are widely used in Oklahoma, Texas, and Kansas for wintering all classes of cattle, but calves make especially good use of this rather uncertain feed supply. Extensive snows, wet weather, or prolonged dry weather all make this source of feed rather unreliable, but a stocker program remains the most profitable one for a man with such pasture to sell through cattle.

Good wheat pasture alone will produce acceptable stocker gains, as shown by Table 80. Although the subject of hormones is discussed in detail elsewhere, note the response obtained from stilbestrol on both the silage and pasture programs. Other implants may be expected to produce similar responses if the use of stilbestrol is prohibited. Wheat pasture is also excellent feed for cows, especially if they are fall-calving cows, but the main disadvantage of this program is that when the wheat pasture fails to materialize or hold up because of adverse weather conditions, cows must either be sold or moved, or fed on roughages that will have a very high cost if purchased or a high sale value if homegrown. Usually a large area is affected by the same weather pattern, which accounts for the generally high cost of feed when purchases are necessary.

Wheat pasture poisoning, also referred to as grass tetany when it occurs on certain grass pastures in some parts of the world, can present real problems to the man who runs cattle on rapidly growing wheat pasture. The condition follows a rapid course, with usually a lapse of only 6 to 10 hours between onset and coma which, unless the condition is treated, is nearly always followed by death. Initially, affected animals appear nervous and excited. The third eyelid flickers, and there is twitching in the muscles

Table 80

	Sorghum Silage + Protein Supplement[b]	Sorghum Silage + Supplemental Ration[c]	Wheat Pasture	Wheat Pasture + Supplemental Ration[d]
Response of Stilbestrol-Implanted Stocker Heifer Calves on Sorghum Silage or Wheat Pasture[a]				
Number of heifers	10	10	10	10
Initial weight (lb)	472	474	471	477
Final weight (lb)	553	667	667	701
Total gain (88 days) (lb)	81	193	196	224
Daily gain (all calves) (lb)	0.92	2.19	2.22	2.54
Daily gain (lb)				
Controls	0.87	1.87	2.03	2.58
12 mg stilbestrol implant	0.95	2.50	2.42	2.51
Daily supplemental feeds[c,d]				
Mixed feed	—	9.64	—	9.34
Cottonseed meal	1.50	1.00	—	—
Mineral and salt	Free choice	Free choice	Free choice	Free choice

[a] Oklahoma Miscellaneous Publication 79, 1967.

[b] 1.5 lb cottonseed meal daily.

[c] Supplemental ration consisting of 77 percent ground milo, 8 percent molasses, and 15 percent chopped alfalfa hay fed at a level consumed by cattle on wheat pasture when provided free-choice plus 1 lb cottonseed meal daily as supplemental protein.

[d] Supplemental ration described in footnote (c) provided free-choice. No additional protein supplement was fed.

of the extremities, followed by tetany, contractions, and convulsions. Rapid breathing and a pounding heart immediately precede coma and death. Cows are most often affected, especially those that are nursing calves, but other classes of cattle can be involved and death losses of 5 to 10 percent are not uncommon in stocker calves on wheat pasture.

The exact cause of this condition is incompletely understood. Recent Oklahoma work suggests that the problem in stocker calves is somewhat distinct from the tetany-type of problem seen in lactating cows. It is suggested that the very high levels of digestible protein in the rapidly growing wheat are responsible for rumenal ammonia concentrations that are toxic. Further, the physical characteristics of the rumen contents are such that rumen motility and belching are impaired, resulting in bloat. These views seem reasonable for it has long been noted that, if some form of dry roughage, even cured grass, is accessible, cattle on wheat pasture eat enough of such forages to prevent the problem described. Magnesium supplementation seems to help prevent the condition in cows but not in stockers. Approximately 20 grams of magnesium per day, usually provided in the form of magnesium oxide added to the salt and mineral supplement, is adequate for cows. Commercial supplements are obtainable in the wheat pasture grazing area, and these should be fed according to directions. Affected animals can be treated with intravenous or intraperitoneal injections of calcium gluconate solutions fortified with magnesium and phosphorus. A second treatment may be required in advanced cases.

In the southeastern states there is extensive use of winter oats and fescue, a perennial grass that remains green throughout much of the winter, in wintering cattle. Wheat and rye are also used to a lesser extent. The Louisiana feeding and grazing trials reported in Table 81 illustrate what may be expected of calves on various combinations of rations in this important cow-calf area. This area is turning more and more to stocker or grower programs, to utilize winter pasture, to increase the output or volume of business, and to spread the marketing period for their calves. Note how compensatory gain in the summer was related to rate of winter gain, as previously discussed.

CONCENTRATES FOR STOCKER CATTLE

Whether stocker cattle should be fed concentrates during the stocker program depends on the amount and character of roughage fed and on the subsequent feeding plan. Calves that are not fed silage should ordinarily be fed 2 to 4 pounds of grain daily because they are unable to consume enough dry roughage to gain more than a pound a day. However, calves that are fed a full feed of high grain-content corn or sorghum silage and 1 pound of a protein concentrate or 4 or 5 pounds of legume hay need not be fed grain unless they are to be marketed following only a short feed on finishing rations. Then the feeding of some grain is advisable because the cattle will have a higher finish and sell for a better price when marketed. The effect of such low-level grain additions is shown in Table 82. Newly arrived calves fed on "warm-up" or "back-grounding" rations for only a short period respond especially well to 3 to 4 pounds of grain. Because their total intake of feed is low, owing to prior stress and the likely presence of a low-grade fever, the high energy content of the grain meets their energy requirements where roughage alone would not.

Except when poor to very poor quality roughages are fed, there is apparently little difference between the various protein concentrates for feeding stocker cattle, provided they are fed in sufficient amounts to

Table 81

A Comparison of Six Methods of Wintering Stocker Calves and Effect on Spring Gains (Three Years)[a]

Group Number	1	2	3	4	5	6
Treatment	Coastal Bermuda Hay	Hay + Grain	Wheat Pasture + Hay	Kentucky 31 Fescue Pasture	Fescue + Grain	Fescue and Wheat Pasture
Winter Period						
Number of acres	Drylot	Drylot	15	5	5	10
Number of calves	10	10	10	10	10	11.3
Initial weight, fall (lb)	454	456	456	456	456	456
Weight, spring (lb)	484	570	579	506	587	575
Winter gain, 105 days (lb)	30	114	123	50	131	119
Average daily gain (lb)	0.28	1.08	1.17	0.47	1.24	1.13
Grain fed per calf per day (lb)	—	5	—	—	5	—
Hay fed per calf per day (lb)	11.4	9.1	8.0	9.8	9.8	9.8
Number of days hay was fed	105	105	19	2[b]	2[b]	2[b]
Number of days grazing:						
fescue	—	—	—	105	105	67
wheat	—	—	86	—	—	38
Cost per cwt of gain[c]	$50.13	$22.04	$28.59	$24.72	$19.50	$17.22
Spring Period						
Weight, summer (lb)	626	677	699	648	692	702
Spring gain, 87 days (lb)	142	107	120	142	105	127
Average daily gain (lb)	1.63	1.23	1.37	1.63	1.20	1.45
Both Periods						
Total gain, 192 days (lb)	172	221	243	192	236	246
Average daily gain (lb)	0.89	1.15	1.26	1.00	1.22	1.28

[a] Red River Valley Station, Louisiana Agricultural Experiment Station.
[b] Hay fed when snow occurred.
[c] Feed and pasture cost: Coastal hay, $25 per ton; grain ration, $50 per ton; fescue pasture, $24 per acre; wheat pasture, $21.50 per acre.

Table 82

Effect of Feeding During the Winter a Half Ration of Grain to Calves That Are to Be Grazed the Following Summer

	Missouri[a] Average of 3 Trials		Kansas[a] Average of 3 Trials	
	Grain and Roughage	Roughage Only	Grain and Roughage	Roughage Only
Average initial weight (lb)	342	348	350	349
Average winter ration (lb)				
Shelled corn	3.8	—	4.6	—
Protein concentrate	0.5	—	1.0	1.0
Corn or cane silage	11.2	14.3	18.4	24.0
Legume hay	4.2	4.5	2.0	2.0
Average daily gain (lb)	1.70	0.88	1.89	1.34
Grazed without grain	56 days	56 days	90 days	90 days
Total gain on pasture (lb)	22	57	98	123
Average daily gain on pasture (lb)	0.34	1.02	1.09	1.38
Full fed grain on pasture	112 days	112 days	100 days	100 days
Total gain (lb)	269	276	256	263
Average daily gain (lb)	2.41	2.47	2.56	2.63
Final weight (lb)	912	823	961	917
Shelled corn fed (bu)	36.5	24.3	36.9	26.2

[a] Mimeographed reports of cattle-feeding experiments.

furnish as much protein or its equivalent in nonprotein nitrogen as is present in 1 pound of the oilseed meals. It has been demonstrated at the Fort Hays branch of the Kansas station that 2 pounds of ground barley or wheat, 3 pounds of wheat bran, or 4 pounds of alfalfa hay are as good supplements to a full feed of silage as 1 pound of linseed, cottonseed, or soybean meal.

RECOMMENDED RATIONS FOR STOCKER CATTLE

Stocker cattle are being successfully fed on a wide variety of rations, and it speaks well for this program that it can be adapted to the efficient utilization of so many different roughages. Some recommended rations, based on the more commonly used roughages and concentrates, are found in Table 83.

Table 83

Rations	Calves (lb)	Yearlings (lb)
Expected daily gain for 120–150 days for stockers to be immediately fattened	1.4–1.8	1.5–2.0
1 { Alfalfa or clover hay	5–6	4–5
Corn or sorghum silage	25	35
2 { Clover hay	8–10	6–8
Corn or oats	4–5	5–6
Straw or hulls[a]	2–4	12–15
3 { Protein concentrate	1–1.5	1
Corn or sorghum silage	20–25	20
Straw or hulls	2–3	10–12
4 { Grass hay	10–12	16–20
Corn, sorghum, or oats	4–5	5–6
Protein concentrate	1–1.5	1–1.5
5 { Legume-grass or oat silage	20–25	35–45
Corn, sorghum, or oats	4–5	5–6
Straw or mixed hay	2	2
Expected daily gain for 120–150 days for grass steers or replacement heifers	0.75–1.25	0.50–1.50
6 { Prairie or all-grass hays	10–15	15–20
Protein concentrate	1–1.5	1
7 Legume-grass hays	12–16	15–20
8 Legume hays	10–12	14–18

Recommended Rations for Stocker Cattle

[a] All straw and hulls should be fed according to appetite.

SUMMER MANAGEMENT OF YOUNG CATTLE

The problems of managing stockers on pasture during the summer are relatively simple in comparison with those arising in drylot. With the availability of spring pasture, feeding problems are largely solved because good pasture is an ideal ration for young growing animals. It is nutritious and rich in growth-promoting and protective vitamins.

Pastures to be used by young breeding animals should not be heavily stocked unless there are adequate facilities for feeding harvested feeds if the supply of grass becomes short. Grass alone ordinarily provides a satisfactory ration for growing cattle during the lush growing season. With stocker steers or grade yearling heifers intended for a commercial breeding herd, economy of maintenance usually requires that pasture alone, if it is available, should form the ration for the entire summer, even though the cattle lose some flesh and may not grow quite so rapidly. If the grass fails because of dry weather, pastures should be supplemented by feeding silage, hay, chopped green corn, or Sudan-sorghum green chop. Selling yearling steers when grass fails, because of unforeseen drought in summertime or for other reasons, may be the wiser course unless homegrown roughages are at hand.

SUMMER GAINS OF YOUNG CATTLE ON PASTURE

The gains made during the summer by young cattle on pasture do, of course, vary greatly with the condition of the animals at the beginning of the grazing season, with the kind and amount of forage available, the genetic potential for growth by the cattle, and with whether they were implanted with hormones. Also, gains vary from year to year because of weather conditions, which greatly affect the amount, palatability, and nutritive value of the forage. Occasionally a severe drought and the discomfort caused by flies and oppressive heat may cause a midsummer loss of much of the gain put on during the spring. As a consequence, the gain for the entire season is disappointingly small. However, such years are offset by unusually favorable seasons, when the gains made are almost double those obtained during an average year.

Fig. 69. Crossbred stocker steer calves on irrigated winter wheat in the Texas Panhandle winter wheat belt. These calves are owned by a custom feeder who leases the wheat pasture from a wheat farmer. (West Texas Livestock Weekly.)

Yearling cattle weighing approximately 550 to 600 pounds when turned onto pasture gain at the rate of 1.00 to 1.50 pounds a day—or 175 to 250 pounds for the season—on good permanent pasture such as bluegrass, bromegrass, or the bluestem pastures of Kansas. Rotation pastures that consist largely of grasses and warm-season legumes, thus retaining their nutritive value and palatability throughout the summer, may usually be counted on for about 25 percent more gain per head than is obtained from permanent all-grass pastures during the same season. The greatest advantage of such combination legume-grass pastures is the greater stocking rate and hence greater cattle gains per acre for the season. Per acre gains of stocker cattle on such pastures in a normal year may be expected to reach 500 pounds. Rotation or strip-grazing of these combination pastures can increase cattle gains by 25 to 35 percent over continuous grazing.

STOCKER OR GROWER CONTRACTS

This subject was briefly mentioned in connection with the question of risk reduction for the operator of the stocker program. There is every indication that this method of financing the stocker program will increase in use. It has much to offer both the small crop farmer and small rancher who have neither the capital nor the risk capability to buy calves for themselves and the feedlot operator who specializes in finishing cattle on high-energy rations. It permits both operators to specialize and to feed only one type of ration. Often roughages, including range, can be grown very efficiently in one area and grain in another. The grower contract thus permits both crops to be efficiently utilized where they are grown, vastly reducing the cost of transportation of feeds. Large commercial feedlots, of course, can and often do carry out both steps in the grower-finishing operation.

The program is not new in that, for some years, much of the Kansas bluestem pasture was used for growing yearlings owned by feeders in Iowa and elsewhere. Much of the new interest, though, is in drylot feeding of stockers, and not always in winter as has been customary. Many smaller feeders in the Greeley, Colorado, area, for example, now feed the corn silage produced on their fertile, irrigated fields to stockers owned by one of several large feedlots in the vicinity. Formerly they may have owned and finished the cattle themselves or may even have sold the corn crop itself, as green corn, to the feedlot when this seemed to present the better chance for profit. Feeding the silage or hay crops on the farm where produced, rather than selling them, reduces transportation costs and leaves the manure where most needed. In view of high fuel and fertilizer

Fig. 70. Many early-day cattle ranchers believed that sheep would "poison' or ruin a cattle pasture. Contrary to legend, these steers and sheep are effectively demonstrating co-use of range to increase the production and income on a New Mexico ranch. Prolonged overstocking with either species can permanently damage a pasture, of course. (Western Livestock Journal.)

costs, these two items will surely become more important and should encourage still more of this kind of contract-feeding.

There are several kinds of contracts. The two most common are based either on a fixed cost for the gain, or on a feed cost plus an extra charge for labor and lot rental. Such contracts should be in written form, and the basis for adjusting for death loss and for computing amount of gain should be well understood in advance by both parties.

CHAPTER 12
THE CATTLE
FINISHING PROGRAM

The finishing of cattle for the slaughter markets is one of the most important enterprises of the American agricultural economy. Cattle finishing once was conducted almost exclusively in the corn-growing regions, but important cattle-finishing areas are now located throughout the United States, including Hawaii. The rapid development of this enterprise is illustrated by the fifteenfold increase in feeding in the Texas-Oklahoma Panhandle sorghum belt, while the country as a whole was only doubling the number of cattle fed during the past 25 years. During the 1970s, this same Panhandle and adjoining areas in New Mexico and Kansas continued to be the most rapidly growing cattle-feeding center in the nation, supplanting California and Arizona, which underwent a similar development at the close of World War II. The Platte Valley areas in Colorado and Nebraska, always important feeding areas, were also experiencing large numerical increases in total numbers fed, especially in the early to mid-1970s.

Table 84 illustrates the 1967–1974 adjustments in numbers of fed cattle in the various sections of the country. It is apparent that the states in the sorghum belt have had the greatest increase in numbers fed and marketed, with Texas more than doubling its numbers in this short period. Kansas and Nebraska saw modest increases, but the rest of the states in the western Corn Belt decreased their numbers. All four eastern Corn Belt cattle feeding states had reductions. It should not be overlooked, however, that the combined Corn Belt states still are feeding 52.1 percent or slightly over half of all fed cattle. This is a reduction, however, from 60.2 percent of all cattle fed since 1967.

Table 85 shows that the number of cattle marketed in 1974, per farm or feedlot feeding cattle, remained smallest in the Corn Belt. Also, the larger lots are mainly located in the 8 southwestern and western states. Over 90 percent of the cattle on feed in those states were in lots of 1,000-head or greater capacity. The number of cattle feeders in the Corn Belt, although still large, is decreasing, and the size of lot is increasing. The majority are still farmer-feeders and not custom or commercial feedlot operators such as are found in the southwestern and western states.

Perhaps equally as important as the increases in total numbers fed are the changes in the nature of the business. Although Corn Belt farmer-feeders continue to feed many cattle, commercial feedlots are feeding

Table 84

Number of Fed Cattle Marketed, by Important States and by Regions, with Comparisons Between 1967 and 1974[a]

State or Region	Number Marketed (1000 head)		Percentage Change 1967–1974	United States Rank		Percentage of U.S. Total 1974
	1967	1974		1967	1974	
Western Corn Belt						
Nebraska	3,057	3,355	+10	2	2	14.1
Iowa	4,057	3,097	−31	1	3	13.0
Kansas	1,312	2,240	+71	6	4	9.4
Minnesota	869	864	−1	8	8	3.6
South Dakota	618	585	−6	11	10	2.5
Missouri	688	400	−72	8	12	1.7
Total for region	**10,601**	**10,541**	**−1**			**44.4**
Eastern Corn Belt						
Illinois	1,281	850	−51	7	9	3.6
Ohio	442	386	−15	13	13	1.6
Indiana	496	361	−37	12	14	1.5
Michigan	240	242	+1	17	18	1.0
Total for region	**2,459**	**1,839**	**−34**			**7.7**
Southwest and West						
Texas	1,654	3,899	+136	4	1	16.4
California	2,049	2,002	−2	3	5	8.4
Colorado	1,330	1,892	+42	5	6	8.0
Arizona	658	895	+36	10	7	3.8
Oklahoma	414	566	+37	14	11	2.4
New Mexico	238	355	+49	18	15	1.5
Idaho	365	344	−6	15	16	1.4
Washington	318	301	−6	16	17	1.3
Total for region	**7,026**	**10,254**	**+46**			**43.2**
Total	**20,086[b]**	**22,634[b]**	**+13**			**95.0**

[a] Adapted from USDA *Livestock and Meat Situation*, 1975.
[b] Total for 18 most important states and 95 percent of total cattle fed.

Table 85

Number of Cattle Feeders and Fed Cattle Marketed, 1974, by Feedlot Capacity and by State and Region[a]

State or Region	Feedlot Capacity 1,000 Head and Over			Feedlot Capacity Under 1,000 Head			Average Number Marketed per Feedlot or Farm
	No. of Cattle Feeders	Cattle Marketed (1000)	% of Total Marketed	No. of Cattle Feeders	Cattle Marketed (1000)	% of Total Marketed	
Western Corn Belt							
Nebraska	460	2,025	60.4	14,510	1,330	39.6	224
Iowa	165	387	10.5	31,835	2,710	89.5	97
Kansas	140	1,840	82.1	5,660	400	17.9	386
South Dakota	77	178	30.4	9,123	407	69.6	64
Minnesota	50	69	8.0	10,970	795	92.0	77
Missouri	21	52	13.0	11,979	348	87.0	31
Eastern Corn Belt							
Illinois	55	95	11.2	14,445	755	88.8	59
Michigan	33	65	26.9	1,667	177	73.1	142
Ohio	25	58	15.0	8,175	328	85.0	47
Indiana	23	25	6.9	10,477	336	93.1	34
Southwest and West							
Texas	199	3,814	97.8	1,001	85	2.2	3,249
Colorado	188	1,761	93.1	425	131	6.9	3,086
California	139	1,989	99.4	28	13	0.6	11,988
Idaho	72	333	96.8	502	11	3.2	599
Oklahoma	42	530	93.6	358	36	6.4	1,415
Arizona	41	894	99.9	6	1	0.1	19,042
New Mexico	41	354	99.7	7	1	0.3	7,396
Washington	21	268	89.1	165	33	10.9	1,618

[a] Adapted from USDA *Livestock and Meat Situation*, 1975.

even more, and for them it is usually their only enterprise, as contrasted to the farmer-feeder who continues to look upon cattle finishing as a supplementary enterprise to crop production. Changes that have occurred and some that are still taking place, in no particular order of importance, may be listed as follows.

1. The number of cattle feeders is decreasing, especially in the Corn Belt states.
2. The average number of cattle fed per feeder is increasing.
3. The number of cattle fed on some kind of contractual arrangement is greater, with many nonfarm people in the business.
4. The energy content of finishing rations is increasing, with a corresponding reduction in roughage content. This change is occurring primarily in the larger commercial feedlots.
5. A greater variety of additives is used, particularly antibiotics, vitamins, wormers, hormones, and minerals.
6. The length of the finishing period is being reduced, not because cattle are sold at lighter weights but largely because initial weights are heavier and gains are more rapid.
7. Fewer cattle are bought and sold in central markets.
8. More of the lower grades of cattle, including Holsteins, are being fed, whereas formerly most such cattle were slaughtered off grass or as veal calves.
9. Grain processing is more sophisticated, with much grain now being steam-processed, micronized, popped, or reconstituted.
10. Less labor is required, because of automation, with a corresponding increase in capital investment in machinery.
11. Rations are selected in many instances through least-cost formulations by computers and slight adjustments may be made almost daily.
12. Many new by-product feeds are in use.
13. More nonprotein nitrogen sources, such as urea and ammonia, are being used as protein-concentrate replacements because of the lower costs made possible by this substitution.
14. Liquid protein supplements, usually containing nonprotein nitrogen sources, are being used by an increasing number of lots.
15. The problems of disease control are intensified because of greater stress and concentration of animals in the feedlots, with less individual attention.
16. Air and water pollution problems occur near urban areas as a result of dust and of manure accumulation.

17. High concentrations of birds and rodents are attracted to large feedlots.
18. More and more feeders contract for cattle procurement and sale, nutritional advice, and veterinary service.
19. Hedging price risks in the futures markets of both feeder cattle and fat or slaughter cattle, and even of the anticipated grain supplies, is now a common practice.

In summary, cattle feeding has become big business, and in no sense of the word can it be called a way of life, a term often used to describe general farming and sometimes farmer-feeding of cattle.

REASONS FOR FEEDING CATTLE

Cattle are usually fed for one or more of the following reasons: (1) to obtain better than current prices for farm-grown grains; (2) to market roughages and pasture at a profit; (3) to maintain and improve the fertility of the farmland; (4) to convert purchased resources, namely feed, into a more valuable finished product, namely beef; or (5) to provide an income tax shelter. These reasons may be combined by saying that cattle are fed to obtain the highest net return for feeds, whether farm-grown or purchased. This statement implies that cattle are finished principally on feeds grown on the farm or in the region where the cattle are fed. This is largely true, although there are large-scale feeders and feedlot operators who buy most or nearly all of their feeds and transport them for as many as 1,000 miles. Such men feed cattle for the same reason that other men operate factories, namely, to make a profit by combining raw materials, which are of lower value in their natural form, into a product for which there is a strong demand and which, therefore, can be sold for a much higher price than the cost of the raw materials.

The high cost of freight on both feed and cattle is tending to push cattle feeding back nearer the major feed-producing areas. Strong demand for beef in an area cannot alone long hold a large cattle-feeding business, because this demand can be more economically supplied with beef in the carcass, by truck or rail, especially if freight rates favor such a trend.

ADVANTAGES AND DISADVANTAGES OF THE CATTLE-FINISHING PROGRAM

The cattle-finishing program offers much opportunity for profit, but at the same time results in as many failures, financially speaking, as all the other beef cattle programs combined. This is mainly because (1) feeds used in finishing rations are relatively expensive, (2) feed conversion is less efficient for finishing purposes than for growth, (3) price fluctuations may erase "price spread," one of the sources of profit, and (4) nonfeed costs are not related to cattle prices and are continually increasing.

Very economical gains in the stocker program or very cheap maintenance rations in the cow-calf program can often offset mistakes in judgment at buying and selling time, but no such opportunity presents itself in the finishing program. Sometimes when both stocker and finishing programs are conducted successively on the same farm, all the money made in the stocker program and more is lost in the finishing program because of poor planning or a disastrous decline in prices. For this and many other reasons, a thorough understanding of this program, the riskiest of all commercial cattle programs, is essential.

Advantages of the Finishing Program. The large number of participants is evidence of belief in the following advantages usually ascribed to the finishing program.

1. Because of its intensive nature it affords an opportunity to market at a profit large quantities of both roughages and grains.
2. Large and spectacular profits are occasionally made as a result of favorable price rises during the period of ownership of a drove of feeder cattle.
3. A large volume of manure high in fertility is produced. As prices for commercial fertilizers increase, this could become an important advantage.
4. It is a relatively short-time program, making it possible to turn two or more droves of cattle per year in some types of finishing programs or to finish off a drove of cattle between peak labor requirements in farming operations.
5. There is flexibility with respect to number, weight, and grade of cattle, as well as to length of feeding period and type of ration fed.
6. Death losses are relatively low as compared with some of the other programs such as the cow-calf program.

Disadvantages of the Finishing Program. Even when the cattle finishing program is the obvious choice for best utilization of the pastures, harvested roughages, and grains found on a farm or in the area, certain disadvantages are inherent in this program and should be recognized.

1. The program is speculative, with the result that large amounts of

money are sometimes lost. As discussed elsewhere, some finishing programs carry less risk than others.

2. Skilled buying and selling are required. This accounts for such common sayings as "well bought is half sold."

3. Relatively large amounts of capital or credit must be readily available, to make the necessary periodic purchases of feeder cattle and, in some instances, feed as well. Commercial feedlots bill their custom-feeder clients for feed and yardage on a monthly basis, and so the capital required must be constantly available.

4. Specialized feedlot and feed preparation equipment is required.

5. Skill and knowledge of feeding principles is very necessary in the finishing program.

6. Serious labor conflicts may develop in the farmer-feeder programs, especially with summer finishing programs.

7. The program is rather confining, requiring constant attention, although pushbutton feeding systems minimize such confinement. The program may interfere with weekend absences and long vacations.

8. Problems arise with respect to manure disposal and increased fly and bird populations. Also, odors and dusts are the cause of many lawsuits from urban neighbors.

9. Increasing export demands for feed grains and protein supplements ensure a fairly strong price on feeds; thus feeders cannot always offset lower fed-cattle prices with correspondingly low feed costs.

SOURCES OF PROFIT AND LOSS

A discussion of profits and losses that may result from a cattle-feeding enterprise must include a breakdown of the following two items.

1. *Price spread*—the difference between buying and selling price per 100 pounds. This is also sometimes simply referred to as "margin" or "spread" and is not to be confused with "feeding margin." It is applied to the purchased weight only.

2. *Feeding margin*—the difference between the feed cost of producing 100 pounds of gain and the selling price per hundredweight. Feeding margin applies only to the weight added during feeding, and only feed costs are used in its calculation, excluding nonfeed costs.

The following example illustrates how these two factors function.

1. Price spread
    700 lb yearling purchased @ $36 cwt delivered.
    1,100 lb choice steer sold @ $40 cwt net.
    $40 - $36 (7 cwt) = $28 potential profit due to price spread.
2. Feeding margin
    1,100 lb - 700 lb = lb gain @ feed cost of $32 cwt gain.
    $40 - $32 (4 cwt) = $32 potential profit due to feeding margin.
    $28 + $32 = $60 total returns above cost of feed and yearling feeder.

After paying for the feeder steer, feed, and buying and marketing costs, a total of $60 potential profit remains, from which must be deducted the following additional costs: labor; interest on investment in steers, feed, equipment, and buildings; depreciation on machinery, equipment, and buildings; death loss; veterinary costs; insurance. If, after paying all of these "out of pocket" costs, any of the $60 remains, this is net profit. These incidental costs are referred to as nonfeed costs and range from 8 to 20 cents per head per day in the feedlot. Many cattle feeders lump all of the incidental costs together and hope the price spread takes care of these items.

When portions of the drylot rations or the pasture are so low in quality that they have no sale value, many farmer-feeders simply do not charge the cattle with these feeds because they would not otherwise be used. Examples are corn stalk fields, meadow aftermath, or permanent pasture that cannot be harvested for hay. The same may be said for labor that is available and would otherwise not be utilized.

Except during the occasional periods of excessive stocker and feeder cattle supplies such as occur during the liquidation phase of a cattle cycle, feeder cattle, especially calves, are fed on a minus price spread—that is, the fed cattle sell for less per hundredweight than they cost as stockers or feeders. This places the entire burden for profit on feeding margin. The relative importance of the two sources of potential profit depends on age, weight, sex, and grade of the feeder cattle, the price of feed, and the relationship between the selling price of cattle and the cost or value of feed and other items that contribute to the total cost of gains. These relationships are discussed further in Chapters 13 and 14.

Table 86 gives some comparisons that illustrate that net returns due to favorable price spreads are generally smaller and proportionately less of the total return for the calf programs than for those programs using older or heavier cattle. The table also shows that older and heavier feeder cattle produce less net return from the feeding margin. This merely reflects the principle that older cattle convert feeds to gain less efficiently than do calves. The 3-year averages for all programs show a negative price spread but in a few individual years a positive price spread prevailed. Note in

Table 86

Effect of Price Spread and Feeding Margin on Profits in Various Steer-Finishing Programs[a]

Feeding Year	Dollars per 100 Pounds of Steer					Return Above Feed and Steer Cost per Head[a]
	Price Paid	Selling Price	Feed Cost	Price Spread[b]	Feeding Margin[c]	
Long-Fed Good-to-Choice Steer Calves						
1962–63	$29.53	$23.41	$17.51	$−6.12	$ 5.90	$ 7.17
1961–62	26.56	26.46	15.97	−0.10	10.49	59.76
1960–61	26.48	23.11	16.28	−3.37	6.83	22.80
1960–63 average	27.52	24.33	16.59	−3.19	7.74	29.91
Long-Fed Good-to-Choice Yearling Steers on Pasture						
1962–63	$27.29	$23.11	$19.42	$−4.18	$ 3.69	$ −6.55
1961–62	25.43	26.98	17.10	1.55	9.88	59.30
1960–61	24.89	22.93	16.46	−1.96	6.47	22.29
1960–63 average	25.87	24.34	17.66	−1.53	6.68	25.01
Long-Fed Good-to-Choice Yearling Steers in Drylot						
1962–63	$27.48	$23.23	$20.09	$−4.25	$ 3.14	$−12.88
1961–62	25.61	26.15	17.26	0.54	8.89	45.64
1960–61	25.22	22.45	18.64	−2.77	3.81	3.10
1960–63 average	26.10	23.94	18.66	−2.16	5.28	11.95
Short-Fed Good-to-Choice Yearling Steers						
1962–63	$25.02	$22.08	$19.29	$−2.94	$ 2.79	$ −9.45
1961–62	24.16	24.45	17.52	0.29	6.93	26.80
1960–61	23.72	22.90	16.81	−0.82	6.09	18.70
1960–63 average	24.30	23.14	17.87	−1.16	5.27	12.02

[a] 1964 Annual Report, Farm Business Farm Management Service, prepared by Illinois Agricultural Experiment Station.

[b] Price paid − selling price = price spread.

[c] Selling price − feed cost = feeding margin.

[a] Does not take nonfeed costs into account.

Figure 11 that the 3-year period covered in Table 86 occurred during the ascending phase of a cattle cycle. In other words, calves were being held back for grazing as yearlings, or as heifer replacements to expand old herds or start new ones. Thus the demand for the diminished supply was strong enough to push purchase prices above sale prices of fed cattle, resulting in the negative price spreads. It is doubtful whether price spread will be consistently reliable as a source of profit in the future when feeding good to choice or better cattle. Programs that must rely on favorable price spreads are much more speculative and carry correspondingly greater risk. This is undoubtedly the chief reason why feeders the country over are feeding more calves, fewer yearlings, and almost no 2-year-olds.

Note in Table 86 that in the feeding year 1962–63 the highest-priced feeder cattle fed, namely the long-fed good to choice steer calves, were the only ones that showed a return over steer and feed cost that year. They did this in spite of the greatest negative price spread, and because of a higher feeding margin resulting from a combination of lower cost of gain and high selling price.

Disastrous financial results can occur, as happened in 1973–74 when both price spread and feeding margins were negative. Such results are most likely to occur during the ascending side of a cattle numbers cycle, and especially in the year when the cycle peaks, because then purchase prices paid are high but the finished cattle sold 6 to 12 months later are selling on a breaking market. Unfortunately, it is not possible to predict with complete certainty the exact year when cattle numbers will peak in a cycle, but a careful study of previous cycles and an awareness of current cattle numbers can be used as guides. For a few years following the price breaks such as occur immediately following the cycle peaks, positive price spreads generally may be relied upon.

METHODS OF FINISHING CATTLE

A wide variety of programs are used in finishing cattle for market. Although there is no best method, even for a particular feeding situation, some methods are likely to prove more satisfactory than others when certain conditions prevail.

Successful feeders choose the one program, or combination of programs, that best fits their needs, and they change only when necessary to adjust to changing conditions. Under some conditions a combination of two or more programs offers some special advantages, especially on larger farms or in commercial feedlots. Among the factors to be considered in the choice of feeding program are the following.

1. *Length of feeding period.* Usually cattle that are fed grain or concentrates for fewer than 100 days are called "short-fed" cattle. Ordinarily, less than 25 bushels of corn or similar concentrate will be fed per head to such cattle. On the other hand, if cattle are fed finishing rations for 8 to 10 months, they are spoken of as "long-fed" cattle. These cattle (usually choice or prime steer calves or light yearlings) ordinarily grade high choice or prime after finishing, provided they were of sufficiently good conformation. They will have consumed 75 bushels or more of corn or similar feeds per head. Anything falling about midway between these limits is regarded as a medium feeding period. Heifers usually are fed about 60 days less than steers of corresponding age.

2. *Relative amount of grain or concentrates in the ration.* This may vary all the way from a full feed of concentrates (1.5 to 2 pounds for each 100 pounds of body weight daily) to a limited feed (0.5 to 1 pound per 100 pounds of body weight per day). Cattle are full-fed in order to finish them quickly. It may not always be the most economical way to produce a pound of gain but feeders follow this practice so that there will be sufficient finish on the cattle to ensure satisfactory grade in the carcass and, of course, to market their grain and shorten the feeding period. In calves, full feeding usually follows a period of first feeding on growing rations or a period of grazing in farmer-feeder situations.

   Limited feeding is usually practiced either during a wintering or grazing period in which principally growth, rather than fattening, is desired, or when a heavy feed of roughages plus some concentrates is fed just prior to a period of full feeding. The latter is done in order to cheapen the gains made during the entire feeding program. Roughages and pastures, when adequately supplemented, make economical gains, but, when unsupplemented, often scarcely do more than maintain the weight of cattle. This fact accounts for the practice of limited feeding of supplemental high-energy concentrates. When all feeds are purchased, roughage nutrients, especially energy, generally cost more than concentrate nutrients as well as require more processing and handling labor; hence the very strong trend toward full feeding of high-concentrate rations by commercial feedlots.

3. *Systems of feeding.* Cattle may be fed finishing rations in drylot or on pasture, or both. For example, many cattle, especially calves, may first be wintered on a grower or stocker ration consisting almost entirely of roughage plus, at most, a limited grain feed. They are then fed either a limited feed or a full feed of grain on pasture during spring and summer, after which they are finally finished in drylot on a full feed for about 60 days. This combination of programs furnishes a market for harvested roughages and grains as well as pasture, which in the

Fig. 71. Long-fed cattle such as these near-prime steers assume considerable importance in a few Corn Belt areas but, in general, choice is the grade of preference
of most cattle feeders. Remodeling so as to mechanize even the small-sized
feedlots in this area is under way in order to reduce labor costs. (Corn Belt Farm
Dailies.)

final analysis is one of the primary reasons for feeding cattle, on farms
at least. Cattle may also be full-fed on self-feeders or by hand daily.

4. *Time of year when fed.* For the most part, cattle finished principally in
drylot in the farming regions are fed during the winter months,
whereas those finished on pasture are naturally fed in the spring and
summer or during the pasture-growing season. Most specialized feedyard operators feed cattle in drylot on a year-round basis, and this
trend is developing even in the farm areas, to keep expensive equipment in constant use so as to spread the fixed, nonfeed costs over more
head fed.

THE KIND OF CATTLE TO FEED

Most feeding programs require a certain kind of cattle if they are to be
most successful. Conversely, each type of cattle should be fed according to

a rather well-recognized plan if the cattle are to return the greatest profit. During a period when prices are rising and profits are large, a particular drove of cattle may return a profit almost regardless of the way they are fed, but when the reverse is true, the kind of cattle purchased must be well suited to the plan of feeding if a profit is to be made or, worse, to keep losses at a minimum.

Feeder cattle vary widely in those characteristics that affect their suitability for different feeding conditions and that are therefore important to prospective buyers. Their chief differences are as follows.

1. *Age.* Calves, yearlings, and 2-year-olds.
2. *Condition.* Thin, medium, and fleshy.
3. *Weight.* Light, medium, and heavy. Weight, of course is the result of both age and condition. For example, a calf 9 months old in fleshy condition may weigh as much as or more than a very thin yearling.
4. *Sex.* Steers, bulls, heifers, and cows. Steers may be any age from 4 to 30 months, but feeder heifers are seldom older than 24 months. If older, they probably are pregnant or have already produced calves and are called heiferettes.
5. *Breeding.* Beef, mixed, and dairy. Cattle are said to show beef breeding when they show the characteristics of the Angus, Hereford, Shorthorn, or Brahman breeds. A drove of cattle of mixed breeding usually contains at least some animals that have both beef and dairy ancestry. Sires of such breeds as Charolais, Charbray, Braford, Santa Gertrudis, and Beefmaster, when mated to British breeds of beef cows, produce calves that are near the beef category, but they possess distinctive qualities that perhaps someday will result in a classification of their own, such as simply *exotics.*
6. *Grade.* Fancy, choice, good, standard, and utility. As discussed in Chapter 11, the grade of an animal, theoretically at least, indicates its all-round desirability for the use to which it will be put—in the case of a feeder, its desirability for feeding. The grade of a feeder steer or heifer is determined chiefly by its conformation, quality, and breeding.
7. *Origin.* Some cattle feeders attach some importance to the region or area from which the cattle have been shipped. It is doubtful whether great preference should be shown for a drove of cattle solely because they come from a certain area. Every important cattle-breeding area of the country produces some excellent cattle and some that are only mediocre. Each drove should be judged on its own merit and not on that of another shipment from the same locality. It is true that cattle from certain areas may adjust more quickly to the environmental conditions existing where they are to be fed out, or they may vary in

degree of parasitism or in amount of condition they carry, but the pricing mechanisms prevailing today usually adjust for these differences.

CONTRACT OR CUSTOM-FEEDING

Occasionally a large rancher, packer, or cattle investor enters into a custom-feeding contract with a cattle feeder or feedlot to finish cattle without a change of ownership. The chief reasons why larger feeders and feedlot operators are increasingly resorting to custom-feeding cattle on contract are (1) the risk is almost entirely eliminated for the feedlot operator, in that the owner of the cattle must take the loss if prices should fall or if, for any of several other reasons, the cattle perform poorly, and (2) less capital is required by the feeder, permitting him to use his own often limited capital to enlarge his operation or even lay in a feed supply in advance as a hedge against seasonal price increases. By becoming involved in a contract, the custom-feeder does forego the occasional opportunity to make large profits when prices rise spectacularly, as they do perhaps one year out of 10 or 12; he does this so as to assure a guaranteed income or return from his operation. Some feeders fill only half or a portion of their lots with contract-fed cattle and thus, as it were, play the market both ways. In the areas where commercial feedlots are concentrated, as many as 75 percent of all cattle being fed are being custom-fed, but for the country as a whole it is doubtful whether more than 10 to 15 percent of all cattle being fed for slaughter are under any sort of contract.

KINDS OF FEEDER CONTRACTS

Contracts are essentially of two kinds and the choice of which to use depends on factors too numerous to discuss fully here. The two types are the following.

1. *Gain-in-weight contract.* The feeder agrees to feed a certain ration for a given period of time, or until cattle attain a certain weight. The owner of the cattle agrees to pay a certain amount per pound of gain. The owner also pays for all medicines and veterinary costs, but the feeder may or may not be paid for the weight gains of any cattle that die. The weighing arrangements should be specifically spelled out for the best interests of both parties.
2. *Feed and yardage contract.* The feeder agrees to provide all feed, labor,

and equipment required, and he is paid an agreed-upon price for feed, per ton, as well as a yardage charge, which may either be built into the feed charge or may be a separate item—for example, 5 to 10 cents per head per day. The owner of the cattle specifies the ration, and a minimum length of feeding period is stipulated by the feedlot owner. The owner of the cattle again provides all veterinary service and drugs. Because a feedlot may be feeding several owners' cattle simultaneously, each owner expects an accurate accounting of all feed fed his particular cattle.

Contract forms are obtainable from various sources for both types of contracts, but none will completely cover every situation. Therefore an experienced lawyer should be consulted. Most owners like to add some type of incentive clause that will provide a motive for better than average care or husbandry of his cattle on the part of the feedlot operator.

Contract feeding has never been popular with farmer-feeders because seldom have both parties been satisfied with results of the arrangement. If prices are rising and the cattle return the owner a handsome profit, the feeder is unhappy that he does not own the cattle himself and thereby receive the owner's profits as well as his own. And if prices are falling and the cattle are sold for less than expected, the owner is dissatisfied when he finds himself taking home less money than he was offered for his feeder cattle the previous fall. Because it takes two parties to make a contract and usually only one is interested in such an arrangement, not much contract feeding is done by the average farmer-feeder.

CONDOMINIUM FEEDLOTS

An interesting new development in the farmer-feeder belt is the so-called cattle condominium. In this instance, 6 to 10 or so farmers join together and build a feedlot that may have many of the characteristics of a High Plains custom-feeding lot. A single storage and feed-processing center and feedlot serves all owners. One of the owners may act as the manager. Owners may or may not provide homegrown grains, but they usually do, and silage and hay often are purchased by the group from members who farm nearest the condominium. Each lot owner provides his own cattle, usually enough to fill one or more pens of 50 to 200 cattle capacity. These cattle may be and often are homegrown calves, or they are purchased calves that first were grown to feeder weight on stocker rations on the home farm of the owner. Investments in facilities and equipment for feed storage, grain processing, manure hauling, and cattle working may be shared, and overall costs may be spread over that many more head. Such feedlots also attract more buyers for fed cattle than when the same numbers are fed on a number of scattered farms. The advantages of this plan seem to outweigh the drawbacks and the practice should spread, particularly in the Corn Belt.

INVESTMENTS IN FEEDLOTS

The capital investments in a feedlot for acreage, buildings, corrals, feed storage, feed-handling and feed-processing equipment, manure-handling equipment, transportation, and utilities can be enormous. Expressed on a per head capacity basis, the investment may range from as little as $25 in the semiarid Southwest for a 20,000-head lot to as high as $400 in the Corn Belt for a completely automated lot with environmental control for 200 head. Table 87 shows how both size of lot and degree of confinement and automation affect investments per head of space in a feedlot. Obviously, if more than one drove of steers can be fed per year, then the annual costs of interest and depreciation can be spread over more cattle fed. It is generally believed that a custom feedlot must be at least 70 to 75 percent filled if it is to break even. It should be mentioned that certain cost

Table 87

Investment per Head Space for Shelter, Buildings, Lots, Equipment, and Feed Storage for Various Beef Feeding Systems[a]

One-Time Capacity	Drylot with Open Shed and Paved Lot			Confinement		Open Lot
	Limited Mechanization	Auger— High Mechanization	Fence-line Bunks	Warm	Cold	No Pavement
40	$171					
80	$150	$202				
100	$140	$180	$231			
200		$142	$138			$75
500		$114	$110	$190 to $300	$188 to $250	$74
1,000		$111				$60
2,000						$52
5,000						$39
10,000						$31
15,000						$28

[a] D. E. Erickson, Economic Comparison of Confinement, Conventional Drylot, and Open-Lot Beef Feeding Systems. AE 4250, University of Illinois, 1970.

items are fixed in a farmer-feeder's lot as well, but underutilization is not so serious in this instance. For example, the tractor used to pull the fenceline feed wagon also is used for field work.

The costs of depreciation, maintenance, taxes, interest, and insurance on the feedlot facility usually make up a major portion of the nonfeed costs in operating a feedlot. A number of factors determine the amount of these nonfeed costs. The environment, including temperature, relative humidity, rainfall, and drainage, determines to a large extent the type of buildings, if any, and feedlot floor that are required. The type of ration fed is a large factor. For example, silage and high-moisture grain storage can be very expensive, in terms of investment capital at least. Steam-processing of grain requires high-cost equipment, especially if small numbers are fed. Concrete feeding floors and concrete bunks, and especially slotted-floor confinement barns with manure pits beneath them, also can be expensive, but nevertheless are often an excellent investment in the Corn Belt where seasonally heavy rain and snowfall, coupled with flat terrain and heavy soils that prevent drainage, make them almost essential. Often such investments are necessary to comply with stringent environmental pollution control regulations.

Example data, shown in Tables 88 and 89, for Ohio and California, respectively, show something of the magnitude of these costs and, more important, how size of feedlot affects the per head nonfeed costs. It is unfortunate that more current cost estimates are not at hand, but the comparative costs of farmer-feeder operations and commercial custom lots are still solid and of interest. In the Ohio report the data are shown on the basis of cost per hundredweight gain. It is evident that on the smaller farms the nonfeed costs amount to $10 per hundredweight gain, 16 cents per day, or about $37 per head fed. On the larger farms the nonfeed costs amount to $5.70 per hundredweight gain, 10 cents per day, or about $26 per head. For contrast, the California data in Table 89 show the effect of differences in still larger sizes of lots. The average custom feeder, who bought nearly all his feed and fed few or no cattle of his own, had nonfeed costs of only 7.06 cents per day. The California farmer-feeder's costs were higher in general, but still below all the Ohio categories. It should be mentioned that since California feeders normally feed for shorter periods of time and have a faster turnover rate than those in Ohio, or the entire Corn Belt for that matter, they can reduce their daily nonfeed charges because they are more likely to have the lot full the year around. Their turnover rate was 1.59 for the lots sampled—that is, 1.59 head were fed per space per year. It appears from the California study that the nonfeed cost reduction, owing to increasing numbers fed, levels off at about 26,000-head capacity.

The fact that many farmer-feeders use existing older buildings already

Table 88

Effect of Number of Cattle on Feed on Nonfeed Costs in Ohio Feedlots, 1958–1964[a]

	Average Number of Cattle Fed and Number of Farms			
Cattle Farms	53 30	123 72	238 32	640 34
Nonfeed costs per cwt gain				
Labor	$3.18	$1.80	$1.23	$1.08
Tractor power	0.72	0.38	0.39	0.30
Veterinary and drugs	0.23	0.17	0.13	0.25
Truck, auto, telephone, electricity	0.25	0.29	0.21	0.30
Buildings and equipment	3.51	2.91	2.43	2.18
Taxes and interest	2.11	1.81	1.74	1.59
Total	$10.00	$7.36	$6.13	$5.70
Miscellaneous performance and cost data				
Average total gain (lb)	369	452	469	460
Average number days fed	225	259	261	258
Average daily gain (lb)	1.63	1.72	1.82	1.83
Purchase weight (lb)	581	574	570	551
Sale weight (lb)	950	1,026	1,039	1,011
Building investment per cwt	$24.65	$20.22	$16.07	$14.71
Death loss (%)	1.18	1.33	.86	1.65

[a] Ohio Agricultural Experiment Station Report, 1966.

on the farm, either dairy, horse, or beef barns, means that some nonfeed cost need not necessarily be included for them or, if so, at less than replacement costs, when calculating relative costs of feeding cattle in different areas or in different situations. Remodeling old structures for mechanization may actually cost more than newer, more efficiently designed buildings. The more costly grain storage facilities, such as oxygen-free storage for high-moisture corn or sorghum, in those areas where these crops are grown in surplus, can be offset by lower purchase and transportation costs of grain. No single area of the country has all the advantages or disadvantages when it comes to costs of feeding cattle.

EXPANSION OF COMMERCIAL FEEDYARDS

Most of the increase in numbers of cattle finished in California, Arizona, Colorado, Nebraska, and in the grain sorghum belt is accounted for by the

Table 89

Average Daily Nonfeed Costs per Head Fed in Eighty-One California Feedlots (Cents)[a]

Cost Items	Type of Operation			Numbers Fed				
	Farmer-Feeder	Commercial Feeder	Custom Feeder	Under 4,000	4,000–10,000	10,000–26,000	Over 26,000	State Average
Salaries and wages[b]	3.67	3.10	3.21	4.68	3.52	2.97	3.29	3.27
Taxes, interest, insurance[c]	1.28	1.29	0.82	2.32	1.42	0.93	0.89	1.00
Utilities	0.52	0.45	0.37	0.59	0.51	0.41	0.40	0.42
Gasoline, oil, grease	0.37	0.27	0.24	0.53	0.29	0.26	0.26	0.27
Depreciation	1.41	1.00	0.88	2.59	1.23	0.88	0.93	1.00
Repairs	0.79	0.88	0.81	1.11	0.62	0.73	0.87	0.82
Veterinary fees, medical	0.66	0.29	0.50	0.85	0.41	0.64	0.41	0.49
Nutrition services	0.06	0.02	0.12	0.01	0.05	0.15	0.07	0.09
Legal and accounting	0.07	0.07	0.17	0.05	0.15	0.07	0.16	0.13
Trucking and freight	0.04	0.12	0.08	0.15	0.28	0.02	0.08	0.08
Promotion	0.02	0.05	0.06	0.01	0.04	0.03	0.06	0.05
Other costs[d]	0.22	0.33	0.22	0.28	0.29	0.19	0.26	0.24
Total gross costs[e]	9.11	7.87	7.48	13.17	8.81	7.28	7.68	7.86
Manure credit	-0.43	-0.38	-0.42	-0.43	-0.13	-0.27	-0.54	-0.41
Total net cost per day	8.68	7.49	7.06	12.74	8.68	7.01	7.14	7.45

[a] *Cattle Feeding in California*, Bank of America, Economic Research Department, 1965.

[b] All salaries and wages and payroll taxes.

[c] Property taxes, interest on investment (computed at 6 percent of book value of total investment in feedlot, mill, storage, office, trucks, etc.) and insurance costs.

[d] Other costs included market reports, odor control, rental fees, and livestock commissions.

[e] Death losses are not included. If death losses average 1 percent and loss occurs halfway through feeding period and animal lost weighs 800 lb, is worth 22.5 cents per pound, and feeding period is 150 days, daily cost per head fed for death loss is 1.2 cents. Death loss is a cost that should not be omitted. Because California cattle feeders use lot accounting and charge deaths to the entire lot costs, and because the loss is higher for calves than for yearlings, an average death loss was not included in the analysis.

increase in commercial feedyard operations. Operators of these feedyards carry on primarily a custom or contract feeding operation, seldom owning all of the cattle being fed. As a rule the operators of such feedyards feed their own cattle only when the outlook for a substantially favorable price spread is indicated or when vacancies occur because of insufficient patronage. Low per unit cost and low risk are essential for profit in this program. Therefore the operator must try to keep his feedyard filled to capacity the year around with customers' cattle.

The finishing rations fed in commercial feedyards vary widely, mainly because the crops grown specifically for cattle feed vary, depending on the area, but also because the by-products of commercially grown crops vary still more. Generally speaking, feedyard operators feed rations of two main types, depending mainly on the source and type of roughage fed. The first category of feeders feed primarily all-dry drylot rations consisting of small grains, sorghum, or corn, and hay (often dehydrated or cubed) or cottonseed hulls, and some dried crop by-products such as sugarbeet pulp, citrus pulp, culled dried fruit, or dried surplus truck crops. The second category includes those who rely heavily on green or wet feeds—either fresh chopped alfalfa, corn, or sorghum, and, in some cases, silages made from these crops. Naturally the green feed or silage is

Fig. 72. Commercial feedyards such as this one in the Texas Panhandle sorghum area are responsible for much of the phenomenal increase in numbers of cattle fed in the Southwest. Low investment in feedlot and shelter, combined with commercial-type feed storage and processing, makes low nonfeed costs possible. Choice cattle are common in the feedlots of this area, no doubt a reflection of the proximity of ranches that produce mainly such cattle. (Western Livestock Journal.)

supplemented with varying amounts of concentrates, depending on the level of gain desired. The feeding program of those in the latter category is similar in principle to the stocker program discussed in Chapter 11, at least in the early stages before the cattle are placed on finishing rations containing a higher level of high-energy concentrates. Actually a feedlot may carry on both programs, but the trend seems to be for a lot to specialize in one or the other.

Successful commercial feedyards generally have the following characteristics.

1. They are large enough to permit labor specialization; that is, each man does one particular job during an entire day.
2. Precision feed-processing and mixing equipment and mechanical feed-handling equipment are used, with large numbers on feed making such equipment economical and practical. Feeding may go on 24 hours around the clock.
3. Specialists such as veterinarians, nutritionists, and marketing men are employed, with large numbers on feed again making this practical.
4. Feeds and incidentals are bought in wholesale quantities, often with competitive bids. Least-cost rations will be calculated almost daily.
5. Buyers for customers' cattle buy many cattle direct from the feedyards, partly because of the large numbers available and, in some cases, because of the reputation for "killing well" that cattle from a particular yard may have.

It is doubtful if the commercial feedyard will assume as important a role in the traditional cattle-feeding areas of the Corn Belt as it has in the West and Southwest. Most commercial-sized feedyards presently found in the Corn Belt are not operating on a custom basis. Furthermore, they are often built around the utilization of some special by-product such as the corn cobs and off-shaped kernels produced by a seed corn producer or the cannery waste resulting from the processing of sweet corn or peas. Individual farmer-feeders are increasing the size of their operations, but they will be limited by the roughage-producing capacity of their cropping systems. Naturally such farmer-feeders are as desirous of reducing their fixed, per head, nonfeed costs as are the commercial feedyard operators. Numbers of cattle feeders in the Corn Belt, especially the traditional smaller farmer-feeders, are decreasing, but confinement-feeding is on the increase and indications are that this increase will continue.

CHAPTER 13
THE IMPORTANCE OF AGE
AND SEX IN GROWTH
AND FINISHING

Changes in consumer demand and in production techniques practiced by cattle breeders and feeders alike have been responsible for a gradual shift to younger and lighter slaughter cattle. How far this shift will go is hard to determine with certainty, because there are factors that tend to counter this trend.

At present the average slaughter weight of fed steers is near 1,000 pounds for the country as a whole, but in certain areas steers weighing 875 to 925 pounds are in greatest demand. Heifers generally go to market at least 100 pounds lighter than steers. Naturally, as the carcass weights of slaughter cattle change, the age at which cattle are placed on feed and finally slaughtered changes too. Whereas as recently as 1940 the majority of the cattle on finishing rations were yearlings or older, today well over half are calves. This is especially true in areas such as West Texas, the Oklahoma Panhandle, and southwestern Kansas, where expansion in cattle feeding has been greatest. Even in the traditional cattle-finishing state of Iowa a recent survey showed that one-half of the cattle feeders fed calves. When feed prices are high in relation to feeder cattle prices, preferred purchase weights work up higher, because feedlot operators believe they can buy weight more economically than they can feed it on. Rarely are heavier feeder yearlings and calves priced higher than lighter-weight cattle of the same age groups, but it does happen, and especially when positive price spreads prevail as they sometimes do in the liquidating phase of the cattle cycle.

More heifers are being fed than formerly, also, as may be seen in Table 90, which shows that heifer beef production more than doubled during the past two decades. This reflects the fact that a smaller portion of the heifer calf crop is kept back for replacements or to build new herds, and that those not kept for breeding purposes are being grain-fed rather than sold as slaughter calves or grass-fat heifers. These changes add to the need for knowledge concerning the effect of age and sex of feeder cattle in finishing programs. The ratio of steers to heifers in the feedlot is strongly affected by stage of cattle cycle. When cow numbers are increasing, heifers for sale are decreased. On the contrary, when cow numbers are decreasing, numbers of heifers available for feeding are increased and, also, more

387

Table 90

		Beef Production by Class			Fed Beef	
Year	Total Production[b] (mil. lb)	Steer (%)	Heifer (%)	Cow[c] (%)	Quantity (mil. lb)	Percentage of Total
1946	9,373	52	10	38	3,427	37
1950	9,534	57	9	34	4,446	47
1954	12,963	55	12	33	5,319	41
1958	13,330	60	15	25	6,760	51
1962	15,298	61	20	19	9,896	65
1966	19,694	56	24	21	13,207	67
1970	21,653	59	25	16	16,602	77
1974	23,138	59	22	19	15,712	68

Estimated Composition of Beef Production, 1946–1974[a]

[a] USDA, *Livestock and Meat Statistics,* 1974.

[b] Includes production for farm consumption.

[c] Includes bull and stag beef, which averaged about 2 percent of total annual supply.

are slaughtered as weaners or as grassed yearlings. The combination of more cull cows and excess heifers slaughtered off grass accounts for the drop in percentage of fed beef in 1974.

EFFECT OF AGE ON RATE OF GAIN

A discussion of the effect of age on daily rate of gain really comes down to a comparison of the rate of gain between weaner calves and yearlings that are about a year older. Also, for practical purposes, one should compare the rates of gain between these two ages of feeders when fed for the lengths of time required for them to reach desirable slaughter weight and grade. In other words, calves weighing initially about 400 to 500 pounds at 7 to 8 months of age and fed for 8 to 10 months should be compared with yearlings weighing initially about 600 to 750 pounds at 18 to 20 months of age and fed for 5 to 6 months.

When comparisons are made in this manner, results such as those summarized in Table 91 are obtained. All cattle in the trial received recommended feed additives such as hormones, vitamin A, antibiotics, and minerals, with energy level differences in the rations being achieved by varying the corn and corn silage levels. When all lots of cattle were fed only until it was felt that at least 80 percent of those in a particular lot would quality-grade USDA choice in the carcass, the yearlings outgained

Table 91

Effect of Age and Energy Level on Performance of Beef Steers[a]

	Calves		Yearlings	
Energy Level	Low	High	Low	High
Number of calves	19	18	19	19
Average initial weight (lb)	448	450	745	746
Average final weight (lb)	1,072	1,084	1,145	1,164
Days fed	305	250	182	154
Average daily gain (lb)	2.04	2.53	2.19	2.70
Total gain (lb)	622	633	399	416
Daily feed (lb)				
Ground corn	2.0	11.8	2.0	15.1
Supplement A	2.0	2.0	2.0	2.0
Corn silage	38.8	15.9	51.7	19.4
Feed per cwt gain				
Air-dry feed (lb)	849	654	1,001	748
Total TDN (lb)	484	539	557	626
Feed cost ($)[b]	31.27	33.60	35.02	38.30
Carcass data				
Hot weight (lb)	649	654	679	698
Loin-eye area (sq in.)	11.00	10.39	12.25	11.71
Backfat thickness (in.)	0.55	0.78	0.63	0.54
Number of choice carcasses	9	1	4	5

[a] Adapted from Indiana Agricultural Experiment Station Research Progress Report 113.
[b] Feed costs used were: corn silage, $20 per ton; corn, $2.52 per bushel; supplement A, $160 per ton.

the calves by about 0.15 pound per day, or by 7 percent. The level of energy fed had no influence on relative gains made by calves and yearlings but did, or course, strongly influence rate of gain, within age group.

When profits from favorable feeding margins are contributing most of the profit in a feeding enterprise, rather than favorable price spreads—and they usually do—then the additional 200 pounds plus total gains made by the calves as compared to the yearlings, as in the trial mentioned above, becomes far more important than the slightly higher daily rate of gain made by the yearlings. When choosing between these two most common ages of feeders, the importance of the difference in rates of gain is often overemphasized. After all, unless the higher rate of gain is accompanied by an improved feed efficiency, the only savings are in the slightly lower nonfeed costs resulting from the somewhat fewer days in the feedlot.

EFFECT OF AGE ON ECONOMY OF GAINS

Numerous experiments have shown that young animals are more efficient in feed conversion than are older cattle. The principal explanation is found in the fact that increases in the body weight of young animals are due partly to the growth of muscles, bones, and vital organs, whereas the weight increases of older cattle consist largely of fat deposits. Fat has much less water and a great deal more energy than an equal weight of any other kind of animal tissue. Thus more feed is required for its formation. Other factors contributing to the more economical gains made by young animals are (1) a slightly larger consumption of feed in proportion to body weight and (2) the much greater maintenance requirement of the older cattle, especially when expressed on a daily basis. As a result of their relatively large feed consumption, younger cattle use a smaller percentage of the ration to satisfy maintenance requirements than is used for this purpose by older animals, thus making available a greater percentage of the total feed consumed for the production of muscle and fat.

The more efficient gains made by young cattle become more and more evident as the feeding period progresses. This fact is well illustrated in Table 92, which shows the results obtained at the Nebraska station with steers of various ages fed for a period of 200 days. Whereas in calves the total feed required per 100 pounds of gain during the last half of the period was only 18 percent above that required during the first, in the yearling, 2-year-old, and 3-year-old steers it was 35, 44, and 69 percent

Table 92

Influence of Age on Variation in Economy of Gains Throughout the Feeding Period[a]

	Feed per Hundredweight Gain (lb)			
Period	Three-Year-Old Steers	Two-Year-Old Steers	Yearlings	Calves
First 100 days				
Corn	586	597	534	431
Alfalfa	363	346	304	221
Second 100 days				
Corn	1,282	1,085	916	623
Alfalfa	320	272	219	148
Total 200 days				
Corn	835	798	702	529
Alfalfa	350	314	266	186

[a] Nebraska Bulletin 229.

greater, respectively. These figures bear out the fact that calves may be fed profitably over a longer period than older cattle. If necessary, they can be held for a considerable length of time, awaiting a favorable market, and all the while they will be making fair gains at not too great a cost. Mature steers, on the other hand, are held at great expense. After being on full feed for 150 or 160 days, they gain very slowly and inefficiently.

The fact that feeding cattle to excessive finish is uneconomical is further demonstrated by the data in Table 93. It will be noted that, although the total investment in feed required to finish a calf is higher than for the older steers, the cost of gain is less, making it more likely to realize a profit from the feeding margin, as indicated earlier. It should be mentioned that, in order for slaughter cattle to grade choice in the carcass, they ordinarily must be fed 30 to 50 days beyond the point where costs of gain are just equal to selling price. About 100 pounds of gain must be added beyond this point that will not directly pay for itself, but is essential to achieve the desired higher selling price. Following this same logic, feeding until cattle grade prime is not advised, for then the last 100 pounds of gain will cost far in excess of the selling price, and the premium price for this grade of slaughter beast seldom covers the added feed cost, let alone the added nonfeed costs. Unfortunately, the extra finish will still not guarantee the prime grade in all too many instances because of lack of

Table 93

Effect of Length of Feeding Period and Age on Feed Cost of Gain of Choice Steers Fed in Drylot[a]

Gain	400-lb Calf	640-lb Yearling	840-lb 2-year-old
Hundred pounds			
First	$ 21.59[b]	$ 25.38	$ 25.84
Second	23.50	29.51	31.82
Third	26.61	36.52	40.53
Fourth	30.24	42.49	56.44
Fifth	35.57	54.36	—
Sixth	42.77	—	—
Seventh	53.44	—	—
Total gain (lb)	700	500	400
Final weight (lb)	1100	1150	1240
Total feed cost	$233.72	$188.26	$154.63
Average cost/cwt gain	$ 33.39	$ 37.65	$ 38.66

[a] Adapted from USDA Technical Bulletin 900.

[b] Based on corn at $2.52 per bushel, mixed hay at $30 per ton, and protein supplement at $120 per ton.

marbling and, even worse, may result in price discounts because of poorer yield grades.

A practical question arising in the minds of cattle feeders relative to age of feeder cattle, is whether to buy the earlier and thus heavier calves from a rancher or whether to buy the later and therefore lighter and younger calves. An Oklahoma study sheds some light on this question, and the pertinent results are shown in Table 94. The calves were all half-brothers, but there was a 15-day average difference in ages. Despite the fact that the heavier calves outweighed the lighter ones by 70 pounds at the start of the test, they gained equally as well as the lighter ones. The phenomenon of compensatory gain usually seen in lighter cattle did not occur, possibly because all calves had been creep-fed and thus were probably in the same condition or degree of fatness. The advantage held by the lighter calves with respect to feed efficiency would have justified paying an extra dollar per hundredweight for them as weaners. Before a feeder pays a premium for light calves, he should know that they are light because they are young or undernourished and not because of poor genetic potential to gain rapidly and efficiently.

EFFECT OF AGE ON LENGTH OF FEEDING PERIOD

It has been mentioned that young cattle can be fed longer with profit than old cattle. If feeder steers of different ages are started on feed in

Table 94

Effect of Age and Weight of Calves on Subsequent Performance in the Feedlot When Full-Fed[a]

	Weight Group		Difference
	Heavy	Light	Heavy—Light
Number of calves	104	107	
Initial age (days)	219	207	12
Initial weight (lb)	514	444	70
Final age (days)	398	386	12
Final weight (lb)	942	866	76
Gain on test (lb)	428	422	6
Average daily gain (lb)	2.39	2.36	0.03
Dressing percent	63.1	62.4	0.7
Carcass grade[b]	10.7	10.9	−0.2
Feed per pound of gain (lb)	9.80	9.34	0.46

[a] Oklahoma Agricultural Experiment Station MP 67.
[b] Low choice = 10; average choice = 11; high choice = 12.

approximately the same degree of flesh or condition, the time required to finish each age of feeder to the same grade varies inversely with the age of the cattle. When cattle are finished for their grade, they should be marketed as soon as possible. To hold them longer will require both feed and labor for which little will be realized, either from increased weight or improved carcass merit.

On the basis of experiments such as that reported in Table 95, as well as

Table 95

Feedlot Performance and Carcass Merit of Steer Calves and Yearlings Fed to Different Market Weights and Grades[a]

Age of Steers	Medium-Fed Calves	Long-Fed Calves	Medium-Fed Yearlings	Long-Fed Yearlings
Number of steers	10	10	10	10
Days on feed	266	392	140	308
Average weight (lb)				
Initial	460	463	688	688
Final	1,057	1,246	1,114	1,442
Average daily gain (lb)	2.25	2.00	3.04	2.45
Average daily feed (lb)				
Corn silage	18.58	17.45	19.10	16.84
Corn, high-moisture	10.89	12.52	16.52	17.92
Soybean meal (50% protein)	1.25	1.25	1.25	1.25
Vitamin A supplement	0.50	0.50	0.50	0.50
Feed per pound of gain (lb) (85% dry-matter basis)	8.13	9.63	7.67	9.67
Feed cost per cwt gain[b]	$30.85	$37.20	$29.95	$38.38
Dressing percent	61.5	65.8	63.4	66.5
Fat thickness over 12th rib (in.)	0.51	1.01	0.60	1.43
Loin-eye area (sq in.)	10.90	12.14	10.38	11.75
Carcass grade[c]				
Conformation	20.3	21.4	19.2	20.6
Quality	19.8	20.8	19.1	20.6
Overall	19.7	20.8	18.9	20.4
Total retailable product, percent of carcass weight	61.5	56.4	67.0	55.2

[a] Adapted from Illinois Cattle Feeders' Day Report.
[b] Feed costs: corn (16% moisture basis), $2.52 per bushel; corn silage, $20 per ton; soybean meal, $120 per ton; vitamin A supplement, $100 per ton.
[c] Low choice = 19; average choice = 20; high choice = 21.

the experiences of practical feeders, it may be stated that 4 to 5 months of heavy feeding are required for yearling feeder steers to be put in choice condition, and from 8 to 9 months for calves. The 1976 reductions in the degree of marbling required for young cattle to attain choice grade will undoubtedly result in reducing the necessary length of feeding period by as much as 2 to 4 weeks. The net effect should be favorable, both from the standpoint of economy of gain and the percentage of retailable product in the carcass. It should, of course, be understood that large numbers of cattle are marketed after a relatively short feed before they are really finished. The increasing demand for leanness in beef cuts may make short feeding more practical, but it must be recognized that with such short-fed cattle, profits must come from favorable price spread. Short-fed cattle do not provide a market for much feed, and a drastic shift to this program by large numbers of feeders would tend to increase the demand for the lower grades of feeders commonly used. Thus feeder cattle costs would be too high to permit the necessary price spreads. Buying warmed-up or half-fat heavier cattle for feeding, in order to shorten the feeding period, is extremely hazardous, for such cattle gain relatively slowly and inefficiently and profits, if made, must come from positive price spreads. Hedging of such cattle on the futures market to ensure a positive price spread is an absolute must.

EFFECT OF AGE ON TOTAL GAIN REQUIRED TO FINISH

There is not a great deal of difference in the increase in weight that must be realized by feeder steers of different ages to attain the same degree of finish. If the cattle are in equal condition when put on feed, the total gain necessary to finish decreases slightly with advanced age. In general this gain is approximately as shown:

Age	Total Gain Necessary to Finish (lb)
Two-year-olds	250–350
Yearlings	350–400
Calves	400–500

When expressed in terms of the ratio of initial weight as feeders to final weight, however, the differences in total gains are significant. Calves double their weight while in the feedlot, yearlings increase in weight approximately 60 percent, and 2-year-olds from 25 to 30 percent. Obviously these percentages are greatly affected by the condition of the steers when started on feed. The older the feeder, the smaller the fixed nonfeed costs per head. However, if a feedlot is kept constantly filled with cattle the year around, nonfeed costs will still be lower for calves because some items of cost, such as interest, are related to size of animal or investment per animal. If profits are being made from feeding margin rather than price spread, then the age of cattle that requires the greatest amount of gain to reach desired grade is favored.

EFFECT OF AGE ON TOTAL FEED CONSUMED

The age of cattle of comparable feeder grades has comparatively little effect on the total amount of feed required to attain a given degree of finish, provided all ages are fed rations with comparable concentrate:roughage ratios. The longer feeding period of the younger animals tends to make up for their smaller daily consumption of feeds, so that by the time they reach the desired finish they have eaten about as much grain, protein concentrate, and roughage as older cattle would have eaten in a shorter time. Consequently the age of the cattle has relatively little effect on the number of animals that should be purchased to utilize a given amount of feed, assuming, of course, that the feed supply is equally well suited for finishing calves, yearlings, and 2-year-old steers.

EFFECT OF AGE ON QUALITY OF FEEDS USED

Although age of cattle has little effect on the amounts of grain, protein concentrate, and roughage required to attain a good or choice finish, it greatly influences the possibility of limiting the grain ration to something less than a full feed and supplying more roughage to take its place. Such a practice may be fairly satisfactory with long yearlings and 2-year-olds, but not with calves, which lack the capacity for large amounts of bulky feeds. Two-year-old steers may be fed large amounts of silage or given free access to good-quality hay without depressing their consumption of grain below the amount required to produce a satisfactory finish, whereas such a practice with calves would be almost certain to reduce the grade of the finished calves.

With the advent of acceptable high-energy rations—that is, rations with less roughage and a higher caloric density than formerly—feeding programs for calves are being used in which they consume as much energy as older cattle, expressed on the basis of energy or calories per hundred pounds of body weight. The result is that calves can be fed even more efficiently, as compared with older cattle, than formerly. This entire subject needs more research, but if the feeding of high-energy rations to calves does not result in excessive fat deposition and therefore does not reduce cutability or yield of lean retail cuts, then we shall surely see the trend toward feeding more calves than yearlings accelerated still further.

EFFECT OF AGE ON PASTURE UTILIZATION

Because of their limited capacity for feed, short yearlings finish less rapidly on pasture than do older cattle. (There are relatively few calves available for grazing in the spring because most of the calves born the previous year are yearlings by then.) Fresh grass is a bulky feed with a high moisture and low energy content. Therefore yearlings scarcely get more digestible nutrients from pasture than they need for maximum growth, leaving little for the improvement of condition. Two-year-old steers, on the other hand, are almost full-grown and nearly all the nutrients consumed above maintenance requirements are available for the production of fat. For this reason older cattle are more popular than yearlings in the better grazing areas of the United States and in Argentina and New Zealand where many cattle are sold for slaughter directly off grass without the feeding of any grain. Even when grain is fed on pasture, yearlings finish less satisfactorily than 2-year-olds because they tend to eat too much of the green pasture forage and less of the high-energy grain. This does not mean that yearlings make less gain per acre on pasture than 2-year-olds, but their gain is in the form of growth rather than fat.

EFFECT OF AGE ON CAPITAL INVESTMENT

Although feeder calves usually cost 5 to 10 percent more, per hundred-weight, than yearling feeders of the same grade, they cost much less per head because of their lighter weight. This is a factor of great importance when feeder cattle are high in price. For example, a 400-pound calf at 40 cents a pound costs $160, whereas a 700-pound yearling at 36 cents costs $250. Thus $25,000 would be needed to buy 100 of the yearlings, but only $16,000 would be needed for the same number of calves.

Calves, on the other hand, tie up capital in cattle and feed for a longer period, as they must be fed for a longer time. They also require a slightly higher capital investment in buildings and equipment, because they require somewhat better shelter than do older cattle.

EFFECT OF AGE ON FLEXIBILITY OF FEEDING AND MANAGEMENT

Young cattle have a marked advantage over older cattle with respect to suitability for a greater variety of feeding programs. Calves may be put on feed immediately or carried on roughage and pasture for 5 to 12 months before being started on a finishing ration. Yearlings, too, may be roughed through the winter but are less satisfactory than calves for such a feeding plan because their gains are relatively slow and expensive. Because calves have a long growing period, they may be held beyond the marketing date originally selected more safely than older cattle, if such a change in feeding plans seems advisable.

EFFECT OF AGE ON CARCASS COMPOSITION

From an industry standpoint, the age at which to slaughter beef cattle is when maximum muscling has occurred but before excessive fattening renders carcasses wasteful and feed efficiency poor. Unfortunately not all parts of the animal body mature or reach maximum development at the same age, and the problem is made more difficult because there are wide variations between breeds of cattle, and even within breeds, in this respect. In other words, the chronological age at which cattle reach physiological maturity is not constant or fixed for the species. Some of the British breeds reach the point where yield of lean cuts begins to decline, on a percentage basis, considerably sooner than do the dual-purpose or dairy breeds. Thus there is a genetic basis for this variation, which means that, through selection, age of maturity can be changed within a breed. This change, if desired, can be made faster by a systematic crossbreeding program, using sire breeds that are either fast or slow in maturity rate.

It would seem that the trend among breeders would be toward slower maturity and thus larger cattle. Certain items of nonfeed costs that are determined by the number of cattle fed rather than weight or size—such as labor, taxes, and grazing fees on leased land—discourage the use of smaller, rapidly maturing cattle. One problem associated with deciding when to stop feeding a drove of cattle in order to ensure maximum cutability or meatiness and yet not stop too soon is knowing body composition of the live animal. Kansas workers have conducted an interesting study with 64 half-brother Angus steer calves that sheds some light on this question. The pertinent data are shown in Table 96. One group of 8 steers was slaughtered at the start of the test when the calves weighed 351 pounds. Then, after feeding on a finishing ration for 56 days

Table 96

Effect of Age and Length of Feeding Period on Performance and Body Composition of Beef Steers Fed a Finishing Ration (0–224 Days)[a]

Group	Age (days)	Days Fed	Daily Gain (lb)	Slaughter Weight (lb)	Carcass Weight (lb)	Carcass Grade	Yield Grade[b]	Percent Trimmed Cuts[c]	Total Bone (lb)	Longissimus dorsi Weight (lb)	Longissimus dorsi Percentage Fat
1	240	0	—	351	188	Good	2.25	51.7	19.6	3.0	5.6
2	296	56	2.35	447	255	Good	2.28	51.6	21.3	4.0	6.8
3	324	84	2.42	493	298	Good	2.35	51.4	23.3	4.5	7.9
4	352	112	2.28	525	328	Good	3.02	50.0	25.5	5.0	11.5
5	380	140	2.37	631	391	Good	3.24	49.5	29.5	5.5	13.3
6	408	168	2.38	682	431	Good	3.12	49.8	30.7	6.0	13.8
7	436	196	2.50	785	488	Choice	3.70	48.4	32.7	6.4	20.1
8	464	224	2.32	835	522	Choice	3.84	48.1	31.9	6.8	23.9

[a] Kansas Agricultural Experiment Station Bulletin 507, 1967.

[b] Yield grade or cutability score, determined by using a USDA equation that accounts for variations in fat thickness, percentage kidney, pelvic, and heart fat, carcass weight, and area of rib eye or *Longissimus dorsi*. Larger values mean lower yield of edible product.

[c] Estimated percentage of carcass weight in boneless, closely trimmed retail cuts from the round, loin, rib, and chuck.

and each 28 days thereafter, further groups of 8 steers were slaughtered. Skeletal, organ, and muscle weights and muscle areas were obtained as measures of growth. Ether extract (fat) within the muscle was chemically determined to measure degree of finish or quality, the principal determinant of carcass grade.

Although this study represents only one breed, and evidently an early-maturing one at that, the data show that growth occurred rather steadily to about 14.5 months of age. Other data not shown indicated that the muscles of the round region followed this pattern, but note in Table 96 that the loin or rib-eye muscle was still increasing in weight when the test ended. As skeletal and muscle growth tended to decelerate, fat deposition within the muscle increased rapidly as reflected in the improvement in grade. The reduction in yield grade is largely a result of fat deposition outside the muscle—that is, on the outside of the carcass (bark)—and in depot fat regions such as the kidney area. The percent edible portion of the carcass decreases steadily with fat deposition.

The Kansas workers concluded that the optimum slaughter age for their test steers was 14.5 months at a weight of 785 pounds, and after a feeding period of 196 days. Feed efficiency was not reported but, based on other data, this age also corresponds closely with the age at which cost of gains increases rapidly. On the side of feeding for a slightly longer period it should be mentioned that a certain amount of apparent excess fat cover is required to prevent excessive shrink in the packer's cooler and to ensure longer shelf-life for the retailer.

DECIDING ON THE AGE OF CATTLE TO FEED

It should be apparent after the foregoing discussion that age is often a very important factor in the purchase of feeder cattle. For example, calves are unsatisfactory for a short feed and should not be bought by the feeder who plans to market his cattle after feeding them only 4 or 5 months. Similarly, yearling steers should not be bought in the fall by the man who expects to carry his cattle into the late summer or fall of the next year, because their gains will be very expensive if they are fed that long.

Two-year-old steers are much better than calves or yearlings for utilizing large quantities of roughage such as silage and sorghum fodder because they have greater digestive capacity. On the other hand, yearlings are better than either calves or 2-year-olds for utilizing stalk fields and late fall pastures, because they are old enough to use such feeds to advantage yet not too old to make gains in the form of growth while on rations that are insufficiently abundant or nutritious to promote the formation of fat.

Some men buy cattle of the age and weight that can most easily be made ready for market by the time they expect the highest prices for finished cattle the following year. This is ideal in theory, but often too many cattlemen make the same "guess" as to when the market will be high, and the increased number of cattle then going to market at the same time causes unsatisfactory prices after all. As a rule, the better plan is to purchase cattle of the age that is able to utilize the available feeds to best advantage and that will be in condition to sell advantageously when the feed supply is gone or when selling is necessary to avoid a labor shortage in the fields. Table 97 contains data collected under average farm conditions in a Corn Belt state, which can be used to estimate potential profit or loss from a feeder program using steers of various ages as well as heifer calves. Actual feeder and feed costs existing at a certain location or time may be substituted for those used in the table to calculate necessary selling or break-even price to pay all costs.

IMPORTANCE OF SEX IN CATTLE FEEDING

On the basis of sex differences, feeder cattle may be either steers, heifers, cows, or bulls. From the standpoint of numbers, steers are more important than the other three classes combined and are the only kind of cattle to be found in fairly large numbers throughout the year. Table 90 shows the composition of the annual United States cattle slaughter by class or sex condition. It will be seen that steers make up over half of the total and that two-thirds of the cattle now slaughtered are classed as fed cattle. It is a safe assumption that a considerably larger percentage of the steers are fed before slaughter than are the other classes. The trade in feeder heifers assumes considerable volume only during the fall and early winter months. Breeding-age bulls are usually as valuable right off grass as after feeding awhile. Consequently these bulls are usually sold at the end of the breeding season to be used as ground beef or processed meats.

Yearling heifers are better suited for a short rather than long feed. Many such heifers have probably been bred before being sold and are likely to show unmistakable signs of pregnancy if fed longer than 90 to 120 days. Even if evidence of pregnancy is not a question, it is seldom profitable to feed heifers for as long as steers of the same condition and quality because the market does not require female beef to be in such high condition as steer beef.

The most desirable heifer carcasses, weighing 425 to 550 pounds, are produced by animals weighing 700 to 900 pounds, showing just enough finish to grade high good or low choice. Heifers of this description sometimes sell for nearly the same price as steers, and loads of mixed

Table 97

Economic Data on Various Ages and Sexes of Feeder Cattle Fed to Grade High Good to Low Choice in the Carcass[a]

Item	Steers			Heifer Calves
	Calves	Yearlings	2-Year-Olds	
Average purchase weight (lb)	420	600	860	400
Days on farm	363	294	163	293
Average daily gain (lb)[b]	1.6	1.7	1.9	1.5
Average total gain (lb)	580	500	310	440
Average sale weight (lb)	1,000	1,100	1,170	840
Feed per hundredweight gain				
Corn (bu)	7.7	11.0	13.0	8.0
Supplement (lb)	40	59	61	47
Silage (lb)	193	261	462	173
Hay (lb)	294	307	251	410
Cost[c]	$27.99	$38.26	$44.55	$30.70
Total feed per head				
Corn (bu)	44.7	55.0	40.3	35.2
Supplement (lb)	232	295	189	207
Silage (lb)	1,119	1,305	1,432	761
Hay (lb)	1,705	1,535	778	1,804
Financial statement				
Purchase cost[d]	$159.60	$216.00	$344.00	$120.00
Value of feed	162.33	192.28	138.09	135.08
Nonfeed cost[e]	36.30	29.40	16.30	29.30
Total cost	$358.23	$444.58	$498.39	$284.38
Break-even price	$ 35.82	$ 40.42	$ 42.60	$ 33.85

[a] These data were constructed from several hundred Farm Business Farm Management Service Cooperator records in Illinois in a recent year and are representative of feeding and management programs used on many Corn Belt farms.

[b] The gains shown are calculated on the basis of purchase pay weights and actual sale weights and thus include shrink on both ends.

[c] Based on 1975 average feed costs and includes all feed fed, including that required to recover shrink: corn, $2.52 per bushel; supplement, $120 per ton; corn silage, $20 per ton; and hay, $30 per ton.

[d] Based on 1975 laid-in purchase costs per hundredweight of $38, $36, $40, and $30 for steer calves, yearling steers, 2-year-old steers, and heifer calves, respectively.

[e] Nonfeed costs of 10 cents per day cover labor, interest on cattle and feed, depreciation of equipment, veterinary costs, taxes, and miscellaneous out-of-pocket costs.

yearling steers and heifers often sell without sorting. Animals in choice condition at this weight, if started on feed as calves, are 12 to 15 months old when slaughtered and there is little likelihood of the females' being pregnant.

Heifer calves finish a little earlier than steer calves. This, together with the fact that heifers need not be so well finished as steers, means that heifers are ready for the market 6 to 10 weeks before steers started on feed at the same time.

The total gains made by heifers while on feed are somewhat smaller and more costly than those made by steers when all are fed the same length of time, as shown in Table 98, because of the earlier maturity and slower rate of growth of heifers. But this point really is of little concern in practical heifer feedlot enterprises because steers are usually fed several weeks longer than heifers. Thus any advantage they may have over heifers in rate and economy of gains at the time the heifers are marketed is likely to disappear during the remaining 8 to 10 additional weeks they are normally continued on feed, when their gains are both slow and expensive relative to those made earlier. Indeed it frequently happens that heifers make larger average daily gains and also cheaper gains during a period of 5 or 6 months than do steers during a feeding period that is 2 or 3 months longer, as shown in data from the Illinois station in Table 99.

A recent study by Tennessee workers, using Angus calves purchased in auction sales and all approximately the same in weight, shows that grade of calves changes some of the relationships mentioned earlier. In Chapter 14 the effect of grade on performance is discussed in detail, but note in Table 100 that heifers of varying feeder grade tend to grade more nearly alike as slaughter cattle than do steers. The result of this difference is that lower grades of heifers are often more profitable than lower-grade steers, because the spread between grades of feeder steers is narrower than that for feeder heifers. Note in Table 100 that rib-eye area, expressed as square inches per hundredweight of carcass, favored the heifers in all cases. This measurement is often used as an indicator of overall meatiness and cutability. The apparent muscularity of the heifer carcass is offset by a greater thickness of outside fat.

A more important factor than the relative rate and economy of the gains made by steers and heifers is their relative price when purchased and marketed. In the Tennessee study the heifers sold for a price much nearer the purchase price than did the steers. Thus they had much less negative price spread to overcome in order to show a profit. The standard-grade heifers showed a slight positive price spread, whereas the choice steers were faced with a $5.03 negative price spread.

Steers usually are preferred to heifers for roughing through the winter and grazing on pasture the following summer, because they grow more and therefore make larger gains on nonfattening feeds. Moreover, yearling heifers, after a summer on pasture and a short period of grain feeding in the drylot, command lower prices than steers because of the competition from the larger number of grass-fat heifers that are marketed

Table 98

Comparative Feeding Qualities of Steer and Heifer Calves When Both Are Fed the Same Length of Time

| | Minnesota | | Missouri | | | |
| | | | Full-Fed in Drylot | | One-half Grain Ration in Winter; Full-Fed on Grass in Summer | |
	Steer Calves	Heifer Calves	Steer Calves	Heifer Calves	Steer Calves	Heifer Calves
Days fed	217	217	182	182	322	322
Initial weight (lb)	451	449	359	358	358	354
Average daily gain (lb)	2.35	2.27	2.16	1.94	1.84	1.64
Average daily ration (lb)						
Shelled corn	12.1	12.1	8.0	8.0	8.3	8.0
Protein concentrate	2.0	2.0	1.1	1.1	1.2	1.1
Alfalfa hay	1.6	2.1	3.2	3.2	1.8	1.9
Corn silage	7.6	8.9	8.7	8.9	5.5	5.5
					(Pasture)	(Pasture)
Feed per cwt gain (lb)						
Shelled corn	517	535	369	410	452	483
Protein concentrate	85	88	53	58	65	69
Alfalfa hay	69	93	115	162	97	118
Corn silage	322	391	400	450	297	336
Return above cost of feed	$9.43	$2.60	$8.30	$8.16	$38.43	$21.14
Dressing percent	—	—	57.3	59.1	59.5	60.3

Table 99

	Iowa (Average of 3 Tests)		Illinois	
	Steer Calves	Heifer Calves	Steer Calves	Heifer Calves
Number of days fed	240	170	200	140
Average initial weight (lb)	415	406	379	379
Average daily gain (lb)	2.18	2.18	2.35	2.56
Average daily ration (lb)				
Shelled corn	10.7	9.1	10.1	9.3
Protein concentrate	1.5	1.5	1.5	1.4
Alfalfa hay	4.9	5.4	2.0	2.0
Corn silage	—	—	8.1	8.2
Feed per cwt gain (lb)				
Shelled corn	492	433	428	363
Protein concentrate	65	65	63	54
Alfalfa hay	226	247	85	78
Corn silage	—	—	343	319
Date marketed	July 18	May 6	July 13	May 14

Comparison of Steers and Heifers When Marketed as Required Finish Is Attained[a]

[a] Mimeographed reports of calf-feeding experiments.

in the fall. Short-fed steers do not experience serious competition from grass-fed steers, as most of the grass-fed steers are sold for further feeding rather than for immediate slaughter.

EFFECT OF PREGNANCY

The once rather prevalent practice of breeding yearling feeder heifers, either accidentally or by intent, has become relatively uncommon. Ranchers, by using more fencing and better range management practices, can more easily graze their yearling heifers apart from the cow herd and thus prevent breeding. The problem of the bred heifer is most common in cattle bought in auction sales in the farm states, where the practice of year-round beeeding is rather prevalent. Therefore late-weaned heifer calves and short yearlings not intended as replacements are often unintentionally bred. Any heifer being sold in weekly auctions and weighing over 550 to 600 pounds may be suspected of being pregnant and should be bought accordingly. If bred heifers will be no more than 5 to 6 months into the

Table 100

Performance of Feeder Cattle of Different Grades and Sexes[a]

	Steers			Heifers		
	Choice	Good	Standard	Choice	Good	Standard
Number of animals	30	30	30	30	30	29
Average days on feed	203	203	204	179	179	179
Average weight and gain per head (lb)						
Initial weight	479	488	492	485	481	488
Final weight	892	917	916	807	830	821
Total gain	413	428	425	321	349	333
Daily gain	2.04	2.12	2.10	1.79	1.95	1.86
Financial statement						
Feed cost per cwt gain ($)	16.10	15.70	16.37	16.50	15.74	16.26
Initial value per cwt ($)	28.58	27.78	25.18	27.45	25.24	23.05
Final value per cwt ($)	23.55	23.38	22.53	23.75	23.67	23.23
Return per head over feed cost ($)	4.65	9.76	11.63	4.27	19.29	22.96
Final on-foot slaughter grade[b]	12.5	11.7	10.8	12.4	12.1	11.4
Carcass data						
USDA grade[b]	12.3	12.1	10.9	11.7	11.7	10.9
Hot carcass weight (lb)	544	553	545	495	502	495
Dressing percent	60.2	59.5	58.8	60.8	60.0	59.9
Rib-eye area (sq in.)	11.22	11.23	10.97	10.88	11.06	11.41
Rib-eye area per cwt carcass (sq in.)	2.06	2.03	2.01	2.20	2.20	2.31
Marbling score[c]	6.1	5.7	5.2	5.6	5.5	5.1
Fat thickness (in.)	0.50	0.42	0.37	0.53	0.48	0.46
Shear force, 1-in. core (lb)	15.51	15.34	15.17	15.24	14.59	15.49

[a] Tennessee Agricultural Experiment Station Bulletin 381, 1964.
[b] Average good = 10; high good = 11; low choice = 12; average choice = 13.
[c] Traces = 3; slight = 4; small = 5; modest = 6; moderate = 7.

gestation period when ready for slaughter, they should be suitable for feeding. However, if they are as much as 3 to 4 months along when started on feed, they will be very near calving or may even calve by the time they reach the intended slaughter weight. Selling pregnant heifers on grade and yield would of course completely eliminate any justification on the part of packers for buying slaughter heifers at a lower price to protect against possible advanced pregnancies and a reduced dressing percent. Chapter 21 includes a discussion of MGA, a synthetic hormone approved for feeding to heifers, which reduces the disturbing effects of the normal estrus cycle seen in feedlot heifers. The use of this hormone may well eliminate the practice of breeding feedlot heifers in order to quiet them.

Unfortunately, at present there are few if any alternatives for aborting feedlot heifers that are approved by the U.S. Food and Drug Administration. Repositol, an injectible form of diethylstilbestrol, is no longer obtainable. The side effects of its use were rather undesirable, in any case, with uterine prolapse and retained placenta being rather common.

Calving out feedlot heifers, especially if they are nearly ready for slaughter, is seldom a satisfactory solution of the pregnant heifer problem. Because of the excess fat accumulated in the vaginal region, severe dystocia is almost a certainty. Severe tearing and bruising of the vaginal lining almost certainly will result in infection and, even worse, posterior paralysis and uterine prolapse are common occurrences. The retained placenta is another problem and extremely good nursing care is needed to save the heifer. It seldom is worthwhile to attempt to have the heifer rear her calf if it is saved. Calving out the pregnant feedlot heifer may seem to be the humane thing to do, but feeders should resist the temptation and, rather, should sell the fattened pregnant heifer to a packer buyer on a grade and yield basis.

SPAYED VERSUS OPEN HEIFERS

One of the principal objections to feeding heifers is the disturbance caused by their coming in heat. When a carload or more of heifers are fed together, hardly a day passes without one or more animals being in this condition, so that the herd is frequently in a state of excitement and unrest. Obviously such conditions are not conducive to rapid and economical gains.

Spaying is sometimes used to avoid the disturbances caused by in-heat heifers. More heifers are not spayed mainly because of the extra cost and the failure of the market to pay sufficient premium for them to cover the cost. The principal advantages claimed for spaying are a more tranquil disposition in the feedlot and a somewhat higher price for the heifers when they are marketed. However, spayed heifers seldom have made as large or as economical gains or have attained as high a finish as open heifers in feeding experiments where they have been compared. Apparently, the removal of the ovaries retards growth and development of young heifers in much the same way that castration retards growth and development of male calves.

Because spayed heifers gain less rapidly than open heifers, spaying cannot be recommended on the basis that it will lessen activity in the feedlot and thus result in faster gains. As a matter of fact, disturbances caused by in-heat heifers usually are most noticeable during the first few weeks of the feeding period and tend to become less frequent as the heifers approach market finish. Consequently, any advantage of spayed heifers must lie in the certainty that they are not pregnant and therefore may be purchased for a long feeding period without danger of advanced pregnancy by the time they are ready for market. Use of the feed additive MGA serves the dual role of an estrus suppressant and a growth stimulant and is strongly recommended over spaying.

FEEDLOT PERFORMANCE OF BULLS

Undoubtedly the research with hormone-like compounds during the past decade in the feeding of steers and heifers has been responsible for revived interest in investigating the performance of slaughter bulls. It has long been observed that bulls outgain steers and do it more efficiently, but for several reasons the feeding of young bulls has not been an accepted practice in America as it has in Europe. Price discrimination by packer buyers, in part justified by a lower dressing percent, lower carcass grade, and a mandatory label of "bullock" or "bull" by government graders, has undoubtedly been the principal deterrent. Feeders themselves have been convinced that the disposition or temperament of bulls was a big disadvantage in the feedlot. The results of a number of recent experiments have caused breeders and feeders alike to reexamine the validity of the practice of castration. The three studies to be reviewed here indicate something of the potential, in terms of increased supply of acceptable lean beef produced at high efficiency, that may follow if the practice of castration is abandoned.

Wyoming workers, in cooperation with Safeway Stores, Inc., and Armour and Company, compared Angus bull and steer calves of similar genetic background. The experiment included a consumer acceptance study, which sets this study apart from most others bearing on this subject. The pertinent data are shown in Table 101. In brief, the Wyoming

Table 101

	Bulls	Steers	Difference Bull-Steer
Number	19	19	
Weaning weight (lb)	353	330	23
Initial feedlot weight (lb)	422	391	31
Slaughter weight (lb)	963	877	86
Gain per day on feed (lb)	2.21	1.97	0.24
Dressing percent	63.7	62.9	0.80
Carcass gain per day of age (lb)	1.22	1.13	0.09
Area of rib eye (sq in.)	13.4	10.5	2.9
Area of rib eye per cwt carcass (sq in.)	2.32	2.07	0.25
Fat thickness at 12th rib (in.)	0.34	0.67	−0.33
Estimated yield of retail cuts (%)	50.3	46.3	4.0
Percent cooler shrink	3.19	3.71	−0.52
Carcass grade	Good	Choice	

Effect of Castration on Production Traits and Selected Carcass Measurement[a]

[a] Wyoming Agricultural Experiment Station Bulletin 417.

workers concluded that bulls, when compared with steers, weaned heavier and gained faster in the feedlot, and produced meatier carcasses with less trimfat, but graded a full grade lower because of less marbling, coarser texture, and darker color. Consumers preferred bull chucks, but steer steaks.

The findings in a Nebraska study with Angus cattle closely parallel those of the Wyoming study, but heifers were included in this test, and the design permitted the evaluation of feed efficiency differences as well as the effect of length of feeding period because the cattle were fed for two different lengths of time. The Nebraska workers found that bulls out-gained both steers and heifers and did so more efficiently. The bull carcasses contained more lean and less fat, but were tougher and graded lower. When comparable retail prices, adjusted for quality grade, were applied to the retail-trimmed cuts of all carcasses, the value of carcasses less feed costs favored bulls over steers and steers over heifers.

It should be realized that at the retail level, bull beef is seldom likely to be obviously labeled as such. However, if a packer requests that a U.S. Department of Agriculture grader grade the bull carcass, the grader is required to label it "bull" if mature, or "bullock" if young, which means that the packer will probably have to sell the carcass at a substantial discount. Nongovernment-graded young bull carcasses might sell at the same price as steer or heifer carcasses or possibly even higher because of their greater muscling and trimness.

Table 102

Effect of Sex on Production and Carcass Traits[a]

	Bulls		Steers		Heifers	
Days Fed	211	255	211	255	219	259
Number	25	24	27	24	18	8
Average initial weight (lb)	435	425	432	409	413	403
Average daily gain (lb)	2.22	1.95	1.78	1.64	1.62	1.45
TDN per pound of gain (lb)	5.14	6.05	6.42	6.85	6.58	7.08
Carcass grade	G+	G+	C−	C−	C−	C−
Tenderness[b]	13.2	17.0	9.8	10.2	10.5	12.8
Rib-eye area (sq in.)	11.4	11.7	9.9	9.9	10.0	9.4
Trimmed and boned cuts (%)	56.1	54.7	53.4	51.4	50.3	49.8
Retail value of carcass less feed cost	$238	$232	$191	$179	$168	$153

[a] Animal Science Department, University of Nebraska.
[b] Warner-Bratzler Shear Values—Larger numbers indicate more pressure required to cut a 1-inch core of the loin-eye muscle.

If a feeder, and especially the smaller farm cow-calf man, can develop a suitable local market for fed bulls—such as a small custom slaughterer who has a trade for freezer beef—he might be advised to feed his male calves as bulls. They should be started on feed early with creep feed and then fed for maximum gain on a fairly high-energy ration so that they reach a weight of 1,000 to 1,200 pounds at 13 to 15 months of age. Best results will be obtained if such bulls are sold on a carcass yield basis.

The role of the newer European breeds in this feeding plan for bulls has not been investigated in the United States, but it is likely that the bulls would need to be fed to undesirably heavier weights before reaching the desired finish. In Europe, most bull calves are slaughtered as intact males but at still younger and lighter weights, and the resulting calf beef is preferred by the European consumer.

FEEDING MALES WITH INDUCED CRYPTORCHIDISM

The "short-scrotum" method of castration (referred to in Chapter 10), also called induced cryptorchidism, results in male calves that gain similarly to bull calves up to weaning time and on through to slaughter weight. Rate of gain and feed conversion are improved over those of normally castrated males to about the same extent as may be expected from the use of hormones, either in implanted or oral form. Data in Table 103 show that, in addition to the improved feedlot performance, carcasses resulting from males castrated by the short-scrotum method were better in yield grade but poorer in quality grade. Hormone-treated steers outgraded short-scrotum steers but produced carcasses with lower cutability, as evidenced by poorer yield grades. Since feed efficiency strongly favored the short-scrotum steers, and since the average price received per head was also greater, in spite of the lower price per hundredweight of carcass, the profit picture obviously was improved by use of this new technique.

As with bulls, there are some feedlot management and marketing problems with steers castrated by this method. Such steers retain their libido and, unless they are fed in relatively large lots, excessive riding may occur, especially if strange cattle are added to a pen. As for marketing problems, at present such males would likely be discriminated against as weaners if sold through normal channels, and when sold for slaughter through normal channels they may not sell as well as steers, on foot at least. If sold on a grade and yield basis and if they are not stamped "stag," "bullock," or "bull" by a USDA grader, they should sell acceptably well. Since marbling requirements were recently lowered slightly, most short-scrotum males should now grade choice. Apparently it matters little how early in life the castration occurs.

Table 103

Influence of Induced Cryptorchidism and Diethylstilbestrol (DES) on Feedlot Performance and Carcass Merit in Male Calves[a]

	Treatment				
Item	Steered at Birth	Steered at Birth + DES	Steered at 60 Days	Short-Scrotum at Birth	Short-Scrotum at 60 Days
Days fed	252	218	252	218	218
Average daily gain (lb)	2.17	2.66	2.26	2.91	2.69
Final weight (lb)	918	1,081	964	1,118	1,063
Feed:gain ratio	7.11	6.65	6.92	6.10	6.19
Number grading choice	7	5	7	0	0
Number grading good	1	4	2	8	9
Average yield grade	2.57	2.44	2.67	2.00	2.00
Value of carcass per head	$420.84[b]	$455.07	$440.43	$468.13	$451.00

[a] Adapted from Eastern Oregon Experiment Station Report, 1974.

[b] Based on warm-carcass prices of $73.50 and $71.50 per hundredweight for choice and good carcass, respectively.

THE "BULLER" STEER

Reference has been made to the excessive riding engaged in by stags and short-scrotum steers because of retained libido. Another kind of problem steer, called a "buller" and a variety of other terms, is the one that arouses the riding instinct in his pen mates. In large, commercial feedlots—and smaller ones, for that matter—where cattle coming from many sources are mixed together and fed as one group, the problem of the buller steer can become an important one. Some researchers believe that a constituent in the urine of the ridden steer arouses the olfactory senses of some of its pen mates. This urinary constituent is probably estrogenic in nature. Satisfactory physiological treatments have not yet been devised that would suppress the excretion of the compound and thereby eliminate the continual mounting. Certain commercial preparations are available for applying to the rump of the buller steer to mask its odor, but none is altogether satisfactory.

As many as 10 percent of the steers in a single lot can be so affected, but 2 to 3 percent is probably a realistic average for the industry. Some feedlot managers simply remove the buller steers and pen them together, whereas other managers pen them with heifers. Neither method is entirely satisfactory, especially in custom lots, for either practice complicates ownership identification, feed and yardage charges, and the like. Immediate slaughter might be the least costly alternative until research finds a better way.

FEEDING COWS

Large numbers of cows are marketed each fall by farmers and ranchers. Many of these cows have been suckling calves all summer and are very thin; consequently, they would seem to be satisfactory for feeding, particularly if they can be bought for only about half the cost per pound of choice feeder calves and yearlings.

Although feeder cows often appear to be worth the money when they are bought in the fall, they frequently prove to be a rather sorry bargain by the time they are marketed. Many misfortunes may happen to a drove of cows during the feeding period, which will reduce the profits materially. In the first place, they are mature animals and consequently all gains are in the form of fat and are very expensive. Gains made by cows are often 50 to 100 percent more costly than those made by calves or yearlings. Such costly gains require a relatively large price spread to prevent a financial loss. However, the price of finished cows is seldom very high, so that large price spreads are the exception rather than the rule. In fact, with ordinary price conditions, very skillful buying and feeding are necessary to realize a profit on a cow-feeding venture. These points are well illustrated in Table 104.

Another objection to buying cows for feeding is that many of them have been bred some 3 to 5 months before being marketed. Thus some of them will calve before they are finished, and a considerable percentage of those that do not will be sufficiently advanced in pregnancy to be seriously penalized in price when they are sold. Men who buy cows to feed frequently change their plans upon discovering that most of the cows appear to be pregnant and decide to keep them until fall and finish them after the calves are weaned. Then, however, the grain already fed yields little return, because the finish put on during the winter is largely lost during the suckling period. Seldom do all the cows prove to be in calf, and the barren ones may be sold directly off pasture the following fall, but if sold at this time of year they may not bring much more per pound than they cost a year earlier. More likely they will bring even less for, as dry cows, they will then be fatter and less desirable for boner or cutter cow purposes because the yield of lean meat will be much lower.

In summary, the feeding of cows must be considered to be among the most speculative of all cattle-finishing programs and one that only the most experienced feeder should attempt. Research is under way at the

Table 104

	Gains and Feed Consumption of Feeder Cows While Being Fattened for Market				
	Nebraska 66 days		Missouri 111 days	Illinois 154 days	
	Shelled Corn	Ground Ear Corn	Shelled Corn	1st 42 Days	Last 112 Days
Average initial weight (lb)	1,006	987	1,004	713.5	
Average daily gain (lb)	1.34	2.09	2.24	1.32	2.44
Average daily ration (lb)					
Corn or grain mix	18.5	26.1	9.9	—	14.4
Protein concentrate	—	—	1.2	1.0	1.0
Corn silage	—	—	41.3	40.0	21.7
Legume hay	8.9	8.1	6.7	2.0	2.0
Feed per cwt gain (lb)					
Corn	1,384	1,246	441	490	
Protein concentrate	—	—	55	47	
Corn Silage	—	—	1,843	1,244	
Legume hay	666	389	302	93	

Ohio Experiment Station in which 2-year-old heifers are allowed to early-wean their first calves, after which the young and still growing cows are fed out for slaughter. This is an entirely different matter from feeding mature cows, and the data, when released, will be of great interest. It may be recalled that the most effective selection tool for replacement females is that of evaluating the first progeny of all females and then selecting those young cows for keepers that demonstrated a breeding value well above average for the herd. The females that are culled while still young might well be most efficiently marketed as fed "heiferettes" or even simply as fed heifers.

CHAPTER 14
THE IMPORTANCE OF GRADE AND BREED IN FEEDER CATTLE

The questions of age and sex of cattle discussed in the preceding chapter are only two of the many decisions that confront the man who is in need of feeder cattle. He also must decide whether to buy crossbred animals or those of one breed, and then which cross or which breed, and whether to buy higher-grade feeder steers that may cost upwards of 40 cents per pound, or nondescript cattle at a considerably lower figure.

At most seasons of the year, feeder cattle of all kinds are available in fairly large numbers at any of the larger central markets and auction sales. Certain markets or areas have a reputation as sources of unusually good selections of cattle or a certain breed, cross, or grade. If a large number of cattle of a given kind are to be bought, it is usually best to go to that market or area where the supply is likely to be largest. Consultation with a reputable order buyer will help one select the area where certain grades of feeders are most likely to be available and the time of year when large numbers are likely to be on hand. If only a load or two are wanted, the nearest market usually is able to furnish them at the lowest cost, freight and other expenses considered.

KIND OF CATTLE TO BUY

There is no hard and fast rule as to which kind of cattle is best to feed. Many factors must be taken into account in deciding this question. Without doubt the strength of the demand in relation to the supply, both for the feeder animals and for the finished beeves, is likely to be an important factor in determining the profit made from the feeding operation. Lower-grade cattle may prove just as profitable, or more so, as choice or prime feeders if they are purchased when the supply of such animals is greater than the demand and if sold at a market that has a strong demand for good-grade fat cattle. On the other hand, the supply of prime, well-finished steers is often so small that they enjoy a market all their own. The profit made on such cattle may then be relatively large, even though the price paid for them as feeders seemed at the time unreasonably high compared with quotations for plainer cattle.

414

The conditions under which the cattle are to be fed and the way they are to be handled should receive first consideration in deciding which kind of cattle to buy. For a long feed in which a liberal allowance of grain will be furnished, only the better grades of feeders should be purchased. Under such conditions, cattle of good breeding and with a reputation for above-average feedlot performance and high cutability will be choice or better when marketed and will command the highest market price. Standard and common feeders, on the other hand, can never be made into "market toppers," no matter how long they are fed nor how excellent the ration. To feed such cattle large quantities of expensive feed is likely to prove unprofitable. As a rule, they should be fed on high-energy rations for a comparatively short time.

KIND OF CATTLE BEING FED

As mentioned in Chapter 11, grade in a stocker or feeder steer is determined by his predicted grade as a slaughter animal. Feeding finishing rations to a choice or higher-grade feeder for a longer time than is usually required to finish him to his corresponding slaughter grade will generally not be profitable, because only a small percentage of this grade of feeder can be upgraded. Also, a few animals may not live up to expectations, thus not making the predicted slaughter grade after the usual length of feeding period.

On the other hand, lower-grading feeders can usually be upgraded one or two grades. There has been a substantial increase in the feeding of the lower grades of feeders, especially in western and southwestern feedlots where it has always been popular, but the practice is not confined to these areas, as many cattle feeders in Colorado, the Corn Belt, and elsewhere are eager to take advantage of the possibilities for upgrading.

Table 105 shows the percentages of the various grades of carcasses in the total United States beef production over a recent period of 20 years. It is interesting to see that prime grade makes up a very small portion of the total and that it appears to have reached a level below which it may not drop further. Choice grade has increased, good has remained level, and standard grade, added only a few years ago, is about equal to the total decrease in commercial and utility grades. These changes largely reflect a growing demand for grain-fed beef. The majority of the carcasses found in the three lower grades consist of cow and bull carcasses. The new grade standards will surely increase numbers of both the choice and standard grades with an offsetting reduction in number of good grade carcasses.

Table 105

Percentage Composition of the United States Beef Supply, By Grades (1946–1966)[a]

Years	Prime	Choice	Good	Standard	Commer-cial[c]	Utility	Canner, Cutter
			Percentage of Total Beef by Grade[b]				
1946–1948	6	28	19		17	17	14
1949–1951	7	34	19		15	15	11
1952–1954	5	34	18		16	13	13
1955–1957	4	33	21		16	13	13
1958–1960	4	36	27		14	10	9
1961–1963	3	47	18	11	4	9	8
1964–1966	4	48	17	9	4	8	10

[a] USDA, *Livestock and Meat Situation*, May 1967.

[b] Data are 3-year averages from 1946 through 1966; included are graded beef and estimate for ungraded beef. Recently less than 50% of the beef supply has been graded; thus data since 1966 are limited and not readily available.

[c] Includes standard and commercial grades from 1946 through 1960.

MARKET CLASSES AND GRADES OF SLAUGHTER CATTLE

Official U.S. Department of Agriculture standards have been established for slaughter cattle and beef carcasses just as for stockers and feeders. The slaughter grades, sometimes called "on foot" or "live" grades, are based on the predicted carcass grade of an animal after slaughter. Color of fat and lean, interspersion of fat and lean (marbling), and age as determined by bone condition are important factors in determining carcass grade that cannot be accurately assessed by examination of the live animal. For this reason, slaughter and carcass grades do not always correspond, but experienced buyers seldom miss their evaluations to any great degree.

Figure 73 shows rear and side views of the more common grades of slaughter steers. Slaughter heifers should have corresponding conformation and degree of finish. Table 106 lists the various grades of slaughter cattle.

Government grading of beef carcasses is done at the request of the packer and thus is a voluntary program. Most government agencies such as the Army and Navy, and many chain stores as well as restaurants, airlines, and hotels, buy their beef on the basis of government grades. Any beef entering interstate commerce must be government-inspected and labeled but not necessarily government-graded. Government-graded beef can be identified by the purple shield, bearing the letters "USDA" and the

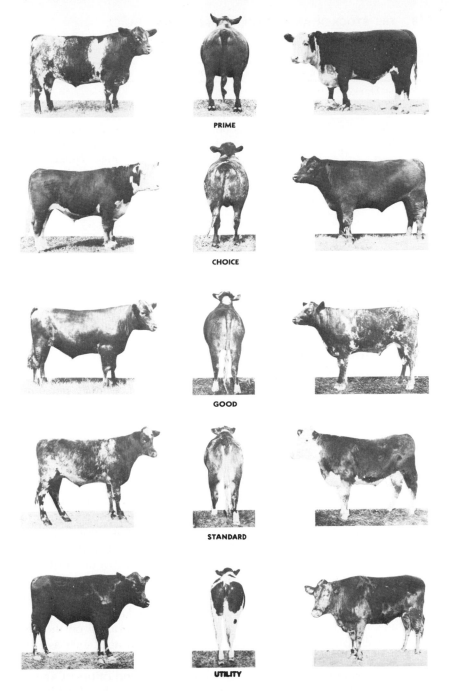

Fig. 73. Steers illustrating the USDA grades of slaughter steers. It is important to be able to judge when a steer is finished for his grade, because economy of gain and length of feeding period are closely related. (USDA.)

Table 106

Market Class	Market Grade
Steers	Prime, choice, good, standard, commercial, utility, cutter, canner
Heifers	Prime, choice, good, standard, commercial, utility, cutter, canner
Cows	Choice, good, standard, commercial, utility, cutter, canner
Bulls	Choice, good, commercial, utility, cutter, canner
Stags	Choice, good, commercial, utility, cutter, canner
Calves	Prime, choice, good, commercial, utility, cull

Market Classes and Grades of Slaughter Cattle[a]

[a] Based on USDA, *Livestock and Meat Situation.*

name of the grade, stamped on practically all retail cuts. Almost all major packers have their own system of labeling the respective grades of beef as well. These are referred to as "in-house" grades and do not always fit the same categories as the Department of Agriculture grades.

Recent federal legislation requires that all packing plants in the United States, whether selling in or out of their respective states, meet inspection standards for health of cattle and plant cleanliness equivalent to those heretofore required of plants that were subject to federal inspection for interstate shipment of beef. Department of Agriculture personnel will be responsible for inspection in this situation unless state regulations are in effect. This may often be the case; if so, regulations or standards must be at least equivalent to or stricter than the U.S. standards. In this case, state department of agriculture inspectors, or their equivalent, will perform the inspections. These inspectors, whether federal or state, are not to be confused with USDA graders.

The specific grade assigned to a live slaughter animal by a grader has been determined by its relative excellence with respect to conformation, finish, quality, and maturity or age. There is now some question whether conformation plays a positive role in carcass retail value. Finish refers to the degree of fatness and the quality and distribution of the fat. The latter factor is associated with palatability, tenderness, and quality of the individual cuts of meat. Quality in the live animal refers to the overall symmetry and smoothness of the animal as well as the refinement of head, hide, and bone. The degree of maturity is appraised on the basis of the animal's physical characteristics associated with age, such as size of head and bone and even the length of tail. In the carcass, on the other hand, hardness and color of bone are the principal indicators of age. Youthfulness and finish are both believed to be associated with palatability. Preliminary research data indicate that tenderness and palatability are heritable. It is conceivable that, in the future, selection for these traits may result in cuts that are tender and palatable without the necessity of feeding to the degree of finish found in high choice or prime cattle.

Since more and more cattle are being sold and bought on a grade and yield basis, it would be helpful if feeders could evaluate the degree of marbling, the principal determinant of carcass grade, in the live animal. This is difficult to do in every instance, but there are some indicators that can be used. Breed has an effect, and it is generally conceded that Angus cattle marble earliest, with the other British breeds intermediate and dairy and dual-purpose breeds last. Crosses between these breeds or types will usually be intermediate between the parent breeds. Age is another factor; the older the animal, the greater the marbling. Length of feeding period on high-concentrate rations is also highly correlated with degree of marbling. Within a breed or type, exterior finish, as seen in the cod or udder regions, brisket, about the hooks and pins, and along the back and loin edge, can also be used with some degree of reliability.

Feeders usually attempt to feed cattle just long enough to bring the majority of a drove of cattle barely into the desired grade. Cattle are graded by full grades on the rail, and if they are bought by grade, there is usually no premium for carcasses in the middle or upper third of the grade. Progeny-testing programs help to identify sires whose offspring tend to marble earlier than the average of the breed.

A few packers are paying premiums for cattle that yield-grade better than average or below 3.0—for instance, as much as $2.00 to $3.00 premium per hundredweight of carcass, for each yield grade better than average. Thus a 600-pound carcass with a yield grade of 2.0 could gross $12.00 to $18.00 more as a result of the yield-grade premium. Needless to say, such packers rightfully may be expected to apply discounts for poor yield-grading cattle. The size of the premiums and discounts is usually negotiable, just as is the base price.

The data in Table 107 give some idea of the USDA grades of slaughter steers and heifers being produced in the High Plains feedlots. Any particular packer attempts to buy the grade of cattle that can be most readily sold for a profit; thus it must be recognized that the data merely represent one packer's purchases from the feedlots in the region. However, if there were a great supply of prime or standard and lower grades of cattle in the area, it is likely that they would be included in larger numbers in the data presented. Note that more than 90 percent of all cattle slaughtered by this plant graded good and choice. It is conceivable that the plant slaughtered more lower-grading cattle but chose not to have these graded because they might sell more advantageously with "house brands" than with USDA grade labels. Also, much of the lower-grade beef may not have been U.S.-graded because it was going to be boned and sold as ground beef.

Table 107

Characteristics of Fed Cattle Slaughtered by a Major Texas Panhandle Packer, by Sex, March–July, 1971[a]

	Sex		
Item	Steers	Heifers	Average[b]
Carcass grades (%)			
Prime	4.9	0.3	4.4
Choice	53.3	45.5	52.6
Good	40.4	48.7	41.2
Standard	1.0	0.1	0.9
Commercial	0.1	5.4	0.6
Utility and lower[c]	0.3	0.0	0.3
	100.0	100.0	100.0
Yield grades (%)			
1	1.2	0.7	1.1
2	34.7	38.4	35.1
3	38.9	52.2	40.1
4	17.6	8.6	16.8
5	7.6	0.1	6.9
	100.0	100.0	100.0

[a] Adapted from Texas Agricultural Experiment Station MP-1069, 1972.
[b] Weighted average.
[c] This category also contains carcasses classified as stags, condemned, and bulls.

The fact that at least one-third of all graded carcasses yield-graded 2.0 or better speaks well for the high cutability of the cattle being purchased by this packer and for the feeding industry in the locale of the packer. A sampling of a packer's kill in the Corn Belt would show considerably more yield-grade 4 and 5 carcasses than in this Texas sample. It is an established fact that the newer commercial feedlots in the Southwest feed many "Okie" cattle that would grade only standard and good as feeders. The fact that the packer slaughtered mostly cattle that graded good and choice in the carcass tells us that the feeder cattle were probably upgraded at least a full grade.

IMPORTANCE OF BODY TYPE

Cattle feeders differ in their opinions as to which body type of steer within a breed or class of cattle is most profitable to feed. Some prefer small,

Fig. 74. Colorado-bred good to choice Angus-Charolais crossbred calves, which are acquiring a reputation as rapid, efficient gainers. Combining heavier weights with faster feedlot gains reduces nonfeed costs through more rapid turnover. (Western Livestock Journal.)

compact, low-set, body-type 1 or 2 feeders, believing that they finish more readily, have a higher dressing percentage, and produce more shapely carcasses than the conventional type of steer. Others favor large, growthy, body-type 4 or 5 cattle with plenty of scale or size, believing that such feeders make faster and more economical gains and return more profit because of their possibly being more efficient and because they believe they have higher cutability in the carcass. Numerous tests have been conducted to obtain information on these points, but few have disclosed consistent differences between the two types. Although the large-type steers usually have gained at a somewhat faster rate, especially if all steers are fed the same length of time, seldom has there been a significant difference in feed efficiency, as seen in Table 108. It should be noted that carcass grade did not differ among the types of steers but, as the cattle tended to be larger of frame or body type, fat thickness tended to be thicker and this is reflected in the lower yield grade. The heavier carcasses also would contribute to the lower yield grade. No doubt the heavier weaning weights of the larger-type steers were partially the result of more condition and this undoubtedly accounts for the poorer feed efficiency in these steers in the feedlot. Had all steers been slaughtered at the same

Table 108

Item	Body Type				
	1	2	3	4	5
Initial weight (lb)[b]	466	537	591	604	660
Final weight (lb)	880	995	1,044	1,043	1,129
Daily gain (lb)	2.73	3.01	2.98	2.89	3.09
Feed/lb gain (lb)	6.0	6.0	6.2	6.4	6.8
Feed cost/lb gain (¢)	21.8	22.0	22.6	23.4	23.6
Fat cover	0.50	0.63	0.65	0.76	0.79
Carcass grade	choice	choice	choice	choice	choice
Yield grade	3.0	3.4	3.6	4.0	4.3
Return per head ($)[c]	14.00	31.00	29.00	35.00	38.00
Gross return per cow ($)[d]	185.00	219.00	230.00	238.00	251.00

Influence of Body Type of Angus Weaner Steer Calves upon Feedlot Performance and Returns to Cattle Feeder and Breeder (153 Days)[a]

[a] Adapted from Wisconsin Agricultural Experiment Station reports, 1973.

[b] All steers were within a month of the same age at the start and end of the trial at which time they averaged 15 months of age.

[c] Initial value ranged from $39 per hundredweight for the lighter calves to $31 for the heavier calves.

[d] Gross return per cow equals carcass value minus feed and yardage charges.

weights as the body-type 1 steers, one can be certain that the larger-type steers would have gained more efficiently, for they would have been fed a much shorter period and the carcasses would have had high cutability although they would not have graded choice.

At the other extreme, if all steers had been fed until they weighed the same as the body-type 5 steers, then the larger-type steers would have excelled in average daily gain, feed efficiency, and yield grade. The point is that, if one knows how long and to what weights to feed each body type of steers in order for them to reach choice grade, then they are equally satisfactory from the cattle feeder's standpoint. If the breeder owns calves all the way to final slaughter weight, then the larger-type cattle are somewhat favored; but, as mentioned elsewhere, the added maintenance cost of the larger-type cow must be subtracted from the higher sale value of the calf when sold.

The Ohio station has a study still in progress with several breeds of cattle, including Charolais. Their preliminary results indicate that all the extra income from the larger-type finished offspring will be required to offset the extra maintenance costs of carrying a larger cow. The Ohio workers tentatively conclude that the energetic efficiency of beef cattle is not influenced by body type per se if one considers the entire life cycle of cattle, and if offspring, regardless of type, are slaughtered when they arrive at the same body composition or degree of fatness.

IMPORTANCE OF GRADE

Apart from overall type or size within a breed, as just discussed, is the matter of feeder grade. In Chapter 11, which deals with the stocker program, it is noted that there is a wide range in grade, reflecting differences ranging from the ultimate in beef type, as seen in the British breeds, to common and inferior at the other end of the scale. Some old and much new research casts considerable doubt on the suitability of the present stocker and feeder grading system as a means of sorting cattle into groups with predictable performance.

Performance-testing programs generally have brought to light the fact that conformation grade is not highly associated with performance in the feedlot nor with carcass merit. Some researchers go so far as to say that they are negatively correlated—that is, the higher the grade, the poorer the performance in the feedlot and the poorer the cutability or yield of trimmed cuts. Perhaps a more middle-of-the-road approach is to say that within each grade there are excellent and poor performers, and that performance records on the sires of stockers and feeders or on their half-siblings are a better guide to their ability than the conformation or USDA grade assigned to them.

An Iowa study summarized in Table 109 illustrates that the lower or plainer grades of cattle can be profitable. The reason is that there is usually a wider spread in the purchase price between grades of feeders than there is in the selling price of the same cattle when sold for slaughter. In the Iowa study the choice and good grade steers were Herefords, the standard steers were of mixed breeding comparable to the "No. 2 Okies" that are so popular in the Southwest and West, and the common steers were Holsteins. Unfortunately, all lots were fed the same length of time. If the standard and common steers had been sold 50 days earlier, the advantage they showed would have been even greater. In practice, lower-grading cattle should not be fed longer than 120 to 150 days. The question that must be asked, though, is whether the upgrading of the plainer cattle that resulted would have occurred on a shorter feed. Figure 75, which diagrams this part of the Iowa study, shows that the choice grade steers actually slipped away from average choice to low choice, but the lower grades were upgraded by two grades. It should be mentioned that the entire reduction in selling price necessary to break even that is shown by the plainer cattle can be accounted for by their lower purchase price.

Table 109

Cattle Fed 198 Days Market to Market (18 Head per Group)	Feeder Grade			
	Choice	Good	Standard	Common
Feedlot Performance				
Purchase weight (lb)	762	763	760	824
Slaughter weight (lb)	1,255	1,292	1,258	1,359
Average daily gain, market to market (lb)	2.49	2.67	2.52	2.70
Feed consumed per steer per day (lb)[b]	21.9	23.0	22.2	24.6
Feed required per cwt gain (lb)	879	861	881	911
Carcass Measurements				
Average carcass weight, shrunk (lb)	776	793	757	800
Dressing percent	61.8	61.4	60.2	58.9
Federal carcass grade	C−	C−	C−	G
Rib-eye area per cwt carcass (sq in.)	1.5	1.5	1.5	1.5
Fat cover per cwt carcass (in.)	0.09	0.09	0.06	0.02
Predicted retail yield of carcass (%)	67.1	66.6	68.7	71.4
Financial Values				
Cost per steer				
Original cost ($)	190.50	179.30	168.72	158.21
Feedlot costs (feed + 3 cents per pound of gain) ($)	97.61	102.16	98.60	109.68
Shipping costs to packing plant ($)	2.51	2.58	2.52	2.72
Total cost to packing plant ($)	290.62	285.04	269.84	270.61
Purchase price per pound (cents)	25.0	23.5	22.2	19.2
Feed cost + 3 cents nonfeed cost per pound of gain (cents)	19.8	19.5	19.8	20.5
Shipping cost per pound to packing plant (cents)	0.2	0.2	0.2	0.2
Break-even price per pound at packing plant (cents)	23.2	22.1	21.4	19.9

Feedlot Performance, Economic Returns, and Carcass Evaluation of Four Feeder Grades of Yearling Steers[a]

[a] *Iowa Farm Science,* 20:22–23.

[b] Feed consumed included 5 lb mixed hay and 1 lb supplement per steer per day plus balance of ration in rolled shelled corn. Feed prices: corn, $1.12 per bushel; hay, $20 per ton; supplement, $100 per ton.

The results obtained in the Iowa study may not hold for many years in the future. As more and more cattle feeders change over to lower grades, the relative cost of these grades of feeders will surely increase because of the increased demand. It can also be said with certainty that the supply

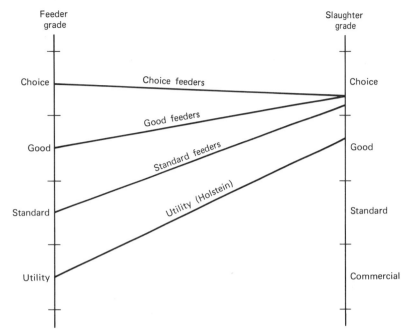

Fig. 75. Lower grades of feeder steers usually can be upgraded in the feedlot, whereas higher-grade feeders seldom can be and, in fact, often are reduced in grade during the finishing process. (Iowa State University.)

will be reduced, because the breeder of these plainer cattle will be unsatisfied with the lower price he has been receiving and will upgrade his cattle. The straight Holsteins, which of course come from dairies, will continue to be available, although even here changes are occurring, as many dairymen are now breeding only their very best cows to dairy bulls, preferring to use beef bulls on the average and lower producers. The resulting crossbred offspring will probably grade good as feeders and thus will cost the feeder more than straight Holsteins.

EFFECT OF BREED

There have been numerous tests with the objective of comparing breeds with respect to feedlot performance and carcass merit, and, as mentioned in Chapter 8 in connection with choice of breed for cow-calf programs, drawing conclusions from a single test is unwise to say the least. A study of the many tests does, however, establish a pattern and the exhaustive study conducted by Tennessee researchers and summarized in Table 110 can

Table 110

Effect of Type and Breed of Feeder Cattle on Performance, Palatability, and Carcass Composition[a]

Breed and Type	Hereford	Angus	Brahman	Brahman X[b]	Santa Gertrudis	Holstein	Jersey
Number fed	32	29	22	10	12	24	25
Performance data							
Average initial weight (lb)	334	338	305	340	368	291	219
Average final weight (lb)	885	865	835	911	893	909	791
Feeding period (days)	310	315	362	303	298	294	368
Average daily gain (lb)	1.84	1.76	1.50	1.90	1.91	2.16	1.56
24-hour live shrink (%)	5.6	6.2	5.2	5.1	6.1	6.9	6.1
Dressing percent	62.2	63.4	62.4	62.4	62.8	59.7	57.5
Feed per cwt gain (lb)	869	909	936	858	892	776	959
Carcass data							
Carcass grade	G+	C−	St+	G−	G−	St	St
Fat thickness (mm)	18.5	19.3	11.5	14.3	16.5	9.1	10.8
Kidney fat (%)	4.1	5.0	3.7	4.3	4.1	4.5	6.5
Rib-eye area (sq in.)	9.4	9.5	9.2	9.9	9.7	8.8	8.0
Loin steak evaluation							
Shear values[c]	4.9	5.6	6.3	5.9	5.4	5.2	5.0
Flavor[d]	7.4	7.3	6.8	7.3	7.3	7.2	7.4
Cooking loss (%)	25.4	25.9	26.6	28.5	27.1	24.2	22.7

[a] Tennessee Agricultural Experiment Station, *Journal of Animal Science*, 22:702, 1001–1008, 1963.
[b] Brahman sires on either Hereford or Angus cows.
[c] Warner-Bratzler value on 0.5-inch core.
[d] Extremely poor = 4; excellent = 9.

serve as a reliable evaluation of breeds with respect to the traits investigated.

Calves were obtained from various sources at an average age of about 5 months and were full-fed a high-concentrate ration until they either weighed 900 pounds or reached 20 months of age, whichever occurred first. British and Zebu types did not differ significantly in any production traits. The dairy type, when both dairy breeds were combined, was inferior in all traits except gain. Angus steers dressed higher than all other breeds and graded highest as well. However, the Angus along with the straight Brahman and Jersey cattle were less efficient in feed conversion. With respect to carcass traits, Brahman carcasses scored lowest on tenderness and flavor, with all other breeds scoring fairly even with one another. The carcasses of the British breeds were fatter and graded higher, but the higher grade did not result in greater tenderness and flavor.

Before leaving the discussion of the effect of conformation and quality on feedlot performance, some mention should be made concerning some of the problems encountered in feeding crossbred cattle that have wide variations in genetic background and size at maturity. A trial conducted at the Illinois station and summarized in Table 111 is somewhat unusual because it uses neither a weight-constant nor a time-constant termination point, but rather, length of feeding period is based on estimates of the slaughter weights at which choice grade would be achieved by cattle of varying mature size, growth potential, and tendency to marble.

All calves were produced on the same pastures and all nursed Hereford dams and thus were of comparable nutritional background. After weaning, calves were gradually worked up to a 90 percent concentrate ration. All calves in a breed group were slaughtered when it was estimated that most of the group would grade choice, conformation not being considered. At the slaughter weights chosen, there were no significant differences in performance, and except for the lower conformation score in the Jersey crosses which slightly lowered their final carcass grades, carcass grade likewise was unaffected. Because yield grade is adversely influenced by increases in carcass weight and fat thickness, the lighter Jersey-cross carcasses yield-graded better than those of British beef cattle, even though the latter had better conformation and carcasses of intermediate weight. The Holstein-cross carcasses graded choice despite lower conformation scores but, at the weights slaughtered, were too heavy and had too much bark to yield-grade well. Although the steers in this study varied extremely in mature size, shape and body type, and genetic potential to gain, when fed just long enough to reach equal condition or body composition as indicated by marbling, there were generally no great differences in feedlot performance or carcass merit among the various breeds and crosses.

In summary, in comparing the merits of the plainer grades of feeders

Table 111

Feedlot Performance and Carcass Merit of Steers as Influenced by Breed and Crossbred Matings When Fed to an Estimated Constant Body Composition[a]

	Breeds and Breed Crosses			
Item	Hereford × Hereford	Angus × Hereford	Jersey × Hereford	Holstein × Hereford
Feedlot Performance				
Average initial weight (lb)	493	511	455	538
Slaughter weight (lb)	1,065	1,023	902	1,198
Number of days fed	201	179	160	229
Average daily gain (lb)	2.85	2.87	2.79	2.88
Feed/lb gain (lb)	6.83	7.30	6.74	7.63
Carcass Characteristics				
Conformation score[b]	21.2	20.8	17.5	20.2
Marbling score[c]	10.9	11.6	11.7	11.7
Quality grade[b]	18.7	19.2	18.9	19.1
Final carcass grade[b]	18.7	19.1	18.3	19.1
Tenderometer ratings[d]	13.9	13.7	15.6	14.4
Carcass Yield-Grade Data				
Hot carcass weight (lb)	676	655	538	765
Loin-eye area (sq in.)	10.9	11.3	10.3	11.4
Loin-eye area/cwt (sq in.)	1.62	1.73	1.91	1.51
Backfat (in.)	0.73	0.71	0.32	0.67
Backfat (in./cwt)	0.11	0.11	0.06	0.09
Kidney and pelvic fat (%)	2.4	3.1	3.3	2.9
Yield grade	3.9	3.8	2.7	4.0
Retail Product Production				
Age at slaughter (days)	449	426	424	490
Hot carcass/day of age (lb)	1.51	1.54	1.27	1.56
Retail product/day of age (lb)	0.88	0.94	0.78	0.95
Feed/lb retail product (lb)	18.7	18.1	18.4	19.0

[a] Adapted from Illinois Cattle Feeders' Day Report, 1975.

[b] Score of 17 = average good; 20 = average choice grade.

[c] Score of 10 = modest−; 11 = modest; 12 = modest+.

[d] Higher values indicate less tenderness.

with feeders of choice or better grades, the following statements seem appropriate.

1. Lower-grade feeder cattle are usually the offspring of larger parents of the Brahman, Holstein, or any of a half-dozen European exotic breeds crossed on British breeds and thus can be expected to outgain straight British-bred feeders.

2. Lower-grade feeders are usually thinner in condition, owing to their tendency to grow larger and mature later and to the fact that their area of origin is likely to provide a lower plane of nutrition. Thinner cattle should, again, gain more rapidly than fleshier cattle.

3. The lower grades of feeders are often 2 to 3 months older than straight British-bred feeders when offered for sale.

4. In practice, lower-grade feeders are usually fed for shorter lengths of time and on higher-energy rations, both of which contribute to higher average daily gains from purchase to sale weight.

5. When finished for slaughter, lower-grade cattle usually are upgraded from their feeder grade by a full grade, whereas straightbred and crossbred British cattle seldom grade as well in the carcass as predicted when graded as feeders.

6. The lower grades of cattle are much more likely to make a profit for the feeder than for the breeder.

7. As British-bred cattle acquire more bred-in performance and the price spread between grades of feeders becomes narrower, the lower grades will lose much of the profit advantage they now enjoy in many instances.

CHAPTER 15
ENERGY IN THE
FINISHING RATION

The principal difference between finishing rations, or those fed to cattle being readied for slaughter, and rations that are fed to cows or to stockers is in their energy content. Improvement in condition or finish is the principal aim in finishing programs, and this is brought about by feeding a ration that supplies energy in excess of the amount needed for maintenance and growth—because, of course, growth occurs simultaneously with fattening in younger cattle, especially in the calf-finishing programs.

Corn, sorghum, and barley are the principal high-energy feeds used in finishing rations, but in some localities other sources of energy, such as molasses, are of some importance. In this discussion the term "grain" is generally used in speaking of "high-energy" feeds, because its meaning is more fully understood. High-energy feeds are also quite generally referred to as carbonaceous concentrates because of their high carbohydrate content, in contrast to protein concentrates that are so termed because of their above-average protein or nitrogen content. It should be mentioned that, generally speaking, protein concentrates, especially the oilseed meals, are also quite high in energy content.

During the 1960s the beef cattle nutritionists probably devoted more time and research funds to study of the energy portion of the ration than to that of any other nutrients. The relative increase in cost of metabolizable energy supplied by roughages, combined with increasing labor and machinery costs for processing roughage, has been responsible for much of the interest in high-concentrate finishing rations.

Because the ruminant's digestive system is uniquely adapted to the utilization of high-bulk, high-fiber feeds, it was formerly believed that rations containing mainly grain should be fed only near the end of a finishing period and only after feeding a stocker or growing ration for some time. Few digestive disturbances occurred under this system because the shift from low or medium to high energy level in the ration was made gradually, perhaps over a 3- or 4-week period.

Recent research in steam processing and rolling of grains has resulted in high-concentrate rations with the "bulky" characteristic of rations that contain some roughage. These rations are readily eaten and, although fewer pounds are consumed, their greater caloric density results in a higher energy intake. The rate of gain is not consistently improved, but fewer pounds of feed are required per pound of gain.

Year-round feeding of finishing rations has become commonplace, meaning of course that many cattle are on finishing rations during the hot summertime. It has rather recently been discovered that the fiber content of rations consumed during hot weather affects maintenance requirements materially. Apparently digestion and metabolism of the energy in crude fiber increases body heat production more than does the utilization of concentrates. This is wasted energy in the sense that it produces neither growth nor gain. And the dissipation of this extra heat itself, because of the increased respiration rate and restlessness, requires further energy. Hence the decreased maintenance requirement and improved efficiency when higher-energy rations are fed in summertime. For these reasons it is quite common for the rations fed in southwestern feedlots in the summer to contain as little as 5 percent roughage as contrasted to the more typical 20 percent. Corn Belt feeders have been slower to take advantage of this new knowledge. Because of the higher humidity and lower wind velocity in this region, feeder cattle in the Corn Belt would no doubt respond even more to a reduction in ration fiber content in summertime.

Apart from their feeding qualities, high-concentrate rations lend themselves to mechanical processing and handling. In addition, the manure disposal problem is lessened, as fewer pounds of manure are produced because of reduced feed intake and lower bulk in the ration.

High-concentrate rations seem best suited to yearlings rather than calves, and to the lower grades of feeders. There is a tendency for younger cattle of the British breeds to fatten too early and at too light a weight on high-concentrate rations. This is especially true of lower-set, more compact types and is even more of a problem with heifer calves than with steers as they often may reach adequate finish at weights no higher than 750 to 800 pounds when fed on high-concentrate rations. Thus there still seems to be a place for the combination stocker-feeder programs in areas where both roughages and grain are produced. This may well mean that the higher grades of feeders will be fed where corn and sorghum silages are economically produced, and the "Okie" type of cattle will be fed mainly in the parts of the country where high-concentrate rations are most feasible.

FULL FEED DEFINED

Most of the cattle finished for market are given a full feed of grain during the latter part of the feeding period. On some farms and in most commercial feedlots, cattle are placed on a full feed as quickly as possible, perhaps within 2 or 3 weeks after they are put into the feedlot; on others a full feed of grain is still only being fed during the last month or two before the cattle are sold for slaughter.

Opinions differ as to what constitutes a full feed of grain. Most feeders consider cattle to be on a full feed of grain when they are fed all the grain they will clean up in an hour or so if fed twice daily. Yet, if they are also fed all the good-quality legume hay and corn silage they want, they can be induced to eat still more grain by limiting the roughage to what they will clean up readily.

It is therefore highly desirable that the term "full feed" be defined more accurately than by saying simply that it is the amount of grain that cattle will eat. The term "full feed" may be applied to the roughage part of the ration as well as the grain. If the grain and hay were limited but the cattle were fed all the corn silage they would eat, we would then say they were given a *full feed of silage*. However, unless otherwise indicated, the term refers to a full feed of grain. The amount of grain that cattle will eat varies much too widely for the term to have much value to the inexperienced feeder who is attempting to feed his cattle in accordance with approved practices.

A much better way to define the term is to speak of the amount of grain they *should eat* if they are to be regarded as full-fed cattle. With this thought in mind we shall define a *minimum full feed of grain* as being 2 pounds daily per 100 pounds live weight, including the grain in any corn silage that is fed. Anything over this amount is, of course, a full feed without qualification.

HOW MUCH FEED WILL CATTLE EAT?

The previous discussion as to what constitutes a full feed of grain has brought out the fact that the amount of grain or roughage eaten can be controlled by increasing or decreasing the other components of the ration. If cattle are being fed principally to utilize roughage, the grain is fed in limited amounts in order to induce the cattle to consume large quantities of roughage. But if the purpose of feeding is to market as much grain through cattle as possible, it is roughage that is limited. This leads to such questions as how much total feed cattle will eat and what the replacement value of roughage is in terms of grain.

The approximate consumption during different parts of the feeding period may be obtained by assuming the total air-dry feed to be 3.2, 2.8, 2.4, and 2.0 percent of the live weight for the first, second, third, and last quarters of the feeding period, respectively. This is on the assumption that calves are fed grain for approximately 240 days, yearlings 180 days, and 2-year-old steers 120 days. For cattle fed a shorter time the estimates for the last quarter should be disregarded; and for a feeding period of only one-half the usual length, only 3.2 and 2.8 percent will apply. Commercial feedlot nutritionists, aware of the reduction in total ration consumption, per unit of live weight, as the feeding period progresses, use a step-up system of increasing the energy content of the ration. This is done by reducing the roughage portion of the ration; for example, the concentrate:roughage ratios during the feeding period may be 5:5, 7:3, 9:1, and 9.5:1 for the four quarters, respectively. This technique ensures that daily intake of energy is maintained in spite of a reduction in total feed intake.

In discussing a full feed of grain and the total feed consumed daily, no mention was made of the amount of protein concentrate fed. This item of the ration has been intentionally omitted from the discussion because the intake of 1 to 2 pounds of protein supplement will not be at the expense of other ingredients, but rather will be consumed in addition to a full feed of grain and roughage. Consequently, it may virtually be ignored in estimating the amount of grain required for a full feed and the total air-dry feed required to satisfy the appetites of full-fed cattle.

REPLACEMENT VALUE OF ROUGHAGE

It appears from the data in Table 112 that grain and air-dry roughage are interchangeable on a pound-for-pound basis for cattle that are fed at least half of a minimum full feed of grain. The replacement value of silage in terms of grain is more difficult to determine because it varies greatly in moisture content. However, assuming that the air-dry roughage equivalent of silage is 33.3 percent of the weight fed, grain and silage will replace each other on an air-dry basis within the limits of practical error.

A SIMPLE METHOD FOR ESTIMATING FEED REQUIRED TO FINISH CATTLE

The factors developed in the preceding paragraphs in regard to the feed consumption of finishing cattle are valuable in estimating the amount of grain and roughage required to finish a given drove of cattle. These factors are also useful in checking the ration at different stages of the feeding period to see whether grain and roughage are being fed in about

Table 112

Replacement Value of Grain and Roughage in the Daily Ration of Feeder Cattle

Replacement Made	Average Weight During Experiment (lb)	Average Daily Ration			Average Total Air-Dry Feed per Day (lb)[a]	Time Fed (days)	Air-Dry Feed per 100 Pounds Live Weight (lb)
		Grain (lb)	Hay (lb)	Silage (lb)			
Grain replaced by hay							
Nebraska							
Heavy feed of grain	1,028	16.4	9.7	—	26.1	140	2.54
Medium feed of grain	1,039	14.0	12.6	—	26.6	140	2.56
Light feed of grain	1,042	12.1	15.5	—	27.6	140	2.65
Ohio							
Full feed of corn	923	13.3	3.2	13.7	22.0	240	2.38
¾ feed of corn	914	10.1	6.2	14.1	21.7	240	2.38
½ feed of corn	889	6.7	8.7	14.2	21.1	240	2.37
Grain replaced by silage							
Kansas							
Full feed of barley	936	14.0	—	17.8	21.2	180	2.27
⅔ feed of barley	916	9.4	—	33.6	22.8	180	2.49
⅓ feed of barley	909	4.7	—	41.8	21.4	180	2.36
Iowa							
Full feed of corn	900	13.2	1.1	17.7	21.4	175	2.38
½ feed of corn	862	6.4	1.1	32.3	20.4	175	2.37

[a] Air-dry weight of silage assumed to be 40 percent of weight of silage fed.

the required amounts. The following example shows how these factors are used.

Given: 30 steers, weighing 700 pounds when purchased, which are to be fed to choice finish. How much shelled corn or sorghum and hay will be needed?

Solution: Estimated time to be fed, 150 days
   Estimated average daily gain, 2.8 lb
   Calculated final weights 700 + (150 × 2.8) lb = 1,120 lb
   Calculated average weight (700 + 1,120)/2 = 910 lb
   Total feed consumed daily per 100 lb live weight = 2.6 lb
   Total feed required daily per head, 2.6 lb × 9.10 = 23.7 lb
   Total feed required daily by 30 steers, 30 × 23.7 = 711 lb
   Average grain eaten daily per 100 lb live weight by cattle on full feed, 2.0 lb
   Average grain eaten daily by 910-lb steer, 910 × 2.0 = 18.2 lb
   Average grain eaten daily by 30 steers, 30 × 1.82 = 546 lb
   Average hay eaten daily by 30 steers = 711 − 546 = 165 lb
   Average hay eaten daily per steer = 165/30 = 5.5 lb
   Estimated total grain needed for 150 days = 150 × 546 = 81,000 lb = 1,462 bu corn
   Estimated hay needed for 150 days = 150 × 165 = 24,750 lb = 12.5 tons

Table 113

Average Amounts of Feed Consumed by Full-Fed Cattle During Different Parts of the Feeding Period[a]

Quarter of Feeding Period	Feed per 100 Pounds Live Weight			Concentrate: Roughage (approximate ratio)
	Total Air-Dry Feed (lb)	Grain (lb)	Air-Dry Roughage (lb)	
First	3.2	1.8	1.4	55:45
Second	2.8	2.1	0.7	75:25
Third	2.4	2.2	0.2	90:10
Last	2.0	1.9	0.1	95:5
Av. entire period	2.6	2.0	0.6	80:20

[a] Average length of grain feeding period: 2-year-old steers, 150 days; yearlings, 200 days; calves, 250 days.

Similarly, if it is desired to know how much grain and hay will be needed during the last month of the feeding period, a close estimate may be made by assuming that approximately 2 pounds of grain and hay combined will be eaten daily per 100 pounds live weight. About 1.9 pounds of this amount will be grain and 0.1 pound will be hay (see Table 113). Multiplying these quantities by the estimated weight of the cattle and dividing the results by 100 will give the approximate average daily consumption of grain and hay during the last month of the feeding period.

The amount of silage required in this example can be determined by increasing the daily hay consumption by 3 or by dividing it by the factor 0.33. Because of the palatability of silage, the grain in the silage will be consumed without reducing the regular grain consumption to any great extent.

NUTRIENT REQUIREMENTS OF FEEDER CATTLE

The simple rules of thumb just discussed for estimating total feed and energy requirements, and those for protein requirements discussed in Chapter 17, simply serve as guides in the formulation of rations and in determining the total feed requirements of a drove of feeder cattle. For those wishing to formulate rations more exactly, or for those wishing to feed a complete mixed ration, Tables 114 and 115 give the National Research Council requirements for feeder calves, yearlings, and 2-year-old cattle. As in the previous NRC tables, ration requirements are shown both on a daily requirement basis and a percentage composition basis. Morrison's standards are shown in Table 116.

STARTING CATTLE ON A FEED OF GRAIN

Feeder cattle coming directly off grass or a stocker ration consisting solely of roughages undergo a rather serious physiological shock if they are abruptly started on a high-energy ration. This is mainly because the rumen microfloral population consists of specific organisms that are adapted to the use of feeds high in crude fiber. Rumen pH drops drastically as a result of the high level of volatile fatty acids produced from fermentation of soluble sugars and starches present in the high-concentrate ration, retarding the growth of certain microflora and thus causing a serious imbalance in the microfloral population. Upwards of 3 weeks are required to reestablish a normal status in the rumen, and feed intake and weight gains are seriously reduced. If the shock just referred to is imposed upon the shock of weaning and exposure to the variety of stresses and viruses found in saleyards and trucks or trains, death losses in starting calves on feed can run as high as 5 to 10 percent.

To further complicate matters, comparatively few of the feeder cattle shipped into the feedlots have been fed any grain. Consequently, they must be taught to eat it. In the farm feedlots this can be done easily by putting about a pound of grain per head in the feed trough and feeding 4 to 6 pounds of hay or silage on top of it. In eating the leaves and chaff from the bottom of the trough, the cattle will also pick up the grain and will quickly learn to eat it. If it is desired to get the cattle on a full feed of grain rather promptly, the grain should be increased by about 1 pound per head daily until they are eating 1 pound of grain per 100 pounds live weight, after which time increases of 1 pound a head every third day for 2-year-old steers, 0.5 pound every third day for yearlings, and 0.25 pound every third day for calves are recommended. All this time, of course, the cattle should have as much roughage in the form of hay or silage or both as they will eat. If all the cattle do not appear hungry when they are fed, the hay or silage should be sufficiently reduced to ensure that all animals come to the feed trough promptly and begin eating when the grain is fed.

If the grain ration is increased gradually as suggested above, 2-year-old cattle will attain a minimum full feed of grain (2.0 percent of their live weight) about 30 days after being started on feed, yearlings in about 40 days, and calves in about 50 days. Further increases, of course, should be made as the cattle increase in weight and become better adjusted to a heavy feed of grain.

Starting cattle on feed in commercial feedlots that have their own feed processing plants is another matter. In this instance the roughage portion of the ration—usually chopped hay, cottonseed hulls, or some other bulky by-product feed—is mixed with coarsely prepared grain, molasses, and a fortified and medicated protein supplement and offered free-choice several times daily. The first feed offered may contain 50 to 60 percent roughage, with reductions being made at weekly intervals or so until the roughage content is down to 5 or 10 percent. The feedbunks should never be completely empty when this method is used.

For cattle that are to be fed on self-feeders, complete mixed starter rations such as those fed in commercial feedlots will serve well, if the feed is not too bulky to feed down in the feeders. Some farmer-feeders use a bulky grain such as oats or barley or ground ear corn as their starter concentrate and additional ground cobs are sometimes used. If cattle are being fed on pasture, they voluntarily consume an adequate amount of roughage to prevent digestive difficulty.

Table 114

Nutrient Requirements of Feeder Cattle (Daily Nutrients per Animal)[a]

Body Weight (kg) (lb)		Average Daily Gain (kg) (lb)		Daily Dry Matter per Animal (kg) (lb)		Total Protein (kg) (lb)		Digestible Protein (kg) (lb)		Energy ME[b] (Mcal)	Energy TDN[c] (kg) (lb)		Calcium (gm)	Phosphorus (gm)	Carotene (mg)	Vitamin A (1000 IU)
Finishing Steer Calves																
150	330	0.90	2.00	3.5	7.7	0.45	0.99	0.30	0.66	9.9	2.7	5.9	21	15	19.5	7.8
200	440	1.00	2.20	5.0	11.0	0.61	1.34	0.41	0.90	13.4	3.7	8.1	23	17	27.5	11.0
300	660	1.10	2.40	7.1	15.6	0.87	1.91	0.58	1.28	19.0	5.3	11.7	26	19	39.5	15.8
400	880	1.10	2.40	8.8	19.4	0.98	2.16	0.62	1.36	23.5	6.5	14.3	25	20	49.0	19.6
450	990	1.05	2.30	9.4	20.7	1.04	2.29	0.67	1.47	25.1	6.9	15.2	21	21	52.0	20.8
Finishing Yearling Steers																
250	550	1.30	2.85	7.2	15.8	0.80	1.76	0.51	1.12	18.8	5.2	11.4	29	20	40.0	16.0
300	660	1.30	2.85	8.3	18.3	0.92	2.02	0.92	2.02	21.7	6.0	13.2	29	21	46.0	18.4
400	880	1.30	2.85	10.3	22.7	1.14	2.51	0.73	1.61	26.9	7.4	16.2	28	23	57.0	22.8
500	1,100	1.20	2.65	11.5	25.3	1.28	2.82	0.82	1.80	30.0	8.3	18.3	26	26	64.0	25.6

ENERGY IN THE FINISHING RATION 439

Finishing Two-Year-Old Steers

350	770	1.40	3.10	10.3	22.7	1.14	2.51	0.73	1.61	26.4	7.3	16.1	30	24	57.0	22.8
400	880	1.40	3.10	11.3	24.9	1.25	2.75	0.80	1.76	28.9	8.0	17.6	30	25	63.0	25.2
500	1,100	1.40	3.10	13.4	29.5	1.49	3.28	0.95	2.09	34.3	9.5	20.9	30	30	74.5	29.8
550	1,210	1.30	2.85	13.7	30.1	1.52	3.34	0.97	2.13	35.1	9.7	21.3	30	30	76.0	30.4

Finishing Heifer Calves

150	330	0.80	1.75	3.5	7.7	0.45	0.99	0.30	0.66	9.9	2.7	5.9	18	13	19.5	7.8
200	440	0.90	2.00	5.0	11.0	0.61	1.34	0.41	0.90	13.4	3.7	8.1	21	15	27.5	11.0
300	660	1.00	2.20	7.3	16.1	0.89	1.96	0.59	1.30	19.5	5.4	11.9	23	18	40.5	16.2
400	880	0.95	2.10	8.7	19.1	0.97	2.13	0.62	1.36	23.2	6.4	14.3	23	19	48.5	19.4

Finishing Yearling Heifers

250	550	1.20	2.65	7.6	16.7	0.84	1.85	0.54	1.19	19.8	5.5	12.1	27	20	42.0	16.8
300	660	1.20	2.65	8.6	18.9	0.95	2.09	0.61	1.34	22.4	6.2	13.6	27	20	48.0	19.2
400	880	1.20	2.65	10.7	23.5	1.19	2.62	0.76	1.67	27.9	7.7	16.9	30	24	59.5	23.8
450	990	1.10	2.45	11.0	24.2	1.22	2.68	0.78	1.72	28.7	7.9	17.4	24	24	61.0	24.4

[a] Adapted from *Nutrient Requirements of Beef Cattle*, Subcommittee on Beef Cattle Nutrition, National Research Council, 1970.

[b] ME (metabolizable energy) requirements for growing cattle were calculated from net energy values for maintenance and gain, as published in appropriate tables, for the weights and rates of gain shown.

[c] TDN (total digestible nutrients) requirements were calculated from metabolizable energy by assuming 3.6155 Kcal of ME per gm TDN.

Table 115

Nutrient Requirements of Feeder Cattle (Nutrient Concentration in Ration Dry Matter)[a]

Body Weight		Average Daily Gain		Daily Dry Matter per Animal		Total Protein (%)	Digestible Protein (%)	Energy		Calcium (%)	Phosphorus (%)	Carotene (mg/kg)	Vitamin A (1000 IU)	
(kg)	(lb)	(kg)	(lb)	(kg)	(lb)			ME[b] (Mcal/kg)	TDN[c] (%)				(kg)	(lb)
Finishing Steer Calves														
150	330	0.90	2.00	3.5	7.7	12.8	8.6	2.82	78	0.60	0.43	5.5	2.2	1.0
200	440	1.00	2.20	5.0	11.0	12.2	8.1	2.67	74	0.46	0.34	5.5	2.2	1.0
300	660	1.10	2.40	7.1	15.6	12.2	8.1	2.67	74	0.37	0.27	5.5	2.2	1.0
400	880	1.10	2.40	8.8	19.4	11.1	7.1	2.67	74	0.28	0.23	5.5	2.2	1.0
450	990	1.05	2.30	9.4	20.7	11.1	7.1	2.67	74	0.22	0.22	5.5	2.2	1.0
Finishing Yearling Steers														
250	550	1.30	2.85	7.2	15.9	11.1	7.1	2.61	72	0.40	0.28	5.5	2.2	1.0
300	660	1.30	2.85	8.3	18.3	11.1	7.1	2.61	72	0.35	0.25	5.5	2.2	1.0
400	880	1.30	2.85	10.3	22.7	11.1	7.1	2.61	72	0.27	0.22	5.5	2.2	1.0
500	1,100	1.20	2.65	11.5	25.4	11.1	7.1	2.61	72	0.23	0.22	5.5	2.2	1.0

Finishing Two-Year-Old Steers

350	770	1.40	3.10	10.3	22.7	11.1	7.1	2.56	71	0.29	0.22	5.5	2.2	1.0
400	880	1.40	3.10	11.3	24.9	11.1	7.1	2.56	71	0.27	0.22	5.5	2.2	1.0
500	1,100	1.40	3.10	13.4	29.5	11.1	7.1	2.56	71	0.22	0.22	5.5	2.2	1.0
550	1,210	1.30	2.85	13.7	30.2	11.1	7.1	2.56	71	0.22	0.22	5.5	2.2	1.0

Finishing Heifer Calves

150	330	0.80	1.75	3.5	7.7	12.8	8.6	2.82	78	0.51	0.37	5.5	2.2	1.0
200	440	0.90	2.00	5.0	11.0	12.2	8.1	2.67	74	0.42	0.30	5.5	2.2	1.0
300	660	1.00	2.20	7.3	16.1	12.2	8.1	2.67	74	0.31	0.25	5.5	2.2	1.0
400	880	0.95	2.10	8.7	19.2	11.1	7.1	2.67	74	0.26	0.22	5.5	2.2	1.0

Finishing Yearling Heifers

250	550	1.20	2.65	7.6	16.8	11.1	7.1	2.61	72	0.36	0.26	5.5	2.2	1.0
300	660	1.20	2.65	8.6	19.0	11.1	7.1	2.61	72	0.31	0.23	5.5	2.2	1.0
400	880	1.20	2.65	10.7	23.6	11.1	7.1	2.61	72	0.28	0.22	5.5	2.2	1.0
450	990	1.10	2.45	11.0	24.3	11.1	7.1	2.61	72	0.22	0.22	5.5	2.2	1.0

[a] Adapted from *Nutrient Requirements of Beef Cattle*, Subcommittee on Beef Cattle Nutrition, National Research Council, 1970.

[b] ME (metabolizable energy) requirements for growing cattle were calculated from net energy values for maintenance and gain, as published in appropriate tables, for the weights and rates of gain shown.

[c] TDN (total digestible nutrients) requirements were calculated from metabolizable energy by assuming 3.6155 Kcal of ME per gm TDN.

Table 116

Morrison's Feeding Standards for Finishing Cattle[a]

Class of Cattle and Weight (lb)	Dry Matter (lb)	Digestible Protein (lb)	Total Digestible Nutrients (lb)	Calcium (gm)	Calcium (lb)	Phosphorus (gm)	Phosphorus (lb)	Carotene (mg)
			Requirements per Head Daily					
Calves finished for baby beef								
400	9.6–12.1	1.05–1.15	7.4– 8.6	20	0.044	15	0.033	25
500	11.3–13.8	1.14–1.26	8.8–10.2	20	0.044	16	0.035	30
600	13.2–15.8	1.26–1.37	10.2–11.8	20	0.044	17	0.037	35
700	14.8–17.5	1.39–1.52	11.6–13.2	20	0.044	18	0.040	40
800	16.7–19.3	1.52–1.68	12.6–14.4	20	0.044	18	0.040	45
900	17.7–20.3	1.64–1.82	13.5–15.5	20	0.044	18	0.040	50
Finishing yearling cattle								
600	15.0–17.6	1.18–1.32	10.7–12.3	20	0.044	17	0.037	35
700	16.5–19.1	1.36–1.52	12.7–14.3	20	0.044	18	0.040	40
800	17.8–20.4	1.52–1.68	14.1–15.9	20	0.044	19	0.042	45
900	18.9–21.7	1.64–1.82	15.4–17.2	20	0.044	20	0.044	50
1,000	20.0–23.0	1.71–1.91	16.0–18.0	20	0.044	20	0.044	55
1,100	21.0–24.0	1.76–1.96	16.5–18.5	20	0.044	20	0.044	60
Finishing 2-year-old cattle								
800	19.6–22.2	1.46–1.62	14.1–15.9	20	0.044	20	0.044	45
900	20.7–23.5	1.53–1.78	14.6–17.4	20	0.044	20	0.044	50
1,000	22.0–25.0	1.65–1.85	16.5–18.5	20	0.044	20	0.044	55
1,100	24.0–27.0	1.70–1.90	17.0–19.0	20	0.044	20	0.044	60
1,200	24.0–27.0	1.70–1.90	17.0–19.0	20	0.044	20	0.044	65

[a] Taken by permission of The Morrison Publishing Company, Ithaca, N.Y., from *Feeds and Feeding*, 22nd edition, by Frank B. Morrison.

Fig. 76. A heavy feed of corn silage with supplements and limited grain during the preceding winter cheapened the overall cost of gains on these steers being full-fed in drylot during the summer for the fall market. (Corn Belt Farm Dailies.)

Obviously the weight of the cattle will increase more rapidly than their capacity for feed. Consequently, it is usually necessary to reduce the roughage from time to time in order to maintain the level of grain consumption at at least 2.0 percent of the live weight. (Calves, because of their rapid growth, may increase in size sufficiently to permit appropriate increases of grain without reducing the roughage below the amount being eaten when they attained a minimum full feed.) However, reducing the roughage to a weight below 0.25 percent of the live weight is seldom justified, since further decreases in the roughage level fed will not result in further increases in concentrate intake. In calculating roughage consumption, the cobs (20 percent) in corn-and-cob meal and the estimated dry weight of the cobs and stover in corn silage should, of course, be considered as air-dry roughage. On a dry basis, corn silage consists of about 50 percent grain and 50 percent cob and stalk.

MEDIUM-ENERGY VERSUS HIGH-ENERGY RATIONS

Farmer-feeders who grow corn for silage often wish to know what kind of performance they may expect when cattle are fed some grain along with a full feed of silage and the necessary supplement. As mentioned in Chapter 11, stocker calves may be expected to gain from 1.5 to 1.75 pounds per day when fed corn silage and supplement alone for 150 to 180 days. Such a program produces more gain, per acre of corn harvested, than any other cattle program. There are times, however, when a farmer-feeder who both grows or backgrounds cattle and finishes cattle in two separate programs may wish to compromise and use some silage and some corn in combination in a single program.

There are abundant experimental data pertaining to the effect of varying levels of corn silage in the ration, and consequently in the level of energy, and the data generally are in agreement, especially when the experimental cattle used are similar in age, sex, and feeder grade and when lengths of feeding period do not differ widely. The data from 17 such steer calf experiments conducted by five of the Corn Belt stations have been combined in an interesting and informative manner by Minnesota workers. They applied a regression analysis on the data, adjusting all initial and final weights to 512 and 1,112 pounds, respectively. Mean values for daily gains and feed requirements were applied to the adjusted 600 pounds gain and uniform nonfeed costs were applied, on a daily basis, to the lengths of feeding period required to attain the calculated slaughter weights. Thus by combining feed and nonfeed costs it is possible to compute the costs of gains, as influenced by level of silage in the ration. Only four silage levels are shown in Table 117. Current feed and nonfeed costs may be applied to the data if they deviate widely from those used in the table. The data may be summarized as follows. Daily gain decreased and total pounds of dry matter required per unit gain increased as ration silage level was increased but feed costs of gain were reduced as silage level was increased. Generally, as corn or sorghum prices reach lower levels, higher-concentrate rations are favored because the lower grain replacement value of the stalk portion in silages will not be high enough in value to offset the longer feeding period required. This item is much more important to a commercial feedlot, which keeps its pens full year-round, than to a farmer-feeder who feeds only one drove of cattle per year.

Cattle fed to the same weight, with varying silage levels in the ration, usually grade alike, but the cattle fed on higher-silage rations may dress 1 to 2 percent lower unless silage level is cut back for several weeks toward the end.

From the standpoint of the farmer-feeder, finishing programs using 15 to 30 pounds of silage and grain equal to about 1 percent of body weight usually have the following advantages.

1. More live weight and carcass beef can be produced per acre of corn fed.

Table 117

Performance of Steer Calves Fed Several Levels of Corn Silage (A Summary of Seventeen University Experiments)[a]

	Percentage of Corn Silage Dry Matter in Ration Dry Matter			
Item	20	40	60	80
Average daily gain (lb)	2.49	2.36	2.17	1.91
Average daily feed (lb DM)				
Corn grain	11.7	8.8	5.5	2.1
Corn silage	3.1	6.4	9.6	12.1
Supplement	0.9	0.9	0.9	0.9
Total	15.7	16.1	16.0	15.1
Daily silage intake at 32% DM (lb)	9.7	20.0	30.0	37.8
Feed per 100 lb gain (lb DM)				
Corn grain	470	373	253	110
Corn silage	126	274	442	631
Supplement	36	37	41	47
Total	632	684	736	788
Feed per 600 lb gain (lb DM)	3,792	4,104	4,416	4,718
Feed cost per 100 lb gain[b]	$ 31.01	$ 30.62	$ 29.84	$ 28.66
Feed cost per 600 lb gain	$186.06	$183.72	$179.04	$171.96
Days to gain 600 lb	241	254	276	314
Nonfeed cost, total[c]	$ 43.29	$ 45.04	$ 47.98	$ 53.08
Total cost per 100 lb gain	$ 38.22	$ 38.13	$ 37.84	$ 37.51

[a] Adapted from Minnesota Cattle Feeders' Report, 1974.

[b] Based on corn silage at $20.00 per ton, corn at $2.52 per bushel, and supplement at $120 per ton.

[c] Based on $11.00 per head fixed charge plus $0.134 per day.

2. More profit is usually made per steer and per acre of corn fed.
3. Retail cutout is improved because of less waste fat trim.
4. A greater number of steers can be fed per acre of corn, making it possible to practice economies of scale.

COMBINED STOCKER AND FINISHING PROGRAMS

A question that many cattle feeders ask is whether a combined stocker-feeder program in which only corn silage and supplement are fed during the stocker program, followed by a shorter finishing period on a higher-energy ration, would not combine the economy of silage feeding with superior carcass merit. The Illinois station conducted such a study, and

Table 118

Effect of Length of Heavy Silage Feeding Period on Performance and Carcass Merit of Feeder Steers (10 Steers per Lot)[a]

Heavy Silage Feeding Phase				
Days on heavy silage	0	112	168	224
Total gain in period (lb)	—	210	272	372
Average daily gain (lb)	—	1.86	1.68	1.66
Average daily and total ration in period (lb)				
Corn silage	—	34.1–3,819	35.8–6,014	34.1–7,638
Soybean meal	—	1.5–168	1.5–252	1.5–336
Feed cost per cwt gain in period[b]	—	$23.46	$26.67	$25.95
Gain per ton of silage (lb)	—	110	91	95
Gain per acre of corn fed (lb)[c]	—	1,980	1,638	1,710
Grain Feeding Phase				
Days on full feed of grain	196	112	105	63
Average daily gain in period (lb)	2.48	2.71	2.05	1.54
Average daily and total ration in period (lb)				
Corn, cracked shelled	12.7–2,489	14.6–1,635	13.9–1,459	12.8–806
Soybean meal	1.5–294	1.5–168	1.5–157	1.5–94
Corn silage	15.9–3,116	18.7–2,094	18.8–1,774	14.0–882
Feed cost per cwt gain in period	$33.29	$34.20	$42.90	$53.06
Combined Silage-Grain Feeding Phases				
Total days to finished weight	196	224	273	287
Average final weight (lb)	1,063	1,095	1,066	1,047
Average daily gain (lb)	2.48	2.29	1.79	1.63

the results are shown in Table 118. Heavy steer calves were fed on silage rations for either 112, 168, or 224 days and finished on a full feed of cracked shelled corn, soybean meal, and limited corn silage, to about 1,050 pounds and estimated choice live grade. Another lot was fed a full feed of corn from the outset as a control. As in the studies summarized in the Minnesota report, the higher-energy rations produced the most rapid gains, when the entire program is considered. Costs of gain varied little, however. The cattle on the medium-energy or high-silage program graded higher in the Illinois test, in contrast to other tests reviewed. This may be explained by the fact that all cattle were fed to approximately equal weights and, as the lot fed only silage for the longest period gained

Total feed consumed per steer				
Corn, cracked shelled (bu)	44.5	29.4	26.0	14.4
Total corn (bu)[d]	50.7	41.2	41.6	31.4
Corn silage (tons)	1.56	2.96	3.90	4.26
Soybean meal (lb)	294	336	409	430
Corn acreage utilized per steer[c]	0.64	0.53	0.54	0.42
Steers fed per acre of corn	1.56	1.89	1.83	2.38
Steer gain per acre of corn fed (lb)	810	971	896	1,116
Feed cost per cwt gain[b]	$33.29	$29.86	$33.84	$31.55
Dressing percent	62.7	61.0	61.8	60.2
Carcass grade	G+	G+	C−	C
Fat cover over 12th rib (in.)	0.63	0.59	0.63	0.39
Return over steer and feed cost[e]	$32.39	$38.41	$54.98	$66.24
Return over steer and feed costs per acre of corn fed	$50.53	$72.59	$100.61	$157.65

[a] Adapted from Illinois Agricultural Experiment Station, Illinois Research Reports.
[b] Feed costs used, per ton: corn silage, $20; soybean meal, $120; shelled corn, $90.
[c] Corn silage yield estimated at 18 tons or 80 bushels per acre.
[d] Includes corn in silage, assumed to equal 15 percent of the weight of the silage.
[e] Based on $36 stocker steer cost and carcass prices ranging from $60 to $67 per hundredweight.

somewhat more slowly, they were older when slaughtered, especially in comparison with the lots on full feed from the start and those fed heavy silage for only 112 days. Marbling score, the chief determinant of carcass grade, is highly related to age, and even though the slower gaining steers had a quarter-inch less back fat, their marbling score was significantly higher.

Profit per steer was higher for the higher-silage steers, mainly because of higher grades and higher selling price in the carcass, and profit per acre of corn fed was even more favorable for the lots on a heavy feed of silage. These findings perhaps are most important for farmer-feeders, who feed cattle in the first place mainly to provide a better market for their crops and who feed only one drove of cattle per year. Nonfeed costs on these farms are not increased appreciably by extending the feeding period from 7 months to 10. In a commercial lot this additional 90 days could add from $10 to $20 in nonfeed costs per head to the overall cost of finishing a steer and, in the large-volume operation, this is especially important.

A point not to be overlooked in favor of the medium levels of energy in the finishing ration is cutability of carcass. The British breeds, especially, have for 50 years or more been selected for early maturity and, if fed the high-energy rations from an early age onward, they often reach choice grade at 850 to 950 pounds. Thus they become inefficient and produce wasty carcasses at weights that the industry cannot afford to live with. Lower grades of cattle such as Okies, exotic crossbreds, or Holsteins can make use of the higher-energy rations more advantageously, as discussed in Chapter 14.

In summary, it appears that corn silage can satisfactorily make up one-third to one-half of a complete, mixed ration fed to finishing cattle, especially by farmer-feeders, or it can be used as the principal ingredient in the ration for 150 to 200 days before finishing on a high-energy ration.

COMMERCIAL FEEDLOT HIGH-ENERGY RATIONS

The principles demonstrated in the discussion of energy levels in the rations fed by farmer-feeders in general may not necessarily hold when it comes to choosing the preferred formula for commercial feedlot rations. Several differences, however, should be noted. The commercial feeder prefers to feed—and in fact almost must feed—a complete mixed ration. His automatic feed-processing and mixing plant and his unloading trucks or wagons are all geared to handling and delivering to the bunk, rations in completely mixed and proportioned form. As already mentioned, this type of feedlot often may feed a lower-grading kind of steer that responds more satisfactorily to high-energy rations, especially in terms of type and amount of finish or cutability produced. Another difference is that the commercial feeder feeds for a shorter period than his counterpart in the farm situation, and this applies especially to heifer feeders.

The Arizona station, noted for its nutrition research in commercial feedlot rations, has compared one lot of cattle that were fed so-called 1954 rations with other lots fed modern commercial feedlot rations that contained 80 or 90 percent concentrates. The rations and performance data are shown in Table 119. The rations for all lots contained 40 percent roughage for the first 56 days.

The obvious differences between the modern and the 1954 rations are

Table 119

Comparison Between 1954 Ration and Modern 80 and 90 Percent Concentrate Rations[a]

Type of Ration	80 Percent Concentrate		90 Percent Concentrate			1954 Ration	
Number of days fed	56	84	56	42	42	56	113
Ration ingredients (lb)							
Ground alfalfa hay	20.0	5.0	20.0	10.0	2.0	25.0	20.0
Cottonseed hulls	20.0	15.0	20.0	15.0	8.0	15.0	—
Steam-processed milo	44.7	63.3	44.6	58.9	73.6	—	—
Dry-rolled milo	—	—	—	—	—	49.3	73.8
Cottonseed pellets	4.5	5.5	4.5	5.0	5.0	4.5	—
Molasses	5.0	5.0	5.0	5.0	5.0	5.0	5.0
Tallow	4.0	4.0	4.0	4.0	4.0	—	—
Dicalcium phosphate	0.5	0.5	0.5	0.5	0.5	0.3	0.3
Urea	0.8	0.6	0.8	0.5	0.6	0.4	0.4
Salt	0.5	0.5	0.5	0.5	0.5	0.5	0.5
Ground limestone	—	0.6	—	0.6	0.8	—	—
	100.0	100.0	100.0	100.0	100.0	100.0	100.0
Vitamin A premix (gm)	10	10	10	10	10	—	—
Stilbestrol	+	+	+	+	+	—	—
Performance data							
Number of steers	16		16			16	
Average initial weight (lb)	626		633			628	
Average daily gain (shrunk 5%) (lb)	3.0		2.9			2.2	
Average daily feed (lb)	23.7		23.3			20.5	
Feed per cwt gain (lb)	786		813			924	

[a] Adapted from Arizona Cattle Feeders' Day Report.

in method of grain processing, additions of tallow, vitamin A, and hormone, and reduction in roughage content, at least with respect to the 90 percent concentrate ration. Protein, calcium, phosphorus, and fiber content were similar; yet the modern ration with 80 percent concentrate produced 32 percent more gain on 17 percent less feed than did the 1954 ration. This and many other experiments clearly show that feeder cattle can perform satisfactorily on very high- or even 100-percent concentrate diets. Feed efficiency is high, when expressed as pounds per pound gain, but the relative costs of concentrate or roughage energy usually determine the least-cost energy level. Farmer-feeders who can grow silage or hay crops efficiently generally prefer to feed at least 15 to 20 percent roughage, but commercial feedlots that buy all their feeds will feed more highly concentrated diets. The reduction in days required to reach slaughter weight when higher-energy diets are fed, plus slightly higher dressing percentages, also favors use of such feeding programs by commercial lots.

Before closing the subject of level of concentrate or energy in the ration, it should be mentioned that there is a very real problem with respect to liver condemnations at slaughter time, often amounting to more than 50 percent, when all-concentrate or very high-concentrate rations are fed. The economic loss is substantial. A highly inflamed rumen wall condition, known as rumen parakeratosis, also occurs and surely interferes with absorption of nutrients from the rumen. Feedlot performance by steers with the abscessed livers has been shown to be reduced substantially, possibly by as much as 10 percent reduction in rate of gain and feed efficiency. The physiology of these two conditions is not well understood, but they are believed to be associated with the lowered pH in the rumen. Including 75 milligrams of the tetracycline antibiotics per day in the rations of cattle fed all-concentrate rations reduces the incidence of abscessed liver. Founder and chronic bloat also may occur in 10 to 15 percent of cattle fed such rations, depending on the degree of husbandry with which the cattle are started on feed. Cattle on such rations should never be allowed to get hungry, and consequently some kind of self-feeding must be practiced. Commercial feedlots often provide feed in their fenceline bunks as frequently as four times daily, whereas farmer-feeders use self-feeders.

CHAPTER 16
CARBONACEOUS CONCENTRATES AND THEIR USE IN THE FINISHING RATION

With respect to source, concentrates are of two kinds: (1) the grains of farm-grown crops and (2) commercially manufactured feedstuffs. With respect to composition, concentrates may be divided into two groups: (1) carbonaceous feeds, or those containing a relatively high percentage of energy and a low percentage of protein, and (2) nitrogenous feeds, or those especially rich in protein. In general the farm-grown grains belong in the carbonaceous class, whereas a majority of the commercial feedstuffs are nitrogenous.

Carbonaceous feeds, also referred to as carbohydrate feeds, as their name implies, are rich in chemical compounds that contain a relatively large amount of carbon. The more important of these compounds are starches, sugars, and fats. All these materials, when consumed by animals, are utilized chiefly for maintenance, muscular energy, and conversion to body fat, which is stored in various parts of the body but especially in the superficial layers of muscle found just beneath the skin. Carbonaceous concentrates are thus of prime importance in finishing cattle for the market and generally make up the major portion of finishing rations.

COSTS OF FARM GRAINS

Prices paid for feed grains by commercial cattle feeders, or prices used in calculating costs by farmer-feeders who may produce their own grain supply, vary cosiderably from area to area. As with most commodities, the prevailing price for a feed grain is a matter of supply and demand and transportation or freight costs. Government price support programs also are a factor when they are in operation, especially in years of average or smaller than average acreages and yields. Feed grain prices in the western Corn Belt and in the Sorghum Belt are well below the national average. This circumstance undoubtedly has contributed greatly to the phenomenal increase in cattle feeding in Nebraska, Kansas, western Iowa, Colorado, and the Panhandle area. One might project that the far western states will reduce their feeding operations and that the southern states will not assume an important role as a cattle-finishing area because of the noncompetitive feed prices in these areas. It is more economical to move surplus cattle to surplus feed areas, but partially offsetting this apparent saving is the cost of moving the fed cattle or carcass beef to consumer centers at either end of the country. Drastic shifts in freight rates can quickly alter the relative competitive position of certain areas, favorably or unfavorably.

Other factors such as export markets for grain, climate, labor costs, feeder cattle supply, and related factors, of course, all enter into this question but, because feed costs amount to about 85 percent of all costs in the cattle-finishing program, relative feed prices must be given a high priority when assessing regional suitability for cattle feeding.

IMPORTANT ENERGY SOURCES

The important carbonaceous concentrates used in cattle feeding in the United States are corn, grain sorghum (milo), oats, barley, wheat, beet pulp, and molasses. Figure 77 shows the contribution made by the various feed grain crops to the total concentrate feed supply consumed by all farm livestock in the United States. Note that total consumption has steadily increased until 1973 and that corn makes up over one-half the total supply. It will be recalled that feeder cattle consume about one-fifth of the

Fig. 77. Feed grains and by-product feeds consumed by livestock and poultry in the United States from 1960 through 1975. Beef cattle consume about one-third of all the concentrate feeds fed annually.

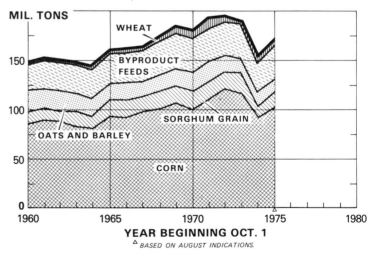

corn, two-thirds of the sorghum, one-third of other grains, and 10 percent of the by-product protein concentrates consumed by all livestock and poultry (see Table 6). Although the use of some of these feeds is limited to certain sections, a brief discussion of all of them will be attempted, along with a discussion of other feeds that are less commonly used.

CORN

In corn (*Zea mays*) is found the explanation of the remarkable development attained by the beef cattle industry of the United States, especially in earlier years and of course specifically in the Corn Belt. Other countries have cattle of as good or better breeding; other regions have as good or more favorable climate; other nations have more and better grasslands; but no other country has such an abundant supply of corn available for cattle feeding. As mentioned in Chapter 2, United States corn production consistently amounts to some 5 to 6 billion bushels annually, of which 25 percent is consumed by beef cattle.

Corn is unsurpassed as a feed for finishing cattle. It is highly palatable, being eaten readily by cattle of all ages. It meets the requirements for a high-energy feed, being high in digestible carbohydrates and fat—the two nutrients that normally furnish most of the energy in finishing rations. Corn is easily stored and easily prepared for consumption. It is produced in great abundance and, consequently, is readily obtainable. In fact, it may be said that corn combines all the essentials of a valuable cattle feed, save one. It is noticeably deficient in protein. For this reason corn should not be fed alone, but should be supplemented with feeds that are high in protein, such as the oilseed meals and alfalfa hay or commercially prepared protein concentrates.

In all sections where corn is extensively grown it should, and normally does, form the major portion of the concentrate of the finishing ration. It is seldom possible for Corn Belt feeders to use other carbonaceous concentrates to any great extent without materially increasing the cost of the ration or lowering its efficiency, or possibly both. Grain sorghums approach corn in this respect in the southern and western fringes of the Corn Belt and in the Sorghum Belt itself—that is, the Panhandle of Oklahoma and Texas, and adjoining parts of Kansas, Nebraska, Colorado, and New Mexico.

NEW VERSUS OLD CORN

In view of the large amount of corn placed in storage on farms or in commercial elevator storage, some of which is carried over for 2 or more years, the question arises concerning the feeding value of such corn in comparison with corn that has been recently harvested. Old corn contains less moisture than new corn and consequently is a more concentrated feed. Assays of corn of different ages do show a loss of carotene with increase in storage time. Feeding tests such as those summarized in Table 120, comparing corn stored for one or more years with recently harvested corn, show no significant difference between old and new corn for finishing cattle.

Another test, comparing still older corn with new corn, was conducted at the Ohio station. They fed steers for 84 days on either the old or new corn, both hammermill-ground, and quality mixed hay, soybean meal, and

Table 120

Relative Value of Old and New Corn for Finishing Cattle (210-Day Feeding Period)[a]

	4-Year-Old Corn	3-Year-Old Corn	2-Year-Old Corn	1-Year-Old Corn	New Corn
Corn data					
Relative hardness of kernels (%)	164	160	174	150	100
Moisture (%)	10.4	10.9	12.4	12.9	16.7
Vitamin A (IU per pound)	821	809	869	1,330	1,720
Crude protein (14% moisture basis) (%)	9.8	8.5	8.6	8.3	8.0
Cattle data					
Average initial weight (lb)	735	736	732	732	738
Average daily gain (lb)	2.20	2.15	2.10	1.97	2.15
Average daily ration (lb)					
Shelled corn (14% moisture)	16.3	15.0	16.1	15.7	14.9
Protein supplement	1.0	1.0	1.0	1.0	1.0
Alfalfa hay	4.0	4.0	4.0	4.0	4.0
Feed per cwt gain (lb)					
Shelled corn (14% moisture)	741	696	766	799	693
Protein supplement	45	47	48	51	47
Alfalfa hay	181	187	190	203	186
Shrinkage (%)	2.0	1.4	3.2	2.3	1.9
Dressing percent	63.7	63.8	63.9	63.5	62.8

[a] Iowa AH Leaflet 160.

Table 121

Comparison of Old and New Corn, with Supplemental Vitamin A, for Finishing Steers[a]

	1960 Corn		1954 Corn	
	Vitamin A	Control	Vitamin A	Control
Average daily gain (lb)	2.11	1.94	2.08	1.98
Average daily ration (lb)	18.8	18.4	19.1	18.5
Feed per cwt gain (lb)	893	952	915	938
Feed cost per cwt gain	$17.61	$18.66	$18.02	$18.40
Carcass grade	G+	G+	G+	G+
Dressing percent	61.0	61.8	61.2	61.8
Blood vitamin A (mcg/100 ml)	39.4	26.1	41.7	26.2

[a] Ohio Farm and Home Research, 1962.

minerals. The steers in one lot fed each type of corn received additionally 20,000 IU of vitamin A. As may be seen in Table 121, the old corn produced results comparable to those of new corn. What appeared to be a gain response to supplementary vitamin A in both treatments was apparently not real as the higher dressing percentages of the unsupplemented steers offset the differences in liveweight gains. The old corn contained only 0.08 milligrams of carotene per pound compared with 0.49 for the new corn. Apparently the hay provided all the vitamin A activity required by these cattle.

The field sheller or corn combine is undoubtedly responsible for a great increase in the feeding of high-moisture new corn, especially if storage facilities such as the airtight Harvestore are available. Such high-moisture grain undergoes fermentation if stored in large quantities and has all the characteristics of silage. For this reason it is discussed more thoroughly under the topic of high-moisture corn.

VALUE OF SOFT CORN

Often a field of late-planted corn fails to mature before the first killing frost, and about once in 8 or 10 years an unusually early frost damages a considerable percentage of the crop over a large portion of the Corn Belt. Frost of course kills the green, growing stalk and halts the translocation of the food nutrients from stalk to ear, and the transfer of water from ear to stalk where it normally evaporates through the leaves. As a consequence the ears fail to dry out, and they contain an abnormally high percentage of moisture, as much as 40 percent not being uncommon. Such corn is commonly termed "soft."

Soft corn is also immature and is not to be confused with mature corn that has a high moisture content. Because soft corn is unmarketable, it is retained on the farm and fed. The feeding value of such corn has long been a matter of controversy, and no one will deny that it is inferior to sound corn. However, results obtained at experiment stations where the two feeds have been compared indicate that soft corn is a better feed for finishing cattle than is commonly supposed. On the basis of dry-matter content, soft corn is as efficient as mature corn in producing gains. However, the gains from soft corn are somewhat less rapid because its lower density results in a smaller consumption of dry matter in comparison with mature corn.

Certain fungi, primarily *Gibberella zea* and *Fusarium moniliforme,* can be something of a problem in soft, frosted corn. If corn is left in the fields until it dries sufficiently to store as shelled corn, these fungi can cause ear and kernel rots or molds that will reduce feed intake still further. These molds are not toxic to cattle, but the effect on intake is a problem in feeder cattle. Cows and heifers in early pregnancy—in September and October, for instance—may abort if allowed to run in frosted corn fields because of the mycotoxins and estrogens produced by the fungi. Swine are much more affected by these rots and molds and will scarcely consume any amount of the moldy corn.

Corn that is abnormally high in moisture will freeze hard during severe weather, so that cattle can scarcely eat it. For this reason soft corn should, if possible, be fed during the fall and early winter. To obtain as great a consumption of corn as possible, the roughage allowance should be limited to 4 or 5 pounds daily per 1,000 pounds live weight. Silage is less satisfactory than hay as roughage for cattle fed on soft corn.

SOFT EAR-CORN SILAGE

One problem connected with the utilization of soft corn is the matter of storage. This difficulty can be avoided by picking the ears soon after the first killing frost and making them into ear-corn silage, which is superior to the same corn fed directly as ear corn, as shown in Table 122. Making silage of the entire stalk is not always successful because the frosted leaves often will have been blown off, but it still may be the best way to salvage a field of soft corn.

Table 122

Value of Soft Corn for Finishing Cattle[a]

Two-Year-Old Steers, Fed 80 Days	Soft Corn		Mature Corn from Previous Year
Form of Corn Fed	Ear Corn	Ear-Corn Silage	Ear Corn
Average percent moisture	35.82	59.65	13.61
Average daily gain (lb)	3.20	3.35	3.52
Average daily ration (lb)			
Ear corn	31.48	—	25.59
Linseed meal	1.89	1.89	1.89
Ear-corn silage	—	45.43	—
Alfalfa hay	2.70	2.35	2.70
Feed per pound gain (lb)			
Ear corn	9.84	—	7.27
Linseed meal	0.59	0.56	0.54
Ear-corn silage	—	13.56	—
Alfalfa hay	0.84	0.70	0.77
Corn (dry matter) per pound gain (lb)	6.42	5.34	6.28
Dressing percent	60.0	60.7	60.6
Cattle and hog gains per acre (lb)	247.0	292.0	—

[a] Illinois Bulletin 313.

HIGH-MOISTURE EAR-CORN SILAGE

The results of an early, pioneering experiment with the feeding of soft corn, harvested as ear-corn silage, are reported in Table 123. The favorable response to the feeding of corn harvested in this form leads one to ask why this method of harvesting and feeding corn need be limited to those occasional years when corn fails to mature.

If high-moisture ear-corn silage will produce the same amount of gain in cattle as will fully matured ear or shelled corn from an equal area of land, there would seem to be several advantages to be gained by harvesting the corn as mature, ground ear-corn silage. First, it permits the harvesting of a considerable portion of the corn crop during favorable autumn weather. Second, it stores the corn in a suitable form for feeding to cattle of all ages without further preparation, eliminating the expense and extra handling of feed involved in grinding or shelling. Third, it is

Table 123

Ear-Corn Silage for Finishing Calves[a]			
Equal Areas of Corn Harvested and Fed	Ear-Corn Silage	Ear-Corn Silage	Ground Ear Corn
Date harvested	Sept. 4–9	Sept. 24–25	Oct. 25–30
Moisture in grain when harvested (%)	53.4	37.5	Under 25
Average daily gain (lb)	2.23	2.12	2.10
Average daily ration (lb)			
Corn component	29.4	27.3	13.3
Cottonseed meal	1.6	1.6	1.6
Alfalfa hay	2.0	2.0	2.0
Oat straw	—	—	2.9
Total days feed from area	3,018	3,153	3,329
Calculated total gain to be credited area (lb)	6,721	6,686	6,987
Selling price per cwt	$15.70	$15.65	$15.35
Dressing percent	61.1	61.4	60.4
Return per bushel of corn obtained from the area allowed to mature	$1.19	$1.11	$1.07

[a] Illinois Experiment Station, Mimeographed Report, 1928–29, Calf Feeding Experiments.

possible to pasture stalk fields while they are still green and palatable, or the green stover may be made into green stover silage or stalklage. This early harvest will enable the farmer to seed winter wheat or rye for fall and winter pasture in regions where this is a feasible agronomic practice.

The method for making ear-corn silage is relatively simple. The ears, with or without the attached husks, are gathered with a mechanical picker without husking rolls and are hauled to the silo to be ground or chopped and blown into the silo. Some ear-corn harvesters have ear-corn grinding attachments that permit field grinding. It is unnecessary to pack the silage in the silo, because the weight of the corn and the absence of much bulky material will cause the silage to settle almost as rapidly as it is made. However, it may be necessary to have someone in the silo to see that the silage is evenly distributed, unless a mechanical spreader is used. Otherwise the cut corn tends to pile up in the center while the husks and cobs accumulate along the walls.

If the corn is overly ripe and contains insufficient moisture to ensure thorough packing and fermentation, a small amount of water should be added through a pipe or hose inserted into the mouth of the blower. The filled silo should be sealed either by running sufficient green stover or other forage through the cutter to make a layer approximately 3 feet deep on top of the ear corn, or by covering it with a plastic silo cap or fine limestone. When ear-corn silage is sealed in this manner, there is little spoilage. The airtight or oxygen-free silo is of course ideal for storing such silage, but heavy-duty bottom unloaders are then required.

Because ear-corn silage is heavy in proportion to its volume, a relatively small silo will hold sufficient silage to feed a carload of cattle, even though they receive little other feed. If the silo is of the conventional upright type and also quite large in diameter, a small drove of cattle will not require enough silage daily to prevent spoilage in warm weather. A silo 12 by 40 feet holds approximately 120 tons of ear-corn silage, or enough to finish 40 to 50 head of cattle during an average-length feeding period.

Research work by the Iowa and Indiana stations is reported in Table 124. These studies deal with corn that was mature to the extent that it contained only 32 percent moisture or less. These studies indicate, as did the Illinois studies with immature ear-corn silage, that the dry matter of high-moisture ear corn is superior to that of mature dry ear-corn when stored as ear-corn silage, because faster gains were produced on less dry matter. A probable explanation is that the cob portion of the ear-corn silage is more efficiently utilized by the rumen microflora as a result of the fermentation it undergoes in the silo. On a dry basis, ear-corn silage consists of approximately 20 percent cob. It should be noted in Table 124 that dry-matter intake was not increased in the high-moisture corn lots. In fact, dry-matter consumption was apparently reduced. Consequently, it cannot be said that the corn is more palatable, although it would seem so upon observing the apparent relish with which cattle eat it.

Experimental data on the feeding value of ensiled high-moisture shelled corn are discussed in Chapter 24, which deals with preparation of feeds. The subject of ensiled ear corn is introduced here only because it is closely related to the subject of ensiled soft ear corn, which appropriately belongs here.

ARTIFICIALLY DRIED HIGH-MOISTURE CORN

Extensive data are unavailable concerning the effect of artificial drying on the feeding qualities of corn. For farmers who have no silo for storing high-moisture corn but who wish to practice field shelling before their corn reaches a safe storage level of 16 to 18 percent moisture, artificial drying seems a practical alternative. A problem presents itself in the heavy corn-growing areas because the capacity of the flow-type dryers commonly operated by the elevators or by farmer-feeders themselves is insufficient to

Table 124

Value of Mature High-Moisture Ear-Corn Silage Compared with Dry Ground Ear Corn for Finishing Cattle[a]

Experiment Station	Indiana		Indiana		Iowa		
Days on Test	117		126		119		
	High-moisture	Dry ground	High-moisture	Dry ground	High-moisture	High-moisture	Dry ground
Moisture and additives (%)	32.2	17.7	32.5	15.5	31	31 + stilbestrol	14.5
Number of animals	10	10	36	35	36	18	36
Sex	Steers	Steers	Heifers	Heifers	Steers	Steers	Steers
Initial weight (lb)	958	960	467	466	804	805	802
Pounds gain	299	272	279	275	354	327	364
Average daily gain (lb)	2.56	2.23	2.21	2.18	2.98	3.17	3.05
Ground ear corn per day (lb)[b]	20.6	22.1	11.88	13.45	20.0	20.4	22.9
Total feed per day (lb)[b]	23.3	25.8	18.43	20.22	24.3	24.7	27.2
Ground ear corn per cwt gain (lb)[b]	807	951	555	617	675	643	750
Total feed per cwt gain (lb)[b]	953	1,111	860	927	819	778	889
Feed required, as % of dry corn	85.7	100.0	92.8	100	92.1	87.5	100.0

[a] Adapted from Indiana and Iowa Feeders' Day Reports.
[b] All high-moisture corn values converted to equivalent moisture for comparison.

Table 125

		Drying Temperature		
Item	Unheated Air	180°F	220°F	260°F
Number of steers	18	18	18	18
Average initial weight (lb)	713	717	737	755
Average final weight (lb)	1,144	1,164	1,189	1,198
Average daily gain (lb)	3.88	4.03	4.05	3.99
Daily intake (lb)	23.0	22.6	22.6	21.8
Feed per pound gain (lb)[b]	5.93	5.61	5.57	5.47

Feedlot Performance of Holstein Steers Fed Heat-treated Corn (111 Days)[a]

[a] Adapted from Illinois Cattle Feeders' Day Report, 1973.

[b] Corn silage was fed in small amounts only during the first 9 days and this silage is not included in feed efficiency calculations. Whole shelled corn was mixed with a complete pelleted supplement (50% C.P.) in the ratio of 7:1 throughout the trial. No roughage was fed after 9 days on feed.

keep up with the large amount of corn that suddenly needs drying when field shelling commences in earnest. Either farm-size or larger portable "batch" dryers ordinarily relying on forced air, heated to temperatures ranging from 180°F upward, are being used to take care of this emergency. The data shown in Table 125 indicate that drying 24 percent moisture content mature corn at 180°, 220°, or 260°F did not reduce its feeding value for cattle. Such high temperatures slightly reduce the feeding value of such corn for swine and poultry, however, and wet corn millers would impose heavy penalties on corn dried at the highest temperature tested because considerable discoloration occurred, along with some popping. It is interesting to note the excellent gains and feed conversion made by the yearling Holstein steers on virtually an all-concentrate ration. The supplement fed contained aureomycin and held liver abscesses to a negligible level. Few farmer-feeders feed their corn exclusively to cattle. Therefore the drying method should be carefully considered if there is a shift to more and more field shelling of corn with high moisture content.

EFFECT OF VARIETY OF CORN ON FEEDING VALUE

During 1930–1940, when hybrid corn was rapidly displacing the open-pollinated varieties, the feeding value of the different hybrids in relation to open-pollinated corn was widely discussed. Some hybrids were claimed to be so hard that they were eaten with difficulty and consequently the kernels were swallowed with only a little chewing. Other varieties were believed not only to be softer but also to possess an aroma and flavor that made them more palatable.

A discussion of this topic is unimportant today, because open-pollinated corn has been replaced by hybrid corn. Plant breeders have continued to do research in breeding "tailor-made" hybrids for specific purposes. For example, special hybrids have been developed that produce more forage tonnage and therefore are ideally suited for silage, although such "silage hybrids" generally fail to produce more beef gains per acre, primarily because of the lower yield of grain. Other hybrids have been produced that are 50 percent higher in protein content than ordinary corn. Feeding trials such as the one reported in Table 126 indicate that protein supplements may be eliminated from finishing rations in the future if plant breeders improve the yield of the high-protein corn hybrids. Hybrids have also been developed that contain twice the usual amount of

Table 126

Ration	Lot 7 Regular Corn	Lot 8 Regular Corn Plus Soybean Meal	Lot 9 High-Protein Corn
Value of High-Protein Corn for Finishing Cattle (175 Days)[a]			
Protein content of corn (%)	8.8	8.8	13.2
Initial weight (lb)	928.0	924.5	926.5
Final weight (lb)	1,244.5	1,310.0	1,319.0
Total gain (lb)	316.5	385.5	392.5
Average daily gain (lb)	1.81	2.20	2.24
Average daily ration (lb)			
Shelled corn	13.1	13.4	14.5
Soybean meal	0.3[b]	1.5	0.3[b]
Corn silage	24.9	25.1	25.0
Feed eaten per cwt gain (lb)			
Shelled corn	726.0	607.0	644.0
Soybean meal	18.0	70.0	15.0
Corn silage	1,377.0	1,138.0	1,114.0
Feed cost per cwt gain ($)	30.85	28.19	26.68
Selling price per cwt ($)	33.50	34.00	34.25
Return over cost of cattle and feed ($)	−16.16	3.58	12.44
Total corn consumed per steer, including corn in silage (bu)	51.8	52.7	56.0
Return realized per bushel of corn fed ($)	1.37	1.75	1.90

[a] Illinois Cattle Feeders' Day Report.

[b] Fed to equalize the protein fed in Lot 8.

oil, and there are other hybrids that contain high levels of sucrose or sugar in the stalks.

Commercial corn varieties are now available that are higher in one of the amino acids that is essential for nonruminants, namely lysine. Feeding trials with swine and poultry indicate that *high-lysine* varieties of corn are more efficiently converted to gain by these species. These corn varieties, generally referred to as *Opaque-2 corn,* also have been tested in cattle-feeding trials but the results have been highly inconsistent as to any advantages, especially when the rations contained considerable roughage. When rations are very high in concentrate, the passage rate from the rumen is apparently so rapid that the digestion pattern is more character-istic of nonruminants, with the protein in the ration escaping normal microbial degradation. The amino acids available for postrumenal absorp-tion might thus more nearly meet the exact needs of a young, rapidly growing ruminant than they would if all or most of the dietary protein or nitrogen were first converted to microbial protein, as occurs in higher-roughage rations where passage rate is slower. This concept is offered by some nutritionists as an explanation for the superiority of high-lysine corn over regular corn when used in high-concentrate rations, although it is of no advantage otherwise. Further research is needed to refine ration formulation to take advantage of higher-lysine corn. Yields per acre are equal to or slightly lower than those for regular corn.

Another new kind of corn still under active research is known as *waxy corn* or waxy maize. In waxy corn, the starch differs from that of regular corn, with 100 percent of the starch being in the form of amylopectin. In regular corn, 25 percent of the starch is in the form of amylose and the remainder is amylopectin. The data are still too variable for drawing firm conclusions concerning the performance of waxy corn. At this time yields per acre are somewhat below those of regular corn. Waxy corn is unsuitable for wet milling and cannot be sold through some commercial channels usually available to farmer-feeders who may wish to sell surplus corn.

The most promising new corn hybrids still under investigation, for cattle feed at least, would seem to be those that are above average in both protein and oil content. Such corn already exists and certain lines contain 10 to 12 percent protein and 6 to 8 percent oil. Improvement in yields is eagerly awaited by cattle feeders.

GRAIN SORGHUMS (MILO)

There has been a great impetus in the growing of the grain-type sorghums with the development of sorghum hybridization in the 1950s.

Fig. 78. Hybrid grain sorghums are increasing in importance as a source of high energy concentrate for finishing rations.

The increase in sorghum acreage can be explained by three reasons: (1) corn, wheat, and cotton acreage allotments in the 1960s stimulated interest in grain sorghum as a replacement crop; (2) great progress in developing combine-type hybrid sorghums with yields of 150 to 200 percent more grain than nonhybrid varieties; and (3) the increase in acreage brought under irrigation in the regions particularly adapted to sorghum growing and now known as the Sorghum Belt.

Sorghums of all kinds are still being grown to some extent throughout the southern half of the country, but the major portion of the acreage grown exclusively for grain, often called milo by cattle feeders, is located in the Panhandle area of Texas and Oklahoma with some overlap in western Kansas, Nebraska, and eastern New Mexico and Colorado.

The newer hybrids not only yield higher with respect to grain, but they are also adapted to harvesting with the combine because of their shorter stalks. A problem that continues to occupy the attention of plant breeders is that of late maturity. Each year thousands of acres are damaged by frost before maturity is reached. The development of open-panicled or "sprangly-headed" hybrids has helped to reduce the moisture in the grain so as to improve the storage qualities, but high moisture is still a major problem.

A discussion of recommended varieties and hybrids has little point here because so much variation exists among sections of the country with respect to rainfall or irrigation water supplies, length of growing season, soil fertility, and drying conditions. Local or state crop specialists should be consulted for recommendations as to suitable sorghum varieties and hybrids and latest improvements, which are being made each year. Yields of 150 bushels or 4 to 5 tons and more per acre are not uncommon in modern grain sorghums, and yields consistently compare favorably with corn.

In chemical composition, grain sorghums in general are quite similar to shelled corn. An exception is protein content, which is higher in grain sorghums—roughly 11 percent compared with 9 percent in corn. The two feeds are approximately equal with respect to total digestible nutrients or energy content, and both are low in calcium and phosphorus. The carotene content of grain sorghums is very low. Fortification of sorghum rations with carotene or vitamin A concentrate is essential in finishing rations that do not also contain at least 4 to 6 pounds of excellent-quality legume hay or an appreciable amount of good silage. High-carotene hybrids, called *yellow endosperm sorghum,* appear promising in the developmental stages and may eventually offset the lack of this nutrient in grain sorghums.

Sorghums should always be processed for feeding to cattle because they are hard-seeded and much of the sorghum grain otherwise passes through the digestive tract in an incomplete stage of digestion. Swine do not do a good job of recovering sorghums from voided cattle feces. The effect of preparation on the feeding value of sorghums is discussed in Chapter 24.

An experiment at the Kansas station, reported in Table 127, made a comparison of grain sources in a finishing ration for heifers, using straight sorghum, a combination of sorghum and corn, straight corn, and a combination of sorghum, corn and wheat. All grains were ground without heat treatment and were full-fed with 2 pounds of alfalfa hay and 1 pound of protein supplement daily. Some prairie hay was fed for the first 56 days. The general conclusions were the following.

1. The straight ground sorghum grain ration produced significantly slower gains.
2. Additions of corn to the ration improved rate of gain and feed efficiency.
3. Addition of wheat did not affect rate of gain but did improve feed efficiency.
4. Mixtures of grain seemed to be more acceptable to the animals over a longer time.

Table 127

Feedlot Results for Finishing Heifers with Sorghum Grain, Sorghum Grain and Corn, or Sorghum Grain, Corn, and Wheat (112 Days)[a]						
Number of heifers per lot	10	10	10	10	10	10
Grain (%)						
Sorghum	100	75	50	25	—	33
Corn	—	25	50	75	100	33
Wheat	—	—	—	—	—	33
Average initial weight (lb)	607.5	608.5	607.0	608.5	607.0	610.0
Average final weight (lb)	897.0	936.5	945.0	946.5	935.0	936.0
Average daily gain (lb)	2.58	2.93	3.10	3.02	2.93	2.91
Average daily ration (lb)						
Grain	17.0	17.6	17.5	16.2	16.2	16.6
Supplement	1.0	1.0	1.0	1.0	1.0	1.0
Alfalfa hay	2.0	2.0	2.0	2.0	2.0	2.0
Prairie hay	1.7	1.7	1.7	1.7	1.7	1.7
Feed per cwt gain (lb)						
Grain	658	600	565	538	553	570
Supplement	39	34	32	33	34	34
Alfalfa hay	77	68	65	66	68	69
Prairie hay	67	59	56	57	59	59

[a] Kansas Livestock Feeders' Day Report.

5. Carcass grade and carcass characteristics were not affected by type of grain in the ration.

The steam-processing and rolling of milo has improved its feeding value so that it compares quite favorably with corn and even surpasses barley, at least with respect to rate and cost of gain. The Arizona test summarized in Table 128 shows that such processing enhances the nutritional properties of milo but of barley considerably less. Data in Chapter 24 show that improved digestibility is the major reason for this effect. The inference is that with the modern processing methods used especially by the commercial feedlots in the Southwest, milo is an excellent cattle-finishing feed that is highly competitive with corn and even excels the other grains commonly used today.

Those feeders on a smaller scale who cannot justify the investment in steam-processing equipment can still obtain satisfactory results from finely ground milo, especially if fed with small amounts of any type of silage, cottonseed hulls, or ground hay.

Table 128

Comparison of Milo and Barley in Dry-Rolled and Steam-Processed Form When Fed to Finishing Steers (140 Days)[a]

Type of Grain	Milo		Barley	
Method of Processing	Dry-Rolled	Steam-Processed	Dry-Rolled	Steam-Processed
Number of steers	10	10	10	10
Average initial weight (lb)	559	551	567	567
Average daily gain (lb)	2.90	3.20	2.90	3.11
Average daily feed (lb)	23.4	24.7	20.6	22.0
Feed per cwt gain (lb)	848	812	749	743
Feed cost per ton	$49.62	$49.62	$57.29	$57.29
Feed cost per cwt gain	$21.04	$20.15	$21.46	$21.28
Major ration components (%)				
Ground alfalfa	5.0		5.0	
Cottonseed hulls	15.0		10.0	
Milo or barley	68.4		74.9	
Cottonseed pellets	4.5		3.0	
Molasses	5.0		5.0	
Urea	0.6		0.5	
Minerals and salt[b]	1.5		1.6	
Total	100.0		100.0	
Crude protein (%)	11.50		11.40	
Calcium (%)	0.48		0.35	
Phosphorus (%)	0.33		0.35	
Crude fiber (%)	10.40		10.20	

[a] Adapted from Arizona Cattle Feeders' Day Report.
[b] Vitamin A was added to both rations.

The problem of high moisture content in sorghum grain can be solved, as it is with corn, by ensiling it at about 30 percent moisture, either as whole grain or as head chops. This subject is discussed further in Chapter 24.

Plant breeders have produced a milo that is resistant to bird damage in the field, one of the important production hazards of this crop. Mississippi investigators fed such bird-resistant milo in a comparison with regular milo in silage form. The silages were cut about 18 inches above ground level, to yield a product high in grain content—about 50 percent—and they also were fed in rolled and unrolled form. The results from the bird-resistant milo silages were disappointing, with significant reductions in performance in spite of greater daily feed intake. (Other workers have reported that the elevated tannic acid content of bird-resistant milo made it unpalatable to cattle; in fact, this is offered as the explanation for the bird resistance that is observed.) Rolling the high-grain-content silages improved feed utilization in both kinds of milo.

As mentioned elsewhere, tannic acid has been investigated as a potential feed additive for the purpose of "protecting" soluble carbohydrates and proteins from microbial degradation in the rumens of full-fed finishing cattle. Apparently the naturally occurring tannic acid at the levels present in the bird-resistant milo did not serve this function in the Mississippi studies because volatile fatty acid determinations failed to reveal any differences in the rumen function in cattle fed the two kinds of silage. Present varieties of bird-resistant milo appear unsatisfactory for cattle-finishing rations, in silage form at least.

OATS

Formerly oats were not important in beef cattle feeding, but after the loss of the market for oats for feeding horses they were used rather extensively for cattle and still are in some parts of the country. For finishing purposes oats are too high in crude fiber and too low in digestible nutrients to be fed alone, but when mixed with other grains such as corn or sorghum, which are higher in energy and less bulky, they give very satisfactory results. The tough, fibrous hull and small kernel make it advisable to grind or roll oats for older cattle, as they swallow a considerable percentage of the kernels with insufficient chewing if the oats are fed whole. Grinding oats is usually inadvisable for calves, unless the grain with which the oats are fed is also ground; in this case grinding ensures a better mixture of the two feeds. Both whole and ground oats are more palatable than corn for calves. When oats are included in finishing rations at levels of 20 to 30 percent, the need for a roughage source is met by the oat hulls. In many farmer-feeder situations, especially in the northern Corn Belt where high yields of heavy oats can be grown, oats are used in grower rations. When fed to stocker calves, 2 bushels of oats are approximately equal to 1 bushel of shelled corn, and oats are approximately equal to ground ear corn, pound for pound, when fed at a level not exceeding half of the concentrate portion of the ration.

Oats may constitute as much as one-half or two-thirds of the grain ration while the cattle are becoming accustomed to a full feed. Because oats contain less energy and are bulkier than corn, their use at this time lessens the danger that some of the steers will overeat and become foundered. However, the oats should gradually be reduced as the feeding progresses, until they form only 20 to 30 percent of the grain ration, or they may be eliminated altogether, during the last third of the finishing period if another roughage source is included in the ration. Obviously the relative prices of corn and oats should be considered carefully in determining the extent to which oats should be used in the finishing of cattle for market. As already mentioned, feeding large amounts is unprofitable unless the price of oats per bushel is less than half that of corn.

Oats are somewhat higher in protein and mineral content than corn, and for this reason they are especially valuable for breeding stock. When used in mixtures with other grains, they are highly regarded by herdsmen who fit cattle for show and sales. For such animals the oat hulls are in no way objectionable, since they give bulk to the grain ration.

BARLEY

Large amounts of barley are used in cattle feeding in the northern and northwestern states and in Canada. At these latitudes corn often fails to mature, and barley, with its shorter growing season, is raised in its place. Barley is also grown to a limited extent along the western edge of the Corn Belt, where a grain crop is desired that will mature before the arrival of hot dry weather.

Barley, like oats, has a kernel surrounded by a tough, heavy hull that materially lessens its digestibility and makes it somewhat unpalatable unless it is processed. The kernel itself is rather hard and flinty in texture and does not "chew up" as easily as corn, making the grinding, crushing, or rolling of barley all the more necessary if satisfactory results are to be obtained.

In chemical composition barley falls midway between oats and corn. Because of its higher protein and mineral content it is a somewhat better-balanced feed than corn, but its lower percentage of fat and greater amount of fiber lessens its energy value. Many comparisons have been made between shelled or ground corn and ground barley as feeds for finishing cattle. Some of the results of these studies are shown in Table 129.

Barley may vary considerably in quality, which may be highly important as shown by the daily gains of three lots of steers fed shelled corn, native barley, and northern-grown barley by the Illinois station. The gains were 3.08, 2.81, and 3.35 pounds per head, respectively. Because weather conditions for the production of high-grade barley are almost opposite those required for a good grade of corn, it is not surprising that results vary widely in studies of the relative value of these two grains.

Although feeding tests show considerable variation in the relative value

Table 129

	Steer Calves[a]		Yearling Steers[b] (4 Years)		Steer Calves[c]	
Barley versus Corn for Finishing Cattle	Ground Shelled Corn	Ground Barley	Ground Shelled Corn	Ground Barley	Shelled Corn	Ground Barley
Daily gain (lb)	2.11	1.96	2.02	2.00	2.34	2.21
Feed per cwt gain (lb)						
Grain	570	613	379	384	476	476
Protein concentrate	47	51	—	—	64	68
Corn silage	—	—	—	—	414	431
Hay	176	199	841	824	119	114

[a] Kansas Cattle Circular 36A.
[b] Oregon Bulletin 528.
[c] Minnesota Bulletin 300.

of corn and barley for finishing cattle, most of them indicate that ground barley is less palatable than shelled corn, that it has a tendency to cause bloat, and that barley-fed cattle usually sell for less than corn-fed cattle when marketed. In view of these facts, the feeding of barley by itself is confined largely to those sections where barley is extensively grown and consequently is relatively cheap.

In the Corn Belt proper, barley should always be fed mixed with corn or corn and oats to lessen the frequency of bloat. In such mixtures barley has a feeding value fully equal to that of corn, but when fed alone it is only about 90 percent as valuable as corn on a pound basis. Or it may be said that 5 bushels of barley have about the same feeding value for finishing cattle as 4 bushels of shelled corn.

ROLLED BARLEY WITHOUT HAY

Barley contains about 15 percent hulls and thus is said to contain its own built-in roughage supply, as does ground ear corn. It is not surprising, in view of increasing costs of processing roughages, that commercial feeders especially have become interested in using rolled barley and fortified supplement as a complete feed for finishing cattle. The South Dakota station compared a mixture of 80 percent rolled corn and 20 percent ground alfalfa hay with dry-rolled barley in finishing yearling cattle. A supplement was designed that equalized the fiber, protein, calcium, and phosphorus fed to all cattle. Results are summarized in Table 130. In

Table 130

Rolled Barley Compared with Corn and Alfalfa Hay Rations[a]

	Trial 1 (154 Days)		Trial 2 (204 Days)	
Basal ration (%)				
Rolled shelled corn	80	—	80	—
Ground alfalfa hay	20	—	20	—
Dry-rolled barley	—	100	—	100
Number of steers	31	31	40	40
Average initial weight (lb)	798	798	700	695
Average daily gain (lb)	2.22	1.97	2.47	2.37
Average daily ration (lb)				
Basal ration	19.8	15.5	23.1	20.1
Supplement[b]	1.0	2.0	1.0	2.0
Hay[c]	—	0.2	0.3	0.3
Total	20.8	17.7	24.4	22.4
Feed per cwt gain (lb)	938	897	961	950
Carcass grade	C−	G+	C	C
Condemned livers	7	8	3	14

[a] Prepared from data in South Dakota Agricultural Experiment Station Bulletin 539.
[b] Supplements contained stilbestrol, vitamin A, and minerals plus enough protein to equalize the rations when fed at the rates shown.
[c] Hay used in getting cattle safely on full feed.

brief, the corn-alfalfa ration resulted in slightly more rapid gains but with about equal efficiency. As a result of this and other tests by this station, they recommend that barley be supplemented with about 10 percent hay for most satisfactory performance. Their recommendations are intended largely for farmer-feeders, but the all-barley ration should not be ruled out for the commercial feedlots in or near the barley-growing areas.

A special strain of barley, containing 40 percent amylose starch compared with 20 percent such starch in regular barley, has been tested in both growing and finishing rations by the Montana station. Slightly better results were obtained from *"Hi-amylose" barley* in the growing rations, feeding barley at the rate of 1 percent of body weight daily, than from regular barley. However, results were comparable when the barley content of the rations was increased to 80 percent in the finishing phase of the study. Further studies are under way.

WHEAT

When prices of feed grains work upward, wheat prices often appear competitive with the more commonly used feeds. This situation may occur as often as once in 4 or 5 years. Many tests have been conducted at various experiment stations, and, not unexpectedly, such major wheat-producing states as Oklahoma, Kansas, Texas, and Nebraska have been most active in the field. These researchers found that wheat was a more satisfactory feed when used as a partial substitute for other grains. Table 131 was prepared after reviewing the recent tests with wheat, especially those conducted at the Fort Hays, Kansas, Agricultural Experiment Station.

Results secured at the Kentucky and Kansas stations indicate that ground wheat may successfully be fed alone if it is spread evenly over a rather liberal feed of corn silage. Evidently, the silage reduces the sticky nature of the ground wheat by supplying more bulk and by compelling the cattle to eat the wheat more slowly. Yearling steers performed surprisingly well at the Illinois station on a self-fed all-concentrate ration consisting mainly of whole wheat. Apparently the whole kernels provided adequate bulk to the ration. Many intact kernels passed through the digestive tract, but conversion was not significantly improved by grinding the wheat.

Table 131

Estimated Break-even Prices for Corn, Milo (Sorghum), Barley, and Wheat

Price of Corn (bu)(56 lb)	Value of Milo (ton)	(cwt)	Value of Barley (bu)(48 lb)	Value of Wheat (bu)(60 lb)
$0.86	$ 30.00	$1.50	$0.70	$1.00
1.01	35.00	1.75	0.82	1.17
1.15	40.00	2.00	0.93	1.33
1.30	45.00	2.25	1.05	1.50
1.44	50.00	2.50	1.16	1.67
1.58	55.00	2.75	1.27	1.82
1.73	60.00	3.00	1.40	2.00
1.87	65.00	3.25	1.51	2.15
2.02	70.00	3.50	1.63	2.33
2.16	75.00	3.75	1.75	2.50
2.31	80.00	4.00	1.87	2.67
2.45	85.00	4.25	1.98	2.83
2.60	90.00	4.50	2.10	3.00
2.74	95.00	4.75	2.21	3.17
2.88	100.00	5.00	2.33	3.34
3.03	105.00	5.25	2.45	3.50
3.17	110.00	5.50	2.56	3.67
3.32	115.00	5.75	2.68	3.84
3.46	120.00	6.00	2.80	4.00

Nebraska workers, in 1974, compared several varieties of wheat as partial or complete replacements for corn in 90 percent concentrate rations fed to feedlot heifers. They found, interestingly, that varieties of wheat differ in their suitability as corn replacements. One variety, Centark, could be substituted at any level, but Trader and Scoutland varieties could be satisfactorily used to replace only one-third of the corn. Otherwise feed intake was not maintained and incidence of abscessed liver increased. These results are mentioned only to illustrate that wheat varieties are not entirely alike as feed sources, and area or state agronomists and nutritionists should be consulted for recommendations with respect to varietal differences.

RYE

Rye is usually regarded as a feed for swine and is ordinarily fed to beef cattle only when its price is considerably below that of corn. Rye grains are small and hard and, of course, should be ground. Like wheat, rye gives much better results when fed with other grains, in which case ground rye may be considered equal to ground wheat.

BEET PULP

In the manufacture of sugar from beets, great quantities of wet beet pulp result as a residue of the sugar-extraction process. Thousands of cattle and tens of thousands of sheep and lambs are finished annually on this material in the beet-growing sections of the West. Because of the large amount of moisture that it contains—approximately 90 percent—wet beet pulp should be regarded as a diluted carbonaceous roughage similar to corn silage and is more fully discussed in connection with that subject in Chapter 20.

A considerable quantity of the beet pulp produced at the sugar refineries is dried with waste steam to produce dried beet pulp, which may be bagged and sold as plain beet pulp or mixed with molasses to produce dried molasses beet pulp. Approximately 100 pounds of the plain dry pulp are obtained from each ton of sugarbeets processed. Four hundred pounds of molasses are mixed and dried with about 1,600 pounds of dried pulp to make a ton of dried molasses beet pulp.

In chemical composition dried beet pulp is a carbonaceous concentrate. Because of its bulk it produces slightly faster gains when mixed with corn or ground barley than when fed alone. Mixtures of one part dried beet pulp and two parts corn, or equal parts of dried pulp and shelled corn, were equal in all respects to a full feed of shelled corn in three tests conducted at the Nebraska station. Plain dried pulp gave as good results as dried molasses pulp when fed with ground barley in two tests made at the Colorado station.

Usually the cost of beet pulp is considerably higher than the price of corn, milo, or barley; consequently, it is not fed to beef cattle to any extent except in the vicinity of sugar refineries where it is purchased direct from the mills or the sugar companies themselves may feed the pulp in lots of their own. Frequently small amounts are used in fitting cattle for show. One quart of dried beet pulp moistened with 1 quart of diluted feeding molasses or plain water and allowed to swell overnight will yield about 2 quarts of soft, succulent feed, which may be fed alone or mixed with other concentrates.

MOLASSES

A large quantity of low-grade molasses is produced each year as a by-product of the sugar-refining industry. Frequently it is shipped in tank cars or river barges to the locality where it is to be fed or mixed with other ingredients in a complete feed.

There are four kinds of feeding molasses—cane, beet, corn, and wood. Most of the cane molasses is made in the South where it is commonly called "blackstrap." Beet molasses is produced principally in the sugarbeet areas of the West. Corn molasses is available in relatively small amounts for feeding purposes. It is a by-product of the corn milling industry in the manufacture of corn sugar. Wood molasses, also called holocellulose molasses, is a by-product of the pressed board or hardboard industry.

Formerly cattle feeders showed a strong preference for cane molasses, claiming that the beet variety was less palatable and much more laxative. More extensive use of beet molasses, however, has convinced feeders that there is little difference between the two kinds. The use of corn molasses in beef cattle feeding has been too limited to permit a definite statement about its feeding value. However, in three tests where it was used at the Illinois and Nebraska stations, no superiority over blackstrap was disclosed. Because of its high viscosity, corn molasses is difficult to handle during cold weather. Indiana studies have recently shown that wood molasses is equal to cane molasses in every respect studied except free-choice palatability. When included in mixed rations, even this problem disappeared.

Feeding experiments indicate that 1 or 2 pounds of molasses usually can be supplied to full-fed cattle without appreciably reducing the consumption of the other feeds. In theory this intake of additional nutrients should result in increased rate of gain, but in only a few of the experiments where small amounts of molasses have been fed has the increase been significant. In fact, in 10 out of 25 experiments where not more than 5 pounds of molasses were fed daily, the average gains were no larger than those of the check lots. As molasses increased the cost of the ration and on the average lessened instead of increased the selling price of the cattle, the use of small quantities for the purpose of accelerating gains and producing a quick finish appears unjustified. (See Table 132.)

Molasses has a definitely useful place in cattle feeding when grain is scarce. By feeding molasses in amounts large enough to replace one-third to two-thirds of the usual grain ration, a sufficient number of cattle may be finished to utilize the farm roughage supply to best advantage. Although cattle fed only grain during such a year probably will return a higher profit per head than those fed a limited amount of grain and considerable molasses, the larger number it is possible to feed by using molasses will probably return larger profits to the individual farmer and to the industry as a whole.

Feeders who make large profits on molasses-fed cattle during a year of feed scarcity should not be misled as to its feeding value in terms of corn or sorghum. When fed to the extent of one-third or more of the grain ration, molasses is approximately 70 percent as efficient as shelled corn or sorghum in producing a given amount of gain. Another way of expressing the value of molasses is to say that 75 pounds of molasses are equal to 1 bushel or 56 pounds of corn or about 60 pounds of sorghum. Therefore, unless 75 pounds of molasses (or 6.5 gallons, as 1 gallon of molasses weighs 11.7 pounds) can be bought for less than a bushel of corn or 60 pounds of sorghum, it is not an economical replacement for large amounts of these grains. Furthermore, when the lower selling price of the molasses-fed cattle is considered, the actual economic advantage of the molasses may disappear altogether.

A substitution of 3 to 5 percent of molasses for grains is commonly made in complete mixed cattle rations fed in many commercial feedlots. The chief purpose of such molasses additions is to reduce the dust problem associated with finely pulverized feeds. If the molasses costs no more than grain per hundredweight, it is a good addition. Even higher levels up to 10 percent may be profitably used in preconditioning or starter rations. Such rations are highly palatable to calves that are unaccustomed to eating anything other than forages. Molasses serves as a good binder if used at the rate of 5 to 10 percent in the making of certain range supplements in the form of pellets or cubes.

Table 132

Value of Small Amounts of Molasses for Finishing Cattle

Age of Cattle	Calves		Yearlings		2-Year-Olds		All Tests	
Number of Tests	12		6		7		25	
	Check	Molasses	Check	Molasses	Check	Molasses	Check	Molasses
Average molasses fed (lb)	—	1.4	—	2.6	—	3.0	—	2.2
Average grain eaten (all lots) (lb)	9.9	10.1	13.8	12.1	15.9	13.5	12.5	11.6
Average grain eaten (unrestricted lots) (lb)	9.9	10.1	15.1	14.8	17.6	15.2	12.6	12.1
Average daily gain (lb)	2.02	2.15	2.53	2.42	2.79	2.72	2.36	2.38
Water drunk daily (lb)	54.0	56.3	—	—	48.8	47.2	52.3[a]	53.6[a]
Feed consumed per cwt gain (lb)								
Grain	491	471	547	496	567	496	509	484
Protein concentrate	98	92	77	80	108	110	96	95
Corn silage	333	311	201	256	840	852	553	549
Hay	90	86	215	215	118	125	129	137
Molasses	—	55	—	124	—	115	—	88

[a] Average of 17 comparisons made of water consumption.

SUGARBEETS

Sugarbeet pulp has already been discussed and the use of beet tops is discussed in Chapter 9. In recent years, as new areas of irrigated land have been brought into production and the acreage of sugarbeets has increased, questions are being raised concerning the feeding value of the fresh beets themselves. The Arizona station conducted trials at their Yuma branch for 2 years to evaluate fresh beets when fed as an addition to a good 85 percent concentrate finishing ration. Table 133 contains the pertinent data. The beets were topped and then dug daily and chopped through a conventional field chopper just prior to being fed. The Arizona workers concluded that cattle receiving the beet-supplemented rations performed comparably with those on the control rations, with no adverse effect on carcass quality. The beets had a feed replacement value of $10 to $14 per ton. Obviously if large quantities are to be fed, specialized harvesting and processing equipment must be developed.

MISCELLANEOUS CARBONACEOUS CONCENTRATES

From time to time, especially when the price of grain is high, cattle feeders become interested in possible substitutes that are ordinarily used only in

Table 133

Effect of Feeding Chopped Sugarbeets in Addition to an 85 Percent Concentrate Ration[a]

	Without Sugarbeets	With Sugarbeets
Number of steers	32	32
Average initial weight (lb)	704	692
Average final weight (lb)	1,009	1,013
Average daily gain (lb)	2.51	2.66
Average daily mixed ration (lb)	19.4	15.5
Average fresh beets per day (lb)	—	21.2
Average beet dry matter per day (lb)	—	4.2
Total dry matter per day (lb)	19.4	19.8
Mixed ration per cwt gain (lb)	812	622
Wet beets per cwt gain (lb)	—	839
Dry beets per cwt gain (lb)	—	168
Total dry ration per cwt gain (lb)	812	790

[a] Adapted from data in Arizona Cattle Feeders' Report. (Two trials at the Yuma station.)

the locality where they are produced. Among such feeds are dried citrus pulp and dehydrated potatoes.

Dried Citrus Pulp.   The rapid increase in production of canned and frozen citrus fruit juices since the early 1940s has made available large quantities of orange and grapefruit pulp for livestock feeding. Much of this pulp is fed in fresh form in the vicinity of canning establishments, but an appreciable amount is dried and bagged for use elsewhere. Since the canning and freezing of citrus juice is an expanding industry, the present production of dried pulp may be but a fraction of the future supply.

Dried citrus pulp compares favorably with ground ear corn in total digestible nutrients but is much lower in digestible protein. Although it has a slightly bitter taste it is eaten readily by cattle and has proved to be a good concentrate for finishing cattle when properly supplemented with a protein concentrate. In two Florida trials, steers full-fed dried pulp without any corn consumed about 1 pound of pulp per 100 pounds live weight but gained more slowly than steers fed snapped corn or a mixture of snapped corn and dried citrus pulp. (See Table 134.)

Dehydrated Potatoes.   A considerable percentage of both Irish and sweet potatoes grown commercially during an average year are cull tubers that are too small or misshapen to be sold for human consumption. Many of them are fed raw to cattle and hogs in the vicinity where they are grown, but some are dried and sold as dehydrated potatoes, which are a much more valuable feed than the fresh form. Dehydrated sweet potatoes are regarded more highly than dried Irish potatoes, because they are more palatable. Neither type is of much economic importance as a cattle feed except in potato-growing areas, but they may be purchased by cattle feeders elsewhere when grain is scarce and abnormally high in price.

Dehydrated potatoes, like dried citrus pulp, compare favorably with corn in total digestible nutrients, but are noticeably lower in protein. They are less palatable than corn and when fed as the only concentrate are not eaten in sufficient amounts to produce a satisfactory rate of gain. In two tests made at the Oklahoma station, steer calves full-fed dried sweet potatoes ate considerably less feed and gained more slowly than the check lot fed ground shelled corn. However, a third lot fed equal parts of ground corn and dried sweet potatoes consumed more concentrates and made faster gains than the check lot. Even in this third lot the substitution of dried potatoes for half of the corn ration reduced the net profit the first year because the dried potatoes cost $19 a ton more than corn, and the second year because the potato-fed cattle sold for 50 cents a hundred less than those fed ground shelled corn. (See Table 134.)

Table 134

Value of Dried Citrus Pulp and Dehydrated Sweet Potatoes for Finishing Cattle

	Florida (Average 2 Trials)			Oklahoma (Average 2 Trials)		
	Ground Snapped Corn	Dried Citrus Pulp	Ground Snapped Corn, 2 lb Dried Citrus Pulp	Ground Shelled Corn	Dried Sweet Potatoes	Ground Corn, Dried Sweet Potatoes
Initial weight (lb)	608	590	607	507	506	508
Final weight (lb)	892	850	883	862	825	877
Total gain (lb)	284	260	276	355	319	369
Average daily gain (lb)	2.37	2.17	2.30	2.14	1.92	2.22
Average daily ration (lb)						
Ground snapped or shelled corn	10.6	—	2.0	11.3	—	6.2
Corn substitute	—	7.9	9.0	—	9.0	6.2
Protein concentrate	2.9	2.9	2.9	1.5	1.9	1.7
Sorgo silage	—	—	—	9.5	8.8	9.5
Hay	5.8[a]	5.7[a]	5.8[a]	1.0[b]	1.0[b]	1.0[b]
Feed per cwt gain (lb)						
Ground snapped or shelled corn	455	—	87	528	—	279
Corn substitute	—	365	322	—	478	279
Protein concentrate	124	135	127	71	100	79
Silage	—	—	—	445	478	430
Hay	245[a]	264[a]	250[a]	59[b]	53[b]	45[b]

[a] Carpet and Bermuda grass hay.
[b] Alfalfa hay.

Table 135

Effect of 4 Percent Fat Addition to Barley or Milo Rations Containing 11 Percent Crude Protein (112 Days)[a]

Type of Grain	Barley		Milo	
	Average Daily Gain (lb)	Feed/Cwt Gain (lb)	Average Daily Gain (lb)	Feed/Cwt Gain (lb)
No fat	3.05	742	2.61	812
4 percent fat	3.32	690	2.97	710
Improvement (%)	8.8	7.0	13.8	12.6

[a] Arizona Cattle Feeders' Day Report.

FATS IN FINISHING RATIONS

Fats and oils have a net energy value approximately 2.50 times that of carbohydrates. In recent years fats have accumulated in surplus quantities because of closer trimming of pork and beef carcasses in the packing plants, and the advent of synthetic detergents with a resultant decrease in demand for fats by the soap industry. Research with poultry has amply demonstrated that if fats can be bought at prices not much higher than those paid for carbohydrates, more economical broiler rations can be formulated with the use of such fats.

Aside from the energy value of fats, the settling of dust arising from finely ground components of the ration such as ground alfalfa hay, dehydrated alfalfa, or finely ground grain is an important factor. The

Table 136

Effect of Level of Animal Fat in Finishing Steer Rations[a]

Fat Levels (%)	0	5	10
Number of animals	16	16	16
Average initial weight (lb)	774	770	776
Average daily gain (lb)	2.01	2.57	2.27
Feed per pound gain (lb)	11.8	9.0	9.9
Carcass grade	G+	C−	C−
Dressing percent	61.4	61.4	60.5
Separable fat, 9th, 10th, 11th rib (%)	27.8	32.1	30.2

[a] Bowman, et al., *Journal of Animal Science*, 16:833.

Table 137

Value of Various Energy Sources Compared to Corn in Feeder Rations with Ration Restrictions[a]

Feedstuff	Value Compared to Corn (%)	Ration Restriction (maximum %)
Corn	100	100
Animal fat	160–180	5
Barley	88–90	100
Beet pulp, dried	88–95	50
Hominy feed	95–98	20
Millet	90–100	50
Milo	85–95	100
Molasses	70	5
Oats	88–94	25
Rye	80–85	20
Wheat	100–105	40
Wheat bran	65–80	10
Wheat middlings	70–85	20

[a] Nebraska Beef Cattle Report, 1972.

lubrication provided for feed grinders, mixers, augers, and pelleting dies, where used, lengthens the life of expensive feed-processing equipment.

Arizona workers recommend 4 percent fat in finishing rations fed in summertime particularly. When this level of fat is used, energy level in the ration can be increased and fiber level can be reduced. Arizona workers believe this largely accounts for the fact that cattle in their summertime tests maintain the same conversion rates as their winter-fed cattle, contrary to the usual 10 to 15 percent reduction in feed efficiency in summer-fed rations not containing fat. One of their many experiments with 4 percent fat added to an 80 percent milo or barley ration fed in summertime is summarized in Table 135.

The level of fat to include in finishing rations has also been studied extensively. Nevada workers compared 0, 5, and 10 percent levels of fat in a medium-energy ration and concluded that a 10 percent level was probably somewhat high but that 5 percent improved performance significantly, as may be seen in Table 136. It is an accepted practice for commercial feedlots to include 2 to 5 percent fat in all their rations, depending on costs of the fat.

Table 137, prepared by Nebraska Cooperative Extension personnel, aptly summarizes the comparative energy values of the wide range of carbonaceous concentrates fed to feeder cattle. Included are guidelines concerning the practical upper limits for including them in mixed rations.

CHAPTER 17

PROTEIN REQUIREMENTS AND PRINCIPAL PROTEIN CONCENTRATES FOR FEEDER CATTLE

One of the first discoveries made by early investigators of animal feeding was the need of the animal body for the complex organic compounds called proteins. The distinctive characteristic of proteins is that they contain nitrogen, an element indispensable to all animal life. Nitrogen is necessary for the building and repair of nearly all the tissues that make up the animal body. Muscle, the skin and its modifications, and the connective tissues consist almost entirely of protein materials that have been elaborated and built up from the nitrogenous compounds in the feeds consumed by the animal. A second significant feature of protein feeds is that they are rich in phosphorus, an element that is very necessary for the development of bone.

Apparently, it is not possible for ration protein in excess of immediate rumen bacterial and body requirements to be stored as protein to satisfy future needs. Instead, after the immediate needs for protein for maintenance and growth are satisfied, the nitrogen from the excess is excreted in the urine and lost, except for whatever fertilizer value it may have. The nonnitrogenous fraction of the excess ration protein can be utilized as energy in meeting maintenance requirements, or it can be converted to fat and stored, as happens when carbohydrate or other feed components are consumed in excess of maintenance requirements. Since protein feedstuffs are relatively expensive, it is evident that their use for the production of fat and energy is not economical. It is therefore important for the cattle feeder to know the protein requirements of feeder cattle of different ages and under different conditions, in order to supply the optimum amount of this nutrient without feeding it in excess of the need. It is estimated that, for each 1 percent of excess protein included in the ration, feed costs of gain are increased by 0.5 to 1.0 cent per pound. On the other hand, for each 1 percent deficiency in protein content in the ration, feed costs may be increased by 4 to 5 cents per pound. Thus it is obvious that it is more costly to err on the low side of the protein requirement than on the high side.

PROTEIN REQUIREMENTS OF FEEDER CATTLE

In Chapter 7 it is shown that protein requirements for beef cattle are generally expressed on a quantitative rather than qualitative basis. That is, protein is protein to a ruminant, provided that the sources are of equal digestibility, and it matters little if the protein is deficient in certain amino acids. It is further shown that nonprotein nitrogen sources, such as urea, can be used to satisfy at least part of the protein needs of cattle. It should be reiterated that research under way in the area of "protecting" high-quality proteins so they escape rumen microfloral degradation may show how dietary protein requirements can be reduced through improving the efficiency of protein utilization, by presenting the essential amino acids in absorbable form to the lower' digestive tract. At present, none of the products or compounds being tested have been approved by the Food and Drug Administration for use as feed additives.

In the nutrient requirement tables in Chapters 9, 11, and 15, protein requirements are expressed in four ways: (1) percentage of total protein in the ration, (2) percentage of digestible protein in the ration, (3) daily requirement of total protein, and (4) daily requirement of digestible protein per head. There are still other ways of expressing protein requirements for feeder cattle, some of which are actually more convenient to use in practice than are the tables referred to above, although they may be somewhat less exact.

METHODS OF EXPRESSING PROTEIN REQUIREMENT IN FINISHING RATIONS

Total Protein Requirement.   According to the latest recommendations of the National Research Council (see Chapter 15), finishing rations should contain between 11.1 and 12.8 percent total protein, or they should furnish between 1.0 and 2.8 pounds of total protein daily, depending on the age, sex, and weight of the animal. When computing daily total protein requirements for a drove of feeder cattle it is often more convenient, although possibly less accurate, to express them on the basis of amount required daily per 1,000 pounds live weight of animal. In this situation 2.5 to 3 pounds may be used as the requirement. Actually calves require somewhat more protein per unit of weight than do older cattle, but because they usually consume more feed per unit of weight, the percentage protein in the ration does not need to be increased as much for calves as one might think when protein is expressed as a percentage of the ration.

Fig. 79. Legume-grass mixtures such as this one, consisting largely of red clover and timothy, produce hay that can supply appreciable amounts of protein in the rations of beef cattle. (University of Illinois.)

Digestible Protein Requirement.   The digestible protein in a ration is, of course, that portion of the total protein that is actually digested and made available to the bacterial flora in the paunch or to the animal itself. Naturally it is less than the total protein figure, and the extent to which it is less depends on the digestibility of the ration and, specifically, the protein in the ration. As shown earlier, digestible protein for cattle represents approximately 60 percent of the total protein in high-roughage rations and 75 percent of the total protein in more concentrated rations such as those fed to finishing cattle.

Fiber content of a protein concentrate or of a ration is a fairly good guide for evaluating the digestibility of the protein in the ration, but this rule is not infallible. Digestible protein requirements, when expressed as percent digestible protein in the ration, range from 7.1 to 8.6 percent, with the younger cattle again having the higher requirement. When requirements are expressed as daily requirements of digestible protein, 0.66 to 1.8 pounds are required daily per head, or approximately 2 to 2.5 pounds per 1,000 pounds of live weight. Protein requirements expressed as a percentage of the ration are most often used by commercial feedlots because they feed complete mixed rations, usually formulated by a nutritionist who uses least-cost formulations to adjust for even slight variations in week-to-week prices of purchased protein concentrates. They most often use the total protein requirements, because purchased concentrates only show total protein in the guarantees and not digestible protein.

Daily Protein Concentrate or Legume Roughage Equivalent.   This expression refers to the supplemental protein needed to balance the grain or carbonaceous concentrate portion of the finishing ration. Corn, for example, when full-fed contributes only about two-thirds of the protein required by cattle being fed finishing rations. If the roughage portion of the ration is nonleguminous, such as stalk portion of corn or sorghum silages and cottonseed hulls, the protein content of the ration will not be improved by the roughage. Consequently all finishing rations consisting of carbonaceous concentrates and nonlegume roughage require supplemental sources of protein. Generally speaking, 2 pounds of a conventional high-protein concentrate (35 to 50 percent total protein) are required daily to balance finishing rations if no leguminous roughage is included. By keeping the daily intake of protein concentrate constant, the percentage protein content in the total ration will be automatically reduced as the feeding period progresses because the daily intake of grain will increase with increasing weight of cattle. Each pound of air-dry legume roughage such as alfalfa or clover hay reduces the protein concentrate requirement by about 0.25 pound. If legume silage is fed, it should be reduced to an air-dry basis by dividing the amount consumed by 3 in determining its contribution of supplemental protein. Legume green chop can be reduced to an air-dry basis by dividing the amount fed by 8. Roughages consisting of mixtures of grasses and legumes generally supply only half as much supplemental protein as straight-legume roughages. Consequently each pound of such air-dry roughage contributes the equivalent of 0.125 pound of protein concentrate. Seldom is it feasible to depend upon legumes for meeting the total supplemental protein needs because the large amount of roughage fed reduces grain intake so greatly that gains and feed efficiency are poor.

Ratio of Protein Concentrate to Grain.   Protein concentrates are often mixed with the grain portion of the ration prior to feeding—for example, when self-feeding or when feeding with automatically metered mixing equipment. In these circumstances it is desirable to know what proportions of protein concentrate and grain to mix in order to ensure correct daily consumption of protein concentrate. In other words, a given ratio is maintained between the weight of grain and the weight of the protein concentrate used. The advantage of such a method is its simplicity. The feeder is much more likely to know how much grain he is feeding than he

Table 138

Ratio Between Nitrogenous Concentrate and Corn for Full-Fed Steers				
Ration	Period	Two-Year-Olds	Yearlings	Calves
Corn-nonlegume roughage	1st third	1:6	1:5	1:4
	2d third	1:8	1:7	1:6
	Last third	1:7	1:6	1:5
Corn-legume roughage	1st third	None	None	1:10
	2d third	None	1:10	1:8
	Last third	1:10	1:8	1:7

is to know how much the steers really weigh. Its chief disadvantage is that the same ratio is likely to be maintained throughout the feeding period. The ration is therefore apt either to be low in protein at the beginning or higher than need be at the end. This problem, however, can be easily overcome by gradually widening the ratio as the feeding period progresses. By the time about two-thirds of the feeding period is complete, roughage or pasture intake will have been reduced to such an extent that more supplemental protein will be needed, and thus the ratio will narrow again. Based on such a plan, the ratios in Table 138 will be found fairly satisfactory.

If ground ear corn is fed instead of shelled corn, the ratio of protein concentrate to grain should be narrowed somewhat, but if grains that are higher in protein content, such as sorghum, wheat or barley, are used, then the ratio should be widened. When full-feeding grain to cattle on legume pasture, the ratio may be widened still further, even to 1:15 during the early pasture season, narrowing to 1:10 or 1:12 as the pasture matures. The latter two methods of expressing protein requirements are most useful to the farmer-feeder who feeds homegrown roughages, often leguminous, and who may feed in self-feeders.

METABOLIZABLE PROTEIN AND UREA FERMENTATION POTENTIAL

Iowa workers, using the terms "metabolizable protein" and "urea fermentation potential," have proposed a method for expressing protein requirements and protein values for feeds, and particularly supplements, that attempts to adjust for the well-established concept that nonprotein nitrogen (NPN) sources such as urea are not uniformly utilized under all circumstances. Through carefully controlled research, they showed that

NPN can actually have negative protein equivalent values when added to certain kinds of rations but, on the other hand, may be equal to preformed proteins when added to still other kinds of rations. It is well recognized that using the formula 6.25 × nitrogen content of a feed or ration to express its total protein content is at best only an estimate of its true protein value. Feed composition tables show crude protein or total protein values derived by this same technique and thus are subject to some of the same criticism. The same applies to comparative evaluations of two feeds or rations compounded to contain equal levels of crude protein, whether based on book values or kjeldahl-determined nitrogen.

In practice, at the farm or feedlot level, the foregoing criticisms are important only if considerable NPN is used in the rations. Since most commercial supplements, both dry and liquid, contain one-third to two-thirds of their protein equivalent in NPN form, and many feeders use such supplements, a better understanding of the proposed Iowa plan for evaluating proteins and protein requirements is in order.

As mentioned in Chapter 7 and earlier in this chapter, only the protein that reaches the absorption sites in the small intestine in the form of amino acids can be used to meet the maintenance and production needs of the ruminant. Microbial protein, synthesized from ration protein and NPN in the rumen and broken down to amino acids in the abomasum, is included in this pool of protein that is available for metabolism by the host animal after absorption as amino acids. Other ration protein which may have escaped bacterial action in the rumen to be finally digested in the abomasum and small intestine makes up the remainder of the protein pool. "Protecting" protein, as discussed elsewhere, is one way of preserving high-quality protein so that it can be more efficiently used by the ruminant after bypassing microbial action in the rumen.

The Iowa workers have estimated the amount of this mixture of absorbable bacterial and ration protein that is needed to meet the animal's need for maintenance and for certain rates of growth. They assumed that 60 percent of the absorbed protein, in the form of amino acids, is metabolized at the cell level. They speak of this absorbed amino acid combination as metabolizable protein (MP), and it constitutes only one portion of the protein requirement as given in their requirement and feed tables. (Other researchers are now establishing which of the specific amino acids are most or first limiting, and they have definitely shown that methionine heads the list.)

The second element of the Iowa method of expressing protein requirements, called the urea fermentation potential (UFP), refers to the amount of urea or NPN that can be synthesized into bacterial protein, to become finally a contributor to MP. They have estimated that only 40 percent of the NPN is converted to bacterial protein under ideal conditions, with the remainder of the nitrogen being lost as ammonia or passing down the tract beyond the site where the microflora can use the nitrogen for bacterial synthesis. The amount of fermentable energy present in the rumen, along with the actual amount of NPN, determines the UFP of the ration nonprotein nitrogen. Negative UFP values are obtained when the ration is very low in readily fermentable energy or high in preformed protein. This explains why NPN supplements perform poorly when fed as supplements to high-fiber forages.

The Iowa method appears to be more precise than methods that use the 6.25 × N formula to express protein requirements or values. Many refinements remain to be made, and actual determinations of MP and UFP values must be made on a large number of feeds and combinations of NPN and preformed protein. For further study of the method and for mathematical examples of its use, the reader is referred to Iowa Experiment Station publication AS-378 (1972) entitled "New Concepts in Protein Nutrition." Other research papers on the method will be forthcoming.

SOURCES OF PROTEIN

While all ordinary feedstuffs contain some protein, the amount furnished by the cereal grains, usually the principal component of finishing rations, is so small that other feeds containing a relatively high percentage of protein must be added if satisfactory results are to be obtained. Because of their nitrogen content, protein feedstuffs are spoken of as "nitrogenous feeds," and are divided on the basis of fiber content into nitrogenous concentrates and nitrogenous roughages.

The concentrates consist principally of the by-products that result from the milling of cereal grains and from the extraction of oil from seeds that have a high percentage of fat. The more commonly known feeds of this class are linseed (flax) meal, cottonseed meal, soybean meal, gluten (corn) meal, corn gluten feed, and wheat bran.

The nitrogenous roughages are represented by the different legume hays and silage and are, of course, entirely farm-grown except for dehydrated legumes. Clover, alfalfa, soybean, cowpea, and lespedeza are the principal legume hays used in cattle feeding. Dehydrated alfalfa, usually fed to cattle in pellet form, remains a roughage though ground and pelleted. In connection with nitrogenous roughages, green legume forage such as alfalfa, trefoil, red, white, alsike, and sweet clover pastures should be mentioned. Steers with access to such grazing obtain a large percentage of the protein needed from these pasture crops.

Ideally, the major portion of the protein needed by beef cattle should

be furnished in the form of legume roughage grown on the farm where
the cattle are fed, because of the price relationships shown in Table 139.
When the price of energy in grain is high—as is the $2.52 per bushel for
corn used in the calculations in Table 139—the energy value of the
roughages, and even of the protein concentrates, is high enough to lower
the net cost of the protein they contain to very economical levels. Thus
when grain costs are high, even though protein concentrates may seem
comparatively high-priced also, they actually are cheaper than when
energy costs are low. This also makes it possible to realize one of the
purposes for which cattle are kept, namely to furnish a means of
marketing legume crops without losing much of the nitrogen that the

Table 139

Net Cost per Pound of Crude Protein in Common Protein Sources for Feeding Beef Cattle

Protein Sources	Protein (%)	TDN (%)	Cost per Ton[a] ($)	Value of TDN per Ton[b] ($)	Net Cost of Protein (¢/lb)
Natural Proteins					
Alfalfa hay	18	56	40.00	62.70	0.0
Alfalfa, dehydrated pellets	17	53	70.00	59.35	3.1
Brewers' dried grains	26	67	82.00	75.05	1.3
Corn gluten feed	22	72	80.00	80.65	0.0
Corn gluten meal	60	79	180.00	88.50	7.6
Cottonseed meal	41	65	130.00	72.80	7.0
Distillers' dried grains	27	77	90.00	86.25	0.7
Linseed meal	35	62	135.00	69.45	9.4
Meat and bone meal	50	65	145.00	72.80	7.2
Peanut meal	49	76	160.00	85.10	7.3
Safflower meal	42	57	145.00	63.85	9.7
Soybean meal	45	76	120.00	85.10	3.9
Soybeans, whole	38	90	150.00	100.80	3.7
Wheat middlings	16	67	90.00	75.05	4.7
Wheat shorts	17	74	110.00	82.90	8.0
Nonprotein Nitrogen					
Biuret	230	0	190.00	76.15[c]	0.3
Urea	281	0	110.00	77.95[c]	0.1

[a] Assumed supplement costs prevailing in winter 1975.

[b] Total digestible nutrients (TDN) value of 5.6 cents per pound was calculated using $2.52/bu for corn and 80% for TDN content of corn.

[c] Corn was mixed with the nonprotein nitrogen sources so that the mixture had a protein equivalent content equal to that of soybean meal. Value of the added corn was subtracted to arrive at final cost.

crops secure from the air. This especially applies to the farmer-feeders in the Midwest, but of course is not applicable to commercial feedlots, which must buy all or most of their feeds.

Despite the large amounts of high-protein roughages produced in this country, they usually fall far short of being enough to balance the enormous tonnage of grains, straw, stubble, and low-protein mill feeds that are used annually in meat, egg, and milk production. Swine and poultry are unable to use legume roughages to any extent; high-protein supplements must therefore be included in their rations regardless of price, if a satisfactory level of production is to be maintained.

PROOF OF THE NEED FOR A PROTEIN CONCENTRATE

In discussing the matter of establishing the need for a protein concentrate under practical feeding situations, the type of roughage fed is of primary importance. For this reason research data are presented that take the different types of roughage into account.

When the Roughage Is Leguminous.   Many feeders consider that cattle receiving a ration of grain and legume hay or silage have little or no need for a protein concentrate. However, most experimental feeding trials show that the addition of a small amount of an oil meal to such a ration usually results in a noticeable increase in the average daily gains. Whether the use of such material proves to be financially profitable depends on the relative costs of the protein concentrate and the feeds that it displaces or saves, as well as on the amount of premium that highly finished cattle command on the market. Under normal conditions the use of a protein supplement is not justified during the first half of the feeding period while the cattle are consuming large amounts of legume roughage. During the last half, however, a small amount of protein concentrate usually is advisable, since the amount of hay or silage eaten at this time is seldom enough to furnish the amount of protein required to maintain the proper ratio between protein and carbohydrates for the most effective action of the rumen bacteria. The ration at this time also will be passing through the tract much faster because it is more concentrated or lower in roughage content; thus inadequate bacterial protein may be synthesized. The supplements fed to calves receiving grain and legume roughages are more apt to be a good investment than those fed to older feeders, mainly because the calves' protein requirements are higher and because they will be fed longer and will respond to the increased palatability of the ration to a greater extent.

When Only Part of the Roughage Is Leguminous.   Unless nearly all of the roughage portion of the ration consists of a good grade of legume roughage, the feeding of a nitrogenous concentrate is usually advisable. Sometimes an unusual demand for the common protein feedstuffs forces their price so high that their use materially increases the cost of gains, but the increase in selling price that results from the better condition and finish of the cattle is usually sufficient to increase the net profit.

Obviously, less protein concentrate is needed where clover or alfalfa makes up a considerable part of the roughage portion of the ration than where carbonaceous roughages, such as corn, sorghum silage, or prairie hay predominate. Table 140 illustrates this type of feeding situation. In all of the experiments reported, the amounts of clover and alfalfa eaten were small. This situation usually exists when corn or sorghum silage is fed, but often it does not exist when legume hay and corn or sorghum stover or oat straw are the roughages used. Should the consumption of legume hay equal 6 or 8 pounds per day, the amount of protein concentrate needed would be appreciably less than the quantities indicated in the table.

When Grain Is Full-Fed on Legume Pasture.   In the Corn Belt many feeder cattle are full-fed ground ear corn, usually in self-feeders, on legume or legume-grass rotation pastures during the peak grazing season. Table 141 shows the effect on rate and cost of gain, and daily consumption of feed, of adding a protein concentrate to a full feed of ground ear corn. Whenever protein concentrate was included, more corn and less pasture was consumed. The feeding of protein concentrate throughout the summer, rather than during the late summer only, proved profitable because of the higher rate of gain and a slightly lower cost per hundredweight of gain.

DIFFERENT AMOUNTS OF PROTEIN CONCENTRATE COMPARED

In the early years of experimental beef cattle feeding, the pratice was to feed 2 to 3 pounds of protein concentrate per head daily to 2-year-old steers after they were on full feed, even though they were fed considerable legume hay. For example, the average daily ration of a drove of 2-year-old steers fed at the Indiana station during the winter of 1910–1911 was approximately 23 pounds of shelled corn, 3.33 pounds of cottonseed meal, and 10 pounds of clover hay after the first 60 days. Such amounts of protein concentrates were soon found to be too large for the most economical gains, and they were gradually reduced. In fact, the results of many feeding experiments later indicated the need for less and less

Table 140

Need for a Protein Concentrate When Only Part of the Roughage Is Legume Hay

	Indiana (Calves) (Average 3 years) Corn Silage and Clover Hay		Minnesota (Calves) Corn Silage and Alfalfa Hay		Indiana (2-Year-Olds) (Average 3 years) Corn Silage and Soybean Hay	
Roughages Fed						
Supplement Fed	None	Cottonseed Meal	None	Linseed Meal	None	Cottonseed Meal
Average daily gain (lb)	1.78	2.17	2.02	2.32	2.35	2.46
Average daily ration (lb)						
Shelled corn	10.8	9.7	13.3	13.5	14.3	12.7
Protein concentrate	—	1.5	—	1.9	—	2.3
Corn silage	9.0	10.2	4.8	4.3	24.1	23.6
Legume hay	2.8	2.9	1.8	1.3	4.4	4.5
Feed per cwt gain (lb)						
Shelled corn	626	467	660	580	611	518
Protein concentrate	—	72	—	80	—	95
Corn silage	510	484	215	187	1,030	961
Legume hay	163	138	91	57	198	183
Total	1,299	1,161	966	904	1,839	1,757

Table 141

Performance of Yearling Steers Self-Fed Ground Ear Corn on Legume-Grass Pasture with or without Supplemental Protein[a]

	Protein Supplement[b]	No Protein
Number of steers	15	15
May 7–July 30, 1957, 84 days		
Average initial weight (lb)	691.3	696.7
Average daily gain (lb)	2.83	2.17
Average daily feed consumption (lb)	17.11	13.65
Cost per cwt gain ($)	16.85	14.67
July 31–August 27, 28 days		Protein Added[b]
Average initial weight (lb)	928.7	879.3
Average daily gain (lb)	2.33	2.95
Average daily feed consumption (lb)	13.45	23.55
Cost per cwt gain ($)	14.40	18.25
Summary—entire 112-day period (May 7–August 27)		
Average final weight (lb)	994.0	962.0
Average daily gain (lb)	2.70	2.31
Average daily feed consumption (lb)	16.20	16.13
Cost per cwt gain ($)	15.12	15.87

[a] Illinois Cattle Feeders' Day Report.

[b] Soybean meal added in the ratio of 1 : 12.5.

protein concentrate in the ration of finishing cattle. Today less than half the amount fed in the Indiana experiment mentioned previously is recommended for mature steers fed a liberal amount of legume hay.

Results of other experiments with younger cattle are shown in Table 142. It has, of course, long been known that young cattle require more protein concentrate in proportion to their weight than do older cattle. However, feeding experiments show that there is little difference in the requirements per head among cattle of different ages when all are fed appropriate amounts of roughage of the same type, that is, legume or nonlegume. Knowing this has greatly simplified the feeding of cattle because the same thumb rules for supplying protein may be applied to cattle of all ages.

The preceding discussion and the information presented in the accompanying tables indicate that a rule for feeding protein concentrates to full-fed cattle need take into account only one factor—the amount of legume roughage consumed daily. Since 4 pounds of legume hay contain approximately the same amount of digestible protein as 1 pound of high-quality protein concentrate, the following simple rule should be sufficiently

Table 142

Value of Different Amounts of Protein Concentrates for Feeder Calves (Fed in Drylot)

Critical Feeds	Oklahoma[a] Average of 4 years (Calves) Cottonseed Cake Alfalfa Hay			Ohio[b] Average of 2 years (Calves) Mixed Supplement Mixed Hay, Corn Silage			Kansas[c] Cottonseed Cake Alfalfa Hay			
Average Protein Concentrates per Day, lb	0.5	1.0	1.5	0.8	1.6	2.4	0.5	1.0	1.5	2.0
Average daily gain (lb)	1.98	2.08	2.17	1.99	2.20	2.21	1.98	2.06	2.07	2.12
Feed per cwt gain (lb)										
Concentrates	588	593	579	609	593	599	515	519	539	550
Dry roughage	51	48	46	85	77	77	101	97	97	94
Silage	346	331	316	351	319	317	462	442	438	432
Profit per head ($)	22.64	26.45	31.74	17.55	25.25	24.81	7.79	9.17	6.62	6.28

[a] Oklahoma Bulletin B-428.
[b] Ohio Bimonthly Bulletins 179 and 186.
[c] Kansas Circular 105.

accurate for practical feeding operations:

> To cattle fed no legume hay
> Feed 2 pounds of protein meal per day,
> But for each pound of legume hay you feed
> One-fourth pound less of meal they'll need.

The application of this simple rule to full-fed cattle nearly always results in their getting enough digestible protein to meet the requirements given in the requirement tables presented earlier.

PROTEIN SUPPLEMENTATION OF HIGH- OR ALL-CONCENTRATE RATIONS

Increasing numbers of commercial feedlots and even farmer-feeders are feeding rations that contain only a few pounds of roughage or, in extreme instances, no roughage at all. Two things result from this practice. Total daily feed intake is reduced and the ingesta pass through the gastrointestinal tract with much greater rapidity. The latter event undoubtedly changes the pattern of microbial activity in the rumen to the extent that less microbial protein is synthesized and more feed protein escapes microbial degradation to be digested in the abomasum and the small intestine. A third consideration is that rumen pH is lowered as ration concentration increases, and the higher acidity decreases the solubility of soybean protein and increases the incidence of rumen parakeratosis and abscessed liver.

An experiment conducted at the Illinois station is summarized in Table 143, which shows how these three phenomena may be interrelated. The data are useful in evaluating the adequacy of several levels of ration protein over a wide range of concentrate:roughage ratios. When energy was limiting, as in the two highest silage levels, there was little response to added protein. However, when most or all of the silage was replaced by corn, responses to higher protein levels became highly significant, demonstrating a positive linear relationship between ration energy level and protein requirement. All lots of cattle were fed until it was estimated that about 80 percent of the steers would grade choice. Obviously the faster-gaining steers on the high- or all-concentrate rations arrived at slaughter weight after fewer days on feed. This means they were younger and no doubt accounts for a slightly lower quality-grade in the steers receiving both more corn and more protein. This slight reduction in grade is more than offset by the accompanying reduction in nonfeed costs which are closely related to days on feed. Without doubt, most of the high good

Table 143

Interaction Between Energy Level and Protein Requirement in Feeder Calves[a]

	Percentage Protein				Corn Silage[b] (%)			
					100	67	33	0
Protein Level[c] (%)	9	11	13	15	High-Moisture Corn[b] (%)			
					0	33	67	100
Number of steers[d]					40	40	40	40
Days fed					234	209	167	167
Daily gain (lb)	9				2.58	2.66	2.90	2.96
	11				2.64	2.82	2.98	3.12
	13				2.51	2.65	3.06	3.42
	15				2.69	2.89	3.32	3.50
Daily feed (lb)[b]	9				19.0	19.4	19.6	16.3
	11				19.8	19.8	19.5	16.0
	13				19.2	19.1	18.3	16.3
	15				19.2	19.9	18.7	17.0

Feed/gain ratio (lb)				
9	7.35	7.30	6.76	7.52
11	7.52	7.04	6.54	5.13
13	7.63	7.19	5.99	4.76
15	7.14	6.89	5.62	4.85
All Protein Levels Pooled				
Average daily gain (lb)	2.60	2.75	3.06	3.26
Daily feed consumption (lb)[b]	19.3	19.6	19.0	16.4
Feed/gain ratio (lb)	7.4	7.1	6.2	5.1
All Energy Levels Pooled				
Average daily gain (lb)	2.77	2.89	2.91	3.10
Daily feed consumption (lb)[b]	18.6	18.8	18.2	18.7
Feed/gain ratio (lb)	6.6	6.4	6.2	5.9

[a] Illinois Feeders' Day Report, 1970.
[b] Level of air-dry feed included in basal ration.
[c] A 1:9 mixture of urea and soybean meal was fed to bring ration protein level to desired level. Corn levels were adjusted accordingly to keep rations isocaloric in protein level comparisons.
[d] Angus × Hereford steer calves, 469 lb average initial weight.

steers would have graded choice if evaluated according to recent USDA grade changes.

The maintenance requirement for protein is relatively constant on a day-to-day basis; thus, if total ration intake is reduced when concentrate level is increased, the protein supplied by the grain and roughage, if fed, is also reduced. This partly explains the response to added protein in the high-grain lots. When the added ration energy results in greater daily gains, obviously this further increases the daily need for protein, an important part of gains, in calves at least.

Along with the ongoing Iowa studies in the metabolizable protein method previously discussed, the foregoing experiment and others like it still in progress are stimulating further study in the area of ration protein-energy relationships. To date, such research has been much more extensive with poultry and swine than with feeder cattle. Nevertheless, the new findings offer possibilities for further refinement of the nutrient requirements for stocker and feeder cattle.

EFFECT OF PROTEIN WITHDRAWAL

The Ohio station published some exciting data in 1972–1973, suggesting the possibility that supplemental protein might be eliminated altogether from finishing rations after cattle reached 800 pounds. Many other stations as well as feedlots followed with studies of this apparent cost-saving technique, but the results generally did not support the Ohio findings. In fact, in later trials the Ohio workers themselves found conflicting results showing that removing the protein supplement resulted in reduced gains and higher feed costs. There are several plausible explanations for the apparent discrepancy in results. In the original experiment the Ohio researchers used fleshy, early-maturing calves that carried more finish at 800 pounds than most of the feeders used by the feedlots and many of the other experiment stations; thus Ohio calves apparently did little growing and mostly fattening. In addition, the shelled corn in the initial experiment analyzed somewhat higher in protein content than much of the corn fed in other instances and the diet employed a higher level of silage, or lower energy level, than that used by many other investigators. Refinement of this technique is under way.

PROTEIN CONCENTRATES AS SUBSTITUTES FOR GRAIN

Sometimes cottonseed or soybean meal is cheaper than, or at least as cheap as, grain in certain unusual circumstances. At such time cattlemen

Table 144

Value of Feeding an Excess Amount of Protein Concentrate to Replace Part of the Corn Ration of Finishing Cattle

	Illinois (2-Year-Old Steers) Soybean Meal			Illinois (Calves) Cottonseed Meal		Oklahoma[a] (Calves) Cottonseed Cake	
Average Protein Concentrate per Day	2.3	3.9	6.4	1.6	4.2	2.0	7.0
Average daily gain (lb)	2.75	2.93	2.89	2.44	2.57	2.23	2.24
Feed per cwt. gain (lb)							
Shelled corn	595	534	442	461	324	451	225
Protein concentrate	85	113	220	67	162	88	312
Total concentrate	680	647	662	528	486	539	537
Dry roughage	150	141	143	82	317	515	514
Silage	—	—	—	333	78	—	—
Net return per head ($)	14.08	18.72	19.42	29.52	28.15	20.67	13.68

[a] Average of three tests.

are likely to feed large amounts of these concentrates in an attempt to cheapen the ration by replacing part of the grain. A study of Table 144 will disclose that protein concentrates fed in excess of the amount needed for their protein content will as a rule replace their weight of grain in producing a pound of gain. The energy of the protein concentrate is used for the same purposes as any other source of energy, as already mentioned.

Occasionally one still hears about "protein poisoning." The data in Table 144 surely demonstrate that such a possibility is remote, at least when the protein concentrate being fed is entirely of plant origin. Under certain conditions there is danger of toxicity from excess consumption of protein concentrates containing high levels of urea, a nonorganic form of nitrogen. This is discussed later in this chapter in relation to another subject.

PRINCIPAL PROTEIN CONCENTRATES USED IN CATTLE FEEDING

Almost all of the common protein concentrates that are used in cattle feeding are by-products of the cereal and vegetable-oil milling industries. Bran and gluten meal are obtained from the cereal mills, whereas cottonseed meal, linseed meal, and soybean meal are by-products of the oilseed-processing industry. Figure 80 gives a breakdown of protein concentrate utilization in livestock feeding. The percentage of any one protein concentrate source that is fed straight, as contrasted to being fed as an ingredient in a mixed supplement, is difficult to determine. Estimates vary, but probably in the neighborhood of 60 percent of all protein concentrates fed to cattle are fed as a part of a commercially prepared supplement. This percentage is increasing because the protein concentrate is being used as a convenient carrier for numerous essential additives such as vitamins, antibiotics, and hormones. Figure 80 does not show how much of the concentrate supply is used by beef cattle, but 20 to 25 percent is a close estimate.

Nonprotein nitrogenous materials such as urea are serving as economical extenders of the protein concentrate supply for ruminants by their use as partial substitutes for the oil meals. A review of the characteristics, processing methods, and comparative value of the principal protein concentrates used in cattle rations should assist feeders in making the proper choice from among the various concentrates available. However, changing price conditions (due largely to variations in supply), quality of the remainder of the ration being fed, and differences in processing

Soybean Meal Equivalent

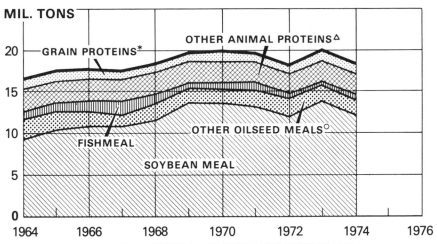

* GLUTEN FEED AND MEAL, BREWERS' AND DISTILLERS' DRIED GRAINS.
△ TANKAGE, MEAT MEAL, AND MILK PRODUCTS.
○ COTTONSEED, LINSEED, PEANUT, AND COPRA MEALS.

Fig. 80. Use of high-protein feeds by livestock in the United States, 1964–1974.

methods used, all tend to make it unwise to set forth hard and fast rules as to the relative value of protein concentrates.

SOYBEANS AND SOYBEAN MEAL

Although soybeans were almost unknown in many parts of the United States before 1920, they now constitute one of the major crops, especially in the Corn Belt and the Delta section of the old Cotton Belt, as shown in Figure 81. Approximately 1.5 billion bushels of soybeans are harvested annually in the United States. About one-third of the crop is exported as beans or by-products. About 90 percent of the domestically used supply is sold to milling companies that extract the valuable oil and sell the residue to commercial feed companies and feeders as soybean meal. The remainder of the soybean crop is processed for industrial and human food purposes.

Soybean oil is a valuable commercial product, being used in the manufacture of paints and varnishes as well as in the preparation of various edible products. In extracting the oil, the beans are finely ground and heated and the oil is either pressed out by mechanical presses or dissolved out with a chemical solvent. If the oil is pressed out, the residue

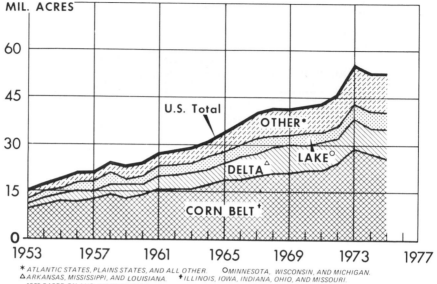

MIL. ACRES

U.S. Total

OTHER*

DELTA△

LAKE°

CORN BELT⁺

1953   1957   1961   1965   1969   1973   1977

* ATLANTIC STATES, PLAINS STATES, AND ALL OTHER.   ○MINNESOTA, WISCONSIN, AND MICHIGAN.
△ARKANSAS, MISSISSIPPI, AND LOUISIANA.   ⁺ILLINOIS, IOWA, INDIANA, OHIO, AND MISSOURI.
1975 BASED ON AUG. 1 INDICATIONS.

Fig. 81. Soybean acreage harvested in the United States, 1953–1975.

is called "old process" soybean meal; if it is dissolved out, the residue is
termed "solvent" or "new process" meal. Two types of presses are used in
making old process meal—hydraulic and expeller presses. Consequently,
old process meal is often call "hydraulic meal" or "expeller meal,"
according to the type of press used in extracting the oil.

The meal is subjected to very high temperatures in the screwlike
expeller presses and as a result has a slightly burnt or "toasted" appearance and flavor not possessed by hydraulic or solvent meal unless it is
given a special "toasting" treatment after the oil has been removed.
Toasted soybean meal is more valuable than untoasted meal for both
swine and poultry but has little if any advantage for beef cattle. In fact,
there is some evidence that the high temperatures used during toasting
depress the digestibility of the protein slightly, but this is no longer a
problem of much practical importance, since most soybeans are now
solvent-processed.

Soybean meal has the highest protein content of any natural feedstuffs
that are available in quantity for beef cattle feeding, namely from 41 to 50
percent. Because of its ready availability and usually comparatively low
price per unit of protein, it is the most common "straight" protein
concentrate purchased by midwestern cattle feeders to supply the protein
needed by feeder cattle. Soybean meal is less important in the Southwest
and West where cottonseed meal is highly competitive in price.

In chemical composition, soybean meal is quite similar to cottonseed meal. Experiments indicate that it is practically equal to cottonseed meal in feeding value. For many years some feeders complained that soybean meal was too laxative for cattle being fed a full feed of shelled corn and legume hay, but with the adoption by milling companies of the solvent process such complaints have almost disappeared. Formerly there was much discussion regarding the relative merits of hydraulic, expeller, and solvent meals and the superior value claimed for the toasted meals. Extensive experiments at several of the Corn Belt stations have disclosed no important advantage of one over the others for finishing cattle. Apparently feeders should buy soybean meal on the basis of the cost per unit of protein content regardless of the method by which it was manufactured.

Soybean meal, like linseed and cottonseed meal, is made and sold as cubes, pellets, or finely ground meal. As a rule, the meal is cheaper and more readily available. However, pea-size cubes or pellets are much better for feeding with shelled corn, and pellets or cubes 1 to 2 inches in size are recommended for use in the range area where the feed is frequently scattered on the ground. The points in regard to meal and size and form of pellets apply to commercial protein concentrates as well.

High-protein soybean meal (50 percent) has assumed considerable importance and is likely to increase in this respect. In the usual solvent-processing methods soybean hulls are virtually all removed, leaving a meal that contains slightly more than 50 percent protein. Usually some of the hulls are added back to the meal so that the protein content is reduced close to whatever the guarantee calls for, usually about 44 percent. The swine and, especially, the poultry feed industries are willing to pay a premium for the lower fiber-content 50 percent protein meals because of their higher energy content. Because of this demand from the mixed-feed industry for the 50 percent meal, and also because soybean processors have developed a fairly good market for their hulls for industrial uses, some processors have discontinued production of the once customary 42 to 44 percent concentrate. The relative feeding value of the higher protein concentrates has not been thoroughly tested with cattle but the results of one Illinois study would seem to make it a safe assumption that these concentrates should be bought on a cost per unit of protein basis, as is the case with other protein concentrates. The lower fiber content does not assume quite the same importance in cattle feeding as it does in feeding poultry and swine because cattle are normally fed higher fiber-content rations and the slightly reduced fiber content of the total ration resulting from the use of a 50 percent protein concentrate is of little importance.

TOASTED OR PROTECTED SOYBEAN MEAL

Current research at Illinois and other experiment stations includes studies of various ways of improving the nutritional value of soybean meal and other concentrates used in cattle rations. One method involves treating high-quality dietary protein with compounds that will "protect" the protein from microbial degradation in the rumen. This should favorably alter the blend or pattern of amino acids presented to the host animal at the absorption sites lower in the digestive tract. The compounds used in the Illinois study, summarized in Table 145, include formaldehyde, glutaraldehyde, glyoxal, and tannic acid. Heavy steer calves were fed an 11 percent protein, all-concentrate ration made up of high-moisture rolled corn and fortified, variously treated soybean meal supplements in a 1:12 ratio (air-dry basis). Performance and feed conversion tended to favor the treated proteins, though not significantly. Concurrent nitrogen balance studies showed improved nitrogen retention and lowered plasma urea nitrogen levels. These results show that the proteins in soybean meal may possibly be better utilized if not subjected to the normal bacterial action in the rumen. Obviously, if supplements were overprotected, not enough nitrogen would be available for normal microbial cellulose breakdown in the rumen. This would be of greater importance if rations contained considerable roughage.

The ultimate in ruminant protein supplements of the future may well consist of enough economical nonspecific nitrogen, to support active microbial action in the rumen so as to derive maximum net energy from forages in the ration, and just enough "protected" high-quality protein or specific amino acids to promote maximum performance on the part of the ruminant. Most of the compounds being investigated as protective agents have not yet been approved for use by the Food and Drug Administration. Certain processing methods—such as encapsulation, pelleting, chemical binding—hold promise as protecting agents, but results of ongoing research must be awaited for recommendations. The principles discussed in relation to this soybean meal study can be applied to other natural protein feeds and perhaps, someday, to the carbonaceous feeds as well.

COTTONSEED AND COTTONSEED MEAL

Despite the fact that the acreage of cotton being grown in the United States is only about half of what it was before the initiation of crop control acreage allotments in the 1930s, cottonseed products remain extremely important to both the finishing and the cow-calf programs. Cotton

Table 145

Effect of Treated Soybean Meal on Performance of Steer Calves Fed All-Concentrate Rations (140 Days)[a]

Item	Level and Kind of Treatment				
	Water 0.125 part	Formalde-hyde 0.125 part 8% solution	Glyoxal 0.125 part 16% solution	Glutar-aldehyde 0.125 part 16% solution	Tannic Acid 0.08 part + 0.105 part water
Number of steers	15	15	15	15	15
Average initial weight (lb)	553	564	545	549	539
Average final weight (lb)	971	1,001	997	984	977
Average daily gain (lb)	3.01	3.12	3.23	3.11	3.13
Feed/gain ratio (lb)	5.5	5.4	5.3	5.4	5.3
Number of abscessed livers	0	0	0	0	0

[a] Illinois Feeders' Day Report, 1970.

production is now concentrated in the Mississippi Delta region, Texas, and in the irrigated areas of Arizona and California. Beef cattle consume about 40 percent of the 1.5 million ton annual production of cottonseed meal and pellets, and nearly all of the 1.0 million tons of cottonseed hulls. The fact that the hulls are produced and processed in the areas where roughages are in short supply makes them especially suitable for providing the bulk or fiber needed to balance the low-fiber grain sorghums used as carbonaceous concentrates. The proximity of the cotton production to the feedlots in the Southwest and West results in tremendous savings in transportation costs.

Protein Content.   Although cotton is not a legume, its seeds contain a relatively large amount of protein, normally about 20 percent. Each ton of cottonseed yields approximately 322 pounds of oil, 931 pounds of meal or cake, 480 pounds of hulls, and 267 pounds of linters and waste material. With the removal of the hulls and extraction of the oil, the protein content of the remaining meal is approximately double that of the whole seed. Although the percentage of protein varies somewhat according to the completeness with which the hulls are removed, the better grades of

Fig. 82. The cotton fields of the South and West constitute an important source of protein concentrate for cattle in the form of cottonseed meal. (International Harvester Company, Chicago, Illinois.)

cottonseed meal contain 41 percent protein and stand at or near the top of the list of high-protein feedstuffs fed to cattle.

About half of all cottonseed is still processed by the expeller process, although a changeover to solvent extraction is taking place in the cottonseed processing industry. The protein and oil content of cottonseed meal usually vary slightly depending on the method used. Experimental work indicates that the solvent method of processing cottonseed results in meals that are slightly lower in feeding value than the expeller meals. This is undoubtedly due to the lower fat and higher fiber content of the solvent meals. Two Oklahoma tests showed a $7.62 per head larger return from the steers fed expeller-processed meal than from those fed solvent meals.

Physical Properties and Form of Cottonseed Meal.   Cottonseed meal should have a rather light yellowish brown color and a pleasant nutlike odor. A dark, dull color signifies a lower-grade product and is due to the presence of an abnormal number of hull particles. Although the finely ground "meal" is the product commonly used in this country, it is by no means the only form in which the material is sold. More and more cottonseed meal is being pelleted by forcing the finely ground meal through steel dies of varying sizes. The resulting pellets range from 0.125 to 0.75 inch in diameter and from 0.25 to 1.5 inches in length. The small pellets are satisfactory for feeding with shelled or coarsely ground corn. The larger pellets, commonly referred to as range cubes, are popular in the range area where cake or cubes are frequently fed to cattle during the winter by scattering them on the dry ground. Often the cottonseed meal is mixed with 10 to 20 percent of its weight of ground alfalfa to obtain a cake that is a valuable source of carotene. Molasses is often added to the mixture in amounts sufficient to produce a pellet that does not crumble during shipment and handling.

LINSEED MEAL

Linseed meal, linseed oil meal, or simply oil meal, as it was once called, is the product that results from the extraction of oil from flaxseed. Flax, like cotton, is not a legume but produces a seed containing a high percentage of protein and oil. When the oil is extracted, the percentage of protein in the residue is further increased. The average protein content of flaxseed is approximately 17 percent, whereas that of the meal is slightly more than twice this amount.

Flax is grown on a much smaller scale than cotton or soybeans in the United States. Altogether only about 200,000 tons of flaxseed are produced annually, over half of which are grown in Minnesota and North Dakota. All flaxseed except that needed for planting is sold to processing plants where the oil is extracted by either the expeller or the solvent method. Almost none of the whole seed is used for feeding purposes.

Many cattle feeders, especially those feeding for the higher market grades, feed some linseed meal in the belief that sleeker haircoats with resultant higher selling prices are obtained. The alleged effective agent, mucin, is a gelatinous material, covering the outer hull of the flaxseed, which is removed from the hull in extraction. While it is true that linseed meal contains mucin, experimental work done at the Iowa station did not satisfactorily prove that either haircoat or carcass quality is affected by the use of mucin.

Physical Properties of Linseed Meal. In appearance linseed meal is grayish brown in color and somewhat coarser in texture than cottonseed meal. Like cottonseed and soybean products, it is made in various degrees of fineness, ranging from finely ground meal to pellets. The pea-size pellet has been increasingly popular and has largely displaced the finely ground product in many sections of the country.

Linseed meal, because of its marked adhesive qualities under pressure, is especially easy to pellet. Consequently, more of it is processed and sold in this form than is the case with cottonseed meal. Practically all of the linseed meal exported is in the form of cake.

GLUTEN FEED

Gluten feed, a by-product of corn, is produced in considerable quantities by corn wet millers. It consists of the outer layers of the corn kernel, which are separated from the starch particles in the wet milling processing of corn. Sometimes the outer hull is separated from the underlying gluten layer, which is then sold as gluten meal, a more valuable product.

Gluten feed contains about 25 percent protein whereas gluten meal has approximately 40 percent. Both these feeds are used more extensively for dairy cattle than for beef cattle.

Gluten feed is decidedly inferior to the oil meals in finishing rations when used as the sole protein supplement, but it can be successfully substituted for up to one-half of the oil meals. Gluten meal, on the other hand, is almost equal to the oil meals but, as is true of gluten feed, the best use of gluten meal is as a partial substitute for the more commonly used oil meals. Equal parts of gluten meal and linseed meal gave results approximately equal to linseed meal alone in finishing calves at the Kansas station. (See Table 150.) Since gluten feed contains less protein than cottonseed or linseed meal, it should be purchased at a correspondingly lower price per ton.

WHEAT BRAN

Large quantities of wheat bran are produced annually by the flour mills of North America. Although comparatively little of this material is used in the finishing of beef cattle, feed prices occasionally are such as to make it the cheapest source of protein available. Because of its bulk and high percentage of fiber, bran is not an especially good feed for cattle being finished for market. Its rather pronounced laxative effect when fed in large quantities is also unfavorable to its extensive use for cattle finishing. On the other hand, these very qualities commend it as a feed for breeding animals and for young cattle intended for the breeding herd. When mixed with the common farm-grown grains, wheat bran adds bulk and lightness to the ration as well as generous quantities of phosphorus, an element greatly needed by pregnant cows and young, growing bulls and heifers.

Bran differs from most of the nitrogenous concentrates already discussed in having considerably less protein. For this reason it must be fed much more liberally than the oil meals to add the same amount of protein to the ration, but the entire protein requirement of the ration should ordinarily not be supplied with bran alone because this amount of bran would make the ration unduly bulky and laxative. Bran contains somewhat more carbohydrate than the oil meals, a fact that should be considered in determining the relative economy of these feeds. For practical purposes, 2 pounds of bran may be said to have the same feeding value as 1 pound of any of the oil meals.

Bran is most extensively used for finishing cattle in the highly specialized cattle-feeding sections of the eastern states where both protein concentrates and carbonaceous feeds must be purchased in rather large quantities. Under such conditions a feed such as bran, which contains both protein and carbohydrates, finds considerably more favor than it does in the Corn or Sorghum Belts where an adequate supply of carbohydrates is available in farm-grown grain.

BREWERS' AND DISTILLERS' GRAINS

In the pre-Prohibition era the expended grains of the liquor industry were commonly fed in the form of wet mashes and "slops" to cattle located near the distilleries. Now, however, they are usually dried, bagged, and sold as brewers' and distillers' dried grains.

Brewers' grains, made principally from barley, contain about twice the protein and fiber but only 90 percent of the total digestible nutrients of the original grain. Because of their bulky nature they are seldom fed to beef cattle but are used principally in the manufacture of mixed feeds for dairy cows. In a test at the Illinois station, brewers' dried grains proved to be a much less valuable source of protein for beef calves than soybean meal. (See Table 146.)

Distillers' dried grains are obtained from corn, rye, wheat, or grain sorghum used in the manufacture of beverage and industrial alcohol. Those from corn and wheat are much higher in protein and total digestible nutrients than those from rye or grain sorghum and consequently are much more valuable per ton as livestock feed.

In the disposal of the distillery slops after distilling off the alcohol, the solid particles of grain are screened out and dried to make distillers' dried grains. The liquid portion, containing the water-soluble nutrients and very fine particles of grain, is condensed by evaporation and dried to form distillers' dried solubles. This product has been widely publicized as an excellent source of the B vitamins, which are so important in the feeding of poultry and swine. As cattle have little need for these vitamins, distillers' dried solubles are seldom fed to beef cattle except when mixed with distillers' dried grains. This mixture is called distillers' dark grains or distillers' dried grains with solubles.

Distillers' dried grains may be regarded as a satisfactory substitute for the oil meals in the rations of finishing cattle if sufficiently large amounts are fed to furnish the proper amount of protein. Feeding tests indicate that 1 ton of distillers' dried grains will replace about 1,500 pounds of soybean or cottonseed meal and 10 bushels of shelled corn if fed at a rate of 2 to 3 pounds per head daily. Feeding larger amounts usually is uneconomical unless the dried grains are no higher in price per pound than shelled corn. (See Table 146.)

OTHER NATURAL PROTEIN SUPPLEMENTS

Mustard meal, sesame seed meal, and crambe meal have all been found to be satisfactory replacements for soybean meal, on a per unit of protein basis, when used as partial substitutes. Some palatability problems have been encountered with mustard meal. These crops are highly specialized and seldom outyield soybeans in areas where soybeans are adapted.

Table 146

Value of Brewers' and Distillers' Dried Grains for Feeder Cattle

	Brewers' Grains Calves, Illinois		2-Year-Old Steers, Iowa			Distillers' Dried Grains and Solubles Heavy Calves, Nebraska	
	Soybean Meal	Brewers' Grains (Barley)	Linseed Meal	Distillers' Grains (Corn)	Linseed Pellets	Distillers' Grains with Solubles	No Protein Concentrate
Protein content of supplement (%)	44.7	31.5	—	—	34.8	25.9	—
Weight of supplement per bushel (lb)	31	18	—	—	—	—	—
Average daily gain (lb)	2.06	1.88	2.36	2.21	1.88	1.68	1.35
Average daily ration (lb)							
Shelled corn	9.2	7.2	10.5	9.2	12.7	10.4	11.3
Protein concentrate	1.3	3.0	1.5	2.0	1.7	2.6	—
Corn silage	7.4	7.4	35.5	32.1	—	—	—
Legume hay	2.0	2.0	1.2	1.4	5.7[a]	5.1[a]	6.0[a]
Total (as is basis)	19.9	19.6	48.7	44.7	20.1	18.1	17.3
Feed/gain ratio (lb)	9.7	10.4	20.6	20.2	10.7	10.8	12.8

[a] Prairie hay fed in Nebraska experiment.

UREA AND OTHER NONPROTEIN NITROGENOUS MATERIALS

As discussed in Chapter 7, rumen bacteria can utilize nitrogen from nonfeed sources such as urea and ammonia in the synthesis of bacterial protein which, in turn, can be digested and used by the ruminant in meeting its own protein requirement. The nonprotein nitrogenous (NPN) substances mentioned are produced synthetically in larger quantities each year and are sometimes in a very favorable competitive position, as to price, with the protein concentrates. Increased demand for urea and ammonium compounds by the fertilizer industry in the mid-1970s has lessened the competitive price advantage of the NPN sources. Since high-priced fossil fuels are used in the synthesis of these NPN compounds, the chances of an improvement in their competitive position to earlier levels are unlikely in the near future. Apparently fertilizer-grade and feed-grade ureas are equally effective in meeting cattle supplemental protein needs, at least in the dry supplements. In liquid supplements, the smaller granules or prills of the feed-grade urea are favored.

Because of its hydroscopic nature, urea is prepared for use as feed in dry form by the addition of substances that prevent caking in storage. Urea contains 46.7 percent nitrogen, but dilution with these substances reduces the nitrogen in feed urea to approximately 42 to 45 percent. Since the crude protein content of a feed is determined by multiplying the percentage nitrogen content by 6.25, the crude protein content of feed-grade urea is 262 to 281 percent. Cottonseed meal is only 43 percent protein; thus approximately 6.1 pounds of 43 percent cottonseed meal are required to furnish as much nitrogen as is present in 1 pound of urea feed. Consequently, if the nitrogen of urea feed were utilized by cattle to the same extent as the nitrogen in cottonseed meal, 1 pound of urea would replace 6.1 to 6.5 pounds of cottonseed or soybean meal as far as the protein of the ration is concerned.

When administered to cattle in solution in the form of a drench, urea is highly toxic; as little as 0.25 pound of urea administered to an adult cow directly into the paunch causes death in 40 to 90 minutes. The toxicity is believed to be the result of rapid conversion of urea nitrogen to ammonium compounds which are absorbed directly into the bloodstream from the paunch. However, if urea is fed to cattle by thoroughly mixing it with the grain ration in liquid or dry form or with silage, no ill effects are noted. Coating the urea particles so as to make them more slowly available also provides some protection against toxicity. Hungry, empty cattle are more subject to urea toxicity.

Early studies of urea as a possible protein substitute indicated that it would be utilized to a higher degree if it were fed with a readily available carbohydrate, such as molasses, so that both nitrogen and a source of energy would be simultaneously available to the rumen bacteria. Feeding trials at the Oklahoma and Iowa stations indicated that the addition of molasses may not be needed by cattle that are fed large quantities of carbohydrates in the form of farm grains. It is possible that molasses favors the utilization of urea by stocker cattle fed mainly low-grade roughages such as ground corn stover and corn cobs, although here the principal value of molasses may be as an appetizer and a source of minerals. It was indicated in Chapters 9 and 11 that NPN is not as well utilized when used to supplement such lower-grade roughages as cured native range, corn stalk fields, or poor-quality hay.

Urea is finding its greatest acceptance in protein supplements fed with high-concentrate complete rations for finishing cattle. An experiment conducted by Oklahoma workers at Panhandle Agricultural and Mechanical College, Goodwell, illustrates that urea makes a satisfactory substitute for the entire amount of cottonseed meal required to bring a high-milo ration up to 12 percent crude protein. The rations were designed to be equal in nitrogen and mineral content. Results after 143 days on feed are shown in Table 147. Note that dehydrated alfalfa additions did not improve this milo ration. Gain and feed efficiency were comparable in all lots and the resulting greater return over feed costs is due to the lower cost of the urea-containing rations. Urea supplied 27.6, 23.5, and 13.1 percent of the nitrogen in three urea-supplemented lots.

The concern with toxicity and poor utilization of urea nitrogen in earlier experiments encouraged development of complex, fortified, urea-containing supplements that would be both palatable and safe when used as recommended. Undoubtedly the various components in the complex urea-containing supplements used in the Indiana trials summarized in Table 148 eliminated the palatability and toxicity problems and enhanced the utilization of the very high levels of urea used. The Iowa station has also formulated an 80 percent protein-equivalent supplement that is being used successfully. These high-urea supplements are usually being fed with rations that contain silage or ensiled high-moisture grain.

There is substantial evidence that the addition of sulfur, in the form of flowers of sulfur or Glauber's salt, to high-urea supplements improves cattle performance. A nitrogen to sulfur ratio of 15:1 is suggested. The sulfur apparently enhances urea conversion to amino acids by the rumen microflora.

Although much research is still being done on the subject of urea, the following paraphrase of statements by Indiana researchers[1] summarizes

---

[1] W. M. Beeson and T. W. Perry, "Formulating and Feeding High Urea Supplements," Purdue University Research Progress Report 249, 1966.

Table 147

Performance of Beef Steers Fed Urea-Containing 12 Percent Protein Rations (143 Days)[a]

Supplement Fed	Cotton-seed Meal	Cotton-seed Meal + Dehy-drated Alfalfa	Urea	Urea+ Dehy-drated Alfalfa	Cotton-seed Meal + Urea + Dehy-drated Alfalfa
Number of animals	23	21	23	22	18
Ration ingredients (%)					
Ground milo	87.50	84.75	96.25	91.50	87.75
Dehydrated alfalfa	—	5.00	—	5.00	5.00
Cottonseed meal	8.10	7.00	—	—	3.20
Urea	—	—	0.98	0.84	0.46
Mineral premix	4.40	3.25	2.77	2.66	3.59
Total	100.0	100.0	100.0	100.0	100.0
Average initial weight (lb)	716	714	714	722	721
Average daily gain (lb)	2.48	2.35	2.38	2.40	2.34
Average daily feed consumed (lb)	20.0	19.7	19.8	19.8	19.5
Feed per pound gain (lb)	8.06	8.39	8.31	8.23	8.35
Return per steer over feed costs	$6.08	$8.44	$13.51	$14.98	$11.60

[a] Adapted from data in Oklahoma Cattle Feeders' Day Report.

the current recommendations concerning the essentials for optimum utilization of high-urea supplements.

1. There should be a readily available source of energy such as molasses or grain.
2. Adequate levels of calcium and phosphorus must be supplied. High-urea rations are usually deficient in both calcium and phosphorus.
3. Special attention should be given to supplying the proper level of trace minerals, especially cobalt and zinc.
4. Sulfur may become a limiting factor for the synthesis of the amino acids methionine and cystine by rumen microorganisms. Experimental evidence indicates that the nitrogen:sulfur ratio should be no wider than 15:1.
5. Dehydrated alfalfa meal should be used as a source of unidentified factors for microsynthesis of protein. High-urea supplements formulated to supply 90 percent or more of the nitrogen from urea should

Table 148

Comparison of Purdue Supplement A and High-Urea Supplements (Yearling Steers, Initial Weight, 640 Pounds; 184 Days)[a]

Supplement Designation	Supplement A	Purdue 64	Purdue 80	Purdue 96
Protein equivalent (%)	32	64	80	96
Ingredient formula				
Macronutrients per 1,000 lb				
Soybean meal (lb)	650	—	—	—
Cane molasses (lb)	140	140	140	140
Dehydrated alfalfa meal (lb)	140	510	407	306
Urea (42% nitrogen) (lb)	—	210	279	347
Bonemeal (lb)	52	104	130	155
Salt (lb)	18	36	44	52
Total	1,000	1,000	1,000	1,000
Micronutrients per 1,000 lb				
Cobalt carbonate (gm)	2	4	5	6
Zinc oxide (gm)	625	1,250	1,563	1,865
Vitamin A (IU, millions)	10	20	25	30
Vitamin D (IU, millions)	1.5	3	4	5
Ration fed and performance				
Supplement per head per day (lb)	2.0	1.0	0.8	0.67
Corn silage (lb)	16.9	16.9	17.0	17.0
High-moisture ground ear corn (lb)	12.5	14.5	13.9	14.8
Average daily gain (lb)	2.39	2.32	2.29	2.32
TDN per cwt gain (lb)	524	548	542	553
Cost per pound gain (cents)	14.6	14.3	14.0	14.2

[a] Adapted from Indiana Cattle Feeders' Day Report.

contain 36 percent or more of dehydrated alfalfa meal for maximum performance on a wide variety of rations.

6. To improve palatability, 3.5 percent salt should be added to high-urea supplements.

7. Fortification with the proper levels of vitamin A and feed additives, such as growth stimulants and antibiotics, is essential to meet recommended daily allowances and to balance the ration completely.

8. Urea should be free-flowing and mixed homogeneously throughout the formula.

9. The maximum intake and desired level of the supplement to be fed should be made clear to the person who will be using the supplement. Mixing the supplement with grain, silage, or total ration is recommended so that each animal obtains the correct amount of supplement.

10. The highest-quality ingredients should be used in a high-urea supplement, avoiding filler feeds such as ground corn cobs, oat hulls, or screenings.

There is no conclusive evidence that supplements that contain urea or similar nitrogenous materials are superior to those that do not. Therefore premium prices should not be paid for such supplements. In fact, because these nitrogen sources are usually cheaper, the feeder should expect to buy them for less than would be paid for supplements containing oil meals alone as protein sources.

Biuret, a nitrogenous material produced by heating urea, is less toxic than urea and thus shows promise as a supplement, but at present it is more expensive, per unit of nitrogen. *Ammonia*, when mixed with molasses, apparently is satisfactory as a nitrogen source but cases of supersensitivity in steers have been reported by Kansas and Oklahoma workers when such supplements were used. *Diammonium phosphate* shows promise as a combined nonprotein nitrogen and phosphorus source.

ALFALFA MEAL

Dehydrated alfalfa meal made from leafy alfalfa cut in the prebloom stage of maturity contains 18 to 22 percent protein and may therefore be regarded as a protein concentrate. Because the meal is usually ground extremely fine, it cannot easily be mixed with coarse feeds and is much too fine and dusty to be fed alone unless molasses or fat is included in the formulation. However, when formed into pellets about the size of corn kernels it is an excellent protein supplement to feed with shelled or ground corn. As its protein content is only about half that of cottonseed or soybean meal, approximately twice as much must be fed to supply a given amount of protein.

The demand for dehydrated alfalfa meal for poultry and swine feeding is so great that it is usually too high-priced to make it as economical a source of protein as linseed, cottonseed, or soybean meal for beef cattle, at least on the basis of protein content alone. Feeding tests conducted at the Nebraska station suggest that dehydrated alfalfa may have a much higher feeding value than its protein content would indicate. (See Table 149.)

Table 149

Value of Dehydrated Alfalfa Pellets as a Protein Concentrate for Beef Cattle[a]

	Finishing Yearling Steers					Finishing Steer Calves		
	1.5 lb Soybean Meal	1 lb Soybean Meal, 1 lb Dehydrated Alfalfa Meal	0.5 lb Soybean Meal, 2 lb Dehydrated Alfalfa Meal	3 lb Dehydrated Alfalfa Meal	1.5 lb Dehydrated Alfalfa Meal	Linseed Pellets	Dehydrated Alfalfa Pellets	No Protein Concentrate
Average daily gain (lb)	2.32	2.52	2.62	2.71	2.47	1.88	1.98	1.35
Average daily ration (lb)								
Ground ear corn	17.6	18.2	17.8	18.4	18.7	—	—	—
Shelled corn	—	—	—	—	—	12.7	11.8	11.3
Protein concentrate	1.5	2.0	2.5	3.0	1.5	1.7	3.3	—
Corn silage	11.3	11.8	11.6	11.9	11.5	—	—	—
Prairie hay	—	—	—	—	—	5.7	5.0	6.0
Feed per cwt gain (lb)								
Corn	757	720	678	677	756	673	596	838
Protein concentrate	64	79	95	110	61	91	166	—
Corn silage	487	466	442	438	465	—	—	—
Prairie hay	—	—	—	—	—	306	254	441
Dressing percent	61.6	63.7	60.8	62.5	59.8	59.6	59.4	[b]

[a] Nebraska Cattle Progress Reports 190 and 194.
[b] This lot was not finished sufficiently for slaughter.

Alfalfa meal made from green leafy alfalfa that was dehydrated immediately after cutting contains abundant carotene. Alfalfa meal makes a real contribution as a source of highly available phosphorus. It should be noted that dehydrated alfalfa pellets had a much lower feed replacement value in the second experiment with calves than in the first experiment with yearlings. Many researchers speak of as yet unidentified valuable components in dehydrated alfalfa meal, especially when used in concentrates high in nonprotein nitrogen.

MIXED SUPPLEMENTS

A question in the minds of most smaller-scale cattle feeders concerns the economy of commercial mixed supplements. Undoubtedly the fact that such supplements are a convenient way to supply all needed additives to rations of grain and roughage has much to do with their increasing use. As a rule they are pelleted and bagged in convenient 50-pound bags, or delivered in bulk with special discounts. Use of the protein supplement as the carrier for hormone-like growth stimulants, antibiotics, vitamin A, or trace minerals, all of which may be desirable ration ingredients under certain conditions, makes this a popular and usually economical method of feeding the supplements.

In addition to these built-in conveniences in commercial mixed supplements, it is sometimes claimed that the mixture or variety of protein sources improves the quality of the supplement. It is sometimes further claimed that the 32 to 35 percentage of protein present in many mixed feeds is in reality more efficient than the much larger amount of protein in the oil meals fed alone, owing to the lack of certain essential amino acids in the ration when a single protein supplement is fed. This argument fails to recognize that the protein compounds eaten by cattle are broken down and resynthesized by the rumen microflora before digestion and assimilation. Consequently the "quality" of protein fed appears to be unimportant in beef cattle rations. This view is supported by results of experiments conducted at the Wisconsin and Kansas stations, in which none of the protein mixtures used proved significantly superior to linseed meal. (See Table 150.) Exceptions may occur in the case of "protected" proteins.

LIQUID PROTEIN CONCENTRATES FOR FEEDER CATTLE

It was mentioned earlier that liquid supplements have not proved to be competitive, as to performance, with dry supplements, either for dry beef

Table 150

Protein Mixtures versus Single Protein Feeds for Feeder Cattle

	Wisconsin[a] (Average of Three Trials)		Kansas[a] (Average of Three Trials)						
	Linseed Meal	Linseed Meal, ½; Cottonseed Meal, ½	Cottonseed Meal	Cottonseed Linseed Meal	Corn Gluten Meal	Cottonseed Meal, ½; Linseed Meal, ½	Cottonseed Meal, ½; Corn Gluten Meal, ½	Linseed Meal, ½; Corn Gluten Meal, ½	⅓ Each, Cottonseed, Linseed, Corn Gluten Meal
Average daily gain (lb)	2.41	2.41	2.18	2.29	2.20	2.33	2.23	2.36	2.35
Feed per cwt gain (lb)									
Shelled corn	343	354	425	418	407	416	420	402	404
Protein concentrate	83	73	45	43	45	42	44	42	42
Corn silage	504	505	368	350	344	338	352	354	356
Legume hay	114	115	91	87	91	85	90	85	85

[a] Mimeographed reports of cattle feeding experiments.

cows or for stockers. In fact, costs generally have been higher and performance has been poorer in such instances. Most researchers explain the negative results by the fact that the liquid supplements usually contain their supplemental nitrogen in the form of nonprotein nitrogen, and such nitrogen is not as well utilized when the remainder of the ration consists mainly of low-quality roughages. The liquid supplements approach equality with dry supplements that are high in NPN when added to corn or sorghum silage rations, either as the silage is made or later when fed. The fact that large tonnages of liquid supplements are nevertheless fed to such cattle is doubtless due to the convenience of feeding a low level of supplement with little labor. Suppliers generally fill the self-feeding tanks at convenient locations, and cattlemen have only to check their cattle periodically.

Liquid feeds for feeder cattle on medium- to high-energy rations seem to be a different matter for, in many trials, fortified liquid supplements high in NPN have been satisfactorily substituted for dry supplements, also high in NPN. It should be mentioned, however, that there also have been negative results from some comparisons, especially when the dry supplements contained natural protein. Probably this is due to the great variety of such liquid supplements. Nearly every commercial firm that makes and sells such supplements claims special, unique merits for its own formula. The ingredients used vary greatly, both as to quality and source. Molasses of course is not a fully standard product and a wide variety of NPN and phosphorus sources are used. At low temperatures, viscosity can present problems, and settling out of solids and instability of some ingredients during storage are special problems with liquid feeds, which also may explain the great variation in experimental and practical results.

Under practical feedlot conditions, liquid feeds are generally fed at a rate of about 2.0 pounds but ranging from 0.5 to 4.0 pounds per head daily, mixed in mixer wagons or trucks with the remainder of the ration. This amounts to about 100 to 200 pounds of liquid feed per ton of complete feed. Other feeders may simply spray or sprinkle the liquid supplement onto the ration in the bunk as a top-dressing. Free-choice feeding from lick tanks is not widely used for full-fed cattle. Anyone planning to mix liquid supplements should acquaint himself with patents that are in effect covering usage of this technique.

Active research is under way to improve liquid supplements by the addition of products that enhance the utilization of the NPN being employed. Distillers' solubles and fish solubles are examples of additives that have proved successful, and there undoubtedly will be others. Solubility and stability are especially bothersome problems with additives used for this purpose.

RECYCLED ANIMAL WASTES

Research done recently and still in progress has established the nutrient possibilities for reusing or recycling a wide variety of animal wastes as feed for beef cattle. Broiler- and layer-house litter, consisting of bedding materials and feces, is being used as a protein supplement substitute in some states. It varies widely in nutrient content, depending on such things as type of bird (layer or broiler), type of litter, ventilation of house, density of birds on the floor, frequency of collection, and so on. Much of the nitrogen is in the form of uric acid and thus may be readily lost into the atmosphere as ammonia. Broiler-house and layer-house litter protein-equivalent values of 28.7 and 11.9 percent, respectively, have been reported, with protein digestibility as high as 77.8 percent. Obviously, with such values prevailing, poultry-house litter has potential as a protein supplement. Its energy value can be substantial also, depending on wasted feed content. Most research on this kind of material is coming from states where the poultry industry is concentrated, such as Arkansas, Virginia, and Georgia.

Dried, activated sewage sludge obtained from municipal sewage plants contains single-cell microbial protein and nonprotein nitrogen of several kinds. These have been used experimentally with limited success. High mineral content and great variability are problems. Questions concerning a great variety of residues, and even the question of pathogens, need to be researched before use of this material can be recommended.

Both dried and fresh rumen contents have been studied and do not appear promising as sources of supplementary cattle feed.

Liquid materials from the oxidation-ditch manure pits located below slotted-floor swine- and cattle-feeding confinement sheds have been successfully used. The 4 percent solids in the cattle-shed ditch liquids contain about 40 percent protein and protein-equivalent materials. Such liquids have been successfully used in fresh form to replace one-third to one-half of the protein supplements used in cattle finishing rations. These oxidation-ditch liquids also have been added to silage, at ensiling time, to provide both needed moisture and supplemental protein. Before recycling such materials, an analysis should be obtained because they vary greatly in content. For instance, total mineral matter is often high enough to cause imbalances. The liquid material can be added directly to complete feeds being fed by means of mixer wagons or trucks.

Scrapings of fresh manure and bedding from feedlot floors or slabs and from pits below slotted floors have been used successfully as a silage additive and this recycling practice seems very promising. The ensiling process consists of mixing fresh slab and pit materials with ingredients such as corn and sorghum whole plant chopped at silage or drier stage, corn stalks, and even high-moisture corn and sorghum. The final mix should contain at least 50 and preferably 60 to 65 percent moisture if conventional upright or trench and bunker silos are used. Materials as low as 35 to 40 percent in moisture content can be stored in oxygen-limiting silos. The ensiling process renders the materials palatable and free of pathogens and parasites. Fresh manure and feeding-floor scrapings have been treated with formaldehyde before mixing with a complete feed. This treatment, by acting as a mold inhibitor, preserves the mixed feed for as long as a week and thus reduces the labor of daily mixing or ensiling.

There are great possibilities for using the animal waste recycling techniques to reduce protein supplement costs and at the same time reduce the problem of waste disposal. Further research, especially with the engineering problems involved, will make some of the successful experimental procedures feasible under practical or feedlot and farm conditions. At present, Food and Drug Administration approval has not been granted for unlimited use of the recycling techniques. The main concerns are in the areas of antibiotic, vermifuge, pesticide, pathogen, and hormone residues in the animal wastes that might be refed.

CHAPTER 18
DRY ROUGHAGES AND
THEIR USE IN
FINISHING RATIONS

Roughages differ from concentrates principally in their fiber content. Most concentrates are very low in fiber, with few of the common ones having more than 10 percent. Roughages, on the other hand, have a large amount of fiber, particularly when in the nonvegetative or mature stage. Hay averages about 28 percent and straw approximately 38 percent of fiber when cut and harvested at the appropriate stage and time.

Fiber consists largely of cellulose, hemicellulose, and lignin, all complex, relatively insoluble compounds that form the walls of plant cells. In young, immature plants the cell walls are comparatively thin and the percentage of fiber, and especially lignin, is relatively small. With the approach of maturity, however, the cell walls thicken, and there is a great increase in the fiber content. More important, with approaching maturity the nondigestible lignin portion increases fastest, lowering the available nutrient content still further.

In chemical composition, fiber is a carbohydrate—that is, it is essentially like starches and sugars. However, because of its relative insolubility it is only partly utilized as a food nutrient by domestic animals. Thus the first important step in the digestion of crude fiber, at least for the nonlignin portion, is the softening of the fibrous tissue that occurs when it absorbs large quantities of water in the rumen. The fiber is then partly converted to less complex, nutritionally available compounds—principally volatile fatty acids—through fermentation brought about by the microorganisms in the rumen. Because of the capacity and structure of the digestive organs and their symbiotic relationship with rumen microflora, cattle are more efficient utilizers of roughage than nonruminant farm animals. The methods by which roughages are broken down in the rumen are discussed in greater detail in Chapter 7.

CLASSIFICATION OF ROUGHAGES

Roughages, like concentrates, may be classed as carbonaceous or nitrogenous, depending on the percentage of protein that they contain. Carbonaceous roughages include hay and pasture from the grasses, straw from the cereal grains, and stalks and leaves of corn and the sorghums. Nitrogenous roughages include the forage from legume crops.

Roughages may also be divided into dry roughages and green or succulent materials. For dry roughages the plants are cut when almost mature and are allowed to cure—that is, to lose moisture through evaporation until the roughage contains only 15 percent or so of moisture. For green roughages, immature green crops are pastured by the animals, or the freshly cut green material is fed to cattle before it withers. In silage the freshness or succulence of green roughages is preserved by storage immediately after cutting in a wide variety of silos. ·

FUNCTION OF ROUGHAGE IN THE FINISHING RATION

Roughage plays three main roles in the cattle finishing ration: (1) it contributes to the total nutritive value of the ration, (2) it supplies bulk, and (3) it serves as a source of minerals and vitamins.

To Provide a Portion of the Required Nutrients. Although most of the gain made by cattle in the feedlot is credited to the concentrate portion of the ration, the part played by the roughage component should not be overlooked. On farms where cattle are fed mainly for the purpose of utilizing unmarketable roughages, the efficiency with which the roughage is used in the production of gains is the factor that often determines the success of the feeding venture.

In finishing rations, roughage is most important during the early part of the feeding period. It is then that the cattle have the greatest appetite for such material, and large quantities can be fed with little likelihood of causing the cattle to overeat or go off feed. Except for short-fed cattle, which, of course, should be got on a full feed of grain in the shortest time possible, roughage may well compose at least half of the ration for the first 4 to 6 weeks. Starting with a ration composed almost entirely of roughage, concentrates should be added gradually until the cattle are on a full feed of grain by the end of approximately 6 weeks, depending on the age and condition of the cattle and relative cost of roughage nutrients. After this the concentrate content of the ration should range from 80 to 100 percent for steers that are to be full-fed until they reach choice grade.

Although the ratio maintained between grain and roughage at different stages of the finishing process varies widely in practical feeding operations, the ratios given in Table 151 are fairly representative of feeding programs in the Corn Belt area in this respect. Less hay would be fed than would be used if hay were scarce and had to be purchased, as it must be by most

Table 151

Approximate Ratio of Grain to Roughage at Different Stages of the Feeding Period

Division of Feeding Period	Ratio of Grain to Roughage (Air-Dry Basis)			
	Large Amount of Roughage Available	Amount of Roughage Limited	Long Feeding Period (Over 200 days)	Short Feeding Period (Approximately 90 days)
1st third	1:2	2:1	1:1	3:1
2nd third	2:2	4:1	2:1	6:1
Last third	2:1	6:1	5:1	9:1
Average for entire period	2:2	4:1	4:1	6:1

commercial feedlots, and this of course explains to a large extent why operators of custom lots design rations that are considerably lower in roughage.

To Furnish Bulk to the Ration.   Owing to the great capacity and peculiar structure of their digestive systems, cattle are capable of consuming and utilizing a considerable amount of roughage. It is being demonstrated daily that cattle can exist on an exclusive concentrate diet, but feeding experiments as well as practical experience demonstrate that only the experienced feeder who, for economic reasons, is almost forced to reduce the roughage content of finishing rations, should attempt the feeding of all-concentrate rations.

Roughage has formed the principal, if not the only, feed of cattle under natural conditions for countless generations. Through processes of natural selection cattle have developed a digestive system, with its population of microorganisms, that functions best when the organs are moderately distended by bulky materials. Cattle naturally only partially masticate their food while eating. When swallowed the food goes into the paunch where it absorbs large quantities of water and begins the process of fermentation through the action of the rumen microflora. It is then regurgitated into the mouth in balls or boluses weighing about 0.25 pound and thoroughly chewed by the resting animal before it is again swallowed for further digestion. Roughage is essential in this process of rumination and also is responsible for the secretion of more saliva, which buffers the rumen contents and thus prevents excess acidosis. Cattle fed on an exclusive concentrate diet spend comparatively little time chewing their cuds. Hence the grain that they eat, if unprocessed on the one hand or ground too finely on the other, is imperfectly masticated and therefore not so thoroughly digested as it would be if part of the ration consisted of roughage.

In furnishing bulk to the ration, roughage tends to lighten or dilute the contents of the alimentary tract so that the particles of concentrates are completely exposed to the rumen microflora and their enzymatic action. Although it is not yet established scientifically, it appears that a certain minimal amount of fiber, or at least bulk, is required to prevent the irritation of the rumen walls. When rations containing less than 10 percent of roughage are fed, rumen parakeratosis may occur in a majority of the animals. Abscessed livers, generally condemned in packing plants by federal inspectors, also are common. Fortunately, although again scientific explanation is lacking, including at least 75 mg of the tetracycline antibiotics in the daily ration prevents the liver malady.

To Furnish Minerals and Vitamins.   A much larger concentration of minerals and vitamins occurs in the leaves and stems of plants than in the seeds. Consequently roughages are a better source of calcium, trace minerals, and vitamins A and D than are the farm grains. Cattle fed an abundance of high-quality roughage seldom show symptoms of mineral or vitamin deficiency, whereas such symptoms are occasionally seen in cattle that are fed heavy grain rations with limited amounts of low-grade roughage or no roughage at all. The roughage portion of the ration also is responsible for the slower passage rate of the ingesta. This enables the microflora to make greater use of nonprotein nitrogen (NPN) substances and to synthesize greater quantities of the B vitamins.

THE AMOUNT OF ROUGHAGE TO FEED

The role of roughage in the finishing ration has received much attention in recent years, mainly because it is the main deterrent to the design and operation of economical automatic feeding systems. The automation of the processing and feeding of concentrates is relatively simple and economical. Thus if roughages could be entirely eliminated from the finishing ration, greater economies, especially in the category of nonfeed costs, could be effected.

One can find published results of successful feeding of 100 percent concentrate rations, but these are most likely to be small experimental lots of cattle. Apparently variations in level of feedbunk management and in amount and kind of fortification in the supplement fed with the high-grain rations account for the failures seen in some larger feedlot operations.

Barley and ear corn, two highly important finishing feeds, apparently have sufficient built-in roughage content in the form of hulls and cob so that it is comparatively easy to design acceptable rations based on either of these concentrates. Test results from the North Dakota and South Dakota stations show that 1 to 4 pounds of hay added to barley rations increased rate of gain about 0.20 pound daily but did not reduce costs of gain.

Shelled corn and milo are another matter, as they are both quite low in fiber. Observations in the commercial feedlots of the Southwest and West suggest that the lowest practical level of roughage in milo rations is in the neighborhood of 6 to 7 percent. When tallow is included in the high-concentrate ration at the level of 3 to 5 percent, slightly higher levels of roughage—up to 15 percent—can be expected to produce the same results as lower roughage content rations, with the bonus of fewer physiological disturbances.

When less than 5 percent roughage is used in finishing rations, it apparently matters little what the roughage is, as the overall fiber and energy content is not greatly changed regardless of which roughage is fed. It is desirable to leave this low level of roughage in its bulkiest form consistent with its ability to fit into an automated system. For example, shredding of hay is preferred to extremely fine grinding in this situation, provided the feed mixer and unloading wagon can properly handle the shredded hay.

Grain processing methods that leave the grain in flake form are preferred in rations with a very low roughage content. Apparently the bulky nature of flaked grains produces the same physiological responses seen when a small amount of roughage is included in a ration that is initially low in roughage.

Another point with respect to all- or high-concentrate rations is the comparative dressing percentage of the finished cattle. In trials comparing levels of roughage, when gains and feed efficiency are calculated on the basis of carcass gain rather than liveweight gain, the higher-concentrate rations produce more favorable results. Cattle on such rations also shrink less in shipment and thus should probably be sold on a grade and yield basis if at all possible.

ROUGHAGES COMMONLY USED IN FINISHING RATIONS

Roughage, especially in the form of hay, is grown and harvested throughout the country. The roughage crops vary in species, quality, and yield, depending on rainfall, soil fertility, and choice of the farmer in some instances (see Fig. 83). The great variety of roughages used in finishing rations may be divided into two broad classifications—leguminous and nonleguminous. The leguminous roughages are usually considered to be a good source of protein, minerals, and vitamins but only average to low in energy or total digestible nutrients. On the other hand, the nonleguminous roughages, such as corn or sorghum silages, used in finishing rations are generally good to excellent sources of energy, especially from a cost standpoint. But the nonleguminous roughages are poor sources of proteins, minerals, and vitamins, and rations using this group of roughages exclusively must be supplemented with rather completely fortified supplements.

LEGUME ROUGHAGES

According to Morrison,[1] legume roughages have the following advantages over other roughages as a livestock feed.

1. They lead in yield of palatable hay per acre.
2. They are the richest of all common forages in protein content.
3. Their protein helps correct the deficiencies in the proteins of cereal crops.
4. Legume forages are the highest in calcium content among all farm-grown feeds.
5. Legume forage excels in vitamin A (carotene) value.
6. Field cured legume hay is rich in vitamin D.
7. Legume forage is rich in other vitamins besides vitamins A and D.
8. Legumes increase the yield and protein content of grasses growing in a legume-grass mixture.
9. Legumes are highly important in maintaining soil fertility.

Cutting at the proper stage and curing under good conditions are essential for the above qualities to be present in hay or silage made from leguminous forages, as discussed later in this chapter.

Alfalfa Hay.   Alfalfa is the most important tame forage crop grown and is the standard by which all others are judged. The development of wilt-resistant and winter-hardy varieties has caused almost universal adoption of this crop as the preferred forage, with its closest competitor among the legume forages being red clover. The alfalfa weevil, which is difficult to

[1] F. B. Morrison, *Feeds and Feeding*, 22nd edition.

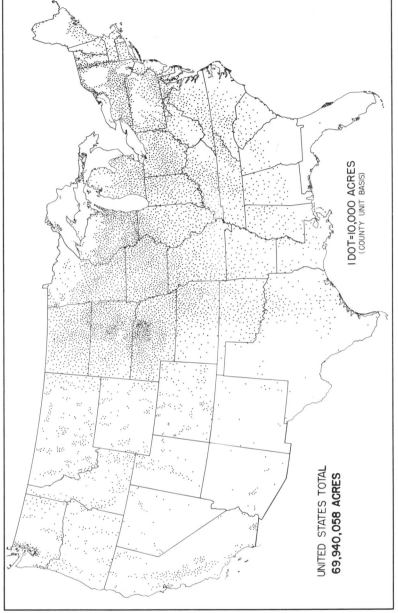

I DOT=10,000 ACRES
(COUNTY UNIT BASIS)

UNITED STATES TOTAL
69,940,058 ACRES

Fig. 83. Hay is cut throughout the United States, with the possible exception of Florida. The hay acreage in the United States has declined to some extent during the last two decades. (U.S. Department of Commerce.)

Fig. 84. One-man hay-baling systems, making use of automatic bale throwers to load bales on an attached trailing wagon, are ensuring that the conventional small, square bale will continue to be used by many cow-calf producers and farmer-feeders. (John Deere.)

control and is rapidly advancing eastward from the alfalfa-growing regions in the Southwest and western Corn Belt, is of great concern to agronomists and cattle feeders alike, as it threatens the loss of this most important roughage source for cattle feeders. Entomologists are searching for methods of controlling this pest but, in the mid-1970s, results are not highly encouraging.

Often the market price of alfalfa hay is 20 to 30 percent above that of clover, indicating its superiority in the minds of most feeders. This superiority is due to its higher protein content and freedom from dust and mold. Western-grown alfalfa is seldom damaged by rain and usually retains its bright green color and does not spoil in the stack or bale. Hay made farther east, on the other hand, is often subjected to rain, which leaves it discolored, dusty, and sometimes moldy. Obviously such hay is less palatable and less nutritious than hay put up under ideal weather conditions.

Occasionally alfalfa hay is criticized because of its somewhat laxative nature when fed in large quantities and because it sometimes causes acute bloating. Both these effects are more pronounced in calves than in older cattle. The problem is lessened when alfalfa hay is fed with ground ear corn instead of shelled corn or in combination with a nonlegume roughage source.

The superiority of alfalfa over clover hay is shown in the feeding trials

Fig. 85. An excellent field of alfalfa being cut with a self-propelled swather-crimper, the machine most likely to make a major contribution in ensuring quality in hay or haylage. (John Deere.)

summarized in Table 152. When protein is definitely limiting and no protein concentrate is fed, as in the Illinois and South Dakota trials, alfalfa hay is the better legume hay. When much of the necessary supplemental protein is provided by protein concentrates, as in the Wisconsin trials summarized in Table 153, then alfalfa and clover hay are comparable.

Clover Hay.   Ordinarily, the term "clover hay" is used to refer to red clover or, more likely, a mixed hay in which red clover, a legume, is mixed with timothy or some other grass. Red clover and timothy make up the backbone of most rotation mixtures in cropping systems used in the Corn Belt that call for leaving the legume-grass seeding down for only one or two years. When longer rotations are used, alfalfa and brome are apt to be substituted, but unless the alfalfa weevil problem is solved soon, perennial clovers may possibly replace alfalfa in these mixtures. Clover hay or mixed clover-grass hay is preferred by some feeders to alfalfa hay because it is less laxative. Since a protein concentrate is often fed, even when a legume roughage is used, because of its palatability factor, the lower protein content of mixed hay is not serious. Clover or mixed hay is more palatable than alfalfa hay. When used along with silage it tends to cause a more normal state in the digestive tract. Obviously the relative proportion of clover and grasses affects the value of this type of hay with respect to

Table 152

	Illinois Experiment Station (2-Year-Old Steers— 126 days)		South Dakota Experiment Station (Yearling Steers— 91 Days)	
	Ear Corn Corn Silage Alfalfa Hay	Ear Corn Corn Silage Clover Hay	Corn Silage Alfalfa Hay	Corn Silage Clover Hay
Average daily gain (lb)	2.38	2.05	2.49	2.29
Average daily ration (lb)				
Ear corn	16.1	16.0	—	—
Corn silage	25.3	26.7	58.3	58.1
Legume hay	4.3	4.0	4.0	3.5
Feed per cwt gain (lb)				
Ear corn	675	783	—	—
Corn silage	1,065	1,302	23	25
Legume hay	181	195	1.6	1.5

Comparison of Clover and Alfalfa Hay for Feeder Cattle

protein, mineral, and vitamin content. Red clover is a biennial crop—that is, it produces a seed crop in the summer of the second year and a large proportion of the stand dies during the ensuing fall and winter, leaving a clover-grass mixture composed mostly of grass.

Miscellaneous Legume Hays. Red clover is only one of several different kinds of clover, all of which are used for hay to some extent. Alsike clover is usually considered more a pasture than a hay crop because of its short growth and consequently low yield. However, when cut and harvested it produces an exceedingly fine-textured hay that is greatly relished by cattle. In chemical composition it has slightly more protein than red clover but is somewhat lower in energy.

Mammoth clover resembles red clover in general appearance but is taller, coarser, and somewhat later-maturing. Although it yields heavily, the hay is often rather coarse and unpalatable. It contains the lowest percentage of protein found in any of the more common legume crops used for hay.

Lespedeza was grown extensively as a hay and pasture crop in the southern and southeastern states in the 1940s and 1950s, but less is being grown each year because of low yields. The improved annual varieties, Kobe and Korean, grow to a foot or more in height and yield about 2 tons of hay per acre. Lespedeza hay cut at the right stage of maturity and properly cured is an excellent roughage. However, if it is allowed to

Table 153

Comparison of Clover and Alfalfa Hay When a Limited Amount of Protein Supplement Is Fed (Steer Calves— Average of 4 Trials)[a]

Average Time Fed, 161 Days	Lot 1 Clover Hay	Lot 2 Alfalfa Hay
Average daily gain (lb)	2.19	2.22
Average daily ration (lb)		
Corn	8.5	8.9
Cottonseed meal	1.5	0.9
Legume hay	5.3	5.3
Corn silage	28.0	28.0
Feed per cwt gain (lb)		
Corn	394	407
Cottonseed meal	70	42
Legume hay	250	242
Corn silage	1,292	1,264

[a] Wisconsin Agricultural Experiment Station, mimeographed reports of calf feeding trials.

become overly ripe, the stems become extremely tough and wiry and are neither palatable nor easily digested. Such overripe hay is sometimes deceptive as to its quality because it often has a good green color and contains an abundance of leaves. *Serecia lespedeza*, a perennial variety, is much lower in value than the annual varieties and can hardly be recommended for use in finishing rations. The lespedezas compete with alfalfa and clovers on the more acid and less fertile soils, and they make their maximum growth during the warm summer months when the cool-season grasses are dormant. For these reasons, lespedeza is more likely to find a place as a pasture forage than as a hay crop.

Sweet clover is not a satisfactory hay crop, because it is generally too coarse and woody to make a high-grade, palatable hay. Its coarse, rank growth makes it exceedingly difficult to handle and adds to the labor of baling it or putting it into the stack or mow. With favorable soil and weather conditions a crop of fair quality hay may be secured in the fall from sweet clover sown in oats or wheat the previous spring. Such hay is quite similar to alfalfa in color and texture and cattle eat it readily. Close clipping of the first-year growth in late September is unlikely to cause serious damage to the stand.

Birdsfoot trefoil and *crown vetch*, both nonbloating legumes that have been highly promoted as species that are excellent for use in pasture mixtures with grasses, are unsuitable for hay crops because of their growth habits and low yields.

TAME OR INTRODUCED NONLEGUME HAYS

Hays made from grasses may differ to some extent in palatability and feeding value in comparison with legume hays but, generally speaking, tame grass hays are similar to each other in chemical composition regardless of species. Factors such as stage of cutting, curing methods, and yield affect the value of grass hays more than does the species. Thus it is advisable to grow the adapted and recommended grasses for a given locality, cutting them for hay when they are in their most nutritious state.

It has been well demonstrated that mixtures of legumes and grasses produce a greater tonnage of higher quality forage than do grasses seeded alone when grown in areas where such mixed seedings are adapted. It seldom pays to grow straight-grass meadows except, of course, in areas such as those where likelihood of erosion precludes growing anything but perennial grass meadows, or where rainfall is apt to limit yields of tame meadows to an unprofitable level.

Grass hays, if cut at any but the very early stage, resemble the better-quality cereal straw in composition, being low in protein, phosphorus, and carotene. Early-cut grass hay from well fertilized meadows, on the other hand, may equal or approach the composition of legume or mixed hay.

Timothy and Orchard Grass.   Many years ago timothy was the standard roughage for finishing cattle, but today most up-to-date feeders apologize for its use. From the standpoint of production, timothy has several advantages. It thrives reasonably well on a wide variety of soils; it is a perennial and thus does not require frequent reseeding; the hay is very easily cured and is usually quite free of dust and mold. Practically its only positive feature apart from its brightness and quality is its nonlaxative properties. Steers or calves receiving this roughage are seldom troubled with scours. In total digestible nutrients, timothy is somewhat lower than the common legume hays, and its protein content is less than one-third and one-half that of alfalfa and clover, respectively.

Orchard grass is a recommended grass for pasture and hay in many parts of the Corn Belt and the cooler sections of the southeastern states. It yields well and, when used as a pasture crop, provides both early and late grazing, thereby extending the grazing season. Hay is often made from surplus spring growth of this excellent grass. The Tennessee station conducted a trial in which they compared orchard grass hay containing 10.1 percent crude protein with alfalfa hay containing 14.3 percent.

Table 154

Orchard Grass Hay versus Alfalfa Hay in Finishing Rations for Yearling Steers (86 Days)[a]

	Corn, Cottonseed Meal, and Orchard Grass Hay	Corn, Cottonseed Meal, and Alfalfa Hay
Number of steers	14	14
Average initial weight (lb)	787	800
Average final weight (lb)	1,004	1,051
Total gain (lb)	217	251
Average daily gain (lb)	2.52	2.92
Average daily ration (lb)		
Ground ear corn	18.7	20.3
Cottonseed meal	2.0	1.5
Hay	4.7	4.7
Feed per cwt gain (lb)		
Ground ear corn	742	695
Cottonseed meal	79	51
Hay	186	161
Total	1,007	907
Slaughter grade	G+	G+

[a] Tennessee Agricultural Experiment Station H-72-7-3.

Supplemental protein was reduced in the alfalfa lot to equalize crude protein intake. Performance was good on orchard grass, but it did not measure up to alfalfa hay in either rate or cost of gain (see Table 154).

Brome Grass. This grass grows well in the northern half of the Corn Belt and is high-yielding on average to fertile soil. It makes heavy first-cutting yields but regrowth in the warm months is poor. Fall regrowth is good and this cool-season grass combines very well with alfalfa, a warm-season species. Seeding failures are more common than with some other species because of the chaffy nature of its seed.

Tall Fescue. This high-yielding grass is well adapted to the poorer to average soils of the southern half of the Corn Belt, and to the higher rainfall belt that extends from eastern Oklahoma to the Atlantic Coast, but it is not considered to be an excellent hay crop. If harvested for hay, the heavy spring growth should be cut before seed-head formation and even then it is not highly acceptable. Alfalfa and clovers should be maintained in tall fescue seedings, but this is difficult because fescue is highly competitive and crowds out other species. The hay made from this grass is quite suitable for dry beef cows if it is of good quality, but it is much less satisfactory in finishing programs. Grass breeders are making progress in developing strains of tall fescue that rate higher in both acceptability and digestibility.

Reed Canarygrass.   This grass is adapted to both wet and dry soils. It grows rank and is coarse and fibrous. Only the early growth should be harvested for hay, and that before heading. Extra care must be exercised to obtain successful seedings, for seed quality is generally poor.

Prairie Hay.   Prairie hay made from native grass meadows in the Plains states resembles timothy in chemical composition and general value as a cattle feed. It is usually much cheaper than timothy, however, especially on the farms and ranches where it is produced. In the western states great quantities of prairie hay are made, and in such areas prairie hay may well be combined with alfalfa, especially in feeding yearling steers. It should be remembered that prairie hay varies greatly in the species of grasses and other plants that it contains. Occasionally it has a large percentage of needle grass or other material that irritates the mucous lining of the cattle's mouths.

   Prairie hay is nonlaxative and does not tend to produce bloat. This fact together with its freedom from mold makes it a popular roughage to feed cattle at stockyards and on the show circuit. Table 155 shows that prairie

Table 155

Value of Prairie Hay for Feeder Cattle					
	Nebraska Bulletin 93 84 Days			Nebraska Bulletin 100 Average of 3 Experiments	
	Ear Corn, Prairie Hay	Ear Corn, Prairie Hay 50%, Alfalfa 50%	Ear Corn, Alfalfa Hay	Shelled Corn, Prairie Hay	Shelled Corn 90%, Linseed Meal 10%, Prairie Hay
Daily gain (lb)	1.20	2.01	2.06	1.51	2.18
Feed per cwt gain (lb)					
Corn	787	470	460	1,171	808
Protein concentrate	—	—	—	—	90
Hay	1,516	1,047	1,075	521	387

hay, when properly supplemented, has a place in the finishing ration. In the future, prairie hay will be used even more for cows and stockers than presently, as less and less roughage is included in finishing rations.

Hays made from the cereal crops and from Sudan or Johnson grass have a feeding value for finishing cattle comparable to timothy or prairie hay and usually are also better utilized by breeding cattle or stockers.

FACTORS AFFECTING HAY QUALITY

Even though hay usually makes up considerably less than one-fourth of the total feeds fed to finishing cattle, the quality of hay, within a kind of

Table 156

Value of Quality Hay When Fed with a Full Feed of Ground Ear Corn to Feeder Cattle[a]

	Trial 1 (259 Days)		Trial 2 (224 Days)	
	Poor Hay	Good Hay	Poor Hay	Good Hay
Number of steers in lot at start	12	10	14	14
Number of steers in lot at close	12	10	14	14
Average weight at start of test (lb)	636	630	472	475
Average weight at close of test (lb)	1,049	1,132	889	965
Average daily gain (lb)	1.60	1.94	1.86	2.19
Average daily ration				
Ground ear corn (lb)	13.6	15.0	11.1	12.4
Supplement (lb)	1.5	1.5	1.5	1.5
Hay (lb)	1.9	3.3	2.1	2.8
Minerals (oz)	2.0	1.9	1.3	1.3
Salt (oz)	0.5	0.6	0.6	0.7
Feed per cwt of gain (lb)				
Ground ear corn	852.0	774.0	598.0	569.0
Supplement	94.0	77.0	80.0	68.0
Hay	117.0	169.0	112.0	127.0
Minerals	8.0	6.0	4.0	4.0
Salt	2.0	2.0	2.0	2.0
Ground ear corn plus supplement				
(lb)	946.0	851.0	678.0	637.0
Total (lb)	1,073.0	1,028.0	796.0	770.0
Dressing percent	59.97	63.04		

[a] Ohio Research Bulletin 732.

Table 157

			Full Bloom and $^1/_{10}$	
Stage of Maturity	Bud	$^1/_{10}$ Bloom	Bloom[b]	Full Bloom
Composition data				
Percent protein	21.3	20.5	19.6	18.4
Digestible dry matter (%)[c]	63.2	61.1	59.5	57.8
Percent leaves	54.9	52.9	52.8	51.1
Yield data				
Average number of cuttings per year	5.0	4.3	4.0	3.7
Dry matter yield, tons per acre	4.18	4.39	4.32	4.35
Protein yield, tons per acre	0.89	0.90	0.85	0.80
Digestible dry matter, tons per acre	2.64	2.68	2.57	2.51

Effect of Maturity at Cutting on Composition and Yield of Alfalfa (3 Years)[a]

[a] Nebraska Beef Cattle Progress Report.

[b] First cutting at full bloom, remainder at $^1/_{10}$ bloom.

[c] Determined by *in vitro* technique.

hay, may materially affect performance. Data such as those shown in Table 156 effectively demonstrate this point.

There is considerable disagreement as to when hay should be cut to obtain the greatest feeding value. It is generally agreed that early cutting favors higher protein content, finer texture, less fiber, and higher digestibility. Many believe, on the other hand, that late cutting results in greater tonnage, more total digestible nutrients, and usually more favorable weather for field curing of the first cutting. The last point is

Table 158

Effect of Method of Curing and Storage on Quality Factors in Hay[a]

Method of Curing and Storage	Hours in Swath	Leaf Loss (%)	Dry Matter Loss (%)	Storage Dry Matter Loss (%)	Dry Matter Left to Feed as Standing Crop (%)
Field cure—rain damage	108	60	40	5	60
Field cure—no rain	54	35	20	5	79
Barn dried—no heat	29	25	20	10	81
Barn dried—heat added	29	25	15	3	85
Wilted silage	8	15	20	18	83

[a] USDA Bureau Animal Industry-Inf. 142.

Table 159

Digestible Protein Content of Alfalfa and Grasses in Relation to Maturity[a]

	Stage of Growth	Digestible Protein[b] (%)
Alfalfa	Immature	17.0
	After bloom	5.4
Kentucky bluegrass	Before heading	15.0
	After bloom	2.7
Orchard grass	Before heading	13.0
	After heading	4.9
Timothy	Pasture stage	13.9
	In seed	2.2
Mixed grasses	Immature	10.3
	At haying stage	4.7

[a] California Extension Service Circular 125.

[b] On 15 percent moisture basis.

undoubtedly true, but data such as those shown in Table 157 disprove the others.

Method of curing and storage also may materially affect hay quality. Data on this subject are plentiful and those in Table 158 are typical. Naturally the amount of rainfall and cloudy weather affects curing time; therefore this factor is a greater consideration in some areas than others.

Table 159 shows the effect of stage of maturity upon digestible protein content, and Table 160 shows the effect of method of harvesting and storing upon carotene content.

The method of cutting hay is of concern today, as several kinds of

Table 160

Carotene Content of Roughages When Cut, Stored, and Fed[a]

Roughage	Carotene (μg/gm dry matter)		
	Cut	Stored	Fed
Early grass	354	341	140
Barn-dried hay	361	190	20
Field-cured hay	211	66	9
Late grass	216	238	84

[a] Cornell Feed Service No. 35.

machines are available for cutting and field-conditioning hay. The Kansas station conducted a trial with weaner heifer calves fed limited rolled sorghum grain in addition to alfalfa hay, processed by various methods, according to appetite. The various treatments applied to the hay were as follows.

1. Control—raked and baled.
2. Crushed—crushed with one smooth steel roll and one spiral-grooved rubber roll, raked, and baled.
3. Rotary cut—cut, lacerated, and windrowed in one operation by means of a 12-foot trail-behind twin-rotor mower, and baled.
4. Swathed-crimped—swathed and crimped by means of a 12-foot self-propelled windrower with crusher-crimper attachment, and baled.
5. Wafered—cut with a flail-type cutter, field-dried to about 15 percent moisture in windrows, and wafered with a commercial wafering machine.

Results of the Kansas experiment are shown in Table 161. Rotary-cut hay proved to be least desirable, undoubtedly because of leaf and particle shedding at baling time. Simultaneous swathing and crimping of the hay produced a maximum response, equal to the more expensive wafering process. The excellent performance by the heifers on the swathed-crimped hay, aside from the reduced curing time and labor saved, undoubtedly accounts for the rapid rise in popularity of the self-propelled swather-crimper in the commercial hay-growing areas where hay is often sold on an analysis basis.

OTHER NONLEGUMINOUS DRY ROUGHAGES

Corn stover, once a fairly important roughage used in finishing the 2- and 3-year-old steers commonly fed in the past, is no longer used except as bedding, or possibly as silage for dry cows. Other stovers are used to a small extent in areas where grown.

Sorghum Stover. The stover of the grain sorghums is essentially like corn stover in chemical composition and feeding value. However, because the stalks are somewhat finer-textured, a greater percentage of the stalk is ordinarily eaten. When fed alone or with other dry carbonaceous roughage, sorghum stover is not very satisfactory for finishing cattle. When combined with legume hay or silage, much better results are secured. Even when legume hay constitutes half or more of the total roughage ration, some nitrogenous concentrate should usually be fed as all stovers

Table 161

Performance of Weaned Heifer Calves Fed Various Field-Conditioned Alfalfa Hays (93 Days)[a]

	Control	Crushed	Rotary Cut	Swathed-Crimped	Wafered
Number of heifers per lot	10	10	10	10	10
Initial weight (lb)	438	441	442	443	442
Average daily gain (lb)	1.10	1.18	1.05	1.30	1.28
Average daily ration (lb)					
Alfalfa hay	11.8	13.1	11.3	11.9	13.0
Rolled sorghum grain	3.5	3.5	3.5	3.5	3.5
Feed per cwt gain (lb)					
Alfalfa hay	1,072.7	1,110.2	1,076.2	915.4	1,015.6
Rolled sorghum grain	318.2	296.6	333.3	269.2	273.4
Total feed required per cwt gain (lb)	1,390.9	1,406.8	1,409.5	1,184.6	1,289.0
Feed cost per cwt gain[b]	$32.00	$31.48	$32.81	$27.19	$29.00

[a] Adapted from Kansas Agricultural Experiment Station Bulletin 460.
[b] Feed costs: alfalfa, $30 per ton; rolled sorghum grain, $5.00 per cwt.

are low in protein. Sorghum stover, like corn stover and the grass hays, is more valuable in rations for cattle other than those on finishing rations.

Cereal Straws.   Of the straws of the common cereal grains, oat straw is the most valuable as a cattle feed. Oat straw is, of course, a carbonaceous roughage and should be fed to finishing cattle only when other constituents of the ration furnish the necessary protein. Feeding a small quantity of oat straw tends to reduce the bloat and scouring occasionally noticed in cattle receiving a heavy feed of alfalfa hay or barley. Also, added to a heavy silage ration it satisfies, at a very low cost, the desire of the cattle for a dry roughage material.

Barley straw is somewhat inferior to oat straw for feeding purposes, principally because most varieties of barley are bearded, and the beards sometimes cause the cattle's mouths to become sore.

Wheat straw has the lowest feeding value of any of the common cereal straws and should be regarded as an emergency feed for finishing cattle, to be fed only when ordinary roughages are scarce and high-priced. Satisfactory gains were obtained in an experiment at the Nebraska station using yearling steers that were fed a full feed of shelled corn, cottonseed cake, and equal parts of wheat straw and alfalfa, hay. Oklahoma trials indicated that wheat straw was equal to cottonseed hulls in a calf-finishing ration containing 15 percent of experimental roughage as well as 15 percent alfalfa. The roughages were all ground through a ⅜-inch screen, and thus differences in palatability may have been obscured.

Corn Cobs.   Corn cobs have long been regarded as an important roughage for grain-fed cattle when fed in the form of ground ear corn. Their value was believed to lie chiefly in the volume and bulk which they gave the ration rather than in the digestible nutrients supplied. Cobs are hard and woody and will not be eaten readily—a fact that has caused cattlemen to consider them almost worthless as a source of feed nutrients. Nevertheless, digestion trials have shown that cobs contain 45 pounds of digestible nutrients per hundredweight, or about the same as oat straw. Cobs have usually saved both grain and roughage in producing 100 pounds of gain in experiments comparing ground ear corn with shelled corn. This shows either that cobs have a definite feeding value or that their presence in the ration favors the utilization of the other components.

Experiments at the Ohio, Iowa, and Nebraska stations, in which ground cobs were fed in addition to those present in ground ear corn, show clearly that corn cobs are a valuable roughage material for finishing cattle. (See Table 162.) In the Ohio and Iowa tests, where both corn and hay were fed according to appetite, the added cobs replaced grain rather than roughage in both the daily ration and in feed consumed per hundred

Table 162

Value of Corn Cobs as a Feed for Finishing Cattle

	Ohio Mimeo. Series 52 Calves and Yearlings (Average 3 Trials)			Iowa Mimeo. AH Leaflet 165 Yearlings (Average 2 Lots Each)			Nebraska Bulletin 396 Yearlings (Average 3 Trials)		
	Shelled Corn	Ground Ear Corn	Ground Ear Corn + Cobs	Shelled Corn	Ground Ear Corn	Ground Ear Corn + Cobs	Shelled Corn	Ground Ear Corn	Ground Ear Corn + Cobs
Average daily gain (lb)	1.92	1.92	1.85	2.38	2.43	2.23	2.42	2.35	2.13
Average daily ration (lb)									
Shelled corn	12.4	11.1	9.4	15.2	14.3	13.0	13.5	13.5	13.1
Corn cobs	—	2.5	4.2	—	2.7	5.4	—	3.1	6.2
Protein concentrate	2.0	2.0	2.0	1.5	1.5	1.5	1.5	1.5	1.5
Hay	4.3	4.3	4.2	5.1	3.6	3.6	—	—	—
Sorgo silage	—	—	—	—	—	—	27.3	16.7	6.1
Total daily consumption of air-dry roughage (including cobs) (lb)	4.3	5.8	8.4	5.1	6.3	9.0	8.2[a]	8.1[a]	8.0[a]
Ratios of corn to roughage (including cobs)	2.9:1	1.6:1	1.1:1	3.4:1	2.6:1	1.4:1	1.6:1	1.7:1	1.6:1
Feed eaten per cwt gain (lb)									
Shelled corn	631	552	478	657	651	576	559	574	613
Corn cobs	—	129	220	—	123	240	—	131	291
Protein concentrate	103	103	108	64	66	67	62	63	70
Hay	221	223	226	219	163	158	—	—	—
Sorgo silage	—	—	—	—	—	—	1,126	712	286

[a] Sorgo silage reduced to 15 percent moisture basis.

pounds of gain. But in the Nebraska experiments, where the roughage fed was sorghum silage, the addition of cobs had little effect on the consumption of corn on a shelled basis but greatly reduced the consumption of silage. These results indicate that corn cobs are a better companion roughage for low-grade hay, which is eaten in small amounts, than for highly palatable hay and silage. It is possible that the additional bulk imparted to the Ohio and Iowa rations by the ground cobs resulted in more complete digestion of the corn and protein concentrate.

Cottonseed Hulls. In the cotton-growing areas, huge quantities of cottonseed hulls are produced in the processing of cottonseed at the oil mills. The 500 pounds of hulls resulting from processing 1 ton of cottonseed add up to 1.3 million tons of hulls produced annually. As they are low in protein and high in fiber, their feeding value would appear to be low compared with most other roughage materials. They are bulky and obviously cannot be economically transported far from their source. Thus they are usually fed in Texas and the western states where cotton production is most highly concentrated.

Fig. 86. Cottonseed hulls are palatable and bulky and make an excellent roughage component of starting rations. High transportation costs usually preclude use of this by-product roughage at distances very far removed from the oil mills where cottonseed is being processed. (National Cottonseed Products Association.)

Table 163

Comparison of Alfalfa and Two Types of Cottonseed Hulls in Finishing Rations (140 Days)[a]

Roughage Source	Alfalfa	Regular Cottonseed Hulls	Delinted Cottonseed Hulls
Alfalfa in ration (%)	20	5	5
Hulls in ration (%)	—	15	15
Number of steers	16	16	16
Average initial weight (lb)	591	595	589
Average daily gain (lb)	2.70	3.02	2.88
Average daily feed (lb)	20.1	23.7	23.8
Feed per cwt gain (lb)	746	786	832
Feed cost per ton	$54.58	$53.68	$53.68
Feed cost per cwt gain	$19.87	$20.48	$21.66

[a] Adapted from Arizona Cattle Feeders' Day Report.

The Arizona station has obtained excellent results from the substitution of cottonseed hulls for chopped alfalfa hay in their 80 percent concentrate ration, especially when they included 4 percent tallow in the ration. Table 163 gives the results of one of their studies, and it will be noted that the steers on regular hulls consumed the most feed and gained fastest, indicating possibly that the hulls were responsible for a more desirable physiological status in the rumen. The fact that the hull rations were not so efficiently converted to gain indicates a lower available energy value, and thus the price paid for hulls probably must be at least one-third lower than that for good alfalfa hay. Feeders who feed complete mixed rations like to include some hulls, because hulls require no processing and make the ration bulkier, and cattle seem to stay on feed better.

A comparative trial conducted by North Carolina investigators, reported in Table 164, shows the relative values of several lower-grade roughages

Table 164

Low-Quality Roughages in Rations for Growing Calves (125 Days)[a]

	Cottonseed Hulls	Oat Straw	Cotton Burrs	Corn Cobs
Number of calves	22	22	22	22
Average initial weight (lb)	506	514	499	505
Average daily gain (lb)	1.78	1.92	1.58	1.85
Feed intake per head per day (lb)	16.4	16.5	16.0	17.3
Feed per cwt gain (lb)	922	862	1,013	937

[a] Adapted from North Carolina State College A.I. Report 73.

when included in rations containing considerably less energy than the ration used by the Arizona workers. The rations used by the North Carolina group contained 61.5 percent of the roughages being compared, so that differences in roughage value are more likely to be expressed. Oat straw appeared to be slightly superior to hulls or cobs. Cotton burrs were definitely inferior. If these roughages were fed at lower levels, such as 10 or 15 percent, it is doubtful if it would matter which one were fed. Therefore, when low-grade roughages are used with high-concentrate rations, the one available at lowest cost should be used, provided that any required processing is comparable in cost. Hulls mix readily with the remainder of the ration, whereas the others, especially cobs and straw, require grinding or chopping, a rather expensive operation.

Whole Cottonseed. Whole cottonseed contains about 25 percent hulls and 25 percent crude protein and for these reasons is a satisfactory source of both roughage and supplemental protein when the price for whole seed is low. Ordinarily the value of the oil in cottonseed, for food and industrial uses, makes the price of whole seed prohibitive for livestock feed. Data such as those in Table 165 show that whole cottonseed can satisfactorily replace alfalfa hay or a combination of hulls and meal. In fact, gains were significantly higher and more economical than in either of the other two treatments. The higher energy content of the whole-seed ration, owing to the 1.5 percent higher fat content, is no doubt responsible for the superior performance. No scours or other problems were noted in the whole-seed lot.

Soybean Hulls and Flakes. As swine and poultry nutritionists and feed manufacturers have formulated rations that are higher and higher in energy content, and thus lower in fiber, they have demanded a soybean meal with little hull content. Consequently a potentially important roughage source has become available for beef cattle feeding in the form of soybean hulls. Soybean hulls contain about 10 percent crude protein, 40 percent nitrogen-free extract, and 35 percent crude fiber, a composition not greatly different from that of the better grass hays. Studies have shown that flaking the soybean hulls increases the digestibility of the dry matter of the hulls from about 60 percent for ground raw hulls to 70 percent for flaked hulls. Undoubtedly the heat treatment involved in the flaking process is responsible for the improved digestibility. Soybean hulls are an example of a roughage with high fiber content that has a rather high digestibility, the reason being that the major portion of the fiber is cellulose, which is digestible by beef cattle if it is not highly bound by indigestible lignin.

The Ohio station conducted two trials with soybean flakes, also called

Table 165

Alfalfa Hay, Cottonseed Hulls, and Cottonseed as Roughage and Protein Sources in Milo Calf-Finishing Rations (3 Years)[a]

Item	Alfalfa Hay	Cottonseed Hulls Cottonseed Meal	Cottonseed Hulls Cottonseed
Ration ingredients (%)			
Steam-rolled milo	74.5	66.9	63.9
Cane molasses	5.0	5.0	5.0
Ground alfalfa hay	20.0	—	—
Cottonseed hulls	—	20.0	15.0
Cottonseed	—	—	15.0
Cottonseed meal	—	7.0	—
Sodium chloride	0.5	0.5	0.5
Limestone	—	0.6	0.6
Total	100.0	100.0	100.0
Number of calves	35	34	36
Number of days fed	149	159	143
Average initial weight (lb)	556	555	554
Average final weight (lb)	977	964	972
Average daily gain (lb)	2.82	2.61	2.94
Average daily feed (lb)	21.5	22.4	22.6
Feed/gain ratio (lb)	7.6	8.5	7.6
Feed cost/lb gain (¢)[b]	20.6	22.7	20.1

[a] New Mexico Agricultural Experiment Station Bulletin 571, 1970.

[b] Costs of complete rations fed were: $54.06, $53.65, and $52.86 per ton for alfalfa, cottonseed hull, and cottonseed rations, respectively.

soybran flakes. The results are shown in Table 166. In one trial they substituted 5 pounds of flakes for 7.5 pounds hay-crop silage and 2.5 pounds hay, or, in other words, they substituted the flakes for a like amount of air-dry roughage. In the other trial they substituted 4 pounds of flakes for 4 pounds of ground ear corn. The Ohio workers concluded that the flakes more than replaced the feeding value of the hay and even proved to be more valuable than ear corn. Thus toasted soybean or soybran flakes have a value comparable to ear corn, oats, or barley, and yet they have the bulky characteristic of a roughage.

Rice Hulls. In some of the southern states, notably Arkansas, Louisiana, and Texas, large quantities of rice hulls result from the rice milling industry. The composition of rice hulls is comparable to that of low-quality

Table 166

Feeding Value of Soybean Flakes Fed Beef Heifers (137 Days)[a]				
Type of Substitution	Roughage Replacement		Ear-Corn Replacement	
Feeds Fed	Hay, Silage, Corn	Hay, Silage, Corn, Soybran Flakes	Hay, Silage, Corn	Hay, Silage, Soybran Flakes
Number of heifers	16	16	24	24
Average initial weight (lb)	458	468	455	458
Average daily gain (lb)	1.51	1.78	1.38	1.52
Average daily ration (lb)				
Ground ear corn	4.0	4.0	4.0	—
Soybran flakes	—	5.0	—	4.0
Hay crop silage	15.0	7.5	15.0	15.0
Mixed hay	5.0	2.5	5.0	5.0
Salt (oz)	0.5	0.5	0.5	0.5
Minerals (oz)	0.4	0.4	0.4	0.4
Feed per cwt gain (lb)				
Ground ear corn	258	223	287	—
Soybran flakes	–	279	—	261
Hay crop silage	988	419	1,078	977
Mixed hay	329	140	359	326
Salt	2	2	2	2
Minerals	2	2	2	2

[a] Adapted from Ohio Agricultural Experiment Station Series No. 122.

hay, and they thus would appear to offer possibilities as an economical source of roughage, since there is little other commercial outlet for rice hulls. The Arkansas station conducted a trial with yearling steers fed a grain mix consisting of 80, 10, and 10 percent of ground shelled corn, crimped oats, and cottonseed meal, respectively, plus vitamin A, minerals, and salt, plus either prairie hay or rice hulls according to appetite. The data are summarized in Table 167. Performance was comparable in all respects and, with the lower price prevailing, the rice hull steers produced gains at $2.55 lower cost per hundredweight gain. Neither bloat, scours, nor other ill effects resulted from feeding rice hulls as the only roughage source.

In other, more recent Arkansas trials, it was found that rice hulls were comparable to Sudan grass hay when roughages made up only 20 percent of the total ration, but rice hulls were about 10 percent lower in value than

Table 167

Performance of Steers Fed Rice Hulls or Prairie Hay as Roughage Sources (84 Days)[a]

Roughage Fed	Prairie Hay	Rice Hulls
Number of steers	8	10
Average initial weight (lb)	711	692
Average daily gain (lb)	2.93	2.86
Average daily ration (lb)		
Grain mixture	20.6	17.7
Prairie hay	3.4	—
Rice hulls	—	4.3
Total	24.0	22.0
Feed per pound gain (lb)	8.5	7.9
Feed cost per cwt gain[b]	$19.79	$17.24
Carcass grades	1 C, 7 G	1 C, 9 G

[a] Adapted from *Arkansas Farm Research*, Vol. 12, No. 4.

[b] Feed costs per ton: prairie hay, $26; rice hulls, $8.

the grass hay when roughage level was increased to 40 percent. At the higher hull intake level, diarrhea, bloat, and excretion of bloody mucus were noted, no doubt caused by the very high silica content of such rations.

Peanut Hulls.   In the Southeast, the most important peanut-growing area, huge tonnages of peanut hulls are available at low cost. In this area, cottonseed hulls also are available, as is Coastal Bermuda grass in pellet form. When corn is grown for finishing cattle, it is sometimes still harvested as snapped corn—that is, in ear form with the shucks left on. Thus several roughage sources are available as substitutes for the more conventional roughages used in cattle feeding. Georgia workers conducted an experiment with yearling steers fed 80 percent concentrate rations and several of these southeastern roughage sources. The data are summarized in Table 168. The steers on Coastal Bermuda pellets and a combination of oyster shells and cottonseed hulls consumed significantly less feed. Otherwise the performance of all steers was virtually identical and it was concluded that all roughages tested were satisfactory in the 80 percent concentrate rations fed. Price, availability, ease of handling, and adaptability to mechanical feeding were listed as items that would determine the preference for one or the other among all the roughages tested.

Table 168

	Kind of Roughage				
Item	Snapped Corn	Coastal Bermuda Pellets	Cotton-seed Hulls	Peanut Hulls	Cotton-seed Hulls + Oyster Shells
Comparison of Various Roughages for Finishing Steers—112 Days (2 Years)[a]					
Ration ingredients (%)					
Ground shelled corn	0	70	70	70	82
Ground snapped corn	92	0	0	0	0
Supplement	8	10	10	10	8
Roughage	0	20	20	20	10
Total	100	100	100	100	100
Number of steers fed	28	28	28	27	28
Average initial weight (lb)	771	770	767	764	762
Average daily gain (lb)	2.54	2.50	2.78	2.58	2.54
Average daily feed (lb)	27.7	22.6	26.6	25.4	22.3
Feed/gain ratio (lb)	10.9	9.1	9.6	9.9	8.8
Average carcass grade	C−	C−	C−	C−	C−

[a] Georgia Coastal Plain Experiment Station Report, Tifton.

MISCELLANEOUS ROUGHAGE SUBSTITUTES

In many cattle-feeding areas, roughage nutrients are costlier than concentrate nutrients. Add to the higher cost of roughage nutrients the processing and storage costs, and the problem of feeding by automated methods, and one can see why feeders and researchers alike are casting about for every conceivable substitute for roughages in finishing rations.

Oyster Shells.   Several stations, notably Nebraska, have studied the use of about 2.5 percent hen-size oyster shells as a substitute for the entire roughage in finishing rations. They have achieved reasonable success, and their results have been confirmed by some, though not all, stations. The high calcium level fed when oyster shells are used apparently must be compensated for by also increasing the phosphorus level in order to keep the calcium-phosphorus ratio in balance. The scabrous character of the shells serves the same physiological function as the "roughness" quality of hays.

Sand.   The addition of 2 percent sand to a completely mixed high-concentrate ration improved performance of steers at the Iowa station. In six experiments the improvement was small but consistent, with a 5 percent improvement in both rate of gain and feed efficiency. Sand did not improve conventional rations containing some hay; thus sand apparently serves the same function as the oyster shells mentioned above.

Sodium Bentonite.   This substance, an inert colloidal clay, has been tested as a roughage replacement, fed at the rate of about 0.5 pound per day. Its use resulted in slight reductions in feed intake and daily gain in most controlled experiments reviewed; hence it cannot be recommended at this time.

Poultry Litter.   Broiler-house litter has been successfully used as both a roughage and a nitrogen source in feeding trials at the Arkansas and Virginia stations. Such litter contains upward of 30 percent protein equivalent and, depending on type of bedding used, may contain 15 to 30 percent crude fiber. Wood shavings apparently result in a less desirable litter, for cattle feeding at least, than cottonseed hulls or peanut hulls which are also sometimes used as litter.

Poultry litter can satisfactorily make up at least 25 percent of the total ration. As the nitrogen is in the form of nonprotein nitrogen, it cannot be expected to be utilized quite as well as natural protein nitrogen. The litter should be ground in a hammermill before mixing. Commercial cattle feeders in the broiler-producing areas of the Southeast are making quite extensive use of litter as an ingredient in their complete feeds.

Cattle Manure.   Cattle manure itself is being tested, and it appears that if the manure from cattle on grower rations is dried and mixed in a complete high-concentrate ration at the level of 5 to 15 percent, it serves as a good substitute for roughage. Manure from confinement cattle-feeding facilities where high-concentrate rations are used is less suitable as a roughage replacement because it is very low in fiber and high in ash or minerals. As mentioned elsewhere, it does supply considerable supplemental nitrogen and single-cell microbial protein.

MISCELLANEOUS DRY-ROUGHAGE TREATMENTS

Several promising methods for improving hay quality or nutrient value are being investigated, although the research has not advanced to the stage where findings can be applied to field application.

Chemically Treated Low-Quality Roughages.   Most low-quality hays and crop residues such as straw, corn cobs, corn stalks, and milo stubble are high in fiber content with low digestibility, and therefore are low in metabolizable energy or total digestible nutrients. Nebraska workers have successfully treated such forages with sodium hydroxide, added at the rate of 4 percent or 4 pounds of chemical per 100 pounds of dry residue. They have reported that when corn and milo combine-tailings were so treated and fed to beef calves, gains were equal to those obtained with a corn silage ration. Treated corn stalks and poor-quality grass hays did not respond as well due to low acceptance of the treated roughage. Nebraska workers also have successfully treated higher-moisture materials, such as husklage, with 3 percent sodium hydroxide, resulting in a product that was 80 to 90 percent as valuable as corn silage. At present the treatment costs about $10 per ton of husklage. They are now testing combinations of sodium hydroxide and calcium hydroxide and they believe a 3:1 combination of the two chemicals appears the most promising.

Preservation of High-Moisture Hay.   Hay quality is affected by moisture content at baling or stacking time but, unfortunately, it often happens that weather conditions simply are not ideal for curing hay. The molding and heating that occur cause reduced intake, lowered nutrient content, and even abortion in some instances. Two compounds that offer promise as preservatives of high-moisture hay are propionic acid and anhydrous ammonia. Both act as fungicides and of course the ammonia, if retained in the hay, adds a supplemental nitrogen source. Indiana researchers, working with 32 percent moisture, second-cutting alfalfa hay, added propionic acid at four levels ranging from 0.4 to 20 pounds per ton, and anhydrous ammonia at the level of 20 pounds per ton. The bales treated with propionic acid, 6 bales for each treatment, were stored under roof while the ammonia-treated bales were left outside and covered with plastic for 48 hours. The results are shown in Table 169.

Both treatments reduced the temperature rise after baling, which is favorably reflected in improved digestibility of dry matter as compared with the nontreated control hay. Crude protein apparently was not affected by level of propionic acid but, not surprisingly, the ammonia treatment increased the crude protein content. The apparent protein content increases resulting from adding propionic acid are perhaps a reflection of losses in carbohydrates during curing and not an actual increase in protein. All treatments but the lowest level of propionic acid decreased dry weight loss, and both ammonia and the 20 pound propionic acid level improved dry matter digestibility. These promising results have encouraged further research, and several manufacturers already have field applicators available for applying chemical preservatives to high-

Table 169

Effect of Propionic Acid and Anhydrous Ammonia Treatment on Compositional Changes in Alfalfa Hay Baled at 32 Percent Moisture[a]

Treatment (lb/ton)	Maximum[b] Temperature (°F)	Crude Protein (%)	Dry-Matter Digestibility (%)	Dry Weight Loss (%)
None	115	16.5	60.5	15.5
Propionic acid				
0.4	113	16.6	61.8	16.7
4.0	100	16.8	62.2	13.2
10.0	97	16.8	61.0	11.7
20.0	91	16.8	65.0	7.6
Anhydrous ammonia				
20.0	90	21.9	66.1	9.9
Baling time[c]	—	14.3	70.5	—

[a] Indiana Beef Forage Research Day Report, 1975.
[b] Temperature was 90°F immediately after baling.
[c] Samples collected at time of baling.

moisture hay as it is being baled by both conventional and big balers and as it is being stacked by the newer field stackers.

Big-Package Hay-Making.   The conventional small, rectangular or round bale, weighing 50 to 100 pounds, is still the most common kind of bale used for hay, but high labor costs, especially for loading, transporting, and moving such bales into storage, have become almost prohibitive. Recent engineering advances have led to development of new equipment for accumulating, packaging, and handling hay in larger units that are more labor-efficient. These larger hay packages range from bales weighing from 600 to 1,300 pounds to stacks weighing up to 8 tons. Twine may or may not be required in the balers, and all of the hay packages are designed for outdoor, unprotected storage.

Most major hay-harvesting equipment manufacturers and many smaller specialty firms now have models of big-package baling machines on the market. Many of them have had only limited testing under a wide variety of field conditions, and modifications are being made almost annually, making it advisable for anyone contemplating purchase and use of these newer machines to become informed about their potential. Extension and resident agricultural engineers in all important hay-producing states have done field testing, and they publish their test results. Field demonstrations by both university personnel and industry representatives are available. Most of the published extension materials include costs per ton of hay harvested and some include wastage data as well. Unfortunately, on smaller farms producing less than 100 tons of hay, purchase of these newer hay-harvesting machines is economically unwarranted. Wastage may be as high as one-third under field storage conditions unless the very best of bale or stack management is practiced. Purdue workers found that feeding large hay packages in well-managed racks reduced wastage from 39 percent to 3.7 percent as compared to simple self-feeding without racks.

Harvesting of hay as big packages by custom operators is now available to many smaller-scale cattlemen, and the use of this service probably will increase because of the high initial cost of owning one's own equipment. Custom operators generally charge a fixed cost, per bale or stack, and thus are likely to operate at maximum speed. The result often is underweight bales and stacks, poorly made and put up at too high a moisture level. Weather damage often occurs and molding in the center of the bales or stacks is common. The choice of bale or stack location also may not be most convenient for the cattleman. Obviously not all operators perform in this manner, but this points up some of the problems of this promising new hay-making method. Much will be learned in the next few years that will determine whether this is only a passing innovation or a real breakthrough. Nutritionally speaking, the net result will likely be a slight to large reduction in feeding value of good-quality roughages, but an overall improvement in nutritional quality of some of the lower-grade roughages.

CHAPTER 19
SILAGE AS A FEED
FOR BEEF CATTLE

The first silage experiment reported in a United States government publication (1875) was that of Professor Manly Miles of the University of Illinois. The studies involved cornstalk and broomcorn silages stored in pit silos and were conducted in cooperation with Funk Farms, Shirley, Illinois. (This farm, incidentally, is still involved in corn production, as seed producers.) Extensive use of silage in beef cattle feeding did not begin until about 1910, however. Before that time it was generally believed that silage was a feed mainly for dairy cows, and it was thought that its succulent nature would produce a marked diarrhea, an undesirable condition for steers on full feed. But the results obtained by using it at agricultural experiment stations and in feedlots of progressive cattlemen were so favorable that the feeding of silage to feeder cattle soon became a common practice throughout the Corn Belt.

By no means has the use of silage been confined to the Corn Belt. Although it is true that corn was the first and is still the principal crop used in silage making, many other crops are ensiled in those parts of the country where corn cannot be grown successfully. In the arid West and Southwest the sorghums as well as corn furnish an enormous tonnage of silage, especially where grown by benefit of irrigation. Large amounts of silage are also now made from alfalfa, oats, field peas, clover, cowpeas, soybeans, rye, Sudan-sorghum hybrids, and other farm crops in various sections of the country. Indeed, it appears likely—if, in fact, it is not already true—that silage will be the principal harvested roughage used in beef production both in the Corn Belt and throughout the United States.

ADVANTAGES OF SILAGE

There are distinct advantages that may accrue from harvesting and feeding forage crops as silage. Listed in no particular order as to relative importance, these advantages are as follows.

1. A greater yield of crop or nutrients may be obtained, mainly because the entire plant is harvested but also because fewer losses occur owing to rain damage and field-harvesting losses, which may occur in making hay.

2. The crop is more nutritious if harvested as silage because the plant can be harvested at the stage of maturity when nutritive value is at its peak.

3. Silages are more palatable and intake is greater and more consistent compared with other forages, making it easier to keep cattle on feed.

4. Silage is more adaptable to mechanized harvesting methods and to feeding with automated equipment.

5. Silage is more conveniently added to a complete mixed ration than is hay.

6. The crop is removed early enough in some instances so that double-cropping may be practiced in some areas of the country.

7. Weeds, including seed, are ensiled and consumed and most weed seeds lose their viability during fermentation in the silo.

8. A complete feed may be mixed and stored at ensiling time. Preservatives, where needed, are easier to add to silage than to hay.

9. Harvesting time can be extended by planting varieties of crops, both within and between species, having varying maturity dates.

10. In larger operations, harvesting and storage costs are comparatively lower.

11. More cattle can be fed on a given acreage because of higher yield of crops, resulting in economies of scale that are especially important to smaller farms.

DISADVANTAGES OF SILAGE

There are instances in which the disadvantages inherent in harvesting and feeding a crop in silage form will make it inadvisable to use this technique. Some of the disadvantages are the following.

1. Initial investment in harvesting and storage equipment is high, especially if less than 200 to 300 tons are harvested annually.

2. Spoilage during ensiling and in storage may run high—up to 25 to 35 percent in some instances.

3. Spoilage after the silo is opened may increase losses further if silage is fed in warm weather, especially if only a few cattle are fed. Some spoilage may even occur in the feedbunk if ad lib feeding is practiced.

4. Silage may create some undesirable effects on the environment, such as odors associated with fluids seeping from the silo, and numerous flies and rodents.

5. Bunker and trench silos, especially when empty, are unsightly to many viewers.

6. Nitrous oxide poisoning in tower silos may pose danger to man.

7. Silages that are too wet and acidic or too dry and moldy will not be consumed readily. These two extremes in moisture level also result in high fermentation losses.

8. Cattle fed on high-silage rations are sometimes discriminated against on the market because they purportedly dress lower.

In spite of the possible disadvantages, the use of silage in the three kinds of beef cattle programs is on the increase, and as long as grain prices and rangeland and pasture remain relatively high-priced, harvesting the entire plant as silage, whatever the crop, will be a technique used by cattlemen who are most conscious of the feed costs associated with their particular enterprise.

PRINCIPLES OF MAKING SILAGE

The making of silage involves choosing the harvesting time that results in the greatest yield of digestible nutrients, and then processing and storing the crop in such a way that losses are lowest. It is well to remember that a silage cannot be better than the crop that was harvested to make it. In fact, it is unfortunately true that much silage is considerably less valuable than the original crop from which it was made. This is due, first of all, to excessive losses during fermentation and from top spoilage and, second, to unfavorable aroma and conditions that reduce consumption to low levels. An understanding of the chemical changes that forage undergoes during ensiling should serve to emphasize the important steps necessary in making good silage. These chemical changes occur in the following order, with much overlapping.

1. Respiration of the plant cells of the chopped forage particles continues for a few hours, consuming much of the trapped oxygen and producing acetic acid, carbon dioxide, and heat as end products. Plant enzymes contained in the forage are also active during this phase.

2. Lactic and still more acetic acids are produced by anaerobic bacteria, provided that sufficient soluble carbohydrate material is available for the fermentation to proceed. This process continues for several days to several weeks, depending on the rate of acid production. When a pH of about 4 is reached, fermentation practically stops, and the silage undergoes little further change. The length of time required to reach the desired pH determines, to a large degree, the amount of energy

lost because of fermentation. At the end of this period, temperatures within the silage mass return to about 85°F and the pH remains at about 4 unless further breakdown occurs in the next step of the silage-making process.

3. If the proper pH is not reached because of insufficient soluble carbohydrate in the forage, butyric acid–producing bacteria become active, further reducing the energy content of the silage. A foul-smelling, slimy, unpalatable silage results from this type of fermentation. Furthermore, the butyric acid–producing organisms may attack the proteins, converting them to volatile fatty acids and ammonia.

In summary, there are two important essentials in making the highest-quality, most palatable silage with minimum nutritive losses from excessive fermentation: (1) pack the finely chopped forage well so as to exclude oxygen; (2) if grain is not a part of the silage plant, add some type of soluble carbohydrate material to promote acetic and lactic acid production and to inhibit the activity of butyric acid–producing bacteria.

EXTENT OF LOSSES IN MAKING SILAGE

For convenience of discussion, silage losses may be divided into (1) seepage losses, (2) fermentation losses, and (3) top spoilage. Extent of seepage loss depends on the amount of moisture present (moisture within the plant itself and added moisture in the form of rain or snow), on pressures exerted, and on whether the excess moisture, containing soluble materials, may drain or seep from the silo. Seepage losses can be eliminated or reduced by (1) wilting freshly cut, and preferably crimped, legume and legume-grass forages to approximately 30 percent dry matter while in the freshly cut swath or in a loose windrow, or (2) by cutting corn and sorghum forages only after they have matured sufficiently to contain above 28 to 30 percent dry matter, or (3) by adding an absorbent material such as ground ear corn, ground corn cobs, or even dry manure from feedlot floors. Seepage losses of nutrients are not as high, on a percentage basis, as the apparent loss in weight, as most of the liquid seeping from a silo is water. Nevertheless, seepage losses of actual nutrients may run as high as 10 percent of the dry matter, mostly in the form of soluble carbohydrate material and valuable silage acids, in large tower silos where much pressure is exerted. In trench, bunker, or stack silos, seepage losses are usually lower, especially if rainwater is diverted. Under ideal conditions no seepage need occur. The possibility of seepage should be allowed for by locating the silo so as to provide drainage into a lagoon or other catchment basin, to prevent stream pollution.

The extent of fermentation losses, usually the greatest source of loss in other kinds of silage besides corn and sorghum, may go as high as 25 percent of the total dry matter, but ranges from 5 to 10 percent in most silages. The extent depends on availability of a soluble carbohydrate for bacterial fermentation or conversion to the volatile fatty acids, lactic and acetic, or the addition of mineral acids themselves, and the effectiveness with which air is excluded from the silage.

Top spoilage losses vary greatly and are the most apparent of all losses because they are easily observed. The extent of these losses, which may range from practically zero to as high as 20 to 25 percent, depends on how well the fresh material is packed, how fine the forage is chopped, whether rain or snow falls on the surface of the silage, how much is fed off the exposed surface each day, and whether a covering material of some sort is used.

Field losses, due to loss of dry leaves as a result of overmaturity or excessive wilting, may be an important loss in legume silages in some situations.

Total losses from all sources may be expected to run from an average of 10 to 15 percent for corn or sorghum silages stored in good tower silos to an average 15 to 25 percent for straight legume or legume-grass silages. Use of oxygen-free storage or airtight silos will reduce losses by at least one-half, while total losses may go as high as 35 percent in poorly located and constructed surface silos. Figure 87 shows the effect of moisture content and type of silo on the magnitude of dry-matter losses in legume-grass silage.

PRESERVATIVES

Preservatives, additives, or conditioners are added to legume and grass silages for one or all of four reasons: (1) to hasten the natural production of lactic and acetic acids so as to lower the pH and prevent undue fermentation losses, (2) to accelerate the lowering of the pH, thus preventing undue heat losses and the more adverse forms of fermentation, (3) to correct certain nutritional deficiencies that may be inherent in the silage, and (4) to improve the odor and palatability of the silage.

Preservatives may be divided into two broad categories: first, the carbonaceous or carbohydrate materials such as molasses and corn and, second, the mineral acids. Certain yeasts and enzyme products are commercially available, but not enough controlled experiments have been conducted to evaluate these products as yet. Some farmers' experiences appear favorable. Good silage can be made, and much is made, without

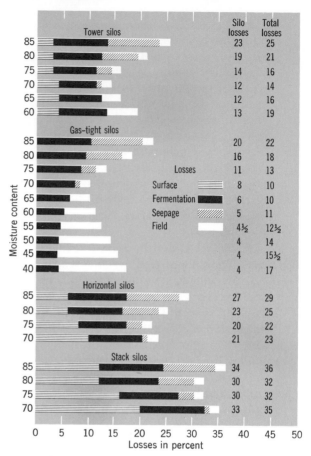

Fig. 87. Effect of moisture content and type of silo on dry-matter losses in legume-grass silages. (A. T. Hendrix, Agricultural Research Service, USDA.)

preservatives, but those who are inexperienced in making silage, especially grass and legume silages, will do well to investigate their use.

The economy of preservatives is difficult to determine for each and every farm. The carbonaceous preservatives can be justified without question because they are recovered in the silage as a carbonaceous concentrate with approximately 85 percent of their energy or feeding value being retained. On the other hand, such preservatives as sodium metabisulfite, while doing an excellent job of reducing losses in the silo and of improving palatability in some cases although not all, make no further contribution to the ration. Application of sodium metabisulfite and similar material presents a mechanical problem because the amount added is extremely small.

Table 170

Rates of Preservative Application with Different Forages			
	Suggested Pounds per Ton of Crop		
Kind of Crop	Molasses[a]	Ground Ear Corn[b]	Sodium Metabisulfite
Grasses	40–60	100–150	5
Cereals	25–50	75–100	5
Legume-grass mixtures	50–100	125–150	7.5
Legumes	75–100	150–200	10
Corn and sorghum	None	None	None

[a] If dried molasses preservatives are used, reduce amounts by 50 percent.

[b] If ground shelled corn is used, reduce amounts about 10 percent.

Table 170 gives recommended levels of preservatives when the material is cut with a direct-cut field chopper. If wilting is practiced, the use of preservatives is questionable. Additional amounts of cob, up to 200 pounds, are often used to reduce seepage of high-moisture silages. Corn or other high-energy feeds may be added up to 40 percent of the weight of the silage in order to balance a finishing ration with respect to energy content.

Numerous experiments have recently been conducted in which 0.5 percent of either limestone or urea or both have been added to fresh-cut whole-plant corn silage as the silo was filled. These 10-pound-per-ton additions raise the pH from about 3.5 to about 4.25 and increase the protein content on a dry basis from about 7.5 to about 9 percent. Silage intake has usually increased along with a 5 to 8 percent increase in rate of gain, and with improved feed efficiency of about the same magnitude. Some further bonuses are a reduction in acid erosion of concrete silo walls, a higher carotene content, and an economical source of supplemental protein. Responses have been greatest when grower or medium-energy finishing rations have been fed. Liquid nonchemical preservatives high in molasses and urea or ammonia and other minerals are available commercially and serve satisfactorily as preservatives, and also supply additional nitrogen and minerals. As these are patented processes, users should be aware of the legal implications of attempting to use similar formulations containing readily available ingredients at possibly lower costs.

SEALING THE SILO

Many farmer-feeders make no attempt to seal their silos but allow a natural seal of rotted and moldy silage to form at the top, through which little air gains access to the silage below. This is a wasteful practice because the amount of spoiled silage, which must be discarded when the silo is opened, may easily represent 10 to 15 percent of the total weight of green forage stored in tower silos, and considerably more in bunker or trench silos. This loss of feed may be greatly reduced by leveling the surface of the silage, packing the silage thoroughly to exclude the air from the top layer, and covering it with a sheet of plastic weighted down with 4 or 5 inches of earth or limestone, or even old tires. Ordinary agricultural limestone makes an especially good seal on uncovered silos because of the tight, impervious crust that it forms after being wet by a heavy rain. Well-packed silage covered with a plastic seal should not shrink in weight more than 3 to 5 percent because of top spoilage.

KINDS OF SILOS

The most common kind of silo found in the Corn Belt is the upright or tower silo, 16 to 20 feet in diameter and 40 to 60 feet high. Concrete staves are the most popular building material, although many monolithic concrete silos and a few glazed tile silos erected many years ago are still serviceable and are being used.

The metal silo is increasing in number, especially the glass-coated steel silo called the Harvestore, which has the feature of being airtight. As a result of the controlled fermentation, the ensiled forage loses fewer nutrients than when stored in conventional upright and especially horizontal silos. The chopped forage is blown into the silo through an opening in the roof and is removed at the bottom by a specially designed bottom unloader. A valve in the top attached to an inflatable "breather bag" under the roof serves to maintain the same pressure within the silo as exists on the outside.

Other manufacturers now market metal silos of various designs, but in general the airtight feature is found in all of them. The method of unloading is the major difference. A big advantage of the bottom-unloading silos is that they may be refilled from the top while feeding from the bottom, allowing for season-long harvesting. A disadvantage of bottom-unloading is the more expensive bottom-loader with its higher-powered motor.

Upright silos made of metal, concrete, or similar material are high-priced, and there have been various attempts to provide less expensive structures for the storage of silage. Perhaps the most common of the less costly forms is the trench silo, which consists of a long straight-sided ditch, 10 to 30 feet wide, constructed on a 5- to 10-degree slope in order that surface water seeping into the trench may drain out at the lower end. An important advantage of the trench silo, besides its inexpensive construction, is the low cost of filling it. No special equipment is required at the silo, since the trucks or unloading wagons that are used to haul the crop from the field harvester are driven into the silo, where they dump and spread the chopped forage with little or no hand labor. The loads also may be dumped into the silo from the side of the silo, using a tractor with manure blade to level and pack the silage.

Driving the truck or tractor and wagon over the silage during filling assists in packing the material, but further packing with a tractor alone is recommended. The principal disadvantages of the unwalled trench silo are that the trench tends to widen each year through the crumbling of the walls and that it is sometimes difficult to maintain roads to the silo over which the silage may be hauled to the cattle during bad weather.

Except in regions of light rainfall, the walls and floors of trench silos should be made of concrete to prevent serious erosion damage by heavy rains during the spring and summer when the silos are empty. Bunker silos, consisting of either concrete or wooden walls and a concrete floor at ground level, are equally satisfactory, and mud usually is not such a problem. This kind of horizontal silo is adaptable to level sites lacking satisfactory natural slopes.

Many farmer-feeders and especially larger feedlots are having fairly good results from storing silage in large, carefully made stacks in the open, with or without retaining walls, without protection to either the tops or sides of the piles. The losses encountered from spoilage in these bunker

Fig. 88. Large-volume trench, bunker, or surface silos and automatic loading equipment, coupled with unloading-mixer wagons and fenceline bunks, reduce nonfeed costs for larger feedlots. (Western Livestock Journal.)

Fig. 89. Filling a bunker silo, using a self-unloading truck. Proper leveling and packing reduce spoilage losses. (John Deere.)

or surface silos are claimed to be less than the annual interest and depreciation charges on a conventional silo. This method of storage is commonly used for preserving pea vines, cannery refuse, beet tops, and other silage materials of low feeding value, but corn and sorghum silages are also being so stored by very large commercial feedlots. The surface silo may well prove to be the best method of preserving grasses and legumes because, as some feeders claim, spoilage costs are lower than silo costs and seepage losses are less of a problem. Labor costs of feeding from surface silos are lower because tractor-mounted scoops or mechanized front-mounted loaders are used to fill automatic unloading trucks or wagons.

Large pits, resembling gravel or sand pits with no concrete used in their construction, may be found in the Southwest and West where rainfall is low. These make very economical and satisfactory silos.

COSTS OF MAKING SILAGE

The total costs of making and feeding silage may be grouped under the following items: (1) depreciation of silo, filling, and feeding equipment; (2) interest on investment in silo and equipment; (3) maintenance and repair of silo and equipment; (4) taxes; (5) silage losses, including field losses; (6) net cost of preservative; (7) labor cost of filling silo and feeding silage.

Such costs naturally vary considerably, depending principally on the type of silo and size of operation. The larger capacity silos have a real advantage in that fixed costs, such as those for filling and feeding machinery, can be spread over more tons of silage or larger numbers of cattle fed. Table 171 shows how size and type of silo affect estimated silo costs and cost of silage losses per ton.

The estimated total cost per net ton fed—that is, the estimated total cost from standing crop to feedbunk per ton actually fed—is shown in Table 172. Some operators do this job more cheaply, but some also have higher costs, mainly because losses are greater and inefficiencies in operating expensive silo-filling equipment are more frequent. Silo-filling equipment is owned cooperatively in many areas, and neighbors often trade work at silo-filling time. The difficulty with this arrangement is that the desirable time for harvesting a crop may arrive simultaneously on all farms in the neighborhood. The custom-harvesting of silage crops is on the increase and may be the only economical way for the farmer-feeder who uses less than several hundred tons of silage to justify using silage at all.

SILAGE IN THE FINISHING RATION

In all cases where cattle are given a liberal amount of grain, any silage fed should be considered a part of the roughage ration. Its use should therefore be in accordance with the well-recognized rules for the feeding of roughage discussed in Chapter 18.

Because silage often contains some corn or other concentrate and because it is finely chopped, feeders who are inexperienced in its use often make the mistake of treating it as a concentrate. Beginning with a small amount of silage, they gradually increase the allowance per day throughout the feeding period when, as a matter of fact, the reverse of this procedure usually should be followed. Because silage contains only about 40 to 50 percent corn or other concentrate when converted to a dry basis, it is impossible for steers to consume any great amount of additional grain and silage at the same time. The logical procedure is to feed the maximum amount of silage during the first part of the feeding period, decreasing the amount gradually, so as to permit a larger consumption of concentrates.

EFFECT OF SILAGE ON GRAIN CONSUMPTION

Except when fed in very small quantities, the use of silage lessens slightly the consumption of grain. This probably because well-preserved silage is

Table 171

Estimated Cost per Ton Capacity of Silos[a]

Type of Silo	Average Initial Cost ($) per Ton Capacity		Estimated Life (yr)	Annual Cost[b] per Ton (% of initial cost)	Annual Cost ($) per Ton Capacity—Silo Only		Estimated Silage Losses (%)	Annual Silo Costs[c] per Gross Ton ($)	
	100-Ton	200-Ton			100-Ton	200-Ton		100-Ton	200-Ton
Tower									
Wood stave	12.00	10.00	25	8.7	1.04	0.87	14	1.74	1.57
Monolithic concrete	12.00	9.00	35	7.4	0.89	0.67	14	1.59	1.37
Concrete stave	13.00	10.00	30	7.8	1.02	0.78	14	1.72	1.48
Clay-tile stave	22.50	17.50	45	6.5	1.46	1.04	12	2.06	1.64
Galvanized steel	17.50	13.50	40	6.7	1.17	0.90	12	1.77	1.50
Gastight steel	32.00	25.10	40	7.0	2.24	1.76	5	2.49	2.01
Horizontal									
Unlined trench	1.00	0.80	5	26.0	0.26	0.21	23	1.41	1.36
Wood wall trench	6.00	4.50	15	11.5	0.69	0.52	20	1.69	1.52
Concrete trench	7.60	7.00	25	8.5	0.65	0.60	20	1.65	1.60
Tilt-up horizontal	6.00	5.20	25	8.5	0.51	0.44	20	1.51	1.44
Wood wall bunker	7.50	5.00	20	9.7	0.72	0.38	20	1.72	1.38
Miscellaneous									
Welded wire, paper	1.00	0.80	3	36.8	0.47	0.38	20	1.47	1.38
Plastic film envelope	1.50	1.40	1	103.5	1.56	1.45	5	1.81	1.70
Stack	0.05	0.04		103.5	0.05	0.04	33	1.70	1.69

[a] A. T. Hendrix, Agricultural Research Service, USDA.

[b] Includes depreciation, maintenance, interest on investment, and taxes.

[c] Includes spoilage loss with green forage crop valued at $5 per ton. New cost data would be desirable but, in their absence, doubling the cited costs would approximate the costs in the mid-1970s.

Table 172

Summary of Estimated Costs of Harvesting, Storage, Storage Losses, and Feeding of Silage by Different Methods per Net Ton[a] (Silo Capacity, 200 Tons)[b]

Type of Silo	Silo Cost per Year, Net Ton ($)	Storage Losses per Net Ton ($)	Filling Costs per Net Ton ($)	Silo, Storage Losses, and Filling Costs per Net Ton ($)	Estimated Feeding Costs per Net Ton ($)	Total Costs, Standing Crop to Feedbunk ($)
Tower						
Wood stave	1.01	0.81	2.65	4.47	1.05	5.52
Monolithic concrete	0.78	0.81	2.65	4.25	1.05	5.30
Concrete stave	0.91	0.81	2.65	4.38	1.05	5.43
Clay-tile stave	1.18	0.68	2.69	4.45	1.05	5.50
Galvanized steel	1.02	0.68	2.69	4.30	1.05	5.35
Gastight steel	1.85	0.26	2.69	4.41	1.09	5.50
Horizontal						
Unlined trench	0.27	1.49	2.69	4.46	0.48	4.94
Wood wall trench	0.65	1.25	2.59	4.48	0.08	4.56
Concrete trench	0.75	1.25	2.59	4.58	0.08	4.66
Tilt-up horizontal	0.55	1.25	2.59	4.39	0.08	4.47
Wood wall bunker	0.48	1.25	2.59	4.32	0.08	4.40
Miscellaneous						
Welded wire, paper	0.48	1.25	3.00	4.72	0.48	5.20
Plastic film envelope	1.52	0.26	2.53	4.32	0.48	4.80
Stack	0.06	2.46	3.58	6.10	0.48	6.58

[a] A. T. Hendrix, Agricultural Research Service, USDA.
[b] Forage crop estimated value $5 per ton in field. (See footnote c in Table 171).

more palatable than most dry roughages and thus competes more actively for the appetite of the steer, and because the concentrate present in the silage naturally tends to replace part of that in the grain ration. The use of corn silage in moderate amounts decreases the consumption of the corn fed as concentrate by approximately 10 percent of the weight of the silage fed, whereas the shelled corn content of the wet silage usually is 15 to 17 percent of the weight of silage eaten. Consequently the feeding of silage in moderate amounts usually results in a somewhat larger total consumption of corn than is realized from a nonsilage ration. This is illustrated in Table 173. Naturally the feeding of silages other than corn or sorghum will not increase grain intake but may in fact reduce it slightly.

AMOUNT OF SILAGE TO FEED

Silage varies so much in moisture content that it is impossible to lay down definite rules as to the amount that should be fed. Some men put the crop into the silo when it is comparatively green, whereas others wait until it is fairly mature. Naturally the greener silage contains more water. Also, it frequently happens that silage in a given silo varies greatly in moisture content from top to bottom. This is particularly likely in a large silo that requires several days to be filled or in a silo containing silage made from rather dry forage to which considerable water was added. Unless the flow of water is carefully regulated, some of it filters down through the silage, leaving the material at the top only moderately moist while that near the bottom is saturated. Consequently, any recommendation as to the amount of silage to feed should be regarded only as an estimate.

The best index to follow in determining how much corn or sorghum silage to feed is the appetite of the cattle. Knowing the approximate amount of grain that should be consumed at a given stage of the feeding period, one may feed as much silage as will not decrease appreciably the consumption of grain. Less than this amount may, of course, be fed but the use of greater amounts is almost certain to result in smaller daily gains and lower finish.

The slightly slower gains resulting from increased silage intake may still be more profitable, especially for the farmer-feeder, who must think first of how to sell an acre of crop for the greatest profit. And, as has been mentioned, gains per acre are usually positively related to silage level in the diet. The new grade standards approved in 1976, requiring lower levels of rib-eye marbling for choice grade, mean that higher silage levels almost certainly will be used by farmer-feeders. This proposition is

Table 173

Effect of Silage on the Consumption of Grain (Two-Year-Old Steers)

Indiana Experiment Station	Number of Trials Averaged	Average Daily Consumption		Corn Present in Silage (lb)[a]	Total Corn Consumed (lb)
		Shelled Corn (lb)	Corn Silage (lb)		
I. { Shelled corn, cottonseed meal, Clover hay, corn silage	10	14.2	25.1	3.8	18.0
{ Shelled corn, cottonseed meal, Clover hay	10	17.1	—	—	17.1
II. { Shelled corn, cottonseed meal, Corn silage	5	14.5	29.9	4.5	19.0
{ Shelled corn, cottonseed meal, Clover hay	5	17.5	—	—	17.5

[a] Assuming that 15 percent of weight of wet silage is corn.

somewhat different as related to commercial feedlots, for they buy roughage as well as grain, and economy of gain may not be as important to the feedlot operator as rate of gain and turnaround of droves of cattle through the lot.

Although from the standpoint of the cattle there is no minimum amount of silage to feed, there are certain practical objections to using this material in small quantities. The principal items in the cost of silage, as mentioned earlier, are the labor, machinery, and storage charges. These items decrease rapidly per ton as the amount of silage made increases. Consequently silage is a much cheaper feed when put up in large quantities. Also, considerable labor is involved in the feeding of silage compared with most other roughages, but this labor decreases per ton with the amount of silage fed. Thus if silage is to prove an economical feed, it must be fed in relatively large quantities. Except in special situations, if silage is used it should furnish 75 to 100 percent of the roughage on an air-dry basis.

Owing to the high water content of corn silage, inexperienced feeders find it somewhat difficult to determine its dry roughage equivalent. Because of the great variation in moisture this can only be approximated, but for practical feeding purposes it may be roughly estimated by dividing the weight of fresh silage by 5. This means that cattle fed 30 pounds of silage are consuming approximately the same amount of air-dry roughage as cattle eating 6 pounds of hay. The remainder of the dry weight of the silage is grain. Dividing the pounds of average corn silage fed by 3 converts it to an air-dry basis. Thus the 30 pounds mentioned earlier contain 10 pounds of air-dry feed (30 ÷ 3 = 10) and 4 pounds of grain (10 − 6 = 4).

FEEDING SILAGE TO CALVES

Some feeders who have found corn or sorghum silage a satisfactory feed for yearling steers consider it too bulky for finishing calves. However, in some respects it is more useful for feeding animals of this age than for older cattle because calves are fed for a much longer time and thus have greater need for variety and succulence in their rations. By starting calves on nearly a full feed of silage, 12 to 15 pounds a head, and adding grain and protein concentrate as the calves grow and attain greater capacity, the calves are gradually got up to a full feed with almost no risk of their going off feed or becoming foundered.

On most farms the silo is empty by April or May, and the change to dry roughage or to still lower roughage levels results in an increased consumption of grain at the time of year when the appetite of full-fed cattle is usually slowed by the onset of hot weather. Should the supply of silage last until calves are finished, the amount should be reduced to 4 or 5 pounds a day with the arrival of summer or about June 1.

The proper amount of silage for calves is a somewhat disputed question. The amount to feed depends greatly on size and age of the calves and length of the feeding period. The Illinois station had excellent success feeding 8 pounds of silage and 2 pounds of legume hay per head daily for the first 6 or 7 months and 6 pounds of silage and 2 pounds of hay thereafter. However, in experiments in which different amounts of silage were compared, equally good gains were secured by feeding 16 pounds of silage and 3 pounds of hay during the first 140 days and approximately half these amounts thereafter. The calves fed in this manner maintained their feed consumption and rate of gain much better during the summer than the other lots. Consequently this plan of feeding appears to be satisfactory for calves that are to be carried into late summer and fall. These roughage levels were fed with a full feed of grain, of course.

VALUE OF A DRY ROUGHAGE WITH SILAGE

Although satisfactory results may be expected from a ration consisting of shelled corn, a protein concentrate, and corn or sorghum silage, feeding experiments indicate that the ration is improved by adding a small amount of dry roughage such as hay or straw. The amount of such roughage eaten is usually only 2 or 3 pounds a day, but it appears to have a noticeable effect on the rate of gain, especially if the dry roughage fed is clover or alfalfa hay. These legumes are rich in minerals and vitamins, both of which are likely to be deficient in a ration in which corn or sorghum silage is the only roughage. Inasmuch as calves have greater need for minerals and vitamins than do older animals, their response to the addition of a legume hay is usually greater. If no legume hay is available, calcium should be supplied in the form of ground limestone.

If the dry roughage fed is straw, hulls, or average to low grade hay, it may be kept before the cattle with little danger of their eating enough to interfere with their consumption of grain or silage. However, if it is good-quality legume hay, it should be fed at the rate of 2 or 3 pounds a head, preferably about noon after the morning's feed of grain and silage has been eaten. Table 174 summarizes the results of several tests which demonstrate the value of small amounts of dry roughage.

Table 174

Value of a Dry Roughage with a Heavy Silage Ration

Number of Trials Averaged	Rations Compared	Daily Gain (lb)	Feed per Pound Gain				
			Shelled Corn (lb)	Cotton-seed Meal (lb)	Legume Hay (lb)	Oat Straw (lb)	Corn Silage (lb)
4 Indiana (two-year-olds)	Shelled corn, cottonseed meal, corn silage	2.37	5.97	1.14	—	—	12.66
	and						
	Shelled corn, cottonseed meal, clover hay, corn silage	2.45	5.74	1.12	1.41	—	11.16
2 Illinois (calves)	Shelled corn, cottonseed meal, corn silage	2.03	5.09	1.02	—	—	7.49
	and						
	Shelled corn, cottonseed meal, alfalfa hay, corn silage	2.20	4.79	0.96	1.53	—	4.64

FEEDING MOLDY AND FROZEN SILAGE

There is seldom any trouble from feeding small quantities of moldy silage to stocker or feeder beef cattle; if the silage is altogether moldy, they will simply reject it. Moldy grain is another matter and should not be fed unless blended in at less than a 5 percent level.

During extremely cold weather a considerable amount of frozen silage is encountered in regions where temperatures drop to 0°F or below, especially in upright silos where little feed is removed each day. Silages above 70 percent in moisture content freeze more solidly than drier silages. This frozen material should be removed from the walls of the silo as soon as it is possible to knock or pry it loose. If the pieces are small and not numerous, they may be piled in the center of the silo after the morning's feed, where they will often thaw before night. With a large quantity of frozen silage, however, or with the temperature much lower than freezing, this method is impractical. Instead, the frozen silage should be piled just outside the silo where it can be watched and fed as soon as it is reasonably well thawed. If left longer, it becomes moldy and unfit for use after thawing.

The presence of small quantities of frozen silage in the ration is unlikely to cause trouble, but the feeding of large amounts is highly inadvisable. Not only is such material unpalatable and difficult for the cattle to eat, but it is likely to cause serious digestive disorders. Excessive scouring is one of the common aftereffects of feeding frozen silage.

The question is often asked as to how long one must wait after filling a silo before it is safe to begin feeding from it. Many successful feeders begin feeding immediately, thus eliminating the top spoilage that occurs with waiting. However, the silage in the upper layer is usually quite warm and lacks the typical mildly acid taste of properly fermented silage; consequently it is not readily consumed by the cattle. It can be mixed with older silage from another open silo if cattle fail to eat it well. Waiting a week or so ensures that most of the fermentation has occurred, the silage will have cooled off, and cattle will start off on the new and strange—to them, at least—feed with far greater relish.

EFFECT OF SILAGE ON SHIPPING AND SALE

Some feeders hold the opinion that silage-fed cattle undergo a much greater shrink between feedlot and market than cattle fed dry roughage. For this reason they sometimes remove the silage from the ration a few days before shipping and supply a dry roughage in its place. However, records kept of the actual shrinkage undergone by silage-fed and nonsilage-fed cattle, respectively, show that this practice is unwise because the silage-fed cattle in reality shrank less than hay-fed cattle. It was true that some silage-fed cattle lost more weight in transit, but they were more inclined to take on a greater fill after they were unloaded. Silage cattle have a reputation at the market for being unusually good "drinkers." Whereas this kind of fame probably does not work to their advantage when it comes to selling price, it does indicate that their owners get good weights. When sold and weighed at the feedlot, silage cattle may have a slight advantage because of greater fill. Selling on grade and yield eliminates the question entirely.

In former years when dry roughage was the orthodox feed for finishing cattle, the few silage-fed steers received at the market were regarded with considerable suspicion. It was feared they would resemble grass-fattened cattle in having a low dressing percentage and dark-colored lean of rather poor quality. Comparisons of the carcasses of such cattle with those of animals finished on dry roughage disclosed no important differences in amount or quality of finish traceable to silage. As the number of silage-finished cattle increased, packers and shippers bought them as a matter of course and frequently found them superior to any dry-roughage-fed cattle included in their purchases. Naturally any discrimination against such cattle soon disappeared, and they were bought on the basis of their actual worth.

Today a large percentage of grain-fed steers received at large midwestern livestock markets have been fed silage, and a buyer seldom stops to inquire whether a given drove has had silage. If the cattle show unusual fill, the buyer may suspect that they have been fed silage, but whether this is the cause of their condition makes little difference to him. He estimates their dressing percentage, judges their worth on the rail, and tries to buy them on that basis.

EFFECT OF SILAGE ON PROFITS

No one ration or plan of cattle feeding is more profitable than another under all conditions. However, if an attempt were made to classify rations on the basis of the net profit realized from their methodical use over a period of years, those containing moderate and fairly liberal amounts of silage probably would be at the top. This is partly because silage-fed cattle have a smaller feed bill than those fed only dry roughage. Then, as a rule, silage-fed cattle develop more uniformly, and a smaller percentage are affected by founder and bloat. Because of their attractive appearance when marketed they frequently sell higher than cattle of similar quality that were fed for the same length of time, but on rations not containing silage.

It is not without significance that nearly all Corn Belt experiment stations have for many years used a ration containing corn silage as a check or standard with which new and little-tried feed combinations are compared. This practice is the result of obtaining better returns from feeding corn silage than from feeding dry roughages in literally scores of experiments made over a period of more than 70 years.

If cattle feeders in the areas where high yields of corn or grain sorghums can be obtained with relative economy fail to use silage made from these feeds rather heavily in their finishing programs, they are ignoring probably their greatest competitive advantage over feeders in other parts of the country who must rely almost exclusively on high-concentrate rations. In fact, taking advantage of this economical roughage source may be the one factor that will keep Corn Belt feeders in a sufficiently competitive position to remain an important segment of the cattle feeding industry.

POISONOUS GAS IN NEWLY FILLED SILOS

It has long been known that gases collect above the silage in tower silos during or immediately after filling. These gases, mostly carbon dioxide, have not been found particularly dangerous although carbon dioxide can cause suffocation. Within the last few years, however, a more ominous situation has been recognized, and numerous fatalities from inhaling poisonous gas have been reported among men working in silos, especially in the main corn-growing states. The toxic gas has been identified as nitrogen dioxide. It seems to develop most readily in corn silage grown in years when drought damages corn that has had heavy applications of nitrogen fertilizer. Under these conditions the plants evidently absorb more nitrates from the soil than they are able to use. If the corn field is weedy and the weeds are chopped with the silage, the chances of toxic gases are greater because weeds are higher in nitrate content. The excess nitrate is released during the ensiling process and, through a series of chemical reactions, is changed into nitrogen dioxide or nitrous oxide gas. Symptoms of "silo fillers' disease" are severe coughing and burning or choking pains in the throat and chest. The condition usually arises as a person enters a silo during filling or within 48 hours after filling. A physician should be consulted if such symptoms occur.

Scientists from the U.S. Department of Agriculture recommend the following safety precautions in filling all tower silos.

1.   Run the blower for 10 minutes before going into a partly filled silo during the silo-filling process. Always keep the blower running while you are inside.
2.   Be alert for irritating odors. Nitrogen dioxide is heavier than air and collects near the surface of the silage. The gas also tends to settle in the silo chute and around the base of the silo in enclosed, attached feed-mixing rooms.
3.   Watch for yellowish brown fumes and chlorine or bleachlike odors—they are a sign of nitrogen dioxide gas. If the silo is dark, use a flashlight so that you can see.
4.   Keep children and animals out of the silo and away from it during filling.
5.   After the silo is filled, wait at least a week before going inside if the blower cannot be operated. Do not let children stay near the silo. If necessary, use a temporary fence to help keep them safe.

COMPARISON OF VARIOUS SILAGES FOR BEEF CATTLE

To the man who lives in the oldest cattle-feeding area, the Corn Belt, the term "silage" usually means ensiled corn. However, outside the region where corn can be successfully grown, various other crops are used for silage purposes. Since the tonnage of corn silage made and fed greatly exceeds that made from other silage materials, it is convenient to compare these other kinds of silage with corn silage as a standard.

CORN SILAGE AS THE PRINCIPAL FEED FOR FINISHING CATTLE

A full feed or varying high levels of corn silage, supplemented with an appropriate amount of protein concentrate and grain, results in economical gains at low cost, as discussed under the subject of limited grain rations in Chapter 15. If the silage is made from high-yielding corn, thin cattle make a noticeable improvement in condition during the first half of the feeding period on silage alone, with no additional corn. Usually it is important to feed as little as 2 or 3 pounds of hay per day, or no hay at all, to obtain the consumption of as much silage as possible. This, of course, means that 1 to 2 pounds of protein concentrate must be fed daily, as well as a mineral supplement.

Cattle with sufficient quality to sell near the top of the market, if they have the necessary finish, usually return more profit if they are fed a high-concentrate ration during at least the last half of the feeding period. Heavy or long yearling steers are more satisfactory for such a plan of feeding than younger cattle, as they attain a satisfactory finish for their grade in a shorter time. Gains made by cattle on limited grain rations are relatively low after 150 to 180 days, depending upon the age of the cattle.

AT WHAT STAGE SHOULD CORN BE ENSILED?

It has generally been thought that the best-quality silage is produced by cutting the corn when the kernels have hardened to the point where the interior has a stiff doughlike consistency, but a large portion of the stalk and most of the leaves are still green. Silage made from immature corn is likely to be sour and "sloppy." Moreover, the cutting of very green corn entails a waste, since the total amount of food nutrients in the corn plant continues to increase until the ears are fully mature. On the other hand, it is difficult to make silage from corn that is almost ripe. Unless sufficient water is added to soften the dry stalks and leaves and permit packing them so firmly that all air is excluded, silage made from overripe corn is very likely to mold.

A measured area of corn ensiled in the undented or "roasting-ear" stage from August 12 to 14 at the Ohio station proved to be much lower in feeding value than an equal acreage ensiled September 5 to 8, when the corn was well dented. Samples taken of the corn at the time of harvest showed that 17 percent less dry matter per acre was obtained from the early-harvested corn. Moreover, considerable juice leached out of the silo after this corn was stored. As a result only 755 pounds of gain per acre were obtained from the immature silage compared with 940 pounds per acre from the well-dented corn. The cattle fed the immature silage were not so well finished and were valued 25 cents per hundred lower than the steers fed the riper silage.

In another study the Ohio station—along with numerous others, it might be added—investigated the possibility of allowing corn to mature beyond the dent stage (the usual time for cutting corn for silage) to permit the crop to store the absolute maximum of nutrients before harvesting. The corn plant normally continues to transfer nutrients to the kernels until the grain moisture content has been reduced to about 30 percent. However, when the plant has reached this stage of maturity, only the upper leaves are green and the moisture content of the entire plant, including ears, has been reduced from the normal silage-cutting moisture level of 70 percent to less than 50 percent. Obviously this drier silage is higher in energy content on an as-fed basis. Thus nutritionists wanted to know if cattle might consume more nutrients per day when such drier silage was fed. If so, rate of gain should increase, and this, combined with a known slightly higher nutrient yield per acre, should result in more beef gains per acre of corn harvested as silage.

The results of the Ohio test are shown in Table 175. The table also supplies data that can be used for another comparison between all-silage, silage-and-corn, and all-corn rations. To balance and equalize all of the rations with respect to protein and minerals, all lots were fed equal amounts of hay, soybean meal, and minerals, in addition to the silage and/ or corn. The heifers eating the more mature silage consumed more silage dry matter and gained faster in the comparison between lots on high silage rations, but required considerably more feed because of poorer utilization of the more fibrous stalk portion of the ration. Thus it is doubtful whether the slight increase in nutrients stored in the more mature plant is enough to offset the depressed utilization that results from late cutting. The chances of spoilage are increased, of course, by the drier silage. Thus there appears to be no real reason for cutting corn at later than dent stage or drier than 45 to 50 percent moisture.

Table 176 shows data from an Illinois study in which the effect of stage of maturity and dry-matter content are related to ear and leaf content of a ton of corn forage. Unfortunately the study was not carried to the more advanced stages of maturity being used by some feeders today. With today's high yields and the drier or more mature stage of cutting now used, much corn silage is being fed today that contains 6 to 7 bushels of corn grain per ton of silage as fed.

As corn matures, the plant portion lignifies and digestibility is adversely affected as it approaches the very mature stages. After a certain point, the depressed digestibility and the leaf loss from the mature plant, as it undergoes weathering, more than offset the deposition of nutrients in the grain as the plant matures. Related studies by Indiana workers are shown in Table 177. They planted a modern corn hybrid on May 14 at the rate of 24,000 plants per acre and fertilized with 210 and 90 pounds of nitrogen and potassium, respectively, per acre. They began harvesting entire-plant material from a measured area 115 days after planting and from equal areas about every 28 days thereafter. They ensiled enough material in oxygen-free bags to conduct digestion trials with steers. Ear-plant separations were made at each cutting date to determine the ear content of the material as harvested.

Digestibility was highest in silage harvested at the earliest dates, then dropped about 3 percent and held steady until February when the moisture content was only 19.7 percent. The most important finding was that the greatest tonnages of digestible dry matter were obtained between the moisture levels of 66.7 and 39.2 percent, which is a rather wide range,

Table 175

A Comparison of Corn Silages Harvested at Two Stages of Maturity (98 Days)[a]

Ration Plan / Stage of Corn Silage	Silage[b]		Silage + Ear Corn		Ear Corn	
	Normal	Mature	Normal	Mature		
Number of heifers	14	14	14	14	14	13
Average initial weight (lb)	480	482	482	491	489	453
Average daily gain (lb)	1.54	1.66	1.95	1.97	2.12	2.33
Average daily ration (lb)						
Corn silage	26.3	16.6	10.0	7.7	—	—
Silage dry matter	7.5	9.2	2.9	4.3	—	—
Alfalfa hay	3.0	3.0	3.0	3.0	3.0	3.0
Ground ear corn	—	—	8.0	8.0	12.9	14.3
Soybean meal	1.0	1.0	1.0	1.0	1.0	1.0
Salt	0.04	0.05	0.04	0.04	0.05	0.05
Minerals	0.04	0.04	0.04	0.04	0.04	0.04
Feed per cwt gain (lb)						
Corn silage	1,703	1,005	512	389	—	—
Silage dry matter	489	559	147	216	—	—
Alfalfa hay	194	181	153	152	141	129
Ground ear corn	—	—	409	404	610	612
Soybean meal	65	60	51	50	47	43
Salt	2	3	2	2	2	2
Minerals	3	2	2	2	2	2
Total air-dry feed per cwt gain (lb)	802	910	779	848	802	788

[a] Adapted from Ohio Agricultural Experiment Station Series No. 129.
[b] Dry matter content of silages: normal, 28.7 percent; mature, 55.6 percent.

Table 176

Ear-Corn Content and Leaf-Stalk Hay-Equivalent in One Ton of Corn Forage at Various Stages of Development[a]

Dry-Matter Content of Forage[b] (%)	Ears[c] (bu)	Leaves and Stalks[d] (lb)	Dry-Matter Content of Forage[b] (%)	Ears[c] (bu)	Leaves and Stalks[d] (lb)
15	0.2	308	24	3.1	312
16	0.5	309	25	3.4	312
17	0.8	309	26	3.8	313
18	1.1	309	27	4.1	313
19	1.5	310	28	4.4	314
20	1.8	310	29	4.8	314
21	2.1	311	30	5.1	315
22	2.5	311	31	5.4	315
23	2.8	312	32	5.7	315

[a] Illinois Bulletin 576.

[b] In the absence of dry-matter determinations, dry-matter content of forage may be estimated on the basis of stage of development, as follows:

Ears beginning to form	15%	Early dent	25%
Kernels forming	17	Well dented	28
Early milk	20	Kernels hardening, most leaves green	30
Late milk	23	Kernels hardening, fewer leaves green	32

[c] On the basis of 15 percent moisture, 70 pounds of ears per bushel.

[d] Hay-equivalent value, on the basis of 15 percent moisture.

suggesting that the question of when to harvest corn silage is determined more by how dry the corn can be for harvesting as silage and still keep the spoilage rate low while allowing normal fermentation.

The type of silo to be used then becomes an important determinant in choosing the optimum moisture level. Seldom can one expect low spoilage rate and desirable fermentation at moisture levels below 60 to 55 percent in conventional upright or horizontal silos, but oxygen-limiting silos can be used with material at moisture levels as low as 35 to 25 percent. In the Indiana study, digestibility at these lower moisture levels was still reasonably good, but leaf loss and stalk deterioration in the field were high when corn was harvested so late in the season, as evidenced by the 80 to 90 percent ear content of corn harvested in late December and late January. Machine-harvesting of such dry material also would result in much ear loss as well.

Table 177

Effect of Stage of Maturity on Yield of Digestible Dry Matter per Acre of Corn Harvested for Silage[a]

Date of Harvest	Days Following Planting	Moisture Content (%)	Dry Matter Content (%)	Yield of Dry Matter per Acre				
				Total Plant (tons)	Digest-ibility (%)	Ear Content (%)	Digestible Dry Matter (tons)
Sept. 7	115	75.9	24.1	5.7	74.8	53.9	4.3
Oct. 5	143	66.7	33.3	7.0	71.4	57.0	5.0
Nov. 2	171	47.5	52.5	7.2	72.4	53.7	5.2
Nov. 30	191	39.2	60.8	6.4	72.0	55.7	4.6
Dec. 28	227	38.9	61.1	3.9	71.3	81.9	2.8
Jan. 25	255	25.0	75.0	3.5	72.2	93.3	2.5
Feb. 22	283	19.7	80.3	3.5	68.4	92.5	2.4

[a] Indiana Research Progress Report 331.

The Oklahoma Extension Service makes the following recommendations as to the most desirable time to cut forage crops for silage.

Corn: when grain is in the dent stage.

Sorghums: when grain is in the dough stage.

Alfalfa: one-fourth to one-half bloom stage.

Cereals: soft dough to hard dough stage.

Vetch: full bloom stage.

Sudan: early heading stage.

Grasses: bloom to early heading stage.

Weather conditions, of course, often determine the time of making silage. During periods of severe drought it may happen that the corn must go into the silo with the base of the stalk and the lower leaves brown and dry and the ear and upper portion of the stalk still green. Whereas first-class silage cannot be made under such conditions, to delay longer would permit the drying up of the entire plant.

VARIATION IN FINENESS OF CHOP

Although the subject of degree or fineness of chop and type of chopper for corn silage might more appropriately be discussed in Chapter 24, which deals with feed preparation, the subject is introduced here because the trials to be described also include data on effect of stage of maturity on feeding value of corn silage. Fineness of chop is obviously related to degree or firmness of packing in the silo, hence to the rate of fermentation and possibly even the end products of fermentation. There is also the possibility of an interrelationship between degree of maturity and fineness of chop of corn silage with respect to quality or feeding value of the final product.

The Michigan station conducted a comprehensive experiment in which corn from the same field was chopped on three dates—in mid-September, mid-October, and mid-November—and stored in two separate silos on each date. The corn in one silo was chopped fine ($\frac{3}{8}$ inch) and the other was chopped medium fine ($\frac{1}{2}$ to $\frac{3}{4}$ inch). The six silages were fed to heavyweight calves for 180 days. In addition to the respective silages, the calves received about half a feed of ground shelled corn and supplement mixture (1 percent of body weight daily). In addition to supplemental protein the supplement contained appropriate levels of calcium, phosphorus, antibiotic, vitamins A and D, salt, and growth hormone. The data of greatest interest are contained in Table 178.

right

SILAGE AS A FEED FOR BEEF CATTLE   583

Table 178

Comparison of Fine- and Medium-Chopped Corn Silage Harvested in September, October, and November (180 Days)[a]

	Sept. 13		Oct. 17		Nov. 14	
Agronomic data						
Harvest date	Sept. 13		Oct. 17		Nov. 14	
Average dry matter (%)	28.2		48.2		59.6	
Adjusted yield of silage, tons[b]	17.3		15.2		13.5	
Bushels corn per ton of silage[b]	4.3		4.9		5.5	
Fineness of chop	Fine	Medium	Fine	Medium	Fine	Medium
Pounds dry matter per cubic foot of silage	13.4	11.1	12.0	10.0	11.9	11.0
Change in volume from fine to medium chop (%)		−16.9		−16.9		−8.0
Cattle performance						
Number of steers	18	18	18	18	18	18
Average initial weight (lb)	539	538	539	538	538	539
Average daily gain (lb)	2.89	2.85	2.78	2.63	2.78	2.69
Average daily ration (lb)						
Corn silage	33.0	32.7	19.2	19.3	17.1	18.2
Corn + supplement	8.33	8.17	8.10	8.02	8.04	7.81
Total[c]	19.5	18.5	19.0	18.5	19.5	19.1
Feed per cwt gain (lb)[c]	678	650	689	703	702	710

[a] Adapted from data in Michigan Beef Cattle Day Report, 1967.
[b] 30 percent dry matter.
[c] 85 percent dry matter.

There was no frost prior to the September harvest date, but both of the other cuttings were repeatedly frosted. The November cutting was also subjected to a 9-inch snowfall. Water was added to the November cutting but not to the others.

The total tons of dry matter produced per acre were lower with each successive cutting, no doubt because of leaf loss after the frost and progressive leaching of soluble stalk components. The increase in bushels of corn grain per ton of silage with increasing maturity is a reflection of the loss of stalk weight already mentioned and a slight increase in corn grain weight itself. The fine-cut silage packed better, as reflected by the 8 to 17 percent greater density in the silo. This means that about 10 percent more fine-cut silage can be stored in the same silo space as medium-cut silage.

The pooled data for the harvest date comparisons show that the cattle fed September-cut silage gained 0.17 and 0.13 pound per day faster than those fed October and November-cut silage, respectively. The feed required per hundredweight gain was 664, 691, 706 pounds for the three lots, in order of harvest date. Carcass grades and selling price also favored earlier dates. With respect to harvest date, the September-cut silage was superior in every respect, with each month's wait being increasingly less desirable.

The pooled data for fine versus medium cutting show that differences in cattle performance were small but slightly in favor of fine cutting. The later the harvest date, the greater the advantage from fine cutting. When the possibly greater power requirements for fine cutting are considered, chopping corn silage finer than one-half inch at normal harvesting time seems unwarranted.

EFFECT OF CORN YIELDS ON VALUE OF SILAGE

It would seem that silage made from hybrid corn yielding 80 to 100 bushels per acre would have a much higher feeding value per ton than the silage made from open-pollinated corn yielding considerably less. Also it would seem that silage made during a dry year would have a low feeding value because of the low yield of grain. However, corn yields have a much greater effect on the yield per acre of silage than on the grain content and hence the feeding value of silage per ton (see Table 179). Although the yields of both dry corn and silage at the Illinois station have varied widely from year to year, the corn content per ton of silage has remained rather constant, especially for the hybrid corn varieties.

Silage made and fed at the Nebraska station during the extemely dry

Table 179

Effect of Yield of Corn on Corn Content of Silage[a]

	Yield of Corn per Acre (bushels)[b]	Yield of Silage Corn per Acre (tons)	Corn per Ton of Silage (bushels)
1. Open-pollinated corn			
1925	72.9	8.84	8.24
1926	44.7	8.71	5.13
1927	68.5	14.73	4.65
1928	78.9	12.62	6.25
1929	48.3	9.70	4.98
Average	62.7	10.92	5.85
2. Hybrid corn			
1946	92.2	13.29	6.64
1947	66.9	10.51	6.37
1948	106.5	16.68	6.39
1949	91.0	12.63	7.20
1950	52.0	9.35	5.38
Average	81.7	12.49	6.40
Percentage increase	30.3	14.4	9.4
3. Dry years			
1934 (dry year)	35.0	6.38	5.49
1935 (normal rainfall)	57.1	10.25	5.57
1936 (dry year)	33.9	7.51	4.50

[a] Illinois Agricultural Experiment Station, unpublished data.
[b] 14 percent moisture.

years 1934 and 1936 produced gains on stocker calves and yearlings as large as those secured in previous tests with normal corn silage. The corn from which the silage was made grew stalks 5 to 7 feet high. Only a few ears filled, and those were immature and watery. Part of the corn was relatively immature when cut and part of it required water as it went into the silo. The yield of silage per acre was very low, but apparently many of the feed nutrients that would have gone into the ears were still present in the stalks and leaves and thus were available to the cattle.

Numerous studies have shown that fertilizers, which increase tonnage of silage, also increase bushel yields per acre to about the same degree. In general, then, silage made from highly fertilized fields has a feeding value, per ton, equal to probably little higher than ordinary silage. The 125- to 150-bushel-per-acre yields, which are becoming common on good Corn Belt livestock farms and in the irrigated sections, result in silages that are higher in grain content, probably because the high planting rates produce relatively light stalks but many ears. New research data are needed on the interaction of row spacing, planting rate, fertilizer level, and variety of corn on grain:stalk ratio in silages made therefrom.

VARIETIES OF CORN FOR SILAGE

Although certain corn hybrids yield an enormous tonnage of silage owing to their rank growing tendencies, such silage is not especially valuable for finishing cattle because of the low percentage of grain present. Silage made from such "silage hybrids" may be suited to dairy cattle that need a highly succulent ration, but the ordinary corn varieties that produce a high yield of sound grain make a much more satisfactory silage for finishing cattle. Not only does it require smaller amounts of additional concentrates when fed to animals that are being finished, but it also produces more growth and development when fed to heifers and stockers that are being carried through the winter.

The Minnesota station conducted a trial with stocker calves in which they compared regular corn silage with silage made from a "high-sugar" hybrid corn and with silage made from one of the popular new sorgo-Sudan hybrids. All lots were fed a full feed of the particular silage, limited ground ear corn, supplement, and about 2 pounds of hay. Separate metabolism trials showed that the two corn silages had similar digestibility and TDN values, but that the sorgo-Sudan silage was 20 percent lower in this respect. The feeding trial data reported in Table 180 indicate, however, that regular corn silage was significantly superior to high-sugar silage, and even more so to the sorgo-Sudan silage. As yields were comparable, and all low for that matter, regular corn again appears to be the best type of corn to grow for silage.

Corn breeders are developing hybrids with grain that is higher in both protein and oil content. As yet, yields are below those of regular corn varieties, but in feeding trials with both stockers and feeders at the Illinois station, protein supplements were successfully omitted from the ration. Thus if yields can be made comparable, these hybrids appear promising indeed. It should be mentioned that the high-lysine corn varieties recently developed, which, being higher in this essential amino acid, are especially valuable in formulating swine and poultry rations, cannot be expected to have higher value for cattle because ordinarily individual amino acids are

Table 180

Performance Data on Calves Fed Regular Corn Silage, High-Sugar Corn Silage, or Sorgo-Sudan Silage (133 Days)[a]

Silage Fed	Regular	High-Sugar	Sorgo-Sudan
Agronomic data			
Wet yield per acre, tons	7.8	10.6	9.7
Dry matter (%)	36.2	26.0	28.7
Dry matter yield per acre (lb)	5,646	5,490	5,555
Price per ton, as fed[b]	$9.63	$7.50	$7.94
Value per acre, as fed	$75.28	$73.20	$74.06
Feedlot data			
Number of calves	18	17	18
Average initial weight (lb)	492	500	495
Average daily gain (lb)	1.86	1.68	1.37
Average daily feed (lb)			
Silage	9.0	8.7	7.9
Supplement	1.2	1.2	1.2
Ground ear corn	3.8	3.8	3.8
Brome-alfalfa hay	2.0	2.0	2.0
Total	16.0	15.7	14.9
Feed cost per cwt gain	$13.24	$14.26	$17.00
Silage dry matter per cwt gain (lb)	424	458	516
Gain per acre of silage fed (lb)	1,332	1,199	1,077

[a] Adapted from Minnesota Beef Cattle Feeders' Day Report.
[b] Assume a price of $8 per ton of 30 percent dry matter silage, of 1.33 cents per pound dry matter. Prices for silages fed were adjusted accordingly, assuming the same value for each pound of dry matter.
[c] 88 percent dry matter basis.

not required as such. However, as mentioned earlier, if the limiting amino acids, such as methionine, could by some means be "protected" from microbial degradation, these special hybrids could become more valuable, even for ruminants.

Waxy corn or *waxy maize,* which is discussed in Chapter 16, contains endosperm starch that is entirely in the form of amylopectin in contrast to the starch of regular dent corn, of which 25 percent occurs as amylose and 75 percent as amylopectin. In laboratory tests, waxy corn has shown higher digestibility values and a different fermentation pattern, favoring propionic acid production. In some feedlot trials, waxy maize has been superior to regular corn, especially when natural protein supplements such as soybean meal were used. Other trials have been negative, indicating that more definitive research should be awaited.

An experimental *brown-midrib hybrid corn* is under study that appears promising because it contains only about 4.5 percent lignin in the stalk portion of the plant, as compared to 6.8 percent for regular corn. In a

Table 181

Item	bm_3	Normal
Effect of Brown-Midrib Mutant (bm_3) on Composition, Digestibility, and Intake of Corn Stover Silage Fed to Cattle[a]		
Fiber (%)	64.8	68.1
Lignin (%)	4.5	6.8
Digestibility (%)		
Dry matter	55.9	49.3
Fiber[b]	60.4	52.7
Daily feed intake (% of body weight)	2.06	1.47
Daily gain (lb)	2.39	1.17

[a] Indiana Beef-Forage Research Day Report, 1975.
[b] Cell wall constituents.

preliminary study using corn stalk stover, Indiana workers obtained the results reported in Table 181. Note the improvement in rate of gain and feed intake in the brown-midrib stover silage lot. This improvement is no doubt due to the reduced lignin and increased digestibility of total dry matter, and especially the fiber portion. In another trial with entire-plant silages, differences were smaller as the fiber content of both silages was lower and the digestibility advantage applied only to the stalk portion of the diet. The practical problem of plant lodging must be solved and grain yields remain low, but the finding of a mutant gene that can significantly lower the lignin content of this widely and abundantly available forage source is encouraging indeed.

VALUE OF FROSTED CORN FOR SILAGE

The silo offers the best way to utilize corn that has been frosted before the ears are mature. If put into the silo immediately after the frost, the silage has practically the same feeding value as it would had the corn been ensiled just before the frost. If, however, the corn is allowed to stand until the leaves are withered and dry, many of the leaves are blown away and the quality of the silage is greatly reduced. A heavy rain on frosted corn also causes much damage by leaching out some of the soluble feed nutrients. It is usually necessary to add some water to frosted corn to make it pack well in the silo.

SORGHUM SILAGE

The various forage sorghums, such as Tracy, Sart, Atlas sorgo, and many new hybrids, are quite satisfactory silage crops and are grown extensively

Fig. 90. A two-row self-propelled chopper can harvest up to 500 tons per day of a forage sorghum crop. Grain content of the silage made from some varieties of forage sorghum may be as high as 40 percent, on a dry-matter basis. (John White, New Mexico State University.)

for this purpose in the semiarid and irrigated parts of the Southwest. The hybrid forage sorghums now in use far surpass those used prior to the 1960s. Yields of silage per acre are usually larger for forage sorghums than for corn, especially in years of low rainfall during the growing season or when both are grown on irrigated acreage. Should early frosts occur, less damage is suffered by the sorghums than by corn because the leaves are not so easily lost.

Silage made from the forage sorghums is somewhat lower than corn silage in both protein and fat and is higher in crude fiber. With few exceptions, feeding experiments comparing corn and sorghum silages point to the superiority of corn silage. Although the difference is not great, it is sufficient to indicate that corn is the more valuable silage crop in areas where corn can be grown competitively, except during unusually dry weather. Under irrigation conditions, the added yield of sorghums as silage makes an acre of forage sorghum equal to an acre of corn.

It is rather difficult to make direct comparisons between corn silage and silage made from the newer hybrid forage sorghums and sorghum-Sudan hybrids because each station doing the testing usually uses a different hybrid or cross, generally bred or selected for the particular climatic

Table 182

Comparisons Between Corn, Sorghum, and Sorghum-Sudan Silages Fed to Finishing Cattle[a]

Type of Silage	156-Day Trial		184-Day Trial		
	Corn	Sorghum-Sudan	Corn	Oat	Sorghum
Average initial weight (lb)	730	738	451	450	450
Average daily gain (lb)	2.76	2.19	2.12	1.70	1.66
Average daily ration (lb)					
Silage	48.4	43.0	25.7	22.4	24.5
Ground ear corn	—	—	5.9	5.7	5.7
Cracked shelled corn	12.7	12.7	—	—	—
Supplement	2.5	2.5	1.9	1.9	1.9
Feed per cwt gain (lb)					
Silage	1,755	1,962	1,213	1,315	1,481
Ground ear corn	—	—	279	333	342
Cracked shelled corn	212	268	—	—	—
Supplement	90	113	90	112	115
Feed cost per cwt gain[b]	$15.33	$17.63	$13.92	$14.97	$15.78

[a] Adapted from data in South Dakota Beef Cattle Field Day.

[b] Costs of silage, per ton: corn, $8; sorghum-Sudan, $7.93; oat, $6; sorghum, $6.

conditions prevailing in the state where recommended. Table 182 shows the results of tests conducted in South Dakota. The results are similar to, but not necessarily the same as, those that have been obtained in other sorghum-growing states. It is apparent that, as yields of both corn and sorghum varieties or hybrid number have improved through breeding and better cultural practices, the grain content of corn silage has improved faster than the grain content of sorghum silage, in relation to yield of silage expressed as tons per acre. Therefore it appears that more grain must be added to forage-sorghum silages, and especially the sorghum-Sudan hybrid silages, than formerly if they are to be competitive with corn silage. This means that the spread in price or value between corn and sorghum silages is widening with improved yields. Farmers who do not actually test their forages for moisture content as they are filling silos nearly always underestimate the moisture in forage sorghums. Therefore the sorghums are usually put up wetter than they should be, and this difference in moisture content alone accounts for much of the variation in performance between steers fed sorghum and those fed corn silages.

HAY-CROP SILAGE

This term refers to silage made from alfalfa, clover, bromegrass, timothy, and other crops that are ordinarily grown, either in pure stands or in mixtures, for hay. Other designations for this kind of silage are legume-grass silage, grass silage, or merely alfalfa, red clover, or bromegrass silage, if one of these species comprises most or all of the freshly cut forage. Whether it is better to harvest these crops as silage rather than cut them for hay depends on a number of factors. In general, the advantages of ensiling over harvesting for hay are as follows.

1. The crop is less seriously damaged by rains during the harvest period.
2. A much smaller percentage of the leaves is lost; leaves are the most valuable portion of the forage.
3. Considerably more protein is obtained per acre, mainly because fewer leaves are lost, but partly because some crops, principally the grasses, are cut earlier for silage than for hay and thus the protein content is higher.
4. The carotene present at cutting time is preserved much better in silage than in hay.

Fig. 91. Direct-cutting of alfalfa for hay-crop silage. Such direct-cut silage is best stored in bunker or trench silos as the high moisture content will result in excess seepage in upright silos. (John Deere.)

5. Hay-crop silage is somewhat more palatable and digestible than hay; consequently it may compose a larger percentage of the ration of finishing cattle without seriously reducing the rate of gain.
6. Silage is more suited to mechanized feeding operations.

The principal disadvantage of using legumes and grasses for silage rather than for hay is the higher cost of harvesting. This added cost is represented chiefly by the expensive field harvesters, the silo-filling equipment, and the silos themselves. The use of surface or horizontal silos reduces the cost considerably. If, by chance, silage-making equipment is already available for making corn silage it can, of course, be used for making hay silage with little additional expense for the pickup head.

MAKING HAY-CROP SILAGE

Much more care and judgment are needed in making high-quality hay-crop silage than in making corn silage. Cutting the forage when it is either too ripe or too green, allowing it to lie in the swath or windrow too long before being ensiled, or failing to add the proper amount of preservative, if required, to the forage, may result in unpalatable silage of low feeding value.

Two types of field cutters are available. One type mows and chops the forage in a single operation, usually called direct-cutting. The other type picks up the forage from the windrow after it has been cut with a mower and windrowed with a side-delivery rake, or has been cut and swathed with a swather-crimper. If the first type of harvester is used, the forage should be cut when somewhat riper than is advisable for the second type, because little moisture is lost by the forage during handling. Cutting the forage when too green with this harvester results in a wet, "sloppy" silage that is unpalatable and of low feeding value because of its high moisture content. If the moisture content is unduly high (75 to 82 percent), a considerable percentage of the feed value is lost in the juices that seep through the doors and walls of the upright silo or from the ends of trenches and bunkers. Free water in a silo, whether or not seepage occurs, is to be avoided because it is highly destructive to the silo, first, through the great hydrostatic pressure exerted on the walls and, second, through the constant corrosive action of the silage acids on the walls of concrete and metal silos. Treating concrete silo walls with epoxy coatings protects the silo, but constant checking and maintenance are required.

Most authorities agree that the best quality of conventional hay-crop silage is made from forage that contains 65 to 70 percent moisture or 30 to 35 percent dry matter at the time it is put into the silo. If it is desired to

Fig. 92. One method of adding ground shelled corn as a preservative to direct-cut alfalfa. (University of Illinois.)

harvest the crop when the moisture content is higher, the forage should be allowed to lie in the swath or windrow to permit evaporation of the excess water. When the forage is mowed, raked, and chopped with a field cutter from the windrow without waiting for it to dry, on a warm sunny day the moisture content drops 3 or 4 percent between the time the forage is cut and the time it is stored in the silo. Mowed forage, lying in the swath, loses about 5 percent moisture per hour on a sunny day.

Color of hay-crop silage can be used as an indicator of digestible protein content. South Dakota workers found that when hay-crop silage underwent a fermentation that resulted in above-average temperature rises, the digestibility of the protein was reduced correspondingly. This may not be detected in a routine chemical analysis for total nitrogen content. Thus, if the haylage routinely becomes brown or chocolate-colored upon storage, the producer should examine his technique. Possibly he needs to add a high-energy preservative. Such dark-colored haylage may also clue the feeder to add a protein supplement for optimum performance.

HAYLAGE

The term "haylage" is in common usage among researchers, farmers, and cattle feeders alike. The product actually is little different from legume or grass-legume silage except in moisture content. Most haylage is harvested with a swather-crimper and left in the swath until it has dried down to 40 to 50 percent dry matter. Haylage has been called "wet hay" or "dry silage." Apparently because of its higher dry-matter content it is generally voluntarily consumed at a higher rate than silage made from the same crop, when intake is expressed on a dry-matter basis. Haylage also is not so acid as silage and has a less pungent, less pronounced butyric acid odor. Because of its higher intake, it apparently can be used as a partial or complete substitute for protein supplement, as seen in Table 183. It can also be seen from the "level of haylage" study in the table that this product can be substituted for the corn in the ration when corn is high-priced.

The storage of haylage presents some special problems, as it is fluffy and difficult to pack. It is best stored in an airtight metal silo, but can be successfully stored in conventional concrete, open-top silos if the walls are in perfect condition and if the top is sealed with plastic and green chop, limestone, or soil for ballast. Note that the studies covered in Table 184 show no real differences between haylages stored in the two types of upright silos. Also note that haylage is not quite comparable to hay but somewhat superior to conventional legume-grass silage made from the same crop.

RECONSTITUTED HAY

Some cattle feeders who have an airtight silo built into an automated feeding system and who thus are not equipped to feed hay will, after the silo has been emptied of conventional haylage, sometimes refill it during the winter or early spring with chopped hay to which water is added. By doing this they need not change rations during the feeding period for a particular drove of cattle. Such chopped hay apparently undergoes fermentation similar to haylage.

FEEDING VALUE OF HAY-CROP SILAGE

Several experiments have compared the feeding value of silage made from grasses and legumes with that of hay made from the same field and also with corn silage grown and harvested the same year. Extensive tests

Table 183

Haylage as a Protein and Roughage Source in Cattle-Finishing Rations[a]

Type of Study	Level of Supplement		Level of Corn			Level of Haylage		
Length of Feeding Trial	182 Days		168 Days			112 Days		
Initial weight of all cattle (lb)	557		760			525		
Number of cattle per lot	40	40	40	40	10	10	10	10
Average daily gain (lb)	2.32	2.29	1.96	2.62	1.68	2.06	2.20	2.30
Average daily feed (lb)								
Haylage[b]	9.2	9.2	16.1	9.7	14.7	11.6	8.1	6.1
Corn[c]	11.6	11.2	7.6	14.7	3.1	6.2	9.3	10.3
Supplement A	—	1.0	1.5	1.5	—	—	—	—
Feed per pound gain (lb)	9.0	9.5	12.4	9.9	10.6	8.7	7.9	7.1
Feed cost per cwt gain[d]	$28.45	$31.09	$35.13	$34.81	$21.43	$21.99	$24.55	$25.43

[a] Adapted from data in Indiana Cattle Feeders' Day Report.
[b] 87 percent dry-matter basis.
[c] 15.5 percent moisture.
[d] Feed prices used: haylage, $30 per ton of 85 percent dry-matter material; corn, $90 per ton; supplement A, $140 per ton.

Table 184

Effect of Type of Storage on Making Haylage[a]

Days Fed and Initial Weight (lb)		196 Days—525 Pounds				168 Days— 760 Pounds	
Type of Roughage	Chopped Hay	Hay-lage	Hay-lage	Silage		Hay-lage	Hay-lage
Type of Storage	Mow	Gas-tight Silo	Open Silo	Open Silo		Gas-tight Silo	Open Silo
Number of cattle per lot	20	20	20	20		40	40
Average daily gain (lb)	2.12	2.08	1.92	1.88		2.26	2.32
Average daily feed (lb)							
Roughage[b]	5.8	5.8	5.6	5.7		12.5	13.3
Corn[c]	11.4	11.7	10.9	11.2		11.2	11.1
Supplement A	1.0	1.0	1.0	1.0		1.0	1.0
Feed per pound gain (lb)	8.5	8.9	9.1	9.8		11.3	11.4
Feed cost per cwt gain[d]	$30.90	$33.33	$33.57	$35.08		$33.70	$33.15

[a] Adapted from Indiana Cattle Feeders' Day Report.

[b] 87 percent dry-matter basis.

[c] 15.5 percent moisture.

[d] Feed prices used, per ton: roughage, $30; corn, $90; supplement A, $140.

were made at the Michigan and Pennsylvania stations. At each station three lots of cattle were fed to compare corn silage, alfalfa silage, and alfalfa hay, respectively. At the Pennsylvania station all lots of cattle were fed the same amounts of ground ear corn, while the silages and hay were fed according to appetite. In the Michigan tests no grain was fed to the corn silage lots during the first half of the test, and about one-third of a full feed during the last half. Grain was fed to the legume hay and silage lots in amounts to make these cattle gain as much as those fed corn silage. Results of these two series of experiments are given in Table 185.

It will be noted that rations based on silage made from alfalfa or other legume crop required the addition of as much as 1 pound of grain per day per hundredweight in order to produce gain and finish comparable to that produced by a full feed of good quality corn silage and hay plus protein supplementation.

MISCELLANEOUS SILAGE MATERIALS

Almost any green material can be ensiled successfully if it is tightly packed in a well-built silo so that no air pockets remain and if sufficient

Table 185

Comparison of Corn Silage, Legume Silage, and Legume Hay for Finishing Cattle

	Pennsylvania (Average 3 Trials)			Illinois (1 Trial)			Michigan (Average 4 Trials)		
	Corn Silage	Alfalfa Silage	Alfalfa Hay	Corn Silage	Alfalfa Silage	Alfalfa Hay	Corn Silage	Alfalfa[a] Silage	Alfalfa[a] Hay
Average initial weight (lb)	627	627	630	852	854	852	491	493	488
Average daily gain (lb)	2.18	2.15	2.06	2.87	2.55	2.45	1.81	1.93	1.81
Average daily ration (lb)									
Corn	12.6[b]	12.6[b]	12.6[b]	16.3	15.5	16.5	2.0[c]	9.2	9.2
Protein concentrate	1.5	—	—	2.3	1.0	1.0	1.6	1.0[d]	1.0[d]
Silage	15.1	20.1	—	22.0	21.7	—	28.0	25.0	—
Legume hay	—	—	8.3	2.0	2.0	6.8	2.0	—	9.5
Feed per cwt gain (lb)									
Shelled corn	581[b]	586[b]	609[b]	568	610	671	111	473	506
Protein concentrate	68	—	—	80	39	41	87	9	10
Silage	688	938	—	766	853	—	1,532	1,293	—
Alfalfa hay	—	—	399	70	79	279	136	—	504
Dressing percent	60.3	60.0	59.8	59.3	59.5	59.4	—	—	—

[a] A mixture of alfalfa and red clover fed in two trials.
[b] Ground ear corn fed in Pennsylvania trials.
[c] Average for entire period but fed during only last half.
[d] Average for last 30 to 60 days, during which time it was fed.

carbohydrate material is present to promote fermentation. Naturally some crops are better suited for silage purposes than others. In addition to having a sufficiently high carbohydrate content to produce enough organic acids to arrest the action of putrefying bacteria before they cause decomposition, a good silage material must have a physical texture that produces an abundance of fine particles to ensure close packing. All plants that have an abundance of leaves in proportion to the amount of coarse, woody stems are good silage materials in this respect. A silage crop should also produce a heavy yield per acre. Otherwise the money or time spent in cutting, hauling, and land rental will be so great that the cost of making the silage will be uneconomical.

Few of the silages discussed below have been used extensively in either practical or experimental beef cattle feeding. However, their yield and other qualities make them suitable silage crops for those regions where they can be successfully grown.

Sudan Grass.   Sudan grass belongs to the same group of plants as the grain sorghums. It differs from the sorghums principally in being finer-textured and in having a comparatively light yield of seed. When seeded in rows or broadcast in fertile soil it grows 6 to 8 feet tall and yields an enormous tonnage of green forage per acre. Sudan grass intended for silage should be planted in rows to encourage better development of heads and leaves and greater seed production. It should be cut when the seeds are in the dough stage, although very good silage can be made at almost any stage of maturity. In the southern states where the growing season is long, both the first and second growths of Sudan grass are available for silage. The newer sorghum-Sudan hybrids are intermediate between Sudan and forage sorghum in feeding value, but at least 50 percent higher-yielding.

Oat and Pea Silage.   A mixture of oats and field peas is often grown for forage purposes in the northern United States and in Canada. If the supply of such forage is greater than needed for pasture, its successful harvest for other purposes is often a perplexing problem because oats and peas do not usually mature at exactly the same time. The silo offers a satisfactory way of harvesting such a crop. The carbohydrates of the oats ensure successful preservation of the pea vines which, if ensiled alone, might spoil because of their high protein content. Silage made from oats and peas is both a palatable and nutritious feed for beef cattle, as shown by Table 186.

Oat Silage.   Oats are grown throughout the country, but they seldom return a satisfactory profit to the farmer, because of their low yield and

Table 186

Oat and Pea Silage for Finishing Cattle[a]			
Full Feed of Grain Plus:	Prairie Hay	Oat and Pea Silage Prairie Hay	Sunflower Silage Prairie Hay
Average daily gain (lb)	1.79	2.48	2.06
Feed per pound gain (lb)			
Grain mixture	5.87	4.06	5.49
Hay	9.24	0.96	1.15
Silage	—	13.24	15.00
Linseed meal	0.55	0.40	0.48

[a] University of Alberta, Canada, Mimeographed Report.

low market price. Their yield is often greatly reduced by the sudden advent of hot dry weather a few days before they are ripe enough to cut. This hazard can be avoided, in part at least, by cutting the oats while they are still green and making them into silage. Oats that are to be ensiled should be cut when the kernels are in the dough stage.

In the northern and northwestern states, where oats are grown much more successfully than corn, oat silage may well take the place of corn silage in beef cattle rations. Extensive tests at the Illinois station and others show that oat and other cereal silages used in finishing rations have a feeding value comparable to that of hay-crop silage but somewhat below that of corn silage (see Table 187). Perhaps even more important than the increased income from the oat crop as a result of harvesting as silage is the improvement in stand obtained from the legume-grass seeding that usually accompanies the oat crop. A good clipping of hay from the new legume-grass seeding during the first fall is not uncommon when the oat nurse crop has been removed as silage instead of by the usual combining procedure.

Millet Silage.   Formerly several varieties of millet were commonly grown by Corn Belt farmers as emergency hay corps. Millet is still occasionally sown in cornfields where the stand of corn has been ruined by floods or insects too late in the season to warrant replanting. Millet is a rank-growing crop and produces a high yield of hay per acre. The quality of hay is poor, however, because of its coarse texture and the numerous hard seeds found in the heavy heads.

Experiments show that ensiling millet greatly improves its value as a feed. In the silo the seeds absorb sufficient water to make them soft and more easily eaten by cattle. Because little labor is involved in growing and harvesting millet, and because of the high yields and the fact that it matures in time to allow it to be seeded after a crop of wheat or early oats

Table 187

Oat Silage for Finishing Cattle

| | Two-Year-Old Steers | | | | Yearlings | |
| | University of Alberta, Canada (Average of 3 Trials) | | Illinois Experiment Station | | Illinois Experiment Station | |
	Oat Silage (lb)	Sunflower Silage (lb)	Oat Silage (lb)	Corn Silage (lb)	Oat Silage (lb)	Hay-Crop Silage (lb)
Average daily gain (lb)	2.48	2.28	2.88	3.08	2.09	2.29
Feed per cwt gain (lb)						
Grain	408	525	568	507	535	504
Protein concentrate	37	41	81	72	54	4
Hay	129	161	70	65	—	—
Silage	1,300	1,562	608	622	1,126	1,158

Fig. 93. The income from an oat crop, when harvested as silage and fed to stockers or cows, may be doubled as compared with the usual harvesting methods. (University of Illinois.)

is removed, millet silage can be produced at very low cost per ton. In feeding value, however, it is distinctly inferior to corn silage, at least when fed as the principal ingredient in the ration.

Canning Refuse.   The pea vines, cobs and husks from sweet corn, and beet tops that accumulate around large canning factories are quite generally used for finishing beef cattle. Frequently these materials are allowed to accumulate in huge stacks, which in time become "natural silos" through the exclusion of air from the interior by the decomposition of the material at the surface. Not only the top but also the sides of the pile may be spoiled for a depth of 2 to 3 feet, depending on the nature of the material and the care with which it was stacked.

Some farmers living in the vicinity of canneries realize the feeding value of this refuse and are willing to buy it and store it in modern silos. Often it is returned to the owners of the land on which the crops were grown in partial payment for use of the land by the canneries, or it is returned in partial payment for the canning crop purchased.

Cattle feeding is quite general in the vicinity of canneries, both because of the large amount of cheap feed available and because the supply of manure helps keep the soil at maximum productivity. Owing to the wide variation in silages resulting from the different crops, stages of cutting, and methods of storing, no definite rules for feeding can be given. In general, better returns are secured by feeding them in combination with grain and hay than by feeding them alone.

Wet Beet Pulp.   In the sugarbeet-growing areas of Colorado and other beet-producing states, large quantities of wet beet pulp are used in finishing cattle for market. Whether it is fed directly from the mill or from a silo, its feeding value compares favorably with corn silage. In feeding trials at the Colorado station 1 ton of corn silage replaced 3,980 pounds of wet beet pulp, but required 43 pounds more beet molasses, 33 pounds more cottonseed cake, and 74 pounds more alfalfa hay.

Potato Silage.   Silage made from surplus or cull potatoes compared favorably with corn silage as a feed for beef calves in tests of these feeds at the Colorado station. Each ton of potato silage replaced 2,642 pounds of corn silage, 42 pounds of barley, and 7 pounds of linseed cake, but required 476 pounds more alfalfa hay.

CHAPTER 20
FINISHING CATTLE ON PASTURE

In the farmer-feeder operation, the contribution of pasture, as such, to the finishing ration may vary all the way from being the only feed used to very little. The first circumstance occurs when older feeder cattle, such as long yearling or older steers or slaughter cows, are sold for slaughter directly off grass. The degree of condition attained naturally is seldom sufficient from such slaughter cattle to grade above high standard or good. At the other extreme are cattle that are fed finishing rations on pasture at very heavy stocking rates with the roughage sources consisting of the small amount of pasture forage they may eat. As pastures are seldom used by commercial feedlots for finishing cattle, the material in this chapter is applicable only to farmer-feeders, some of whom are found in all parts of the United States.

In the traditional Corn Belt farmer-feeder area and in some irrigated sections of the West, pasture makes its greatest contribution in the finishing programs that are conducted during the summer. In this circumstance various combinations of pasture and concentrates are used, depending on the relative amounts of pasture to be harvested and sold through feeder cattle and the amounts of grain or concentrate available, the time at which cattle are to be sold, and the availability and skill of the labor being used. In parts of the Southwest and South, where winter small grains are grown, the feeding of grain on pasture is more likely to occur in fall, winter, and spring when the forage is most nutritious and abundant.

With the steady rise in tillable land values in recent decades, less and less pasture is planted if the soil can grow cash crops successfully. Thus many farmer-feeders are forced into winter or year-round drylot feeding, particularly in the prairie sections where all the land is tillable. For the grain farmer who must keep most of his land in cultivated crops, winter feeding has certain advantages, as will presently be discussed. But for that man who, because of the large size of his farm or the presence of rolling or timbered land, has a considerable area of pasture, summer feeding on pasture will always have an appeal.

ADVANTAGES OF SUMMER FEEDING ON PASTURE

The main points in favor of summer feeding on pasture may be summarized as follows.

1. Summer gains on pasture are usually cheaper than winter gains in drylot because (1) less grain is eaten per pound of gain; (2) pasture is a cheaper form of roughage than harvested hay or hay crop silage; (3) less labor is required in feeding and caring for cattle, as the labor involved in feeding roughage is eliminated, and cattle fed grain on pasture are usually fed by means of self-feeders, whereas hand feeding is often used in drylot.

2. Ordinarily larger daily gains are secured in the summer than in the winter over feeding periods of equal length. Cattle are likely to be more comfortable in summer than in winter because weather conditions are more ʼuniform and feedlots are dry. In addition, summer rations are on the whole superior to winter rations, as explained in the next paragraph.

3. Cattle fed on pasture during the summer usually receive a better-balanced ration than cattle fed during the winter in drylot. Fresh-pasture forage is an excellent source of protein, minerals, and vitamins, and cattle fed on pasture are not often deficient in these important nutrients.

4. No investment in buildings to afford shelter is required.

5. The manure produced is spread on the fields or pastures by the cattle themselves, avoiding loss of fertility through leaching and heating, and saving much labor.

6. Summer-fed cattle are commonly marketed during the late summer and fall when well-finished cattle are usually higher priced than at any other season of the year. Winter-fed cattle, on the other hand, are marketed in the late winter and spring when prices paid for fed cattle are relatively lower.

DISADVANTAGES OF SUMMER FEEDING ON PASTURE

Summer feeding of cattle on pasture has certain disadvantages, which are more likely to apply to the small farmer with a quarter-section of land or less than to the man who owns several hundred acres.

1. The land used for pasture, if tillable, may return a larger gross cash income if planted in crops.

2. An adequate supply of feed in the form of pasture is uncertain, owing to the possibility of unfavorable weather. Winter killing or a late freeze after germination may result in complete failure of tame pasture. Drought, flood, and insect or hail damage are always a threat.

3. Grains, especially corn and milo, are relatively high-priced during summer and early fall.

4. The farmer has less time to devote to cattle in summer than in winter.

5. Flies and extremely hot weather may cause cattle much discomfort.

6. If permanent pastures are used, the manure is dropped on the same ground year after year and not on the cropland that needs it most.

7. Shade and water are hard to provide in temporary and rotation pastures.

8. If spring purchases are made, feeder cattle are scarce, high in price, and the grade and weight desired are difficult to obtain.

9. Summer pasture feeding programs are not so well adapted to use of laborsaving equipment such as may be used in drylot, unless cattle are self-fed.

10. Weather damage to grain in feeders or bunks is possible because of rain and windstorms, which are more prevalent in summertime.

11. Stream pollution is more difficult to control, especially if the cattle have direct access to a stream on the farm.

DRYLOT VERSUS PASTURE FOR CORN BELT SUMMER FEEDING

A large number of cattle are fed in drylot for the late summer and fall markets. However, a majority of these cattle are purchased in the fall and will have been fed considerable grain by the arrival of spring in some type of stocker program. Consequently they usually carry too much flesh to be fed on pasture during the summer and therefore are kept in the drylot until they are ready for market. Only steers of strictly choice grade with body-type conformation scores of 4 or 5 justify such a long feeding period. However, if they have sufficient conformation and finish to sell near the top of the market, they usually are more profitable if carried to heavier weights because higher prices are paid for choice and prime steers during late summer and fall. Only a small portion of the fed steer supply meets these criteria today.

It is generally agreed that cattle in good, thrifty feeder condition in the spring will make faster and more economical gains when fed on pasture than in the drylot, as will be noted in Table 188, for the yearling trials. Note also that older steers, however, do not benefit from the pasture environment. It is often claimed that these advantages are more than offset by the lower prices received for pasture-fed cattle when marketed. This opinion is well supported by numerous feeding experiments in which

Table 188

Comparison of Drylot and Pasture for Summer-Fed Steers

	Full-Fed in Drylot				Full-Fed on Pasture			
	Daily Gain (lb)	Shelled Corn per Day (lb)	Concentrate[a]	Hay	Daily Gain (lb)	Shelled Corn per Day (lb)	Concentrate[a]	Hay
2-year olds								
Illinois	2.12	19.7	927	411	2.00	19.9	992	—
Kentucky (3-year-average)[b]	1.81	10.1	608	908	2.07	10.1	531	—
St. Joseph stockyards	3.08	18.2	671	288	2.76	18.6	763	112
Average 5 trials	2.13	13.6	684	685	2.19	13.8	670	20
Yearlings								
Kansas	1.76	11.5	706	242[c]	2.10	13.0	667	55[c]
Ohio	1.88	12.8	787	401[c]	2.36	12.7	625	—
Ohio	1.92	15.0	840	331[c]	2.13	15.0	758	—
Missouri	1.85	13.4	804	121	2.18	15.5	780	—
Missouri	2.32	12.9	611	229	2.20	12.3	613	—
Missouri	2.46	15.4	680	135	2.13	12.7	653	—
Nebraska (average of 3 trials)	2.19	16.1	743	187	2.31	15.7	686	112
Illinois	2.45	14.1	618	204	2.34	12.3	483	58
Average of 10 trials	2.12	14.3	728	222	2.24	14.1	665	45

Feed per Cwt Gain (lb) column headings appear over Concentrate and Hay.

[a] Grain plus protein concentrate.
[b] Mixed legume-grass pasture used in this experiment.
[c] Includes some silage reduced to dry roughage equivalent.

prices paid for the lots fed on pasture have usually been 25 cents to a dollar less per hundredweight than for those fed in drylot.

Packer buyers defend these prices by saying that cattle fed on pasture yield less beef in proportion to their live weight, that their carcasses have a higher shrinkage, and that the beef is poor in color, with the lean being too dark and the fat being yellow, instead of white as found in cattle fed in drylot. Although none of these claims except that relating to the fat color has been supported by slaughter and carcass studies made by impartial investigators, the fact remains that market buyers look with disfavor upon pasture-fed cattle, even though they show good quality and finish.

Grass-fed cattle can be distinguished while alive from drylot cattle by their rough, dry, sunburned hair and the greenish color of their feces. Removing the cattle from pasture about 2 weeks before marketing will correct the color of the feces, but even a month of drylot feeding usually does not improve their hair sufficiently to escape some price discrimination. There is no improvement in appearance of the hair unless the drylot adjoins a barn or shed as protection from the sun. Selling on grade and yield of course will eliminate the possibility of price discrimination for pasture-fed cattle.

Although the financial statements on comparable droves of cattle frequently show a larger net return over feed costs for drylot than for pasture feeding, it should not be assumed that this system of feeding is

Fig. 94. Full-fed steers self-fed on rotation pasture. Whether to self-feed on pasture in the summer, rather than in drylot, is a matter for each farmer to decide after weighing the advantages against the disadvantages. (Corn Belt Farm Dailies.)

necessarily better under all conditions. Such financial statements seldom take into account the relative farm costs of harvested roughages and pasture, the relative amounts of labor used in drylot feeding, or the investment in the feedlot itself. All these items are very much in favor of the pasture-fed cattle.

All things considered, the net difference between pasture and drylot feeding is not great. In the last analysis, the amount of available pasture is the factor that usually determines whether cattle purchased in the spring or carried through the winter in stocker condition are fed on pasture or in the drylot. If the topography of the farm or the crop rotation results in a large acreage of pasture, the cattle, in all likelihood, will have been purchased mainly to utilize the otherwise surplus pasture. Thus, in the Corn Belt at least, feeding on pasture is usually the result of a system of farming rather than a separate project undertaken because of advantages peculiar to the enterprise itself. Often the farm that was suited to a cow-calf operation or that was formerly operated as a dairy farm is the one that is best suited to a pasture feeding program because fences, water, and pasture are already established.

SELECTING CATTLE FOR SUMMER FEEDING

Cattle to be finished during the summer may be purchased any time between September or October of the preceding year and the date when it is desired to turn them on pasture. Purchases made in the fall consist largely of calves and light yearlings, which will make considerable growth during the winter before the period of heavy feeding begins. Spring purchases, on the other hand, are often more mature steers or yearlings with sufficient flesh to ensure their being in choice slaughter condition after a feeding period of 4 or 5 months.

Regardless of their age and time of purchase, cattle to be fed during the summer should be selected with considerable care. Cattle from drought areas that are light but healthy and of good to choice grade usually work best in this program, especially if summer feeding is preceded by a stocker program in the winter.

In the Winter Wheat Belt, literally millions of calves are bought in the fall for grazing the wheat fields from fall to spring. Unfortunately, moisture conditions necessary for adequate growth on the wheat to graze all of these calves happen only occasionally. Thus many of the calves are prematurely sold in the Wheat Belt from January to March, and even in good years most of them are sold by May 1. These calves all make ideal cattle for feeding in summertime. Steers are usually more suitable than heifers for summer feeding, mainly because so many of the heifers will be in calf, but heifers should not be ruled out, for they are cheaper and finish earlier and ahead of the steers, and often sell at only a little below steer prices.

Steers bought in the spring for summer feeding should have a fair amount of flesh; otherwise it is difficult to get them into choice grade by the end of the grazing season. In selecting steers with plenty of flesh, one should avoid cattle that have been "warmed up" on corn or, still worse, poor-doing steers cut from droves that were fed grain all winter. Most experienced feeders prefer cattle that have never had any grain. While each spring thousands of half-finished, corn-fed steers are sent back to the country for further feeding, the gains made by such animals are usually more expensive than those made by green cattle unaccustomed to a heavy grain ration.

CORN BELT PASTURES FOR FINISHING CATTLE

Before the mid-1930s bluegrass was the pasture most commonly used in the summer feeding of steers. Even though legume-grass rotation pastures have taken over as the most important pasture for summer feeding, bluegrass is still considered by many to be an especially valuable forage for grain-fed cattle because it does not tend to cause scouring or bloat. It is ready to graze much earlier in the spring than are most of the legume forages, and its firm sod withstands trampling much better than most other pasture crops.

During the spring and early summer, bluegrass is palatable and nutritious, but after ripening its seed in midsummer it becomes more or less dormant, especially during a dry season. At this stage it is not so palatable, and cattle getting a full feed of grain eat comparatively little of the grass.

It is this tendency to go dormant that reduces the carrying capacity of bluegrass and lessens its value in a summer feeding program. True, with the coming of fall rains and cooler weather, it starts growing again and often furnishes considerable grazing during September and October. The common practice is to remove the cattle to the drylot before this growth begins, however, lest the new, green grass interfere with the grain consumption. The fall growth of bluegrass pastures is utilized to greater advantage by newly purchased stocker calves or yearlings.

Along the western and northern borders of the Corn Belt, bluestem, bromegrass, and orchard grass pastures are used to some extent for summer-fed steers. All these pasture forages are similar to bluegrass in composition, and similar gains may be expected from them. Bromegrass, orchard grass, canary grass, and even tall fescue, in combination with alfalfa and clovers, are now being grown extensively in the Corn Belt where they have given much better results than straight grass pastures. The advantage comes from the greater carrying capacity made possible by the greater forage production. Some tests also show that protein supplements may be reduced when cattle are being full-fed on these good legume-grass pastures, not only because the legumes are high in protein themselves but because the grasses also are made higher in protein by the atmospheric nitrogen that is fixed in the soils by the leguminous plant nodules.

Many longtime cattle feeders look with disfavor upon straight legume pastures for grain-fed cattle, believing that they produce both scouring and bloat. However, comparisons of these forages with bluegrass have shown that legume pastures are capable of producing so much more gain per head and per acre that they must be rated as valuable pasture crops for cattle despite these objections. Both red clover and alfalfa, either alone or in mixtures, have given much larger and more economical gains than bluegrass in experiments summarized in Table 189. Fairly good results have been obtained with sweet clover, but the short grazing season of the second year's growth makes it necessary to transfer the cattle to other pasture or to the drylot about the middle of August.

TURNING STEERS ON SUMMER PASTURE

Better results are secured if grass or legume-grass pastures are allowed to make a fairly good start before the cattle are turned onto them. The first growth that appears is high in moisture and very low in energy content. By April 20 to May 1 bluegrass is usually mature enough to be used in the central part of the Corn Belt. In the famous bluegrass sections of Virginia and Kentucky, and in Iowa the grazing season opens about 10 days earlier and later, respectively.

An exception to these dates are steers that have received a heavy feed of grain during the winter. Better results are usually obtained if such cattle are turned onto the pasture in late March or early April as soon as the grass begins to grow. In this way they become accustomed to the grass so gradually that they are not apt to lose their appetite for corn. When full-fed cattle are turned onto a heavy growth of grass their grain consumption temporarily falls off, sometimes to only half of what it was in the drylot. Although the cattle usually regain their appetite for grain by the end of the third or fourth week, gains made during such a transition period are far from satisfactory.

Steers should never be turned onto straight-legume or legume-grass rotation pastures before the pastures are 6 to 8 inches high. Grazing too

Table 189

Value of Legume Forages for Steers Full-Fed on Pasture

	Nebraska May 5–Nov. 3 182 Days			Illinois May 4–Oct. 23 172 Days			Illinois May 6–Nov. 15 193 Days		
Yearling Steers Initial Weight 550 to 650 lb	Drylot	Alfalfa Pasture	Native Pasture	Blue-grass Pasture	Red Clover Pasture	Sweet Clover Pasture	Brome-grass Pasture	Alfalfa Pasture	Mixed Clover Pasture
Date turned on pasture	—	May 5	May 5	May 5	May 5	May 5[b]	May 6	May 6	May 13
Date turned off pasture	—	Sept. 22	July 28	Oct. 23	Oct. 23	Oct. 23	Sept. 23	Sept. 23	Sept. 23
Days on pasture	—	140	84	172	172	172	140	140	133
Steers per acre	—	3	2	2	2	2[b]	2.4	2.4	2.4
Days in drylot	182	42	98	None	None	None	53	53	53
Average daily gain on pasture (lb)	2.67[a]	2.70	2.49	2.16	2.47	2.27	2.25	2.44	2.28
Average daily gain in drylot	2.19[a]	2.60	2.51	—	—	—	2.57	2.36	2.27
Average daily gain total period (lb)	2.41	2.68	2.50	—	—	—	2.34	2.42	2.28
Feed per cwt gain (lb)									
Shelled corn	659	605	661	658	663	648	574[d]	553[d]	609[d]
Protein concentrate	—	65	141	45	44	44	25	24	27
Alfalfa hay	289	29	18	—	—	—	58	57	62
Hog gains per steer (lb)	—	—	—	67	67	58	—	—	—

[a] Average daily gain in drylot before and after July 28.
[b] Spring seeding of sweet clover used after August 7, one steer per acre.
[c] Clover mixture: approximately 80% sweet clover and 20% red clover.
[d] Weight of ground ear corn.

early results in reduced forage production through the remainder of the grazing season. Such pastures usually reach the desired height by April 15 to May 10 in the midwestern cattle-finishing areas, with earlier dates in the South and the West.

BLOAT ON LEGUME PASTURES

Use of legume and legume-grass rotation pastures has resulted in an increase in the incidence of pasture bloat. This type of bloat is often fatal and is not to be confused with the troublesome, but less dangerous, common feedlot bloat. Great economic losses are incurred during some years because of deaths and poor performance of bloated animals. Accurate data are difficult to obtain, but the figure often used is $100 million lost due to bloat per year in the United States alone. The annual loss from bloat is estimated at one death per 200 head of cattle. Of even greater importance, however, is the loss resulting from failure to use these nutritionally more valuable pastures instead of grass pastures simply because of the cattle feeder's fear of bloat.

The causes of legume bloat are still under investigation, but the picture is becoming clearer. In almost all cases the paunch becomes markedly distended on the left side, because of the formation of a stable frothy mass in the left dorsal portion of the rumen. Trapped within this mass is a large quantity of rumen gas, mostly carbon dioxide and methane, which cannot easily be eliminated by the usual belching process. Gas continues to be produced faster than it can be expelled, and pressure builds up until the animal dies of apparent suffocation. The actual cause of death is not fully established, but suffocation currently seems the most plausible explanation.

The important discovery that appears to be the best explanation for the real problem, the stable froth, was made by a group of Wisconsin investigators. They believe that certain plants, notably alfalfa, are bloat-producing because they contain critical amounts of a plant enzyme called pectin methyl esterase or PME. The enzyme, when present in sufficient quantity, reacts with pectin, a carbohydrate always present in forages, and changes it to pectic acid and alcohol. Pectic acid reacts with calcium, a valuable mineral in legumes, in the rumen to produce the sticky, foamy mass that traps the rumen gas, preventing its escape and finally resulting in the fatal buildup of pressure.

A number of antifoaming agents are available, and many of them are effective if administered at a sufficient and a constant rate. These agents, called surfactants, generally act by reducing the surface tension of the otherwise stable froth, permitting the gas to escape at a rate fast enough to prevent serious consequences. Among the products being used for this purpose are animal fats, lecithin, vegetable oils, mineral oils, detergents, turpentine, silicones, glycerol, and—one that looks especially promising— poloxalene, a synthetic polymer. The effectiveness of this product was first announced by a Kansas group and it has now been approved by the Food and Drug Administration for use in livestock feeds as a bloat preventive.

A test conducted at Kansas compared two methods of administering poloxalene with a third lot receiving no treatment, using steers that were grazing immature alfalfa pasture. Results are summarized in Table 190. One group received poloxalene from a molasses-salt block containing 30 grams per pound of block. No other salt was provided. Another group received poloxalene in a self-fed ground milo preparation to which "Bloat Guard," a commercial premix containing 53 percent of the active agent, was added at the rate of 10 grams per pound of milo. This group and the control group also received molasses-salt blocks to which no poloxalene was added. The steers were rotated between treatments every 14 days to allow for the possibility of anatomical differences in the steers.

The poloxalene in the molasses-salt block most successfully prevented bloat. No steers died, but many severe cases of bloat were produced, so that there was no question that bloat-producing forage was available.

Bloat is also common in cattle fed in drylot on almost any combination of alfalfa hay and barley. Several stations have tested poloxalene in the prevention of bloat under these circumstances and here, too, it apparently is effective so long as at least 2 grams of the compound are consumed daily per 100 pounds live weight.

Some old and some new recommendations for preventing and controlling bloat through management other than medication are as follows.

Table 190

Effectiveness of Poloxalene for Controlling Bloat in Steers Grazing Immature Alfalfa (42 Days)[a]

Treatment[b]	Poloxalene in Feed	Poloxalene Block	Control
Average number of animals	4	4	4
Number of bloat ratings[c]	336	336	336
Cases of bloat	48	20	98
Highest bloat score	3	2	3+
Average bloat score[d]	0.69	0.54	1.38

[a] Adapted from Kansas Livestock Feeders' Day Report, 1967.

[b] All 12 animals received each treatment, being rotated at 14-day intervals.

[c] All 4 animals on each treatment were rated twice daily for 42 days.

[d] A scale ranging from 0 for no bloat to 4 for severe bloat was used.

1. Permit cattle to fill on grass or hay before first turning out on pasture.
2. Allow cattle free access to dry forage or a grassy pasture adjoining the legume pasture, or mow a strip and leave the clippings in the field as a source of dry roughage.
3. Leave cattle on the legume pasture 24 hours a day rather than penning them at night.
4. Use pasture mixtures that provide no more than 50 percent of legume forage.
5. Mix additional cob or oats with ground ear corn or other grains fed on legume pasture.
6. Avoid cattle with genetic potential for high susceptibility to bloat. Never use a bloater bull or save daughters of bloater cows. Angus cattle from some herds are more susceptible than cattle from other breeds.
7. Provide poloxalene in self-fed blocks or liquids in pastures with a history of bloat problems.

TREATMENT OF BLOAT

There are almost as many methods for the treatment of bloat as there are recommendations for preventing it. The presence of bloat is easily recognized by a pronounced swelling of the left flank. So long as the distention causes the animal no great discomfort, there is no need for alarm, as recovery usually occurs with treatment. However, the bloated animal should be kept under observation, because the condition occasionally worsens very quickly.

If bloating persists or proceeds to a more advanced stage, treatment by medication or otherwise many be necessary. Treatment consists in administering compounds that will break down the stable froth that is preventing the gas from escaping. As already mentioned, numerous products are available from commercial sources. Also, either 1 pint of mineral oil, or 2 ounces of aromatic spirits of ammonia, or 2 ounces of turpentine, diluted with 1 pint of cold water often brings relief when used as a drench.

A simple way of relieving acute bloat in animals that are readily caught and not critically advanced in bloat is to force the escape of the gas through a 6-foot length of smooth $\frac{1}{2}$- or $\frac{5}{8}$-inch rubber hose, one end of which is inserted far back into the animal's mouth and carefully pushed down the esophagus and into the paunch. With a little practice there is no difficulty in getting the end of the hose over the trachea and into the esophagus.

If the formation of gas proceeds despite the remedies administered or after use of the stomach tube, it is necessary to tap the paunch to permit the gas to escape. This is done by means of a trocar, a sharp-pointed instrument encased in a cannula or sheath. The point of the trocar is placed over a hollow spot on the animal's left side equidistant from the last rib, the hipbone, and the transverse processes of the lumbar vertebrae. The handle is then struck sharply with the palm of the hand in the direction of the animal's right knee. Then, while the cannula is kept in place, the trocar is withdrawn, permitting the gas to escape slowly. This procedure is a last resort and should only be used in extreme cases.

SMALL GRAIN PASTURES

Opinions differ as to the best way to use winter small-grain pastures. In the Winter Wheat Belt of Kansas, Oklahoma, and Texas, calves are most often grazed without supplemental feeding except during inclement

Fig. 95. Steers, winter-grazing a Bermuda pasture that has been overseeded with rye, in eastern Oklahoma. Oats, wheat, and ryegrass also are used in much of the Southeast for the purpose of providing excellent green forage from dormant Bermuda pasture from fall to spring. (Oklahoma Farmer-Stockman.)

weather. Because of the low dry-matter content of such pastures, some operators add a source of dry roughage or even a high-energy feed to increase the total intake of nutrients—and, it is hoped, to improve performance. Furthermore, if only small grain forage is available, much protein is lost because the protein content may run as high as 25 percent—much higher than required by calves or lactating cows, for that matter. This excess protein, after conversion to ammonia, is incriminated by some investigators as the main cause of wheat pasture poisoning, a result of the direct absorption of large quantities of ammonia from the rumen.

The Oklahoma station conducted a test in which half of the calves grazed wheat pasture alone and the other half had free access to a high-concentrate mixture consisting of 77 percent ground milo, 8 percent molasses, and 15 percent chopped alfalfa hay. The results are summarized in Table 191. The calves that were offered supplemental feed consumed 10 pounds per head daily but gained only 0.61 pound more per head daily than calves consuming no supplemental feed. This added gain would not offset the cost of the supplemental feed. Possibly a lower level of supplemental energy would have been economical.

In the Southeast, winter small grains are often grown specifically for winter pasture. The question arises as to how best to utilize these pastures and at the same time produce a steer ready for slaughter. The Georgia station studied this problem at its Coastal Plain station at Tifton, and the results of the test are shown in Table 192. They used five management systems as follows.

1. Drylot—on ground snapped corn, cottonseed meal, and Coastal Bermuda hay.

Table 191

Response of Heifer Calves to High-Energy Feed on Wheat Pasture (114 Days)[a]		
	Wheat Pasture Only	Wheat Pasture + High-Energy Feed
Number of heifers	20	20
Average initial weight, 11/21/67 (lb)	481	482
Average final weight, 3/14/68 (lb)	637	708
Total gain (lb)	156	226
Average daily gain (lb)	1.37	1.98
Feed consumed per head (lb)		
Hay (during snow cover)	135	90
Concentrate feed	—	1,216

[a] Oklahoma State University, Animal Science Research Progress Report, Miscellaneous Publication 80.

Table 192

Systems of Utilizing Winter Grazing in Finishing Yearling Steers[a]

	Drylot Feeding	Drylot + Pasture	Oats Pasture	Rye Pasture	Limited Oats Pasture
Number of steers	10	10	10	10	10
Total days fed	147	147	223	223	223
Days maintenance feeding	—	—	76	76	76
Days in drylot	147	76	35	49	35
Days on pasture	—	71	112	98	112
Average initial weight (lb)	714	716	710	716	730
Average final weight (lb)	1,024	1,061	1,094	1,084	1,104
Average daily gain, finishing period (lb)	2.10	2.35	2.42	2.39	2.44
Average daily ration (lb)[b]	25.36	14.45	6.69	10.00	14.74
Ground snapped corn	17.73	10.08	3.75	6.07	12.66
Cottonseed meal	2.50	1.29	0.60	0.90	0.60
Coastal Bermuda hay	5.13	3.08	2.34	3.03	1.48
Acres grazing per lot	—	7.00	10.00	10.00	5.00
Feed cost per cwt gain	$24.04	$20.28	$18.65	$22.06	$20.36
Average dressing percent	60.2	59.2	58.3	58.8	59.2
Average carcass grade	G	G−	G	G−	G

[a] Adapted from data in Georgia Agricultural Experiment Station Circular N.S. 31.
[b] Based on 147-day finishing period.

2. Started in drylot on above ration, then transferred to grazing plus limited ground snapped corn.
3. Fall maintenance feeding followed by winter grazing and limited feeding on oat pasture.
4. Same plan as 3, except rye pasture was used.
5. Same plan as 3, except grazing was for only 5 hours daily and ground snapped corn was full-fed.

The pastures were seeded in October after 100 pounds of actual nitrogen was applied to the soil per acre. A charge of $40 per acre was made for the pasture. Grazing began in mid-December and continued for 112 days. A short, 35-day, drylot feeding period followed for steers on treatments 3, 4, and 5 to bring them to comparable slaughter weights. All of the pasture treatments produced more rapid and more economical gains, with the most profitable being the lot on oat pasture continuously plus limited grain and hay. Carcass grades were comparable for all treatments. Other Georgia studies with feeding steers on oat pasture are reviewed in Chapter 24.

It appears that systems making full use of the small-grain pasture with only enough supplemental feed to provide intakes at the level of about 1 percent of body weight offer the most profitable way to utilize such pastures in finishing programs. Short periods of drylot feeding at the end of the grazing season in the spring are recommended.

DRYLOT VERSUS PASTURE FEEDING ON FESCUE-LEGUME PASTURES

Around the western, southern, and eastern fringes of the Corn Belt and well down into the southern and southeastern states, tall fescue-legume pastures are the most prevalent of all improved pastures. Since this also is important cow-calf country, there is a large supply of weaner calves and light yearling cattle available for feeding for slaughter purposes. Nearly all of the experiment stations in these areas have conducted studies on the feasibility of keeping larger numbers of the locally produced calves and feeding them out for slaughter on the abundant, nutritious improved permanent spring pastures and a limited feed or a full feed of grain, also available in the area or nearby.

The data shown in Table 193 summarize a 2-year study conducted by Arkansas workers. North Carolina, Tennessee, Louisiana, and Alabama stations have reported studies that cover a wide variety of steer management systems on such pastures, but space does not permit discussion of them here. The Arkansas study with fleshy yearling cattle shows that

Table 193

Finishing Beef Steers on Fescue-Clover Pasture Compared to
Drylot Finishing (Early April–Mid-June)[a]

Item	Drylot	Fescue-Clover Pasture
Days fed	70	70
Initial weight (lb)	763	668
Final weight (lb)	975	1,000
Average daily gain (lb)	3.05	3.35
Carcass grade	good+	choice −
Daily feed consumption (lb)		
Concentrate	22.4	22.0
Cottonseed hulls	7.5	—
Total feed per head (lb)		
Concentrate	1,557	1,531
Cottonseed hulls	512	—
Total carcass value per head ($)	227.11	232.34

[a] Arkansas Experiment Station Report, 1969.

pasture-fed cattle gained about one-third pound faster and produced carcasses grading slightly higher than those fed in drylot, but the pasture forage did not sufficiently reduce concentrates required per unit gain to give the pasture a very high value per acre. Limited feeding of the grain would probably have produced a higher value in the fescue-clover pastures.

SUPPLEMENTATION OF STEERS ON SORGHUM-SUDAN PASTURES

Sorghum-Sudan hybrids, a favorite warm-season annual forage crop, are grown on extensive tillable acreages in the southern Corn Belt, the Southwest, and the southeastern states. This fast-growing, productive annual crop is successfully used by cow-calf men and by producers who grow out lightweight calves for sale or placement on feed in the fall. Chemical analysis of the dry matter in this forage suggests that it offers possibilities as a feed for lightweight yearling cattle intended for the slaughter market.

The Alabama station conducted a trial through three growing seasons with short yearling cattle, to determine whether pasture alone or pasture plus a limited feed of a high-energy feed would be feasible options in the utilization of the sorghum-Sudan crop. They seeded a modern hybrid on a properly fertilized, prepared seed bed in early May. Their rotationally

Table 194

	Grazing Alone[b]	Grazing + Supplement[b]
Effect of Energy Supplementation on Performance of Steers Grazing Sorghum-Sudan Pasture (Average of 3 Years)[a]		
Item		
Average days on test[c]	77	77
Animal performance		
Average stocking rate (head/acre)	2.8	2.9
Average initial weight (lb)	673	669
Average final weight (lb)	756	789
Average daily gain (lb)	1.10	1.57
Total steer gain/acre (lb)	210	335
Supplemental feed/day (lb)	0	7.0
Feed/lb extra gain (lb)	0	14.4
U.S.D.A. carcass grade	high std.	low good
Dressing percentage	53.6	56.3

[a] Adapted from "Supplementation of Steers Grazing Sorghum-Sudan Pastures," Alabama Agricultural Experiment Station Circular 188, 1971.
[b] Unweighted averages.
[c] Early June to mid-August.

managed grazing trials began in early June and ended in mid-August, as an average for the 3 years. Each year, half of the steers were hand-fed an 80 percent concentrate feed at the rate of 1 percent of body weight daily. Results are summarized in Table 194. They obtained excellent gains early in the growing season and, in one year, were able to carry as many as 5 steers per acre for about 45 days. Gain per acre averaged only 210 and 335 pounds for unsupplemented and supplemented steers, respectively, however, because gains were poor in July and August. In all years, neither total gain per steer nor carcass grade was satisfactory.

These studies and others reviewed indicate that sorghum-Sudan hybrid pastures are generally unsatisfactory for steers intended for the slaughter market. Cool-season perennials have produced as much or more gain with considerably lower crop production costs. The added daily gain and gain per acre made by fed steers generally has been insufficient to pay for the cost of the concentrate fed. Carcass grade in general has been too low to command slaughter prices as high as the steers would have brought as feeder steers. It appears that such pastures are best utilized by lactating cows and lightweight stocker calves.

DAY VERSUS NIGHT GRAZING

Some farmers who full-feed their cattle during the summer confine them in the drylot overnight and turn them onto the pasture during the day.

The object of this procedure is to induce the cattle to eat as much grain as possible by holding them in the feedyard.

As cattle that are continually on pasture spend more time grazing at night than during the day, especially in hot weather, it would appear that confining them in the lot during the day, where they have access to water and shade, and turning them onto the pasture at night might be a better procedure.

These two methods of using pastures were compared at the Illinois station. A third and fourth lot, fed continuously in drylot and on pasture, respectively, were also included in the test. Grazing at night proved to be much the best plan of utilizing the pasture. Observations of the cattle showed that those confined in the drylot during the day and turned onto the pasture at night usually grazed steadily for about 3 or 4 hours after being turned onto pasture, whereas the cattle pastured during the day usually sought the protection of an artificial shade soon after being put into the field. The main difference noted between the cattle pastured at night and those left on pasture continually was that the night-pastured cattle spent much time during the day lying in their well-bedded shed, while the cattle continually in the pasture stood most of the time under their sunshade. (See Table 195.)

Table 195

A Comparison of Night and Day Grazing for Full-Fed Cattle[a]	Confined in Drylot	On Pasture at Night	On Pasture During Day	On Pasture Continually
Average daily gain (lb)	1.74	2.20	1.99	2.00
Average daily ration (lb)				
Shelled corn	12.3	13.6	12.4	13.7
Soybean meal	1.0	1.0	1.0	1.0
Clover hay	4.7	1.9	1.8	0.2[b]
Pasture, acres per steer	—	0.25	0.25	0.50
Feed per cwt gain (lb)				
Shelled corn	706	617	625	686
Soybean meal	55	45	49	49
Clover hay	272	86	93	11
Pasture days	—	11	12	24
Return per head above feed costs ($)	14.00	25.72	20.58	16.89

[a] Illinois Mimeographed Report.
[b] Hay fed in drylot 6 days before marketing.

GRAIN RATIONS FOR PASTURE-FED STEERS

Either corn or milo may be fed to cattle on pasture. Their concentrated form makes them combine well with a bulky roughage such as pasture. Ground ear corn is too bulky and unpalatable for cattle fed on nonlegume pastures but is better than shelled corn for cattle on sweet clover or alfalfa because the cob particles tend to reduce the prevalence of scours and bloat (see Table 196).

Owing to the relatively high percentage of protein in young grasses or legumes, the need for a nitrogenous concentrate is not so urgent with pasture-fed cattle as it is with those in drylot. Except for slightly higher daily gains resulting from a larger consumption of the more palatable ration, there is no important advantage in feeding protein supplements to full-fed steers that have access to abundant high-quality green pasture. However, bluegrass, bromegrass, and all legume forages contain a much lower percentage of protein in midsummer than they do in the spring. Moreover, they are less palatable and consequently are consumed in smaller amounts. For these reasons a protein concentrate should always be fed with the grain on grass pasture beginning about July 1, and on legume pastures about August 1, unless unusually favorable weather conditions delay the maturing of the forage beyond the usual time.

Many feeders feed protein supplement from the start of the grazing period to make certain that they do not wait until the pasture is too dry before starting it. In such cases, and especially if the pasture is legume-grass or straight legume, feeders may widen the usual 10–12:1 ratio of ground ear corn to protein concentrate to 12–15:1 during at least the first 2 months of the grazing season.

The success achieved by including several additives with the supplements fed to finishing cattle being fed in drylot has tempted many feeders who feed grain on grass to use such fortified supplements. The rate of gain is almost always increased to some extent, but when the cost of the fortified supplement is added to the cost of grain also being fed, the cost per pound of gain invariably comes out higher, rather than lower as one might expect from an increase of gain. Hormone-like growth-promoting implants and phosphorus are apparently the only two additives that can be relied on to reduce cost of gain in steers being fed on pasture, assuming that protein is adequate, of course.

TURNING HALF-FINISHED CATTLE ONTO PASTURE

Cattle that have received a fairly liberal ration during the winter, so that they are half-finished or better at the opening of the grazing season,

Table 196

Shelled versus Ground Ear Corn for Cattle Full-Fed on Pasture

	Mixed Pasture[a]			Bluegrass[b]		Alfalfa[b]	
	Shelled Corn	Ground Shelled Corn	Ground Ear Corn	Shelled Corn	Ground Ear Corn	Shelled Corn	Ground Ear Corn
Average weight (lb)	660	667	667	703	709	705	708
Days fed	190	190	190	133	133	133	133
Average daily gain (lb)	2.47	2.40	2.19	2.38	2.08	2.31	2.32
Average daily ration (lb)							
Shelled corn	16.6	15.9	15.3[c]	15.0	13.2[c]	12.8	11.4[c]
Protein concentrate	—	—	—	1.0	1.0	1.0	1.0
Alfalfa hay	2.2	2.4	2.3	—	—	—	—
Feed per cwt gain (lb)							
Shelled corn	673	663	654[c]	599	635[c]	554	491[c]
Protein concentrate	—	—	—	42	50	43	43
Alfalfa hay	90	98	105	—	—	—	—

[a] Nebraska Mimeographed Circular 140.
[b] Illinois Mimeographed Report.
[c] Shelled basis.

should be finished in a drylot rather than on pasture. To turn such animals onto pasture will probably result in a marked decrease in grain consumption for the first 3 or 4 weeks while they are becoming accustomed to pasture. This decreased grain consumption, together with the "washy" character of the spring pasture, results in very moderate gains for the first 4 to 6 weeks. In fact, it is not unusual for cattle under such conditions to show an actual weight loss when weighed 2 or 3 weeks after leaving the drylot.

If it is necessary to put half-finished cattle on pasture because of a scarcity of dry roughage, they should be turned onto pasture early in the spring when the grass first starts to grow, and their dry roughage should be continued until the pasture is fairly mature. In this way the cattle become accustomed to the change in their ration very gradually and are unlikely to go off feed (see Table 197). Actually a better choice of steer management program when roughage supplies are short is simply to omit roughage completely, as discussed in Chapter 14.

GREEN CHOP IN DRYLOT FINISHING RATIONS

Modern methods of harvesting forages for silages are being adapted to daily cutting of forage for feeding in drylot in the fresh state as green chop. This system of utilizing legume-grass or legume forage is especially suited to operators of feedyards and to large farm feedlots for the following reasons.

1. Carrying capacity of pasture is doubled because of more complete utilization of the crop.
2. The quality of the forage is higher by at least one government grade, over the same forage harvested and fed as hay, if harvested at the proper stage, and it usually can be.
3. The forage is more uniform in quality.
4. Ration changes and adjustments can be made more accurately and conveniently.
5. Less fence, shade, and watering equipment are required.
6. The cattle expend less nonproductive maintenance energy because walking about the pasture is eliminated.
7. Fewer cases of bloat occur, although green chop is not a sure cure for bloat.
8. Flies are easier to control in drylot than on pasture.
9. The process of checking cattle for disease, bloat, and other problems is simplified.
10. Weeds and insect damage to pastures are easier to control.

Table 197

Drylot versus Pasture for Finishing Half-Finished Cattle[a]

Place Finished	First Trial		Second Trial		Third Trial		Average 3 Trials	
	Drylot	Pasture	Drylot	Pasture	Drylot	Pasture	Drylot	Pasture
Weight in spring (lb)	880	880	870	875	827	809	859	855
Average daily gain (lb)								
First month	2.06	0.66	2.16	1.61	2.37	1.34	2.20	1.20
Second month	1.90	2.05	1.58	1.41	1.35	1.66	1.61	1.71
Third month	1.01	1.00	1.38	1.56	1.35	1.38	1.25	1.31
Total—90 days	1.65	1.24	1.71	1.53	1.69	1.46	1.68	1.41
Grain eaten per day (lb)								
First 10 days	14.90	8.22	14.55	10.27	16.00	14.50	15.15	11.00
Total—90 days	17.04	11.84	16.17	14.46	16.74	15.92	16.65	14.07
Grain per pound gain (lb)								
First month	7.60	13.94	7.06	6.62	6.96	11.02	7.21	10.53
Total—90 days	10.28	9.55	9.46	9.45	9.89	10.88	9.88	9.96

[a] Indiana Bulletin 142.

11. Puddling and packing of pasture soil is lessened but not eliminated.

12. Haircoats are in better condition in cattle fed in drylot.

Problems arise in this method of feeding, some of which are serious enough to cause some feeders who have tried green chop to abandon the plan. The disadvantages of green chop are as follows.

1. Prolonged wet weather makes harvesting difficult or impossible.

2. Labor requirements are increased, especially in smaller operations. One study found that 0.4 man-hour of labor was required per ton of fresh forage harvested.

3. Equipment costs are high; machinery costs amounted to $1.75 per ton in one study. If silage-making equipment is already on hand, however, this extra item of expense is small.

4. Large numbers must be on feed—at least 100 head or more—to keep per-head costs of labor and machinery low.

5. Scouring is prevalent with this type of feed.

6. Forage quality of roughage consumed may become poor if the crop becomes fibrous because of advanced maturity. Steers on pastures selectively graze only the new growth, thereby consuming a more nutritious roughage.

7. A system of feeding green chop is confining, in that fresh forage must be chopped daily.

Results of the Iowa experiments reported in Tables 198 and 199 show what may be expected from steers being managed under two different programs for utilizing pasture forage. In general, the contribution to the ration by pasture forage was about twice as great when it was harvested as green chop instead of pasture, in a steer program in which no grain was fed for 120 days followed by a 75-day full feed of grain in drylot for both lots. (See Table 198.) Profits per steer were no greater, but more steers could be fed per acre of pasture.

In a comparison between steers full-fed on pastures and steers fed limited grain plus pasture forage harvested as green chop, the pasture made a substantial contribution toward total feed needs when harvested and fed as green chop. Profits per steer fed were noticeably higher than when the pasture was grazed, especially if supplement was not fed. (See Table 199.)

FINISHING CATTLE ON GRASS ALONE

Comparatively few cattle are grazed without grain in the Corn Belt with the expectation of selling them for slaughter in the fall. There is grassland

Table 198

Comparison of Pasture and Green Chop When Steers Were Grazed or Fed
Green Chop Without Grain from May to September (120 Days) and Full-Fed in
Drylot for 75 Days (Average of 2 Years)[a]

	Grazing		Green Chop (Soilage)	
	No Supple- ment	Supple- ment	No Supplement	Supplement
Number of steers	16	15	16	15
Average initial weight (lb)	778	782	807	801
Average final weight (lb)	1,183	1,182	1,186	1,176
Average daily gain (lb)	1.86	1.84	1.95	1.97
Average feed consumed per steer daily				
Pasture, brome-alfalfa (acres)	0.85	0.85	—	—
Clippings, brome- alfalfa (lb)[b]	—	—	58.2	55.0
Ground ear corn (lb)	8.4	7.8	7.7	7.1
Hay, brome-alfalfa (lb)[c]	0.4	0.4	0.9	0.9
Supplement (lb)	0.6	1.8	0.4	1.8
Feed cost per cwt gain ($)	20.00	22.10	19.70	21.40
Selling price per cwt ($)	23.16	23.45	23.36	23.10
Margin per steer over feed costs ($)	22.13	16.79	25.25	14.83
Estimated beef per acre (lb)	192	161	283	256
			1.5 tons hay	1.5 tons hay
Returns per acre of pasture ($)	41.00		46.42	

[a] Iowa Miscellaneous Publication AH 693.

[b] Daily consumption during pasture season was 79 pounds per steer in the first year and 90 pounds in the second.

[c] Daily consumption during finishing period was 3 pounds per steer in both years.

in many parts of the country, notably the Flint Hills section of Kansas and Oklahoma, where older steers will reach choice grade on grass. Other cattle programs are proving more profitable, so that this system of finishing cattle is passing out of the picture in the United States, but it is quite common in Australia, New Zealand, and Argentina. The most common practice where no grain is fed on pasture is to buy heavier cattle in the spring or at the start of the grass season and carry them to fall, or the end of the best grazing, then sell them as long 2- or even 3-year-old

Table 199

Comparison of Pasture and Green Chop When Steers Are Full-Fed on Pasture or Limited-fed Grain and Green Chop in Drylot for 135 Days (Average of 2 Years)[a]

	Grazing plus Full-Feed of Corn		Green Chop plus Limited Corn[b]	
	No Supple-ment	Supple-ment	No Supple-ment	Supple-ment
Number of steers (2 yr)	16	16	16	16
Average initial weight (lb)	814	820	820	816
Average final weight (lb)	1,123	1,162	1,125	1,130
Average daily gain (lb)	2.32	2.56	2.33	2.39
Average daily feed consumed				
Pasture, brome-alfalfa (acres)	0.25	0.25	—	—
Clippings, brome-alfalfa (lb)	—	—	51.6	49.2
Ground ear corn (lb)	20.8	20.1	13.6	13.4
Hay, brome-alfalfa (lb)	—	—	0.1	0.1
Supplement (lb)	—	2.0	—	2.0
Feed cost per cwt gain ($)	23.40	23.60	20.00	22.70
Selling price per cwt ($)	24.18	24.88	24.58	24.18
Margin per steer over feed costs ($)	21.29	28.41	35.30	25.52

---

[a] Iowa Miscellaneous Publication AH 693.

[b] Ground ear corn—5 pounds per steer daily during first month, 10 pounds second month, 15 pounds third month, and a full feed of ground ear corn thereafter.

steers ready for slaughter. A few long-aged Mexican and Canadian steers are exported to the United States just for this purpose each year.

Choice-quality 2-year-old steers often attain enough finish on excellent pasture alone to grade low to average choice as slaughter cattle. However, most feeders who graze steers of this age and weight prefer to feed for 30 to 50 days in drylot to improve the condition sufficiently for the steers to grade still higher. Ordinarily the 10 to 15 bushels of corn or sorghum required are more than paid for by the higher selling price received.

CONSUMER ACCEPTANCE OF GRASS-FED BEEF

During periods when there is either an overabundance of stocker and feeder cattle in relation to demand from cattle feeders or when grain and concentrate prices are high in relation to fat cattle prices, packers are able to buy fleshy weaner calves or yearlings, or even young cows, at prices that are attractive enough to cause them to shift their slaughter from fed cattle to "grass cattle." To sell this added tonnage of beef, which ordinarily is available only in small quantities and for fairly short periods of time, intense advertising campaigns are conducted by the retailers, and especially chain stores, extolling the virtues of the lean, low-fat content beef which they can offer at prices below those of fed beef. Such beef is labeled with a variety of special names, with each chain store having its distinct descriptive label, but it seldom is assigned a USDA grade. Often this lean beef is called "baby beef," which it certainly is not, at least in terms of the long-accepted definition that baby beef is from young animals fed grain for 90 to 150 days. Much of the imported beef from Australia and New Zealand is correctly termed "grass beef," but it usually is from more mature cattle, sometimes even from 3-year-old steers, and thus is carrying considerably more marbling. Much of this imported beef, if from steers and not cows, is equivalent in quality to medium good to low choice grade, whereas much of the so-called grass-fed baby beef of domestic origin would only grade standard.

Nutritionally speaking, the American grass-fed beef from young cattle is usually a good buy for it is high in protein and low in fat. Because it is higher in water content, cooler shrink—especially when the meat is displayed in unpackaged, retail-cut form—is higher and the meat shrinks somewhat more when broiled. Many consumers, who are unfamiliar with this kind of beef, wonder about its suitability as a replacement for the conventional fed beef that grades low choice on the average. Table 200 contains data from one trial conducted by Kansas workers in relation to this subject. Only 10 cattle per treatment were slaughtered and they came from various sources: The grass-fed steers were obtained off grass at the end of summer, the short-fed cattle were from a commercial feedlot, and the longer-fed steers came from an experiment station feeding trial. Nevertheless the carcasses were no doubt fairly representative of cattle available to the packing industry in the fall of the year. Not surprisingly, the carcasses did grade differently and cuts from the grain-fed cattle were more tender and were rated more acceptable by a taste panel. The data also show that the grass-fed beef was leaner and the carcasses, and presumably also the retail cuts, were lighter in weight. The data show, however, that the grass-fed beef was reasonably well accepted, but a logical question is whether consumers, largely adapted to grain-fed beef, would continue to accept the grass-fed beef should it become available at competitive prices on a year-round basis.

LIMITED GRAIN RATIONS FOR PASTURE-FED CATTLE

Feeding a limited grain ration to cattle on pasture is unusual. Unless they are moved from the pasture to the drylot each day for feeding, some of the animals may not be at the bunks when the grain is fed and therefore

Table 200

Acceptability and Carcass Traits of Grass-fed, Short-fed, and Conventionally Fed Beef[a]

Item	Number of Carcasses			Carcass Weight (lb)	Fat Thickness (in.)	W-B Shear Test[b]	Meat Acceptability Score[c]		
	Choice	Good	Standard				Day 0	Day 3	Day 5
Grass-fed steers	0	5	5	501	0.19	10.70	5.88	5.65	5.71
Steers fed finishing rations									
70 days	1	9	0	533	0.36	7.54	6.80	6.85	6.98
150 days	5	5	0	615	0.53	7.01	7.36	7.33	7.59

[a] Kansas Feeders' Day Report, 1975.

[b] Pounds shear force required to slice across a beef cut with Warner-Bratzler Shear Test apparatus.

[c] Days = number of days in cooler. Score of 1 = extremely undesirable; 9 = extremely desirable.

will not get any. The more common practice is to feed no grain at all during the first 2 or 3 months while the pasture is palatable and nutritious, and to supply a full feed during the late summer and fall when grazing conditions are less favorable. This plan has the advantage of utilizing the pasture when it is at its best and of supplying abundant digestible nutrients during the period when cattle on even a limited feed of grain will eat relatively little of the dry, unpalatable pasture. Another advantage of deferring the feeding of grain to the last half of the summer is that much labor is saved.

It has already been said that the feeding of some form of dry feed on legume pastures reduces the tendency to bloat. Ground ear corn or oats, fed in limited amounts, is quite effective for this purpose and at the same time serves as an energy source and as a moisture-absorbing agent in the digestive tract. This absorbent effect slows the passage of the feeds consumed, which conceivably should improve digestion and absorption of nutrients. Thus, if the pasture consists largely of succulent legumes, the limited feeding of ground ear corn or oats, which contain considerable fiber, seems justified, although admittedly this recommendation is not based on extensive research.

North Carolina workers have extensively studied the use of limited grain on fertilized Coastal Bermuda grass pastures using the technique of adding 10 percent animal fat to mixed rations to inhibit intake to about 1 percent of body weight daily. Results of a 4-year study are summarized in Table 201. Grazing trials ran from May 1 to mid-October for about $5\frac{1}{2}$ months. Implanted, wormed yearling steers were provided either energy alone, protein alone, both energy and protein, or no supplemental feed on the pastures. The pastures actually were understocked and the surplus hay was clipped. The steers were slaughtered at the end for carcass data and a complete economic analysis of the programs, as shown in Table 201.

Unsupplemented steers gained 1.15 pounds per day for the $5\frac{1}{2}$-month grazing period. Supplemental protein increased gains by 0.25 pound per day while either energy or energy plus protein supplementation, consumed at the rate of about 1.25 percent of body weight daily, increased daily gains to 2.16 and 2.38 pounds, respectively. The 10 percent stabilized fat added to the supplemental energy rations did not quite restrict voluntary feed intake to the intended level of 1 percent of body weight daily. Only about 5 pounds of supplemental energy concentrate were required per pound of gain, although the grass of course supplied additional nutrients. The steers receiving no supplemental energy on grass produced only standard carcasses while most of those receiving about a half-feed of concentrates graded good with a few grading choice. There are commercial supplements that contain inhibitors that can be added to self-fed mixtures to restrict intake on pasture to about half a full feed.

Table 201

Performance of Steers Provided Limited Concentrates on Coastal Bermuda Grass Pastures, May 1 to Mid-October (4 Years)[a]

		Treatment		
Item	Pasture Only	Pasture + Ground Shelled Corn + 10% Fat Free-Choice	Pasture + 1.5 lb Soybean Meal/Day	Pasture + Ground Shelled Corn + 10% Fat + 10% Soybean Meal Free-Choice
Average days on test	163	163	163	163
Average initial weight (lb)	682	671	677	671
Average final weight (lb)	869	1,025	906	1,043
Average daily gain (lb)	1.15	2.16	1.40	2.27
Average stocking rate (steers/acre)	2.38	2.38	2.38	2.38
Average concentrate consumed/day (lb)	0	10.7	1.5	10.9
Concentrate/lb gain (lb)	0	5.03	1.07	4.79
Beef gains/acre (lb)	440	845	538	893
Hay clippings (tons/acre)[b]	1.8	3.3	1.8	3.0
Dressing percentage	54.7	55.4	55.4	56.2
Carcass grade	std.+	good−	std.+	good−
Return above cost/steer ($)[c]	34.21	63.13	32.25	60.51
Return above cost/acre ($)[c]	81.24	149.93	76.59	143.71

[a] Adapted from North Carolina Experiment Station Bulletin 430.

[b] Pastures were understocked and amounts of hay clipped are the surplus growth.

[c] Cost of steers at $23/cwt and land rental, fertilizers, and all feed costs accounted for with hay being credited at $25/ton.

AREA OF PASTURE PER STEER

Pastures vary so much in productivity that no definite statement can be made as to their carrying capacities. When cattle receive a full feed of grain, only one-third to one-half as much pasture is required as when the cattle are finished on pasture alone. Some feeders use a minimum of pasture, feeding as many as 80 yearling steers and at least an equal number of hogs on 20 acres. A pasture so heavily stocked, however, becomes so thickly covered with manure that the grass is made unpalatable. This is especially likely in pastures adjoining feeding yards that are used year after year. The heavy growth of grass under such conditions is not always proof that the cattle have an adequate amount of pasture. The same cattle turned onto an equal area, similar as far as growth of forage is concerned but free of objectionable odors, might eat the grass down to the roots within 2 or 3 weeks. As previously mentioned, green-chopping and feeding the forage in drylot overcomes this objection.

Legumes and legume-grass mixtures grown in a 3- to 5-year crop rotation are valued largely for the nitrogen and organic matter they add to the soil, and they should not, if pastured, be grazed too closely or they will be of little benefit to the grain crops that follow. It is much better to stock them with only enough cattle to eat 50 to 70 percent of the forage, leaving the remainder to be returned to the soil. Even permanent pastures such as bluegrass and bromegrass do not remain productive if they are grazed closely year after year. Under such unfavorable conditions the stand becomes thin and the bare spots are gradually taken over by weeds. A weedy permanent pasture is almost unmistakable evidence of overgrazing. It is believed that the stocking rates recommended in Table 202 will prevent overgrazing except during extremely dry years. As a result, permanent pastures become better with the passing of each year and improved pastures add large amounts of both nitrogen and humus to the soil. Stocking rates for the range area vary so much with rainfall and type of vegetation that local specialists should be consulted as to recommended rates.

Table 202

Recommended Rate of Stocking Beef Cattle Pastures in Nonrange Areas (Acres per Head)[a]

	Permanent Grasses Unfertilized		Legumes or Mixed Grasses and Legumes in Improved Pastures	
	Yearlings	2-Year-Olds	Yearlings	2-Year-Olds
Pasture only				
Entire season	1.5	2.25	1.0	1.5
Until Aug. 1, then				
removed to drylot	1.0	1.50	0.66	1.0
Full-fed on pasture				
Entire season	0.5	0.75	0.33	0.5
After July 1	1.0	1.50	0.66	1.0

[a] For soils of average fertility that will produce 75 to 90 bushels of corn or 2.5 tons of hay in an average season.

CATTLE GAINS PER ACRE

Gains secured from improved pastures during recent years by steers without grain have been so high as to disprove the statement that the level lands of the midwestern farm belt are too valuable to be seeded to pastures for beef cattle. (See Table 203.) It is easily shown that an acre of pasture, which puts 400 pounds of gain on a yearling steer each summer, will return as much net profit per year as an acre of 100-bushel corn but prices of both cattle and corn can affect net profits either way.

The most surprising fact disclosed by the pasture experiments summarized in Table 203 is that the productivity of pastures seems to bear little relation to the natural soil fertility. For example, pastures established on thin, eroded soil at the Dixon Springs, Illinois, station in the Ozarks section of the state have been somewhat more productive than those at Urbana in central Illinois on level, brown silt loam. Part of the difference may be explained by the longer grazing season, but the main difference seems to be that the legumes encounter less competition from the grasses on the poor soil and therefore constitute a larger percentage of the available forage.

Much heavier applications of fertilizers, principally limestone and phosphate, are required on the poor soil in preparing it for seeding. Also more labor is usually needed to clear the land of brush, establish and maintain pastures, and prevent reencroachment of brush.

BETTER METHODS OF UTILIZING PASTURE

There are many different methods of utilizing pastures in the finishing of cattle, each of which probably has more or less merit for a given situation. Farmers who have a considerable area of pasture land will do well to study critically their present use of pastures, because pastures can easily be a liability rather than an asset in a cattle-finishing enterprise. In fact, a cattleman of much experience and a close observer of other feeders' methods once said that "inexperienced feeders have lost more money trying to use pastures to save a little feed than pastures have ever made for those few feeders who know how to use them wisely." This is an excellent evaluation and it emphasizes the fact that summer-feeding cattle on pasture is in many respects a more difficult task than winter-feeding in the drylot, so far as profits are concerned.

Apparently one common error in pasture management is an attempt to use the available pasture in a cattle-feeding enterprise that would succeed as well or even better without it, instead of adopting a plan of feeding that

Table 203

Gains Secured per Acre of Improved Pasture When No Concentrates Are Fed[a]

Location	Kind of Forage	Period Grazed	Days Grazed	Gains per Acre (lb)
Central Missouri (Columbia)	Bluegrass	May 6–Sept. 27	144	216
	Bluegrass-lespedeza	May 6–Sept. 27	144	279
Northwest Missouri (Lathrop)	Wheat-lespedeza	Apr. 21–June 26 July 27–Aug. 31	91	313
	Bluegrass-sweet clover	Apr. 27–Oct. 4	160	315
	Bluegrass-ladino	Apr. 27–Oct. 4	160	359
Northwest Indiana (Upland)	Alfalfa-timothy	May 5–Oct. 14	162	264
	Bluegrass	May 5–Oct. 14	162	198
Central Illinois (Urbana)	Bluegrass	Apr. 15–Sept. 20	158	160
	Bluegrass-ladino	Apr. 15–Sept. 20	158	329
	Bromegrass-ladino	Apr. 15–Sept. 20	158	304
	Bromegrass-alfalfa	Apr. 13–Nov. 1	202	342
	Alfalfa	May 6–Sept. 23	140	383
	Sweet clover, 2nd year	May 6–Aug. 2	88	220
	Haas mixture, 2nd year	Apr. 30–Nov. 1	185	416
Southern Illinois (Dixon Springs)	Basic mixture[b]	Apr. 20–Nov. 25	219	277
	Basic mixture + alta fescue	Apr. 20–Nov. 24	218	325
	Basic mixture + bromegrass	Apr. 20–Nov. 24	218	364
	Basic mixture + orchard grass	Apr. 20–Nov. 24	218	337
	Basic mixture + bluegrass	Apr. 20–Nov. 24	218	296
	Ladino, timothy, red clover, alta fescue			564[c]
Florida (Everglades)	St. Augustine grass		365	1,004

[a] Based on data in mimeographed reports of state agricultural experiment stations.
[b] Basic mixture (pounds per acre): ladino 1, timothy 4, redtop 3, alfalfa 4, lespedeza 5.
[c] Three-year average; combined gains of cattle and sheep.

Fig. 96. Fleshy yearling steers, just off brome-alfalfa pasture, being started on green, chopped entire corn plant and a small amount of hay. No more than 90 to 120 days should be required to finish these steers for market, once they have been worked up to a full feed of corn. (Prairie Farmer.)

seems to offer the best opportunity to use the pasture efficiently. This is merely another instance of the necessity of fitting the cattle-feeding program to the available feed supply.

Cattle that are full-fed grain on pasture throughout the spring and summer use too little grass to make this feeding plan suitable for utilizing comparatively large areas of pasture. If the pasture is stocked heavily enough to consume most of the grass—that is, 3 to 4 head per acre—the number of cattle required may be greater than the supply of corn that is available for satisfactory finishing.

Pasture is at its best in spring and early summer and should be utilized at this time if the greatest returns are to be realized from a given area. To stock a pasture in the spring with only the number of cattle that it will carry through the entire season under average weather conditions is a great waste of forage through failure to harvest the pasture when it is most valuable. Too heavy stocking at any time during the grazing season results in little or no gain, since most of the available forage is used for maintenance and little for production. Consequently, it is logical to stock the pasture with the number of cattle that will ensure the consumption of the forage at about the rate it grows at the time of year. This means stocking the field fairly heavily at the beginning, perhaps 3 or more head to the acre, and removing animals from time to time to supplementary pastures or to the drylot where they are given a full feed of grain. Harvesting a portion of the abundant early spring growth in the form of hay or silage is one practical means of balancing forage supply and cattle numbers.

STRIP GRAZING AND SOILAGE

California studies reported in Table 204 give some interesting comparative data on different methods of harvesting an alfalfa crop through beef steers. The strip grazing was controlled by an electric wire to provide a 2- or 3-foot strip across the field once or twice daily. Although oat hay was fed to minimize the danger of bloat, one steer from the soilage or green chop lot was lost from this cause. Despite the one death in the soilage lot, however, bloat was a greater problem in the pasture lots. Strip grazing of

Table 204

Methods of Utilizing an Alfalfa Crop Through Beef Steers[a]				
	Pasture Lots		Drylot	
	5-Day Rotation Grazing	Strip Grazing	Hay	Soilage
Number of steers	9	9	10	10
Days on test	132	132	132	132
Initial weight (lb)	550	565	574	532
Average daily gain (lb)	1.79	1.64	1.45	1.73
Feed consumption per head per day (lb)				
As fed				
Alfalfa	—	—	21.4	69.0
Oat hay	4.7	4.9	—	3.9
Dry basis				
Alfalfa	—	—	18.9	14.1
Oat hay	4.2	4.4	—	3.5
Dry matter per cwt gain (lb)	—	—	1,303	1,017
Beef production per acre (lb)	689	739	856	1,080
Percentage of soiling	64	68	79	100
Percentage of rotational grazing	100	107	124	157

[a] *Journal of Animal Science*, 15:64.

this sort is quite commonly practiced in Great Britain, northern Europe, and New Zealand, where pasture utilization is highly developed.

THE KANSAS PLAN OF UTILIZING PASTURE

The Kansas station has conducted extensive experiments on pasture utilization in finishing yearling cattle for market, with particular reference to the bluestem pastures of that state. Although this grass differs somewhat from rotation pasture, results obtained at the Missouri, Nebraska, and Illinois stations, where the same grazing methods have been used, are so similar to those obtained in Kansas as to leave little doubt that the grazing plan recommended by the Kansas station is well suited to all pasture forages that produce their heaviest growth in the spring and early summer.

The object of the Kansas experiments was to determine the best method of using pasture in the finishing of beef calves purchased in the fall and marketed approximately a year later as fed yearlings. Some of the different methods of feeding and grazing used are shown in Table 205. The following plan almost always proved most profitable.

1. Steer calves were wintered sufficiently well to gain 1.3 to 1.5 pounds a day. A ration consisting of 16 to 20 pounds of Atlas sorgo silage, 2 pounds of legume hay, 1 pound of protein concentrate, and 4 to 5 pounds of shelled corn per head daily was found to be excellent for this purpose. Omitting the shelled corn from the winter ration was advisable with heifer calves but not with steer calves, as steers wintered only on sorghum silage and hay lacked sufficient finish to sell satisfactorily the next fall. A full feed of corn silage, plus supplement and 2 to 3 pounds of hay, also would produce such gains.

2. Steers were grazed on pasture without grain for approximately 90 days, or from May 1 to August 1. Although these were the limiting dates in all of the Kansas experiments, results obtained at the other stations indicate that condition of the pasture should be considered in deciding when to begin and end the grazing period. Nothing is gained by leaving cattle on a pasture that does not furnish enough feed to ensure reasonably good gains.

3. Cattle were full-fed in drylot for 100 days, or from about August 1 to November 15. Various other feeding plans were tried, but none was found to be as satisfactory as this. Steers fed on pasture during the last 100 days gained less rapidly than those removed to the drylot, and they sold for a much lower price. Other plans involving various combinations of pasture and drylot feeding were also inferior to the standard method.

Table 205

The Kansas Method of Utilizing Pasture in Finishing Yearling Cattle[a]						
Comparison of Standard Plan with Variations (3-year averages)	Winter Gain 136 Days	Pasture Gain 90 Days	Gain While Full-Fed Grain	Total Gain	Margin per Head	Corn Fed per Head (bu)
Standard plan	258	98	256	612	$43.54	37
No corn fed in winter	183	123	263	569	40.54	26
Standard plan	270	91	285	646	11.76	39
Full-fed on pasture after May 1	268	—	283	551	.69	43
Full-fed on pasture after Aug. 1	269	93	271	633	4.17	39
Standard plan	240	93	259	592	4.66	37
Fed 60 days on pasture after Aug. 1; last 40 days in drylot	236	93	250	579	3.01	37
Standard plan	231	90	265	586	10.17	38
Winter grain ration discontinued gradually during first 4 weeks on pasture	229	104	256	589	5.87	39
Winter grain ration continued on pasture	219	131	247	597	3.64	46

[a] Kansas Agricultural Experiment Station, Mimeographed Report.

The main reasons for the effectiveness of the Kansas plan are: (1) the cattle are wintered in a way that enables them to make compensatory gains and thus use pasture efficiently the following summer, (2) the grass is grazed at the stage of growth when it is most palatable and nutritious, and (3) when the productive pasture season is over the cattle are removed to the drylot and full-fed for the late fall market, which usually is a good time to market grain-fed yearling steers of choice quality.

INDIANA SUMMER PASTURE FEEDING SYSTEMS

The Indiana station conducted a comprehensive 4-year study to compare a variety of management systems. One system was a comparison of season-long grazing of legume-grass pastures without grain, several levels of limited feeding on pasture, and full-feeding of grain on grass and in drylot. The data are summarized in Table 206. Hereford steer calves were wintered to gain at the rate of 1.00 to 1.25 pounds per day on limited corn silage and supplement. The calves were then assigned to summer treatments on May 6 with one lot of steers remaining in drylot and full-fed an 8:1 mixture of ground ear corn and Purdue Supplement A to slaughter weight. Four lots were grazed without grain for 58 days or to July 2, after which one lot remained on grass alone and three lots were placed on one-third, two-thirds, or a full feed of the above-mentioned concentrate mixture on pasture. On September 18 all lots of steers not already on full feed were worked up to full feed in drylot and fed to slaughter weights. The main objective of this management plan was to maximize utilization of the heavy spring and early summer forage production by feeding no concentrate on pasture during this time. As the forage production, and hence carrying capacity, lessened during the hot summer months, concentrates were to be fed on pasture, but the question to be answered was how much concentrate should be fed. Therefore the various levels indicated were fed until fall or September 17. For final finishing, all cattle were full-fed in drylot to overcome the objection to yellow fat and lower carcass grade.

The steers on pasture during spring without energy supplementation gained an average of 1.76 pounds daily at a stocking rate of 2.39 steers per acre for 58 days. Thus each acre of spring pasture produced 244 pounds of gain during this short maximum forage production period without additional feed except minerals and salt.

During the summer dormancy period of 77 days, from July 2 to September 8, unsupplemented steers gained only 0.19 pound per day with pastures stocked at the rate of 1.47 steers per acre. Thus only 22 pounds of gain per acre were produced during this time. If the pasture had contained more legume such as alfalfa, which is a warm-season crop, the gains would have been more satisfactory, possibly up to 1.50 pounds per day. All levels of energy concentrate fed improved gain and increased stocking rate, in direct relationship to level of energy fed.

After the summer grazing period, when all cattle were on full feed in drylot, gains were inversely related to the summer gains, owing to compensatory gain. The cattle that had been in drylot all the while gained only 1.74 pounds per day during the fall drylotting period but they reached slaughter weight soonest. Length of the fall drylot finishing period and the total amount of feed required were negatively correlated to their summer energy level used on pasture. A comparison of either the total amounts of corn and protein supplement required or the concentrate required per 100 pounds of gain can provide information concerning the contribution made by the pasture. The carcass quality-grades were comparable for all systems, and all were slightly below the desired low choice level. The steers drylotted on full feed continuously were fattest and would have yield-graded poorest.

The overall plan tested in this study has several advantageous features for a farmer-feeder. Calves were bought in the fall when cheapest and were wintered on supplemented corn silage to grow the steers to short yearling stage without excessive condition. Spring pastures were maximally utilized, and then any of the levels of supplemented grain fed on pasture during the summer dormancy period kept the steers gaining efficiently. Finally, finishing the steers in drylot during the fall assured high-cutting carcasses on the market at a desirable comparative price level. The decision as to how much corn to feed on summer pasture could depend on such items as price and availability of corn, labor conflicts in late fall, and need to avoid conflict of owning two droves of steers at one time.

THE DIXON SPRINGS GRAZING PLAN

The University of Illinois, at the Dixon Springs station located in the hilly, eroded section of the southern part of the state, compared several methods of utilizing pastures with homebred calves. One plan that gave excellent results differs from the Kansas plan only with respect to the finishing period during the late summer and fall. In the Dixon Springs plan both yearling and 2-year-old steers were full-fed on late summer and fall pasture, whereas at the Kansas station they were fed in drylot. (See Table 207.) Fescue and ladino clover or alfalfa, in mixed seedings, are excellent forages during August, September, and October, and their value is unutilized if they are not grazed heavily during these months.

Yearling and 2-year-old steers that were marketed directly from pasture in October without being fed any grain were also included in the Dixon Springs tests. The younger cattle usually were sold for return to the country for feeding, but the 2-year-olds almost always were sold for slaughter. The fact that these grass-fat 2-year-old steers sold for as high as $32 per hundred when the top of the market was $35.50, and dressed as high as 59.8 percent, is evidence of the high nutritive value of the improved pastures at the Dixon Springs station.

HYDROPONIC PRODUCTION OF FORAGE FOR CATTLE

Many equipment manufacturers have equipment for growing green feed from sprouted grain—oats being the grain most often used. The seed is

Table 206

Performance of Yearling Steers Grazed Without Concentrate During Spring and then Fed Four Levels of Concentrate During Summer Compared to Total Drylot Feeding[a]

Item		Level of Concentrate			
	Spring	0	0	0	0	Full Feed[b]
	Summer	0	1/3 Full Feed[c]	2/3 Full Feed[c]	0 Full Feed[c]	Full Feed[b]
Initial weight (lb)		636	628	634	640	611
Spring (May 6–July 2) (58 days)						
Stocking rate (steers/acre)		2.39	2.39	2.39	2.39	0
Average daily gain (lb)		1.76	1.76	1.76	1.76	3.40
Summer (July 2–Sept. 18) (77 days)						
Stocking rate (steers/acre)		1.47	1.67	2.08	3.13	0
Weight off pasture (lb)		760	840	905	947	1,027[b]
Average daily gain (lb)		0.19	1.35	2.13	2.87	2.82
Daily concentrate intake (lb)		0	7.2	14.4	20.8	22.7

Spring and Summer Pasture Period

Drylot Finishing Period (Sept. 18 to Slaughter)

Number days to slaughter	125	118	103	85	46
Slaughter weight (lb)	1,113	1,142	1,150	1,132	1,106
Average daily gain (lb)	2.83	2.55	2.43	2.37	1.74
Daily concentrate intake (lb)	23.8	24.1	23.7	23.0	22.6
Total feed required[a]					
Ground ear corn (bu)	37.4	43.2	45.0	45.1	51.9
Protein supplement (lb)	327	378	394	495	454
Hay (lb)	143	142	103	39	226
Concentrate/100 lb gain (lb)	617	662	687	722	825
Carcass information					
Chilled weight (lb)	670	696	700	688	676
Quality grade[e]	6.4	6.5	6.6	6.6	6.4
Backfat (in.)	0.5	0.5	0.6	0.5	0.7
Loin-eye area (sq in.)	11.1	11.9	11.7	11.6	11.9

[a] Adapted from Indiana Beef-Forage Research Day Report, 1975.
[b] This lot was full-fed in drylot throughout the trial.
[c] Full feed of concentrates as determined by ad lib feeding.
[d] Does not include pasture.
[e] 6 = high good; 7 = low choice.

Table 207

Methods of Utilizing Pastures at the Dixon Springs Experiment Station in Southern Illinois[a]

	Sold as Yearlings				Sold as 2-Year-Olds			
	Full-Fed 100 Days on Pasture		Sold Directly off Pasture		Full-Fed 94 Days on Pasture		Sold Directly off Pasture	
	Total	Per-centage	Total	Per-centage	Total	Per-centage	Total	Per-centage
Winter period, drylot (days)	138	39	140	44	292[b]	40	292[b]	42
Pasture, without grain	119	33	181	56	336[b]	47	405[b]	58
Full-fed on pasture	100	28	—	—	94	13	—	—
Total	357	100	321	100	722	100	697	100
Gain during winter (lb)	127	28	141	46	303[b]	38	299[b]	42
Gain on pasture, no grain	145	32	164	54	285[b]	36	418[b]	58
Gain while full-fed on pasture	184	40	—	—	206	26	—	—
Total gain	456	100	305	100	794	100	717	100
Initial weight	407		399		396		394	
Final weight	863		704		1,190		1,111	
Shrinkage	40		40		42		83	
Sale weight	823		664		1,148		1,028	
Feed per head								
Corn (bu)	23.6		—		26.9		—	
Protein concentrate (lb)	267		141		394		276	
Corn silage (lb)	3,327		3,466		9,047		9,047	
Hay (lb)	397		350		1,285		1,285	
Pasture (acres)	1.7		1.7		3.3		3.3	

[a] Illinois Agricultural Extension Service Mimeographed Circular.

[b] The 2-year-old steers were carried through two winters and sold at the end of the second summer.

first soaked in water, then placed in trays for germination. As the new plant grows, it extends its roots downward into a liquid that contains nitrogen plus the other essential minerals. Temperature tolerances are very narrow, and maximum light must be provided to supply the energy for photosynthesis. About 6 days are required to develop the plant to feed stage and the entire plant is fed. Fantastic claims are made for the feeding potential of this hydroponically grown green feed.

A comparison between the chemical composition of oats grain and hydroponically grown oats green feed shows some differences. Protein content of the dry matter of the green feed increases from 15 to 21 percent and the mineral and vitamin values nearly all increase, some of them dramatically, in comparison with oats grain. Unfortunately, some changes are deleterious, principally the decrease in nitrogen-free extract, the principal energy source in both feeds, which decreases from an average of 65 percent for oats grain to 43 percent in the green feed. Further, the crude fiber content increases from 11.7 in oats grain to 26.1 percent in the green feed, which obviously lowers digestibility of the latter. About 8 pounds of green feed result from 1 pound of oats grain but when the green feed is corrected to an equal dry-matter basis with the grain, 83 percent of the original dry matter has been lost. One study indicates that oats green feed dry matter contains 72 percent total digestible nutrients (TDN) as compared to 75.7 percent for oats grain; thus, not only is much dry matter actually lost in the growing process, but the resulting green feed is lower in feeding value, at least with respect to energy. The added vitamins provided by the green feed, mainly carotene and the B vitamins, are either already adequate in cattle diets or can be supplied cheaply.

When the initial costs (ranging from $4,000 to $10,000) of equipment for producing fresh, green feed from grains are added to the loss in energy value already referred to, it is estimated that TDN from sprouted oats costs at least 400 percent more than the TDN in the original oats grain, making it very difficult to recommend hydroponic production of green feed for cattle.

CHAPTER 21
MISCELLANEOUS RATION ADDITIVES

Developments in genetics, physiology, ruminant nutrition, and feed preparation techniques have increased the daily gains of cattle by an estimated average of at least 25 percent during the last 25 years. As a result the feeding period is being reduced on the average by 50 to 75 days. More important, from a profit or loss standpoint, is the estimated 20 percent improvement in feed efficiency. Although the price received for the finished cattle may be unaffected, a savings in nonfeed costs due to less time required to carry cattle to the same weight can amount to $10 to $20 per head, depending on size of feedlot. (Chapter 12 explains how the number of cattle fed affects nonfeed costs more than any other item.)

To show that improved feed efficiency is more important than improved rate of gain, consider that when an average feed cost of $30 per hundredweight gain is reduced 20 percent for a yearling steer fed to gain 400 pounds, a $6 reduction in feed cost is made for each 100 pounds of gain. Thus there is a $24 reduction in the feed bill for the steer. Calves fed to gain 600 pounds at a $25 feed cost before taking advantage of recent developements would have their feed costs reduced to $20 per hundredweight gain by the 20 percent improvement in feed efficiency, or a saving of $5 per hundredweight for a total of $30 per head fed. Savings in both categories can, of course, be made on the same steers, but the saving in feed efficiency is in the end the more important.

Several individual additives or management practices may account for half as much improvement as the estimate made above for all of the improvements combined. Unfortunately the benefits from all of the newer improved practices are not cumulative—that is, the individual responses cannot all be added on top of each other. In reality this is good, because it enables a feeder to choose, from among the wide variety, a few reliable additives and practices that are also low in cost. After all, some of the additives are more expensive than others and all of them must be paid for either out of savings in the nonfeed cost category or out of total feed cost savings.

The effect on the end product, the finished carcass, should not be overlooked. Improvement in performance through selection for more rapid and more efficient gains fortunately nearly always also improves carcass yield or cutability. Most feeder-accepted and government-approved additives at least do not reduce carcass merit. Only a very few improve the selling price of cattle. Thus the profits must come from the two sources mentioned above.

Progressive feeders today would not think of feeding cattle without taking advantage of the stimulus supplied by certain ration additives or treatments discovered through recent research. Some of these new developments are so recent as to require further testing before recommendations can be made concerning their use. Those that have been thoroughly tested are discussed below.

HORMONE-LIKE AND ANABOLIC AGENTS

The greatest development in beef cattle feeding and management since the establishment of the need for protein supplementation is the development of practical methods of utilizing synthetic or manufactured hormone-like compounds to enhance growth and feed efficiency. Some of the natural sex hormones such as estradiol, progesterone, and testosterone, singly or in combination, also show promise, but the use of some of these products is still under study.

Diethylstilbestrol, generally referred to as "stilbestrol" or DES, is a synthetic estrogen-like compound that has many of the physiological properties of estrogen, the female sex hormone. Its action on the animal body resembles that of the natural sex hormones, both male and female, in several respects. First, it stimulates growth in immature animals. This growth, as growth nearly always is, is accompanied by economically important improved feed conversion, because it results in a corresponding increase in the protein or lean meat content of the animal or carcass, with a corresponding lessening of external fat or bark.

Stilbestrol is mentioned first because it was the first of this category of compounds to be extensively researched and eventually approved, in 1954, for use in cattle feed. Like all feed additives of this nature, it had to meet rigid standards set by the U.S. Food and Drug Administration (FDA) as to its safety, both for the cattle consuming it and for humans consuming the beef produced by such cattle. Carcasses were analyzed for residues that might prove harmful and, with the routine analytical procedures in use in the early 1960s, detectable harmful residues were not found in the edible portion of the carcass.

Recently, however, more refined analytical techniques have been developed and are now in use that can measure DES residues as low as 0.01 part per billion. Enough cattle have since been found, upon slaughter, with minute detectible residues in the offal, mostly in the liver and lungs, to cause concern on the part of consumers and government regulatory officials alike. This is mainly because tumors have been produced in experimental rats and mice treated with this compound, although the levels fed to the experimental animals exceeded the levels of residues found in certain tissues of the beef carcass by many thousandfold. An amendment to the original act regulating food additives contains a zero tolerance for residues which, of course, permits no detectible residue, on any level. For this reason, by action of the regulatory arm of the FDA, stilbestrol is no longer approved for use by cattle feeders. Other hormone-like compounds to be discussed are approved, but undoubtedly will remain under close scrutiny as well.

Most cattlemen who used DES in their programs used it in oral form or as a ration additive, at the rate of 10 milligrams per day. It also was effectively used in ear-implant form at rates ranging from 12 to 36 milligrams per implantation. The majority of the compounds approved for use at the present time—in the mid-1970s—are administered in implant form.

IMPLANTS FOR SUCKLING CALVES

Three ear-implant materials are available for implanting suckling calves. Studies with these materials were usually made in comparison with untreated calves or with calves implanted with DES. It should be mentioned that DES implants generally improved weaning weights of steers by about 20 to 25 pounds or about 5 percent, and of heifers by about 30 to 35 pounds or about 7 to 8 percent.

Synovex Compounds.   *Synovex-S* implants, recommended only for steers, consist of 200 milligrams of progesterone and 20 milligrams of estradiol benzoate. *Synovex-H*, recommended only for heifers, consists of 200 milligrams of testosterone propionate and 20 milligrams of estradiol benzoate. The active components of the two implants are natural hormones that are chemically identical to those produced by the animal's endocrine glands. They are absorbed, metabolized, and eliminated from the body in the same manner as are the hormones produced by the animal itself. The side effects associated with the use of the banned synthetic hormone-like compounds are not seen in presently approved materials. Such side effects were: excessive riding by both sexes, mammary development in males, and uterine prolapse in females.

Nebraska workers have compared responses in calves to the two Synovex compounds with untreated calves, and the results are reported in Table 208. Calves were implanted when steers averaged 189 pounds and heifers 166 pounds and weights were taken monthly up to weaning at

Table 208

	Weight Gain (lb/head)				
Treatment Group	1st Month	2nd Month	3rd Month	4th Month	Total
Control heifers	46	65	35	37	183
Synovex-H heifers	58	70	37	35	200
Control steers	52	67	31	33	183
Synovex-S steers	55	72	28	37	192
Synovex-H steers	63	70	32	32	197

Effect of Synovex Implants on Steer and Heifer Calves[a]

[a] Adapted from Nebraska Beef Cattle Report, 1972.

about 7 months of age. In both the heifer and steer calves, sizable responses were obtained from the implants during the first 2 months, and the weight advantages were not lost thereafter. The 17-pound advantage from implanting heifers and 9- and 14-pound advantages from implanting steers with the two kinds of Synovex products are smaller by almost half in comparison to responses obtained from DES in many previous trials. The fact that the response was not sustained suggests that reimplantation should be studied. The calves in this study were not creep-fed, and it is likely that the responses to implants would have been greater if additional energy had been available to the calves. Eight rather large individual pellets are necessary to carry the dosage given in the Synovex implantations.

Ralgro.   *Ralgro,* the trade name for zeranol, a resorcyclic acid lactone, is the third implant presently approved for use in calves. It is applied at the 36 milligram level in three 12 milligram ear-implant pellets. Montana workers compared DES- and Ralgro-implanted steer calves with untreated controls and obtained responses, from about 3 months of age to weaning time, of 6 pounds for DES and 11 pounds for Ralgro. In a New Mexico study, using both sexes, DES-implanted calves responded about twice as well as Ralgro-implanted calves. In summary, it appears that the response from Ralgro is about the same or slightly less than from Synovex.

The latest information concerning FDA clearance of the implants just discussed should be obtained before using these products. Heifers or bulls being retained for breeding purposes should not be implanted.

IMPLANTS FOR STOCKER OR GROWING CATTLE

The same three products mentioned above in reference to suckling calves also are presently approved for use in stocker calves. Synovex-S and

Table 209

				Advantage over Controls		
Experi-ment	Number of Animals per Treatment	Initial Weight (lb)	Days Fed	Weight Gain (%)	Feed Intake (%)	Feed Efficiency (%)

Response of Growing Steers to Synovex-S[a]

Experi-ment	Number of Animals per Treatment	Initial Weight (lb)	Days Fed	Weight Gain (%)	Feed Intake (%)	Feed Efficiency (%)
1	25	532	99	13.4	3.0	− 9.1
2	25	445	170	17.9	2.9	−13.8
Average per experiment				15.6	3.0	−11.4

[a] Montana Nutrition Conference, 1973.

Ralgro are recommended for steers and postweaning responses to these implants are greater than for suckling calves. Two trials, summarized in Table 209, were conducted with steer calves fed on roughage rations supplemented with just enough concentrates to produce gains of 1.75 to 2.00 pounds per day. The 15.6 percent improvement in gain with a 11.4 percent reduction in feed required per pound of gain is substantial and compares favorably with responses from DES implants in stocker steers.

Arizona workers conducted two trials with Ralgro for steer calves fed for only 68 days on a growing ration of 25 percent rolled milo and 75 percent ground alfalfa hay. Their results are summarized in Table 210. Ralgro was as effective as DES in this trial with an excellent growing ration. A 27 percent improvement in gain and a 14 percent improvement in feed efficiency were obtained from the Ralgro implant as compared to no implant.

Synovex-H, an approved implant for heifer calves, and DES implants were compared in lightweight weaner heifer calves grazing winter wheat pastures in an Oklahoma study. During a 122-day grazing period, DES

Table 210

A Comparison of Diethylstilbestrol (DES) and Ralgro Implants for Stocker Calves (68 Days)[a]

Item	Control	12 mg DES	36 mg Ralgro
Number of steers	39	39	39
Average initial weight (lb)	421	422	426
Average daily gain (lb)	2.46	3.05	3.14
Average daily feed (lb)	16.3	17.8	17.9
Feed/gain ratio (lb)	6.63	5.84	5.70

[a] Arizona Cattle Feeders' Day Report, 1974.

and Synovex-H heifers gained almost identically, 1.59 and 1.57 pounds per day. Unfortunately no untreated heifers were included in the study as controls.

In summary, it appears that either Synovex-S or Ralgro implants are about as effective as DES for growing steers and Synovex-H is an effective implant for growing heifers.

IMPLANTS FOR FINISHING CATTLE

The presently approved implants have been more thoroughly investigated for finishing cattle than for either suckling calves or stockers. This probably stems from the fact that oral DES was being fed to an estimated 80 percent of all cattle on finishing rations before the compound was banned, whereas, it is doubtful whether more than 25 percent of all calves and stockers received DES in either implant or oral form. Thus a replacement for DES is of greater direct importance to the finishing segment of the industry than to other cattlemen.

Both Synovex-S and Ralgro produce responses in gain and reduction in feed requirements when compared to control or unimplanted steers. Gain responses average about 10 percent, as compared to about 15 percent for DES, and feed efficiency is improved about 6 to 8 percent as compared to 10 to 12 percent for DES. There appears to be some problem with carcass grade, as was observed also in DES-implanted steers, until feeders learn how to manage around the problem. If the steers are fed to the same weights as nonimplanted steers, they show a little less marbling, probably because they are younger when sold, as a result of the increased rate of gain. The answer is to feed the cattle the same length of time as if they had not been implanted, which should produce a satisfactory grade while selling weight will average about 40 pounds heavier. This extra weight will have been added efficiently as well. The 1976 reductions in marbling requirements for the lower end of choice grade will of course alleviate much of this problem.

A few examples of reported trials will illustrate the response to be expected in finishing steers from using presently approved implants. Arizona workers reported a trial in which Synovex-S and Ralgro were compared with several levels of DES. They fed Okie-type yearlings on a 60 percent concentrate ration for 56 days and finished them on a 90 percent concentrate ration for 96 days. Their results are summarized in Table 211.

Based on this study, the three implants would rank in the order of DES, Synovex-S, and Ralgro. Although it is not shown in the table, the

Table 211

Item	24 mg DES Implant	Synovex-S Implant	Ralgro Implant
Effect of Diethylstilbestrol (DES), Synovex-S, and Ralgro Implants on Fattening Steers (142 Days)[a]			
Number of steers	16	15	16
Average initial weight (lb)	628	621	623
Average daily gain (lb)	2.64	2.51	2.40
Average daily feed (lb)	20.4	20.4	19.5
Feed/gain ratio (lb)	7.7	8.1	8.1
Carcass grade	good+	choice−	choice−
Cutability grade	3.4	3.5	3.3
Number of abscessed livers	7	2	9

[a] Adapted from Arizona Cattle Feeders' Day Report, 1974.

researchers reported that, at 112 days into the trial, all gains were about the same, and the final differences seen were the result of a persistent response in the DES steers whereas the effectiveness of the other implants diminished after about 112 days following implantation. Possibly reimplantation with Synovex-S and Ralgro about midway in the trial would have prevented the decline in response.

In connection with reimplantation, it should be mentioned that when Ralgro and Synovex-S and -H were approved, a final date prior to slaughter was established beyond which the compounds were not to be administered, to ensure that tissue residues would be absent in the carcasses. This means that Ralgro and Synovex compounds may not be implanted later than 65 and 60 days, respectively, prior to slaughter.

Feedlot heifers respond to Synovex-H and Ralgro, on a percentage basis, at the same rate as steers but, because their rate of gain is somewhat slower, the response in daily gain and feed efficiency is also somewhat less. No side effects are reported from the use of these implants in feedlot heifers.

Although data are not extensive on this subject, it appears that implantation with any of the approved products during the calf suckling or growing programs does not adversely affect the response obtained from reimplantation during the finishing program. There are data that suggest that it might be beneficial to change from one implant to another kind when reimplanting.

Bulls have not responded to Synovex or Ralgro implants when fed out in the same manner as steers; thus the compounds are not recommended for use in bulls.

IMPLANTING TECHNIQUE

Response to implantation may vary widely because of differences in technique or skill on the part of the person doing the implanting. Steps that may improve the efficiency and speed of this operation are as follows.

1. Restrain the animal securely in a headgate or squeeze chute.

2. Pull the haltered head to either side and anchor the head to prevent upward and downward movement, unless the headgate is of the type that prevents such movement—as most are not. A nose lead may be used in place of a halter but is less humane.

3. Using the ear that is farthest from the front of the headgate, select the implantation site in the middle third of the back surface of the ear, locating the site $1\frac{1}{2}$ to 2 inches from the head. Avoid prominent blood vessels.

4. Position the implanting gun parallel and flush with the ear. Pick up the loose skin with the point of the needle and insert the needle completely, under the skin and toward the head, taking care not to push the needle through the ear cartilage. Positioning the beveled edge of the needle against the ear usually eliminates this likelihood.

5. After inserting the needle full-length, pull the trigger or depress the plunger, slowly withdrawing the needle as the implant pellets are deposited. All pellets should be deposited or should have left the end of the needle by the time it is withdrawn halfway. By all means, the pellets should not be crushed by too-rapid expulsion or they will be absorbed too quickly.

MELENGESTROL ACETATE (MGA)

A synthetic progestin was approved by the FDA in 1967 for oral use in cattle rations for heifers. This synthetic hormone, melengestrol acetate (MGA), suppresses estrus and thus prevents the continual disturbance that results from some heifers being in heat every day when a drove of 40 to 50 or more are fed together. It apparently also is a growth stimulant, as may be seen from the summary presented in Table 212. In trials in which both stilbestrol and MGA were used, it is interesting that MGA produced the greater response. Unfortunately no control groups were fed, hence one cannot tell how much above that of untreated heifers was the response of the two treated groups. The MGA does not produce a response in steers and it may not prove suitable when mixed droves are

Table 212

Response to Melengestrol Acetate (MGA) in Feedlot Heifers (25 Trials)[a]			
	Controls	MGA	Percent Improvement
Average daily gain (lb)	2.14	2.38	11.2
Feed per pound gain (lb)	9.24	8.54	7.6
	Stilbestrol	MGA	Percent Improvement
Average daily gain (lb)	2.16	2.31	6.9
Feed per pound gain (lb)	9.68	9.07	6.3

[a] U.S. Department of Agriculture, Federal Extension Service, Animal Science Report No. 4, 1968.

fed. There are apparently no ill effects from using MGA on heifers which may be put back in the herd for breeding purposes, nor does it adversely affect pregnant heifers. A management problem that may arise is that of heifers coming into heat after withdrawal of MGA at no later than 48 hours prior to slaughter, as prescribed by the FDA. A large percentage of heifers will be in heat within 2 days after withdrawal; so unless heifers are immediately slaughtered, they may be in heat on the day of slaughter and the disturbance may present problems in transporting the heifers or while they are in holding pens at the slaughter house. If they are being sold on grade and yield no problems should result, but if they are sold at a central market the heifers are not likely to be attractively presented. There is no evidence that the carcass would be adversely affected.

RUMENSIN

A compound produced by the mold *Streptomyces cinnamonensis* and called monensin during its developmental stages by Elanco Products Company, was approved by the FDA in 1976 as a feed additive, under the trade name Rumensin. The compound was first used as an effective agent against coccidiosis in poultry and subsequently was found to exert a beneficial effect on feed conversion in beef cattle. The compound apparently is not absorbed from the ruminant gastrointestinal tract, so does not enter into the metabolic systems and thus presents no residue problems. It influences the fermentation patterns in the rumen by shifting the acetic:propionic acid ratio in favor of the highly desired propionic acid. As mentioned in earlier chapters, rations that result in higher propionic acid levels in the rumen are more efficiently utilized, especially at the cell level. Digestibility of the carbohydrate fraction of the ration appears not to be influenced by Rumensin, although there is preliminary evidence that protein digestibility may be improved somewhat. For some reason not yet fully explained, feed intake is slightly reduced when Rumensin is fed, and daily gain is either unaffected or slightly improved. Obviously, then, with less feed being consumed without a reduction in gain, feed efficiency must improve—which has been the case in more than 80 trials conducted to establish the efficacy of Rumensin as a feed additive. Fortunately, no adverse effects have been found in carcass studies and none would be anticipated since rate of gain and shrink were not adversely affected.

Only one of the many Rumensin trials that have been conducted is reviewed here for the sake of space. The Arizona station conducted a study with steer calves carried through both a growing and a finishing phase. Three levels of Rumensin supplementation were used and during the finishing phase one lot of Rumensin-fed steers was implanted with Synovex-S. Thus the researchers could simultaneously study the effects of energy level in the diet and of level of additive, and whether steers would concurrently respond to both hormone-like implants and Rumensin. The growing ration, fed for 115 days, consisted of ground alfalfa hay free-choice and 4 pounds of dry rolled milo per day, serving as a carrier for the additive. The finishing ration, fed for 162 days, contained 80 percent concentrates and 20 percent ground alfalfa hay during the latter two-thirds of the finishing period, and thus would be considered a medium-energy finishing ration. The Rumensin was added to the complete mixed ration at the test levels indicated and was adjusted at about 2-week intervals to maintain the desired level of additive intake.

Results of the complete trial are shown in Table 213. The data from the 40 grams per ton level of Rumensin are not included as no added responses were obtained beyond the 30 gram level. The Arizona workers summarized their findings as follows.

1. Rate of gain was unaffected by any level of Rumensin in a growing ration fed to unimplanted steer calves. Feed intake was reduced at the 30 gram Rumensin level and feed efficiency was improved by about 6 percent over the control calves.
2. Rate of gain was improved by 0.32 pound per day in the calves receiving 30 grams Rumensin per ton of finishing rations when compared to control calves. Synovex-S, when superimposed on the 30 grams level of Rumensin, improved gains an additional 0.28 pound or a total of 0.60 pound over the controls. Feed intake was unaffected by Rumensin alone but increased in the Synovex-S steers. Feed efficiency

Table 213

		Rumensin (gm/ton of diet)		
Item	Control	20	30	30 + Synovex-S
Growing trial, 115 days				
Number of steers	23	24	23	
Average initial weight (lb)	450	456	441	
Average daily gain (lb)	1.82	1.84	1.79	
Average daily feed (lb)	16.4	16.3	15.2	
Feed/gain ratio (lb)	9.0	8.9	8.5	
Feed cost/lb gain (¢)	43.6	42.7	41.5	
Finishing trial, 162 days				
Number of steers	15	18	17	18
Average initial weight (lb)	668	682	663	629
Average daily gain (lb)	1.76	2.00	2.08	2.36
Average daily feed (lb)	18.1	19.2	18.8	20.1
Feed/gain ratio (lb)	10.7	10.0	9.5	8.9
Feed cost/lb gain (¢)	62.9	59.0	56.0	52.5
Carcass data				
Carcass grade	choice−	good+	choice−	choice−
Fat thickness (in.)	0.32	0.33	0.26	0.39
Rib-eye area (sq in.)	11.5	12.4	11.7	10.9
Yield grade	2.4	2.2	2.5	2.8

Effect of Three Levels of Rumensin in Growing and Finishing Rations[a]

[a] Adapted from Arizona Cattle Feeders' Day Report, 1974.

was improved by 11 percent at the 30 grams level and by 20 percent when both Rumensin and Synovex-S were used.

3. For the combined 277-day growing and finishing periods, 30 grams of Rumensin per ton of feed resulted in a 10 percent or 100 pound reduction in feed required per 100 pounds of gain.

4. Rumensin additions alone or in combination with Synovex-S did not affect carcass characteristics.

5. Volatile fatty acid values determined on rumen fluid samples showed an increase in propionate and a decrease in acetate during the growing phase, but during the finishing phase, acetate tended to increase and butyrate to decrease. Total volatile fatty acid production was somewhat increased in this phase.

Rumensin is available in FDA-approved form in dry protein supplements prepared by commercial mixers using a concentrated premix. The FDA approval for Rumensin is for not less than 5 nor more than 30 grams per ton of total ration, or for not less than 50 nor more than 360 milligrams per head per day. The compound has been favorably tested in self-fed blocks and in liquid feeds but use of the product in these forms awaits approval. The same holds true for combinations of Rumensin with antibiotics or other oral growth-promoting substances. The simultaneous use of implants and Rumensin apparently is permitted.

It is estimated that this product will cost about 1.5 cents per head daily when fed at the 30 grams level, and present knowledge suggests that using less than the 30 grams level to reduce costs is not economically feasible.

The unusual reduction in feed intake observed soon after the start of feeding Rumensin may alarm first-time users of this product. It will be observed that feed intake will drop about 20 percent on the second day of feeding and will remain at this reduced level for 3 or 4 days, after which it will gradually increase to a more nearly normal level that still will be about 10 percent below the consumption of non-Rumensin-fed cattle.

OTHER RATION ADDITIVES

VITAMIN A

Preformed or synthetic vitamin A is now recognized as a highly important ration additive. The requirement for this vitamin, both as the natural vitamin and as its precursor, carotene, is discussed in various appropriate chapters, as is the method of administration, whether as a natural ingredient of ration components, as an injection, or as a ration additive. The same practices and principles apply to use of the synthetic forms.

ANTIBIOTICS

The encouraging results obtained with the use of antibiotics in swine and poultry rations led investigators to study the use of these additives in beef cattle rations. Early work with antibiotics for dairy calves affected by digestive disturbances of bacterial origin appeared promising, so it was natural that tests would be made with suckling beef calves. Table 214 shows the favorable results obtained in two such experiments using aureomycin. Many tests in other herds were negative, however, making it appear that disease level and sanitation practices determine whether favorable responses may be expected from antibiotics fed to suckling calves.

Similarly, the continuous feeding of either aureomycin or terramycin in

Table 214

Effect of Aureomycin on Gains of Suckling Beef Calves[a]				
	First Trial		Second Trial	
Aureomycin (daily)	0	24 mg/100 lb body wt.	0	20 mg/calf
Number of calves	7	6	7	8
Average initial weight (lb)	61	60	61	64
Average 80-day weight (lb)	188	203	146	157
Average daily gain (lb)	1.59	1.78	1.46	1.60
Incidence of scouring (days)	38	7	—	—

[a] Indiana Cattle Feeders Report.

the roughage rations of stocker cattle has produced variable results. Calves infected with shipping fever after transport from range to feedlot usually will respond to continuous feeding of 60 to 80 milligrams of antibiotic daily.

Because the response to antibiotics is so inconsistent as far as feeder cattle are concerned, the decision of whether to use this ration additive must be made on an individual farm or feedlot basis.

A test conducted with yearlings by a large commercial feedlot in California, under the supervision of veterinarians, resulted in some interesting data concerning incidence of feedlot disorders, and the effect of the antibiotic aureomycin on their occurrence. The antibiotic was fed in a fortified premix at the level of 500 milligrams daily for the first 28 days, followed by 75 milligrams daily for remainder of the feeding period. Table 215 shows how the number of cases of respiratory diseases and, notably, foot rot were significantly reduced by the use of aureomycin. Average daily gain in the lot receiving the antibiotic was 0.20 pound greater, mainly because of the slower gains of the animals requiring treatment for the various diseases.

Table 215 also provides information on the relative incidence of disorders normally encountered in feedlot cattle. As the steers used came from a large ranch in New Mexico and were delivered direct to the feedlot, there was reduced exposure to viruses normally encountered in marketing channels, such as sale rings and the like. This undoubtedly accounts for the low incidence of shipping or "stress" fever observed, even in the control lot.

Antibiotics for Prevention of Liver Abscess. Abscessed livers, which are condemned by U.S. Department of Agriculture inspectors in packing plants, have become a major problem since the advent and increased use

Table 215

Effect of Aureomycin on Incidence of Disease in Feedlot Cattle[a]	Control	Aureomycin
Number of cattle	681	684
Pneumonia	13	4
Shipping fever	26	19
Tracheitis (necrotic laryngitis)	10	1
Listerellosis	1	0
Foot rot	172	2
Photosensitization	1	0
Encephalitis (nonspecific)	0	1
Digestive disturbances	4	2
Not determined	13	7
Liver condemnation at slaughter	23	20

[a] Abstracted from *Veterinary Medicine*, 52:375.

of high- or all-concentrate rations for finishing cattle. It is estimated that the economic loss from condemned livers now amounts to at least $15 million annually. An even greater monetary loss is the estimated 5 to 10 percent reduction in feed efficiency in feeders so affected. Aureomycin, bacitracin, and terramycin have been approved for some time to be fed continuously at a level of 75 milligrams per head per day. Tylan, long .used in swine rations, is the antibiotic most recently approved by the FDA as a cattle feed additive that is specifically effective in preventing liver abscess. Nutritional aspects of this problem are discussed in Chapter 15.

TRANQUILIZERS

Both natural and synthetic tranquilizing drugs have been used by the medical and veterinary professions to reduce the symptoms associated with hypertension and nervousness. Research workers have explored the possibility of including these drugs at extremely low levels in the rations fed to farm animals. Results are extremely variable but unfavorable in enough trials to question whether to recommend tranquilizers as a feed additive. Individual cattle response varies widely and is unpredictable, causing some cattle to become overtranquilized to the point where they go down in the trucks during shipment, to be trampled and bruised.

GOITEROGENS

Goiterogens such as thiouracil, thiourea, and methimazole are among the compounds reported to produce a quieting effect on cattle. Goiterogens exert a depressing action upon the thyroid gland, resulting in a "hypothyroid" condition. Unfortunately, in most controlled experiments, appetite was reduced to the extent that subnormal performance resulted.

DRIED RUMEN CONTENTS

Veterinarians have used rumen contents, prepared by special drying procedures that supposedly do not destroy the rumen microorganisms, to bring cattle back on feed after illness, infection of the digestive tract, or other rumen disorders. Heavy use of sulfa drugs or antibiotics in the treatment of rumen disorders reduces the normal rumen microflora to abnormally low levels. Thus the administration of specially prepared rumen contents sometimes produces favorable results by more quickly reestablishing the normal microfloral population. Although some cattle feeders have been adding such materials to the rations of healthy cattle at considerable cost, experimental evidence does not justify this ration additive under normal circumstances.

YEAST

From time to time, yeasts of various forms are highly advertised as being valuable ration additives. Extensive research with yeast as a cattle feed, including relatively recent work by the Iowa and Ohio stations, shows that yeast serves no essential function in cattle rations.

The principal value of yeast as cattle feed lies in its high protein content, which is approximately 50 percent. Consequently, it may replace soybean or cottonseed meal or other high protein supplements on a pound-for-pound basis. However, it is usually sold to be fed in very small quantities of only 0.25 or 0.10 pound per day for the reported purpose of seeding a small quantity of the yeast germs in the paunch, where they will multiply rapidly and aid in fermentation and digestion of the feed. Feeding a small quantity of yeast for this purpose from time to time would perhaps be of value were it not that cattle in normal health, being fed a well balanced ration, already have literally billions of the yeast-type of organisms in their digestive tracts. Consequently the addition of a few more produces no noticeable results.

DRUGS

Feeding tonics or drugs to beef cattle to stimulate appetite and improve performance is occasionally practiced by professional herdsmen and fitters in conditioning cattle for the show ring. The drugs most commonly used for this purpose are Fowler's solution and nux vomica. Fowler's solution is an aqueous solution of potassium arsenite (K_3AsO_3). Nux vomica is a drug, made from the poisonous seed of an Asiatic tree, containing several alkaloids, chiefly strychnine. Both of these drugs are highly toxic, and their stimulating effect is achieved by their powerful reaction on the metabolic systems. Obviously the dosage must be gauged with great care or an acute toxemia is produced that may result in death.

Yearling heifers fed nux vomica and arsenic trioxide at the Washington station showed no greater appetites than the control heifers but made somewhat larger daily gains. When the feeding of arsenic trioxide was discontinued the rate of gain rapidly dropped from 2.25 to 1.51 pounds per day, despite the fact that the daily feed consumption did not change significantly during this period. Such behavior in cattle after stopping the feeding of these highly stimulating drugs has been noted frequently by experienced cattlemen.

It is commonly believed that feeding Fowler's solution, and perhaps nux vomica as well, to cattle in appreciable amounts or over long periods adversely affects their fertility. Consequently the use of such drugs is regarded as dishonest and unethical and is strongly condemned.

ENZYMES

Experiments with enzyme additions to barley, high-moisture corn, dry corn, and milo rations by various experiment stations have failed to demonstrate a consistent response to enzymes in the ration. Both cellulytic and proteolytic enzymes have been used and, while an occasional trial may appear to show a favorable response, it is likely that such an effect occurs merely by chance. At present, the use of enzymes in cattle rations seems unwarranted. Some enzymatic products are available as silage preservatives and, although more controlled experiments are needed, some farmer-feeder experiences have been impressively favorable.

GRUB CONTROL AGENTS

Systemic or organic phosphate compounds are effective control agents for cattle grubs and certain kinds of cattle lice. They can be given orally, sprayed, or administered as a back pour-on.

Cattle feeders generally are aware of the need for controlling the external and internal parasites but they expect to see results in terms of improved performance in their cattle in order to justify the expense and inconvenience caused by the use of these control compounds. The South Dakota station conducted a trial with yearling steers raised in the central part of their state, to obtain data on the use of systemic phosphate, thiabendazole (a deworming agent), and aureomycin in finishing rations. The incidence of grubs and internal parasites was not reported, making it difficult to apply their results to all cattle. Knowing the climate prevailing in the area of origin, one would suspect that the incidence of both grubs and other parasites was low.

A summary of the South Dakota data is given in Table 216. Application or dosage rates of the compounds were: systemic phosphate (Neguvon), 0.5 ounce per 100 pounds body weight, as a back pour-on; thiabendazole, a single oral dose of 20 grams per head; and aureomycin, 70 milligrams daily, in the feed. The systemic phosphate and the antibiotic produced significant improvement in rate of gain and feed efficiency while the anthelmintic or "wormer" did not. The latter compound has performed favorably in heavily parasitized cattle originating in environments with higher rainfall where worm infestation is a major problem.

ANTHELMINTICS

Cattle raised in high-rainfall areas such as in the southeastern states and cattle that graze certain high mountain meadows of the West are generally

Table 216

Effect of Neguvon, Aureomycin, and Thiabendazole on Feedlot Performance of Calves (142 Days)[a]

	Neguvon		Thiabendazole		Aureomycin	
	Control	Treated	Control	Treated	Control	Treated
Number of steers	72	72	72	72	72	72
Average daily gain (lb)	1.86	1.97	1.92	1.91	1.86	1.97
Percent change from control		5.9		−0.5		5.9
Average daily feed (lb)	17.3	17.3	17.2	17.3	17.2	17.4
Percent change from control		0		0.6		1.2
Feed per cwt gain (lb)	933	879	896	916	928	884
Percent change from control		−5.8		2.2		−4.7

[a] South Dakota Cattle Field Day Report.

expected to be heavily parasitized with a wide variety of stomach and intestinal worms. The same is often true of farm-sized cow herds throughout the country where cattle of all ages run together and where they are lotted near barns and feedlots for a large part of the winter. Such herds often drink from ponds, dirt tanks, or seasonally active streams. The combination of moist pasture conditions, heavy concentration of cattle around water sources, and mild winters favors a build-up of high populations of internal parasites in grazing cattle.

The Mississippi station grazed yearling heifers of the British breeds on improved pastures in the prairie area of northeastern Mississippi and compared three lots treated with anthelmintics or deworming compounds with a control lot of untreated cattle. They were interested in the degree of infestation, as indicated by fecal worm-egg counts, and in the effect of the three dewormers on cattle gains and changes in worm egg count after deworming. All dewormers were given by drench at the start of the trial and all heifers were grazed together in the same pasture.

The Mississippi workers concluded that, first, the heifers never were seriously parasitized since few egg counts of more than 100 eggs per gram of feces were found. An egg count of 100 is generally considered to be the point at which there should be serious concern about internal parasites. Even with the relatively low level of infestation, however, gain responses were obtained from all treatments although less response was observed in the phenothiazine-treated heifers. When both weight gains and reduction in egg count were considered, thiabendazole and parabendazole were equally effective and both were more effective than phenothiazine. The improvement in weight gain, over 128 days, was adequate in all three treatment groups to more than pay for the cost of dewormers. This study is summarized in Table 217.

An injectible deworming compound called Tramisol has been approved for use in cattle. In commercial feedlots where all incoming calves are run through the chutes and headgates for a variety of treatments, it is very convenient to use this form of anthelmintic. Another common dewormer in use at present is Thiobenzole (thiabendazole) in pellet form, which is orally administered as a feed additive. Arizona workers compared the injectible and oral dewormers in a trial with yearling heifers grazing irrigated alfalfa pasture. The experimental heifers were produced in cow herds that had grazed two different kinds of irrigated pastures—Bermuda and alfalfa—making it possible to study the effect of kind of cow-calf pasture on worm incidence in weaner calves as well. The heifers were allotted into three treatment groups with both sources of heifers represented in each group. As the heifers were weighed at the start of the trial, fecal samples were collected for egg counts. Heifers were either left untreated, injected with 12 cubic centimeters of Tramisol, or fed the

Table 217

		Pheno-thiazine (4.0 oz/head)	Thia-bendazole (0.5 oz/cwt)[b]	Para-bendazole (0.5 oz/cwt)[b]
Weight Gains and Worm Egg Counts in Heifers Treated with Anthelmintics, 128 Days (2 Years)[a]				
Treatment Dosage	Control			
Average initial weight (lb)	443	447	452	473
Average gain (lb)	126	130	140	141
Average daily gain (lb)	0.98	1.02	1.09	1.10
Eggs/gm feces, May 30	37	51	42	26
Eggs/gm feces, July 3	42	39	25	10

[a] Mississippi Agricultural Experiment Station Publication ASC Series 1–8, 1969.
[b] Per 100 lb body weight, administered in a single dose.

prescribed level of Thiobenzole in 1 pound of a complete feed. Results are summarized in Table 218.

The Arizona workers reported that heifers raised on Bermuda pastures, presumably untilled for some years, averaged 125 worm eggs per gram of feces whereas heifers raised on alfalfa averaged only 37 eggs per gram. Ten out of 24 heifers off Bermuda pasture had egg counts over 100, whereas only one heifer off alfalfa had an egg value as high as 100. Tramisol reduced all egg counts to zero whereas, in the lot receiving the Thiobenzole pellets, only 13 out of 17 heifers had counts reduced to zero. It would appear that some heifers ate too little of the dewormer-

Table 218

Item	Control	Thiobenzole	Tramisol
Tramisol Injection Versus Thiobenzole in Feed as Deworming Agents for Yearling Heifers[a]			
Number of heifers	10	17	17
Average weight, Oct. 10 (lb)	594	595	589
Pretreatment[b] count (eggs/gm)			
Average	44	80	68
Range	0–99	3–245	0–333
Posttreatment[b] count (eggs/gm)			
Average	23	4	0
Range	0–96	0–37	0–0
Treatment cost/head ($)	none	1.25	1.44

[a] Arizona Cattle Feeders' Day Report, 1974.
[b] Pretreatment data were taken on October 10; treatment occurred on October 16; posttreatment data were taken on October 27.

containing feed for it to be totally effective and, when cattle are group-fed in feedbunks, that is likely always to be the case. The Arizona workers concluded that the injectible material appeared to be more promising in spite of the slightly higher cost compared with the ingestible compound.

The reduction in egg count in the control or untreated group is similar to what is often observed when cattle are changed to a higher plane of nutrition. Entomologists have no explanation for this phenomenon at present.

FLAVOR AND COLOR

Feed intake, or level of feed consumption, has a profound effect on performance in beef cattle, as gains can only be made from the feed consumed over and above that required for maintenance. Therefore, any addition to the ration that might improve voluntary consumption would presumably be worthwhile, cost considered.

Various natural and synthetic flavor compounds have affected intake in some species of livestock and their effect on the acceptability of food by man is well known. As different colors of feed, artificially applied, have had some influence on intake in other species of farm animals, investigators were led to explore the possibility of improving intake, hence performance, of beef cattle by using certain colors and flavors.

Texas workers used a cafeteria system of feeding a medium-energy, low-palatability ration to calves in two experiments, to determine whether factors such as color and flavor influenced feed intake. The trough or feedbox contained four compartments, one for the control ration and the other for control ration plus the flavor or coloring agent being tested.

Table 219

Influence of Flavor and Color on Acceptability and Intake of Ration[a]					
Flavor Test (15 Steers—24 Days)					
Ration treatment	Control	Sucro	Fenugreek	Sessalom	Total
Daily feed intake (lb)	3.12	4.62	3.49	5.20	16.43
First day's intake (lb)	2.60	7.00	4.10	10.30	24.00
Color Test (15 Steers—21 Days)					
Color used	Control	Green	Red	Blue	Total
Daily feed intake (lb)	3.97	4.16	3.24	3.14	14.51
First day's intake (lb)	7.90	5.80	4.20	3.10	21.00

[a] Adapted from Texas Agricultural Experiment Station Miscellaneous Publication MP-591.

Calves were individually fed by being given daily access to their respective stalls with their compartmented feedboxes, from 7:30 A.M. until 1 P.M. Results are shown in Table 219. Color had no effect on the choices made by the steers, but differences in intake between flavored and control rations suggest a selectivity by cattle for flavored rations in general. The flavor products used contain a variety of organic and synthetic materials. The purpose of showing the data here is only to suggest that flavors may play a future role in cattle feeding, but at present no one flavor or combination appears to be outstanding.

PART V
SPECIALIZED BEEF
CATTLE PROGRAMS

CHAPTER 22
THE BABY-BEEF
AND FAT-CALF PROGRAMS

Cattlemen in the nonrange areas who manage and handle their cows with the view of producing stocker and feeder calves and yearlings to sell to cattle feeders, in direct competition with the ranchers in the range states, are not making full use of all the opportunities available to the nonrange area cow-calf man. In the Corn Belt states, and in much of the remainder of the nonrange area for that matter, grain and harvested roughage are produced on every farm. Markets for fed cattle of all weights and grades are nearby and, in the Southeast and South particularly, weather conditions permit fall and winter calving with a minimum of shelter. Winter small grains, ideal pasture for lactating beef cows, are being grown more extensively in the Southeast and many of the farm cow herds in the Winter Wheat Belt in Oklahoma and Kansas are on a fall- or winter-calving program in order to take advantage of the milk-producing potential of the protein-rich wheat pastures.

For the cow-calf man who has considerable pasture and varying amounts of harvested roughage and grain, there is a real opportunity to increase the size of his business by feeding out his own home-raised beef calves. Two specialized programs that are proving profitable in such situations are the baby-beef and the fat-calf programs. The principal differences between them are the age and weight at which the slaughter calves are sold and the beef characteristics found in the cow herds. Both programs call for maximum growth rate from birth to market to the extent that the calves "never have a hungry day in their lives," as someone has said.

In general the baby-beef program is more suitable for the heavier grain-growing areas, whereas the fat-calf program is better adapted to the southern fringes of the Corn Belt and the whole of the Southeast, the South, and the Southwest.

BABY-BEEF AND SLAUGHTER FAT CALVES DEFINED

A "baby beef" slaughter animal is usually defined as a beef calf 8 to 15 months old, weighing 650 to 950 pounds, and carrying enough condition

669

to merit a quality grade of at least low choice in the carcass. Slaughter "fat calves" are calves that weigh 500 to 750 pounds at ages of 6 to 10 months. These calves usually show considerable bloom and condition, yielding carcasses that grade in the upper end of good grade and better. The beef conformation or "beefiness" of the fat-calf carcass is less pronounced than in the baby beef. Such calves are usually sold directly from the cow or "off the teat"—that is, without a postweaning feeding period. These two kinds of slaughter cattle are not to be confused with the so-called grass beef sold at low prices in some years, especially those years when supplies of such cattle far outstrip demand from cattle feeders. Packers then buy them at prices as low as those paid for slaughter cows, for processing mostly into boneless cuts and ground beef.

OPERATION OF THE BABY-BEEF PROGRAM

It is estimated that more than half of the feeder cattle fed out in the grain-growing states are of native origin—that is, they are produced in herds not too distant from the feedlots where they are eventually finished. Because grain and harvested roughage of average to good quality are generally produced either on the farm that maintains the cow herd or nearby, it seems only logical that the cow-calf man in these sections should feed out his own home-raised calves. This partially offsets the low volume of business that is one of the disadvantages of maintaining a cow herd in these areas, because at least twice as many pounds of calf can be sold. Furthermore, it is very easy to buy calves of the same breeding and weight in the weekly auction sales in the vicinity if additional calves are desired to increase the volume of the finishing program.

Consumer preference studies show that the cuts from well-finished young animals of good beef breeding are always in strong demand. Animals that yield this kind of carcass are never plentiful but they are especially scarce in summer and early fall, resulting in strong prices for baby beef during these months.

QUALITY OF ANIMALS TO USE

Since baby beeves are sold at a relatively young age and yet are beyond the "milk fat" stage, the cows and bulls used in this program must have at least choice beef conformation. Although growth rate is important, early maturity or early finishing characteristics must also be present along with beefiness and thickness. Heavy-milking cows are essential because the most profitable baby beeves make more than half their total gain while nursing their dams. The extra finish or bloom on the calves at weaning time is not lost, as is the situation in the typical range cow-calf operation, because the calves immediately go onto at least a partial feed of grain, if not on full feed. Only heavy-milking cows of choice or better grade of the major British beef breeds, with pure or nearly pure breeding, should be used if the baby-beef program is to succeed. By all means, cows showing much dairy breeding should be avoided. Such cows are better utilized in a fat-calf program, as will be discussed later. Purebred beef bulls of average size and scale, with exceptional thickness or muscling, should be used and the so-called easy-keeping characteristic should be present. The crossbred calves resulting from crossing sires of the so-called exotic breeds on British beef cows are not satisfactory because they are still noticeably in the growing stage and thus do not grade well at the weights desired for baby beef.

BEST SEASON FOR CALVING

The smaller farm herds in the Corn Belt are usually fed and sheltered in such a way that winter or early spring calving may be practiced. This practice is most satisfactory if heavyweight baby beeves are to be sold about 15 months later, during the months of highest prices for such slaughter cattle, namely late spring, summer, and early fall.

Table 220 shows the advantage of early calving in a Missouri station test comparing winter calves with spring calves. For these calves to be considered genuine baby beeves they would have had to be fed until the spring months, but the data still show that early calving resulted in faster gains and better utilization of the concentrates fed. Note also that creep-feeding of the nursing calves did not cause the late calves to make up the deficiency in summer gains.

Farther south, where winter small-grain pastures are possible, still earlier and even fall calving is recommended, with the fed calves being sold in early fall the next year at about 10 months of age.

DEVELOPMENT OF HOME-BRED CALVES

Cow-calf men who follow the baby-beef plan have a choice among three recognized methods of developing their calves. First, the calves may be allowed to run with their mothers without grain until weaning time, when they are weaned, removed to the drylot, and started on feed. When

Table 220

Effect of Winter or Spring Calving on Calf Gains and Feed Requirements[a]		
Calving Season	January–February	March–May
Nursing period (creep-fed)		
Number in each lot	12	14
Average initial weight (lb)	165 (Mar. 13)	157 (June 5)
Average weaning weight (lb)	583 (Oct. 3)	507 (Dec. 3)
Average gain to weaning (lb)	418	350
Average daily gain (lb)	2.05	1.89
Average daily ration (lb)		
Shelled corn	2.70	2.79
Cottonseed meal	0.34	0.35
Average total feed consumed (lb)		
Shelled corn	548 (9.7 bu)	498 (8.9 bu)
Cottonseed meal	68.7	62.4
Feed required/cwt gain (lb)		
Shelled corn	131.7	146.6
Cottonseed meal	16.5	18.4
Postweaning period		
Average final weight (lb)	752 (Dec. 3)	658 (Feb. 12)
Average gain after weaning (lb)	169	151
Average daily gain (lb)	2.77	2.13
Average daily ration (lb)		
Shelled corn	11.7	12.1
Cottonseed meal	1.6	1.6
Alfalfa hay	3.7	2.4
Average total feed consumed (lb)		
Shelled corn	715 (12.7 bu)	859 (15.3 bu)
Cottonseed meal	94	108
Alfalfa hay	226	168
Feed required/cwt gain (lb)		
Shelled corn	423	568
Cottonseed meal	56	71
Alfalfa hay	134	111
Total feed requirement/calf (lb)[b]		
Shelled corn	1,263 (22.6 bu)	1,357 (24.2 bu)
Cottonseed meal	163	170
Alfalfa hay	226	168
Total	1,652	1,695
Feed/gain ratio (lb)	2.81	3.38

[a] Missouri Agricultural Experiment Station Bulletin 652.

[b] Includes only creep feed and feed fed in 2-month postweaning period.

Fig. 97. Home-bred calves of choice or better quality are preferred for the baby-beef program. (American Shorthorn Association.)

handled in this way, home-raised calves are at least comparable to western-bred calves as far as weight, condition, and other qualities are concerned.

The second method of handling home-bred calves is to creep-feed them on a good grain mixture beginning when they are 3 to 4 months old and continuing until weaning time. This feeding may be done by the use of creeps when the calves are running with their dams, or the calves may be kept in a separate lot or pasture and hand-fed, with the cows being turned in for nursing during the night. Calves fed in this manner during the suckling period are much heavier and fatter at weaning time than western-bred calves, hence may be marketed after a much shorter feeding period. Usually such animals are in choice slaughter condition when 12 to 14 months old, when they weigh 800 to 900 pounds. This is the kind of calf that the meat retailer has in mind when he speaks of the superior quality of "native baby beef."

The third method of developing home-bred calves represents ultra-baby-beef production. This plan requires calves that are born early, preferably from September to February or at least by the middle of March, so that they are sufficiently mature and well finished to be marketed directly off the cows. If backed up with the right kind of early-maturing ancestry and if fed to the limit of their ability to utilize feed, such calves weigh 600 to 750 pounds when 9 to 10 months old. Though the beef from such animals lacks the color and flavor usually associated with more mature, choice beef, meat markets are able to dispose of it to customers whose chief considerations when purchasing meat are tender-ness and absence of waste fat. People accustomed to veal or calf beef often will readily accept this youngest of all baby beeves.

Each of the three methods of developing calves has its advantages and disadvantages. The method that is best suited to a particular farm depends largely on the type of the individuals in the cow herd, the feeds available for the cows after calving, the equipment for taking care of young calves in cold weather, the milking qualities of the cows, and the intended date and outlet for marketing the finished calves.

BABY-BEEF PROGRAM FOR LIGHTWEIGHT NONCREEP-FED CALVES

In some areas cow-calf producers may choose to wean calves early because of poor gains resulting from poor pasture conditions or for a variety of other reasons. In any case, many lightweight calves are available at times. For example, in the Southeast the runs of such calves are large in July and August after pastures have stopped growing because of forage dormancy or drought. At the same time, demand from stocker-calf buyers is weak because the fall wheat pastures in the Panhandle area are not yet ready. Thus the question arises as to the suitability of these calves for feeding on a finishing ration to slaughter condition weighing 500 to 650 pounds. Such calves would produce carcasses in the same weight range as heavy fat calves off the cow.

The Louisiana station conducted an experiment bearing on this question. Predominately British breed steers and heifers weighing just over 300 pounds were dewormed, implanted, and placed on an all-concentrate ration during a 14-day preliminary period, after which the objective was to determine the length of feeding period required for such calves to achieve carcass grades of good to choice on a finishing ration containing adequate energy to promote fattening as well as growth. The pooled steer and heifer data on feedlot performance and carcass data are shown in Table

Table 221

Effect of Length of Feeding Period on Feedlot Performance and Carcass Grade of Light-Weight Weaner Calves[a]

Item	Days on Feed				
	70	84	98	112	126
Number of calves	12	12	12	12	12
Average initial weight (lb)	328	327	330	314	331
Average final weight (lb)	501	528	561	582	634
Average daily gain (lb)	2.47	2.39	2.36	2.39	2.37
Average daily feed (lb)	11.20	11.77	12.29	12.05	12.17
Total feed (lb)	784	989	1,205	1,350	1,533
Feed/gain ratio (lb)	4.4	4.9	5.3	5.1	5.2
Dressing percentage	58.9	58.1	58.8	59.7	60.0
USDA grade	std.+	low−	good	good+	good+

[a] *Louisiana Agriculture* 12(4):3.

221. Note that average daily gains were well over 2 pounds and that gains still had not begun to decrease after 126 days on feed. The excellent feed/ gain ratio is also noteworthy, with the conversion rate continuing to the end at about 5 pounds of feed per pound of gain. While the data do not support the necessity of feeding beyond 112 days because dressing percentage, carcass grade, and feedlot performance appeared to have reached a plateau, neither do they discourage feeding for at least 126 days and possibly even several weeks longer. Eventually, however, the calves would become too fat to fill the requirements of the retailer who prefers to market a finished, lightweight carcass that requires little trimming.

When the Louisiana workers compared the data for steers and heifers fed for 126 days, they found that heifers gained 2.43 pounds daily compared to 2.32 pounds for steers but the heifers were slightly less efficient, requiring 5.1 pounds of feed per pound of gain compared to 4.9 pounds for steers. Carcass grades were identical, but steers outdressed the heifers, 59.3 to 58.9 percent. They concluded that lightweight calves of either sex are suitable for this highly specialized baby-beef program, utilizing a very high-concentrate ration for 112 to 126 days. In addition, because heifer calves usually cost considerably less per hundredweight than steers, they should prove to be more profitable, especially if sold on a grade and yield basis.

The 1976 reductions in USDA marbling requirements would certainly ensure that calves fed as those were in the Louisiana experiment would grade at least low choice. At the same time, because gains and feed conversion were so favorable, even up to 126 days, there would be no reason to shorten the feeding period. The greater the gain that can be profitably put on each calf, the lower the nonfeed or fixed costs per calf fed.

BABY-BEEF PROGRAMS FOR HEAVIER-WEIGHT CALVES

For those cow-calf producers who wish to maintain ownership of their calves after weaning and sell the calves at heavier weights after further feeding, there is the question of which feeding program to use immediately after weaning. The question really is, should the calves continue to be grown for a while on a growing or stocker ration before finally being finished on a high-concentrate ration, or should they go immediately onto the finishing ration.

The Oklahoma station conducted a 194-day trial with straightbred Angus steer calves that provides some answers to this question. Half of their weaner calves were fed a grower ration for 76 days before changing to a high-concentrate finishing ration for 118 days, while the other half

Table 222

Feedlot Performance and Carcass Data of Steers Fed on Two Programs for Producing Baby Beef[a]

Item	Grower Ration Steers	Finishing Ration Steers	Difference, Finishing − Grower
Number of steers	45	46	
Feedlot data			
Average initial weight (lb)	435	430	−5.0
Ration concentrate:roughage ratio[b]			
Grower phase	16:84	88:12	
Finishing phase	88:12	88:12	
End of grower phase			
Final weight (lb)	603	646	+43.0
Change in wither height (in.)	2.55	3.17	+0.62
Average daily gain 1st period (lb)	2.22	2.84	+0.62
Feed/gain ratio in period (lb)	6.74	5.52	−1.22
End of finishing phase			
Final weight (lb)	982	980	−2.0
Average daily gain, final period (lb)	3.21	2.83	−0.38
Feed/gain ratio, final period (lb)	6.49	6.30	−0.19
Total for both periods			
Average daily gain (lb)	2.82	2.83	+0.01
Feed/gain ratio (lb)	6.57	6.00	−0.57
Carcass data			
Hot carcass weight (lb)	589	601	+12.0
Rib-eye area (sq in.)	10.8	10.8	—
Average fat covering (in.)	0.73	0.85	+0.12
Carcass grade	good+	choice−	+1/3 grade
Cutability (%)	49.3	48.5	−0.72

[a] Adapted from data in Oklahoma Agricultural Experiment Station MP-87, 1972.
[b] Grower ration contained 84 percent alfalfa plus miscellaneous concentrates; finishing ration contained 78 percent milo, 10 percent other concentrates, 12 percent roughages.

went directly onto the finishing ration and remained on it for the full 194 days. The results of their trial are shown in Table 222.

During the first 76 days, the calves on the high-concentrate ration gained faster, grew more as indicated by wither height, and converted their feed more efficiently, as anticipated. The daily gain values are both high, namely 2.22 and 2.84 pounds for the two groups, which probably reflects the fact that off-truck weights, after a 90-mile haul, were used. This of course also then favorably influences the feed conversion data in both instances but the comparative values for gain and feed efficiency are still valuable in comparing the performance during this period.

The calves that had been on a grower ration for 74 days outgained their full-fed mates for 118 days, when both groups were on the high-concentrate ration, demonstrating the principle of compensatory gain discussed in earlier chapters. Surprisingly, however, they were no more efficient and, in fact, required slightly more feed per pound gain when compared with calves receiving finishing rations from the outset. The combined improved feed efficiency for the latter group of calves is the most significant finding in the study, for the reduction of 0.57 pound of feed per unit gain represents a feed saving of 10 percent. The relative prices of roughage and concentrate would determine whether this difference is significant.

In this study, the slightly heavier carcasses produced by the calves not fed a grower ration reflects a higher dressing percentage, and this, combined with a slightly higher grade, would no doubt mean the carcasses had somewhat higher values per head. Roughage nutrients would have to be considerably cheaper before the higher carcass value of the calves full-fed from the beginning would be offset. In summary, it appears from this study that, under normal feed pricing situations, it is most feasible to place weaner calves intended for the baby-beef steer market immediately on a high-concentrate ration.

VALUE OF CREEP-FEEDING

Experiments conducted on a cooperator's farm, under the supervision of the U.S. Department of Agriculture and the University of Missouri, show conclusively that calves fed grain while suckling their dams weigh off heavier at weaning time but may return no more above feed costs when sold for slaughter at weaning time than calves that are not so fed (see Table 223). Feeding 3 to 6 pounds of grain daily, or an average of approximately 700 pounds per head, produced calves that were 100 pounds heavier at weaning time but, because of lower value per hundredweight because of lower demand from cattle feeders than from the packers, and because of relatively high grain prices, less gross profit was potentially available from the creep-fed calves at weaning time.

After 84 days of full-feeding, at which time the cattle weighed 665 to 770 pounds, the gross profit picture altered in favor of the two droves of calves that received the most creep feed. At this weight, feeder steer demand for the heavier steers caused them to show a greater value, and feed efficiency was still favorable to this point. After another 112 days of full-feeding, or 196 days after weaning, all systems of handling the calves resulted in about the same returns above feed costs. During the latter 112 days, the two lots that consumed the most creep feed as sucklers were most inefficient in feed conversion, no doubt because they were heavier and fleshier at the beginning of this period.

Table 223

Effect of Feeding Nursing Calves on Return Realized When Calves Are Sold at Various Ages[a]

Item	Not Fed	Creep-Fed While Running with Cows	Kept from Cows and Fed Separately	Creep-Fed During Last 4 to 8 Weeks
Sold at weaning time				
Average weight (without shrink) (lb)	490	593	588	522
Value per hundredweight ($)	36.00[b]	35.00	35.00	35.50
Value per head ($)	176.40	205.55	205.80	183.30
Grain fed to date (lb)	—	720	720	180
Value of grain fed to date ($)[c]	—	32.40	32.40	8.10
Gross return/head above feed cost ($)	**176.40**	**173.15**	**173.40**	**175.20**
Sold after 84 days of drylot feeding				
Average weight (without shrink) (lb)	665	770	760	695
Value per hundredweight ($)	38.50[b]	40.00	40.00	39.00

THE BABY-BEEF AND FAT-CALF PROGRAMS 679

Value per head ($)	256.00	308.00	304.00	271.00
Grain fed to date (lb)	760	1,760	1,700	1,050
Hay fed to date (lb)	335	265	245	290
Silage fed to date (lb)	220	135	120	200
Value of feed fed to date ($)[c]	41.75	84.55	81.40	53.60
Gross return/head above feed cost ($)	**214.25**	**223.45**	**222.60**	**217.40**
Sold after 196 days of drylot feeding				
Average weight (without shrink) (lb)	927	1,007	976	947
Value per hundredweight ($)	41.50[b]	43.00	42.50	42.00
Value per head ($)	384.70	433.00	414.80	397.75
Grain	2,535	3,610	3,445	2,840
Hay	720	580	550	655
Silage	500	365	295	470
Value of feed fed to date ($)[c]	129.90	174.80	166.22	142.32
Gross return/head above feed cost ($)	**254.80**	**258.20**	**248.60**	**255.43**

[a] Adapted from U.S. Department of Agriculture and Missouri Agricultural Experiment Station Reports.

[b] All values per hundredweight are estimated and are representative of values prevailing in January 1976 for cattle of the weights and estimated grades indicated.

[c] Feed prices per ton: grain, $90; hay, $30; silage, $20.

OPERATION OF THE FAT-CALF PROGRAM

The nonrange area cow-calf man with a herd of cows somewhat lacking in beef breeding will find that the fat-calf program offers the same opportunities for enlarging his business and for better utilizing his homegrown feed supply and pasture that the baby-beef program offers to the man with the better-bred cows discussed in the preceding section.

Returns come quickly in the fat-calf program, as calves are seldom over 8 or 9 months old when sold. There is a ready market for the 275-to-350-pound carcasses yielded by these calves because they are light, tender, and require a minimum of trimming. Many consumers in the region below a line extending from Washington, D.C., to Amarillo, Texas, are especially fond of this kind of beef. This is fortunate because the area corresponds to the region in which this specialized cow-calf program is best adapted. There is some indication that consumers in this section are gradually changing to a preference for more mature, fed beef, however. This has resulted in some softening in the demand for fat-calf beef.

BREEDING ANIMALS PREFERRED IN THE FAT-CALF PROGRAM

In contrast to the commercial cow-calf and baby-beef programs, something less than ideal beef conformation is quite acceptable in the brood cows used in the fat-calf program. In fact, cows of mixed beef and dairy breeding are preferred because they usually are better milkers than cows of straight beef breeding, although of course there are variations within each group. Cows with some Brahman breeding also work well because of their adaptation to adverse conditions of heat and humidity in hot climates and their inherent growth rate and milk production.

Investigators at the Kentucky station have studied this program extensively and are among its strongest proponents for mixed farming areas such as those of Kentucky. They recommend that the females selected for this program should be healthy, rugged heifers or cows of fair to poor dairy breeding or mixed dairy-beef breeding. Cows should produce 2 to 3 gallons of milk daily at calving and should maintain fair milk production for 8 to 10 months. The better dairy cows are unsuitable because they give more milk than one calf can utilize. High-grade or purebred beef cows usually sell too high and give insufficient milk to justify their use in this plan. The dual-purpose breeds such as Milking Shorthorns and Red Polls are suitable but relatively scarce and high-priced. However, crossbred cows sired by bulls of either of these breeds should be equally suitable.

The Kentucky workers concluded that the bulls best suited to the fat-calf program are purebred beef bulls, larger than average in size and thick

Fig. 98. Well-bred Red Poll cows are exceptionally well suited to the fat-calf program. Calves may be allowed to run with their mothers or, if milk flow is heavy, it may be preferable to allow the calves to nurse twice daily. (Pinney-Purdue Farm, Wanatah, Indiana.)

and meaty in type. Size and the ability to sire large, growthy calves with beef conformation are more important considerations than breed itself.

Unlike the recommendations made for the regular cow-calf programs, herd replacements for the fat-calf program should usually not be made from among the heifers raised in the herd. Continuing to save the heifer calves sired by the purebred beef bulls recommended for use in this program tends to increase the beef breeding in the cow herd with each succeeding generation, resulting in lighter calves owing to the reduced milking ability of the cows. An exception may exist when a man is striving to upgrade his herd in order to have enough quality in the herd eventually to produce high-quality feeder calves or baby beeves.

Herd replacements of the desired breeding can usually be bought in the neighborhood. Buying them at weaning time has several advantages. If heifer calves are bought directly from the breeder, the sire and dams can be evaluated, the cost is less than for older females, the health of the herd and of the heifers can be assessed, and calfhood vaccinations, where indicated, can still be practiced if the calves are not older than 8 months.

Reliable data on the economics of the two cow-calf programs just discussed are not available, but observations by the author would indicate that most of the smaller cow herds in the South, Southeast, and Southwest are more profitable when the fat-calf plan is followed.

BEST SEASON FOR CALVING

Because the fat-calf is best sold for slaughter at weaning time or "off the teat," the period of highest prices plays a larger role in determining the calving date. Figure 99 shows that, in the Southwest at least, the best prices

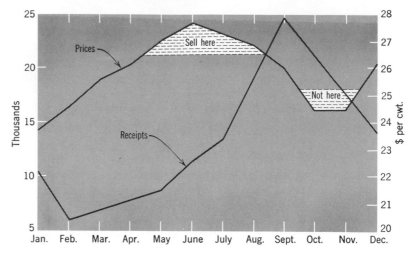

Fig. 99. Effect of season on receipts and average price received for choice slaughter calves. (Oklahoma Extension Circular 619.)

for slaughter calves are received during May, June, and July. Therefore in this region the best time for calving is about 8 months previous to this period, or from September to December. Fortunately the weather is favorable for calving at this time of year in this region, and both cows and calves are quite adequately nourished on the winter oats and wheat pastures grown there.

Local seasonal market demands, winter feed supply, quantity and quality of the labor supply, the weather, and the shelter available are all involved in the choice of calving season but, in general, fall and winter calvings are more profitable in this program than spring calving.

FEEDING THE LACTATING COWS

Cows that calve in the fall or winter present nutritional problems that differ from those previously discussed. Home-grown feeds and pasture must be economically used; otherwise the cost of meeting the rather high nutrient requirements of the lactating cow soon more than absorb the profits possible with this program. If fall and winter calving is practiced, the cows will be dry in summer when pastures are usually of sufficient quality to produce good gains on such cows. Therefore, the feeding program for the lactating cows should be adequate for good milk production and rebreeding, but it need not contain enough energy to maintain cow weight as was advocated for the cow-calf programs in which cows were dry in winter. Some weight loss in the lactating cow should be considered normal. A possible exception is the first-calf 2-year-old heifer that might fail to conceive for the next calf unless she is fairly well nourished.

Tables 224 and 225 show that winter pasture, in this case rye-vetch, is a satisfactory, yet cheaper ration for the lactating cows than native grass and supplement. The calves nursing the silage-fed cows were heavier and

Table 224

Average Cow Data from Different Feeding Systems in the Production of Fall-Dropped Calves[a]

Winter Treatment	Native Grass + Supplement	Native Grass + Silage and Supplement	Rye-Vetch Pasture
Number of cows producing calves	18	17	18
Average cow weights (lb)			
Fall	1,206	1,184	1,156
Spring	1,053	1,005	1,093
Winter weight loss	−153	−179	−63
Average daily winter ration			
Cottonseed meal (lb)	2.5	1.5	
Ground ear corn (lb)	3.0		
Alfalfa hay (lb)[b]			4.0
Oat hay (lb)[b]			12.0
Silage (lb)[c]		51	
Rye-vetch pasture (acres)			1.5
Native grass pasture (acres)	8	5	4
			(summer only)
Yearly feed cost per cow ($)[d]			
Winter supplement	23.59	7.56	6.64
Silage		32.64	
Rye-vetch pasture			20.62
Native grass	22.50	17.50	14.00
Total	46.09	57.70	41.26

[a] Oklahoma Miscellaneous Publication MP-48.

[b] Fed during a 42-day period in January and February when it was necessary to supplement cows on rye-vetch pasture.

[c] Silage (from drought-damaged, immature corn) was available from a self-feeding pit silo, every other day. Amount consumed was estimated from silo measurements at 42 pounds of silage per cubic foot.

[d] The prices used in this study should be adjusted upward by at least 100 percent to be applicable in the mid-1970s, but the systems comparisons are still valid.

Table 225

Average Calf Data from Study of Systems of Management[a] **(All Calves Creep-Fed and Sold for Slaughter)**[b]

Winter Treatment of Dams	Native Grass + Supplement	Native Grass + Silage and Supplement	Rye-Vetch Pasture
Number of calves marketed	18	17	18
Steers	10	9	10
Heifers	8	8	8
Average birth date, October	20th	9th	11th
Average calf weights (lb)			
Birth	77	77	69
End of winter phase, 4/19	462	474	440
Final weight, 5/5	551	574	540
Slaughter data			
Average yield (%)[c]	57.2	57.7	56.4
Average carcass grade (USDA standard)	C−	C−	G+
Average market value per cwt ($)[d]	19.50	19.74	18.63
Total value per calf ($)	107.77	113.27	100.65
Creep-feed consumed per calf (lb)	898	856	590
Creep-feed cost per calf ($)	21.21	20.22	13.97
Cow feed cost per cow-calf unit ($)	46.09	57.70	41.26
Total cow-calf feed cost ($)[e]	67.30	77.92	55.23
Net return over feed cost ($)	40.47	35.35	45.42

[a] Oklahoma Miscellaneous Publication MP-48.

[b] Based on average of steer and heifer data for each lot.

[c] Hot carcass weights shrunk 2.5 percent (minus hide weight). Values based on final Ft. Reno weights.

[d] On-foot market value calculated from yield, grade, and current value of carcass, based on Ft. Reno weights.

[e] Doubling the feed costs, including those for the creep feed, would make these cost data reasonably applicable in the mid-1970s.

fatter at weaning and therefore sale time, but the higher cost of the cows' winter ration was not covered by the additional gross return per calf.

The use of winter pasture also reduced the amount of creep ration consumed by the nursing calves by about one-third. The carcasses were slightly less well finished, but profits were higher despite the lower selling price.

The native pasture used in the Oklahoma tests consisted of grass species that do not deteriorate so badly from weathering as do the pastures found in the Southeast and South. For this reason the native pasture results in the tests mentioned are not highly applicable in the other regions. The feeding of supplemental silage or hay is justified, in fact necessary, in these regions. In addition, winter pastures are not quite so satisfactory in these sections because the higher rainfall results in muddy pastures that are damaged by heavy grazing. A combination of winter pasture for grazing when weather permits, and silage or hay for supplemental feeding when rain and cold weather make grazing inadvisable, seems best. Tables 45, 46, and 47 give the nutrient requirements for cows that are nursing calves; in Tables 45 and 46 the heavy milking cow requirements should be used.

The area where the fat-calf program is recommended also happens to be the area where minerals are generally deficient in the soil (unless corrected by fertilization) and therefore in the plants. A mineral mixture should be available to both cows and nursing calves. A typical adequate mixture consists of equal parts of salt, limestone, and bonemeal. The bonemeal may be replaced by one-half as much dicalcium phosphate.

CREEP-FEEDING IN THE FAT-CALF PROGRAM

Creep-feeding of the calves produced in this program is highly recommended, although it is not always recommended for the commercial cow-calf program mentioned earlier. Creep-feeding is more likely to be profitable if the herd consists of numbers of first-calf heifers or old cows, if drought or mud reduces the forage available as winter pasture, or if there is an appreciable spread between standard- or good-grading and choice-grading calves.

Naturally creep-feeding is successful only if the calves consume the ration offered. The quality of the pasture, location of the creep, stage of lactation of the cows, and the feeds used in the creep ration all affect the amount of creep ration consumed. The effect of the quality of the pasture has been discussed. The best location for the creep naturally varies from farm to farm, but it should be near the place where the cows spend most of their nongrazing time. This spot may be near a shade, the water supply, or the salt and mineral feeders. On farms where the pasture joins the farmstead, creep feeders can be located in a part of the barn or shed. In any case, the creep feeder should be convenient to the calves, yet protected from damaging rain and from other livestock.

The creep ration should be extremely palatable, high in energy, and coarse in texture. Calves prefer either whole grains or coarsely cracked or rolled grains to finely ground feed, and a combination of grains such as corn and oats is generally preferred to a ration consisting of a single grain.

686 **SPECIALIZED BEEF CATTLE PROGRAMS**

Table 226

Comparison of Grain Rations for Nursing Beef Calves

Rations Compared	First Series Average of 3 Years[a]			Second Series Average of 2 Years[b]		
	Shelled Corn	Shelled Corn 8 Parts, CSC[c] 1 Part	Shelled Corn 2 Parts, Oats 1 Part	Shelled Corn 8 Parts, CSC 1 Part	Ground Corn 8 Parts, CSC 1 Part	Ground Corn 8 Parts, CSC 1 Part, Ground Alfalfa, Molasses
Initial age (days)	79	81	81	77	81	85
Initial weight (lb)	222	221	220	216	217	217
Final weight (lb)	501	522	496	536	529	524
Total gain (lb)	280	301	277	320	312	307
Daily gain (lb)	2.0	2.15	1.98	2.29	2.23	2.19
Grain eaten per cwt gain (lb)	177	199	251	187	241	233
Final value per cwt ($)	11.65	12.10	11.60	6.80	6.80	6.70
Feed consumed daily (lb)						
1st 28 days	0.5	0.9	1.0	0.7	1.4	1.4
2nd 28 days	1.8	2.8	3.0	1.9	3.7	2.8
3rd 28 days	4.0	4.7	5.4	4.3	5.7	5.1
4th 28 days	5.1	5.9	6.9	6.5	6.9	7.1
5th 28 days	6.3	7.0	8.4	8.0	9.1	9.0
Total, 140 days	3.5	4.3	5.0	4.3	5.4	5.1

[a] USDA Technical Bulletin 397.
[b] USDA Technical Bulletin 564.
[c] Cottonseed cake.

Fig. 100. A creep-feeder, ideally located near a shady nook where the cows tend to gather. (American Angus Association.)

The protein concentrate, if any, should be in pellet form, so that it will not sift out from the rest of the feed mixture. Grain sorghums or barley should always be rolled or coarsely ground, but oats and corn can be fed whole. Including a molasses feed increases feed intake but does not always result in more profit. Table 226 gives results from two tests in which various grain and supplement combinations were tested. Including a protein concentrate is usually profitable only during the late stages of lactation.

The question of the value of stilbestrol and the antibiotic terramycin in creep-feeding was tested during the last 47 days of the Oklahoma test summarized in Tables 224 and 225. Table 227 shows that both the feeding of 5 milligrams of stilbestrol daily and a combination of 5 milligrams of stilbestrol and 40 milligrams of terramycin daily were beneficial. The use of stilbestrol in oral or implant form is no longer approved, but the trial results in this table are included because other trials mentioned in Chapter 21 show that the approved implants, Ralgro or Synovex-S and Synovex-H, are about as effective as stilbestrol.

There is no one best combination of feeds for the creep ration. Rather, the available supply of homegrown grain should determine the ration used, with supplemental feeds being held to only the necessary minimum.

Table 227

Effect of Stilbestrol or Stilbestrol Plus Terramycin in Creep-Feeding Beef Calves (Last 47 Days on Test)[a]

	Basal Creep Feed	Basal + 5 mg Stilbestrol per Calf	Basal + 5 mg Stilbestrol + 40 mg Terramycin
Number of calves per group[b]	15	15	15
Average calf weights (lb)			
Initial, 4/19	470	473	470
Final, 5/5	563	572	578
Average daily gain	1.97	2.10	2.30
Creep feed consumed per calf (lb)[c]	254	207	229
Creep feed per cwt gain (lb)	273	209	212
Slaughter data[d]			
Yield (%)	57.9	57.7	57.1
Carcass grade	G+	G+ to C−	G+ to C−
Marbling score	3.47	2.97	3.32

[a] Oklahoma Miscellaneous Publication MP-48.

[b] Nine steer calves and 6 heifers per group, 5 calves in each group from each lot of the original treatment.

[c] Concerns only the amount during this phase.

[d] Yield based on hot carcass weight shrunk 2.5 percent (hide off). Marbling score: 1 abundant, 3 moderate, 5 very slight.

Ordinarily the home grown grain should make up at least 90 percent of the ration. Oats, because of their high fiber content, should not constitute more than half of the ration, at least in the final months before sale.

It may be necessary to make the opening into the creep feeder enclosure wide enough to permit entry of a yearling heifer in order for the nursing calves to learn to use the creep feeder. Penning a few calves in the enclosure during the day or night also is effective in teaching calves to use the creep. The addition of wheat bran or a higher proportion of oats—or using oats entirely—in the first feed mixture placed in the feeder is also practiced. This first filling of the feeder should be both palatable and bulky, the latter feature being important to prevent founder in a hungry calf that may overeat at the outset. Swine, sheep, and poultry should by all means be kept away from the feeder.

CHAPTER 23
THE PUREBRED PROGRAM

The aim of this chapter is to give information that will, it is hoped, enable a breeder or prospective breeder of beef cattle to conduct the purebred program more profitably. Most herds are, after all, operated with the intention of making a profit, contrary to the belief of some who think that most purebred herds are only a hobby or a tax write-off. It is true that a number of wealthy men are purebred breeders, but in most instances they too, like most other breeders, are seriously interested in producing the best cattle possible and in making a lasting contribution, to the beef cattle industry in particular and to agriculture in general.

Some of the requirements for success with the purebred program are the following.

1. Keen business judgment on the part of the owner or manager.
2. Location near, or within ready access of, an area with heavy beef cow population.
3. Skill in choosing herd replacements or in making outside purchases.
4. Knowledge of cattle feeding and herd management techniques.
5. Sufficient capital for the necessary investments in cattle, land, equipment, and buildings.
6. Sufficient land to produce all of the pasture and roughage and most of the grain needed.
7. A reputation for honesty and integrity.
8. Dedication, commitment, and awareness that purebred-cattle breeding is a long-time enterprise.

Successful purebred herds are being operated by men who fail to meet all of these requirements, but, in general, if the herd is to be a financial success and achieve its other objectives over a period of years, the above requirements and possibly others must be met. Management and skill can be hired, and in some instances successful herds are owned by men who depend almost entirely on such assistance.

AIMS OF THE PUREBRED BREEDER

Every would-be breeder of purebred cattle should decide what his ultimate aims or goals are as a breeder. It is important also to analyze

689

these aims in terms of the means available—that is, capital, land, climate, and skill or know-how—because these factors determine whether the aims can be achieved. It would be futile for someone living in a remote part of the country—or in a section unsuited to beef cattle because of low soil productivity or extremes in climate, for instance—to aspire to reach the top as a cattle breeder. The odds against success would be too great, regardless of how much capital such a prospective breeder might have. This is not to imply that superior cattle could not be produced, but the venture would almost certainly be a financial failure.

Purebred breeders and their operations can usually be divided into four rather broad categories. Although there is considerable overlapping and breeders are often in the process of shifting from one category to another, most breeders can be placed in one of the following groups.

1. The so-called master breeders. The master breeder is one who makes a lasting contribution toward improving the cattle of the breed of his choice. He not only produces superior individuals or show-ring champions; he produces sires or families of females that will leave their influence on the breed for generations to come. The master breeder usually stays in business for a long time and often passes the intact herd on to the next generation. Although there are notable exceptions, master breeders are often men of means, and the beef cattle industry should be thankful they are. They are usually able to obtain the cows or bulls needed in their breeding programs without undue consideration of cost. If a particularly costly individual proves unsuccessful as a producer, financial ruin does not result as may be the case with the breeder of more limited means. Master breeders usually display their cattle at the large shows which, although they may not be a paying proposition, promote the breed and serve as an educational tool for the industry as a whole.

Master breeders refuse to be swayed by the momentary popularity of temporarily fashionable pedigrees or of a type of cattle at variance with the one that will be most beneficial to both cattle producer and consumer. Master breeders have never been numerous and perhaps never will be.

2. Purebred breeders who furnish the purebred bulls for the commercial cow-calf man and the females for many new purebred breeders. Approximately only 3 percent of all beef cattle in the United States are purebred. A broader genetic base for improvement of all cattle would be established if there were more breeders of purebreds.

The quality of cattle produced by this type of breeder varies greatly, from cattle good enough to suit the master breeder, to cattle that should not be multiplied. The better breeders of this group are the backbone of the commercial beef cattle industry, because the bulls produced in their herds determine the quality of stocker and feeder cattle produced in the commercial herds. The size of herd and financial resources of purebred breeders in this group vary greatly and have little bearing on the success of the program.

One rather consistent characteristic of this program is its location, with the herds usually being situated near a market for large numbers of bulls for commercial use. Certainly not all of the calves produced in such herds should be saved as bulls or replacement females. Rather, a good percentage should be sold as stockers or feeders or fed out on the farms or ranches where they are produced. Breeders in this category often go to the master breeder as a source of herd bulls.

3. Breeders of commercial cattle who maintain a purebred herd that is large enough to produce the bulls and at least some replacement females for their own herds. This group includes the breeders who are gradually converting their herds from a commercial cow-calf program to a purebred herd. These purebred breeders are often found in the range areas where herds are comparatively large and many bulls are needed each year. Many young men use this method of growing into the purebred business and, although the process sometimes takes many years, such herds are usually founded on a sound basis because the breeders are aware of the requirements of the commercial operation.

4. Four-H and FFA projects. Members of 4-H and Future Farmers of America organizations, with projects of one to a dozen cows and heifers, comprise an extremely important category among the purebred breeders. It would be interesting to know what percentage of the present-day adult breeders got their start in one of these organizations. Knowledge and motivation gained while caring for a 4-H or FFA calf project have started many a breeder in business and, although this method of building a herd is slow, only a small initial investment is required and such herds often are very soundly built. Details concerning this program can always be obtained from the local county agricultural agent or from the vocational agriculture teacher.

GETTING A START WITH PUREBREDS

The choice of breed is discussed in connection with the commercial cow-calf program in Chapter 8, and most of the points made there apply also to the purebred program. Personal preference, adaptation to climate, and the feed and pasture supply are the most important considerations in choosing a breed.

Purebred cattle are usually purchased in a different manner than are commercial cattle. First of all, such cattle are usually bought by the head, rather than by the pound as commercial cattle are bought. Values of purebred cattle are, of course, determined by supply and demand, as is true of commercial cattle, but there is no real basis for determining whether the prices asked are reasonable. Predicting the price that may be received for the offspring produced in a purebred program 5 years from now is all but impossible. It is true that purebred cattle prices swing up and down in sympathy with the commercial cattle market, but many other factors enter the picture. For instance, a drastic shift in demand for purebred cattle of certain bloodlines owing to an outcropping of such an inherited undesirable trait as dwarfism can quickly reduce values to the level of commercial cattle prices.

Another difference in method of purchase between purebred and commercial cattle is source. Among the more common sources of purebred cattle are the following.

1. Direct purchase from a breeder by private treaty. This purchase may apply to an entire herd or to a few head and is the best, although not the most economical, method of buying purebred cattle.
2. Production sale by larger breeders. The prospective buyer can obtain much knowledge concerning the performance of the sale cattle by studying the sires, dams, and close relatives of the offerings. It is also possible to observe the conditions under which the cattle were developed if the sale is held at the breeder's farm or ranch. Herd records kept through participation in herd improvement programs sponsored by breed associations or the extension service are extremely valuable in evaluating sale cattle from a single herd. Such records help single out the animals that excelled in comparison with their herd mates, and these should be the cattle with the most desirable genes.
3. Consignment sale of cattle produced by the members of state, district, or local associations. This is perhaps the least desirable source of purebred cattle, because little information other than pedigree and the individuality of the animals themselves is available. Cattle sold in such sales are often highly fitted because of the competitive aspects of this method of selling. The high condition of such cattle not only serves to cover up weaknesses in conformations, but also may impair the fertility and shorten the productive life of many cattle.
4. A father's or neighbor's herd is often the source of 4-H and FFA project heifers and is to be highly recommended if the quality is sufficiently high. Project heifers can of course be obtained from the above sources as well, but costs will be higher.

Although the auction method of selling purebred cattle may be a profitable way of merchandising them, it is a questionable method of buying as far as starting a purebred herd is concerned. Sound judgment as to values is often laid aside in the competitive atmosphere of an auction sale. "Bargains" all too often turn out to be someone else's mediocre or cull surplus females, fitted for sale to be bought by the beginner. If a large number of cattle are to be purchased, the buyer would do well to seek the advice and help of a highly reliable fellow breeder or breed association field man.

A highly trained, experienced herd manager or herdsman is often a real asset when beginning a purebred program because of his ability to select and purchase the foundation stock. A new breeder, with little or no experience, who personally buys his first breeding cattle may often pay exorbitant prices for cattle that are considerably below foundation quality. After making his initial investment he may feel the need for an experienced manager, but the financial outcome of the program is already under a real handicap. Such programs are usually doomed to failure from the outset.

Which quality of animal to buy depends largely on the breeder's goals and on the available finances. The average breeder may logically pay only moderate prices for his females but he should be willing to invest whatever is necessary to obtain a sire with superior genetic potential or breed by artificial insemination to top-quality bulls.

Production-testing information is even more important in the purebred program than in the commercial cow-calf programs. Many new purebred breeders are demanding performance records on animals being purchased to start new herds. Such records do not guarantee that the new herd will be a financial success but they help.

BREEDING SEASON

The purebred breeder has a few additional problems to consider in establishing his breeding season as compared with commercial breeders. He must not only be concerned with feed supply, available shelter and labor, and marketing time. If a purebred breeder plans to show his animals, or if he wishes to attract buyers who do, he must consider the age classifications that are fairly standard in shows throughout the country. January 1, May 1, and September 1 are the principal base dates for such classifications, and the following classification is representative of those found in open class shows:

Two-year-old—calved between January 1 and August 31, two years previously.

Senior yearling—calved between September 1 and December 31, two years previously.

Junior yearling—calved between January 1 and April 30 of the previous year.

Summer yearling—calved between May 1 and August 31 of the previous year.

Senior calf—calved between September 1 and December 31 of the previous year.

Junior calf—calved after January 1 of the current year.

In several of the breed-association-sponsored shows there is a further breakdown of the three main groupings into two each, with additional base dates of March 1, July 1, and November 1. This applies especially in the yearling and calf classes. These classifications are subject to change, but the trend has been to lower the maximum ages given above and to offer a wider variety of classes for the younger ages. A new development is that of dividing all cattle of each sex into approximately equal numbers per class. This means that there may be several classes for cattle born between March 1 and May 1 but only one class for all cattle born between May 1 and September 1. Since most calves are born in the spring months, this method places most cattle into groups that are close in age. This system will tend to encourage the breeding of cows when they are ready to breed, without worrying whether the calf may come too late to qualify for a particular show class.

Under most circumstances it is highly desirable to bunch calvings just as soon after the base dates as is practical. Since six base dates are coming into use, this point is less important because no calf would be more than 2 months younger than its competitors. In practice each breeder must decide what his calving plan is to be. Many successful purebred herds follow a calving season that is based entirely on the same factors used in a commercial program, and still others, especially if artificial insemination is used, breed cows to calve the year around. Larger calf crops result from year-round breeding, but naturally the management of the cow herd is then less clear-cut, and overall feed and labor requirements are greater. Estrus synchronization gives the purebred breeder a new tool that virtually enables him to tailor his calving season to fit perfectly with his individual management requirements.

SUMMER FEEDING OF THE PUREBRED HERD

Most purebred cows calve in the spring or fall but, as just mentioned, some herds may calve the year around. The summer feeding program is determined by the calving season and is therefore discussed under six headings.

Spring-Calving Cows. Because an ample milk supply for the suckling calf is highly desirable in order to ensure maximum development and growth, an adequate supply of nutritious pasture is essential for the cow herd. Stocking rates slightly below those recommended for commercial cows are in order as a guarantee against the hazards of drought.

Some breeders, especially in the nonrange areas, separate the calves from the cows during the day in order to hand-feed the calves and to feed the cows a small feeding of grain when they are penned with the calves at night for nursing. First-calf heifers and old cows especially benefit from such treatment. The economy of this practice depends on method of selling, and it is doubtful whether it pays unless calves are sold at or soon after weaning or are to be fitted. Certainly it is not essential for the proper nourishment of the mature cow herself. It should be mentioned that this practice facilitates both hand mating and breeding by artificial insemination.

Creep-feeding calves that are nursing cows in summer is generally recommended for purebred calves to ensure bloom and maximum development. This feed should probably be included under the category of advertising expense because, nutritionally speaking, calves, like their mothers, do not require the added feed. Feeding the calf extra grain during the nursing period is more justifiable than feeding the cow, however. Creep rations are further discussed in Chapter 22.

Fall-Calving Cows. These cows are dry during the summer, so that there is an opportunity to effect some savings at this time. It is never justifiable to feed grain to dry bred cows on pasture in the summertime. Possible exceptions may be valuable but very old or lame cows or cows that are to be sold in an auction sale in the fall or winter. Supplementing the pasture with legume or mixed hay may be advisable, especially if pastures consist entirely of grasses and if drought reduces the total supply of forage. If the pasture is entirely legume, feeding some grassy hay reduces the danger of bloat. Cows that calve in the fall are actually in a better state of nutrition at calving time, even if pasture is short and poor in quality, than cows that calve in the spring. Problems such as poor milk production and calving difficulties caused by poor nutrition seldom occur in fall-calving cows.

Replacement Heifers. Heifers that have definitely been chosen as herd replacements need only good pasture during the summertime, whether they are being bred as yearlings or as 2-year-olds. Feeding extra feed can more easily be done during the wintering period if it is believed necessary to ensure maximum development. In the summer, economy can be practiced without fear of detracting from the herd's appearance.

Sale Heifers.   Heifers that are to be sold in the fall or winter as open yearlings or as bred yearlings or two's are more favorably received if fed some grain or high-energy concentrate while they are on summer pasture. A limited feed—that is, about 1 pound per 100 pounds of body weight daily—of a bulky concentrate mixture results in gains of approximately 2 pounds per day and a definite improvement in condition. A 30-to-60-day drylot feeding period in the fall brings the heifers into sale season with adequate condition without impairing their breeding value.

Using large amounts of oats in the summer and fall rations of sale heifers is recommended. A protein concentrate at this time is of doubtful value if pasture or hay contains considerable legume.

If fall sale heifers are handled in drylot during the summer, a ration consisting of at least 50 percent roughage should be fed and, as mentioned, oats are desirable for summer feeding. Barley or ground ear corn are also bulky and therefore acceptable, but shelled or ground shelled corn or grain sorghum should be fed sparingly. A low-protein, high-molasses, vitamin-and-mineral-fortified concentrate is often used in the fitting rations fed in drylot to ensure adequate intake of total ration and to protect drylot-fed cattle against vitamin A or mineral deficiencies.

Sale Bulls.   Most bull sales are held in late winter and early spring, and grain feeding of yearling or older bulls on grass the previous summer is less important than it is for fall sale heifers. If, however, the breeder wishes to attract buyers who may want to show the bulls, grain feeding from calfhood is essential. Most buyers of bulls to be used on commercial cows—and this applies especially to the ranchers in the rougher range areas—would rather have their new bulls in less than show condition. They find that such bulls are more fertile and more apt to range with the cows when turned out to pasture during breeding season. If grain is fed on pasture, bulky feeds such as oats or ground ear corn are preferred. It is doubtful whether it is economical to maintain all sale cattle, and especially bulls, in high condition at all times with the hope that the occasional visitor will make a purchase. Rather, the feeding program should usually be designed so that most of the sale cattle are in attractive sale condition only when seasonal demand is high.

The Herd Bull.   The mature herd bull needs no additional feed when running with the cow herd on good summer pasture. Yearling bulls are sometimes used in the nonrange areas, and the daily feeding of 6 to 10 pounds of oats, ground ear corn, or similar bulky concentrate is recommended to promote growth and development. The same supplemental feeding may prolong by several years the useful life of old bulls that fail to graze enough to maintain their condition.

During the summer the water, salt, mineral, and shade needs of purebred cattle should be given due attention. Often, because some grain is fed, minerals are overlooked.

WINTER FEEDING OF THE PUREBRED HERD

The problem of winter feeding will also be discussed on the basis of the nutritive requirements of the various age and sex groupings. The cow herd often consists of both wet and dry cows. Dry cows can be wintered economically in the same manner as commercial dry cows—that is, on a full feed of legume hay or, better still, mixed hay, or silages made from the same crops. A partial feed of corn or sorghum silage, grass hay, or cereal straw free-choice, and a pound of protein concentrate per day is used extensively in the nonrange areas. If cured-range grass is used, supplementation with 2 pounds of a vitamin and mineral-fortified protein concentrate results in good calf production but some loss in weight on the part of the cows. This may detract from the cows' appearance to some extent, but investment in winter feed can be kept low.

Open yearling or bred yearling and 2-year-old replacement heifers attain larger mature size if 2 to 4 pounds of a farm grain or high-energy concentrate are fed with the better roughages during the winter. Again, however, it is doubtful whether the increased feed cost can be recovered in terms of larger or heavier calf crops.

Weanling heifer and bull calves should definitely be fed for at least normal growth during wintertime with possibly some improvement in condition, especially if the calves have not been sorted into sale and replacement groups. A full feed of good mixed hay or silage plus 4 to 6 pounds of almost any grain mixture is adequate. If prairie hay or other nonlegume hay or corn or sorghum silage is used, additional protein is needed. One pound per head of any of the oilseed meals is recommended in such cases.

Winter feeding and management of the herd bull are discussed in Chapter 9. If annual production sales are held during the winter, additional condition is often desired in herd bulls simply for display purposes. Although this practice cannot be advocated from a health standpoint, there is no doubt that prospective buyers like to see the sires of sale animals carrying considerable condition.

SALE CATTLE

Winter rations or, for that matter, any ration that is going to be fed to full-fed sale cattle should meet the nutrient requirements for feeder cattle on finishing rations as given in Chapter 15. However, there is a growing consensus that the purebred industry as a whole would benefit if sale cattle were sold without fitting—that is, without being fed to high condition. Someone must pay the added cost of the unnecessary fitting ration; hence it either cuts into the profits of the original breeder or increases the initial cost to the new breeder. Even more important, the reproductive lives of such animals are shortened, and much research data would indicate that fitting of young females reduces their milk production for life. Founder and cattle fed "off their feet and legs" are all too common in purebred herds. Such cattle can only be salvaged at commercial prices.

Many special fitting rations are used in purebred herds. More frequent feedings, special preparation methods such as steamrolling the grain to ensure a bulkier, dust-free ration, cooking the grain portion of the ration, using molasses and beet pulp as appetite stimulators, and many other special adaptations were used by old-time experienced herdsmen to increase ration consumption and to maintain appetite over a long feeding period. The economy of such special treatment depends on the added price received for sale cattle. Undoubtedly the larger breeders who wish to attract buyers from among other purebred breeders are going to continue the practice, but the breeder who sells only to the commercial breeder can reduce his feed and labor investments by maintaining his cattle in less condition by use of simple and practical home-grown rations. Fortunately there is a decided trend toward fitting cattle for show in more natural flesh condition—no doubt a result of discrimination against cattle with a tendency to mature and fatten too early. The emphasis on trimness and muscling in show cattle, and especially in steers, has been responsible for the use of more practical rations in fitting cattle.

SELECTING THE SHOW HERD

The selection of animals for the show herd should be made as early as possible. For most larger breeders the show season begins about August 1 and continues through the fall and winter. Because it takes 4 to 6 months to bring an animal from good breeding to show condition, the candidates for the show herd should be chosen no later than late winter or early spring. For calves that will be shown in the calf classes, selections should be made while calves are still nursing so that they can be creep-fed if desired. If more than one candidate is available for a given class, two should be fitted so that a substitute will be on hand should accident, sickness, or unsatisfactory development disqualify the more promising individual. Substitute animals are especially desirable in the younger classes, where the growth rate is rapid and marked changes in conformation are common.

An important factor in selecting a string of show cattle is uniformity. If all individuals in a show herd are of the same general type and show a uniform gradation in size from the youngest to the oldest, they present the best possible appearance while stalled and when shown in the group classes. Uniformity of color, although of minor importance, is not without value in the group classes.

GENERAL CARE AND MANAGEMENT

In fitting cattle for show or sale, every reasonable precaution should be taken to ensure their comfort and well-being because they make optimum gains and development only when they are quiet and contented. Comfort implies protection from extreme weather, from annoying parasitic insects, and from disturbance by other animals.

In hot weather the cattle should be kept in the barn during the day to prevent their hair from being sunburned and becoming dry and stiff. Moreover, the cattle will be more comfortable in the barn and will gain more rapidly than if left in the pasture throughout the day. If possible, roomy box stalls, well supplied with clean bedding or sand, should be provided, so that the cattle are inclined to lie down and rest except when they are eating. Beginning in June or July the barn should be sprayed to control flies, and the use of fans or even air-conditioning is sometimes justified in the more humid parts of the country.

Show cattle need a reasonable amount of exercise to stay sound in their feet and legs. Exercise also tends to stimulate the appetite and helps keep cattle on feed. Young animals usually exercise sufficiently if turned together at night in a small pasture or lot. Older cattle often stand around or lie down, especially if their feet are tender, and should be made to exercise by walking them about a mile each day.

THE FEEDING OF SHOW CATTLE

The outstanding difference between show animals and those in the breeding herd is a difference in condition. Cattle entered in an exhibition are presumed to be as nearly perfect as it is possible to make them. They are expected to show their muscling, growthiness, soundness, and breed and sex character to the best advantage. Show animals that are noticeably lacking in finish invite the criticism that they do not have the ability to finish at a reasonably young age, or that they are "hard doers." On the other hand, extreme condition is detrimental to breeding animals, as few will deny. It is a healthy sign that judges are tending to accept show cattle with less finish than traditionally has been demanded. In the end, a continuation of this trend will be for the betterment of the cattle, from the health standpoint, and for the viewing public, which can then more accurately judge the true quality of the animal.

In general, the feeding of show cattle is similar to the finishing of steers for market, except that the length of feeding period is shorter for show cattle and the ration contains only about 60 to 75 percent concentrates. Bran also is usually fed in rather generous amounts, not only for its protein content but also because its bulkiness exerts a regulatory effect on the digestive tract.

The grain ration of show cattle is almost always coarsely ground or rolled, particularly during the latter part of the feeding period. Such preparation has a desirable effect on uniformity of consumption from week to week and also makes it easier to mix such feeds with other materials.

Show cattle are commonly hand-fed twice daily. Supplying a relatively smaller amount of feed in several feedings is much better than feeding a larger quantity at one time. So far as possible, the amount fed should be no more than will be eaten within about an hour or two. Leftover feed soon becomes stale and unpalatable and should be promptly removed from the feedbox.

If rather large numbers of cattle are being fitted for sale, self-feeders can be used quite satisfactorily. Including more of the bulky feeds such as oats, bran, ground ear corn, or barley makes this an entirely safe practice and saves labor just as it does in a commercial operation.

The following grain mixtures (in pounds) are examples of the many that are recommended for cattle being fitted for show:

I.	Cracked corn	40
	Crushed oats	35
	Wheat bran	15
	Linseed pellets	10
II.	Crushed barley	40
	Crushed oats	25
	Wheat bran	20
	Molasses feed (dry)	15

III.	Shelled corn	40
	Whole oats	30
	Linseed pellets	10
	Molasses feed (dry)	20

Some fitters moisten each batch of the mixed feed with just enough molasses diluted with an equal volume of water to remove dust and dampen the feed. Cooked barley is sometimes used for the same purpose, as is wet beet pulp, at the rate of about 0.25 pound per feeding.

Show cattle on a full feed of grain are sometimes turned into a short grass pasture each night during summer and fall. Animals that have the run of a clean pasture are unlikely to be troubled by foot rot. In addition to the pasture, show cattle should be fed a limited amount of bright mixed or straight grass hay.

After frost when pastures are no longer green and after winter weather arrives, the day and night procedure may be reversed with the cattle spending the day outdoors and the night inside. Having the cattle spend some time in the open promotes general health and longer, healthier haircoats. During the summer and fall, hair growth can be encouraged by dampening the hair with a sprayer each morning after the cattle have cleaned up their feed.

FINE POINTS OF FITTING, GROOMING, AND SHOWING

Such subjects as care of feet and legs, grooming, washing, breaking to lead, care of horns, clipping of head and tails, and selection of equipment for fitting and showing are discussed in detail in excellent publications obtainable for the asking from the various breed associations, whose addresses are listed in the Appendix, and from the extension livestock specialists in each state. Generally speaking, problems encountered in developing, fitting, and showing beef cattle are quite specific in nature. It is likely that consultation with an experienced herdsman, field man, extension service representative, or vocational agriculture teacher is the quickest and best solution to individual problems.

LETTING DOWN FITTED CATTLE

Fitted cattle that are being retired from a show herd or that have been purchased at a sale need special attention when they are to be let down in condition. Simply turning such cattle out to pasture or placing them on an all-roughage ration is not recommended. Rather, there should be a gradual reduction in concentrate allowance. Raising the fiber content of the grain ration by increasing the oat content is also often practiced. In any case, the feeding of some grain is recommended .for as long as 3 months while the animal is gradually losing its excess fat. Increased exercise such as results when cattle are penned on pasture at all times also assists in letting down fitted cattle.

THE FINANCIAL ASPECT OF PUREBRED CATTLE

The purebred breeder's financial problems are quite different from those of the man who handles only commercial cattle. In the first place, the size of his investment in animals is largely a matter of his own choosing, with "the sky as the limit." This is true even though he maintains only an average-sized herd, for what he lacks in numbers can easily be made up in quality and breeding if he is willing to pay the price. Nonfeed costs also are extremely variable, with investments in barns and labor often being 2 to 3 times those for commercial cattle.

Purebred cattle values vary greatly from one herd to another, because there are wide differences in individual merit and in the popularity of the bloodlines represented. They also vary greatly from one period to another, largely in relation to the degree of prosperity enjoyed by the men who handle market cattle. When prices of commercial cattle are high, there is a strong demand for purebred bulls to produce steers and heifers for the market. This demand results in increased prices for purebred cattle.

The close correlation usually existing between slaughter steer prices and prices for purebred cattle may be seen in Fig. 101. During the 21-year period 1900–1920, purebred prices, as determined from published reports of all important public sales of Shorthorn, Hereford, and Angus cattle, averaged approximately 3.25 times the average value per head of native beef steers on the Chicago markets. During the early years of this period some purebred sales were held at which scarcely better than beef prices were realized. Later, however, literally scores of sales averaged around $500 while the offerings from the more noted herds averaged $1,000 and upward.

Figure 101 shows that purebred cattle prices skyrocketed during World War I but declined immediately afterward to about the same level they had occupied before 1914. A similar rise in prices occurred during World War II, but instead of receding after the war, prices climbed higher and higher.

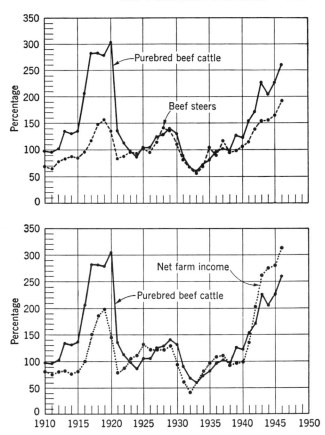

Fig. 101. The relationship of prices of purebred beef cattle to (*a*) prices of beef steers at Chicago and (*b*) net income of farm operators of the United States. Index number 1935–1939 = 100. (University of Minnesota.)

In the 1960s and early 1970s, purebred prices continued to rise and fall with commercial cattle prices but, because of some occasional fantastically high prices, sale averages are becoming less meaningful in terms of what the average purebred breeder may expect. If anything can be concluded it is that the previously mentioned 3.25-times increase for purebred over commercial prices has been reduced to perhaps 2.50 for most purebreds produced in small- to average-sized herds.

Some observers believe that, within some of the breeds, the domination of the show and sale rings by a few large breeders makes the small-scale breeding of purebreds of those breeds a rather discouraging prospect. In the end, if the "little" man does drop out, the resulting reduction in number of breeders could work serious harm on these breeds.

IMPORTANCE OF DEPRECIATION CHARGE IN PUREBRED HERDS

The relatively large amount of money invested in purebred cattle exerts an appreciable influence on the financial outcome of the enterprise. Interest and depreciation charges and death risks, particularly, are affected. Depreciation is an item that may be almost disregarded in the financial aspect of a herd of grade cows, because the cows are usually sold before age affects their market value. In a herd of purebreds, however, depreciation is of considerable importance, because animals bought at high prices are eventually sold for beef. Occasionally the loss suffered from the depreciation on a single animal is sufficiently large to wipe out the profits accruing from the rest of the herd for a full year. Losses from depreciation are particularly heavy during periods of financial depression when prices are falling rapidly. These, however, are to a certain extent covered by the appreciation or increase in values enjoyed during periods of prosperity when prices are advancing. Unfortunately, comparatively few breeders keep many cattle on hand when prices are rising, whereas barns and pastures are likely to be full when values are falling. Moreover, there is a pronounced tendency for new men to enter the purebred business after prices have reached a high level, and a majority of them have nothing to sell until the arrival of a period of depression, which inevitably follows inflation and overexpansion. The losses suffered on herds so founded are sometimes very great, occasionally necessitating the sale of the cattle for only a fraction of their initial cost. A tax expert should be consulted by breeders of valuable animals in order to choose the best plan for computing taxable income, depreciation, and deductions.

OPERATING COSTS IN PUREBRED HERDS

Operating expenses per animal are greater for purebred than for grade cattle. The purebred breeder, to make progress and succeed, must keep his herd in fairly good condition to ensure maximum development of the young stock and to make a favorable impression on prospective purchasers. This means a larger outlay for feed and labor. Further calls for labor result from the general practice of hand mating, from the necessity of keeping the bull and heifer calves separated, from the need for more or less individual feeding and care in order to give each animal the best possible chance to develop, and from the desire of most breeders of purebred cattle to keep the barns and farmstead neat and clean and open for inspection by visitors at any time.

Usually the building and equipment charge per animal is much higher in purebred herds than in grade herds. Although a certain amount of extra equipment beyond that needed for ordinary cattle is highly desirable, elaborate barns and costly equipment are by no means essential to the raising of animals good enough to meet any requirement. A sufficient number of fairly large, roomy box stalls and well drained lots to permit proper separation of animals on the basis of age and sex constitutes the principal item of equipment needed for the satisfactory management of a herd of purebred beef cattle.

Insurance against death losses is available for highly valuable cattle. It is quite expensive but must be added as a necessary item of operating cost for breeders who cannot afford such losses. Insurance is usually used mainly for herd bulls and purchased females.

COST OF SELLING PUREBRED CATTLE

The marketing cost of purebred cattle is appreciably higher than that of grades. Because there is no established market for breeding cattle, each breeder is forced to find his own market, which he does through various kinds of advertising. Newspapers, livestock journals, billboards, and letterheads announce to the reading public that he has stock for sale. Reproduced photographs, show herds, and public displays of prize ribbons and trophies bear witness to the excellence of his herd as a whole. Even a certain percentage of the cost of an imposing barn, the expense of painting the fences, and the cost of keeping the breeding animals in higher condition than necessary to ensure their maximum usefulness should properly be charged to advertising. Returns on money spent in this manner come from a quicker sale or from higher prices for surplus stock rather than from lower operating costs. Again, a tax expert should be consulted because many of these expense items are deductible from income under certain conditions.

The beginning breeder should be watchful lest his selling costs go out of proportion to the prices received. To consign a $300 animal to a sale is poor policy when expenses, including advertising, catalogs, auctioneer, stall rent, freight, expenses of attendant, and the like, amount to $50 or more per head. It is likely that the net return would be greater if the sale were made at home at little more than beef prices.

RECEIPTS FROM PUREBRED CATTLE

Receipts from a herd of purebred beef cattle are very irregular in both amount and season. During periods of prosperity every surplus animal is taken by eager buyers who show little tendency to haggle over prices. During less favorable times even the more noted breeders must resort to peddling tactics in order to make sales. At such times it is common for a large percentage of bull calves to be castrated, finished, and sent to market along with the plainer heifers and those cows that have shown only mediocre breeding ability. Although such sales are trying to the individual breeder, they may be of great benefit to the breed as a whole, as they eliminate the poorer animals from breeding herds.

Price fluctuations have been the downfall of a great many cattle speculators who attempted to get into the purebred business by buying a readymade herd at peak values. However, for the man who builds his own herd from the progeny of a few carefully selected cows of proven merit, price fluctuations, although by no means welcome, are unlikely to prove disastrous.

PART VI
SPECIAL PROBLEMS IN
BEEF PRODUCTION

CHAPTER 24
THE PROCESSING OF FEEDS
AND METHODS OF FEEDING

Probably no question connected with the feeding of beef cattle gives the feeder more serious concern than that of choosing the method of feed processing or preparation. It is small wonder that he is confused, for every medium of communication is used to praise the virtues of every conceivable type of feed grinder, crusher, roller, shredder, cooker, popper, micronizer, and the like. Many of the claims made by the manufacturers are in direct contradiction to one another, and obviously all cannot be correct in all situations. The importance of convenience too often is given more emphasis than it deserves, and the economics of a processing method is sometimes almost entirely overlooked.

Both old and new research findings indicate that altering the form in which certain feeds are fed may affect performance and therefore the subject of processing is an important one. Much of the new research is centered on the influence of kind and level of roughage fed along with the variously processed grains. For example, one form of processing may work wonders if the grain is fed with silage but have no influence whatever in all- or very high-concentrate rations. The method of offering feed to cattle also may favorably affect the financial outcome of a feeding enterprise, especially if a labor saving can be made without too great a capital investment in equipment.

Common goals of any processing method or system of offering feed to cattle are the following.

1. Improved feed conversion or efficiency.
2. Increased feed intake, usually accompanied by faster gain.
3. Improved carcass grade, quality, and cutability.
4. Reduction in wastage of feed at the trough or feeder.
5. Lower transportation and storage costs.
6. Less labor through mechanical handling.
7. Low energy requirements for processing.
8. Improved health of cattle through reduction of digestive disorders.
9. Reduction in harvesting costs such as occurs with field shelling of corn.
10. Maximized utilization of processing equipment and storage such as

might result from reconstituting dry grain in an oxygen-free silo several times per year.

In general, feeds should be fed with the least processing necessary. It is easy to overprocess feeds without necessarily enhancing their value but at the same time adding labor, equipment, and storage costs that must all be recovered from improved efficiency or lowered profits. Consultation, both with nutritionists and engineers, is indicated if an appreciable investment in a new feed processing and handling facility is contemplated. This may be even more important if remodeling an existing plant is being planned.

PREPARATION OF GRAINS

The topic of grain preparation or processing concerns most feeders more than does roughage preparation, simply because proportionately more grain is fed. An exception is the large commercial feedlot that uses complete rations in which the roughage, even though it makes up only a small portion of the ration, is blended with the grain in processed form. In this situation roughage preparation is an important factor, but the average farmer-feeder who relies on hay or silage has few decisions to make of this kind.

In discussing the many aspects of grain preparation, more emphasis will be given to preparation of corn than to other grains because, first, corn is the most important feed fed to beef cattle that requires some form of preparation and, second, more experimental work has been done with this feed. For feeders in the milo area, processing this grain is equally important, of course.

Corn is fed to beef cattle in the following well-recognized forms: (1) dry shelled corn, (2) ground, crimped, or rolled dry shelled corn, (3) high-moisture shelled or processed corn, (4) ground ear corn, and (5) high-moisture ground ear corn.

DRY SHELLED CORN

After mechanical corn pickers came into common use in the 1930s, most corn was harvested and stored as ear corn. Progressive feeders noted that, if some of the dry ear corn were shelled and fed, either in whole or processed form, beef cattle performance was improved, in comparison with ground ear corn, with respect to both gain and feed efficiency. This was not surprising because the 20 percent roughage contained in the ground ear corn was corn cob, which had a lower feed value than the hay or silage used as roughage when the shift was made to the shelled corn. In the change to shelled corn to take advantage of the better cattle performance, feeders noted that considerable whole-kernel corn could be found in the droppings. However, because most farmer-feeders were running hogs behind their feeder cattle, they were not greatly concerned about what they regarded as low digestibility of the shelled corn on the part of cattle.

In the 1950s a series of developments brought about the use of various kinds of grinding or processing equipment to alter the form of shelled corn (which by then had almost entirely replaced the use of ear corn in cattle feeding). First, low-cost electricity and farm tractors with power take-off units became available on every farm to make the grinding of grain convenient at reasonably low energy cost. Then field-shelling of corn was increasingly practiced and grain dryers, also powered by low-cost electricity, made possible the safe storage of large quantities of shelled corn in convenient unloading bins. Third, manufacturers of grain-processing machinery were extremely active in developing efficient multipurpose, farm-sized grinders; thus Corn Belt cattle feeders, the majority of whom were also feeding hogs, purchased such grinders for processing shelled corn to be used in both swine and cattle rations.

Whole Versus Ground or Cracked Shelled Corn.   Data to support the shift from shelled to processed corn were somewhat limited. Trials that showed an advantage for processed corn generally were trials with yearling cattle fed rations containing considerable roughage. Also most of the feeding trials related to corn processing in the 1950s and 1960s were concerned with method of processing, fineness of grind, and so on, and not with whether processing was required in the first place. It was only after large commercial feedlot feeders and their nutritionists began to

Table 228

	Cracked Corn	Whole Corn
Cracked Versus Whole Shelled Corn for Yearling Steers Fed All-Concentrate Diets (82 Days)[a]		
Item	Cracked Corn	Whole Corn
Number of steers	12	12
Average initial weight (lb)	728	717
Average final weight (lb)	1,018	1,072
Total gain/steer (lb)	290	355
Average daily gain (lb)	3.44	3.99
Feed/gain ratio (lb)	5.29	4.84

[a] Illinois Cattle Feeders' Day Report, 1971.

Table 229

Whole or Ground Shelled Corn Fed with Two Roughage Levels to Yearling Steers (150 Days)[a]

| Item | Level of Hay after 56 Days | | | |
| | 0 | | 10% | |
	Ground Corn	Whole Corn	Ground Corn	Whole Corn
Number of steers	50	50	50	50
Average initial weight (lb)	647	648	649	645
Average daily gain (lb)	2.74	2.94	3.07	3.16
Daily corn intake (lb)	19.3	19.2	19.6	18.8
Feed/gain ratio (lb)	7.9	7.4	7.7	7.2
Corn/cwt gain (lb)	702	653	640	596
USDA carcass grade[b]	11.7	12.1	12.1	12.3
USDA yield grade	2.9	3.1	3.1	3.0
Number of abscessed livers	22	4	6	2

[a] Illinois Cattle Feeders' Day Report, 1971.

[b] A carcass grade score of 12 = low choice; 11 = high good.

reduce the roughage content of rations, for reasons already discussed in Chapter 15, and after energy costs skyrocketed, that researchers began to reinvestigate the old question of whether the feeding qualities of corn were enhanced by processing.

Two trials with yearling steers conducted at the Illinois station are summarized in Tables 228 and 229. In the first trial, either cracked or whole shelled corn was full-fed along with a fully fortified pelleted protein supplement, with no roughage whatever. The increased rate of gain and improved feed efficiency in this 82-day trial effectively demonstrated that the processing of shelled corn, for feeding to yearling cattle in an all-concentrate ration, was not only unnecessary but, in fact, detrimental.

In the second trial, also with yearling cattle but fed for 150 days, coarsely ground and whole shelled corn were compared. One of the two lots of steers on each kind of corn received 10 percent chopped hay in the ration throughout the trial while the other lot received no hay or roughage after 56 days on test. Initially all rations contained 40 percent chopped hay, which was gradually reduced to the 10 percent level by 56 days and then removed entirely from the rations of half the steers. The results again show the advantage of shelled or unprocessed corn in the all-concentrate ration. Even when 10 percent hay was included, processing still was not indicated. Carcass quality and yield grade were unaffected and, interestingly, incidence of liver abscess was markedly reduced in the steers finished on the whole corn all-concentrate diet. Possibly more of the whole corn bypasses the rumen and thus does not undergo degradation, resulting in reduced rumen acidity, which has been suggested as a predisposing condition for liver abscess. The whole corn, being of larger particle size, also may provide a "roughness factor" which other investigators suggest is an agent responsible for prevention of liver abscess.

Ohio investigators, in a comparison of whole or shelled corn with processed corn, have reported that starch digestion was unimproved by the crimping of shelled corn, whether fed with an all-concentrate ration or with one containing 20 pounds of silage per day. They also found that digestibilities of ration dry matter, organic matter, and protein all were higher in an all-concentrate ration than when the ration contained corn silage. The Ohio workers concluded that the favorable responses obtained from shelled corn rations in all- or high-concentrate rations cannot be attributed to differences in overall ration digestibility and, further, with reference to the processing of grain, crimping—often used as a selling point for corn processing equipment—does not in itself improve digestibility.

Flaked Shelled Corn.   If corn is to be processed to enhance its bulk or roughness capability, several promising methods have been developed for this purpose. The process is usually known as flaking but it also may be called steam-rolling in some parts of the country. Flaking, as most

Fig. 102. Flaked corn, showing two degrees of thickness, with the thicker flakes at left. The bulky nature of the thin flakes ensures more nearly complete digestion and greater efficiency when used in high-concentrate finishing rations. (Colorado State University.)

commonly used in commercial feedlots and especially in Colorado, consists of steam-cooking or heating the shelled corn in a chamber for 15 to 30 minutes at atmospheric pressure and 200°F. The moisture content is raised to about 20 percent in this process. The corn is then rolled or flaked, the thinner the better, as shown in Table 230. The Colorado workers used a constant cost figure for all three kinds of processed corn, but in reality the processing cost varies. It is fairly obvious that mill capacity is affected by thinness of flake. In any case, there was no question that the bulkier, thin flake—and, for that matter, even the thick flake— were superior to finely ground corn. Undoubtedly the bulky nature of the thinner flake is involved in maintaining the normal physiological status of the rumen. The high-moisture flakes must be dried if not fed immediately. Figure 102 shows the extent to which the thin or flat flake is bulkier than the thicker flake.

The high initial cost of the equipment for processing grain by this method makes it prohibitive for the farmer-feeder with fewer than 500 head of cattle. Several manufacturers are working to develop smaller-sized

Table 230

Performance of Heifers Fed Flaked Corn of Different Thickness or Finely Ground Corn (163 Days)[a]

Method of Processing	Thin Flake ($^1/_{32}$ in.)	Thick Flake ($^1/_{12}$ in.)	Finely Ground ($^1/_4$-in. screen)
Number of animals	14	13	13
Average initial weight (lb)	485	483	490
Average daily gain (lb)	2.82	2.70	2.65
Average daily ration (lb)			
Corn	12.41	12.66	12.83
Fortified supplement	0.78	0.82	0.82
Beet pulp	1.49	1.63	1.79
Chopped alfalfa hay	0.90	0.93	0.95
Corn silage	5.58	5.82	5.52
Feed consumed per pound gain			
(air-dry basis) (lb)	6.14	6.66	6.88
Feed cost per cwt gain	$15.60	$17.00	$17.74
Dressing percent[b]	64.21	63.71	63.25
USDA carcass grade			
Choice	13	12	12
Good	1	1	1

[a] *Colorado Farm and Home Research*, 17:4.

[b] Dressing percent = hot carcass weight less 2% cooler shrink divided by actual final weight less 4% shrink multiplied by 100.

units that are efficient and yet not priced out of reach of the smaller feeder. A mill with continuous grain processing is rather easily adapted for use by the very large feedlots that both process and feed for 24 hours around the clock.

Extruded Shelled Corn.   Colorado workers have developed yet another method for increasing the bulkiness of shelled corn. The grain is forced through an orifice by an auger-like rotor, producing "ribbons" that break into flakes of irregular shape and size as they exit from the extruder. As in the other processing methods involving steam or dry heat, the heat generated by the friction applied to the grain gelatinizes the starch of the corn. Cattle fed on rations containing either whole, flaked, or extruded corn and a medium level of roughage all gained at about the same rate but feed efficiency was improved by 10 percent by both flaked and extruded corn. Other research is under way to find ways of further enhancing the physical character of the extruded product.

Micronized Shelled Corn.   A newer heat treatment for processing dry grain is called micronization. The method has been researched more with milo than with corn and, in fact, much milo used in the large southwestern feedlots is now processed by this method. Feedlot data are presented in the section dealing with the processing of milo.

Micronization consists of passing whole dry grain in a continuous process over an oscillating or vibrating steel plate. Gas-fired infrared generators, located above the plate, heat the grain by microwaves to about 300°F for a few seconds as the grain flows across the sloping, oscillating steel plate. As the grain leaves the plate it drops between steel rollers which apply extreme diagonal and parallel pressure upon the grain by means of spiral grooves. The flakelike end-product, in addition to having been heat-treated, is quite bulky, weighing only about 25 pounds per bushel.

Popped Shelled Corn.   Air-dry grain ranging from 10 to 14 percent in moisture content is heated to 700° to 800°F for 15 to 30 seconds. Most of the kernels explode or pop, and the resulting product contains only about 3 percent moisture. The starch becomes gelatinized and some moisture must be added back before the popped grain is rolled prior to feeding. The final product is very high in bulk and is best used in all- or very high-concentrate rations. Roasting in a dry-heat whole-soybean roaster, commercially known as a "Roast-A-Tron," at 275° to 300°F is another method used to process corn. The end-product has a pleasant aroma and an oily, puffed, and slightly caramelized appearance. Results of a feeding trial using corn processed by this technique are presented in Chapter 16 along

Fig. 103. Ground high-moisture shelled corn can be economically and efficiently stored in trench silos if they are covered. This silo was covered with an 8-inch layer of silage, a sheet of black plastic, and a layer of used tires. Note the self-contained Ensiloader and mixer truck. (Progressive Farmer and Ladd Hitch, Jr.)

with a discussion of the effect of several drying temperatures. As in most other heat treatments, feed efficiency was significantly improved by about 10 percent.

Mills for Grinding Corn.   Farmer-feeders still may wish to resort to the hammermill, burr mill, reel-type knife mill, or roller mill for processing their dry corn, especially if they are feeding a complete mixed ration with some kind of processed roughage. Research does not indicate that one type of mill is either better or less effective than another, so the mill that grinds the corn most economically, consistent with convenience and adaptation to automation, is the one to use. If moist feeds such as silage, haylage, or green chop are not a part of the ration, it would seem that coarse processing is preferred, to provide as much bulk as possible. If on the other hand, moist feeds are fed and the grain is fed mixed or spread on top of the moist feed, then fine grinding is preferred.

HIGH-MOISTURE SHELLED OR PROCESSED CORN

The mechanical harvesting of corn with a combine, as high-moisture shelled corn, is a development that introduces some new questions for cattle feeders. As discussed in Chapter 19, corn has been harvested as high-moisture ear corn for many years, but in those circumstances it was to salvage a frosted crop of immature corn. The change to field shelling is prompted by agronomic aspects of corn production rather than for the purpose of harvesting a special kind of cattle feed.

Table 231 offers a concise explanation for one of the chief reasons for the change to field shelling. If corn is field shelled at approximately 30 percent kernel moisture, there is a reduction in harvesting loss of about 5 bushels per acre, compared with conventional harvesting at about 18 percent kernel moisture. The kernel losses due to inefficient cylinder shelling noted in the table have since been greatly reduced by more efficient cylinder and concave design in field-shelling combines. Farmers who grow corn as a cash crop also are using field shelling, but they usually prefer to wait until the moisture content of the corn is only 20 to 24 percent. The corn is then dried on the farm with forced air, either heated or unheated, in batch or bin driers, or it may be delivered to elevators in wet form to be dried in large-capacity continuous-flow driers.

Table 231

Field Losses in Corn Harvested with a Picker-Sheller at Different Moisture Levels[a]

Observation	Percent Kernel Moisture Content When Harvested			
	35	30	25	18
Loss expressed as bushels per acre				
Ear (machine and detached)	0.8	0.8	2.7	5.4
Shelled corn (snapping rolls and separation)	2.6	1.6	2.3	2.1
Cylinder (on cobs)	5.2	2.4	2.4	0.4
Total	8.6	4.8	7.4	7.9

[a] Illinois Cattle Feeders' Day Report, 1959.

Table 232

Summary of Cattle Feeding Trials Comparing Dry and High-Moisture Shelled Corn Processed by Several Methods[a]

Comparison	Dry Corn			Mois-ture Content (%)	High-Moisture Corn		
	Daily Gain (lb)	Corn Intake[b] (lb)	Feed/Gain Ratio[b] (lb)		Daily Gain (lb)	Corn Intake[b] (lb)	Feed/Gain Ratio[b] (lb)
Processed at feeding time							
Illinois—3 trials							
Dry cracked vs. high-moisture rolled	3.03	—	6.7	28.3	2.86	—	6.4
Percentage change					−5.6%		−4.5%
Iowa—3 trials							
Dry rolled vs. high-moisture rolled	2.11	—	6.8	25.3	2.30	—	6.0
Percentage change					+8.6%		−10.6%
Michigan—1 trial							
Dry rolled vs. high-moisture rolled	3.54	—	8.0	32.0	3.33	—	7.2
Percentage change					−5.9%		−9.8%
Minnesota—2 trials							
Dry rolled vs. high-moisture rolled	2.75	—	5.2	28.0	2.73	—	5.1
Percentage change							−2.1%

Nebraska—3 trials							
Dry rolled vs. high-moisture rolled	2.75	—	7.6	25.0	2.78	—	7.3
Percentage change					+1.1%		−4.0%
Processed before ensiling							
Indiana—5 trials							
Dry rolled vs. high-moisture ground	2.45	7.7	6.0	28.7	2.38	6.5	5.9
Percentage change					−6.9%	−15.6%	−1.2%
Nebraska—2 trials							
Dry rolled vs. high-moisture ground	2.52	—	8.2	25.0	2.29	—	8.7
Percentage change					−9.1%		+6.7%
Reconstituted before ensiling							
South Dakota—5 trials							
Dry rolled vs. high-moisture rolled	2.52	12.2	9.0	—	2.65	11.7	8.3
Percentage change					−5.2%	−4.1%	−7.5%
Indiana—4 trials							
Dry rolled vs. high-moisture rolled	2.43	7.8	5.8	22.1	2.48	7.6	5.7
Percentage change					+2.1%	−2.3%	−2.6%
South Dakota—4 trials							
Dry whole vs. high-moisture whole	2.85	19.9	8.4	—	2.76	18.9	8.3
Percentage change					−3.2%	−5.0%	−1.6%

[a] Adapted from Report to North Central Regional Livestock Specialists' Conference, B. A. Weichenthal, 1973.
[b] Converted to the same dry-matter basis.

Cattle feeders who have silos and who practice field shelling prefer to harvest their corn at about 25 to 30 percent moisture content, storing it as shelled corn in either airtight or conventional silos. If large pits or trenches are used—as they are by the commercial feedlots in Colorado, Oklahoma, Kansas, and Nebraska and even in the Corn Belt where bunkers are used—the shelled corn is ground as it is stored, to reduce spoilage and to make it easier to mix with silage in self-unloading mixing wagons or trucks without having to process or grind at that time.

Many experiments have been conducted to compare ground or rolled high-moisture shelled corn with processed dry shelled corn. Results of a number of these tests are summarized in Table 232. Note that feed efficiency was improved in by far the majority of the studies and gains were slightly reduced in most of the lots on high-moisture corn. Feed intake was reduced in all high-moisture corn lots, when expressed on an equal dry-matter basis. Thus it is apparent that the rapid increase in this method of harvesting and feeding corn is the result of improved feed efficiency and of improvements in harvesting efficiency, rather than its effect on feed intake or rate of gain.

The same comparative advantage for high-moisture shelled corn, in terms of improved feed efficiency, prevailed regardless of whether the corn was rolled before or after ensiling. This was true for reconstituted corn also, but the reduction in corn dry-matter intake was smaller than in ordinary high-moisture corn. (Reconstituted corn is ordinary dry corn to which enough water is added before ensiling to raise the moisture level to about 25 percent, to ensure fermentation.)

High-moisture corn is slightly more efficiently utilized if fed rolled than if fed whole, as shown in Table 233. When considerable roughage is contained in the ration, rate of gain also is improved by rolling the high-moisture corn. If other than oxygen-limiting storage is used for storing high-moisture shelled corn, rolling or grinding should be done before ensiling to ensure better packing and exclusion of oxygen; otherwise considerable spoilage can occur.

Acid Treatment of High-Moisture Shelled Corn.   In order to reduce the storage cost of high-moisture shelled corn, organic acids are employed as sterilizing agents to inhibit mold and bacterial growth. Such acid-treated corn may then be safely stored in low-cost storage facilities such as concrete-floored open sheds, or even in piles on the ground if the area has low rainfall. Several commercial acid-treatment preparations are available and more are under study. Acetic and propionic acids combined in a 60:40 solution are commonly used, and special applicators are available commercially. The caustic nature of some of the acids used causes excessive corrosion and rapid deterioration of metal mixing bins and

Table 233

Whole Versus Rolled High-Moisture Corn with Two Levels of Roughage[a]

Annual Report	Whole High-Moisture			Rolled High-Moisture	
	Daily Gain (lb)	Feed/ Gain Ratio (lb)	Mois- ture Content (%)	Daily Gain (lb)	Mois- ture Ratio (%)
With low levels of roughage					
Minnesota 1967	2.34	755	35.0	2.39	686
Minnesota 1968	2.58	704	34.4	2.57	656
Illinois 1970	2.84	603	26.0	2.90	568
South Dakota 1969	2.44	890	27.4	2.27	929
South Dakota 1970	3.30	638	29.7	3.10	646
South Dakota 1971	2.68	872	21.9	2.75	841
South Dakota 1971	2.62	898	21.9	2.65	876
Nebraska 1973	3.26	620	25.0	3.05	640
Nebraska 1973	2.15	870	25.0	2.04	920
Average	2.69	761		2.64	751
Percentage change				−1.9%	−1.3%
With higher levels of roughage					
Minnesota 1967	1.32	378	35.0	1.36	367
Minnesota 1968	1.39	358	34.4	1.55	324
Michigan 1967	2.73	663	26.0	2.62	708
Ohio 1970	2.39	825	27.0	2.32	808
South Dakota 1970	3.16	739	29.7	3.33	696
South Dakota 1970	2.01	929	28.0	2.06	920
South Dakota 1970	2.08	979	28.0	2.28	922
South Dakota 1971	2.70	819	25.7	2.70	796
Average	2.22	711		2.28	693
Percentage change				+2.7%	−2.5%

[a] Adapted from Report to North Central Regional Livestock Specialists' Conference, B. A. Weichenthal, 1973.

feeder wagons, metal augers, and even silo unloaders. This should be taken into account before deciding to convert to acid treatment as a means of preserving high-moisture corn.

The data summarized in Table 234 show that acid treatment results in corn that is, at best, no better than dry corn and, more likely, in corn that is about 5 percent less valuable for feeder cattle. The averages shown are for nine trials conducted in the Corn Belt states, some of which were conducted before techniques had been refined and, in one or two instances, the acid levels used may have been inadequate for preservation.

Table 234

Annual Report	Treatment	Untreated High-Moisture Corn		Treated High-Moisture Corn	
		Daily Gain (lb)	Feed/ Gain Ratio (lb)	Daily Gain (lb)	Feed/ Gain Ratio (lb)

Untreated Versus Acid-Treated High-Moisture Corn in Finishing Rations[a]

Annual Report	Treatment	Daily Gain (lb)	Feed/ Gain Ratio (lb)	Daily Gain (lb)	Feed/ Gain Ratio (lb)
Indiana 1972	A[b] & P[c]	2.51	613	2.50	607
Indiana 1973	A & P	2.24	644	2.01	675
Indiana 1973	P	2.24	644	2.18	632
Iowa 1972	P	2.35	590	2.02	690
Illinois 1972	A & P	2.91	510	2.93	513
Nebraska 1973	A & P	3.05	640	2.85	680
Nebraska 1973	A & P	3.26	620	3.24	620
Nebraska 1973	A & P	2.04	920	1.86	1,000
Nebraska 1973	A & P	2.15	870	1.94	1,020
Average		2.53	672	2.39	715
Percentage change				−5.5%	+6.4%

[a] Illinois Cattle Feeders' Day Report, 1972.
[b] Acetic acid.
[c] Propionic acid.

Research is still under way with other acids and acid combinations, and forthcoming results should be watched. Some researchers hoped that the energy value of the added acids would itself be recoverable as an energy source to the cattle but preliminary data do not support such hopes. Cost of the acids being used in the mid-1970s was 10 to 20 cents per bushel, depending on corn moisture level.

Reconstituted Corn.   Grain that is harvested in dry form can be reconstituted to high-moisture form by adding water at whatever level is necessary to obtain a final moisture content of 25 to 30 percent. The grain is then stored whole, in oxygen-limiting storage, after adding the moisture, for at least 20 days of fermentation prior to feeding. Usually the grain is rolled as it is being unloaded and the feeding value of the fermented grain is comparable to that of grain harvested and stored as high-moisture corn. The main reason for using this procedure is to provide a year-round supply of high-moisture grain without the need for the sizable investment in oxygen-limiting storage for a year's supply of feed. Dry grain might be stored in lower-cost storage or, in the case of large commercial lots, purchased as needed and then processed to high-moisture stage by refilling the oxygen-limiting storage at intervals as needed.

When a farmer-feeder contemplates changing to field-shelling his corn as high-moisture corn, he should use that opportunity to reexamine his entire feeding system so as to take full advantage of automation and thus reduce the handling and storage of his feeds to the simplest system possible. For example, when high-moisture corn is stored in an oxygen-free unit, the corn must be stored as shelled corn or else the heaviest-duty bottom-unloader must be used. Placing a roller or burr mill at the mouth of the bottom-unloader and tying it in with an auger-bunk feeding system eliminates the handling of the corn entirely. Metered devices may be used to add silage and supplements into the auger system as the corn is being removed from the silo and processed, thus eliminating the need for a mixer of any kind. Belt-feeders, with lower power costs, are now available.

For larger-scale feeders who use mobile mixing wagons or trucks to feed in fenceline bunks, a short elevator may be used to deliver the corn and other ingredients directly from the upright or horizontal silo into the vehicle instead of into a bunk, again doing away with the need for labor.

It is possible to design literally scores of combinations of feed storage and handling systems. Most Land Grant universities have agricultural engineering extension specialists on their staffs who may be consulted for advice, and of course manufacturers of the components of feeding systems also provide consulting services.

GROUND EAR CORN

Although more than a score of comparisons have been made between shelled corn and ground ear corn, it appears impossible to rate one above the other without making a number of qualifications. Most of the early experiments showed that shelled corn was somewhat superior in respect to feed intake, rate of gain, and degree of finish produced, but several more recent tests have raised doubts with respect to these points. For instance, in the experiments reported in Table 235, calves and yearling cattle gained at practically the same rate on shelled corn as on ground ear corn. In all of the Ohio tests, substantially fewer bushels of corn were eaten per hundredweight gain by the steers fed ground ear corn.

In another Iowa test not reported in the table, ground ear corn produced significantly less gain than shelled corn with only 1 pound of linseed meal, but slightly larger gain with 2 pounds of protein supplement, suggesting that the ground ear corn rations need a somewhat higher level of protein supplementation.

Table 235

Comparative Value of Shelled and Ground Ear Corn for Feeder Cattle

| | Calves | | | | Yearlings | | | |
| | Ohio (Average 3 Trials) | | Nebraska | | Nebraska (Average 4 Trials) | | Iowa | |
	Shelled Corn	Ground Ear Corn	Shelled Corn	Ground Ear Corn	Shelled Corn	Ground Ear Corn	Shelled Corn, 240 Days	Ground Ear Corn, 120 Days; 25% Ground Ear Corn, 75% Ground Shelled Corn 120 Days
Average daily gain (lb)	2.00	2.10	2.52	2.48	2.59	2.53	1.94	1.98
Average daily ration (lb)								
Corn	9.7	11.2	10.9	12.8	18.3	22.3	12.1	12.0[a]
Protein concentrate	1.8	1.8	1.1	1.3	—	—	1.7	1.7
Corn silage	7.1	7.1	—	—	—	—	11.5	11.5
Hay	2.7	2.8	3.9	3.9	5.2	4.0	1.5	1.5
Feed per cwt gain								
Corn (bu)	8.7	7.6	7.7	7.4	13.5	13.2	11.2	10.9
Protein concentrate (lb)	88	83	43	51	—	—	88	86
Corn silage (lb)	267	260	—	—	—	—	593	581
Hay (lb)	280	274	155	158	212	164	79	77

[a] Grain portion only.

Ground ear corn is superior to shelled corn for cattle that are full-fed on legume pastures, such as sweet clover and alfalfa, because the dry particles of cob tend to overcome the tendency of these forages to cause scours and bloat. However, shelled corn appears to be much better on bluegrass and other nonlegume pastures, because larger quantities of grain are eaten and therefore greater gains are secured. Shelled corn is less badly damaged by rain than is ground corn and is better suited for feeding outdoors and in self-feeders, which may have rain or snow blown into them during severe storms.

COMPARISON OF METHODS OF GRINDING CORN

The question of which type of mill to use for grinding either shelled or ear corn receives an unwarranted amount of discussion among cattle feeders. It is true that small differences in feed intake can be demonstrated, especially over short periods of time, if corn is coarsely ground.

Table 236

Comparison of Three Methods of Grinding Ear Corn for Full-Fed Yearling Steers in Drylot (56 Days)[a]

	Burr Mill	Hammermill	Reel-Type Knife Mill
Number of steers	10	10	10
Average initial weight (lb)	947	941	948
Average final weight (lb)	1,068	1,059	1,069
Average total gain (lb)	121	118	121
Average daily gain (lb)	2.17	2.11	2.17
Average daily feed consumed (lb)			
Ground ear corn	17.1	17.4	17.1
Cottonseed meal	1.5	1.5	1.5
Mixed hay	3.6	3.6	3.6
Feed consumed per cwt gain (lb)			
Ground ear corn	790.1	828.4	790.1
Cottonseed meal	68.8	70.4	68.8
Mixed hay	168.6	172.9	168.6
Feed cost per cwt gain ($)[b]	35.02	36.49	35.02
Cobs refused per day per steer (lb)	0.22	0.82	0
Average fineness moduli	4.33[c]	4.20	4.30

[a] Illinois Feeders' Day Report.

[b] Feed prices used: ground ear corn, $2.52 per bushel; cottonseed meal, $120 per ton; mixed hay, $30 per ton.

[c] Flour has a fineness modulus of 0 and particles larger than 0.375-inch are given a score of 7.

Such differences are not only small, but they may be more than offset by slower mill capacity and larger power requirements. Data such as those in Table 236 show that the type of mill used in grinding corn is not of great importance. In the Illinois test, where the degree of average fineness was the same for the corn ground by all three mills tested, steers tended to sort out the larger particles of cob in the hammermill-ground corn, causing slightly lower actual consumption of ground ear corn and slightly slower gains. If feedbunks are not perfectly leakproof or if they are exposed to strong winds, there is appreciable loss of the finer particles of corn.

Hammermills are available with screen sizes that grind ear corn at various degrees of fineness. Minnesota workers compared the feeding value of ear corn that was ground through $\frac{1}{4}$-, $\frac{1}{2}$-, and $\frac{3}{4}$-inch screens and fed to heavy yearling steers. There was no apparent effect on level of feed consumption but average daily gains, for 89 days, were 2.33, 2.52, and 2.51 pounds for the three screen sizes, respectively. Thus the ear corn ground through the two coarser screens was more efficiently utilized, with feed/gain ratios of 8.8 versus 8.3 pounds, or about 6 percent improvement in the coarsely ground lots. Mill output also was greatly affected by screen size, with 189, 381, and 445 pounds of ground ear corn per minute being produced by the three screen sizes, respectively, meaning that energy requirements are much lower for coarse grinding.

HIGH-MOISTURE EAR CORN

Many of the advantages mentioned with respect to high-moisture shelled corn also prevail when it comes to harvesting and feeding high-moisture ear corn. Most important, harvesting date is 2 to 4 weeks earlier, resulting in less ear loss. A major disadvantage is the relative cost of storing the cob portion, a low-grade roughage, in expensive silos—especially if airtight silos are used. The cob portion can supply considerable nutrients, especially when fed in ensiled form. Michigan studies showed that 7 percent fewer bushels of corn were required per hundredweight gain when high-moisture ear corn was fed in place of cracked shelled corn.

A more direct comparison may be made between dry and high-moisture ear corn. Table 237 shows that, in 14 different tests, ground high-moisture ear corn was about 10 percent more efficient than dry ground ear corn. Gains were slightly improved also, in contrast to ground high-moisture versus ground dry corn in shelled form.

High-moisture ear corn should be ground before going into the silo. Conventional silos with ordinary top-unloaders are quite satisfactory for high-moisture ground ear corn, although the airtight silos have more

Table 237

Summary of 14 Experiments Comparing Feeding Value of High-Moisture versus Dry Ground Ear Corn[a]

Station	Days on Feed	Animals per Lot	Percent Moisture		Daily Gain (lb)		Concentrate per 100 Pounds Gain (lb)		Percent Stimulation over Dry Corn	
			High-Moisture Corn	Dry Corn	High-Moisture Corn	Dry Corn	High-Moisture Corn[b]	Dry Corn	Gain	Efficiency
South Dakota	97	18	40	15	2.13	1.78	790	878	19	11
Indiana	117	10	32	18	2.47	2.34	866	988	6	14
Indiana	117	10	32	18	2.56	2.33	807	951	10	17
Indiana	126	36	32	15	2.14	2.18	555	617	-2	11
Iowa	119	36	31	15	2.98	3.05	675	750	-2	11
Iowa	56	36	38	14	3.34	3.24	471	528	.3	12
Indiana	133	36	37	24	1.86	1.94	634	660	-4	4
Colorado	112	8	53	14	2.41	2.52	433	483	-5	10
Michigan	147	10	31	18	1.97	1.53	754	973	29	23
Colorado	140	8	55	15	2.08	2.25	519	564	-8	8
Iowa	175	6	30	14	2.12	2.31	726	805	-8	10
Iowa	175	6	30	14	2.40	2.32	667	634	3	-5
Ohio	119	21	36	12	2.15	2.04	685	803	5	15
Michigan	203	13	23	19	1.65	1.60	734	725	3	-1
Average	131	18	36	16	2.30	2.24	663	740	3	10

[a] Adapted from Michigan Beef Cattle Day Report, 1960.
[b] Corrected to same moisture content of dry corn.

built-in convenience. If the airtight bottom-unloading silo is used, the heavy-duty forage unloader is required. This may be a convenience if the silo is used during part of the year for forage storage.

PREPARATION OF OATS

The question of processing oats for cattle is discussed to some extent in Chapter 16. Because the saving effected by grinding oats for calves usually does not exceed 5 to 10 percent, grinding oats for animals of this age is unlikely to be profitable unless oats are high-priced.

PROCESSING BARLEY FOR FEEDER CATTLE

The use of barley in finishing rations is more likely to occur in farm feedlots rather than in commercial feedlots, excepting some of the larger lots in the Northwest. Steam-flaking equipment for processing barley is therefore a rarity, but dry-rolling equipment is made in sizes to fit the needs of such farmer-feeders.

Dry Barley.   A series of trials with various processing methods for both dry and high-moisture barley was conducted by workers in Montana, where this grain is highly important. Table 238 summarizes two trials, one

Table 238

	Trial 1		Trial 2	
Item	Whole Barley	Dry-rolled Barley	Whole Barley	Dry-rolled Barley
Number of steers	9	9	8	8
Average initial weight (lb)	853	853	675	680
Average final weight (lb)	1,101	1,127	1,014	1,050
Average daily gain (lb)	2.12	2.34	2.51	2.74
Average daily ration (lb)[b]				
Protein supplement	0.91	0.91	0.91	0.91
Barley	17.9	16.6	16.8	15.3
Corn silage	1.0	1.0	—	—
Barley straw	—	—	2.2	2.2
Feed/gain ratio (lb)[b]	9.35	7.90	7.92	6.72

Whole and Dry-Rolled Barley for Finishing Steers[a]

[a] Adapted from 16th Montana Livestock Feeders' Day Report, 1973.

[b] Dry-matter basis.

Table 239

Comparisons of Barley Processing Methods for Feeder Cattle (Average of 5 Trials)[a]

Item	Form of Barley		
	Reconstituted High-Moisture	Combined High-Moisture	Dry
Number of steers	35	35	36
Days fed	138	132	132
Average initial weight (lb)	774	776	770
Average final weight (lb)	1,071	1,081	1,056
Average daily gain (lb)	2.16	2.32	2.15
Average daily ration (lb)[b]			
Protein supplement	0.89	0.89	0.89
Barley	13.4	14.6	13.2
Corn silage	2.4	2.4	2.4
Total	16.7	17.9	16.5
Feed/gain ratio (lb)[b]	7.74	7.75	7.60

[a] Adapted from Montana Research Progress Report, 1971.
[b] Dry-matter basis.

with calves and one with yearlings, fed whole and dry-rolled barley rations. In both trials, with rations containing about 90 percent barley, rolling the grain increased the rate of gain about 10 percent and improved feed conversion by about 18 percent.

High-Moisture Barley. Another series of trials at the Montana station compared dry barley, barley harvested at high moisture levels, and reconstituted high-moisture barley. Barley is more difficult to reconstitute than other grains because dry whole barley absorbs only about 10 percent moisture in the usual routine procedure used for reconstituting grains. The Montana workers added water to whole barley and left it standing in a truck overnight, after which it was wetted again while being rolled at reconstitution and silo-filling time. This procedure increased the moisture level to 28 percent and produced a normally fermented high-moisture grain. The averaged results of the five trials are shown in Table 239.

Apparently barley is not improved by storage at high moisture levels and fermentation to the same extent as are the other feed grains. Feed intake and gain were slightly greater in the grain harvested and stored in high-moisture form, but feed conversion was unimproved and the reconstituted high-moisture barley performed no better than dry barley. Steers on the high-moisture barleys did remain on feed better and chronic bloating was not a problem, as it often is with dry barley.

There are agronomic advantages to harvesting barley at about 40 percent moisture with a forage harvester as barley head chop. The material ferments well and is readily consumed. However, because it contains about 50 percent grain and 50 percent straw, head chop is better for growing out stocker calves than for finishing rations. Montana workers harvested barley in four forms, with respective digestible dry-matter yields in pounds per acre as follows: whole plant silage, 2,774.9; head chop, 2,478.5; high-moisture grain, 1,999.5; and dry grain, 1,739.5.

The steam processing of barley is mentioned in Chapter 16, and results of an Arizona trial comparing dry rolling and rolling of steam-processed barley are presented in Table 128. This and other experiments showed that steam processing increases feed intake and daily gain about 10 percent, but feed efficiency is unimproved when a roughage source other than the barley hulls is included in the formula.

High-moisture barley, stored in airtight silos at 30 percent moisture, was compared with dry rolled barley by Minnesota workers. When only a small amount of hay was included in the ration, the high-moisture barley was consumed at a higher rate, resulting in an increase in gain—but, as was true of steam-processed barley, there was no improvement in feed efficiency. When about 5 pounds of hay were fed per day, there was no increase in either intake or gain, and again feed efficiency was unimproved. Apparently both steam processing and rolling of high-moisture barley result in a large bulky flake that is more palatable to cattle on high-concentrate rations. The resulting increased intake steps up gain but without an improved feed efficiency or reduction in cost of gain. Therefore processing of barley, other than grinding, dry rolling, or crimping, seems unwarranted at present.

PREPARATION OF WHEAT IN FINISHING RATIONS

Ordinarily wheat is priced too high, in relation to corn or milo, to use as an energy source in cattle finishing rations. However, when wheat can be purchased at prices no higher than 8 to 10 percent above those of other grains, it is economically feasible to feed wheat. Grain elevator fires resulting in water, smoke, or heat damage to large quantities of wheat sometimes make this potential energy source available at lower cost. As most of the wheat available for feeding is in dry form, the processing of dry wheat will be discussed first.

Dry-Processed Dry Wheat.   Oklahoma workers compared four methods of processing dry, hard winter wheat by self-feeding 90 percent concen-

Table 240

Feedlot Performance of Steers Fed Dry Wheat Processed by Different Methods[a]

Item	Dry-rolled Milo	Dry-rolled Wheat	Coarsely Ground Wheat	Finely Ground Wheat	Whole Wheat
Number of steers	12	12	12	12	12
Average initial weight (lb)	693	693	688	688	685
Average final weight (lb)	1,058	1,024	1,033	1,003	1,038
Average daily feed (lb)[b]	20.8	18.6	20.8	17.5	23.0
Average daily gain (lb)	3.01	2.70	2.86	2.54	2.87
Feed/gain ratio (lb)	6.90	6.93	7.28	6.89	8.19
Density of grain (lb/bu)	39.9	33.1	44.6	46.4	60.0

[a] Adapted from Oklahoma Experiment Station MP-87, 1972.
[b] Adjusted to 90% dry matter.

trate finishing rations to yearling steers for 122 days, with wheat composing 70 percent of the total ration. Their results are shown in Table 240. A milo check lot was included for comparison.

Processing methods affected feed intake and average daily gain, but only the whole wheat was converted to gain at a significantly lower rate. All rations were equalized for protein content and all contained 5 percent cottonseed hulls and 5 percent ground alfalfa hay. As no differences were seen in carcass traits, it would appear that dry wheat compares favorably with dry-rolled milo when fed dry-rolled or coarsely or finely ground, but not when fed whole. Using different levels or forms of roughage might cause different results.

Steam-Processed Dry Wheat.   Because steam-flaking has produced favorable results with corn and milo, this process also was investigated with wheat by Oklahoma workers. Larger feedlots are likely to have equipment for steam-flaking grains already on hand, and information about its use for wheat would be valuable for times when wheat may be a feasible substitute for milo or corn.

Table 241 shows data collected in a wheat feeding trial with heifer calves in which steam-flaking for two lengths of time was compared with dry-rolling. The 90 percent concentrate rations contained 70 percent hard winter wheat, on a dry-matter basis, and were fed for 150 days to heavy heifer calves. Both steam-flaking treatments significantly increased feed intake and daily gain and improved feed efficiency. The carcasses of heifers fed steam-flaked wheat were fatter and consequently had lower cutability, but if all heifers had been slaughtered at nearly equal weights,

Table 241

	Steam-flaked Wheat for Finishing Beef Cattle[a]		
		Wheat Treatment	
Item	Dry-rolled	Steam-flaked (6 min)	Steam-flaked (12 min)
Number of heifers	12	12	12
Average initial weight (lb)	509	487	500
Average final weight (lb)	872	922	933
Daily feed (lb)[b]	14.8	16.0	16.1
Daily gain (lb)	2.42	2.87	2.89
Feed/gain ratio (lb)	6.12	5.58	5.57

[a] Adapted from Oklahoma Experiment Station Report MP-92, 1974.

[b] Dry-matter basis.

carcass traits no doubt would have varied less. These data show that the conventional steam-flaking procedure used on corn and milo can be successfully used on whole wheat and that 6 minutes of steam treatment are adequate.

Processed High-Moisture Wheat. Wheat can be efficiently combine-harvested at moisture levels higher than normal storage levels, but unless it is stored in an oxygen-limiting silo or treated with preservative acids, spoilage is excessive. Corn and milo, when stored and fed in high-moisture form, are utilized about 10 percent more efficiently than the same grains in dry form. Oklahoma workers harvested both whole, hard winter wheat at 23.4 percent moisture and wheat head chop at 28.0 percent moisture on June 2. They stored some whole high-moisture wheat in an oxygen-limiting silo and treated some with propionic acid as a preservative before storage in a wooden bin. The head chop contained 63.9 percent grain and 36.1 percent nongrain dry matter. The three forms of high-moisture wheat were compared with dry wheat harvested at 11 percent moisture from the same field and fed in rolled form, by feeding yearling heifers for 72 days. The dry and high-moisture whole wheats were rolled as fed while the head chop, harvested with a self-propelled field chopper and cut just below the head, was hammermill-ground before ensiling in an oxygen-limiting silo.

All rations containing whole wheat were compounded to contain 70 percent wheat and 10 percent roughage while the head chop ration

Table 242

High-Moisture Wheat and Wheat Head Chop for Finishing Beef Cattle[a]				
	Form of Wheat and Treatment			
Item	Dry Wheat, Dry-rolled	High-Moisture Wheat, Oxygen-limiting Silo	High-Moisture Wheat, Acid-preserved	Wheat Head Chop
Number of heifers	12	12	12	12
Average initial weight (lb)	690	695	698	697
Average final weight (lb)	882	897	887	866
Average daily feed (lb)[b]	19.6	20.0	19.4	22.9
Average daily gain (lb)	2.67	2.82	2.67	2.31
Feed/gain ratio (lb)	7.38	7.29	7.56	9.94

[a] Adapted from Oklahoma Experiment Station Research Report MP-90, 1973.

[b] Expressed on a 90% dry-matter basis.

contained 95 percent wheat heads and no added roughage. All rations contained approximately equal amounts of protein and other additives. The data are summarized in Table 242.

Heifers consumed more of the ensiled head chop than of the three forms of rolled wheat but, because the head chop contained 36.1 percent nongrain roughage material, gains were significantly slower and more inefficient. The Oklahoma workers suggest that the head chop is more suitable for stocker cattle than for feeders. Acid-preserved rolled wheat was equal to dry, rolled wheat; thus this method of preserving wheat seems feasible.

Reconstituted Dry Wheat. Dry wheat in whole kernel form may be reconstituted to high-moisture condition by adding enough water to restore the moisture level to about 30 percent and storing the wheat in an oxygen-limiting structure for about a 3-week fermentation period. Oklahoma workers fed heifer calves for 129 days on 90 percent concentrate finishing rations containing 70 percent wheat (dry basis) in three forms, namely dry, rolled; whole reconstituted; and whole reconstituted, rolled. Feed intake, on a 90 percent dry-matter basis, averaged 11.9, 16.5, and 12.8 pounds daily for the three processing methods respectively, with the whole reconstituted wheat being consumed at a significantly higher rate. The respective daily gains were 1.67, 2.07, and 2.00 pounds and the feed/gain ratios were 7.15, 7.99, and 6.42 pounds. Thus the heifers on dry, rolled grain gained slower and, most important, feed conversion was more efficient in heifers on reconstituted wheat that was rolled as fed.

PROCESSING OF MILO GRAIN

Milo is the most important energy source for finishing cattle in the High Plains region and the far West and it is assuming some importance in the western fringes of the Corn Belt. Most of the milo is combine-harvested at moisture levels in the range of 12 to 16 percent but increasingly more is being harvested as high-moisture grain and as high-moisture head chop. The question of how to process milo depends largely on the harvesting and storage system used and upon the size of feedlot, for this determines how much grain must be processed daily.

Researchers have long known that dry milo does not quite measure up to its potential, based on its chemical composition. Early studies were concerned primarily with effect of grinding per se, and with fineness of grind and type of grinder. Not surprisingly, feeding value of milo was improved by almost any form of grinding, as compared to feeding the milo whole. Arizona and California workers led the way in developing a steam-flaking process, using mostly batch-type units. Industry has now developed continuous-flow processing units of large capacity that operate 24 hours around the clock to provide freshly steam-flaked milo for feeding throughout the day and night. Not all feeders and feedlots have operations that are large enough to justify the great investment required for such units. Thus research continues in the hope of finding still other methods for processing milo that might improve its nutritional value and at lower costs.

Steam-flaked and Reconstituted Dry Milo.   A field-scale study conducted by Texas workers in which conventional steam-flaking was compared with two forms of reconstituted milo is summarized in Table 243. The cattle consumed less of the reconstituted whole milo that was rolled as fed, whereas equal amounts of the steam-flaked and ground reconstituted milo were consumed. Gain and feed efficiency were about equal in the light steer lots, but both reconstituted forms of milo were superior when fed to the heavier steers. Reconstituting the grain in whole form and rolling it as fed resulted in the most efficient gains and processing costs were lowest.

Oklahoma workers agree with the Texas study just cited with respect to the time when grain that is to be reconstituted should be processed. They ground whole dry milo as reconstituted to 30 percent moisture or ground the high-moisture whole milo as fed after reconstituting. They also included a dry, finely ground milo check lot for comparison of dry and high-moisture milo. When compared to dry, finely ground milo, the reconstituted ground milo resulted in a 11.6 percent faster gain and a 9.0 percent better feed efficiency, whereas the ground reconstituted milo resulted in 1.8 and 3.5 percent decreases in gain and feed efficiency, respectively. In another study these same workers determined that milo that is being reconstituted should be allowed to ferment for at least 20 days. Five- and ten-day fermentation periods resulted in little improvement over dry grain.

Micronized Dry Milo Grain.   A gas-fired infrared burner has been adapted for quickly heating grains, by means of microwaves, to very high temperatures as the grain passes in a continuous process about 6 inches below the heating unit. Dry grain is heated just to the point of popping and is then rolled into a bulky, flaky product. The heat treatment dries the grain to about 7 percent moisture and gelatinizes the starch, as do the other heat treatments, and improves digestibility by about 10 percent and feed efficiency by about 12 percent. Units sized to accommodate medium-

Table 243

Effect of Method of Processing Milo for Two Weights of Yearling Steers Fed High-Concentrate Rations[a]

| | Processing Method | | | | | |
| | Steam-flaked | | Whole Reconstituted | | Ground Reconstituted | |
Item	Small Steers	Large Steers	Small Steers	Large Steers	Small Steers	Large Steers
Cattle performance						
Number of steers	75	75	75	75	75	75
Average initial weight (lb)	662	803	668	807	665	812
Average daily ration (lb)	21.6	22.8	20.7	21.7	21.8	23.0
Average daily gain (lb)	2.83	3.07	2.81	3.23	2.74	3.33
Feed/gain ratio (lb)	7.63	7.43	7.37	6.72	7.95	6.91
Processing costs[b]						
Costs per ton	$1.81		$1.54		$1.93	
Initial investment per head capacity	$8.89		$9.14		$5.18	
Annual cost per head capacity	$2.07		$1.75		$2.20	

[a] Adapted from Texas Experiment Station Miscellaneous Reports.

[b] Based on an assumed 20,000-head, one-time capacity lot with a 2.4 annual turnover rate and an 8% interest charge.

Table 244

Micronized and Steam-flaked Milo in Beef Cattle Finishing Rations[a]

	Preparation Method	
Item	Steam-flaked	Micronized
Number of steers	200	200
Days fed	100	100
Average initial weight (lb)	752	769
Average daily gain (lb)	2.98	3.14
Average daily ration, as fed (lb)	24.2	25.2
Feed/gain ratio (lb)	7.14	7.27

[a] Adapted from Texas Experiment Station Miscellaneous Reports.

sized feedlots feeding 2,000 to 5,000 head are available and processing costs are reportedly as low as $1.50 per ton, depending on daily output.

Texas workers planned a cooperative feedlot trial involving a commercial feedlot, a feed manufacturer, the developer of the micronizing equipment, and extension and experiment station personnel. Commercial-sized pens of cattle were fed on a medium high-concentrate ration containing equal levels of milo dry matter in either steam-flaked or micronized form. Performance of the steers (Table 244) was similar for the two forms of milo, although slightly favoring steam-flaking in the important area of feed efficiency. Small, farm-feedlot-sized micronizing units are under study.

High-Moisture Milo.   High-moisture milo may be harvested and stored in airtight silos or very well packed in conventional silos in several forms. The entire head may be removed with a field chopper and ground through a regular screen along with some of the stem. These heads contain about 70 to 75 percent moisture. Milo also may be harvested when it contains about 30 percent moisture and ensiled whole. Finally, as discussed earlier, water may be added to dry milo to make reconstituted high-moisture milo. The Texas station has extensively studied the feeding of high-moisture milo. In a comparison between high-moisture ground heads and dry ground grain they got comparable gains and a significant improvement in feed efficiency from the high-moisture heads. They have also consistently observed improvement in feed efficiency but little or no change in rate of gain when reconstituted grain was compared with dry grain. It appears that storing milo at about 30 percent moisture in silos

Table 245

Preparation of Milo	Medium Rolled	Finely Ground and (⅜-in) Pelleted
Number of heifers per lot	9[b]	10
Average weights (lb)		
Initial	498	497
Final	826	838
Average daily gain	2.09	2.17
Average daily ration (lb)		
Rolled milo	11.96	
Pelleted milo		11.49
Cottonseed meal + stilbestrol[c]	1.50	1.50
Dehydrated alfalfa meal pellets	1.00	1.00
Sorghum silage	11.58	10.77
2–1 mineral mix	0.10	0.06
Feed required per cwt gain (lb)		
Milo	572	529
Cottonseed meal	72	69
Dehydrated alfalfa pellets	48	46
Sorghum silage	554	496

Rolled Versus Pelleted Milo for Feeder Heifer Calves[a]

[a] Oklahoma Livestock Feeders Day Report.

[b] One heifer removed from data because of founder.

[c] Cottonseed meal fed per head daily contained 10 mg stilbestrol.

offers the same potential for increasing the value of sorghums as it does for corn.

ROLLED VERSUS PELLETED GRAIN

Pelleting or cubing of the grain alone or of the entire concentrate portion of the ration has been given much attention. The Oklahoma study summarized in Table 245 shows that considerably less sorghum was required per unit of gain when the sorghum was pelleted instead of rolled. One explanation may be that all of the smaller particles usually resulting from rolling or grinding sorghum grain were saved by being incorporated in the pellet. Another possibility is that of higher digestibility of the pelleted grain caused by the heating that occurred in the pelleting process. The advantage for pelleting might be even greater if compared with whole or finely ground sorghum grain. No studies have been reported comparing pelleted grain with processing methods such as steam-flaking or micronizing. It is predicted that the latter methods would be comparable in feeding value and pelleting would cost at least 50 percent more.

GRINDING AND CHAFFING HAY

The claim that the feeding value of hay is improved by grinding is not supported by most of the tests comparing long and ground hay. (See Table 246.) Instead, these experiments show that cattle often do not like ground hay, that feeding ground hay decreases the time spent in rumination, and that, because of reduced consumption and less thorough digestion of the ground roughage, gains are smaller and more costly.

The practice of grinding the hay and feeding it mixed with the grain has sometimes resulted in larger gains than when long hays were fed, but a number of experiments yielded opposite results. In few if any of the tests have the advantages been sufficient to pay for the expense of grinding and mixing the feeds.

Apparently the principal advantage of ground hay over long hay is that less is wasted. Cattle fed long hay often pull some of it out of the manger

Fig. 104. Grassy hays, harvested as big bales, can easily be ground or chopped for mixing in complete rations by using high-volume tub grinders. (Haybuster.)

Table 246

Value of Grinding or Chaffing Hay for Feeder Cattle

| | Nebraska (Two Trials) | | Iowa | | | Minnesota | |
| | Shelled Corn and Alfalfa Hay | | Shelled Corn and Alfalfa Hay | | | Ground Shelled Corn and Alfalfa Hay | |
	Long Hay Fed Separately	Ground Hay Mixed with Corn	Long Hay Fed Separately	Coarsely Chopped Hay Mixed with Corn	Finely Chopped Hay Mixed with Corn	Whole Hay Fed Separately	Ground Hay Mixed with Corn
Average initial wt (lb)	611	610	692	690	690	688	689
Average daily gain (lb)	2.20	2.15	2.35	2.40	2.35	2.49	2.52
Average daily ration (lb)							
Corn	13.5	13.8	15.7	15.3	14.8	17.0	17.2
Protein supplement	—	—	1.0	1.0	1.0	—	—
Hay[a]	5.5	3.5	4.6	4.6	4.5	7.0	6.4
Feed per cwt gain (lb)							
Corn	616	642	667	636	629	682	684
Protein supplement	—	—	42	41	42	—	—
Hay	250	162	196	192	194	279	256

[a] Including hay that was wasted.

or bunk and trample it under foot and they usually leave some of the stems uneaten, particularly if the hay is full-fed. Both of these losses are greater if the hay is coarse and poor in quality. Since only poor quality hay is fed with appreciable waste, it is this kind of hay that is most often ground. Even if the waste of the long hay were as much as 25 percent, it is doubtful whether grinding would be profitable, because the coarser portions constituting most of the waste usually are worth much less than the cost of grinding the entire supply of hay fed.

If hay is so scarce and expensive that waste must be prevented more than is possible by using properly built racks and mangers, it should be cut or chaffed by running it through a silage cutter or similar machine that will cut it into 2-to-4-inch lengths. Chaffed hay is much preferred by ruminants to ground hay and is prepared with much less labor, power, and energy.

One of the advantages of chaffed or chopped hay is the low cost of harvesting and storage. If it is chopped from the windrow with a field cutter and blown into the barn where it can be pushed down to the cattle with little or no handling, less labor is required than in harvesting and feeding long or baled hay. Also, approximately twice as many tons of chopped as of long hay can be stored in a given shed or mow.

Hay that is to be chopped in the field must be much drier when stored than ordinary hay or hay that is stored in the bale. Consequently a rather high percentage of the leaves and fine stems are reduced to such fine particles by the chopper and blower that they are lost in harvesting and feeding. Storage of chopped hay with a moisture content much above 15 percent usually results in loss of valuable feed nutrients from extensive "mow burning." There is also the risk of losing both hay and barn from spontaneous combustion. Thus the field chopping of hay can be recommended only when it is to be stored in a barn or bin equipped with forced ventilation. Stored in this manner it may be chopped while still sufficiently tough to prevent loss of valuable leaves and stems because the excess moisture is driven off by the forced ventilation before serious heating occurs.

In the final analysis, the method of harvesting and storing roughage matters little to cattle being fed on finishing rations, as seen in Table 247. After all, roughage makes up only about 15 to 25 percent of the total ration consumed. Thus the method used in harvesting a hay crop which will be fed largely to feeder cattle is determined by the relative cost of the various harvesting methods. The so-called big bales discussed in Chapter 18 are seldom used in finishing programs although they might be ground in tub grinders for mixing in complete rations. Hay in big bales is generally slightly to considerably lower in quality than the same hay stored in conventional square bales.

Table 247

Effect of Method of Harvest and Storage of Alfalfa Hay on Its Value for Feeder Cattle (Average of 3 Trials)[a]

	Stacked Alfalfa Hay	Baled from Wind-row	Chopped from Wind-row	Dehydrated Alfalfa Hay	Molasses Alfalfa Silage
Average initial weight (lb)	696	692	689	694	717
Average daily gain (lb)	2.04	2.11	2.13	2.19	2.09
Average daily ration (lb)					
Grain	15.5	15.6	15.4	15.5	15.8
Soybean meal	1.0	1.0	1.0	1.0	1.0
Hay or silage	3.6	3.9	3.9	3.5[b]	9.5[b]
Feed per cwt gain (lb)					
Grain	773	756	738	724	777
Soybean meal	48	47	46	45	47
Hay or silage	177	187	183	155[c]	453[c]
Average TDN per cwt gain	712	705	689	669	702

[a] Colorado Experiment Station Progress Report.
[b] Includes approximately 0.5 pound of long stacked hay.
[c] Includes approximately 20 pounds of long stacked hay.

EFFECT OF GRINDING AND CHOPPING ROUGHAGE ON PALATABILITY

It may seem that ground or chopped roughage should be more palatable than whole roughage because it can be eaten more easily. Also, all the ground roughage fed is usually eaten, but a noticeable percentage of long hay or stover is often refused. With coarse, stemmy hay and corn stover the refused portion may amount to 20 to 40 percent of the total amount fed.

Feeding experiments show, however, that cattle usually eat more long hay than ground or chopped hay if both kinds are fed according to appetite. For example, in the Nebraska experiments reported in Table 246 it was necessary in both trials to limit the feeding of long hay in order to obtain the desired consumption of corn, but there was difficulty in getting the lots fed ground alfalfa to eat their hay. Also, in palatability tests conducted at the Wisconsin station with purebred beef cows and heifers a marked preference for long hay and coarsely chopped hay (1.8 inch) and a dislike for hay chopped to 0.5 and 0.25 inch was noted. In one of the Wisconsin tests, where all four kinds of hay were available in adjacent mangers, whole hay constituted 71.5 percent of all the hay consumed.

PELLETING OF BEEF CATTLE FEEDS

The pelleting or cubing of roughages or of complete rations by using pressure to extrude or force the finely ground feedstuffs through dies ranging from 0.25 to 1 inch in diameter, is one of the noteworthy developments of the last 20 years. The advantages usually named in favor of this rather costly method of preparing beef cattle feeds are the following.

1. Easier handling because of reduced bulk.
2. Increased feed intake, especially of lower-quality roughages.
3. Improved feed conversion rates.
4. Increased rate of gain on high-roughage rations.
5. Control of ratio of concentrate to roughages at the desired level.

PELLETING HAY

Of all the various components of a beef cattle ration, the roughage portion is most affected by pelleting from the standpoint of storage and labor of feeding. The bulkiness of roughage also is the limiting factor with respect to intake or level of consumption by beef cattle. Thus it appears that pelleting of roughage should offer more promise than the pelleting of concentrates. Several tests have been conducted with pelleted roughages, with conflicting results. The Illinois Dixon Springs station conducted two tests with an all-hay ration for stocker calves with results overwhelmingly in favor of pelleting, as may be seen in Table 248, which summarizes the first test. The hay fed was a first-cutting mixed hay of average to poor quality consisting of two-thirds timothy and one-third alfalfa. Before being pelleted into $\frac{3}{16}$-inch pellets, the hay was finely ground. No molasses or steam was used in the pelleting process and, as the data show, daily consumption of the ration was greatly improved by the pelleting.

The second trial produced similar results with poor-quality alfalfa, with Serecia lespedeza, and again with a timothy-alfalfa mixture. As might be expected, some of the advantage of the more rapid winter gains made by the calves on pelleted hay was lost during the subsequent summer grazing period because of compensatory gain by the lighter cattle. The roughages fed in these tests were only poor to average in quality, but much of the hay fed in any but the mountain and western states is of similar quality.

Even if further tests reveal that pelleting roughages for beef cattle is advantageous, it will be helpful if equipment manufacturers can develop a portable pelleting machine that can chop and pellet the hay out of the swath in the field. The matter of excess moisture in hay for safe storage as pellets also presents a problem that may be solved only by adding drying units to the pelleting machine or in the storage area. The cost of such a complex machine probably would be prohibitive for individual farmer-feeders. For larger commercial feedlots that buy their roughage in the

Table 248

Comparison of Timothy-Alfalfa Mixture When Fed Chopped, Pelleted, or as Long Hay (119 Days)[a]

Method of Hay Preparation	Baled	Pelleted	Chopped
Number of steers	15	15	15
Average initial weight (lb)	421	430	423
Average final weight (lb)	496	636	497
Average daily gain (lb)	0.63	1.73	0.62
Daily feed consumption (lb)	11.0	15.7	10.7
Feed required per cwt gain (lb)	1,732	906	1,722
Feed cost per cwt gain ($)	17.32[b]	13.59	17.22
Gain per ton of feed (lb)	115.5	220.7	116.2

[a] Illinois Dixon Springs Mimeograph.

[b] Feed prices used: baled and chopped hay, $20 per ton; pelleted hay, $30 per ton.

neighborhood, this method of harvesting and feeding roughage shows promise, but it is doubtful whether the average or small feeder will be pelleting roughages in the near future unless he has it done on a custom basis.

WAFERS AND CUBES

Great strides have been made in developing hay-harvesting equipment that processes the hay from the windrow into extruded wafers or cubes. It is estimated that 500,000 tons of alfalfa hay were processed by this method in 1974, much of it in California, Arizona, and New Mexico.

The wafers or cubes range in size from 1.25 inches square to 4 inches but most machines make the smaller cube. The wafering or cubing machines are self-propelled, pulling a trailer bin that holds about 4 tons of wafers. Each machine will handle 500 to 1,000 acres during the season, depending on the number of cuttings. Their rated capacity is usually about 5 tons per hour. In California economic studies it was estimated that the complete cost of wafering was $7.60 per ton as compared with $4.80 for baled hay.

The large 4-inch wafer first made did not prove especially satisfactory, and gains and feed efficiency were unimproved. There was much wastage because cattle dropped the wafers to the ground and into the manure or mud. A large commercial feedlot in California compared smaller 1.25-inch cubes with comparable hay fed in baled form to steer calves as a stocker ration and obtained gains of 1.55 and 0.95 pounds per day for cubes and baled hay respectively. The cube-fed calves required 28 percent less alfalfa. Thus it appears that when cubes make up all or most of the ration, they are more efficient, and if the cost of making the cube is no more than $3 to $4 per ton more than the cost of making hay into bales, the use of cubes may be warranted.

If cubed alfalfa is used in finishing rations, some calves tend to sort or select out the cubes, leaving other calves to eat a higher-concentrate ration. California workers were interested in the effect this selectivity might have on performance of yearlings; hence they fed hay in three forms—whole cubes, ground cubes, and chopped hay. The hay, regardless of form, was mixed with the finishing ration in the ratio of 15:85. Results are shown in Table 249. They concluded that the amount of sorting done was serious enough in the early stages of the feeding period to warrant crushing of the cubes. Most commercial feedlots that now buy their hay in cube form first break up the cubes by passing them through a cube-breaking machine and then mix them with the remainder of the total ration, usually in unloading-mixer trucks or wagons.

Table 249

Form of Hay	Whole Cubes	Ground Cubes	Ground Hay
Influence of Form of Hay on Feedlot Performance of Beef Steers[a]			
Number of animals	36	36	36
Average initial weight (lb)	617	616	629
Average final weight (lb)	945	934	978
Average daily weight gain (lb)	2.94	2.84	2.93
Slaughter data			
Dressing percent	60.18	60.40	60.44
Percent choice grade	64.0	69.0	72.0
Feed intake and utilization			
Daily feed intake (lb dry matter)	17.66	18.09	18.54
Feed per pound gain (as fed) (lb)	6.59	6.97	6.32

[a] Adapted from California Cattle Feeders' Day Report.

Field wafering or cubing, done on a custom basis or by cooperatively owned equipment to keep down investment costs, has enough advantages with respect to handling and storage to give it a place in the already wide variety of methods of harvesting and feeding a hay crop.

PELLETING COMPLETE RATIONS

Increased mechanization of feed handling in cattle feeding has caused interest in complete mixed rations. Roughages in chopped form, either dry or as silage or green chop, are often mixed with the concentrate portion of the ration for feeding with automatic unloading wagons or in various types of conveyor or auger-bunks. Naturally the idea of pelleting the entire ration, where dry roughages are used, offers the possibility of simplifying the feeding operation considerably.

The Illinois Dixon Springs station has done some pioneer work with complete pelleted rations. Table 250 shows the result of a 130-day test comparing a complete ration fed in pelleted and in meal form. The steers fed on pellets gained faster and more efficiently, but if a $6 per ton pelleting charge is made, the economic advantage is lost.

Table 251 is a summary of another Dixon Springs station experiment in which rations with various ratios of concentrate to roughage were compared when fed in pelleted form. Increasing the roughage from 25 to 35 percent did not change performance appreciably, and even when the roughage content of the ration was increased to 45 percent, reduction in rate of gain was not serious.

Table 250

	Pellets	Meal	
Form of Ration	Self-Fed	Self-Fed	Hand-Fed
Comparison of Feeding a Ration as Pellets and as a Meal to Yearling Steers (130 Days)[a]			
Number of steers	18	18	18
Average initial weight (lb)	651	646	653
Average final weight (lb)	1,008	981	969
Average total gain (lb)	357	335	316
Average daily gain (lb)	2.75	2.58	2.43
Average daily ration (lb)			
Concentrate and hay[b]	20.0	21.0	20.4
Corn silage	12.1	11.9	13.0
Feed eaten per cwt gain (lb)			
Concentrate and hay	729	845	840
Corn silage	442	463	534

[a] Illinois Dixon Springs Mimeograph 40-329.
[b] The pellets or meal consisted of 65% ground ear corn, 5% blackstrap molasses, 10% soybean meal, and 20% ground alfalfa hay.

Pelleting the complete ration as a general farm feedlot practice awaits further developments on the part of the processing equipment manufacturers. Some commercial feedlots are pelleting their complete rations with satisfactory results, but capital investments in costly equipment can be spread over a far greater number of cattle in such operations than is possible with farm or ranch feeders.

Table 251

Response of Yearling Feeder Steers Self-Fed Complete Pelleted Rations of Varying Ratios of Concentrate and Roughage[a]

	Lot 1	Lot 2	Lot 3
Number of steers	15	15	15
Pelleted ration (%)			
Timothy-alfalfa hay	25	35	45
Ground shelled corn	65	55	45
Soybean meal	10	10	10
Average initial weight (lb)	703	700	698
Average final weight (lb)	1,145	1,137	1,112
Average daily gain (lb)	2.89	2.85	2.71
Average daily pellet intake (lb)	21.5	22.8	23.0
Pellets required per cwt gain (lb)	745	800	851

[a] Illinois Dixon Springs Mimeograph 40-333.

METHODS OF FEEDING

Two developments during the 1960s that have profoundly affected modern methods of offering feed to full-fed cattle are, first, the perfection of many models of efficient, economical self-unloading mixer wagons and trucks and, second, the gradual shift to high- or nearly all-concentrate rations. Such rations must be available to cattle at virtually all hours of the day or the incidence of founder will be disastrously high. Self-feeding has been commonly practiced by farmer-feeders for decades, using self-feeders for the concentrate portion of the ration with the roughage being offered in racks or bunks apart from the concentrates. When ground ear corn was used, the cob portion often made up the roughage in the ration. Nowadays, since most self-fed rations may be in complete mixed form and contain little roughage, and since equipment is available for delivering fresh feed directly to fenceline bunks or to double-sided bunks by means of augers, the old-fashioned self-feeder that is filled once weekly is less common.

SELF-FEEDING VERSUS HAND-FEEDING

For cattle that are on a full grain ration during practically the entire feeding period in farm-feedlot situations, the self-fed method of feeding has many advantages. On the other hand, self-feeders are impractical for an extended feeding period using less than a full feed of grain.

The principal advantages of using self-feeders are the following.

1. There is a saving in labor per steer fed, on smaller farms at least.
2. Larger daily gains are secured.
3. The cattle are less likely to go off feed.

Of these advantages, the saving of labor is by far the most important. By using self-feeders with large bins or hoppers, it is necessary to feed cattle only once every week or 10 days. Large quantities of feed can be prepared and mechanically delivered directly into the feeders with augers or self-unloading wagons, with a minimum of time and labor. Still more convenient is the practice of grinding directly into the self-feeder, in which case almost all handling of grain is eliminated.

Hand-feeding, on the other hand, requires the preparation of each feed at the time it is fed, or rehandling if the material is prepared in quantity and stored. This saving of labor is of most consequence to the farmer-feeder during the summer when he is busy in the fields with farm work. For this reason, self-feeders are more extensively used in summer than in winter feeding. Carefully conducted feeding experiments indicate that self-feeding usually results in larger gains than hand-feeding. This is usually true even where the hand-feeding is carefully done at regular intervals by experienced feeders. Under ordinary farm conditions, where there is often irregularity in the time of feeding, quantity and character of rations, and so on, the difference in favor of self-feeding is usually greater than the result obtained in experimental trials comparing these two methods.

Next in importance to the saving of labor is the value of the self-feeder in lessening the tendency of the cattle to go off feed, especially on all- or very high-concentrate rations. Cattle that are self-fed show comparatively little tendency to go off feed. They soon learn that the feed is accessible at all times and they eat in a leisurely manner and only enough to satisfy their appetites at the moment.

Care must of course be exercised in putting cattle on a self-feeder. As a rule, the roughage content of the complete ration is high at first and gradually reduced at each filling until the cattle are accustomed to a full feed of grain. This period may vary from a week to 30 or even 50 days, depending on the length of the feeding period.

The principal disadvantages of using self-feeding are the following.

1. It is more difficult to utilize large amounts of farm-grown roughages.
2. Large amounts of high-moisture grain cannot be prepared at one time.
3. A larger investment in equipment is required.
4. There is a slight increase in cost of gains.
5. Cattle are apt to be less carefully observed.

None of these items requires any extended discussion. It is freely admitted that self-feeders have no place on farms where the grain ration is more or less limited with a view to utilizing large quantities of roughage.

If a nitrogenous concentrate is to be fed in a self-feeder along with the corn, the corn may be either shelled or ground, but a better mixture results with ground corn than with shelled. In all instances, of course, milo must be processed. If the moisture content of either ground ear corn, ground shelled corn, or milo is much over 18 percent in winter or 13 percent in summer, molding is apt to occur in the self-feeder, as it will if the feeder is used on pasture where a certain amount of rainwater is almost bound to get into the self-feeder.

It is rather generally believed that hand-feeding results in more economical gains than can be obtained from self-feeding. This may be true where the hand-feeding is carefully done but this advantage is unlikely in the ordinary methods of hand-feeding. As nearly half of the experiments involving hand-fed and self-fed lots show a slightly lower

Fig. 105. Illustrating two types of self-feeders in use on many midwestern farms; one is top-loading, the other is end-loading. The trim middles on the steers suggest that little, if any, roughage is being fed. (Hubbard Milling Company, Mankato, Minnesota.)

feed requirement per pound of gain with self-fed lots, it may be concluded that no material difference exists between these methods of feeding as far as the cost of gains is concerned (see Table 252.)

It is difficult to place a monetary value on good husbandry in the full feeding of cattle, but the old adage, "The eye of the master fattens his cattle," is undoubtedly an important factor in the success or failure of this cattle program.

GROUP VERSUS STALL FEEDING

Occasionally one hears lavish claims made for the stall method of feeding practiced in some European countries. However, reports of attempts to use this method in America do not lend encouragement to its adoption here. Apparently cattle are influenced by "crowd psychology" in much the same way as human beings. Each is interested in what its pen mates are doing; each has a strong desire to enter into the activity engaged in by the rest. In no way is this tendency more clearly shown than in the activity of eating. No sooner does one steer get up and walk to the feed trough than

Table 252

Comparison of Self-Feeding and Hand-Feeding Steers

| | Wisconsin 168 Days | | | Iowa Average of 2 Trials | | | | |
| | | | | First Trial 160 Days | | Second Trial 150 Days | | |
	Self-Fed Corn; Other Feeds Fed Once a Day	Fed Twice a Day	Fed Once a Day	Self-Fed	Hand-Fed	Self-Fed	Hand-Fed
Average daily gain (lb)	2.55	2.51	2.56	3.24	3.06	2.94	2.68
Average daily ration (lb)							
Corn	12.8	12.4	12.4	15.15	14.47	16.14	15.00
Protein concentrate	1.50	1.50	1.50	2.25	2.25	2.25	2.25
Legume hay	2.0	2.0	2.0	1.44	1.44	1.28	1.28
Corn silage	17.1	17.1	17.1	32.47	31.43	29.24	28.74
Feed per pound of gain (lb)							
Corn	5.00	4.95	4.86	4.71	4.76	5.55	5.64
Protein concentrate	0.59	0.60	0.58	0.69	0.73	0.77	0.84
Legume hay	0.79	0.80	0.79	0.45	0.47	0.44	0.48
Corn silage	6.68	6.79	6.66	10.04	10.27	10.00	10.78

others lying nearby begin to do likewise. Apparently animals with sluggish appetites are stimulated to eat through observing the behavior of their hungry mates. It is a well known fact that cattle fed together out of a common trough will eat considerably more than if they are fed separately. A new and interesting field of research, not yet yielding many data with respect to feedlot cattle, is that of animal behavior. A "peck order" very definitely is established in a pen of cattle, with some cattle dominating and intimidating the more timid ones. Removing the "boss" steer or cow is not a permanent solution because another one soon emerges.

DETAILS OF HAND-FEEDING

There are two main considerations in the practice of hand-feeding: (1) methods of feeding or combining the components of the ration and (2) when and how often to feed during a day.

Combining Feeds.   Usually grain and other concentrates should be fed separately from the roughage, except when cattle are fed silage or complete pelleted rations. When silage is fed, the silage should be put into the trough first and the grain and protein concentrate poured over it. Mixing all three together with a self-unloading mixer wagon or truck is ideal. If the grain is self-fed, the protein concentrate should be spread over the silage instead of fed with the grain. When silage is not fed, all concentrates used, including grain, should be mixed together.

Number of Feedings per Day.   Among practical cattle feeders there is wide variation in the number of times the different constituents of the ration are fed each day. Sometimes the cattle are fed only once daily, being given enough of every item to supply them for the next 24 hours. (Should the practice be to offer a surplus so that feed is before the cattle at all times, hand-feeding no longer exists, even though the usual self-feeding equipment is not used.) Although feeding only once a day saves much time and labor, it does not have as good results as the feeding of smaller quantities at shorter intervals. This applies to the grain rather than the roughage, particularly if more than one kind of roughage is used.

Two feedings per day are the usual number, although commercial feedlots may provide fresh feed in the bunks as often as four times per 24-hour day by running their trucks day and night. The advantage is that frequent feedings stimulate the appetites of the cattle by supplying fresh, clean feed that has not been "mussed over." Also, at each feeding time the animals' curiosity is aroused and they come to the troughs to see what they have been given.

When two feedings per day are made they should be given fairly early in the morning and late in the evening, the exact time depending on the season of the year. Seldom are more than two feedings of roughage given per day. If more than one roughage material is used it is a common practice to feed one in the morning and the other in the evening. The less palatable material should be fed in the evening to allow a longer time for its consumption, during wintertime at least. When silage and a legume hay are fed, the former may be fed in the evening and the latter in the morning, as usually more silage than hay will be used; or both silage and hay may be fed in the evening and only silage in the morning. The cooler nights also will prevent the silage from heating in the bunks in the summer if silage is fed in the evening.

If an auger-bunk feeding system is used and the complete system is automated, including the silo unloader when one is used, there are advantages to feeding as often as six times per day or at 4-hour intervals. Less bunk space is required, and research indicates some improvement in feed efficiency, due apparently to a more even intake of feed over the entire 24-hour period. Large commercial feedlots, equipped with self-unloading trucks and all-night lights over the feedbunks, make more efficient use of their equipment, such as trucks and feed mill, by feeding continuously for 24 hours. Each pen may be fed 2 to 4 times per day. For all practical purposes the cattle are self-fed but fresher feed, especially silage, is available if frequent feedings are made.

No firm rules can be given concerning either the best time or the best method of feeding. They necessarily must be determined largely by the location of the feed supply in relation to the cattle and by the amounts of the different feedstuffs used. The more labor involved in feeding a given material, the greater the likelihood that a single feeding per day will give the best results.

CHAPTER 25

BUILDINGS AND EQUIPMENT FOR BEEF CATTLE

Reductions in capital investments in buildings and equipment, and in labor required per unit of beef produced or per head fed, have not in general kept pace with improvements in the feeding and breeding of beef cattle, especially for average- to small-sized units. An exception is found in the fast-developing feedlot areas in the Panhandle region and parts of the West, which have been able to take advantage of the experience gained in the years since 1960.

Recent studies by the U.S. Department of Agriculture show that crop labor requirements per unit of production have been reduced by 34 percent within the last 30 years. The reduction for all livestock enterprises has amounted to only 7 percent, and it is doubtful whether the beef production segment of the industry equals the average in this respect.

An average of 6 to 7 man-hours of labor are required to feed and care for a single feeder steer, carried through a normal feeding program in the Corn Belt area. In the commercial feedlot areas the figure is down to 2 to 3 hours per steer. The largest single labor requirement is for feeding, especially in the stocker and feeder programs; thus any appreciable reduction in labor requirements will occur only when the feeding problem is regarded as a materials-handling problem.

Efficient shed and feedlot layouts are the first step toward reducing labor requirements, followed closely by the use of carefully chosen mechanical feed processing and handling machinery. Either when building a new set of beef cattle buildings and feedlots equipped with the latest in new equipment or when remodeling older improvements, the farmer feedlot operator or the rancher must remember that most of the new capital investment must be paid for out of labor savings, and the amount that can be saved per animal unit is very small. Thus either the capacity of the feedlot must be large, with year-round feeding operations to spread the cost over a large number of head, or the initial investment must be minimal. The great interest in mechanization, with all its possible combinations of convenient feeding systems, may lead some cattlemen to overbuild. Actually in the handling of beef cattle, comparatively little equipment is needed, especially for stockers and feeders. Purebred animals usually are allowed more pretentious shelter and equipment, although even here the investment in equipment need not be large. On the farms of the larger, wealthier breeders, however, it often totals several thousands of dollars, much of it in buildings, apparatus, and devices that make for more efficient herd management and reduce the amount of labor required for its proper feeding and care.

Following is a brief discussion of the more important articles of equipment found useful on beef cattle farms.

SHELTER

The amount and kind of shelter needed for beef cattle is determined almost wholly by the section of the country where the cattle operation is located. Items that influence the need for shelter include the following: minimum, maximum, and mean day- and night-time temperatures; wind velocity; precipitation in the form of rain, snow, or sleet; humidity; local pollution control and zoning regulations; and kind and age of cattle. Thus it is obvious that no single set of recommendations can be made to fit a very large segment of the cattle producers in the country because of the wide variation in the items just mentioned. Space limitations prohibit the presentation of quantities of experimental data that would support the recommendation of one kind of shelter over another. Nearly every state agricultural experiment station is conducting research on this subject. Producers in a given state or geographic region would do well to consult the beef cattle and agricultural engineering researchers in their locality with respect to their specific needs for beef cattle shelter. In addition, many states are still in the process of developing legislation that will conform with federal guidelines relative to environmental pollution control, and until such legislation is passed and enforcement is initiated, it is probably unwise to make large investments in new or remodeled cattle housing facilities.

Comparisons between types of shelter can be misleading if they do not take into consideration the fact that a different type of lot surface is often used. Even on farms, the man with a good cattle shed will have paved lots, while the man with a poor shed or none at all will also run his cattle in unpaved muddy lots. Examples of tests in which the lot surface may have had more to do with results than type of shelter are shown in Table 253. Although the favorable results in these tests cannot entirely be ascribed to the type of shelter, the combinations tested are quite common in the states doing the testing, and thus the results can be applied to a majority of the farms. All of these states are in areas where there is either considerable snow or wintertime rain or where winter temperatures are severe, or both. This being the case, it is quite possible that the feedlot surface is the main source of differences in performance.

Table 253

Effect of Shelter and Feedlot Surface on Cattle Performance in Wintertime[a]

Station	Kansas		Michigan		Kansas		Arkansas	
Cattle Program	Wintering Yearling Steers		Finishing Heifer Calves		Finishing 2-Year-Old Steers		Finishing Steer Calves	
Type of Shelter, Lot Surface	Open Shed, Concrete Lot	No Shelter, Dirt Lot	Barn, Paved Yard	Open Shed, Unpaved Yard	Open Shed, Concrete Lot	No Shelter, Dirt Lot	Open Shed	Bermuda Lot
Number of cattle per pen	20	20	26	26	20	19	39	39
Average initial weight (lb)	716	738	434	436	853	849	—	—
Average daily gain (lb)	1.47	1.16	1.88	1.65	2.33	2.17	1.98	1.66
Average daily ration (lb)[b]	19.6	19.6	17.0	16.6	29.3	29.2	19.6	19.9
Average feed cost per cwt gain	$11.80	$15.07	$15.48	$17.34	$21.63	$23.28	$17.75	$21.50

[a] Compiled from miscellaneous Feeders' Day Reports.
[b] Air-dry basis.

Fig. 106. An example of an inexpensive pole-type cattle shed with bedding or hay storage in back, a compact adjacent concrete paved lot, a nearby large-capacity bunker silo, and airtight high-moisture corn storage combined into an efficient cattle-feeding plant. (University of Illinois.)

Cattle being fed in drylot in summertime, whether on stocker or finishing rations, are much more adversely affected by heat and humidity inside barns or even sheds if poor ventilation prevails than when simple shades or even trees are the only shelter. Fences are also a factor in air circulation, with wooden fences retarding the flow of air more than cable fences.

Although breeding cows and stocker steers on little more than maintenance rations undoubtedly are more sensitive than finishing cattle to extremely low temperatures, it is unlikely that there is any great advantage in housing them in closed barns.

PAVED FEEDLOTS

Muddy feedlots impose a severe handicap on winter-fed steers. Many a man has ordered a truck and shipped his half-finished steers to market for no other reason than to get them out of muddy lots in which they were standing up to their knees. Others have permanently abandoned late winter and spring feeding, acknowledging their defeat in the annual battle against slush and mud.

The best insurance so far devised against possible loss in cattle feeding from bad weather conditions is the paved feedlot. The advantages of paving, briefly stated, are as follows.

1. It adds greatly to the comfort of the cattle, thus promoting greater and more efficient gains.

2. It reduces the labor involved in feeding and handling the cattle.

3. It makes possible the saving of more manure.

4. Losses caused by disease should be lower.

The open lot used by feeder steers does not need to be large. Only enough space is required to permit the cattle to move about freely without unduly disturbing one another. Fifty square feet of floor space per mature animal, 40 for yearlings, and 30 for calves is sufficient, especially if the lot adjoins an open shed affording an equal area per head. The cost of a 5-inch concrete pavement is not excessive and soon pays for itself in the increased pork and manure credits secured, to say nothing of the greater gains made by the cattle or the labor saved in feeding and cleaning. Asphaltic concrete pavings are a questionable investment, especially in the extreme northern and southern states, because of the freezing and softening, respectively, that are encountered in these regions. A 6- to 10-foot concrete ramp should be provided for single-sided bunks. Equally essential is a 10-foot ramp around the water tank. Reducing the feedlot concrete to these areas makes cleaning the feedlot more inconvenient, but the comfort of the cattle comes first as far as paving is concerned.

In locations where there are slopes of 6 percent or greater and the soil is rocky, gravelly, or sandy enough to provide good internal drainage, lots

Fig. 107. Several components of a modern feedlot in the making. Protected water tank with enclosed electric water heater on a concrete slab with apron. In the background, poured-concrete bunk and 8-foot apron; steel cable fence. (Western Livestock Journal.)

may not require paving to prevent undue muddy conditions. Use of earthen mounds is a compromise that may serve to prevent the problem of unpaved lots. Such mounds should be 8 to 10 feet high and have side slopes of 4:1. Capping the mound with 4 to 6 inches of fine limestone will help prevent muddy and slippery spots during rainy days. Obviously there will be run-off from such lots and they can only be used if provision is made for diverting the run-off into lagoons or settling basins.

TOTAL CONFINEMENT BARNS

In the Midwest, where snow, ice, and mud prevail in winter and where high temperatures, day and night, combine with high humidity to cause problems in the summer, a few farmer-feeders are now feeding cattle in total confinement barns with or without controlled environment the year around. A portion or all of the floor is slotted, having concrete slats 6 inches wide laid 1½ inches apart, usually over a manure pit about 6 feet deep.

Fig. 108. Fenceline bunks and automated feeding wagons are replacing conventional feedlot bunks and the scoopshovel in the Corn Belt area. Such modern equipment is common in the larger commercial feedlots of the Southwest and West. (International Harvester Company.)

Fig. 109. Two-sided precast concrete bunks with economical automatic chain and paddle feeder, used in feeding a high-silage ration to yearlings in confinement on an Illinois farmer-feeder's farm. (Prairie Farmer.)

Pits that have no agitation equipment or added liquid in them are generally known as anaerobic pits and are emptied periodically. Other pits have water added to them in order to maintain a solids level of about 4 percent dry matter in the liquid ditch contents, and such pits, generally called "oxidation ditches," have paddle wheels that aerate the mixture and keep the oxygen content high and also keep the liquid flowing continuously in the ditch channels. The agitation, in the presence of free oxygen, ensures aerobic fermentation and reduces odors much more effectively than does the anaerobic ditch. The level of liquid in the ditch must be kept constant; thus some arrangement must be provided for removing liquid almost daily. Liquid may be stored in lagoons and recycled in the ditch after settling to remove the solids, or it may be used to irrigate or fertilize pastures or cropland as removed from the ditch. The liquid has been successfully used as a partial protein supplement to dry rations, as fed, on a daily basis.

An automated feeding system is built into the barn to reduce labor to an absolute minimum. Considerably less floor space is required under these conditions than in conventional barns. At first many such barns were completely enclosed but now most are open-fronted. Performance of cattle housed in these plants is reportedly improved. Most researchers report either no improvement or up to 5 percent improvement in gain and about 8 or 9 percent reduction in feed requirements as compared to the use of conventional sheds. The elimination of bedding and reduction in labor required for feeding and barn clean-up are other positive factors. The high initial capital investment may prevent many farmer-feeders from installing this kind of system. It is estimated that annual operating costs of such equipment may equal 20 percent of initial costs, including interest and depreciation on initial investment.

Fig. 110. Auger-bunks that deliver both roughages and concentrates to cattle at very low power costs are genuine labor savers. Another kind of auger-bunk in which the auger is enclosed in a tube with openings at frequent intervals permits the separate feeding of each component of the ration. (University of Illinois.)

FEEDBUNKS

Portable feedbunks used by beef cattle, if made of wood, should be made of 2-inch material, preferably undressed cypress or oak. The legs should be cut from 4- by 4-inch material and should be both bolted and braced to the body of the bunk. Bunks placed in the open, where cattle can eat from both sides, should be at least 30 inches wide, or preferably 36 inches for mature cattle. The floor of the bunk should be 20 inches from the ground and should have 6-to-10-inch sides, depending on the age of the cattle. Feed troughs built inside a barn or shed should be made adjustable in height to accommodate cattle of various ages. It will be necessary to raise the troughs from time to time if the manure and bedding are allowed to accumulate in the shed.

Development of the self-unloading wagon or truck has been closely followed by a great increase in the use of fenceline bunks. Besides saving labor, this type of bunk, if used with self-unloading equipment, shortens travel routes, makes for easier cleaning of lots, and cuts down on damage to bunks. However, twice the linear footage of bunk space must be provided if such single-sided feeders are used.

Bunk space of 20 to 24 inches per head should be provided if all cattle are to eat at once. This may be reduced if cattle are fed two or more times daily or if bunk self-feeding is practiced. Paved, graveled, or rocked driveways are an integral part of the fenceline bunk feeding system. In all instances, a sloping paved strip should be provided adjacent to the bunks for cattle to stand on. A 4- to 6-inch-high raised concrete strip about 8 to 12 inches wide in front of the bunk itself will prevent cattle from backing up to the bunk to scratch or defecate into the bunk.

CORRALS AND RESTRAINT EQUIPMENT

A set of working corrals for farm or ranch with such accessories as an unloading and loading chute, a squeeze chute with headgate, sorting chute or gate, dipping vat where needed, scales, and holding pens with sorting gate, are important tools of the beef cattle man. The size of the operation, of course, determines the most economical size of such equipment. Plans are available from the U.S. Department of Agriculture and the various agricultural experiment stations for almost any type and size of layout needed. Figure 111 is one example of a plan for an expanding operation. Note that the simplest form of the plan still contains the squeeze, headgate, and loading chute. Figure 112 shows several types of headgates, some hinged for a front-opening chute and others stationary for a side-

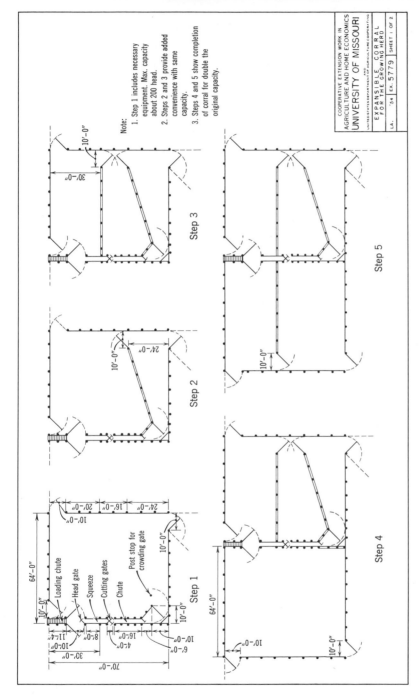

Fig. 111. An expansible corral, which can be built in stages as the herd grows in size up to 200 head. Step 1 includes the essential loading chute, headgate, squeeze, and chute. Steps 2 and 3 provide added convenience with the same capacity. Steps 4 and 5 show completion of the corral, doubling the original capacity. (University of Missouri and USDA cooperating.)

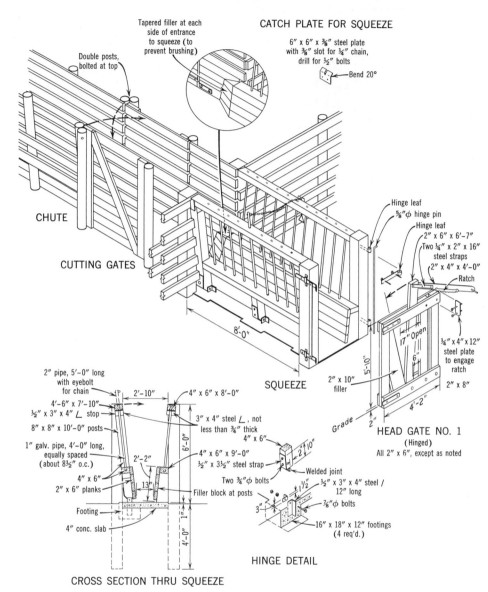

Fig. 112. Squeeze chute with details and three kinds of headgates. (University of Missouri and USDA cooperating.)

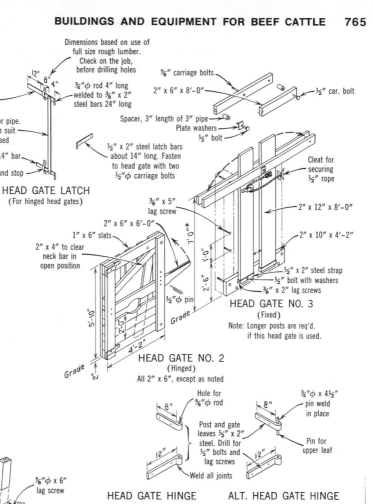

Dimensions based on use of
full size rough lumber.
Check on the job,
before drilling holes

12" 8" 4"

Drill for
⅝" x 5"
lag screw

Steel bar or pipe.
Length to suit
gate used

¾"ϕ rod 4" long
welded to ⅜" x 2"
steel bars 24" long

⅜" x 2" x 14" bar

Steel guide and stop

HEAD GATE LATCH
(For hinged head gates)

⅝" carriage bolts

2" x 6" x 8'-0"

½" car. bolt

Spacer, 3" length of 3" pipe
Plate washers
½" bolt

½" x 2" steel latch bars
about 14" long. Fasten
to head gate with two
½"ϕ carriage bolts

Cleat for
securing
½" rope

⅜" x 5"
lag screw

2" x 6" x 6'-0"

1" x 6" slats

2" x 4" to clear
neck bar in
open position

7'-0"*

1'-0"

2" x 12" x 8'-0"

2" x 10" x 4'-2"

½" x 2" steel strap
½" bolt with washers
⅜" x 2" lag screws

HEAD GATE NO. 3
(Fixed)
Note: Longer posts are req'd.
if this head gate is used.

½"ϕ pin

Grade

2'-6"

Grade

5'-10"

2'-0"

4'-2"

HEAD GATE NO. 2
(Hinged)
All 2" x 6", except as noted

Grade

2"

Hole for
⅝"ϕ rod

8"

8"

¾"ϕ x 4½"
pin weld
in place

Post and gate
leaves ½" x 2"
steel. Drill for
½" bolts and
lag screws

Pin for
upper leaf

12"

12"

Weld all joints

HEAD GATE HINGE

ALT. HEAD GATE HINGE

4" x 6" x 1'-10"
at center of
fixed side

⅝"ϕ x 6"
lag screw

18"

½" x 3" x 4" steel L
12" long

12"

16" x 18" x 12" footings
(4 req'd.)

16"

ANCHOR DETAIL

Note:
Squeeze parts and
all wood in contact
with earth should
be pressure-treated
with creosote or
other preservative.

COOPERATIVE EXTENSION WORK IN
AGRICULTURE AND HOME ECONOMICS
UNIVERSITY OF MISSOURI
AND
UNITED STATES DEPARTMENT OF AGRICULTURE COOPERATING

CATTLE SQUEEZE
AND HEAD GATES

| LA. | '54 | EX. 5789 | SHEET 1 OF 1 |

opening chute. All such devices should combine ease of operation with strong construction, as it frequently is necessary, in confining an animal, to close and fasten a gate in the fraction of a minute. Many firms manufacture satisfactory equipment that can become an integral part of the barns or lots and working corrals found on farms and ranches.

FEEDLOT LAYOUT AND EQUIPMENT

A number of important factors must be considered in planning a feedlot of the commercial type. Some, but not all, of these principles apply in the smaller farmer-feeder operation as well. Assuming that the section of the country is already determined, there are still some local considerations involving site that must be taken into account. For instance, access to a hard-surfaced road or railroad ensures year-round shipment of cattle and feed, to and from the feedlot. Availability of an adequate supply of good-quality water is essential. Some natural drainage is desirable, and a soil type that drains well will reduce mud problems. Electrical power should be economically available. Local and state water and air pollution laws and county zoning ordinances should be studied carefully. Amount and prevailing direction of wind should be considered because of potential odor and dust problems involving the surrounding area. There should be sufficient land available for future expansion if desired.

The number of cattle to be fed has much to do with both the layout and the type of feeding equipment used. Feedlots with less than about 1,000-head capacity may well be fed and managed in pens holding 100 head each, equipped with self feeders. Auger-bunks dividing two lots so that cattle may eat from both sides may also be used. Fenceline bunks and self-unloading trucks may also be used for this size lot and are almost a necessity for lots that are larger in size.

Pen capacity may vary from as few as 50 head up to 200 as a practical maximum, with 100 head being about ideal for heavier cattle and 200 for calves. This number requires a pen 100 by 200 feet to provide the necessary space if the lot is unpaved.

Some additional, smaller lots are convenient for holding cattle for various reasons. A 12-foot-wide working or sorting alley and several catch pens are essential. These should be adjacent to the loading chute. Small hospital pens and perhaps a hospital shed, depending on the climate, are essential. In large lots, working alleys should be separate from feeding alleys or there will be congestion and loss of time. Feeding alleys may range from 12 to 40 feet in width, depending on size of lot. A width of 20 feet permits passage of two vehicles and in large lots this often saves time. Feeding alleys should be paved with concrete if at all possible, and in

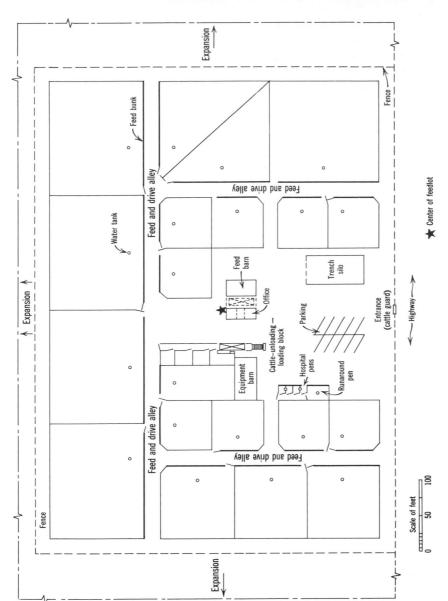

Fig. 113. A proposed layout for a commercial cattle feedlot with a capacity of 1,000 head. (Texas A. and M. University.)

localities with 20 inches or more of rainfall annually such concrete is almost a must. Gates should be on the side of the pen facing the working alley and 12 feet is a good length, especially if it matches the width of the working alley.

Materials for the fence may be treated lumber, $\frac{3}{8}$-to-$\frac{1}{2}$-inch new or used steel cable, or oilfield sucker rod. Cable and similar fencing allows free movement of air but provides no protection from cold winds; thus location dictates the type of material to use. Fences should be at least $5\frac{1}{2}$ feet high. Details concerning post and cable spacing, setting of posts, and so on are obtainable from builders' supply houses or the county agricultural agent. The same is true for recommendations concerning feed troughs, water systems, type of shades, and the like. Local weather conditions have so much to do with the detailed planning of such equipment that local advice should be obtained.

Research is under way to determine the value of all-night lighting and windbreaks. Preliminary evidence indicates that lights are unessential but do reduce stampedes and minor disturbances as well as provide light for overnight feeding if desired. Shelters or windbreaks have proven beneficial in states such as Kansas, Nebraska, and South Dakota where snow and cold winds are common, but such structures are harmful in California, Arizona, and Texas where free circulation of air is to be desired.

In the fully equipped feedlot, arrangements must be made for office space, weighing facilities for large trucks hauling both feed and cattle, equipment storage, repair space for vehicles, and a parking lot for employees. Horses, used by pen riders, must be provided and housed.

Manure disposal and pen cleaning must be planned for when designing

Fig. 114. One man with self-unloading wagon can feed thousands of cattle daily, especially when the ration is fed in complete form—that is, roughage and concentrate combined in one mixture in the desired proportions. (National Cottonseed Products Association.)

a feedlot. About 0.5 ton of manure is produced per month per animal. In large dirt-surfaced lots, manure or scrapings may be bulldozed into mounds and removed only between droves of cattle fed. Concrete-surfaced lots must be hosed off or scraped more frequently.

Details concerning the feed storage and feed processing facilities should be provided by an engineering service specializing in such work. Some states have Agricultural Extension Service engineers who are experienced enough to be of assistance.

CATTLE STOCKS

Cattle stocks, or bull stocks as they are often called, are essentially an item of equipment for the purebred herd. Their important features may be combined with those of an ordinary dehorning chute to make them useful with unbroken cattle, but it is more usual to build them as a separate or detached unit into which cattle are led rather than driven. This fact alone limits their use to animals that are rather easily handled.

The characteristic features of cattle stocks are (1) rollers on either side at

Fig. 115. Control panel for a fully automated feedmill, in the process of being installed. Large feedlots often operate their own feedmills and storage facilities adequate for a year's supply of grain. (Western Livestock Journal.)

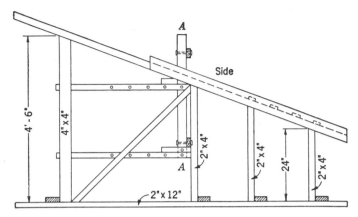

Fig. 116. Diagram of a breeding stock, showing details of construction. (University of Illinois.)

a height of about 4 feet from the floor, made of 5-inch cedar posts or 4-inch steel pipe, for supporting a canvas sling that is placed under the animal to prevent it from lying down while in the stocks; and (2) wooden sills, 6 by 8 inches, that extend along either side at a height 15 inches from the floor, on which the feet may be rested and tied while undergoing treatment or trimming.

Cattle stocks are exceedingly useful for the restraint of animals during the performance of such minor operations as the ringing of bulls, clipping of heads, surgical treatment of lump jaw, and trimming of feet. An adjustable canvas surcingle allows the weight of the body to be taken off the legs, making it practically impossible for the animal to throw itself or lie down, and thus reducing chances of injury to itself or the herdsman.

Commercially made, tilting trimming tables serve the same purpose as the stocks described above. The table is tilted vertically and the animal is fastened to it by the legs while standing. The table is then tilted to a horizontal position with the animal lying on its side. The feet are readily accessible for trimming while in this position. The chances of injury to the animal are lessened with this equipment but, unfortunately, it is quite expensive.

BREEDING CHUTES

As discussed in Chapter 5, breeding chutes are useful in safeguarding the breeding of young heifers to old, heavy bulls. A satisfactory type of chute or stock is shown in Fig. 116. The back portion of each side of the stall is

Fig. 117. A permanent loading chute designed to load cattle into single- or double-decked trucks. The 8-inch concrete step ensures more orderly loading and unloading than does a slotted floor. (Western Livestock Journal.)

surmounted by a 2-by-12-inch plank on which cleats are nailed to prevent the bull's front feet from slipping. The gate or stanchion labeled *A* in the diagram is adjustable to accommodate cows varying in height and length of body.

LOADING CHUTES

A well-designed loading chute is a necessary item of equipment on every beef cattle farm. Stationary chutes should be located in reference to both roads and holding pens to make them accessible to heavy trucks and so that cattle may be quickly loaded and unloaded after the truck has been set. Tests conducted at commercial stock yards indicate that livestock prefer low stairstep risers to a cleated ramp. Whether the advantages of such a chute are sufficient to justify the extra materials and labor required may be questioned by the farmer who handles only a load or two of cattle a year.

A portable loading chute that may be towed behind a truck is often useful in loading or unloading cattle at a distance from the farmstead

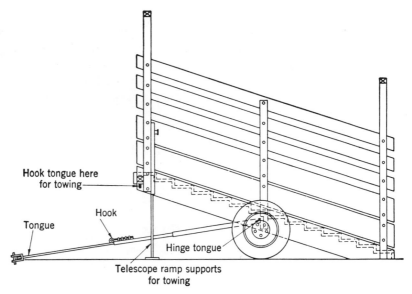

Hook tongue here for towing

Hook

Tongue

Hinge tongue

Telescope ramp supports for towing

Fig. 118. A portable step-ramp loading chute developed by the Union Stock Yards and Transit Company. (Livestock Conservation, Inc., Chicago, Illinois.)

where the stationary chute is built. A sketch of a portable chute with a stairstep ramp is shown in Fig. 118.

LABORSAVING EQUIPMENT

The shortage of labor on livestock farms has caused a wide interest in laborsaving devices that will shorten the time and lighten the labor required to care for the animals. Such equipment includes (1) blowers and conveyors for moving feed, (2) storage of feed and penning arrangements for animals that will require as little movement of the feed as possible, and (3) equipment for cleaning the sheds and yards. First-class equipment is fairly expensive and thus better suited for the large feeder who handles 100 or more cattle than for the farmer who feeds only a load at a time. On the other hand, the small feeder is usually able to store his feeds nearer the cattle than is possible in large operations. Often by remodeling his barns and carefully planning the location of his grain bins and silo, the small-scale feeder can feed and care for his cattle entirely under cover and with a minimum of hand labor.

Cleaning cattle sheds and lots and hauling and spreading the manure have always been regarded as among the hardest and most disagreeable jobs on the farm. They probably have caused many good farm boys to become grain farmers or even leave the farm entirely instead of following in the steps of their cattle-feeding fathers. However, the presence of tractors with hydraulic manure loaders on practically every farm has now eliminated most of the hand labor formerly required in cleaning sheds and feedyards on livestock farms.

CHAPTER 26
THE MARKETING
OF CATTLE

The matter of selling or buying stockers or feeders is discussed in Chapter 11; therefore the discussion in this chapter is confined mainly to the marketing of slaughter cattle.

A rancher, farmer, or cattle feeder with cattle to sell seldom has any difficulty in disposing of them at a price near their actual value. This price may not be so high as the seller had hoped, but with the great variety of marketing agencies generally available today and the competitive bidding they afford, few "steals" are made by buyers of slaughter cattle. In some instances, of course, usually owing to a low population of slaughter cattle, a locality has a limited choice of marketing facilities, and sometimes only one or no convenient outlet is available.

In most areas with a relatively dense livestock population, most if not all of the following avenues are available for disposal of slaughter cattle.

1. Consignment to central or public terminal livestock market.
2. Use of local cooperative shipping and selling association.
3. Consignment to local auction sale.
4. Direct on-the-farm or feedlot sale to order or packer buyers.
5. Sale of cattle direct to packer buyers at concentration yards or at the packing plant.
6. Consignment of cattle to a packer on the basis of carcass dressed weight and grade.
7. On-the-farm sale to cattle dealers or local packers and locker plants.
8. Fulfillment of a futures contract by delivery of cattle.

The choice of the most suitable market is not a simple one, and there are no infallible rules for making such a decision. The factors that should play the most important roles in making this choice are (1) quoted market price per hundredweight, (2) transportation costs, (3) selling expenses, (4) shrinkage, (5) services rendered by the market or its marketing agencies, and (6) weighing conditions.

Still other factors that enter into the choice of a cattle market are convenience, custom, and personal preference. Many cattlemen simply prefer to deal directly with a buyer because they feel that less risk is involved. On the whole, all of the market channels listed previously offer a good market for cattle, but each man with slaughter cattle to sell owes it to himself to make comparative studies of the various markets for the particular cattle he has to sell. The market of choice varies greatly depending on locality. Unfortunately, up-to-date statistics on numbers sold through the various available channels are difficult to find.

Until recent times the majority of slaughter cattle eventually were sold to packers in the large central markets. Today, however, the trend is for more and more slaughter cattle to be sold direct from the feedlot to a packer buyer or at least direct to a country buyer who is buying on order for the killer. Some of the formerly large central markets have changed entirely into auction markets where, of course, fat cattle as well as stockers and feeders are sold. The overall effect is that the central market has declined in importance as a slaughter cattle market and packers fill less than one-third of their needs there.

As already mentioned, a large percentage of slaughter cattle are purchased direct from the feedlot by the packer but the packer buyer may deal directly with a commission man who sells cattle directly from the feedlot or farm rather than in a terminal market. The reason usually given for this practice of direct packer buying is that it assures the packer of the necessary number of certain kinds of cattle required for a full day's kill. Many marketing experts feel that the packers follow this practice principally because buying competition is reduced, especially in areas not served by more than one packer. Another undesirable feature of this method, from the feeder's standpoint at least, is that direct-purchased cattle are usually of better quality than the cattle selling through other channels. The price at the central market is established on the poorer quality cattle offered. Because the direct-purchase price is usually calculated on the basis of central market prices, it can be seen that the overall effect is a lowering of all prices.

A new development that will be watched with interest is the selling of cattle, either live or in the carcass, directly to the consumer through special retail outlets, as is being done by some commercial feedlots and even some larger farmer-feeders. The slaughtering, in such instances, usually is done by a local firm on a custom basis.

The 1976 approval of a U.S. Department of Agriculture requirement that all cattle that are USDA quality-graded must also be yield-graded could greatly influence the cattle feeder's choice of a market system. Undoubtedly a system of premiums and penalties or discounts will be developed to accurately adjust the prices paid, within a quality grade, for cattle with high and low yield grades. It is too early to estimate whether this will increase grade and yield selling. It is certain that sellers and buyers alike must learn to estimate both quality and yield grade in live cattle unless some system of selling is used that calls for paying specific agreed-upon prices for each quality and yield grade. Undoubtedly marketing methods will be changed accordingly as the new grades and the compulsory yield-grading are activated.

WHEN TO MARKET

Various rules for marketing have been stated by successful feeders, but it is doubtful whether they are taken very seriously, even by their originators. Perhaps the one most often heard is "Ship when the cattle are ready, regardless of the condition of the market." However, it would seem from the appearance of many animals received at the yards that a rule more frequently followed is "Ship when the market seems right, regardless of the condition of the cattle." Neither rule is a good one if blindly followed, although each expresses an element of truth that should not be overlooked if satisfactory returns are to be realized.

The approximate time of marketing should be decided at the time the cattle are placed on feed. Only by having a definite feeding plan and length of feeding period in mind can the purchase and selling dates be planned intelligently. As the cattle begin to reach the degree of finish desired, the short-run market trends should be studied in an effort to obtain as favorable a price as possible. However, there is no way of knowing definitely which way the market will go. Even the reputation salesmen and large buyers, who have had long years of experience, occasionally make serious errors in predicting future price tendencies. Nevertheless, the judgment of such men deserves respect, and their advice regarding shipping should be carefully considered. They have at hand much information regarding total supplies, expected loadings, religious holidays affecting consumption, and conditions of the dressed beef trade that is not available to many individual shippers.

In general, the relative prices being paid for a particular weight and grade of cattle are determined by the total supply of such cattle on the major markets in the country. One of the leading livestock journals has prepared a calendar for marketing beef cattle, based on average prices generally received at the major markets throughout the year (see Fig. 119).

SOURCES OF MARKET INFORMATION

Factors affecting supply and factors affecting demand are constantly changing and no one can predict with certainty when they will occur and how great these changes will be. However, to understand and, to some

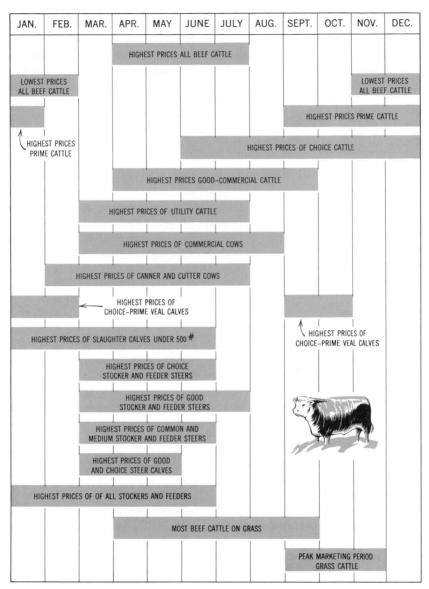

JAN.	FEB.	MAR.	APR.	MAY	JUNE	JULY	AUG.	SEPT.	OCT.	NOV.	DEC.

HIGHEST PRICES ALL BEEF CATTLE

LOWEST PRICES ALL BEEF CATTLE

LOWEST PRICES ALL BEEF CATTLE

HIGHEST PRICES PRIME CATTLE

HIGHEST PRICES PRIME CATTLE

HIGHEST PRICES OF CHOICE CATTLE

HIGHEST PRICES GOOD–COMMERCIAL CATTLE

HIGHEST PRICES OF UTILITY CATTLE

HIGHEST PRICES OF COMMERCIAL COWS

HIGHEST PRICES OF CANNER AND CUTTER COWS

HIGHEST PRICES OF CHOICE–PRIME VEAL CALVES

HIGHEST PRICES OF CHOICE–PRIME VEAL CALVES

HIGHEST PRICES OF SLAUGHTER CALVES UNDER 500 #

HIGHEST PRICES OF CHOICE STOCKER AND FEEDER STEERS

HIGHEST PRICES OF GOOD STOCKER AND FEEDER STEERS

HIGHEST PRICES OF COMMON AND MEDIUM STOCKER AND FEEDER STEERS

HIGHEST PRICES OF GOOD AND CHOICE STEER CALVES

HIGHEST PRICES OF OF ALL STOCKERS AND FEEDERS

MOST BEEF CATTLE ON GRASS

PEAK MARKETING PERIOD GRASS CATTLE

Fig. 119. A calendar for marketing beef cattle. Each weight and grade of slaughter cattle has its period of high and low prices during the year. There are indications that these seasonal fluctuations are narrowing, undoubtedly owing to year-round feedyard operations. (National Livestock Producers.)

extent, anticipate these changes, it is helpful to have up-to-date information about these market factors. Some of the most reliable information is found in U.S. Department of Agriculture (USDA) publications, but even these are open to criticism for they sometimes lead to wrong guesses. The sampling procedure used by USDA data gatherers is constantly under revision and the department makes use of the latest electronic procedures.

The following government publications are among the most useful for making one's own estimates of market trends.

1. *Livestock and Meat Situation Report,* Economic Research Service, USDA, Washington, D.C. 20250. Published six times a year, this report includes historical summaries of data and professional comments on general economic conditions affecting the various livestock and meat markets.

2. *Livestock Slaughter and Meat Production,* Statistical Reporting Service, Crop Reporting Board, USDA, Washington, D.C. 20250. Published monthly, this report gives the slaughter and meat production by numbers and by weights for the previous month.

3. *Livestock, Meat and Wool Market News,* Livestock Division, Consumer and Marketing Service, USDA, Washington, D.C. 20250. Published weekly, this report serves as a current source of information on livestock receipts and slaughter.

4. *Calf Crop Report,* Statistical Reporting Service, Crop Reporting Board, USDA, Washington, D.C. 20250. Published annually in July, this report gives an estimate of the number of calves born during the current year.

5. *Livestock and Poultry Inventory,* Statistical Reporting Service, Crop Reporting Board, USDA, Washington, D.C. 20250. Published annually in February, this report shows the January 1 livestock and poultry inventories.

6. *Cattle on Feed,* Statistical Reporting Service, Crop Reporting Board, USDA, Washington, D.C. 20250. This report, published monthly, shows the placements and marketings of cattle and calves on feed.

The Land Grant university or college in each state, through either the agricultural extension service livestock marketing specialist or the resident staff of the agricultural economics department, prepares livestock marketing outlook materials at certain times of the year. These are distributed by a variety of communications media, usually all the way down to the county level. In addition, most states have specialists in their departments of agriculture who prepare weekly market price quotations and related information. It is easy to be placed on mailing lists to receive reports from these two sources.

Privately prepared publications such as the well-known *Drovers' Journal*, issued weekly in some instances, are excellent sources of market information and statistics. Many cooperative marketing associations have daily radio and television market reports and, in fact, reports may be made several times during the day. Still others have a telephone hook-up with recorded market reports which can be received throughout the 24-hour day.

An excellent example of a producers' organization which provides market information on a subscription basis is that of the American National Cattlemen's Association. They recently organized a market information gathering system under the name Cattlefax. Information supplied by a sizable sample of feeders is interpreted in a central office and then disseminated to subscribing members.

It is evident that there is no shortage of sources of market information. In spite of this, beginning feeders might still be best advised to leave the marketing of their cattle to experienced persons in the marketing agency that seems best able to serve their needs in a given area or circumstance.

THE FUTURES MARKET

Few if any cattle feeders who feed their own cattle know for certain what the selling price of purchased cattle will be in 3, 6, or 12 months, when they have been finished out. Obviously there is a risk as to whether the enterprise will prove profitable. Unless the selling price is high enough to cover the original cost of the feeder animal plus the feed fed and the incidental costs entailed, profit cannot result and, in fact, a loss will occur. Many feeders, in search of a way to reduce the risk, make use of the livestock futures market by hedging their purchase. This procedure was developed long ago for other commodities that also follow volatile price patterns. Space does not permit a detailed discussion of the practice of hedging, and those wishing to use this method should avail themselves of the services of a commodities broker.

A hedge means, simply, that one takes an opposite position or stance on the futures market from the position he is in with respect to the cash market. If one buys feeder cattle in October that will be sold in May as fat cattle, one would sell a May futures contract as a hedge. Then when he actually sells his finished cattle in May, he buys back the contract sold the previous October. It would actually be possible to deliver cattle to fill the contract, but this is seldom done. If cattle prices have dropped during the period of ownership, the cost of the futures contract also will have dropped and so no money should be lost. Obviously if prices rise sharply during the period of ownership, then futures prices also will have risen and the opportunity to make a large profit will have been lost. The purpose of using the futures market is to protect oneself from a disastrous decline in prices. Some buyers and sellers of futures contracts trade in contracts almost daily, as speculators, but this is another aspect of futures trading and should not be confused with a simple hedge. Many lenders of money to be used in financing a feeder cattle operation understandably require that purchased cattle be hedged so as to protect the loan they have made.

Rules and regulations concerning futures trading in commodities, of which feeder and fat cattle are only two examples, are constantly under revision as to size of contracts, commissions, specifications as to grade and weight of cattle involved, and so on. Thus it is imperative that one should obtain the latest information concerning these topics by attending meetings where such matters are discussed or by relying on the advice of a competent broker.

PREPARING CATTLE FOR SHIPMENT

There is much difference of opinion regarding the advisability of attempting to reduce shrinkage in cattle that will be weighed at some market off the farm by changing the ration before shipping. Of course, this is not a problem when cattle are sold direct from the lot, as more and more are today.

Some feeders remove all laxative feeds such as protein concentrates and legume hay or silage a day or two before shipment and supply nonlaxative feeds instead. Others withhold both feed and water on the day of shipment in the belief that the cattle fill better at the market if they arrive hungry and thirsty. Occasionally an unscrupulous shipper salts his cattle heavily on the day of shipment to obtain a heavier consumption of water at the market. Usually nothing is gained from abnormally large fills, however, because buyers quickly detect such cattle and refuse to bid on them, or they adjust their bids accordingly.

Cattle that are to be sold direct and weighed at the feedlot or on the nearest truck scales after loading are ordinarily fed as usual on the day of shipment. Large feedlots cannot afford to manipulate rations so as to get heavier fills because they need the repeat buyers coming to their lots weekly. These they would surely lose after only one bad experience with excessive fills.

There is little information regarding the control of shrinkage in cattle. Yet it is an extremely important subject, particularly when cattle prices are

Table 254

Effect of Changing Feed on Shrinkage of Slaughter Cattle During Shipment[a]

| | Withholding Feed and Water on Day of Shipment | | Replacing Laxative Feeds | | | |
| | Heifer Calves | | Heifer Calves | | Yearling Steers | |
	Fed as Usual	Feed and Water Withheld	Shelled Corn, Clover Hay	Shelled Corn 50%, Oats 50%, Timothy Hay	Shelled Corn, Linseed Meal, Clover Hay	Shelled Corn 50%, Oats 50%, Timothy Hay
Weight out of experiment (lb)	999	998	649	649	1,070	1,095
Length of change period	10hr.	10hr.	6days	6days	4days	4days
Shipping weight (lb)	999	982	666	665	1,069	1,104
Market weight (lb)	963	952	616	624	1,011	1,052.5
Shrinkage (lb)						
On final experiment weights	36	46	33	25	59	42.5
On shipping weights	36	30	50	41	58	51.5
Shrinkage (%)						
On final experiment weights	3.6	4.6	5.1	3.9	5.5	3.9
On shipping weights	3.6	3.1	7.5	6.2	5.4	4.7
Dressing percent	60.6	59.6	57.9	56.0	59.7	58.1

[a] Illinois Mimeographed Report.

high. When cattle are worth $40 per hundredweight, the 40 pounds lost during transit amounts to $16 per steer. Anything the shipper can do to reduce the loss by 5 to 10 pounds per head without noticeably impairing the slaughter value of the cattle is highly profitable. Just as with stocker cattle, some of the shrink is tissue shrink and a reduction of this loss is the goal.

Preliminary studies made at the Illinois station indicate that withholding feed and water on the day of shipment increases rather than reduces shrinkage (see Table 254). Apparently cattle handled in this way are greatly disturbed by the abrupt change in their feeding schedule and spend the day on their feet vainly waiting for feed and water instead of lying quietly at rest. As a result they are tired and nervous when loaded and arrive at market in a fatigued condition. Though shrinkage based on loading weights may be in their favor, the cattle actually have lost considerable weight before they were loaded. This loss may well be 20 to 30 pounds, or the weight of the feed and water that would have been consumed if the regular feeding schedule had been followed.

Other studies made at the Illinois station indicate that substituting oats for part of the shelled corn ration, and timothy or mixed hay for alfalfa or clover hay, is a sound practice because these changes result in less shrinkage (see Table 254). However, these nonlaxative feeds should be introduced into the ration 4 or 5 days before shipment.

RAIL VERSUS TRUCK SHIPMENT

Trucks have almost entirely replaced rail transportation for shipping cattle from within about a 200-mile radius of the central market or slaughter plant, and the large trailer trucks are often used instead of rail for even the longer distances.

The comparative merit of rail and truck transportation for fed cattle is a disputed point. The greater convenience afforded by the truck in loading the cattle at the farm or feedlot, at the hour most agreeable to the owner, and the greatly reduced shipping time have made the truck the favorite method of most feeders who are within easy trucking distance of the market. Since the heavy traffic on main highways and the poor condition of fences along secondary roads make it impractical to drive cattle to the railhead in most communities, they must be trucked from the farm or feedlot to the rail loading point even when they are shipped to market by rail. Once they are in the truck, the cattle can usually be taken directly to market in much less time and at little more expense than would be incurred if they were trucked only to the local station and reshipped by rail (see Table 255).

Table 255

Comparison of Truck and Rail Shipment of Fed Steers				
	Colorado Bulletin 422		Illinois Unpublished Data	
	(3-Year Average)		(Average of 3 Shipments)	
	Truck	Rail	Truck	Rail
Number of cattle	100	100	69[a]	70[a]
Shipping weight (lb)	835	833	1,044	1,052
Market weight (lb)	806	805	1,009	1,009
Shrinkage (lb)	29	28	35	43
Shrinkage (percentage)	3.6	3.4	3.4	4.1
Bruised carcasses	5.5	4.0	10[b]	10[b]
Hours in transit	3.17	7.0	7.1[b]	13.0[b]
Distance shipped (miles)	70	70	135	135

[a] Total cattle in three shipments.

[b] Average of two shipments.

CATTLE PER CAR OR TRUCK

The standard stock car is 8 feet 6 inches wide and either 36 or 40 feet long. The shorter car is billed with a minimum weight of 22,000 pounds and the longer one with 24,400 pounds. The shipper must pay freight on this minimum weight whether his load weighs that much or not. In computing freight charges over and above that for the minimum weight just mentioned, market weights are taken, less 800 pounds deducted for fill. No maximum weights are specified, and the shipper may crowd as many cattle into the car as he wishes. The average load for a 40-foot car is about 22 two-year-old slaughter steers or 28 yearlings. Cattle ship better if the car is comfortably filled, although overcrowding is more objectionable than underloading. If fewer than 15 mature cattle are to be shipped, they had better be partitioned off in one end of the car and the remainder of the space used for some other class of livestock. This will reduce the freight charges and will avoid undue jolting of the cattle by the sudden starting and stopping of the train.

Truckers engaged in hauling livestock have instigated more improvements in recent years than have railroads. There are a few instances where railroads give top priority in use of the lines to freight trains that are hauling large numbers of cars of live cattle and hogs from the Midwest to either coast. This makes possible a considerable reduction in the time from point of origin to destination. Livestock trucks deliver their cargo

Table 256

	Rail Cars[a]		Trucks[b]				
Average Capacity of Single-Deck Rail Cars and Trucks for Transporting Cattle and Calves							
Length (ft)	36	40	14	18	30	36	42
Calves							
350 lb	55	62	18	24	42	50	58
450 lb	46	51	15	20	34	41	48
Cattle							
600 lb	36	40	12	16	27	33	39
800 lb	30	33	10	13	22	26	31
1,000 lb	26	28	8	11	19	22	28
1,200 lb	22	24	7	9	16	19	22
1,400 lb	19	21	6	8	14	17	20

[a] Western Weighing and Inspection Bureau, Chicago, Illinois.
[b] Livestock Conservation, Inc.

with still greater dispatch, with the time generally being saved at both ends of the haul. The net result usually is less shrink.

Truck sizes have increased immensely, with double-decked bodies not uncommon, and this has made trucks more competitive with respect to cost for hauling cattle (Table 256). Express highways with few or no stoplights also result in faster and smoother rides with lower bruise rate. On some of the very long hauls, rest stops for truck-transported cattle, especially for stocker cattle, are required by the buyer, although not by law as is the case with rail-transported cattle.

SHRINKAGE

Shrinkage refers to the loss in weight that occurs between feedlot and market scales. It may be expressed either in pounds per head or in percentage of the weight before shipment. The percentage method is preferred because the amount of weight lost is usually in direct ratio to the size of the cattle.

Shrinkage during shipment is mainly due to excretions of urine and feces and, to a lesser degree, to tissue moisture given off by the lungs in breathing. Some of this loss is regained at the market from the feed and water consumed between arrival time and the time the cattle are sold and weighed. The amount of shrinkage expressed in percentage of the home or loading weight varies greatly between different loads of cattle. The main causes of this variation are (1) length of the journey, (2) condition of the cattle at loading time, (3) the degree of comfort en route, (4) the kind of feeds used, (5) the degree of finish of the cattle, and (6) the fill at market.

Length of Journey. The longer the journey the greater the shrinkage. The loss in weight, however, is not in direct ratio to the distance traveled, because the greatest loss occurs during loading and the first few miles in transit. (See Table 257.)

Condition of Cattle at Loading Time. As far as possible the condition of the cattle should be normal when they are loaded. Tired, hungry, or thirsty animals are in poor physical condition to stand the trip and are slow to recover upon reaching the market. Likewise, cattle that have consumed large quantities of green grass or succulent roughage or have taken too great a fill of water just previous to loading are in poor condition for the journey.

Degree of Comfort en Route. During extremes of hot or cold weather, shrinkage runs unusually high. Badly crowded cars or trucks and slow, rough runs with frequent stops are certain to cause considerable loss in weight.

Table 257

Effect of Length of Haul on Shrinkage of Grain-Fed Cattle Transported by Trucks[a]

Weight Classes	Number of Head	Average Full Weight Out of Lot	Cumulative Percentage of Shrinkage			
			After 25 Miles	After 50 Miles	After 100 Miles	After 200 Miles
Under 1,000 lb	11	954	1.5	2.2	3.1	3.9
1,000–1,099 lb	10	1056	2.1	3.0	3.8	4.1
1,100–1,199 lb	24	1139	1.8	2.6	3.4	4.1
Over 1,200 lb	15	1263	1.9	2.4	3.1	3.6
Group average	60	1122	1.8	2.5	3.3	3.9

[a] Unnumbered report, Chicago Union Stock Yards and Transit Company.

Note: The cattle in this study were hauled in groups of five in a truck that had a scales in one end, on which they were weighed individually after covering the respective distances. No opportunity to fill was allowed before weighing. Initial weights taken in morning before the cattle were fed.

Kind of Feeds Used. Cattle that have been fed large quantities of roughage such as grass and silage commonly lose more weight than those that have received a full feed of grain. Also cattle that have been fed laxative feeds such as soybean meal and alfalfa hay shrink more than cattle fed feeds of a less laxative nature. Contrary to popular belief, silage-fed cattle do not shrink as much as cattle fed dry roughages (see Table 258).

Degree of Finish.   As shrinkage is mainly due to the loss of excrement, it is more closely related to the size of the animal than to the degree of finish. In other words, thin 2-year-old steers lose about the same amount of weight per head during shipment as fat 2-year-olds that are 200 to 300 pounds heavier. However, the shrinkage per 100 pounds live weight is of course much higher for the thin cattle. Grass-finished cattle and cows are likely to have a much higher shrink, expressed as a percentage of the loading weight, than grain-fed cattle, because they have had a more laxative ration and because they are carrying less finish.

The Fill at Market.   The consumption of feed and water after arrival at the market is the most important factor in determining the net amount of shrinkage. This consumption in turn is influenced by several factors, the more important being (1) weather conditions at the market on the day of sale, (2) the length of time the cattle are in the pens before they are sold and weighed, and (3) the condition of the cattle upon arrival.

Smaller fills are obtained during cold, damp weather than on bright, warm days because cattle drink little at such a time and have little appetite for hay after it becomes wet.

The most satisfactory fills result when the cattle reach the market about daylight, which permits them to be penned and fed at about their usual feeding time. As a rule the market is not under way until 9 o'clock so that the cattle have at least 2 or 3 hours in which to eat and drink. Cattle that arrive after the market has opened may be in the pen only a few minutes before being sold and are weighed almost empty. Although the price paid for such cattle is often higher than would have been bid had the cattle taken on the usual fill, the buyer rather than the seller is most likely to profit from the late arrival. After cattle have been in the pens for 4 or 5 hours they stop eating and drinking and begin to lose rather than gain in weight. Consequently, good fills are associated with brisk, active markets, rather than with slow, long-drawn-out trading that runs into the afternoon session.

Cattle that arrive at the market tired and worn out from a long hard journey, or weakened by insufficient water and feed immediately before or during shipment, frequently lie down upon being unloaded and will not eat or drink to any extent until they have obtained some rest. If sold

Table 258

Shrinkage of Cattle During Transit[a]

Class of Cattle	Hours in Transit	Number of Shipments	Number of Cattle	Average Weight at Origin (lb)	Gross Shrinkage		Fill at Market		Net Shrinkage		
					Range (lb)	Average (lb)	Range (lb)	Average (lb)	Range (lb)	Average (lb)	Percent of Live Weight at Origin
Grain-fed											
Nonsilage	Less than 24	4	164	1,303	59–95	67	4–48	16	20–64	51	3.91
Nonsilage	24–36	59	1,853	1,167	47–128	85	19–52	37	18–88	48	4.11
Silage-fed	Less than 24	14	666	1,168	46–128	76	6–97	52	+7[b]–67	24	2.05
Silage-fed	24–36	4	169	1,204	84–121	101	50–64	58	27–75	43	3.57
Grass-finished											
Mixed range	Less than 24	21	1,511	700	19–84	37	1–56	22	+12[b]–71	15	2.14
Mixed range	24–36	17	872	848	27–118	72	−8[c]–55	18	19–114	54	6.37

Average shrinkage of grain-fed cattle in transit less than 36 hours, 3.62 percent.
Average shrinkage of grass-finished cattle in transit less than 36 hours, 3.88 percent.

[a] USDA Bulletin No. 25.
[b] Abnormal load, market weight exceeding loading weight.
[c] Abnormal load, sale weight less than unloading weight.

on the day of their arrival, their shrinkage is much above the average because of their small fill. If given time to recover, or if they are only hungry or thirsty, they probably take on such a large fill that buyers will bid lower or ignore them until the effects of the fill have largely disappeared. Best results are secured when the cattle are only moderately hungry and thirsty when received. In such condition they fill to a moderate extent, but not to the point where they invite unfavorable consideration.

Mud or Dirt on Cattle Hides.   At certain times of the year, usually in late winter or early spring, a percentage of slaughter cattle carry substantial but varying amounts of mud on their hides. A combination of factors may be involved, but usually muddy lots, at a time when alternate thawing and freezing temperatures occur, are the main cause. An equitable system for determining the amount of shrink that should be allowed by the seller or that should be expected by the buyer if the cattle are bought live—that is, not on a grade and yield basis—has not been developed heretofore.

Kansas workers recently conducted a study with 167 slaughter cattle, averaging 1,000 pounds, which were visually rated as to degree of muddiness in one of four categories ranging from clean to very muddy. They then determined the weight of dirt or mud actually adhering to the hair and hides. Obviously, dividing cattle into "muddiness" categories on the basis of a subjective or "eyeball" evaluation leaves room for disagreement, but if all cattle could be sorted into just four categories, it would at least be a step toward arriving at a fair basis for shrinking cattle in relation to the condition of their hides.

Figure 120 shows representative cattle in each of the four degrees of hide condition. The Kansas data showed that the weight of mud carried by cattle in clean, slightly muddy, muddy, and very muddy condition was 0, 5, 12, and 23 pounds, respectively. In reference to 1,000-pound cattle, this translates into shrink percentages of 0.0, 0.56, 1.27, and 2.31 for the four categories respectively. Removing the switch from the tail when cattle first go on feed, especially in slotted-floor confinement units, removes one site for mud and manure deposition and this practice is recommended.

LOSSES SUSTAINED DURING SHIPMENT

Losses resulting from injury or death of cattle in transit are not readily apparent to the casual observer. Because of their size and strength, cattle are less likely to be injured by the rough ride in freight cars or trucks than are swine and sheep. They also withstand unfavorable weather conditions encountered en route better than do the other classes of meat animals.

Clean 0.0 lb. Mud

Slightly muddy 5.7 lb. Mud

Muddy 12.8 lb. Mud

Very muddy 23.2 lb. Mud

Fig. 120. Representative cattle that illustrate four categories of mud deposition on the hair and hide. (Dr. Dell Allen and Dr. Herb Ramsey III, Kansas State University.)

Probably 99 percent of all cattle shipments arrive at the market with no dead or crippled animals. Even so, there is an estimated annual loss of more than $80 million owing to bruises, crippling, and death.

Statistics kept by the Department of Agriculture of all dead and crippled livestock received at the public stockyards of the United States

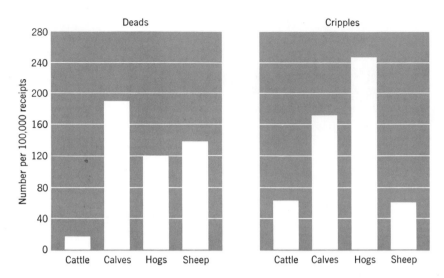

Fig. 121. Number of dead and crippled animals encountered in receipts of the different classes of livestock, as reported by the Bureau of Agricultural Economics for a typical year. (Livestock Conservation, Inc., Chicago, Illinois.)

show that a much smaller percentage of cattle are injured during shipment than of veal calves, swine, or sheep (see Fig. 121). Extensive data are unavailable comparing the losses of cattle when shipped by truck and by rail.

The relatively few dead and crippled cattle that arrive at the market are not the most serious losses for which shippers must pay through lower prices received for their cattle. Unfortunately, bruised animals cannot be detected by the buyer at the time of purchase; consequently he must buy all cattle on the basis that there will be a loss of $1.50 to $2 per head during slaughter as the result of bruises that must be trimmed out of the carcass beef (see Fig. 122). This loss represents 30 to 40 cents per 100 pounds live weight, or $60 to $90 a car or truckload.

An individual shipper can reduce losses from dead and crippled cattle by careful handling at the feedlot and during loading, but he cannot escape the losses resulting from bruising that occurs in transit or at the killer's yards. Consequently, every cattle feeder should give his enthusiastic support to the program of Livestock Conservation, Inc., a nonprofit, educational organization representing railroads, truckers, commission firms, packers, and stockyard companies, whose principal objective is to reduce losses in marketing livestock.

MARKETING COSTS

A knowledge of the items that make up the total marketing expense is highly desirable if the feeder is to estimate accurately the value of his cattle in the feedlot on the basis of current market quotations. Frequently he will need to choose between selling his cattle to a local order buyer or packer at a certain price ,or shipping them to the market on his own account.

Fig. 122. Location and estimated causes of bruises observed on carcasses of beef cattle during slaughter tests conducted by Wilson and Company. (Livestock Conservation, Inc., Chicago, Illinois.)

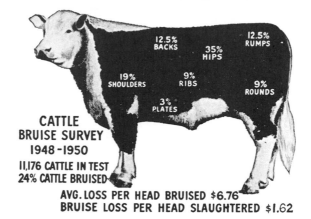

LOCATION OF CATTLE BRUISES

12.5% BACKS
35% HIPS
12.5% RUMPS
19% SHOULDERS
9% RIBS
9% ROUNDS
3% PLATES

CATTLE BRUISE SURVEY 1948-1950
11,176 CATTLE IN TEST
24% CATTLE BRUISED
AVG. LOSS PER HEAD BRUISED $6.76
BRUISE LOSS PER HEAD SLAUGHTERED $1.62

CAUSES OF CATTLE BRUISES

CROWDING BUMPING AND RUSHING   ←66%

TRAMPLING   ←14%

CANE WHIP CLUB   ←10%

HORNED CATTLE   ←3%

OTHER CAUSES   ←7%

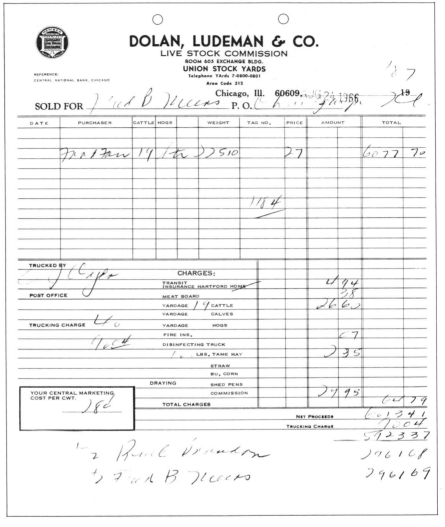

Fig. 123. Account of sale for a truckload of steers sold in Chicago. (Courtesy of Fred Meers, Champaign, Illinois.)

Obviously he must know the approximate marketing cost per hundredweight before he can make an intelligent decision.

Marketing costs are of two kinds, direct and indirect. The direct costs include the cash charges made by the transportation companies and marketing agencies. Indirect costs refer to the weight loss or shrinkage of the animals between feedlot or shipping point and the market.

Figure 123 shows the account of sale for a truck shipment of steers hauled a distance of 130 miles. Only the direct costs are considered in computing the average marketing expense per steer and per hundred-weight shown in Table 259. These costs are made up of (1) freight, (2) commission, (3) yardage, (4) insurance, (5) federal transportation tax, (6) feed, and (7) National Livestock and Meat Board deduction.

Freight.   Freight constitutes by far the largest item of direct marketing expense unless, of course, the cattle are sold at home. Freight charges have risen steadily, but additional services sometimes offset some of these increases. The freight charge varies, of course, with the distance traveled and the weight of the load. In general, freight charges are figured on the sale weight of the cattle, but if only a partial load is hauled the charge may be made on a mileage basis. Although it may appear that freight costs are eliminated by selling at home, the buyer in turn has to pay comparable freight costs; naturally the buyer will pay enough less for the cattle to offset these costs. If he is buying for a local or nearby firm he, of course, could bid more for the cattle, because his freight bill would be less in such a situation.

Commission.   Cattle shipped to a terminal market must be consigned to a livestock commission firm, of which there are many at each of the four or five large markets in or near the Corn Belt. Employees of the commission firm receive the animals from the stockyards company, drive them to the pens, and see that they have feed and water. When the market opens, a salesman, frequently one of the members of the firm,

Table 259

Marketing Expenses for a Truck Shipment of Steers (19 Head, Hauled 130 Miles)

Item of Expense	Cost, Entire Shipment	Cost per Head	Cost per Cwt	Percent of Total Marketing Cost
Freight	$90.04	$4.74	$0.400	58.3
Commission	29.95	1.58	0.133	19.5
Yardage	26.60	1.40	0.118	17.2
Insurance	5.01	0.26	0.022	3.2
Feed	2.35	0.12	0.011	1.5
National Livestock and Meat Board	0.38	0.02	0.001	0.3
Total	$154.33	$8.12	$0.685	100.0

shows the cattle to prospective buyers and, unless otherwise instructed by the owner, finally sells them for the highest offer he has received.

Immediately after the cattle are sold they are driven to the scales where they are weighed and locked in holding pens to await the orders of the purchaser. The weigh ticket is sent by messenger to the office of the commission firm, where the marketing expenses are computed and a draft is drawn on the firm's account for the net proceeds of the sale. The draft, together with a statement of the sale, is delivered to the shipper, if present, or is mailed to his home address or to his bank.

All settlements for freight, yardage, feed, and the like, and the collection of the money from the purchaser of the cattle are made by the commission firm without any trouble whatever to the owner. The charge for all these services is based on the number and, on some markets, the weight of cattle sold. Formerly buying charges for feeder cattle were considerably lower than selling charges, but currently they are the same on most markets. The selling commission varies slightly between different markets and is changed from time to time to meet new business conditions.

Considering the multiplicity of details attended to, the value of the product sold, and the amount of responsibility assumed by the salesman individually and the firm collectively, the commission charges are remarkably low, averaging less than 0.5 to 1 percent of the gross value of fed cattle. It is doubtful whether farmers take home so high a percentage of the sale value of any other major agricultural product as they do in the case of beef cattle.

Commission charges in auction sales are generally based on the gross returns of a sale and average about 3 percent of the gross returns and this covers yardage charges as well. At some sales the buyer reserves the right to refuse the highest bid, in which case the cattle are "passed out" with either no charge or up to 50 percent of the normal sale charge. In a few instances, commission charges are made on a straight per-head basis regardless of the age, weight, or value of the animal, but this system is relatively uncommon except where cattle are sold one head at a time.

Yardage.   A fee of 75 cents to $1.50 per head is charged by the stockyards company at the terminal markets for the use of the pens, watering facilities, scales, and so on. This is the main source of revenue for the stockyards company, and from it must come all funds for cleaning the pens, repairing fences, pavement, and buildings, as well as for interest, taxes, and dividends. Yardage is paid only once, regardless of how long the cattle are held before slaughter or reshipment.

Insurance.   Transit insurance is required if cattle are hauled by public transportation, and it is usually a worthwhile investment. As mentioned earlier, the percentage shipping losses are small; consequently the premium rate is correspondingly low. In addition to transit insurance, at all large markets a charge of 5 to 15 cents per car is made against both shipper and buyer to provide for insurance of the cattle against fire while they are in the yards. In this way a fund is maintained which is adequate to reimburse owners for the full market value of any animals destroyed. The wisdom of providing for such a fund was fully justified in October, 1917, when a fire in the Kansas City Yards destroyed nearly 10,000 head of cattle and calves. Fortunately the insurance fund built up in the preceding years was sufficient to reimburse fully every owner involved. Checks totaling $1,733,779.99 were mailed the day after accredited appraisals had been made.

Federal Transportation Tax. All agencies that haul livestock for the public are required to pay a 3 percent federal transportation tax, based on the transportation charges. Naturally these costs are passed on to the shipper in most instances.

Feed. After being unloaded the cattle are fed hay at the rate of about 10 pounds per head, or 200 pounds per car or truckload. This hay is purchased from the stockyards company at approximately twice the price of hay on the farm. However, the price includes delivery of the hay to the pens and placing it in the mangers before the cattle.

   If corn-fed cattle are held overnight before being sold, they are given a feed of shelled corn. Other feeds are usually available if the owner of the cattle wishes to use them.

National Livestock and Meat Board. A 10-cent-per-head deduction is made by commission firms in yards that cooperate with the National Livestock and Meat Board, a nonprofit organization that promotes the consumption of meat and meat products. Such promotion is conducted through all of the various mediums of communication such as demonstrations, radio, television, and newspapers. Meats research is sponsored in universities and private research organizations throughout the country to investigate subjects such as the value of meat protein in the human diet and palatability factors in meat. The deductions are voluntary and are usually matched with an equal amount by the slaughter plant that is buying the cattle.

   In summary, the total direct marketing costs per hundredweight, exclusive of freight and shrink, amounted to only approximately 30 cents in the example illustrated in Table 259. This figure would be about 60 cents in the mid-1970s, mainly because of increased transportation costs. It is doubtful whether so much selling and service can be obtained for so little cost in any other area of agricultural production.

REGULATORY MEASURES AFFECTING CATTLE MARKETING

A wide variety of federal and state regulations designed to protect livestock producers, processors, and consumers alike and to ensure a free and unhampered competitive market are in force and they directly and indirectly affect all segments of the beef cattle industry. In several instances regulations are in the process of being modified to meet the changing needs of several segments of the beef industry and alterations undoubtedly will continue, over time, as they have in the past. Beef producers should seek to remain current in their information concerning government regulations that relate directly to their operations.

Packers and Stockyards Act.   This act is a federal law, first enacted in 1921 and last amended in 1958, which is administered by the U.S. Department of Agriculture. Its main purpose is to prescribe rules of fair competition and fair trade practices for all firms or persons engaged in livestock marketing and meat packing in interstate or foreign commerce. Included are dealers, stockyards, and auction markets, regardless of size, if they engage in interstate business. Some of the regulations under the act which provide marketing protection are: accurate weights, protection of a stockman's funds, full and accurate accounting, reasonable and adequate facilities at fair and nondiscriminatory rates, and sale of cattle under open competitive conditions. Firms or dealers covered by the act are required to post bond to ensure full payment for cattle sold. Many states have regulations concerning bonding that are even more stringent than those imposed by the federal act. The act requires that the seller or buyer be given a true written account of any sales or purchase transaction, including an itemized list of all charges.

The regulations of this act are enforced by representatives of the Packers and Stockyards Administration of the U.S. Department of Agriculture, who visit markets, packing plants, and buying stations on a periodic but unscheduled basis. Such representatives are based in 15 area offices throughout the country. Anyone desiring to register a complaint to a Packers and Stockyards Administration supervisor should do so, in writing, as soon as possible but no later than 90 days after a transaction has been made.

Federal Meat Inspection Act.   Federal meat inspection legislation was first enacted in 1890, mainly to assure the foreign importers of U.S. meat that the meat was not diseased. In 1891 and 1895, meat inspection was extended to include the inspection of live animals prior to slaughter. The first full-protection act, the Meat Inspection Act of 1906, included both ante-mortem and post-mortem inspection along with other features, most of which remain in effect today. Only plants that dealt in interstate and foreign meat trade were covered.

In 1967 the Federal Meat Inspection Act, also known as the Wholesome Meat Act, was passed. This act effectively extended meat inspection to all plants, including those selling meat intrastate and even local abattoirs and freezer plants. Actually the act specified that states must enact legislation providing inspection standards and procedures at least equal to federal requirements by December 1, 1969, or they must submit to federal inspection of all meat and plants in each state. Most, but not all, states have enacted such legislation. In effect, the Wholesome Meat Act has brought all processors under a uniform and effective meat inspection service. Only farm-slaughtered beef, intended for consumption on the farm and not for sale to anyone, is legally excluded from inspection today.

Obviously the intent of this legislation is to safeguard the health of meat consumers by eliminating diseased and otherwise unfit meat from consumer channels. Sanitary slaughtering and processing practices are assured and use of harmful additives is prevented. False and misleading labeling is made illegal.

All cattle must pass ante-mortem inspection by a veterinarian employed specifically for the purpose. During slaughter, all carcasses and viscera are inspected. If abnormal conditions are found, including severe bruising, the carcass is tagged and given more careful inspection. Certain parts may be trimmed—in instances of severe bruising and grub damage, for example. Carcasses that are passed unconditionally are stamped with a circular imprint, made with edible purple fluid, bearing the plant number and the abbreviated inscription "U.S. Insp'd and P's'd" and imprinted at numerous places on the carcass so that each wholesale cut carries the stamp.

Meat imported from foreign countries as fresh beef must meet the same requirements as domestically processed meat. Inspectors from the United States annually visit the foreign plants previously approved for slaughtering beef intended for the U.S. import trade. Other U.S. inspectors check incoming shipments as they arrive at U.S. ports.

Buyers of slaughter cattle may purchase animals suspected of serious injury or disease on a basis of "subject to inspection," thereby agreeing to pay a specified price if the animal and its carcass pass complete or partial inspection.

The cost of federal meat inspection is borne by the government and the inspection actually is done by personnel of the Consumer and Marketing

Service of the U.S. Department of Agriculture. When states perform their own inspection, the federal government enters into cost-sharing arrangements, usually providing 50 percent of the cost along with technical assistance.

Food, Drug and Cosmetics Act.   This act, passed in 1938, provides for the Food and Drug Administration (FDA) and its regulatory functions and is administered by the Department of Health, Education, and Welfare, and not by the Department of Agriculture as some cattlemen believe. The function of the FDA that directly affects cattlemen is that covering residues in beef carcasses, and even in manure, which might result from using additives such as hormones, pesticides, and even antibiotics and other drugs. Feed manufacturers who process and sell fortified supplements to cattlemen must have approval of the FDA before selling feeds containing additives. The FDA may actually prosecute individuals for selling slaughter cattle containing tissue residues. FDA regulations governing withdrawal times for certain additives prior to slaughter have been mentioned elsewhere.

Animal Disease Control.   The Animal Health Division of the Agricultural Research Service, U.S. Department of Agriculture, is responsible for administering the regulations relating to control and eradication of animal diseases. These include matters related to interstate movement of cattle and to import and export trade of live cattle. The records of meat inspectors at slaughter plants are available to disease control officials in the interest of detection of carrier animals and their source. Federal inspectors also function at stockyards and auction sales and they receive reports from practicing veterinarians. Quarantines are imposed on farms, feedlots, and ranches when cases of certain diseases are detected, by whatever means, until the condition has been corrected in order to prevent spread of the disease. Cattlemen are sometimes caught up in an area quarantine, even though their own cattle may be free of the disease.

In extreme cases it may be necessary to destroy and dispose of all animals in an area, as has happened in local outbreaks of foot-and-mouth disease and rinderpest. Animals found to be infected with brucellosis are required to be slaughtered immediately, but carcasses must not be destroyed. A complete embargo may be imposed on live cattle or fresh meat from countries whose cattle are afflicted with certain diseases.

Space prohibits a more thorough discussion of the numerous federal and state regulations related to movement of cattle, quarantine, immunization programs, indemnities paid for animals destroyed, and penalties for violation of these regulations. All cattlemen owe it to themselves to keep abreast of the current status of the many regulatory measures in effect.

CHAPTER 27
DISEASES OF
BEEF CATTLE

The incidence of disease in beef cattle is low compared with the disease rate of the other important species of livestock. Nevertheless losses do occur and may be of considerable importance in individual herds. The monetary loss from cattle diseases in the United States was estimated at $300 million in a 1965 report of the Animal Health Institute, with values per head having at least doubled since that date. The beef cattle industry cannot afford such a loss.

Data on the incidence of specific diseases in beef cattle are limited and subject to errors in diagnosis on the part of the farmer, rancher, or veterinarian. A rather comprehensive survey covering most of the problems encountered in the beef cattle business, including diseases and parasites, was conducted by the Washington station in cooperation with the Research Committee of the American National Cattlemen's Association. Much of the Corn Belt area, the northeastern, and the upper southeastern states are not included in the survey. However, the 1,588 questionnaires, representing 502,616 head of cattle, supply information that serves as an adequate basis for discussing the incidence of disease among beef cattle.

Table 260 summarizes the data from these questionnaires with respect to incidence of the various nonnutritional diseases encountered. An annual mortality rate of 0.59 percent from nonnutritional diseases and ailments was reported, with 60 percent of all death losses being caused by pneumonia, calf scours, shipping fever, and blackleg, in order of importance.

Ailments of a noninfectious nature and nutritional diseases were tabulated separately in the survey, with results summarized in Table 261. An annual mortality rate of 0.32 percent was attributed to diseases and ailments of a noninfectious nature, with 71 percent of the death losses in this category resulting from bloat, poisonous plants, and urinary calculi, in order of importance.

Fortunately the ailments, both infectious and noninfectious, that occur most frequently in beef cattle yield rather readily to simple treatment. There are, however, a few diseases that are of a serious nature. In addition, minor ailments if not properly treated may quickly develop into serious disorders that may prove fatal. Consequently the beef cattle man should want to become proficient in detecting the symptoms of the **799**

Table 260

Incidence of Beef Cattle Nonnutritional Diseases and Ailments[a]

Disease	Herd Incidence of Disease (Cattlemen Reporting Disease)[b] (%)	Incidence of Disease (% of Total Cattle Afflicted by Disease) (%)	Herds Reporting the Disease in Which the Veterinarian: Diagnosed the Disease (%)	Herds Reporting the Disease in Which the Veterinarian: Treated the Disease (%)
Pink eye	46.2	3.07	3.0	2.1
Calf scours	30.8	1.09	3.9	2.5
Shipping fever	11.7	0.84	2.5	2.2
Foot rot	28.3	0.74	4.8	3.8
Pneumonia	23.2	0.46	7.5	6.0
Warts	21.3	0.33	2.0	1.6
Cancer eye	36.8	0.32	7.2	8.6
Lumpy jaw and wooden tongue	25.4	0.22	5.5	6.2
Brucellosis	8.8	0.18	4.2	2.8
Leptospirosis	0.9	0.17	0.6	0.4
Calf diphtheria	8.1	0.13	3.8	2.8
Sunburned udder	3.5	0.11	0.0	0.0
Prolapse of uterus	13.0	0.10	4.0	4.9
Navel infection	7.4	0.09	1.0	1.1
Blackleg	10.2	0.04	2.6	2.1
Vaginitis	1.8	0.03	0.8	0.8
Brisket disease	0.4	0.01	0.0	0.0
Mucosal disease	0.7	0.01	0.3	0.3
Red water disease	1.3	0.01	0.4	0.4
Tetanus	0.7	0.01	0.3	0.3
Circling disease	0.5	0.00	0.2	0.3
Mastitis	0.4	0.00	0.2	0.2
Retained placenta	0.1	0.00	0.0	0.0
Anthrax	1.1	0.00	0.3	0.3
Pulmonary emphysema	0.06	0.00	0.0	0.0
Rabies	0.3	0.00	0.1	0.1
Vibriosis	0.06	0.00	0.1	0.1
Hardware disease	0.6	0.00	0.1	0.2
Johne's disease	0.2	0.00	0.1	0.0
Diseases not diagnosed	6.7	0.05	0.1	0.1
Other diseases	3.2	0.02	1.4	1.2
Total		8.03		

[a] Washington Experiment Station Bulletin 562.

[b] This column will total over 100% because many cattlemen reported having several diseases in their herds.

Table 261

Incidence of Beef Cattle Noninfectious Ailments and Nutritional Diseases[a]

Disease or Ailment	Herd Incidence[b] (%)	Percent of Cattle Afflicted
Bloat, pasture	20.91	0.39
Vitamin A deficiency	2.58	0.35
Bloat, feedlot	10.01	0.15
Poisonous plants	10.14	0.08
Urinary calculi	10.39	0.07
Grass staggers	3.15	0.05
Salt sick	1.01	0.04
Fluorine poisoning	0.63	0.03
Pine needle abortion	1.32	0.02
White muscle disease	1.64	0.02
Iodine deficiency	2.02	0.01
Phosphorus deficiency	1.45	0.01
Sweet clover disease	0.57	0.01
Alkali disease	0.88	0.01
Oat hay poison	0.69	0.01
Poisons, chemical	0.63	0.01
Anemia	0.94	0.01
Acetonemia	1.13	0.00
Molybdenum deficiency	0.06	0.00
Rickets	1.26	0.00
Oak poisoning	0.19	0.00
Prussic acid poisoning	0.13	0.00
Milk fever	0.38	0.00
Diseases not diagnosed	0.38	0.01
Other diseases	0.69	0.04
Total		1.33

[a] Washington Experiment Station Bulletin 562.

[b] Percent of cattlemen reporting the ailment or nutritional disease.

common diseases and ailments so that proper treatment may be given before the illness makes extensive progress. Perhaps even more important is a knowledge of preventive measures that can eliminate or reduce the problem in the first place.

Treatment should usually be prescribed by a competent veterinarian. Some of the minor ailments, such as scours and bloat, are of such common occurrence as to warrant keeping on hand a supply of prescribed medicines which the herdsman may administer himself.

DISEASES COMMON TO BEEF CATTLE

ANAPLASMOSIS

Anaplasmosis is a serious blood disease of cattle that has not been encountered in some sections of the United States. It is believed that the disease has been present in the southern states for many years, but was not distinguished from Texas fever until that disease had been definitely eradicated. Anaplasmosis resembles Texas fever in being caused by a blood parasite that is carried from animal to animal, mainly by ticks, horseflies, and mosquitoes, but the protozoan that causes anaplasmosis may be transmitted by more than 25 species of insects. This makes control and eradication very difficult. Anaplasmosis has been observed in nearly all of the southern states, but the few outbreaks that have occurred in the North have been confined mainly to shipped-in cattle that presumably were infected when purchased.

The disease is especially severe in mature cattle, resulting in death losses of 30 to 50 percent of the animals infected. Calves and yearlings seldom are visibly affected, but may have the disease in mild form and make a complete recovery. Animals that have the disease and recover, regardless of age, are immune to future infection but usually remain carriers throughout life, and when introduced into clean herds they are a source of infection for other cattle. Unfortunately, several species of deer and other game species also harbor the disease and insect vectors may transmit the disease to cattle.

Anaplasmosis may occur in either chronic or acute form. In the acute form the first symptom is a high temperature of 103°F to 107°F. After two or three days the temperature falls to subnormal, breathing becomes difficult, the mucous membranes are pale and yellowish due to severe anemia, and the muzzle is dry. Urination is frequent, but the urine seldom is bloody. The patient usually is constipated and either blood or mucus, or both, may be voided with the feces. In acute anaplasmosis death usually occurs 1 to 5 days after the first symptoms are observed. In the chronic form the infected animals live longer and a greater percentage recover. Recovery of mature animals is usually very slow. Cows with the disease often become belligerent and hostile when moved or, at the other extreme, they may fall back to the rear of a herd and may even fail to keep up with the herd if driven for several miles. Such a crude test often sorts out most of the sick animals. Laboratory tests of blood samples should be made to verify a tentative diagnosis, and state and federal veterinarians should be notified.

No specific treatment has been perfected for anaplasmosis; consequently treatment varies among veterinarians. Good nursing and care, of course, are vital to recovery. Daily transfusions with 2 gallons of normal bovine blood usually save an infected animal but this treatment is of course expensive. Drugs that stimulate blood formation are sometimes helpful, and tetracyclines have been reported beneficial in the early stages of anaplasmosis. Daily treatment with these antibiotics at the rate of 5 milligrams per pound body weight, in the feed, for 45 days has been used successfully.

A vaccine for anaplasmosis was approved and released for use in 1965. Young cattle are given two treatments at 6-week intervals, and results after testing under commercial ranching conditions appear quite favorable. Isolation of carrier animals and control of insect vectors is advisable when an outbreak occurs within a herd.

ANTHRAX

Anthrax, also called splenic fever, is a highly infectious disease and often fatal. Not only is it transmissible from one animal to another, but it may spread rapidly to all species of livestock and even to man. Historically it has proved to be one of the worst scourges of animal life. This disease existed among European cattle as long ago as the early seventeenth century and has caused enormous losses. Its presence in the United States is, in general, confined to certain local areas of rather low, moist land more or less mucky in character. In such soil the spores of anthrax remain potent for years. Hay grown in such soil may carry the germs of the disease far and wide. Hides have been responsible for outbreaks of anthrax among the cattle in the vicinity of tanneries.

Death from anthrax is very sudden; finding a dead animal in the pasture or lot is commonly the first sign of the presence of the disease. In all cases of sudden death without forewarning symptoms of illness, anthrax should be suspected and the carcass handled with extreme caution until the death can be attributed to some other cause. The characteristic signs of death due to anthrax are (1) a tarry consistency of the blood, which is blackish in color and fails to clot firmly, and (2) an enlargement of the spleen of two to five times its normal size. A positive diagnosis can be made only by examining a blood sample under a microscope for the presence of the specific organism, *Bacillus anthracis*.

When anthrax is suspected a veterinarian should be called without delay. Burning or deeply burying the carcass of the dead animal, and prompt vaccination of all other animals in the herd with antianthrax serum are the best known methods of preventing spread of the disease.

Following a diagnosis of anthrax, animals showing high temperatures should be separated from the others and given serum plus antibiotics, whereas others may be treated with serum and spore vaccine. Those receiving the serum and antibiotics should be revaccinated after several weeks with both serum and vaccine.

BLACKLEG

Blackleg is an acute, infectious disease attacking mainly cattle between 6 and 24 months of age. Man apparently is not susceptible to this disease. The main symptoms are a marked lameness and pronounced swellings over the shoulders, thighs, and flanks, caused by the formation of gas in the subcutaneous tissues. When pressure is applied to these swellings a peculiar crackling sound is heard, which is the characteristic symptom of blackleg. The disease runs a rapid course and almost always terminates fatally in 12 to 36 hours. Like anthrax, it is more or less restricted to certain sections of the country and even to individual fields where the soil is infested with the spores of the causative organism and where outbreaks commonly occur annually unless prevented by vaccination.

Blackleg is common in the western half of the United States and is especially prevalent throughout the Southwest and on the eastern slopes of the Rocky Mountains. Outbreaks have been reported in nearly all of the Corn Belt states, but there the disease is confined to limited areas.

Prevention rather than treatment is the practical method of combating the disease. All young cattle brought into a region known to be infested with blackleg should immediately be vaccinated with blackleg bacterin, made by treating cultures of the blackleg organism with formalin to destroy their infectious properties but not their antigenic value.

Calves born in a region where blackleg has occurred even infrequently during the preceding 10 years should be vaccinated when they are about 6 months old. Usually vaccination is performed with a combination blackleg and malignant edema vaccine during late spring when other operations such as branding and dehorning are done, and often again at weaning time. The cost of vaccination is so small that most buyers of stocker and feeder cattle of unknown origin routinely vaccinate all new cattle upon receipt simply to ensure protection.

BRUCELLOSIS

Brucellosis is also called Bang's disease, contagious abortion, or simply abortion. Abortion of the infectious type is one of the worst problems for cattle breeders. It is caused, usually, by a specific organism, *Brucella abortus*, although other uterine infections of a more general character may cause a number of abortions in a herd of cows.

Apparently infection usually gains entrance through the mouth, having been scattered in the uterine discharge of an aborting animal over grass, hay, and other feed materials, or into the water source. All aborting cows should be segregated from the herd and quarantined until at least 3 weeks after all uterine discharge has ceased, or until a negative blood test has been obtained. Some authorities hold that an infected cow is never rid of the organisms, although she may eventually become sufficiently immune to carry a calf full term. Bulls as well as cows may contract the disease, and again the avenue of infection is thought to be the digestive tract. The prudent breeder, however, refrains from mating a bull that he knows is clean with cows that are suspected aborters, to prevent possible contamination from the genital tract.

The blood agglutination test furnishes a dependable basis for the control and eradication of brucellosis. In this test, which uses about 10 cubic centimeters of blood drawn from the jugular vein, various dilutions of the blood serum are mixed in small test tubes with specially prepared cultures of the *Brucella* bacillus and allowed to stand 24 to 48 hours. If the animal is negative the bacteria remain in suspension, keeping the solution cloudy, but if it is positive the bacteria are precipitated to the bottom of the tube, leaving the fluid above them clear. Samples intermediate between positive and negative are called "suspects," and animals from which these came must be rebled and tested again.

A much faster agglutination test is now in use that appears promising. This test is called the buffered Brucella antigen (BBA) test, also known as the card test. The tail vein may be used as the source of blood to be used in testing for this disease and for others such as leptospirosis. The need for a squeeze chute is thus eliminated and the cattle are not unduly disturbed. More important, results of the test can be obtained within a few minutes. Those animals that react positively—and they are either positive or negative, never "suspect" as in the older test—are branded with a "B" on the jaw and slaughtered. Herds in which positive reactors are found are quarantined until all members of the herd have been retested and found negative.

Another test known as the ring test, applied to milk samples, is impractical for beef herds but is useful in screening dairy herds for possible infected animals.

There is no known cure for brucellosis. Immunization of heifer calves between 2 and 10 months of age with a standardized live vaccine, *Brucella abortus strain 19*, is a reasonably effective immunizing agent. In older animals, immunization may effectively prevent the disease, but it interferes with blood agglutination tests, thus making later blood tests difficult to interpret. For this reason females older than 6 to 8 months are seldom vaccinated. In some states, immunization of heifer calves is being phased out but this leaves the potential hazard of a return of the disease. Bull calves should not be vaccinated. Eradication of brucellosis is usually most successful if pursued on an area plan, operated under joint federal-state arrangements. The numerous individual herds that are tested annually by a state or federal veterinarian and found to be free of reactors in two successive years are designated as "certified herds," and it is these herds, if kept closed or free of outside introductions, that often are no longer being vaccinated. States where cases of the disease are found in less than 1 percent of the herds or less than 0.2 percent of all animals tested are classified as "certified-free" while states in which less than 5 percent of the herds or 1 percent of all cows tested are infected are classified as "modified-certified." Federal and state regulations relating to the transportation of breeding cattle may vary from area to area and are subject to change, as are the various approved state plans for eradication of the disease. Every breeder of beef cattle, therefore, should keep abreast of the latest regulations in effect in his area.

The handling of infected animals or the drinking of milk from infected cows can spread the disease to man in a form known as undulant fever. The disease is not often fatal in man but a long, debilitating illness results.

FOOT-AND-MOUTH DISEASE

Foot-and-mouth disease is not prevalent in either the United States or Canada. It has gained access to the United States on ten different occasions since 1870, but in each instance it was promptly stamped out by the destruction of all infected and exposed animals. Such measures are justified in view of the terrible handicap imposed by this plague upon the livestock industry of the countries of Europe and South America where the disease is constantly present. Foot-and-mouth disease is not transmissible to man.

The disease was introduced into Mexico in 1947 when a shipment of Brahman bulls was smuggled into the country from Brazil, and only through heroic efforts and the expenditure of more than $100 million by the Mexican and United States governments working cooperatively was the disease brought under control. At first an attempt was made to stop the disease by the slaughter of all infected and exposed animals, but it was soon obvious that this plan was impractical, and a program of vaccination and rigid quarantine and inspection was inaugurated. Such plans are in use throughout South America.

Foot-and-mouth disease is seldom fatal, but it causes enormous losses through the loss of weight suffered because of lameness and while the animals' mouths are so sore that they almost refuse to eat. Recovery is very slow, and many of the animals are hopelessly ruined from the standpoint of the feedlot. There is no specific cure for the disease, although vaccination is being used in many parts of the world with reported success. Federal-state agreements for prompt cooperative action in case of an outbreak of foot-and-mouth disease can quickly be put into effect. Indemnities are paid in the eradication programs that follow a diagnosis of the disease. Fresh, chilled, or frozen meat or live cattle may not be imported from countries afflicted with this disease.

GRASS TETANY

Grass tetany is a metabolic or nutritional disorder causing an increasing number of death losses among mature, lactating cows throughout the regions where the cool-season grasses are grown. Years ago this condition was commonly called grass staggers. A similar or related disorder, known by the same names, has caused increased losses in stocker calves grazing mainly winter small grain pastures.

Mature, heavy-milking spring-calving cows turned onto early green growth of predominantly tall fescue, brome, or orchard grass pastures are most likely to be affected. A dead cow, often one of the best in the herd, found in the pasture may be the first indication of trouble. Close daily observation of cows with calves, after first turning out on spring grass, sometimes makes it possible to detect early symptoms of the disorder. Affected cows are in an excitable state, with erect ears, and may even attack a person. They may be blind and, if driven or roped, often go down after the excitement. Other symptoms are grinding of the teeth, trembling, and possibly deep coma followed by sudden death. Milder cases resemble milk fever. "Downer" cows seldom respond to treatment although death may possibly be prevented by injection with calcium gluconate. Early detection by blood analysis, followed by injection with magnesium sulfate, may save afflicted animals. Blood serum magnesium levels of 1.0 milligram or lower per 100 milliliters of serum, compared with a normal level of 2.25 milligrams, are associated with tetany.

A discussion of the agronomic aspects of the grass tetany problem can lead to development of a preventive program. Grass pastures that are beginning spring growth on soils high in nitrogen and potassium and low in magnesium seem to offer the greatest danger. Cloudy, wet, windy days with daytime temperatures between 40° and 60°F and soil temperatures below 50°F are associated with a high incidence of this malady.

Missouri researchers in nutrition, agronomy, and veterinary medicine have prepared a list of recommendations for prevention of grass tetany in cows that seems to cover all possibilities, given the present state of knowledge concerning this perplexing problem. The Missouri recommendations are as follows.

1. Provide supplemental magnesium from October through May. Provide mineral supplements continuously in several convenient locations. Any of the following mixtures is suitable and choice depends on price, convenience, and availability.
   Mineral mixtures—(a) Iodized salt, 30; bone meal or dicalcium phosphate, 30; magnesium oxide, 30; and dried molasses, 10 parts. This supplement contains 18 percent magnesium. (b) Iodized salt, 75; and magnesium oxide, 25 parts.
   Concentrate mixtures (self-fed)—(a) Ground grain, 65; magnesium oxide, 20; salt, 15 parts. (b) Cottonseed or soybean meal, 65; magnesium oxide, 20; salt, 15 parts.

2. Refrain from grazing spring pastures until growth reaches 8 to 10 inches in height, or provide legume hay, ad lib, if grazed earlier.

3. Graze legume or legume-grass pastures first.

4. Graze less susceptible animals on high-risk pastures (heifers, dry cows, cows with calves over 4 months of age, and stockers are less susceptible).

5. Precautionary measures are especially called for in cows with a previous history of grass tetany.

6. Improve the magnesium content of soil on which pastures are grown. Dolomitic limestone is a good and economical source of magnesium.

7. Pastures may be dusted with magnesium oxide at the rate of 25 to 30 pounds per acre at the beginning of the critical 2-to-4-week spring period.

8. Nitrogen and potassium fertilizers should be applied in split applications in spring and early fall, rather than in one spring application.

The winter small grain pastures—wheat, rye, and oats—are at times even more dangerous than the cool-season grasses. Fall-calving cows and thin stocker calves are most susceptible. The magnesium supplements recommended for spring-calving cows on grass pastures are suitable in this case as well. Reduction of the high rumen-ammonia levels that result from grazing lush winter small grain pastures by diluting the high-protein forage intake with dry roughages seems to be an added protection against losses from grass tetany or wheat pasture poisoning in fall and wintertime.

JOHNE'S DISEASE

This disease, known also as paratuberculosis, is believed to be on the increase. It is usually spread by the droppings of infected animals which contaminate pastures, water sources, and even feedbunks. A severe, chronic and intermittent diarrhea is responsible for the ease of contamination, and this diarrhea, accompanied by extreme loss in weight, is the principal outward symptom. Owing to the long incubation period, calves that contract Johne's disease usually do not show symptoms until they are about 2 years old or, in the case of heifers, after they have dropped their first calf. Appetites remain good, temperature and pulse rate are unchanged, but milk production decreases. The animals soon appear unthrifty and emaciated, continuing to scour and waste away until death results in many cases. Because these symptoms are similar to the effects of internal parasites and malnutrition, Johne's disease is rather difficult to diagnose. An interdermal injection of johnin, prepared and used in the same way that tuberculin is used for the tuberculosis test, with resultant thickening of the skin within 48 hours indicates a positive reaction.

No satisfactory treatment is known and satisfactory vaccines are not yet available. Johne's disease has not been reported in man.

LEPTOSPIROSIS

The incidence of this disease is not well determined, but it has been found in 48 states and seems to be spreading rapidly. The disease is spread by urine from infected animals, and the principal point of invasion is the mucous membranes. Because the kidneys are attacked, the urine may be dark red or wine-colored. Abortions may occur when pregnant animals are infected, causing this disease to be confused with brucellosis. Leptospirosis has a short course of 3 to 10 days, with death losses running as high as one-third in calves.

Treatment is of doubtful value, although antibiotics sometimes give relief. A bacterin has been produced, but protection through its use is not always certain. The difficulty is that about 25 species of the *Leptospira* genus have been identified. Vaccines or bacterins for the *L. pomona, L. hardjo,* and *L. grippotyphosa* species have been prepared and often are used effectively in combination. A herd with a leptospirosis problem should institute an annual vaccination program, with all cows and yearling heifers being vaccinated 3 weeks before the beginning of the breeding season. Bred cows and heifers should not be vaccinated. A serological test has been developed for diagnostic work, and it has become a generally recommended practice to use the blood samples drawn for brucellosis testing to screen a herd at the same time for leptospirosis.

A few cases of this disease have been found in man. Until the extent of the danger to human beings is known, the usual antiseptic precautions should be taken in working with cows that abort or show other symptoms that may indicate presence of leptospirosis.

RESPIRATORY DISEASE COMPLEX

Several diseases, including infectious bovine rhinotracheitis (IBR), parainfluenza-3 (PI3), and bovine virus diarrhea (BVD), are virus diseases which, together, are often spoken of as the respiratory disease complex and are frequently associated with shipping or stress fever. Symptoms are high fever, nose and eye discharges, excessive salivation, coughing, lesions in the membranes of the respiratory and digestive tracts, and diarrhea.

Rhinotracheitis or IBR or "red nose" is the most important of the various diseases of this complex, as it has spread throughout the large feedlots of the West where large cattle numbers are congregated. Weight losses are great owing to severe dehydration. The inflammation is usually confined to the respiratory system, and death often results from suffocation and secondary pneumonia.

The second most important of the virus diseases of this complex is *parainfluenza-3*. This virus is widespread in the cattle population but seldom exerts its influence except when calves are stressed. The infection by itself seldom causes death but often leaves the respiratory membranes vulnerable to bacterial invasion. One of the newest of PI3 modified live virus vaccines is administered by nasal spray, and its antibody response generally is developed in 4 or 5 days. Thus the vaccination is not highly effective if administered after calves have made a long, stressful trip to the feedlot. Rather, the vaccination should be a part of a backgrounding program initiated on the farm or ranch where the calves were produced and perhaps even prior to weaning.

Bovine virus diarrhea, the third member of this disease complex, is an infection confined largely to the digestive tract and occurs in calves more often than do the other diseases of the complex. Late winter or early spring is the season of greatest incidence. Symptoms are a rise in temperature to about 106°F and then a drop to nearly normal. There is loss of appetite, nasal discharge, and diarrhea that progresses with the disease until the feces contain much mucus and blood. Ulcers may be found in the nostrils, on the muzzle, lips, and gums, and in the mouth. Sometimes there are congested areas and hemorrhage in the colon. Death losses in a herd may amount to 20 to 50 percent of the young stock, and 100 percent losses have been reported. There is a chronic form of this disease that may leave the calves with elongated hooves and arched backs. Modified live virus vaccines are available for this viral disease. In fact, combination vaccines are now available for all three entities of this disease complex but, to be effective, they must be used before stress occurs, to allow time for development of antibodies. Directions as to time of administration are found on the labels of containers in which the vaccines are marketed. No completely successful control has been developed for any of the diseases in the respiratory disease complex, however. Antibiotics and sulfa drugs are often used but serve only to reduce the danger of secondary infections such as pneumonia.

So far as is known, no danger to man exists with respect to this group of diseases.

SHIPPING FEVER

This disease has been called hemorrhagic septicemia, Pasteurellosis, feed-lot pneumonia, and stress fever, as well as shipping fever. Actually "stress fever" is a fitting name because any sudden change, such as weaning and shipment or change in weather, which puts a stress or strain on the cattle, particularly calves, makes them susceptible to the infective agent. A single causative organism has not been identified; rather, it is believed by most to be a mixture of viruses and bacteria. The viral infections just discussed under the heading "respiratory disease complex" above are included in the broad complex commonly called shipping fever. If the viral infections are arrested, the problem may advance no further and it may be said that the disease has been prevented. The term "shipping fever" is more inclusive, however, and covers the advanced stages of the disease, including even pneumonia and complications leading to death.

Shipping fever spreads easily from animal to animal upon contact. Contaminated stockyards, sale barns, trucks, and cattle cars, and cattle-working equipment such as chutes and scales, are all likely sources of infection. It is not uncommon for a new shipment of stockers or feeders to spread the disease to cattle that are already on feed.

As indicated earlier, shipping fever is one of the big four in causing cattle death losses. Indirectly it may account for additional losses as calves or yearlings that have been weakened by shipping fever often fall victim to pneumonia.

Shipping fever is a respiratory infection that usually appears in cattle within 10 days of arrival after shipment or within 14 days after first exposure. A tired, hang-dog appearance is characteristically an early symptom. Appetites are dull and a mild cough is usually present, becoming more evident when the calves are moved about. Temperatures rise to 107°F in the more serious cases. Since easily recognized symptoms may not always be present, the use of a rectal thermometer to check the temperature of all calves in a drove is the only way to find those really needing treatment. In the usual outbreak of shipping fever many calves recover spontaneously but some—and this varies with the virility of the infection and the amount of stress the calves have undergone—contract pneumonia and die if left untreated.

Because shipping fever is apparently brought on by an impaired condition of the animal, preventive measures rather than a specific cure should be the first concern of the cattleman. Procedures to be avoided are: hard driving, excessive exposure to dust, too much "cowboying" or undue disturbance during loading and unloading, overcrowding in trucks or cars, insufficient bedding, too long a trip without a rest stop, inhalation of diesel truck exhaust fumes, and use of unpalatable rations and unfamiliar watering arrangements upon completion of the journey.

If the cattle arrive during cold weather, especially if it is wet and stormy, adequate shelter should be provided. The cattle should have sufficient time to recover from their trip and become accustomed to their new surroundings before they are dehorned, implanted, subjected to detailed sorting, and so forth. The ration for the first few days should be very palatable and medium to high in energy content. Calves will eat little, so each bite should be nutritious and fortified with antibiotic-sulfa combinations. During favorable weather the cattle may have the run of a rather short pasture; hay, if fed at all, should be fed on the grass.

The use of serums and mixed bacterins after arrival at the final destination is of doubtful value in the prevention of stress or shipping fever, because their effectiveness at this time is questionable in view of the present uncertainty concerning the real cause of the disease. If an outbreak of shipping fever occurs in spite of all precautions or if the buyer had no control over the previous treatment of a shipment of calves, prompt diagnosis and treatment are necessary to keep both weight and death losses low. Immediate treatment with antibiotics or combinations of sulfa drugs and antibiotics on arrival of all calves is sometimes practiced, but unless the calves have had an unusually long and rough trip, such treatment probably is unwarranted. Rather, only those calves showing temperatures above 105°F need such treatment, and the remainder may be cared for in the manner previously described.

High-level feeding of antibiotics and sulfa drugs at the rate of 700 milligrams daily in the form of AS-700 (350 milligrams aureomycin and 350 milligrams of sulfamethazine) for 28 days is effective in reducing the effects of shipping fever. There is a practical problem of ensuring intake at the proper level. Newly weaned calves are unaccustomed to eating concentrates unless they were previously creep-fed, which is usually not the case. The antibiotics, usually contained in a manufactured or commercial pellet or meal, must therefore be fed with a highly palatable feed to ensure uniform consumption of the antibiotic. All too often the calves that are most apt to contract the disease owing to excessive exposure or rough treatment are also the ones that prefer to eat nothing but hay or pasture. Because affected cattle will usually drink water, water-soluble sulfa drugs are being used with some success. There is some evidence that cattle fed antibiotics do not respond as well to injectible antibiotics if they later require such treatment. For this reason, some feedlot managers prefer to wait and then treat only the sickest calves.

There is no evidence that human beings contract shipping fever from exposure to infected animals, although feedlot employees and truckers often have a high incidence of colds, no doubt caused by the same stresses that affect the calves.

PINK EYE

Pink eye is the term commonly applied to an infectious inflammation of the eye, technically called infectious keratoconjunctivitis. It is usually encountered only during the summer months and is more prevalent in cattle on pasture than in those kept in drylot. Frequently the disease persists in a herd for several months, with nearly all of the young animals becoming affected in one or both eyes.

The disease is characterized by an intense inflammation of the mucous membrane of the eye accompanied by tears mixed with pus, which flow down the sides of the face. In its most aggravated form it causes a large grayish yellow ulcer to appear on the cornea, making the eye temporarily blind. The cornea may even rupture, with loss of sight in the affected eye.

All animals in a herd in which pink eye is present should be examined carefully every day, and those that show symptoms of the disease should be segregated, if possible, in a darkened barn and supplied with plenty of fresh water and succulent feed. Antibiotics, in the form of a spray, ointment, or powder, can be applied to the eye daily by almost any caretaker. An injection of cortisone and antibiotic can be given underneath the mucous membrane of the eyelid, but this usually requires a veterinarian. Eye patches, available commercially, or homemade ones cut from a burlap bag and glued over the eye with "sale-barn number" glue, to darken the area will serve about the same purpose as the darkened shelter. The patch may be left on until it wears off, but it should be left open at the bottom for aeration. Faceflies, attracted to the moisture from the tears in and below inflamed eyes, are a genuine nuisance and are believed by some entomologists to be involved in spread of the disease. Control of faceflies is difficult, but efforts may often be rewarding. Use of vitamin A and vaccines is of doubtful value in the prevention or control of pink eye.

PNEUMONIA

Pneumonia is one of the most common diseases affecting young cattle during the winter months. It is especially prevalent during cold, damp weather and among calves that undergo considerable exposure because of poor shelter. The characteristic symptoms are a high temperature of 105 to 107°F, quick, shallow breathing, with dilated nostrils, and a hard, pounding pulse. By applying the ear to the chest, the herdsman can hear the rasping sound made by the affected lung. The animal habitually lies on the side opposite the diseased lung in order to keep the infected organ uppermost. In advanced cases and when both lungs are involved, the animal shows little disposition to lie down. Instead, it takes an unsteady position with head down and forelegs wide apart to make for as much ease in breathing as possible.

The usual treatment for pneumonia is to administer sulfa drugs either alone or in combination with antibiotics. Whereas formerly a high percentage of pneumonia cases resulted in death, a large number are now saved by the judicious use of these wonder drugs. The animal should be kept as quiet as possible because exercise or excitement tends to aggravate the condition by increasing the pulse and respiration rate. Warmth and dry shelter are essential. One should especially be on the lookout for pneumonia in the late stages of all shipping fever cases. Pneumonia is most likely to occur during the third week after a new drove of calves has been received.

SCOURS OR DIARRHEA

An abnormal looseness of the bowels is a common disorder among young calves kept in unsanitary surroundings or fed improper rations. Calves ranging in age from a few days to 3 months are most likely to be affected, although older animals are by no means immune. The condition in reality may not be a disease in itself, but may be the result of an abnormal condition of a portion of the digestive tract that causes the feed to be improperly digested. Prompt treatment should be administered because a calf infected with a bad case of scours derives little nourishment from the milk and feed eaten and consequently loses flesh rapidly because of dehydration and becomes weak and thin.

Most cases of scours are caused by the presence in the digestive tract of harmful strains of *Escherichia coli*, a bacillus that brings about the formation of toxic products. The first step in the treatment of scours is the administration of an oral antiseptic, sulfa drug, or antibiotic, to destroy these organisms. Many biological supply houses manufacture orally administered medications that give good results when used for scours. By giving them according to the directions on the label, along with 1 to 2 quarts of an electrolyte-glucose solution, depending on the size of the calf, the trouble is usually checked. In order for these medicaments to bring the desired result as quickly as possible, the calf's feed intake should be somewhat reduced for 3 or 4 days. Rations of heavy-milking cows may need to be reduced also, to reduce their milk flow and ensure that calves already scouring do not aggravate the problem by overconsumption of milk.

It is important that the newborn calf obtain colostrum early, preferably within 4 to 6 hours of birth, and it may be necessary to milk the cow and hand-feed the calf if the calf is weak. It is a good practice to keep a gallon of colostrum frozen, to be given to weak calves or to calves whose mothers fail to claim them or who give little milk. Colostrum usually can be obtained from a local dairy farmer.

A particularly virulent form of scours, which is highly contagious among the calves of a herd, is characterized by grayish yellow or dirty white feces with a highly offensive odor. Whenever such symptoms are observed, a veterinarian should be consulted immediately because "white scours" is one of the most serious ailments affecting young calves. White scours appears within 3 days after birth, whereas other diarrheas seldom develop before the calf is a week or 10 days old. This helps to differentiate between the two types of scours. The early scours is usually caused by a virus called a reovirus and a vaccine is available to prevent the disease. The vaccine is given orally at birth.

TUBERCULOSIS

Tuberculosis is one of the most serious diseases affecting domestic cattle. It is caused by a specific organism that attacks all parts of the body, but particularly the glands of the lymphatic system. The glands of the neck, chest, and mesentery are most likely to be affected. In advanced cases the disease spreads to all parts of the body, forming huge masses of tubercles that interfere greatly with the normal functions of the vital organs. Not only is the disease highly contagious from one cow to another, but it may be transmitted from cattle to swine through the manure, and from cattle to man through the milk or meat.

Specific symptoms of tuberculosis are difficult to detect in the live animal. In advanced cases the hair becomes harsh and dry and the animal has a general rundown appearance. Such symptoms, however, may be caused by a variety of diseases and cannot be regarded as conclusive. The reliable test for the presence of tuberculosis is the so-called tuberculin test. Tuberculin is a laboratory product which, when injected under the skin, causes a characteristic reaction or swelling within 72 hours.

Because there is no method of treating tubercular cattle, all animals reacting to the test must be sold for slaughter. "Reactors" that are to be salvaged for meat must be shipped to markets where they may be slaughtered under federal inspection.

Tuberculosis has been almost eradicated in the United States by these procedures, but constant surveillance is still necessary. The incidence of the disease is reduced to such a low status that regulations concerning free testing are in the process of change. For example, a "back tagging" system is being used in most states wherein farmers and ranchers voluntarily tag all of their cull and older cows being marketed. This enables the federal inspectors, upon finding an animal with tubercular lesions, to identify the farm or ranch from which the animal came. Obviously the purpose is to find the herds that need testing. Unfortunately, producers seem reluctant to use this voluntary method for detecting infected herds, but in the long run it should be the most economical method for accomplishing complete eradication of this disease.

Federal and state governments cooperate in paying indemnities for cattle slaughtered after a positive reaction to the tuberculin test. Tuberculosis is transmissible to man, and it remains a health problem among human beings as long as a vestige of the disease exists in the cattle herds. Breeding cattle consigned to purebred auction sales must usually be tested for tuberculosis, brucellosis, and in some cases, leptospirosis, within 30 days before the sale. This testing is done to permit entry of animals into accredited areas or across state lines. Anyone planning to sell or buy breeding females, in or out of state, should know his home state's regulations pertaining to these diseases.

VESICULAR STOMATITIS

This disease is not common, but outbreaks sometimes occur in the western states. It resembles both foot-and-mouth disease and the mucosal complex. The fact that it resembles the former makes it imperative that anyone suspecting that his cattle are so affected notify the state veterinarian in the area. The disease has several forms and is caused by a virus that is transmitted by direct contact or by insect bites.

The first symptom of vesicular stomatitis is an acute rise in temperature followed by the formation of reddened patches in the mucosa or lining of the mouth, on the tongue, inside the nostrils, between the toes, and on the teats of fresh cows. These areas quickly develop into vesicles or blisters up to an inch in diameter. They rupture, and a yellowish fluid or serum oozes from the reddened area beneath the scab that forms. Usually healing occurs in 8 to 15 days in uncomplicated cases, after much loss in condition and weight. Death seldom occurs.

Quarantines are usually imposed on all cattle in the immediate vicinity of an outbreak until it can be determined by incubation tests that the disease is not the much more serious foot-and-mouth disease. Horses may be affected. Treatment consists of isolation, to prevent further spread, and the provision of soft, palatable feeds such as bran, silage, or green chop. Dry hay is to be avoided if possible.

VIBRIOSIS

This infectious disease of the genital tract of cattle is fast assuming equal importance with Bang's disease and leptospirosis as a cause of abortion. It appears to be widespread, especially in the eastern parts of the Rocky Mountain states. The disease is caused by a bacterium, *Vibrio fetus,* which is spread in mating with an infected animal. Artificial insemination with semen from infected bulls can also spread the disease. Young breeding cattle seem to be most susceptible.

Infected animals may or may not conceive, but conception is followed by early abortion or, more often, resorption of the fetus. Evidence that this is happening will be the cows' returning to heat several months after they apparently were settled. Abnormally long or short heat periods are also good evidence of the presence of vibriosis. Infected cows or bulls show few other symptoms of the disease; hence any suspected animals should be tested by a veterinarian. Infected bulls should be eliminated, but sexual rest of cows for at least 3 months, plus treatment with antibiotics, will usually cure them. Only a young, virgin bull or semen from tested bulls should be used on the rested females.

A vaccine for vibriosis has been developed by Colorado workers and is commercially available, to be used on heifers in the prevention of this disease. At present, annual vaccination is required, as the vaccine does not produce permanent immunity.

NONINFECTIOUS DISEASES AND AILMENTS

Bloat and vitamin A deficiency are two conditions coming under this heading and have already been discussed in Chapters 20 and 7, respectively. Other ailments in this category, although they may not cause a large number of deaths, can be particularly trying at times. The order in which they are discussed has no bearing on their relative importance.

IMPACTION, CONSTIPATION, AND INDIGESTION

Because cattle often consume large amounts of coarse, dry roughage, they are somewhat disposed to certain digestive disorders during the fall and winter months. Impaction is the term applied to an abnormal accumulation of material in the paunch. Constipation, on the other hand, implies a clogging of the large intestine with hard, dry feces. Indigestion is a more general term referring to both of these conditions, as well as to other digestive disturbances.

The first symptoms of indigestion are usually a loss of appetite and a rise in body temperature. When such conditions are noted, the bedding material should be examined for the amount and character of the recently voided feces. If the feces are scanty or hard and dry, the patient should be given a strong purgative such as 2 pounds of Epsom or Glaubers' salts dissolved in 2 quarts of water. In the case of impaction, as much as 2 or 3 gallons of water should be given, and the left side of the animal should be powerfully kneaded with the fist to bring about a breaking up and softening of the impacted mass. Constipation ordinarily calls for less drastic treatment. Drenching with 1.5 pints of castor oil, 2 quarts of raw linseed oil, or 1.5 to 2 pounds of Epsom salts usually brings relief. For calves these doses should be reduced by approximately one-half.

LUMPY JAW OR ACTINOMYCOSIS

Lumpy jaw is caused by a fungus that attacks the tissues of the throat, the parotid salivary gland, and the bones of the upper and lower jaws. Its presence is indicated by a round swelling, usually quite hard and firmly adherent to the surrounding parts. The swelling gradually increases in size until it finally breaks open to form an abscess that discharges thick, creamy pus and becomes filled with raw, bleeding tissue. If not properly treated, the fungus invades the interior of the jaw bones, causing a loosening of the teeth. Also growths may be formed in the mouth and pharynx, sizable enough to interfere greatly with eating and breathing.

Not all swellings in the neck region are evidence of lumpy jaw. A diagnosis can easily be made by microscopic examination of a sample of the abnormal growth. The fungus will appear to be made up of club-shaped bodies, all radiating from the center of the mass to form a rosette. For this reason the fungus is often called "ray fungus."

Cattle affected with lumpy jaw should be separated from the herd because there is danger of spreading the disease by scattering the fungus on feed that other animals consume. Animals of no more than market value should be promptly sold where they will be slaughtered under proper inspection. Only in advanced cases of the disease are the carcasses likely to be condemned as unfit for food. Valuable breeding animals that are suspected of having lumpy jaw should be examined by a veterinarian and, if found to be diseased, should be placed under his care. The most satisfactory treatment consists of surgically opening the soft center of the lump to allow drainage of the pus. The cavity may then be packed with iodine-saturated gauze. The drainage should be collected and destroyed. If surgical aid is unavailable or if position of the growth makes an operation hazardous, the injection of sodium iodide combined with organic iodide administered orally over an extended period may bring about recovery. Antibiotics also may be helpful in aiding the lump to regress. However, when surgical treatment is possible, it is much to be preferred.

FOUNDER OR LAMINITIS

Founder or laminitis in beef cattle is usually the result of overeating during the early stages of a finishing program. It most frequently occurs among feeder or show cattle that are put on a full feed of grain too rapidly for them to become gradually accustomed to their new ration or among cattle being fed on all-concentrate or high-concentrate rations containing less than 5 to 10 percent roughages.

Treatment of badly foundered cattle is seldom satisfactory, because usually the condition is not observed until the damage to the feet has been done. If it is detected in its early stages, some relief may be had by applying cold packs to the feet or compelling the animal to stand in a pond or stream of cold water. Administration of cortisone and antihistamines is helpful if given early. Such treatment may prevent the growing out of long hooves and the expense and bother of treatment may be justified in valuable purebred cattle. Owing to the intense pain of standing and walking, foundered animals spend a large amount of time lying down, and they go for feed and water only when driven by pains of hunger and thirst. They frequently lose weight, especially if they were carrying considerable finish when the attack occurred. Animals only mildly affected need cause little concern, but those badly involved should be disposed of as soon as it is obvious that their condition is chronic. If possible they should be sold locally, as they are likely to lie down and be trampled if shipped to a distant market. Moreover, such cattle do not command satisfactory prices at terminal markets because of their poor appearance. They are suitable for home consumption as beef, but may not be carrying enough finish to suit all consumers.

FOOT ROT

Foot rot or foul foot is most likely to occur in feedlot and show cattle that are shut up in dirty barns and lots or kept on contaminated pastures during the summer. Occasionally thorns, stones, and corncobs or other bedding materials may become lodged between the toes in a way that sets up inflammation of the skin in the interdigital spaces. As this part of the foot is very sensitive and tender it is easily scratched or cut, allowing infection or attack by the fungus that causes the problem. If prompt attention is not given, the inflammation and swelling may extend well above the hoof and around to the heel.

Treatment consists of cleaning the foot thoroughly and applying a strong antiseptic that destroys the causative organism, *Actinomyces spherophorus*, which has gained access through the skin. Mild cases respond to undiluted creolin or iodine, but if the infection is well advanced, a 20 percent solution of blue vitriol or potassium permanganate is preferred. Bandaging the foot and keeping the bandage well saturated with one or the other of these solutions usually effects a cure within 3 or 4 days. Intravenous injections of sulfas are recommended for unbroken cattle that cannot easily be handled. Organic iodides, administered in the feed or in salt-mineral mixes, may be indicated on farms where previous cases have been numerous.

Foot rot in show cattle can be largely prevented by keeping the feet properly trimmed and by turning the cattle into dew-laden pastures at night throughout the summer and fall. A shallow footbath containing a 30 percent solution of copper sulfate, through which the animals must walk as they enter and leave the barn, is also effective in preventing the disease.

CANCER EYE

Cancer eye or epithelioma is a malignant tumor on the eyeball or eyelid of cattle. The disease is found everywhere but is more prevalent on the ranges of the Southwest where intense sunlight and irritating dust are believed to be at least indirect causes. Hereford cattle are much more susceptible to eye trouble than are other common breeds. Apparently the lack of pigment in the eyelid and eyelashes to shield the eye from intense sunlight contributes to the prevalence of the disease. For that reason Hereford cows with a ring of red skin around their eyes are highly regarded by many ranchers of the Southwest and in other countries such as Australia and Argentina. Many southwestern ranchers will not use a bull from a cow that developed cancer eye. There is strong evidence that there is a genetic basis for such discrimination against the offspring of either bulls or cows that develop this condition.

Usually by the time the cancer is observed it has affected the eyeball or the eyelids to such an extent that treatment, other than surgical, is of little avail. If the cow or bull is of great value it should be treated by a veterinarian, who will partially or completely remove the eye. Less valuable animals should be sold for beef soon after the condition is discovered because, in most serious cases, the entire carcass will be condemned.

URINARY CALCULI

Much trouble sometimes results from the formation of stonelike mineral deposits in the bladder in male cattle, particularly in steers that are finished on grain sorghum in the Southwest. The calculi cause no noticeable discomfort, unless they enter or block the entrance to the urethra and prevent normal discharge of urine. When this occurs the animal is extremely nervous, refuses to eat, lies down and gets up at frequent intervals, and shows other signs of acute distress. Should the bladder rupture, temporary relief may be obtained, but peritonitis caused by the urine in the abdomen soon results in death. If the bladder does not rupture, death results from uremic poisoning brought about by the continued absorption of urine.

The cause of urinary calculi is not known. There is some evidence that it is associated with abnormal mineral metabolism or imbalance. The mineral composition of the stones varies greatly, with common forms being magnesium phosphates and various silicates, indicating that a number of factors are probably involved. Drinking water that is high in mineral content does not of itself cause urinary calculi.

There is no effective treatment of urinary calculi, and cattlemen living in areas where it frequently occurs should attempt to prevent the problem. Including ammonium chloride in pasture supplements or in feedlot rations appears to be one method of prevention. An intake level of 0.75 ounce or 21.3 grams per day has been cleared for use by the Food and Drug Administration. Frequent salting to induce the drinking of large quantities of water is recommended in some parts of the country to reduce the incidence of calculi. Apparently the chloride ion contained in salt replaces the insoluble magnesium carbonate ion in these instances.

WARTS

Warts are small skin tumors that frequently appear on young cattle, especially on the neck, shoulders, and head. They are more often observed during the late winter and early spring, when the skin is in poor condition because of the low sterilizing effect of winter sunlight, and perhaps because of faulty nutrition which lowers the natural resistance of young cattle. It has been demonstrated that warts are caused by a virus that may be transmitted from one animal to another and possibly to other species, including man.

Warts are usually a temporary condition that eventually disappears of its own accord. The disappearance can be hastened by the daily application of salicylic acid and castor oil, which favors their absorption. Absorption can also be accelerated and the spread of the warts to other cattle made less likely by use of a commercially prepared vaccine consisting of finely ground fresh wart tissue suspended in a salt solution to which formalin has been added to destroy the virus.

RINGWORM

This skin condition is prevalent during the winter when cattle may be rather closely confined. It is caused by a fungus and appears as circular patches of roughened, scaly skin over the body but mainly about the head and neck and at the root of the tail. Itching is a symptom easily observed as cattle rub the affected areas, thus further spreading the condition to other parts of the body or to other cattle. Curry combs and brushes can easily spread ringworm through an entire show string of cattle; consequently the condition should receive immediate attention once it is observed.

Treatment for ringworm is simple, consisting of scrubbing the infected areas vigorously with soapy water, then daubing them with iodine, lime sulfur, or prepared medicants available from veterinarians. This treatment should be applied every 3 to 5 days until the condition is remedied. Access to bright sunshine in wintertime often remedies the problem as rapidly as any other treatment. Gloves should always be worn when working with cattle infected with ringworm because it is highly transmissible to man.

HARDWARE DISEASE

This condition carries its descriptive name because it is caused when such objects as nails, pieces of baling wire, or parts of machinery are picked up by cattle, accidentally during eating or through curiosity, and swallowed into the reticulum or honeycomb compartment of the rumen. There the sharp edges or ends of the objects may puncture the stomach wall, passing into the pericardial cavity where they may cause immediate death by injury to the heart, or cause an accumulation of fluids that eventually leads to death.

Symptoms of hardware disease are a rapidly developing, unthrifty condition, swellings in the brisket region, a sound of fluid movement in the heart and lung regions, and obvious pain when the animal moves about. Treatment is seldom successful, and most cases are inoperable. However, when cases involving extremely valuable animals warrant the expense, objects that can be located by fluoroscopic examination are sometimes surgically removable.

Mineral deficiencies, especially of phosphorus, often are responsible for animals' picking up metal objects. Correction of the deficiency and careful removal of all bits of wire and the like from pastures and feedlots reduce the incidence of this problem. The use of magnetic metal arresters in feedmills removes a large part of the metal objects often responsible for the condition in feedlot cattle. Magnets placed in the stomach are of questionable value as a preventive of this condition in beef cattle.

POISONOUS PLANTS AND METALS

The number of cases of sickness and death of cattle caused by eating poisonous plants and other harmful materials is difficult to determine with any degree of accuracy and it varies greatly from area to area. In the midwestern and southeastern sections of the United States the number of plants that are poisonous to livestock is small compared with the number found in the range area. Also, the feed supply is usually sufficient to cause animals to pass by poisonous plants, nearly all of which are quite unpalatable compared with good, wholesome forage. However, it occasionally happens, especially in early spring and during prolonged periods of dry weather, that pastures are so short and bare that cattle begin to eat any vegetation available, especially weeds and shrubs that are still green and succulent.

The majority of poisonous plants are found in moist, shaded regions, especially near ponds and streams. The forage of timbered and marshy pastures should be examined closely each year to see if there are plants that are poisonous to livestock. If any are found, a close watch should be maintained to detect the first tendency of the cattle to eat them.

Treatment of animals that have eaten poisonous plants should be prescribed by a veterinarian. Remedies given by mouth are often of little value in overcoming the toxic effect of poisons, owing to the great volume of the digestive tract in cattle and the mass of stomach and intestine contents that must be reached by the drugs.

Prevention of poisoning from plant sources is a matter of being able to identify the dangerous plants and conditions under which they are toxic. Limited space does not permit a detailed description of all plants that are poisonous to cattle, and only the more common ones can be described here. The local county extension agent and veterinarian usually have information relative to the plants in a given area.

COMMON CROWFOOT

This is a common weedy wild flower of the buttercup family that is found in meadows and pastures. The juices of the plant are extremely acrid, and cattle usually will not eat it except in the early spring when they are first turned onto pasture. Several varieties of crowfoot are found in the central states; nearly all are more or less poisonous in the green state but harmless after being cut and dried. The stems, which vary greatly in height according to species, are smooth, hollow, and much branched. The flowers are small, with pale yellow petals surrounding a prominent seed head.

Symptoms of crowfoot plant poisoning are gastric enteritis and diarrhea with black, foul-smelling feces, together with nervous symptoms such as difficult respiration, slow chewing of cud, and jerky movements of ears and lips, sometimes followed by convulsions and death within a few hours.

ERGOT

Poison from ergot occurs chiefly during the winter and spring when considerable amounts of grain and cured roughages are being fed. Ergot is a fungus-caused plant disease that affects the seeds of some plants, making them hard, black, somewhat curved in shape, and several times larger than natural. Of the grains, rye is most likely to be affected, and bluegrass, fescue, and redtop hays are also highly susceptible to ergot.

Ergotism in cattle is of two types, the spasmodic and gangrenous. In the former the symptoms are muscular trembling, convulsions, and delirium. Abortion is often brought about by this form of ergot poisoning. In the gangrenous type there is a mummification and sloughing off of the extremities such as ears, tail, or feet through degeneration of the small blood vessels supplying those parts. Treatment of animals with marked symptoms of ergotism is of little avail. Destruction is recommended except in the case of valuable breeding animals.

FESCUE FOOT

It is believed by some that this aptly named condition is identical with ergot poisoning, and it does resemble it in many respects. It differs principally in that it is always associated with the grazing of a heavy growth of fescue, usually in the late fall or winter. Some investigators believe that it differs from ergot poisoning because ergot is seldom found in fescue, especially after the grass drops its seed in the fall. In any case, the same crippling condition results from destruction of small blood vessels in the extremities, mainly the hind feet (therefore the name "fescue foot"). Gangrene and actual sloughing of toes, tail, and ears may occur because of the loss of circulation.

Some preventive measures that can be applied are: sowing a mixture of pasture grasses and legumes rather than straight fescue; feeding supplemental concentrate or roughage; mowing the fescue to keep it from growing tall; and rotation grazing, using the fescue no more than half the time. There is no known cure for the condition, but prompt removal of cattle from the pasture if symptoms are noticed, and good nursing, may prevent the loss of an animal if the condition is not far advanced. Badly affected animals should be sold for slaughter before undue loss of condition occurs, because recovery of advanced cases is seldom complete.

FERN OR BRACKEN

The common fern or bracken that grows in moist, shaded spots in woods and along streams often causes poisoning in cattle. The early symptoms are unsteady gait, loss of appetite, constipation, nervousness, and congestion of the eyes. These symptoms are followed by a spreading apart of the legs in an effort to maintain balance, extreme nervousness, and a general loss of muscular control.

JACK-IN-THE-PULPIT OR INDIAN TURNIP

This is a plant of the arum family found growing in wooded areas. It is easily recognizable by its peculiar purple-striped, vase-shaped flowers with large erect stamens and pistils. The poisonous nature of this plant is not yet fully understood, and few data are available concerning the specific symptoms in cattle that have been poisoned from this source. However, there is ample reason for classifying it among the plants dangerous to cattle.

HENBANE AND JIMSON

These weeds, which belong to the nightshade family, are coarse, smooth-stemmed, ill-scented plants that inhabit neglected fields and waste places. They prefer a rich soil and are often found in old feedlots and in fields that have been heavily manured. Poisoning is most likely to occur when good grazing is scarce or when hungry feeder cattle are turned into infested lots upon their arrival at the farm.

The common symptoms of nightshade poisoning are unsteady gait, cramps, convulsions, and loss of consciousness. Respiration is difficult and the pulse is very rapid. Fortunately cattle seldom eat these weeds except when forced to do so by extreme hunger.

WATER HEMLOCK

This plant is one of the deadliest that may be eaten by cattle. The root is the harmful part, and losses are more severe during the early spring when the ground along streams is badly eroded by high water. Also, at this season the roots may be pulled up by cattle grazing the green tops of the plants. Symptoms of hemlock poisoning are intense nervousness, frothing at the mouth and nose, and violent convulsions. Death frequently occurs within an hour after the root of a single plant is eaten.

WHITE SNAKEROOT

White snakeroot is a rather large, coarse perennial weed found in many woodland pastures throughout the central states. It can be identified rather easily by its leaves, which are opposite, from 3 to 5 inches long, broadly ovate, with serrated edges. Each leaf stalk has three main veins extending from the base of the leaf in numerous branches. These veins are prominent on the underside of the leaf. In late summer the white flowers of the plant appear as compound clusters of 8 to 30 flowers.

Poisoning from white snakeroot occurs usually during the late summer and autumn when pastures are dry and brown. Affected cattle show a peculiar trembling of the muscles, which accounts for the name "trembles" that is applied to the disease in some localities. This trembling is most pronounced after exercise and tends to disappear following rest. Constipation, general weakness, loss of weight, and incoordination of the voluntary muscles are associated with the condition.

Cows affected by snakeroot poisoning secrete the toxic principle of the plant in the milk. There is thus great danger that the disease may be transmitted to human beings, calves, and other animals that are fed milk from cows running in pastures infested with snakeroot plants.

NITRATE POISONING

The number of cases of nitrate poisoning has increased during the past decade, caused, it is believed, by the use of heavier application of nitrogen fertilizer, and by heavier or thicker planting rates. A number of climatic factors can be responsible for abnormal mineral metabolism in the plant, resulting in accumulation of nitrates and nitrites, mainly in the stems and leaves. Drought, excessive cloudiness, uneven distribution of rainfall, and self-shading due to a very thick stand of plants, all may cause nitrate accumulation. Corn or sorghum cut for green chop, small-grain pastures, mountain meadow hay, Sudan grass, sorghum, and pigweed all cause difficulty at times.

A nitrate level higher than 0.4 percent of the dry matter of the ration, expressed as potassium nitrate or KNO_3, is cause for concern. Pregnant animals may be affected by even lower levels of nitrate. At present the only satisfactory way to prevent difficulty is to dilute the nitrate-containing feed with a feed lower in or free of the nitrate. Molasses, mixed with or sprayed onto hay, will reduce the danger.

Animals afflicted with acute nitrate toxicity may die in a short while but usually live for several days. Intravenous injections of 4 percent methylene blue at the rate of 100 cubic centimeters per 1,000 pounds live weight are used by most veterinarians in the treatment of cases of nitrate toxicity. The difficulty is an anemia caused by a failure in the formation of normal hemoglobin, resulting in fatal or chronic anoxia brought on by insufficient oxygen in the body tissues.

SORGHUM AND SUDAN GRASS POISONING

Under certain conditions, rapidly growing sorghum, Sudan grass, and sorghum-Sudan hybrids become dangerous feeds for cattle because of their high content of hydrocyanic or prussic acid under certain conditions. Severe drought and frost appear to promote the formation of this poisonous material in these forages, especially in the regrowth that springs up after the main crop has been harvested. The grazing of sorghum stubble fields should be done with extreme caution so that the first signs of illness may be detected and the animals removed from the field. Feeding the harvested fodder or silage causes no trouble, even in frosted fields or fields badly fired from drought. Intravenous injections of sodium thiosulfate and methylene blue are used by veterinarians in treating the condition. The consumption of considerable wild cherry foliage can cause the same poisoning.

LEAD POISONING

Numerous cattle deaths have resulted from licking buildings and fences that were covered with a lead-containing paint. There have even been instances in which a single old paint bucket or paint brush thrown on a rubbish pile has been responsible for the death of several valuable animals.

The symptoms of lead poisoning are general dullness, blindness, convulsive movements of the limbs, champing of the jaws, and violent bellowing, followed by prolonged stupor and death. Lead acts as a toxin that is particularly damaging to the kidneys, resulting in perforation of the glomeruli.

Use of nonlead paints has reduced the incidence of lead poisoning, but when old lead paints have only been covered over by the new, safe kinds, access to the lead-containing paint is still possible. If detected in its early stages, lead poisoning may be arrested by certain injections into the blood-stream. Prompt consultation with a veterinarian is essential in treating this condition.

RADIATION DAMAGE

Although beef cattle are unlikely to be primary targets in the event of the use of nuclear weapons, some cattle would be affected unless provided protection. Naturally, the nearer the target the greater the protection needed because of the correspondingly higher intensity of radiation and rapidity of fallout. Radiation effects are less severe in cattle than in human beings, and some cattle are more resistant than others. It is therefore likely that, although some cattle might receive external burns, most of them would survive a nuclear explosion. Temporary sterility would occur in many cases and permanent sterility in others. However, a cow that was badly burned and scarred in the first experimental atomic explosion at Trinity Site in New Mexico lived to produce 16 normal calves after she was moved to Oak Ridge, Tennessee.

The lean tissue or muscle of an animal apparently does not absorb radiation fallout. Consequently the meat from exposed animals would not become radioactive, even though burned externally, and thus would furnish a ready supply of safe food following a disaster. Such animals should not be fed contaminated feed or allowed access to contaminated water, however, if they are to be kept alive and used for food. Contamination could occur during slaughter unless precautions were taken. To ensure future generations of beef cattle, breeding stock could be fed the feeds that were subjected to the least radiation exposure. Radiation decays rapidly, and most feeds and water would be usable within a few days.

Shelters can be provided that will afford protection for cattle, but naturally in large operations it would not be feasible to attempt to protect all cattle. Priorities might be established, with valuable seedstock and milk cows having top priority. Plans for shelters, as well as estimates of feed and water requirements, are obtainable from the county extension agent. It is to be hoped that protection from radiation damage will never be required, but some attention and thought should be given to this potential problem.

CHAPTER 28
PARASITES AFFECTING
BEEF CATTLE

Internal and external parasites of beef cattle are responsible for rather serious monetary losses in individual areas, and for smaller losses in almost all situations, unless preventive and control measures are applied. Unfortunately, many of the losses are not readily apparent and thus do not seem important to many cattlemen. In 1965 the Animal Health Institute reported that cattlemen in the United States were losing $162 million annually because of cattle parasites. As mentioned elsewhere in reference to loss from disease today, this figure must be at least doubled because of higher values, per head affected, and because of larger cattle numbers. In addition, some effective insecticides are no longer approved for use and few new breakthroughs have occurred; thus the economic loss from parasites is growing. Undoubtedly the judicious use of insecticides is one of the best investments a cattleman can make in terms of net return per unit of investment.

The incidence of recognizable cases of harmful parasitism is difficult to assess. The Washington station survey, discussed in Chapter 27, is the most comprehensive and recent study of this problem. When cattlemen were asked which of the parasites they considered most damaging, they ranked lice, grubs, screwworms, and hornflies at the top, as shown in Fig. 124. Three of these four insects are still the most important ones, but the screwworm has almost been eliminated. Diseases such as anaplasmosis and coccidiosis were included by many cattlemen because parasites are responsible for their transmission (see Chapter 27).

The Washington survey revealed some interesting information concerning the control measures applied by cattlemen. Table 262 indicates that lice, hornfly, and grub control measures were applied by about 40 percent or more of those answering the questionnaire. The use of control measures for intestinal parasites has increased in the South in recent years. Systemic phosphates are now being used for control of grubs, lice, intestinal parasites, and even flies.

Any discussion of control measures for parasites is apt to become outdated because of rapid developments in this area of research. Readers would do well to follow the research work of entomologists and parasitologists through the press releases and other publications of local and state extension specialists, experiment stations, and industrial organizations.

Many parasitic infestations are geographic in nature, and consequently

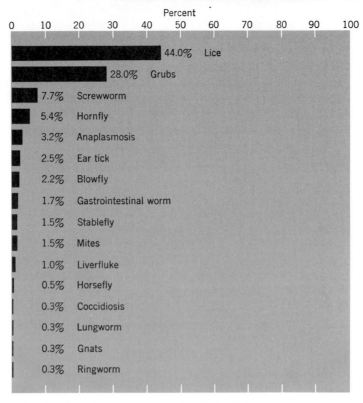

Fig. 124. Most damaging beef cattle parasites in the United States, as indicated by percentage of herds affected. (Washington Experiment Station.)

only a few recommendations concerning use of certain sprays, vermifuges, and the like, are applicable in all instances throughout the country. Suppliers of insecticides, as well as county extension agents and vocational agriculture teachers, should be able to supply information for local conditions in a given area.

COMMON EXTERNAL PARASITES OF CATTLE

FLIES

The hornfly is the most damaging fly among those that are harmful in adult form. This fly feeds on the backs, necks, and at the base of the horns

Table 262

Percentage of Cattlemen Applying Beef Cattle Parasite Control Measures[a]

Parasite	Cattlemen Applying Control Measures (%)
Lice	52.7
Hornfly	41.2
Grubs	37.7
Stable fly	19.3
Blowfly	15.6
Mites	5.8
Ringworm	5.2
Gastrointestinal worms	4.7
Coccidiosis	2.8
Liver fluke	2.5
Anaplasmosis	2.3
Cattle tick fever	1.3
Lungworm	0.8
Bovine trichomoniasis	0.4
Others	2.5

[a] Washington Experiment Station Bulletin 562.

of cattle. Thousands may be found at one time on a single animal, and since each fly feeds several times daily by sucking the blood, the afflicted animal experiences much annoyance and disturbance along with loss in vitality and poor performance caused by loss of blood. Hornflies may be almost eliminated by use of any of the recommended control measures from the beginning of the grazing season and throughout the summer. Note that several chemicals are effective, and these may also be applied by spraying or dipping or by means of back rubbers and dust bags as shown in Table 263. Preliminary investigations indicate that feeding a granular mineral mix containing 5.5 percent Ronnel controls hornflies.

STABLE FLIES

Stable flies are much more difficult to control than are hornflies because they seem to become resistant to many fly-spray materials after a few

Fig. 125. Hornflies can be controlled by periodic high-pressure spraying with any one of several excellent insecticides. If systemic materials are used, other insects, including grubs and even intestinal parasites, are also controlled. (Western Livestock Journal.)

years. This species of fly is most troublesome around barns and feedlots, where it breeds in wet straw, manure, spoiled feed, and other kinds of filth. Stable flies can be greatly reduced by frequently cleaning lots and sheds and by plowing under the manure or piling it at least half a mile from the barns.

Frequent spraying of the legs of cattle with one of several effective chemicals, at levels recommended by the manufacturers, controls stable flies. The same treatment is recommended for horseflies and mosquitoes, which suck blood from the backs and ears of cattle. Various materials may be used, as recommended in Table 263.

Table 263

Materials and Application Methods for Control and Treatment of Cattle Parasites[a]

Parasite	Treatment	Insecticide	Minimum Days from Last Application to Slaughter	Treatment Method	Precautions and Comments
Hornfly	Spray	0.5% Methoxychlor	—	Spray on 2 qt to backline of animal every 3 weeks or as needed.	Grub control can also be effected by Ruelene and Ronnel sprays if high pressures are used.
		0.5% Ronnel (Korlan)	—		
		0.06–0.125% Co-Ral (Coumaphos)	—		
		0.5% Malathion	—		
		0.375% Ruelene	28	Spray thoroughly, no more than 1 gal per animal.	
	Dust bags	3% Ciodrin	—	Place insecticide inside trippled burlap bag, hang so that 4–6 in. of bag touches backs of animals. Place where animals congregate or in walkways to water, mineral blocks, etc.	
	Cable rubbers	5% Methoxychlor	—	Saturate cable rubber with insecticide in No. 2 diesel oil. Recharge at 3-week intervals.	Do not use crankcase oil on cable rubbers.
		1% Ronnel	14		
		1% Co-Ral	—		
		2% Malathion	—		

	Pour-on	8.3% Ruelene (in water) or 9.4% (ready-mix)	28	Apply 1 fl oz per 100 lb body weight up to 800 lb, no more than 8 fl oz per animal. Pour evenly along animal's backline.	Use only prepared solution in hot, humid weather. Keep sparks away from treated animals.
Grubs	Pour-on	8.3% Ruelene (in water) or 9.4% (read-mix) 4% Co-Ral 8% Neguvon (Trichlorfon)	28 — 21	Apply Ruelene at 1 fl oz per 100 lb body weight up to 800 lb, no more than 8 fl oz per animal; Co-Ral and Neguvon at ½ fl oz per 100 lb body weight up to 800 lb, maximum of 4 fl oz per animal. Pour evenly along animal's backline.	Initiate treatments 1–2 months after end of heel fly season. Apply no later than Nov. 1. Also controls horn flies and lice.
	Spray	0.375–0.5% Ruelene 0.375–0.5% Co-Ral (WP) or 0.375% Co-Ral (EC) 1% Neguvon	28 — 14	Wet entire body to skin. Use nozzle pressure of 300 psi and a narrow band spray with not more than 1 gal Ruelene per animal.	Do not treat sick animals or those under 3 months old. Begin treatment after end of heel fly season. Apply no later than Nov. 1. Also controls horn flies and lice.
	Feed additive	0.6% Ronnel 0.26% Ronnel	60 28	Mix with grain or protein supplement so that animals receive 0.3 lb treated (0.6% Ronnel) feed per 100 lb body weight daily, 7 consecutive days, or 0.3 lb treated (0.26% Ronnel) feed per 100 lb body weight daily, 14 consecutive days.	Start treatment after end of heel fly season. Apply no later than Nov. 1. Aids in control of horn flies and lice.

Table 263 (*Continued*)

Materials and Application Methods for Control and Treatment of Cattle Parasites[a]

Parasite	Treatment	Insecticide	Minimum Days from Last Application to Slaughter	Treatment Method	Precautions and Comments
	Mineral block or granules	5.5% Ronnel	21	Feed continuously for not less than 75 days at level of 0.25 lb medicated block or granules per 100 lb body weight per month.	
Lice, Mange, Scabies	Spray	0.5% Methoxychlor 0.06–0.125% Co-Ral 0.5% Malathion 0.5% Ruelene 0.5% Sevin (Carbaryl)	— — — 28 7	Spray thoroughly; repeat after 3 weeks if needed. Apply with high pressure for best results. Increase methoxychlor concentration to 1.0–1.5% for tail louse.	Treatments also effective against horn flies. Ruelene also effective against grubs.
	Pour-on	9.4% Ruelene	28	Apply Ruelene pour-on at 1 fl oz per 100 lb body weight up to 800 lb, no more than 8 fl oz per animal. Pour evenly along animal's backline. Repeat after 3 weeks if needed.	Also effective against horn flies and grubs.
	Dip	0.06–0.125% Co-Ral 0.025% Ronnel	— 56	Immerse animal. Repeat treatment after 2–3 weeks if needed.	Stir insecticide mixture in dipping vat regularly to prevent settling out.

Parasite	Application	Chemical and concentration		Directions	Remarks
Ticks	Spray or dip	0.125–0.25% Co-Ral	—	Immerse or spray thoroughly. Use pressures of 300 psi for spray applications. Repeat as needed.	
Ear tick	Spray	0.75% Ronnel	56	Dust lightly inside ears to reach infested areas, also adjacent head area. Repeat as needed. Use only low-pressure sprays.	Avoid ear damage by using very low pressure sprays. Handle applicators carefully.
	Dust	5% Co-Ral	—		
Screwworm, Blow-fly maggot	Spray	0.5% Malathion	—	Dust wound and surrounding area thoroughly; repeat as needed.	Collect specimens before treating wound. Preserve larvae in 70% alcohol and submit to county agricultural agent for positive identification to assist in eradication program.
	Aerosol	2.5% Ronnel	21	Spray wound thoroughly and wet entire body; repeat as needed.	
	Dust	5% Co-Ral	—		
	Spray	0.25% Co-Ral	—	Spray wound thoroughly.	
		0.5% Ronnel	56		
Horseflies, Stable flies, Mosquitoes	Aerosol	2.5% Ronnel	21	Dilute in mineral oil. Spray 1–2 oz as a mist spray twice daily.	Only temporary control is effected. Vapona and Ciodrin not registered for control of horseflies.
	Spray	1% Vapona (Dichlorvos)		Same as for faceflies.	
		1% Ciodrin		As a spray, 1–2 qt per animal every 2 or 3 days, or as mist spray of 1–2 oz per animal daily.	
		0.6–1% Pyrethrum + 0.5–1% Synergist			

Table 263 (*Continued*)

Materials and Application Methods for Control and Treatment of Cattle Parasites[a]

Parasite	Treatment	Insecticide	Minimum Days from Last Application to Slaughter	Treatment Method	Precautions and Comments
Facefly	Face rubbers	1% Ciodrin	—	Dilute insecticide in No. 2 diesel oil. Saturate rubber.	
		1% Co-Ral	—	Construct rubber to permit animal to rub its face, or force animals to use rubber by installing it in front of mineral troughs, watering troughs, etc.	
		1% Ronnel	14		
	Bait	0.2% Vapona (Dichlorvos)	—	Add bait to mixture of 75% corn syrup, 25% water.	

General Precautions: These control suggestions are for beef cattle only. Other sources should be consulted for control suggestions on dairy animals. Do not treat any animal with systemic insecticide (Coumaphos, Ronnel, Ruelene, Neguvon) while under treatment with another systemic. Do not treat animals under 3 months of age with Coumaphos. Spray animals 3 to 6 months old lightly. Do not treat calves less than 1 month old with Malathion. Do not use Coumaphos with synergized pyrethrums. Do not apply systemics in conjunction with oral drenches or other medications such as phenothiazine or with other organophosphates. Do not spray or pour-on animal for 10 days prior to shipping or weaning. Do not treat sick animals or animals under stress. Recommended levels of insecticide applications and mandatory withdrawal periods before slaughter are constantly changing. Always consult the label on the insecticide container before making applications.

[a] Prepared with the assistance of Dr. Grant Kinzer, Department of Botany and Entomology, New Mexico State University. Revised in 1976 with the assistance of Dr. Leroy Biehl, College of Veterinary Medicine, University of Illinois.

GRUBS

The grub or ox warble is a serious cattle parasite in nearly all parts of the northern hemisphere. It has long been present throughout the United States, although in many localities it is of little economic importance. The degree of infestation appears to vary from year to year even in highly infested areas. This variation leads to the belief that the grub population is affected by climatic conditions, particularly severe cold weather, since the pupa stage of the insect is spent in the ground. The heaviest infestations are found in the Southwest, and it is possible that most of the severe outbreaks of this parasite in feedlots can be traced to feeder cattle coming from that region.

The adult fly, commonly known as the heel fly and somewhat resembling a small black bee, appears during the first hot days of summer. It deposits clusters of eggs on the hairs about the feet and legs of cattle, especially just above the heel. Although the flies can neither bite nor sting, cattle show a natural dread of them and often run from one end of the pasture or lot to the other, holding their tails high, attempting to escape from the flies. Yearlings are more likely to be attacked than calves, and calves more than adults. Because the insects do not fly over water, cattle often protect themselves by retreating into ponds or streams.

The larvae, which hatch in 4 to 6 days, immediately penetrate the skin at the base of the hairs to which they were attached. In one species, the developing grubs then proceed, by no well-defined route, to the neck region where they gather about the esophagus. In another species, the migrating grubs locate mainly along either side of the spinal column or back. No plausible reason has yet been advanced to explain why the grubs congregate only in these two specific areas. During winter, probably in February and March in the Corn Belt but as early as October in the South or Southwest, the larvae leave the neck or spinal column regions and migrate to the back where they form characteristic swellings beneath the skin. There a hole is made through the hide, toward which the larva directs its posterior. In this position it increases rapidly in size, feeding on the pus and mucus caused by its presence. During this time the larva undergoes at least two metamorphoses, becoming larger and stouter until finally, with the arrival of spring, the fully grown grub works its way through the hole in the skin and falls to the ground. There it pupates before being transformed into a mature fly, when it begins the cycle over again.

The numerous holes made by grubs in the backs of cattle greatly lessen the value of the hides, and the beef in the vicinity of the grubby lesions often is slimy and has a bad color. Badly damaged carcasses require extensive trimming in the valuable loin region. Altogether the total economic loss to the country caused by the heel fly is estimated at nearly $115 million per year.

Several recommended treatments are given in Table 263. The initial treatments may begin as early as November in southern and southwestern cattle, even if shipped to the Midwest, but native midwestern cattle may not require attention until February. Cattle raised in the Northwest, North, and Northeast seldom are affected by grubs. Cattle shipped in from the areas where flies may overwinter are affected during the first winter but probably not thereafter. It is hazardous to treat badly infested cattle with systemics if the grubs have migrated to the neck and back regions. Death may result, evidently because of the side effects of absorption of the dead grubs, and possibly the effect on nerves in the spinal region.

SCREWWORMS

The screwworm, once the scourge of the southern and southwestern stockman, is all but a thing of the past. From a pest that caused untold millions of dollars in death losses and much labor and inconvenience in treating infested cattle, the screwworm has dwindled to a few hundred cases per year. The almost complete eradication of this once destructive insect pest is an outstanding example of teamwork and the widespread application of research findings.

Screwworm control was brought about and is being continued in the following manner. Each year billions of male screwworm flies are produced in the laboratory, made sterile by radiation, and then released from airplanes in areas where the female fly is found, including areas where cases are detected in the southern United States and across a belt in lower Mexico. Because the female screwworm fly mates for life with only one male, the majority of females in these areas produce infertile eggs, and the insect's life cycle is broken because only a few new young are produced. As a result, only a few isolated instances of screwworm infestation are now found in the United States, but constant vigilance is necessary to prevent the problem from recurring.

Relatively few sterile male flies are now needed in the States, but a barrier zone several hundred miles wide is still being seeded with such flies each year across Mexico to prevent migration of fertile males back into the United States. Credit for the virtual elimination of the screwworm problem should go to the state and federal governments of both countries, to the various livestock producers' organizations, and to the individual ranchers and farmers concerned. This is one of the finest examples of how all elements of an industry have worked together and strenuously attacked a common problem. It is hoped that eventually the seeding zone can be moved southward into Central America to a point where it can be shortened to less than 200 miles. This obviously would reduce costs and lessen the chances of untreated flies breaking through the barrier.

Treatment methods for screwworm are included in Table 263, although it is doubtful whether they will often be needed. Ordinary blow-fly maggots are occasionally found in open wounds, especially in older, suppurating wounds. Such maggots may be treated by the same procedure used for the screwworm.

FACEFLY

The facefly was first observed in Canada in 1952 and by 1965 had migrated into most of the United States. The facefly originated in Europe and Asia where it was more a household nuisance than a problem with livestock.

The adult facefly, slightly larger than a housefly, is distinguished by its feeding habits. The flies cluster about the eyes, nostrils, and muzzles of cattle, living off the secretions found at these sites. They do not bite or suck but are an annoyance to livestock and cause reduced gains and milk production. Faceflies are also strongly suspected of contributing to the spread of pink eye.

The adult female facefly deposits her eggs in fresh manure where they hatch within 1 day. The yellowish larvae mature in 3 or 4 days and pupate in the soil underneath the manure deposits for 3 or 4 more days, after which they emerge as adult flies, the generation having been completed in 10 to 12 days. Adult flies overwinter in warm or heated buildings, and 10 to 15 generations can be produced in one year, accounting for the large numbers that can result from only a few overwintering adults. Reasonably effective control can be obtained by following the recommendations given in Table 263, but control is made difficult because as little as 10 percent of the local fly population may feed on cattle at one time. Thus control is a continual battle and sporadic sprayings are not very effective.

LICE

Cattle are subject to three species of lice: long- and short-nosed blue sucking lice, and the red biting louse. Although numerous, the biting lice feed mainly on particles of hair and dead skin and are not especially troublesome. The sucking lice, however, cause the cattle much annoyance and, if numerous, sap large quantities of blood. Lice-infested cattle spend much time rubbing themselves against posts, trees, and gates in an effort to relieve the intense itching. Frequently large patches of hair are worn off and the hide becomes hard and callused.

Lice are most numerous during late winter and spring. Cattle on pasture are seldom troubled with them. Treatment should be begun in the fall by dipping or thoroughly spraying with any of the materials recommended in Table 263. Pour-on materials are also effective and easy to apply.

MANGE OR SCABIES

Mange of cattle, also called scabies in one of its forms, is caused by small mites that attack the skin, causing it to become thickened, covered with crusts, and devoid of hair. Usually the rump is the part most likely to be affected, although the back and shoulders also may be involved. In many respects mange of cattle resembles the disease known as scab in sheep, although it is not considered so serious. The sprays and dips recommended for lice also are effective in controlling mange and scabies. When outbreaks of this problem occur in commercial feedlots, quarantines are often imposed because of the greater risk of spreading the condition when so many animals are congregated in small spaces. Regulations regarding movement of animals from these areas are administered by federal veterinarians, thus cattle feeders should acquaint themselves with federal, as well as state, regulations and be alert to changes that may occur on rather short notice.

TICKS

Losses from tick infestation are not so great today as they were in the preceding century when the tick that was the carrier of the dreaded Texas fever caused millions of dollars in losses every year, providing the impetus for government control programs for this and other parasites.

There are far too many species of ticks to allow a detailed discussion here, but several of the more important ones should be mentioned. The ticks found in the Southeast are often the vector or carrier for anaplasmosis in cattle. Elimination of all ticks from this area would go a long way toward eliminating this disease, but other insects also may serve as vectors, and wild animals may also carry ticks, so it is likely that the problem will never be eliminated completely. Other species of ticks serve as vectors for Rocky Mountain spotted fever and tularemia, diseases that affect man.

Spraying or dipping cattle with the preparations recommended in Table 263 at about 30-day intervals will control body ticks. The same treatments that control hornflies and lice also control most kinds of ticks.

A tick calling for special methods of control is the spinose ear tick. This parasite, more commonly called "ear tick," is rather prevalent and can be found in other species of farm animals besides cattle, and in dogs. Few death losses occur and a figure for monetary loss is difficult to estimate. Poor performance, due to the constant annoyance and irritation that cause cattle to scratch at their ears, is the real problem. The tick spends about 6 months in the ear of an animal, then drops to the ground in mature form and lays its eggs on dry debris on or near the ground surface. The eggs hatch within a few days, and the small immature ticks or larvae crawl onto grass or other plants and from there onto the animal. They then migrate through the hair to the inside of the ear, and the cycle is complete.

Several good treatment methods are available and are described in Table 263, but unfortunately all require individual treatment, usually in a squeeze chute. If cattle are being handled individually anyway for some other purpose such as Bang's testing, pregnancy testing, weighing, or feedlot processing, this is a good time to treat.

INTERNAL PARASITES

The situation with regard to the incidence of stomach and intestinal worms in beef cattle is not clear-cut at the present time. Research has shown that cattle are hosts to many species of worms, and parasitologists say they can find worm eggs in nearly every sample of feces examined under the microscope. A problem is that differentiation between the harmful and harmless types on the basis of fecal examination alone is extremely difficult, if not impossible, with presently available methods. The problem is further complicated by the fact that the number of eggs in a fecal specimen does not necessarily indicate the severity of worm infestation, although the correlation is generally high.

The damage or loss resulting from internal parasite infestation is also difficult to assess. Few death losses occur; rather, the losses are in terms of slower gains on pasture or in the feedlot, digestive disturbances that reduce efficiency of feed conversion, or lighter calf weaning weights owing to reduced milk flow in the mother cows. In general, internal parasite damage is higher in areas of high rainfall such as the coastal regions, but other sections such as irrigated pastures or well-watered pastures in the mountain ranges may be just as hard hit.

Symptoms of severe infestation with internal parasites are diarrhea, dead, rough haircoats, anemia, poor performance in general, and lowered resistance to disease. Swellings under the jaw are usually seen in extremely heavy infestations, and younger animals may die if the situation is not alleviated. Worm eggs are expelled in the feces of infested cattle, contaminating the grass with the resulting larvae and thus spreading the infestation to clean cattle.

Phenothiazine, consumed daily at the low level of 1.5 to 2 grams, aids in controlling internal parasites. Such constant low levels of consumption can be maintained by mixing salt or a mineral mixture and phenothiazine in the ratio of 10:1 and feeding it free-choice. For treatment of heavily infested cattle, the oral administration of 20 grams per 100 pounds of body weight, but not over 70 or 80 grams total, as a bolus, capsule, or drench is recommended. Mixing the phenothiazine in the feed is not highly successful because it is unpalatable and because each animal may not receive the correct dosage. The day following such treatment the urine is quite red in color, but this is no cause for alarm.

Thiabendazole, one of the more promising anthelmintics, is administered orally in the form of a drench, bolus, paste, or feed additive at the rate of 3 grams per 100 pounds of body weight. This dewormer is safe and can be given to debilitated animals. This treatment is rather expensive but is used routinely by commercial feedlots that buy feeders originating in the Gulf Coast region. *Tramisol* or tetramisole has been shown to be effective against many of the gastrointestinal parasites and also lungworms. In addition to being available in bolus, drench, or feed additive form, tetramisole has the advantage of being available in injectible form. Animals with very severe parasite infestations should always be retreated with an anthelmintic within 3 weeks.

MISCELLANEOUS PARASITIC DISEASES

TRICHOMONIASIS

Trichomoniasis is a venereal disease of cattle caused by a parasitic organism that infects the reproductive tract of both cows and bulls. A veterinarian may diagnose the disease from the presence of the flagellate organisms in vaginal or cervical smears examined under a low-power microscope. The organism is more difficult to obtain from the sheath of the bull; consequently his infection is more readily confirmed by mating him with a virgin heifer and checking her uterine and cervical smears for the organism at her next heat period.

Infection of cows results in a mild inflammation of the vagina and uterus with an accompanying discharge, which reduces the likelihood of conception or which may cause abortion, especially during the second to fourth month of gestation. Frequently, however, the dead, macerated fetus remains in the uterus, which becomes filled with a thin, grayish white, almost odorless fluid. When this condition occurs, the cow seldom exhibits symptoms of illness, but fails to show the usual signs of approaching parturition.

Bulls infected with trichomoniasis frequently show no visible signs of the disease, although usually there is a mild inflammation of the prepuce and penis during the early stages. The disease is often transitory in cows, but it usually is chronic in bulls, which reinfect the cows each time they are bred. If infected bulls are sold for slaughter and breeding operations are discontinued for at least 3 months, the disease usually disappears from the herd.

APPENDIX

848   APPENDIX

Table A-1
Composition of Feeds[a]

Feedstuff	Dry Matter (%)	Protein Total (%)	Protein Digestible (%)	Metabolizable Energy (Mcal/kg)	TDN (%)	Crude Fiber (%)	Calcium (%)	Phosphorus (%)	Carotene (mg/kg)	Carotene (mg/lb)
Alfalfa, fresh	27.2	19.3	15.0	2.21	61	27.4	1.72	0.31	152	68
Alfalfa hay, prebloom	84.5	19.4	11.7	2.28	63	28.5	1.25	0.23	500	227
Alfalfa hay, midbloom	89.2	17.1	12.1	2.10	58	30.9	1.35	0.22	33	15
Alfalfa hay, full bloom	87.7	15.9	11.4	2.06	57	33.9	1.28	0.20	37	17
Alfalfa hay, mature	91.2	13.6	9.5	2.00	55	37.5	1.33	0.24	27	12
Alfalfa haylage	55.0	17.9	10.7	1.88	52	32.4	1.61	0.38	52	23
Alfalfa meal, dehydrated	93.0	19.2	15.0	2.24	62	26.1	1.43	0.26	173	79
Alfalfa silage	30.4	17.8	11.9	2.02	56	30.4	1.61	0.38	90	41
Alfalfa-brome, fresh	21.6	19.6	14.4	2.24	62	25.3	1.52	0.37	—b	—b
Barley, grain	89.0	13.0	9.8	3.00	83	5.6	0.09	0.47	—	—
Barley, straw	88.2	4.1	0.5	1.01	41	42.4	0.34	0.09	—	—
Beet molasses	77.0	8.7	5.0	3.22	89	—	0.21	0.04	—	—
Beet pulp, dried	91.0	10.0	4.5	2.60	72	20.9	0.75	0.11	—	—
Beet pulp, wet	10.0	9.0	4.0	2.46	68	20.0	0.90	0.10	—	—
Beet pulp with molasses, dried	92.0	9.9	6.5	2.68	74	17.4	0.61	0.11	—	—
Beet tops, ensiled	20.7	12.7	10.0	1.95	54	13.3	2.32	0.20	—	—
Bermuda grass hay	91.1	8.9	4.8	1.56	43	29.6	0.46	0.20	129	58
Bluegrass, fresh	30.5	17.3	12.6	2.60	72	25.1	0.56	0.47	383	174
Bluestem, fresh, immature	31.6	11.0	7.2	2.31	64	28.9	0.63	0.17	219	99
Bluestem, fresh, mature	71.3	4.5	1.7	2.24	62	34.0	0.40	0.11	—	—

Feed									
Brome, fresh, immature	32.5	20.3	15.1	2.46	68	23.9	0.30	0.26	460	209
Brome, fresh, mature	56.1	6.4	3.3	2.35	65	33.0	—	—	—	—
Brome hay	89.7	11.8	5.0	1.59	44	32.0	0.30	0.26	17	8
Buffalo grass, fresh	47.7	9.2	5.7	2.13	59	27.7	0.52	0.16	94	43
Cactus, prickly pear	17.1	5.0	2.8	2.13	59	13.3	6.29	0.08	—	—
Canarygrass, fresh	25.8	13.2	9.1	2.39	66	26.8	0.40	0.30	—	—
Citrus molasses	65.0	10.9	5.6	2.78	77	—	2.01	0.25	—	—
Citrus pulp, dehydrated	90.0	7.3	3.9	2.78	77	14.4	2.18	0.13	37	17
Clover, red, hay	87.7	14.9	8.9	2.13	59	30.1	1.61	0.22	—	—
Corn and cob meal	87.0	9.3	4.6	3.25	90	9.2	0.50	0.31	—	—
Corn cobs, ground	90.4	2.8	0.0	1.70	47	35.8	0.12	0.04	—	—
Corn distillers' grains, dehydrated	92.0	29.8	23.4	3.18	88	9.8	0.10	0.40	2	1
Corn gluten feed	90.0	28.1	24.2	2.96	82	8.9	0.51	0.86	4	2
Corn gluten meal, dehydrated	91.0	47.1	39.2	3.04	84	4.4	0.18	0.44	7	3
Corn grain, No. 2 Dent	89.0	10.0	7.5	3.29	91	2.2	0.02	0.35	25	11
Corn silage, dough stage	27.9	8.4	4.9	2.53	70	26.3	0.28	0.06	15	7
Corn silage, mature	55.0	7.8	4.5	2.57	71	23.0	0.27	0.19	24	11
Corn stover, dry	87.2	5.9	2.2	2.13	59	37.1	0.49	0.09	—	—
Corn stover silage	27.2	7.2	2.9	2.10	58	32.1	0.38	0.19	—	—
Corn, sweet, cannery refuse, ensiled	29.4	8.8	4.9	2.60	72	26.8	—	—	2	1
Cotton burrs	92.0	9.6	2.4	—	45	39.0	1.13	0.12	—	—
Cottonseed	92.7	24.9	15.7	3.29	91	18.2	0.15	0.73	—	—
Cottonseed hulls	90.3	4.1	0.2	1.48	41	47.5	0.16	0.10	—	—
Cottonseed meal, expeller	94.0	43.6	35.3	2.82	78	12.8	0.17	1.28	—	—
Cottonseed meal, solvent	91.5	44.8	36.3	2.71	75	13.1	0.17	1.31	—	—
Fescue hay	88.5	10.5	6.0	2.24	62	31.2	0.50	0.36	29	13
Grama grass, fresh, immature	41.0	13.1	9.0	2.31	64	27.2	0.53	0.19	30	14

Table A-1 (*Continued*)

Composition of Feeds[a]

Feedstuff	Dry Matter (%)	Protein Total (%)	Protein Digestible (%)	Metabolizable Energy (Mcal/kg)	TDN (%)	Crude Fiber (%)	Calcium (%)	Phosphorus (%)	Carotene (mg/kg)	Carotene (mg/lb)
				Dry Basis (Moisture-free)						
Grama grass, fresh, mature	63.4	6.5	3.4	2.10	58	32.7	0.34	0.12	—	—
Grass-legume silage	29.3	11.8	6.0	2.02	56	31.4	0.78	0.28	30	14
Johnson grass hay	91.0	7.6	3.4	2.30	55	33.3	0.81	0.31	17	8
Lespedeza, fresh	25.0	16.4	11.8	2.42	67	32.0	1.35	0.21	50	23
Linseed meal, expeller	91.0	38.8	34.1	2.93	81	9.9	0.48	0.98	—	—
Linseed meal, solvent	91.0	38.6	34.0	2.85	81	9.8	0.44	0.91	—	—
Milk, dry, skim	94.0	35.6	32.1	—	80	0.2	1.34	1.09	—	—
Milk, whole	12.0	25.8	24.2	—	133	0.0	—	—	—	—
Molasses, sugarcane	75.0	4.3	2.4	3.29	91	—	1.19	0.11	—	—
Oat hay	88.2	9.2	4.4	2.21	61	31.0	0.26	0.24	101	46
Oat silage	31.7	9.7	5.5	2.13	59	31.6	0.37	0.30	120	55
Oat straw	90.1	4.4	1.4	1.88	52	41.0	0.78	0.10	—	—
Oats, grain	89.0	13.2	9.9	2.75	76	10.0	0.11	0.39	—	—
Orchard grass, fresh	23.8	18.4	13.5	2.35	65	23.6	0.58	0.55	337	153
Orchard grass hay	88.3	9.7	5.8	2.06	57	34.0	0.45	0.37	34	15
Prairie hay, midbloom	91.0	8.1	4.1	1.84	51	32.1	0.34	0.21	20	9
Prairie hay, late bloom	91.3	6.6	2.2	1.77	49	32.5	0.36	0.13	8	3
Prairie hay, overripe	91.5	4.0	0.5	1.63	45	35.4	0.52	0.08	—	—
Rice bran	91.0	14.8	9.6	2.39	66	12.1	0.07	2.00	—	—
Rye grain	89.0	13.4	10.6	3.07	85	2.2	0.07	0.38	—	—
Ryegrass, Italian, fresh	24.3	16.3	10.1	2.24	62	21.8	0.64	0.41	—	—

Safflower meal, solvent	90.5	49.1	41.3	2.75	76	9.4	0.26	1.83	—	—
Sorghum grain, milo	89.0	12.4	7.1	2.89	80	2.2	0.04	0.33	—	—
Sorghum silage, sorgo	26.0	6.3	1.7	2.10	58	26.8	0.35	0.20	3	1
Sorghum stover, milo, silage	29.4	7.3	2.0	2.06	57	26.3	0.25	0.18	3	1
Sorghum stover, milo, sun-cured	85.1	5.3	1.8	2.06	57	32.6	0.40	0.11	—	—
Soybean hay	89.2	16.3	10.1	1.88	52	32.1	1.29	0.23	27	12
Soybean hulls, flakes	91.3	13.7	8.8	1.63	45	38.9	0.59	0.17	—	—
Soybean meal, solvent	89.0	51.5	43.8	2.93	81	6.7	0.36	0.75	—	—
Soybean seeds	90.0	42.1	37.9	3.40	94	37.9	0.28	0.66	—	—
Soybean straw	87.6	5.5	1.7	1.37	38	44.1	1.59	0.06	—	—
Sudan grass, fresh	17.6	16.8	12.2	2.53	70	30.9	—	0.31	—	—
Sudan grass, hay	88.9	12.7	5.5	2.13	59	28.9	0.56	0.35	—	—
Timothy, fresh, midbloom	28.1	9.6	5.7	2.39	66	33.7	0.50	0.18	—	—
Timothy, hay, late bloom	88.0	8.3	4.1	2.10	58	32.4	0.38	0.22	—	—
Trefoil, birdsfoot, fresh	20.0	15.6	10.7	2.71	61	29.6	1.75	0.22	—	—
Turnips, roots, fresh	9.3	14.0	9.7	3.04	84	11.8	0.64	0.34	—	—
Vetch, hay	88.2	20.0	13.2	2.21	62	28.5	1.36	—	—	—
Wheat, fresh, immature	21.5	28.6	22.2	2.64	73	17.4	—	1.32	520	236
Wheat bran	89.0	18.0	14.0	2.53	70	11.2	0.16	0.41	—	—
Wheat grain	88.0	18.0	14.0	3.15	87	2.9	0.06	1.01	—	—
Wheat middlings	90.0	19.1	13.6	3.00	83	8.9	0.16	0.08	—	—
Wheat straw	90.1	3.6	0.4	1.74	48	41.5	0.17	0.35	—	—
Wheatgrass, crested, fresh	30.8	23.6	18.0	2.68	58	32.6	0.46	0.84	434	198
Whey, dried	94.0	13.9	12.6	—	78	—	0.93	1.54	—	—
Yeast, brewers', dried	93.0	47.9	44.1	2.82	78	3.2	0.14	1.81	—	—
Yeast, Torula, dried	93.0	51.9	47.2	2.89	80	2.2	0.61		—	—

[a] Adapted from feed tables in *Nutrient Requirements of Beef Cattle*, Subcommittee on Beef Cattle Nutrition, National Research Council, 1975.

[b] Dash (—) = no data available.

Table A-2

Composition of Calcium and Phosphorus Supplements[a]

Mineral Supplement	Calcium (%)	Calcium (gm/lb)	Phosphorus (%)	Phosphorus (gm/lb)	Fluorine (%)
Bonemeal, cooked, dehydrated, ground	27.3	124	13.1	59	—
Bonemeal, steamed, dehydrated, ground	30.5	138	14.3	65	0.037
Defluorinated rock phosphate	33.1	150	18.0	82	0.150 or less
Dicalcium phosphate	23.1	105	18.6	84	—
Limestone, ground	33.8	153	0.02	—	—
Monosodium phosphate	—	—	22.5	102	—
Oyster shells, fine-ground	38.1	173	0.07	—	—
Sodium tripolyphosphate	—	—	26.0	118	—
Spent bone black	30.1	137	14.1	64	—

[a] Adapted from feed tables in *Nutrient Requirements of Beef Cattle*, Subcommittee on Beef Cattle Nutrition, National Research Council, 1975.

Table A-3

Estimated Carotene Content of Feeds in Relation to Appearance and Methods of Conservation[a]

Feedstuff	Carotene (mg/lb)
Fresh green legumes and grasses, immature	15 to 40
Dehydrated alfalfa meal, fresh, dehydrated without field curing, very bright green color[b]	110 to 135
Dehydrated alfalfa meal after considerable time in storage, bright green color	50 to 70
Alfalfa leaf meal, bright green color	60 to 80
Legume hays, including alfalfa, very quickly cured with minimum sun exposure, bright green color, leafy	35 to 40
Legume hays, including alfalfa, good green color, leafy	18 to 27
Legume hays, including alfalfa, partly bleached, moderate amount of green color	9 to 14
Legume hays, including alfalfa, badly bleached or discolored, traces of green color	4 to 8
Nonlegume hays, including timothy, cereal, and prairie hays, well cured, good green color	9 to 14
Nonlegume hays, average quality, bleached, some green color	4 to 8
Legume silage	20 to 30
Small grain silages	5 to 20
Corn and sorghum silages, medium to good green color	2 to 10
Grains, mill feeds, protein concentrates, and by-product concentrates, except yellow corn and its by-products	0.01 to 0.2

[a] *Nutrient Requirements of Beef Cattle*, revised 1958, National Research Council (table prepared by the late H. R. Guilbert, Davis, California).

[b] Green color is not uniformly indicative of high carotene content.

Table A-4

Weight Unit Conversion Factors[a]

Units Given	Units Wanted	Conversion Factor	
		Multiply Units Given By	Divide Units Given By
lb	gm	453.59	
lb	kg		2.2046
oz	gm	28.4	
kg	lb	2.2046	
kg	gm	1,000	
kg	mg	1,000,000	
gm	mg	1,000	
gm	μg	1,000,000	
ppm	mg/lb	.45359	
mg/gm	mg/lb	453.59	
mg/kg	mg/lb		2.2046
μg/kg	μg/lb		2.2046
kcal/kg	kcal/lb		2.2046
kcal/lb	kcal/kg	2.2046	
μg/gm	ppm	0	0
mg/kg	ppm	0	0
mg/kg	%		1,000
mg/gm	%		10
gm/kg	%		10

[a] *Joint United States-Canadian Tables of Feed Composition*, Publication 1232, National Research Council, National Academy of Sciences, 1964.

Table A-5

Breed Associations

American Angus Association, 3201 Frederick Boulevard, St. Joseph, Missouri 64501; Lloyd D. Miller, secretary.

American Belted Galloway Cattle Breeders' Association, South Fork, Missouri 65776; Charles C. Wells, secretary.

American Brahman Breeders Association, 1313 La Concha Lane, Houston, Texas 77054; Kirby Cunningham, secretary.

American Charbray Breeders Association, 475 Texas National Bank Building, Houston, Texas 77000; Mrs. Quinta Arrigo, secretary.

American Chianina Association, P.O. Box 11537, Kansas City, Missouri 64138.

American Hereford Association, Hereford Drive, Kansas City, Missouri 64105; H. H. Dickenson, secretary.

American-International Charolais Association, 1610 Old Spanish Trail, Houston, Texas 77025; Dr. J. W. Gossett, secretary.

American Pinzgauer Association, 617 West Rock Creek Road, Norman, Oklahoma 73069; W. A. Livingston, executive secretary.

American Polled Hereford Association, 4700 East 63rd Street, Kansas City, Missouri 64130; Orville K. Sweet, secretary.

American Red Danish Cattle Association, Marlette, Michigan 48453; Mrs. Harry Prowse, secretary.

American Shorthorn Association, 8288 Hascall Street, Omaha, Nebraska 68124; C. D. Swaffar, secretary.

American Simmental Association, 1 Simmental Way, Bozeman, Montana 59715; Don Vaniman, secretary.

Devon Cattle Association, P.O. Drawer 628, Uvalde, Texas 78801; Stewart A. Fowler, secretary.

Foundation Beefmaster Association, 201 Wyandot Street, Denver, Colorado 80223.

Galloway Cattle Society of America, Springville, Iowa 52336; Archie Minish, secretary.

International Brangus Breeders Association, 9500 Tioga Drive, San Antonio, Texas 78230; Roy W. Lilley, secretary.

North American Limousin Foundation, 100 Livestock Exchange Building, Denver, Colorado 80216; Robert H. Vantrease, executive vice president.

Red Angus Association of America, P.O. Box 776, Denton, Texas 76201; Julius C. Todd, secretary.

Santa Gertrudis Breeders International, P.O. Box 1257, Kingsville, Texas 78363; Richard Thallman, secretary.

Dual-Purpose Cattle

American Milking Shorthorn Society, 313 South Glenstone Avenue, Springfield, Missouri 65802; Harry Clampitt, secretary.

American Red Poll Association, 3275 Holdrege Street, Lincoln, Nebraska 68503; Wendell Severin, secretary.

INDEX

Abomasum, role of in digestion, 185
Abortion, caused by, leptospirosis, 809
    vibriosis, 817
  contagious, *see* Brucellosis
  from eating moldy hay, 552
  from grazing frosted corn fields, 456
  induced, in feedlot heifers, 406
  pine needle, incidence of, 801
  as result of, ergot poisoning, 825
    sperm senility, 144
  and vitamin A deficiency in cows, 204
Acetic acid, as corn preservative, 720, 722
  as product of digestion, 185
  in silage-making, 557
  in utilization of ration energy, 186
Acetonemia, incidence of, 801
Acidosis, and ration roughage content, 525
Additive gene effects, and performance of
    crossbred, purebred, and inbred cat-
    tle, 102–104
Afterbirth, *see* Placenta
Age of feeder, effect on, capital investment,
    396
    rate of gain, 388–389
    total feed consumed, 395
  as factor in finishing cattle, 387–400
  and length of feeding period, 392–394
  and quality of feeds used, 395
  related to carcass composition, 397–399
  and total gain required to finish, 394
Alaska, cattle numbers, 21
Alfalfa, carotene content, 853
  as cause of bloat, 612, 613
  in commercial feedlot rations, 385
  compared with cottonseed hulls as rough-
    age source, 545
  composition, 848
  as cubes in finishing rations, 745
  price related to livestock market, 36
  as protein source, 37
  strip-grazing of, 637
  as summer pasture for breeding herd, 284–
    286
  when to harvest for silage, 582
Alfalfa hay, *see* Hay, alfalfa
Alfalfa meal, as protein concentrate, 516–518
Allantois, 164, 166
  during parturition, 174
All-concentrate rations, effect on, dressing
    percentage, 527

  feeding method, 748–749
  rumination process, 525
  protein supplementation of, 495–498
American National Cattlemen's Association,
    survey of diseases and parasites, 799–
    801
Amides, in meeting protein requirement, 186
Amino acids, role of in ruminant nutrition,
    186
Ammonia, production and utilization in ru-
    men, 189–190
  as protein supplement, 516
Ammonium chloride, in prevention of uri-
    nary calculi, 822
Ammonium salts, in meeting protein require-
    ment, 186
Amnion, 164, 166
Amniotic fluid, 165
Anaplasmosis, causes, 802
  efforts to control, 832
  as major parasite problem, 831
  similarity to Texas fever, 802
  symptoms and treatment, 802–803
  ticks as carrier of, 842
Anemia, incidence, 801
  and iron deficiency, 200
  and nitrate poisoning, 828
  as symptom of internal parasites, 843
Angus, as affected by heat and humidity, 293
  carcass traits, 427
  as carrier of dwarfism gene, 92
  comparative adaptability to cow-calf pro-
    gram, 217–219
  in crossbreeding experiments, 110–116
  gestation period, 168
  heritability of polled trait, 305
  heterosis effects compared with other
    breeds, 105–109
  marbling ability, 88, 419
  performance compared to Angus-Holstein,
    117, 118
  registration data, 18–19
  susceptibility to bloat, 614
Angus, Red, registration data, 19
Animal wastes, as protein supplement, 521–
    522
Anthelmintics, as feed additives, 662–665
Anthrax, areas where prevalent, 803
  causes, 803

857